Graduate Texts in Contemporary Physics

Springer
New York
Berlin
Heidelberg
Hong Kong
London
Milan
Paris
Tokyo

Graduate Texts in Contemporary Physics

S.T. Ali, J.P. Antoine, and J.P. Gazeau: **Coherent States, Wavelets and Their Generalizations**

A. Auerbach: **Interacting Electrons and Quantum Magnetism**

T. S. Chow: **Mesoscopic Physics of Complex Materials**

B. Felsager: **Geometry, Particles, and Fields**

P. Di Francesco, P. Mathieu, and D. Sénéchal: **Conformal Field Theories**

A. Gonis and W.H. Butler: **Multiple Scattering in Solids**

K.T. Hecht: **Quantum Mechanics**

J.H. Hinken: **Superconductor Electronics: Fundamentals and Microwave Applications**

J. Hladik: **Spinors in Physics**

Yu.M. Ivanchenko and A.A. Lisyansky: **Physics of Critical Fluctuations**

M. Kaku: **Introduction to Superstrings and M-Theory, 2nd Edition**

M. Kaku: **Strings, Conformal Fields, and M-Theory, 2nd Edition**

H.V. Klapdor (ed.): **Neutrinos**

J.W. Lynn (ed.): **High-Temperature Superconductivity**

H.J. Metcalf and P. van der Straten: **Laser Cooling and Trapping**

R.N. Mohapatra: **Unification and Supersymmetry: The Frontiers of Quark-Lepton Physics, 3rd Edition**

R.G. Newton: **Quantum Physics: A Text for Graduate Students**

H. Oberhummer: **Nuclei in the Cosmos**

G.D.J. Phillies: **Elementary Lectures in Statistical Mechanics**

R.E. Prange and S.M. Girvin (eds.): **The Quantum Hall Effect**

S.R.A. Salinas: **Introduction to Statistical Physics**

B.M. Smirnov: **Clusters and Small Particles: In Gases and Plasmas**

B.M. Smirnov: **Physics of Atoms and Ions**

M. Stone: **The Physics of Quantum Fields**

(continued after index)

Boris M. Smirnov

Physics of Atoms and Ions

With 98 Illustrations

Boris M. Smirnov
High Temperatures Institute
Russian Academy of Sciences
Moscow 127412
Russia
smirnov@orc.ru

Library of Congress Cataloging-in-Publication Data
Smirnov, B.M. (Boris Mikhailovich)
[Fizika atoma i iona. English] Physics of atoms and ions / Boris M. Smirnov.
p. cm.—(Graduate texts in contemporary physics)
Includes bibliographical references and index.
ISBN 0-387-95550-X (alk. paper)
1. Atoms. 2. Ions. I. Title. II. Series.
QC173.S4732413 2003
539.7—dc21 200202663

ISBN 0-387-95550-X Printed on acid-free paper.

Printed in the United States of America.

9 8 7 6 5 4 3 2 1 SPIN 10887080

www.springer-ny.com

Springer-Verlag New York Berlin Heidelberg
A member of BertelsmannsSpringer Science+Business Media GmbH

Preface

This book is devoted to the quantum theory of atoms and ions. It deals with the properties of atoms and ions, interactions between these atomic particles, and collisional processes involving them. The goal is to present and demonstrate the basic concepts of atomic physics and provide the reader with various atomic data. The book is intended for both students and postgraduate students studying atomic physics, and for specialists in physics, chemistry, and the other sciences, using information concerning atomic particles in practice. On one hand, the book outlines the basic concepts of atomic physics and the physics of atomic collisions. At the same time, it contains data which are necessary for the analysis of processes and phenomena in adjacent areas of physics and chemistry. Thus, this book can be used both as a textbook and as a reference book.

The text of the book contains two levels of complexity. In the main text the author tried to use simple and convincing methods to explain the fundamental properties of atomic particles and the processes involving them. Certain detailed aspects are placed as separate problems at the end of the chapters and can be omitted by the general reader.

The basis of the content of the book is the quantum mechanics of nonrelativistic atomic particles. The book consists of two parts. The first part is devoted to the properties of free atoms and atomic ions considered as quantum systems consisting of a heavy Coulomb center and electrons. The concept of a self-consistent atomic field and the shell model of atoms are a central part of the description of these systems. This allows us to give a logical and consistent description of the structure and properties of atoms and ions. At the same time, this gives relatively simple and reliable models for the analysis of the properties of atomic particles. As a result, we have a simple way of describing the basic concepts of atomic physics and obtaining transparent models for solving certain problems. The style of the first part of the book is similar to that in [1]–[4]. This is based on the author's book [5].

The second part of the book deals with the interactions and collisions between particles. The style of this part resembles [6]. In this part the interaction between atomic particles is described, and the analysis is given for bound atom–atom and atom–ion systems, such as molecules, clusters, and solids. Problems of atomic collisions are also considered. However, the author does not try to give a strict description of these problems. The goal of the second part of the book is to demonstrate that specific features of the interactions inside free atomic particles influence the character of the interaction between particles and atomic transitions resulting from interparticle collisions. Then, concentrating on certain principal limiting cases, we can extract those which are important with regard to both methodology and application. In addition, we are free from describing the formalism of the more complicated cases of the theory. Thus, we are well-positioned to choose the problems which are of importance for applications and can be described in a physically transparent way.

For example, resonant collisional processes provide us with the connection between interatomic interaction and collision-induced transitions between atomic states. These processes are characterized by large cross sections, and their analysis is important for the corresponding applied problems. These resonant processes are also of interest from a methodological point of view: Due to large cross sections, they proceed at large interatomic distances where the interaction between particles is relatively weak. This allows us to construct a strict mathematical theory of the resonant processes, where a small parameter is an inverse value of the cross section. On this basis, we obtain a connection between the interactions and the cross sections. Then one can explore the character of competition for the interactions and mixing of certain processes, such as the mixing of a charge exchange process, the processes of rotation of atomic momenta during collisions, and the processes of transitions between states of multiplet structure. As a result, we construct a bridge between the properties of the corresponding atomic particles, the character of their interaction in the molecule, and the rates of transitions which are caused by these interactions during collisions.

Thus, the second part of the book contains separate problems of the interactions and collisions of atomic particles. On one hand, these problems explore the connection between the properties of free atomic particles and the properties of systems of bound atoms, or rates of transitions between states of these particles during collisions. On the other hand, the results of this analysis are of interest for applications. For this purpose, some information in the form of the numerical data in tables, figures, and spectra is included in this part of the book. References indicating the books which contain a more detailed description of certain aspects of the considered problems are given at the end of the book.

Although the analysis of the properties of atomic particles and processes involving these particles is based on analytical methods, this does not contradict the numerical methods which can employ understanding of the problems in more complicated cases. The analytical methods under consideration can be algorithms for numerical calculations in general cases.

References

[1] H.A. Bethe and E.E. Salpeter, *Quantum Mechanics of One- and Two-Electron Atoms*. Springer-Verlag, Berlin, 1957.
[2] H.A. Bethe, *Intermediate Quantum Mechanics*, Benjamin, New York, 1964.
[3] L.D. Landau and E.M. Lifshitz, *Quantum Mechanics*, Pergamon, Oxford, 1965.
[4] I.I. Sobelman, *Atomic Spectra and Radiative Transitions*, Springer-Verlag, Berlin, 1979.
[5] B.M. Smirnov, *Physics of Atom and Ion*, Energoatomizdat, Moscow, 1986.
[6] E.E. Nikitin and S.Ya. Umanskii, *Theory of Slow Atomic Collisions*, Springer-Verlag, Berlin, 1984.

Boris M. Smirnov
Moscow, Russia
October 2002

Contents

Part I

Quantum Mechanics and the Structure of Atoms and Ions

CHAPTER 1

The Development and Concepts of Atomic Physics

1.1 The History of the Creation of Atomic Concepts

The creation of atomic physics was a long process of the study and understanding of some physical phenomena and objects. It is of interest that the character of the investigations of macroscopic objects which led to an understanding that the surrounding world is a matter of which elements are atoms. This understanding opened a new era of physics and science. Therefore, we describe below the history of the creation of atomic concepts. Just as the description of atom structures allowed us to construct quantum mechanics and introduce new concepts that, in principle, gave a new point of view on the surrounding world. But the creation of atomic physics corresponds to the usual development of science, when new concepts and new standpoints for objects and phenomena arise from the accumulation of separate facts and from the study of separate aspects of the problem. Therefore, it is useful to retrace the ways of creation of atomic physics in order to understand its connection with the other fields of physics.

One can follow several lines of the study of matter which have led to the creation of atomic physics. The first is the chemistry and physics of gases which led us to the conclusion that the elements of gases are atoms or molecules and gave methods for their study. The second line reflects the investigations of electrical processes and phenomena. In this way, the prospective experimental methods of the study of matter were developed. In the end, on the basis of these methods, the conclusion was arrived at that atoms consist of electrons and ions, and reliable methods of the investigation of electrons and ions were developed. The third line of investigation of atomic matter is connected with radiation and its interaction with atomic particles. The study of the spectra of atomic particles and the character of the interaction of radiation with atomic particles allowed us to understand the quantum nature of atomic particles. It is necessary to add to this a new understanding of nature, due to the discovery of radioactivity and X-rays. Alongside the principal significance of these phenomena, they gave new methods for the study of matter. Finally, joining all these lines and methods has led to the creation of reliable models of atomic particles. This gave a new understanding of the nature of the surrounding world. Hence, we consider below, in detail, the evolution along each of the above lines.

1.2 The Atomic Concept of Gases

The development of the related fields of physics consisted of many steps, each of which gave a certain contribution to the understanding of the surrounding world. We start with an analysis of the problems of the physics and chemistry of gases which influenced the creation of atomic physics from 1738, when Daniel Bernoulli (Switzerland) worked out a quantitative theory of gases. In 1789 the French scientist, Antoine Laurent Lavoisier, had formulated in his book the concept of chemical elements and established the method of verification of the law of matter conservation in chemical reactions. In 1799 Joseph Louis Prust from France established the law of the definite proportions of elements in chemical compounds. From his point of view, the chemical constituents of a sample are conserved as a result of its formation or decomposition. The following step was made by the English teacher John Dalton who, in 1803, introduced atomic weights. He had formulated the law of multiplied proportions in chemical compounds that was the basis of his theory of matter. According to his conclusion, if the reaction of two substances A and B can result in the formation of different substances, then the masses of A and B in these products correspond to the used masses of the initial substances.

In 1808 Joseph Louis Gay-Lussac (France) suggested the law of combining volumes of gases. According to this law, if two gases A and B are joined to a third gas C, then the ratios of the volumes of A, B, and C, being taken at the same temperature and pressure, must correspond to the ratios of simple integers. This led to some contradiction with the Dalton law. In 1811 the Italian physicist L.R.A. Avogadro overcame this contradiction. Let us consider this contradiction for the example of the formation of water from hydrogen and oxygen. Within the framework of the Dalton law, one elementary particle of water consists of two elementary particles of hydrogen and one elementary particle of oxygen. According to the Gay-Lussac law, which takes into account a number of molecules in a certain volume, one elementary particle of water—a water molecule—is formed from one elementary particle of hydrogen and one-half an elementary particle of oxygen. Avogadro suggested the concept of "elementary molecules" (or atoms), "constituent molecules" (or molecules of elements), and "integral molecules" (or molecules of compounds). He postulated that matter consists of small particles—molecules. From the analysis of the reaction of the formation of water and ammonia he concluded that gaseous oxygen and nitrogen exist in the form of diatomic molecules. The Avogadro law, in the form "at the same temperature and pressure equal volumes of all gases contain the same number of molecules," was of importance for the future. From this it follows that 1 mole of any substance contains the same number of molecules. This value, which equals $6.022 \cdot 10^{23}$, is called the Avogadro number.

Although the Avogadro concept was not accepted by his contemporaries, the idea that matter consists of molecules, and that molecules of compounds include atoms of elements, became widely spread. In 1813 the Swedish chemist Johns Jacob Berzelius introduced the symbols of chemical elements which are still used. Other steps in the atomic concept of matter were to be of importance in the future. Although the atomic concept in gaseous laws was a rather convenient work hypothesis than a new understanding of the nature of matter, it was principally for the development of physic and chemistry of gases. More detailed measurements of atomic masses proved the correctness of the above laws, and the atomic concept was widely used for the analysis of chemical compounds and chemical reactions. Thus, the study of the physics and chemistry of gases led to the creation of the atomic concept of gases which was useful for an explanation of the observed gaseous laws.

The above chain of laws which has followed from experimental research led to a work hypothesis that gases consist of individual particles—molecules. Although this hypothesis was discussed earlier, it can be considered as a guess. But after the establishment of the gaseous laws, the real basis for such an understanding of their nature arose. In order to show the state of the gaseous theory at that time, we present in Table 1.2 the values of the atomic weights of some elements, which were given by Berzelius in 1818, and their contemporary values.

The establishment of the gaseous laws and their usage led to the dissemination of the atomic concepts for the analysis of the various properties of gases. Keeping in mind gases as systems of free atomic particles was profitable for an explanation of the equation of the gas state and such phenomena as diffusion, thermal conductivity and viscosity of gases, propagation of sound in gases, etc. On the basis of the atomic concept, the kinetic theory of gases was created by R.J.E. Clausis (Germany), J.M. Maxwell (England), L. Boltzmann (Austria), etc. This theory uses the distributions of molecules on velocities in gases for the analysis of gaseous properties and transport phenomena. The advantage of the kinetic theory of gases was the cogent

Table 1.1. Atomic masses of elements which were obtained by Berzelius (A_B), and their contemporary values (A).

Element	A_B	A
C	12.12	12.011
N	14.18	14.007
O	16.00	15.999
S	32.6	32.06
Cl	35.47	35.453
Cu	63.4	63.54
Pb	207.4	207.2

argument for confirmation of the atomic structure of matter. In order to understand this, let us consider in detail some elements of the kinetic theory of gases.

The kinetic theory of gases is based on the hypothesis that a gas consists of molecules whose collisions determine various gaseous properties. As an example, let us calculate the pressure of a resting gas within the framework of the kinetic theory of gases. The pressure is the force which acts on a unit area of an imaginary surface in the system. For an evaluation of this force note that if an element of this surface is perpendicular to the x-axis, the flux of molecules, with velocities in an interval from v_x to $v_x + dv_x$ through the surface, is equal to $dJ = v_x f\, dv_x$, where f is the distribution function. Elastic reflection of a molecule from this surface leads to the inversion of v_x, i.e., $v_x \to -v_x$, as a result of this reflection. Therefore, a reflecting molecule of mass m transfers to the surface the momentum $2mv_x$. The force which acts on this surface is the momentum variation per unit time. Hence, the gaseous pressure, the force acting per unit area, is equal to:

$$p = \int 2mv_x \cdot v_x f\, dv_x = m \int v_x^2 f\, dv_x = mN \left\langle v_x^2 \right\rangle, \tag{1.1}$$

where N is the number density of molecules, and the angle brackets mean averaging over the velocities of atomic particles. We take into account that the pressures from both sides of the area are identical.

On the basis of the Maxwell distribution of molecules on velocities, one can connect the mean square of molecule velocity with the gaseous temperature T. This relation has the form

$$m \left\langle v_x^2 \right\rangle = kT, \tag{1.2}$$

where k is the Boltzmann constant. Thus, on the basis of the kinetic theory of gases, we obtain the state equation of an atomic gas in the form $p/N = kT$, which can be written in the form

$$pV = nRT, \tag{1.3}$$

where V is the gas volume, n is the number of moles in this volume, and R is universal constant. Relation (1.3) is a generalized form of the Boyle–Mariott law.

Note that the form of equation (1.2) becomes possible, after J.R. Mayer (Germany) and J.P. Joule (England) proved in 1842–1843 that the mechanical and thermal energies can be transformed to each other with conservation of the total energy, and after W. Thomson (Lord Kelvin, Ireland) introduced the absolute temperature in 1848. Then the right-hand side of equation (1.3) can be connected with the heat capacity of the gas. Indeed, the kinetic energy of molecules of the gas is

$$E = NV \left\langle \frac{mv^2}{2} \right\rangle = \frac{3kT}{2} NV,$$

where NV is the number of molecules in a given volume. This gives, for the heat capacity of a monatomic gas,

$$C_V = \left(\frac{dE}{dT} \right)_V = \frac{3}{2} kNV. \tag{1.4}$$

The heat capacity of 1 mole is equal to $3R/2$ for a monatomic gas. Hence the measurement of the gas heat capacity allows us to determine the universal constant R which is equal to $R = 22.41$ l/mol. Thus the kinetic theory of gases allows us to understand the properties of gases more throughly and led to a generalization of the gaseous laws.

Note that our consideration does not determine the parameters of individual molecules. All the above gaseous laws result from the action of many molecules, and these laws do not allow us to determine the Avogadro constant. This can be obtained only from the analysis of the parameters of kinetic processes. In this connection, is the important discovery of the English botanist Brown in 1827 when he observed a chaotic motion of small pollen grains suspended in water. The chaotic character of motion relates to all types of small particles suspended in liquid. The Brownian movement or diffusion of particles is explained as a result of collisions with surrounding molecules. In rare gases the diffusion motion can be simpler because of the separation of successive collisions with gaseous molecules. The parameter of this motion is expressed through the mean free path of molecules—a distance which a molecule passes through between neighboring collisions with other molecules. If we accept the model of billiard balls for molecules, the mean free path λ is equal to $\lambda = (N\sigma)^{-1}$, where N is the number density of molecules, $\sigma = \pi r^2$ is the cross section of collisions, so that r is the sum of radii of the colliding molecules. Hence, one can evaluate the parameters of individual molecules on the basis of measured diffusion coefficients. This also relates to the thermal conductivity coefficient, viscosity coefficient, and other transport coefficients for other transport processes. Joseph Loshmidt (Germany) was the first who, in 1865, estimated the Avogadro number (at first it was called the Loshmidt value) and the diameter of molecules on the basis of the kinetic theory of gases. In addition, he introduced one more fundamental constant—the Loshmidt number, which is the number of molecules in 1 cm^3 of a gas at temperature 20° C at atmospheric pressure. At first the accuracy of the determination of these values was low. For example, in 1875 Maxwell, on the basis of the diffusion coefficients for twenty pairs of gases, found the Loshmidt number to be $1.9 \cdot 10^{19}$ cm^{-3} instead of its contemporary value of $2.7 \cdot 10^{19}$ cm^{-3}. The precise determination of the Avogadro number was made later, in 1908, by J.B. Perrin (France) who studied the sedimentation equilibrium of small particles, of equal size and mass, suspended in water. These particles were prepared by the centrifuge separation of particles of different sizes. As a result, at that time, Perrin obtained a very precise value of the Avogadro number. Thus, the kinetic theory of gases allowed us to generalize the gaseous laws and to determine the parameters of individual molecules. It was shown that the atomic concept is not only a convenient method for the analysis, but that atoms and molecules are real objects. Therefore, the kinetic theory of gases proved the atomic nature of matter.

Another evolution of the atomic concepts in chemistry has led to the creation of the idea of the chemical valence of elements. In 1863 J.A.R. Newlands (England) postulated the law of octaves according to which the chemical properties of elements are characterized by periodicity. In 1869 D.I. Mendeleev (Russia) suggested the periodical system of elements. This gave impetus to the development of the chemistry of elements and was further useful for the understanding of the atom nature after creation of the atom model. Thus the development of the atomic concepts in gases was twice of importance. First, this concept allows one to create a new description for some properties of gases and therefore was fruitful for the understanding of the gaseous nature. Second, the laws of the chemical transformation of substances resulting from chemical reactions have obtained a simple and strict form on the basis of the atomic concepts of matter.

1.3 The Physics of Electrical Phenomena and Electricity Carriers

The research on electrical processes was of importance for the understanding of the atom's nature. Finally, these investigations showed that the charge carriers—the electron and the ion—are constituents of the atom. We start with the electrical studies which influenced creation of atomic physics from 1705, when F. Hauksbee (England) made the first powerful electrostatic generator that allowed him to observe electrical breakdown in air and also to produce electrical discharges in air which are accompanied by air glowing. This was the first step in the study of electrical gas discharges. Note that gas discharge is a form of passage of electric currents through a gas, and that this current is created by electrons and ions. Hence, the investigation of electric gas discharges must lead to the study of the elementary carriers of electric currents—electrons and ions, which are the constituents of atoms. During the gas discharge investigations Stephen Gray from England discovered the property of conductivity in 1731. In 1734 C.F. de Cisternay Dufay (France) showed the existence of two types of electrification and suggested two types of electrical flows. In addition, he observed that the air conductivity near hot objects was evidence of air conductivity at high temperatures. In 1745 E.J. Von Kleist (Germany) and P. Van Musschenbroek (Netherlands) independently invented an electric capacitor which was called the "Leyden jar." The investigations of American scientist Benjamin Franklin were of great importance. In 1752 he proved on the basis of experiments that lightning has an electrical nature and can be considered as an electrical current propagating in the atmosphere air. He considered this phenomenon as an electrical flow between charged objects. These first studies of electricity flows through the atmosphere showed that electrical conductivity results from the flow of electrical charges. Subsequent research allowed us to ascertain the nature of the carriers of charges that was of importance for the construction of the atom model.

The principal information for the understanding of the structure of atomic matter followed from the investigations of electrolysis by Michael Faraday (England) in 1833–1834. The obtained laws of the electrolysis can be expressed by the formula

$$q = Fm/A, \tag{1.5}$$

where q is the value of electricity passing through the electrolyte, m is the mass of substance which is formed on an electrode, and A is the chemical equivalent of the substance (its atomic weight). The proportionality coefficient is called the Faraday constant which is equal to $F = 9.649 \cdot 10^4$ C/mole. As a matter of fact, this coefficient is the ratio of the charge of an ion which partakes in the electrolysis to its mass. Alongside the established fundamental laws of electrolysis, Faraday introduced new terms such as "ion," "anode" (a way down in Greek), and "cathode" (a way up in Greek) for the analysis of the passage of electricity through matter. In due course these terms migrated to the physics of gaseous discharge and later became general terms in physics and chemistry. On the basis of the subsequent investigation of electrolysis, Stoney (England) estimated in 1874 the value of the elementary charge quanta which are transported during electrolysis. Stoney suggested "electron" as a name for the charge quanta. Note that an electron obtains its name as a result of the electrolys investigations, while electrons are not present in this phenomenon in a free state.

The development of the techniques of gaseous discharge played a key role in the understanding of the nature of electricity carriers. An important technique was the creation of an inductive coil by Rumkopf in 1851 that gave the possibility of obtaining high voltages in a simple way. Alongside this, the development of pump techniques allowed one to work with discharges of low pressure. In 1851 J. Pluecker (Germany) discovered specific rays emitted from the cathode which were later called cathode rays. Just as in a gas discharge of low pressure, electrons emitted by the cathode can be accelerated up to relatively high energies because their collisions with gaseous molecules are rare. Therefore, in future the discharge of low pressure allowed one to study the properties of electrons.

The specifics of cathode rays are such that they turned under the action of a magnetic field and caused the fluorescence of substances on which surfaces they are directed. According to subsequent studies, objects which are found in the way of cathode rays give a shadow. This testified to the propagation of cathode rays along straight trajectories. In 1871 C.F. Varley (England) proved that these rays are negatively charged. In 1876 Eugene Goldstein (Germany) discovered rays of positive charge which are called Kanalstrahlen. Investigating cathode rays resulted from the discharge of a high voltage and a low gaseous pressure, Kruks (England) found in 1887 that cathode rays are forced to rotate blades of the electrometer if the blades are located on the trajectory of the cathode rays. This confirmed the corpuscular nature of these rays. Finally, all these facts allowed us to determine the nature of particles which are constituents of the cathode rays.

Note that the experimental research with charged particles is simpler than with neutral particles because of the possibility of acting by external electric and magnetic fields on charged particles. The experiments with cathode rays led to the discovery of electrons and to the determination of its parameters. This was made by J.J. Thomson (England) in his classical work for the deflection of cathode rays by electric and magnetic fields. These fields are located in a space where accelerating fields are absent, i.e., electrons reach a certain velocity and move with this velocity to a target (see the device scheme in Fig. 1.1). The deflection angle or

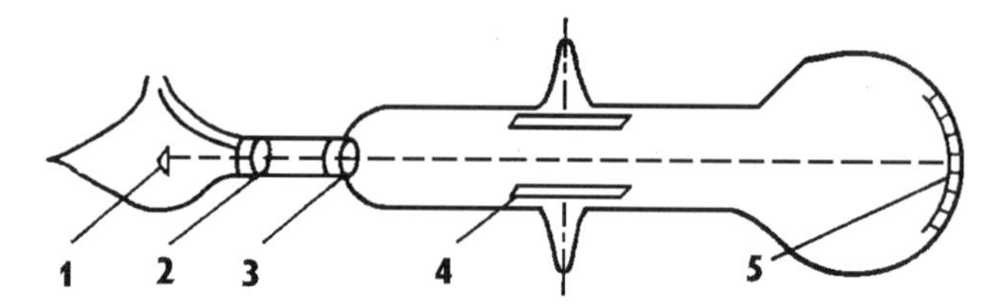

Figure 1.1. The Thomson device for the study of cathode rays. 1, cathode; 2, anode; 3, governed electrode; 4, plates of a declined capacitor; and 5, rule for the measurement of the deflection of a cathode ray.

the displacement distance of a target under the action of a transversal electric field depend on the electron velocity. If the magnetic field is directed perpendicular to the ray and to the transversal electric field, it causes a ray deflection in other directions. The measurement of two deflections allows us to determine both the electron velocity and the ratio of the electron charge to its mass.

Later (in 1907–1911) J.J. Thomson used this method for the study of rays of positive charges. Positive ions which are formed in gas discharges move to the cathode and their action on the cathode causes the emission of electrons. This mechanism can be different depending on the type of discharge. For example, in glow discharge collisions of accelerated ions with the cathode cause the emission of secondary electrons. If the cathode has a hole, part of the ions can be extracted. Using the accelerating electric field and the deflecting electric and magnetic fields, J.J. Thomson determined the ratio of charge to mass for ions of different gases. From this it followed that the mass of the hydrogen ion, which is the lightest one among positive ions, is 1836 times more than the electron mass. The Thomson method becomes the basis of the mass spectrometer operation.

Alongside the measurement of the ratio of charge to mass for an electron, other electron parameters and properties were of interest. In 1883 T.A. Edison (USA) discovered that a heated incandescent filament in vacuum emits a current of negative charges, and in 1887 Heinrich Hertz (Germany) discovered the photoelectric effect when electric current results from the irradiation of a metallic surface. In 1888 Wilhelm Hallwachs (Germany) proved that negative charges are emitted as a result of the photoelectric effect. In 1899 P.E.A. Von Lenard (Hungary) showed for the photoelectric effect, and J.J. Thomson for the Edison effect, that in both cases electrons are emitted. This gave the method of the independent determination of electron parameters. In 1902 P. Lenard discovered that the photoelectric effect has a threshold as a function of the light frequency, and the energy of released electrons does not depend on the light intensity. According to the Einstein theory of the photoelectric effect, which he suggested in 1905, the threshold of this phenomenon corresponds to the photon energy which is equal to the electron binding energy. Then, measuring the energy of released electrons as a function of a frequency of incident light, one can establish the connection between photon and electrical energies which are expressed in electronvolts. According to this definition, an electronvolt is the energy which an electron obtains after the passage of 1 volt. On the basis of measurement of the power of a light source, and the power of the flux of released electrons, one can establish the connection between the mechanical, electrical, and photon energies. In principle, this approach allows one to determine the electron charge and mass.

But the determination of the electron charge was made by another method. In 1909 R.A. Millican (USA) found the electron charge on the basis of measurement of a free fall of small oil drops which move in air in an external electric field. Drops were extracted from a mist, which was formed by spraying oil through a nozzle, and these drops have both positive and negative charge. Under these conditions, the drops conserve their charge and mass for a long time. Because of a small size of the drops, their descent down was observed with a microscope, and the time of passage of an individual drop between two lines of a microscope with a certain distance between them was measured. Hence, the velocity of the descent of a drop can be determined and is connected with the drop radius.

This process was observed between two horizontal metallic plates to which an electric voltage was applied. The velocity of an individual drop was determined, and later an electric field was switched on. The electric field strength was increased until the drop stopped in mid-air. This means that the force on the drop from the electric field equalizes the force from its weight. This position of a drop can last long enough (up to several hours) which gave the possibility of determining a drop charge precisely enough. Thus, the two measuring parameters, the free-fall velocity and the equalized electric field strength, allowed one to determine the drop charge. As a result, a drop charge occurred which was the whole value of an elemental charge. This elemental charge was accepted to be the electron charge.

Measurement of the electron charge removes the deficit of information concerning electrons and ions. Together with the atomic data, this gave the possibility of constructing different models of the atom consisting of electrons and ions. All the atom models before the Bohr model were based on the classical behavior of electrons and hence were wrong. Construction of the realistic atom model must use the radiative properties of atoms which are determined by the quantum nature of atoms. These studies are considered below.

1.4 Blackbody Radiation and the Radiative Properties of Atoms

The quantum nature of atoms was understood on the basis of the analysis of its radiative properties. Let us consider the main steps of this study within the framework of its importance for the problems of atomic physics. The basis of such investigations was the spectral method which was created as early as the seventeenth century. In 1666 Newton had separated the Sun's light into its spectral components as a result of the passage of light through a prism. This method led to a new understanding of the properties of radiation through to the nineteenth century.

In 1802 Wollaston (England) had shown that the spectra of different light sources (Sun, flame, and electric spark) are different. In 1814 Fraunhofer (Germany) separated the dark lines of absorption in the solar spectrum. The principal results were obtained in 1859 by G.R. Kirchhoff (Germany) who postulated the connection between the processes of emission and the absorption of radiation. This connection was expressed in the Kirchhoff law which is as follows: "The ratio of the radiant emittance of a surface to its absorptance is the same at a given temperature and is equal to the radiant emittance of a blackbody at the same temperature." This led him to the concept of blackbody radiation. Establishment of the connection between the processes of absorption and the emission of light led to the creation of methods of spectral analysis. The first success of this method was the discovery in 1860 of two new elements—rubidium and cesium—on the basis of the spectral methods.

Along with the development of spectral methods, a new understanding of radiation due to the blackbody concept was of importance. This followed from the concept of a blackbody irradiation. Developing this concept, J. Stephan (Poland) in 1879 postulated the so-called "Stephan–Boltzmann law" according to which the flux of a blackbody summarized over a spectrum is proportional to T^4, where T is the blackbody temperature. Boltzmann obtained this result from thermodynamic consideration. In 1893 Wilhelm Wien (Germany) gave the spectral dependence for the radiation of a blackbody at long wavelengths, and in 1900 J.W. Strutt (Lord Rayleigh, England) and J.H. Jeans (England) obtained this dependence for the limit of short wavelengths. Considering the electromagnetic field of radiation as a system of oscillators, Max Planck (Germany) tried to connect these two limited cases. In order to escape an "ultraviolet catastrophe," which corresponds to an increase of the spectral power of radiation as the wavelength decreases, he introduced a discrete energy for photons. This refers to the following relation for the energy of an oscillator of a frequency ν:

$$E = h\nu, \tag{1.6}$$

where h is a constant called the Planck constant. As for the spectral distribution of a blackbody, this operation allowed one to describe it for all of the spectrum. But the introduction of this constant meant essentially more because this gave the beginning of quantum mechanics.

Note that the Planck constant and the electron charge allow us to connect electric and spectral units, i.e., electronvolt and cm^{-1}. In particular, there is the following expression for the photoelectric effect:

$$eV = h(\nu - \nu_o),$$

where ν_o is the threshold frequency of the electromagnetic waves, ν is their frequency, and V is the electric potential which stops electrons. Measuring $V(\nu)$ for a given surface, one can connect the above values. This connection has the form

$$1\,\mathrm{eV} = 8066\,\mathrm{cm}^{-1}.$$

The introduction of the Planck constant in physics was of importance for both the creation of atomic physics and quantum mechanics. Along with this, a detailed study of atomic spectra and their analyses was of importance for atomic physics. In 1885 Balmer (Norway) found a simple approximation for the wavelength of a sequence of spectral lines of the hydrogen atom which is given by the formula

$$\lambda = Cn^2/(n^2 - 4), \tag{1.7}$$

where C is a constant of this sequence. Further this formula was represented in the form

$$1/\lambda = A - B/n^2, \tag{1.8}$$

where A and B are constants, and similar dependencies for the spectra of other elements were found. In 1890 J.R. Rydberg gave a generalization for the wavelengths of a series of spectral lines on the basis of the expression

$$\nu = 1/\lambda = A - Ry/(n + \alpha)^2, \tag{1.9}$$

where ν is the frequency of a given transition, n is the number of the line in a given series, A, α are parameters which depend on both the element and on a spectral series, and $Ry = 109,737\ \text{cm}^{-1}$ is the so-called Rydberg constant. In 1908 Ritz generalized this formula by usage of the following expression, for the parameter A,

$$A = Ry/(m + \beta)^2, \tag{1.10}$$

where the parameter β depends on an element and includes a certain set of serials, and m is an integer. In 1908 Pashen found, for the hydrogen case,

$$\nu = Ry(m^{-2} - n^{-2}), \tag{1.11}$$

where m and n are integers. Although all these expressions are empirical, they are based on experimental data and generalize the results of experiments. In addition, due to a high accuracy of spectral measurements, these relations include much information on atom properties. This information was of importance in checking the atomic models.

1.5 Radioactivity and X-Rays

In 1895 W.C. Roentgen (Germany) discovered X-rays, and in 1896 A.H. Becquerel (France) discovered the radioactivity of uranium. Although these great discoveries did not directly influence the creation of the atom model and the quantum theory of matter, they promote the development of physics, including atomic physics. Along with a general value of these discoveries for the understanding of the surrounding world and for the main behavior in it, they were of importance for atomic physics because they gave new instruments for the analysis of atomic particles. Below we reflect on this aspect of these discoveries.

Let us start from the phenomenon of radioactivity. In 1897 E. Rutherford found that the radiation of decaying uranium consists of two parts—soft or alpha-rays and hard or beta-rays. In 1900 A.H. Becquerel showed that beta-rays are like cathode rays, i.e., they are a flux of electrons. Next, it was discovered that uranium decay was accompanied by the formation of helium. In 1908 W. Ramsay and F. Soddy found helium in a remarkable amount in all radium compounds, and in 1909 E. Rutherford and T. Royds ascertained that alpha-particles are double-charged helium ions. In addition, in 1900 P. Villard (France) showed that gamma-rays are formed at some radioactive decays. Gamma-rays are a flux of ultra-short radiation.

Thus, particles of high energy result from radioactivity decay. Therefore, fluxes of these particles can be used for the analysis of atomic matter. In 1911–1913, E. Rutherford and E. Marsden studied the passage of the flux of alpha-particles formed in uranium radioactive decay through thin foils. According to the results, the character of the scattering of alpha-particles did not correspond to the existing models of the atom at that time. For example, in the most popular model of that time, the Thomson model, the positive charge of the atom has the form of a liquid drop in which electrons are floating. Then the oscillations of electrons in this liquid drop determine the observed spectrum of atom radiation. Within the framework of this and similar models one can expect that the scattering of a fast charged particle colliding with the atom occurs mostly at small angles. But in reality scattering at large angles was observed.

On the basis of these data, Rutherford has suggested a new atom model where a point positive charge located at the atom center occupies a small part of the atom volume, and electrons are distributed over all of the atom volume. Then electrons do not partake practically in the scattering of alpha-particles which is determined mainly by the Coulomb interaction between an alpha-particle and a positive ion. Rutherford calculated the differential cross section for the scattering of two charged particles, and now such a process is called Rutherford scattering. In 1913 Geiger and Marsden (England) studied the scattering of alpha-particles on their passage through thin foils of silver and gold in order to test the Rutherford theory, and their results confirmed the Rutherford conclusions. Note that the Rutherford atom model contradicts the nature of a classical atom of this structure. Indeed, such a classical atom has a limited lifetime because of the radiation of moving electrons in the field of the positive ion. But this model which followed from experimental results became the basis for the creation of a realistic atom model by Nils Bohr.

Another great discovery at the end of the nineteenth century, X-rays, also introduced a contribution to the creation of atomic physics. X-rays are formed as a result of bombardment of the anode by electrons in gas discharges of low pressure and of high voltage which are used for the generation of cathode rays. Because of a small gaseous pressure, collisions of electrons with atoms are seldom in such discharges. Therefore the electron energy can reach eV, where e is the electron charge and V is the discharge voltage which is several tens of kilovolts in such discharges. Within the framework of the contemporary standpoint that X-rays are electromagnetic waves, it follows from high-electron energy that the wavelength of X-rays is enough small. In particular, assuming that all the electron energy transfers to a photon which is formed as a result of the bombardment of a surface, we obtain by analogy with formulas (1.4), (1.5) the following relation $E = eV = h\nu$, i.e., the frequency of the generated radiation ν satisfies the relation

$$\nu \leq eV/h.$$

For example, for $V = 50\,\mathrm{kV}$ this formula gives that the wavelength of X-rays, in this case, is more than 0.25 nm. This value is of the order of atomic size and, therefore, at the scattering on condensed substances the photon "fills" their structure. Using this consideration and testing the wave nature of X-rays, Max von Laue (Germany) in 1912 studied the diffraction of X-rays on the crystal lattices of NaCl and KCl. This experiment not only confirmed that X-rays are electromagnetic waves, but also gave a new method for the analysis of crystals. In 1913 W.H. Bragg and W.L. Bragg constructed an X-ray spectrometer on the basis of X-ray scattering resulting from the passage through crystals. This device was of importance for the study of the interaction of X-rays with atomic systems, and with its help some aspects of the internal structure of atoms were understood. Thus the investigations of radioactivity and X-rays have prepared the ground for a detailed study and understanding of the nature of atomic matter.

1.6 The Bohr Atom Model

The discovery of the electron stimulated the development of atomic physics. It was clear that the atom consists of a heavy positive nuclei and electrons that led to the creation of various atomic models constructed on the basis of this concept. The most popular was the Thomson atom model which describes an atom as a positive spherical liquid drop inside which electrons move along ring trajectories. It was postulated that each ring can include a certain number of electrons. Such a model was capable of explaining the chemical and physical properties of atoms. According to this model, atoms with the same number of electrons in the largest rings have similar physical and chemical properties. At a certain number of electrons in a ring, the electron distribution becomes unstable due to interaction with another atom, and this atom can be lost or obey electrons depending on a number of electrons in the rings. In this way was explained the observed valence of atoms. In addition, the vibrational properties of electrons in this atom are responsible for its spectrum. Thus, in spite of the imperfections of this model, the chemical valence of atoms and their physical properties are connected within the framework of this model with the distribution of electrons inside the atom. This was an important step in the understanding of the nature of atomic matter.

Along with the other first atomic models, Nagaoki (Japan) suggested in 1904 a planet atom model which is close to the contemporary standpoint. In this model the positive nucleus occupies the atom center, and the electrons rotate around it along classical trajectories. The charge of the electrons equalizes the nuclei positive charge so that the atom is neutral. But this model was wrong within the framework of the classical laws, because, rotating electrons emit radiation and lost energy.

Hence, such an atom is unstable, and the model was not developed in its initial form in spite of its closeness to contemporary representations. The Rutherford atom model was not a development of this model, but resulted from experimental analysis. Therefore, it differs from the Nagaoki atom model in some detail.

The atom model which was suggested by Nils Bohr (Denmark) in 1913 began a new era in atomic physics. Bohr found the right way to overcome some contradictions which arose because a sum of results for the atom nature cannot be explained by classical considerations. Within the framework of contemporary knowledge, the Bohr postulates of its first atom model can seem to be inconsistent. But with time this model took a more convincing and consistent form, and can explain new observational facts. The main idea of the Bohr model contradicted the classical representations of that time, but this idea just happened to be the advantage of this model. Therefore below we concentrate on this idea.

The Bohr considerations during the creation of its models were as follows. The classical atom models as the Thomson model using three-dimensional parameters—an atom size a, the electron mass m, and an electron charge e. One can construct a value with any dimensionality from these parameters. In particular, a typical frequency of an atom spectrum is of the order of $em^{-1/2}a^{-3/2}$ which corresponds to a real spectrum. Among different atom models we must make a choice in favor of the Rutherford model because it uses additional observational facts. But the Rutherford model does not contain an atom size in its basis. Hence, in order to make a consistent atomic model, it is necessary to add to the Rutherford model one more dimension parameter. Bohr chose as this parameter the Planck constant, and by analogy with the Planck operation requires that the electron momentum expressed in the corresponding units is a whole number for the hydrogen atom. This gives certain stationary states of the electron and leads to a hydrogen atom spectrum which corresponds to the observed one.

These considerations were realized in the following way. The classical atomic electron moves along elliptic trajectories in the Coulomb field of the positive nuclei. According to the Keppler law, the frequency of rotation of the electron on such an orbit ν is connected with the electron binding energy W (i.e., the energy required for removal of the electron to infinity) by the relation

$$\nu^2 = \frac{2}{\pi^2} \cdot \frac{W^3}{me^4}. \tag{1.12}$$

In this formula the charges of the electron and nuclei e are identical, i.e., this formula corresponds to the hydrogen atom. Because of the quantum character of motion, the connection between the binding energy W and the frequency of the orbital motion ν is given by the Plank formula, which Bohr has written in the form,

$$W = \frac{nh\nu}{2}, \tag{1.13}$$

where n is an integer. The numerical coefficient between the above values was chosen such that it can also give the correct spectrum of the hydrogen atom. From formulas (1.12) and (1.13) it follows that

$$W = \frac{me^4}{2\hbar^2 n^2}, \tag{1.14}$$

where $\hbar = h/(2\pi)$.

As is seen, the assumption of the quantum character of electron motion on the basis of formula (1.13) has led to a discrete set of the electron binding energies. For justification of his concept, Bohr used some assumptions which were the basis of the atom description and were also called the Bohr postulates. These Bohr postulates were formulated in 1913 in the following way:

(1) An elemental system with moving electrons around the nuclei emits radiation only during the transitions between stationary states in contrast to the classical electrodynamics when radiation is emitted continuously.
(2) The dynamical equilibrium of this system is submitted to classical laws when classical laws are not valid for transitions between stationary states.
(3) The radiation that resulted from the transitions between stationary states is monochromatic. The change of the energy of system E and the radiation frequency ν are connected by the relation $E = h\nu$. Only in the region of small frequencies does the transition frequency correspond to classical electrodynamics.

The atom Bohr model was of importance both for the creation of a self-consistent atom model and quantum mechanics. It is not without reason that the Bohr postulates are called the "old quantum theory." This means that the Bohr atom model was an intermediate step on the way to the creation of contemporary quantum mechanics.

1.7 Corroboration of the Bohr Atom Model

At the first stage the Bohr model seemed to be semiempirical, and its main advantage was an ability to explain various experimental facts. But with time it was transformed into the harmonious and self-consistent theory. In addition, this model was confirmed by new experimental facts. Let us pause at some of them starting from the study of the X-ray spectra of elements. In reality, the generation of X-rays as a result of bombardment of a surface by an electron beam has the following character. Fast electrons excite the internal electrons of surface atoms, and the radiative transitions of these atoms to the ground state cause the radiation of short-wave photons. C.G. Barkla (England) in 1908 concluded, on the basis of absorption experiments, that the secondary X-rays of various elements can be divided into groups which he called the K-, L-, and M series of X-rays. In 1913 H.G.J. Moseley (England) made more detailed experiments for the analysis of X-rays and obtained spectrograms of X-rays for some elements from calcium to nickel for K- and L- series. He shown that the characteristic frequency of K- and L-radiation is determined by the charge of its positive ion Z and can be written in the form

$$\nu^{1/2} = a(Z - b),$$

where the parameters a, b depend on the series and number of the line. For short-range radiation this formula has the form

$$\nu = \frac{3}{4} Ry(Z - 1)^2, \quad (1.15)$$

where Ry is the Rydberg constant. Thus Moseley concluded on the basis of experiments the existence of an atom fundamental value which increases monotonically from element to the following one. He stated that this value can only be the nuclei charge. Therefore, this analysis allows one to show the internal atom structure.

Developing the Moseley investigations, Kassel (England) in 1914 showed that the Moseley results correspond to the Rutherford–Bohr atom model. Locating electrons in the atom on individual orbit-rings, he concluded that the K_α-radiation corresponds to the transition from the state-ring with $n = 2$ to the ring with $n = 1$, the K_β-series corresponds to the transition between $n = 3$ and $n = 1$, and the L_α-series corresponds to the transition between $n = 3$ and $n = 2$. In this case the combination principle must be fulfilled, which is similar to the corresponding principle for spectral lines and has the form

$$\nu_{K_\beta} - \nu_{K_\alpha} = \nu_{L_\alpha}. \quad (1.16)$$

Experimental data confirmed this law with an admissible accuracy. Thus, the Moseley law and its development corroborated the validity of the Rutherford–Bohr atom model for internal electrons. These investigations exposed the internal atom structure and were of importance for the understanding of the atom's nature.

The most importance for corroboration of the Bohr atom model were the results of the experiment by James Franck and Gustav Hertz (Germany) which they made in 1915. They constructed a specific gas discharge with the use of a discharge tube whose cathode had a hole, so that part of the electrons could penetrate to the other side of the cathode. These electrons were accelerated by a voltage V between the cathode and one more electrode—an anode. Further electrons were gathered by a collector, and the electron current between the cathode and collector was measured. A discharge tube was filled with mercury vapor at a low pressure. At the first stage of study, the authors wanted to lock the electron current to the collector. Then measurement of the ion current would allow one to determine the threshold of the formation of electrons and ions. But this was not realized. Then the electric current versus the voltage between the cathode and anode was measured.

Assuming that the anode does not influence the number of electrons which penetrate through the cathode, one can obtain that, in the absence of collisions in the tube, the electron current is proportional to the electron velocity, i.e., $V^{1/2}$, where V is the electric potential between electrodes. In reality, a stronger elastic scattering of slow electrons on atoms compared to fast ones makes this dependence a stronger one. Let us take the vapor pressure such that the elastic collisions of electrons with atoms are weak. The cross section of the excitation of the resonance

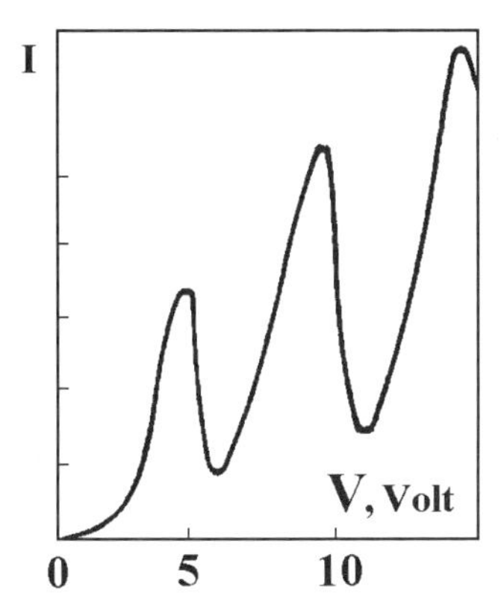

Figure 1.2. The dependence of the collector current I on the accelerating voltage V in the Frank–Hertz experiment.

state of the mercury atom has maximum at energy ε_o which is close to the threshold energy. Assuming the cross section of atom excitation to be more than the elastic cross section of electron–atom scattering, one can obtain an oscillation form for the current–voltage dependence (see Fig. 1.2). The first local minimum occurs at the voltage which leads to the electron energy near ε_o because electrons which reach this energy, lose it as a result of an inelastic collision with a mercury atom. These electrons do not make a contribution to the electron current. The following local minimum occurs at the energy near $2\varepsilon_o$ due to the loss of two electrons. This local minimum is expected to be weaker than the first one.

Although by mistake, the authors identified the measured excitation energy with the ionization potential of the mercury atom, this experiment was of importance because it allowed us to determine the connection between atomic parameters by a new method. The measured value ε_o occurred at 4.9 eV and its error was estimated as 0.1 eV. The measured energy was close to the photon energy of wave length 253.6 nm that was observed in the mercury atom spectrum and corresponds to the excitation from the ground state of the mercury atom. Such a coincidence induced the authors to make a direct experiment for the simultaneous determination of "the ionization potential" and the frequency of photons resulted from this process. The experiment was made in a quartz pipe with a platinum incandescent filament and a platinum anode. Mercury introduced into this pipe was heated to a temperature of 150° C for the formation of vapor. Radiation of vapor was detected by a UV-spectrometer. As the voltage between the wire and anode reached approximately 4.9 eV, photons of a wavelength of 253.6 nm were detected. At lower voltages these photons were not observed. Thus, this experiment established the connection between the electron units of energy (electronvolt) and the radiative ones. Note that the above picture of the process is rough enough and is not universal. Therefore the method of the first experiment was not used for the determination of the atomic parameters. Nevertheless, at that time, this experiment confirmed the correctness of the Bohr atom model because it showed the quantum character of the electron binding energy, and the second experiment gave a direct connection between the electron and photon units of energy.

Another argument in favor of the Bohr model was the experiment by Theodor Lyman (USA) in 1916. He found a new series of spectral lines of the hydrogen atom which was predicted by the Bohr theory and corresponds to transitions from its ground state. This series is called the "Lyman series." The development of the Bohr theory allowed us to explain the splitting of spectral lines in electric and magnetic fields (the Zeeman, Pashen, and Stark effects), the structure of more complex atoms, and other observed facts of the atom structure. Thus, the Bohr model became the turning point in the evolution of atomic physics.

1.8 The Development of the Bohr Atom Model

The Bohr atom model gave a lift to the development of atomic physics, and in this improved form it allowed us to explain the various details of atomic spectra. According to one of the Bohr postulates, the electron rotation momentum in the hydrogen atom is the quantum value which is a divisible quantity with respect to $h/(2\pi) = \hbar$. On the basis of the Bohr model, it explained the Zeeman experiment for the splitting of the spectral lines of atoms in a magnetic field. But this effect was also described by Lorenz on the basis of the classical motion of the electron in a magnetic field. This means that the Zeeman effect has a classical nature. Further, the problem of the quantum character of the electron momentum arose from the analysis of the multiplet structure of atomic spectra. In order to show the behavior of the electron moment, Otto Stern (Germany) suggested in 1921 the experiment scheme in which results were obtained in 1922 together with Walter Gerlach. This experiment was of importance, and we consider it below.

The experiment by Stern and Gerlach was based on the following considerations. If the atom magnetic moment is the quantum value, its projection onto the magnetic field direction is a discrete value. Then, if an atom with a magnetic moment moves in a nonuniform magnetic field, the force acted on the atom, due to nonhomogeneity of the magnetic field, depends on the moment projection. Creating this force to be directed perpendicularly with respect to the motion of an atomic beam, one can separate the beam into several beams. If a plate is placed in the way of the beam, the trace of the beam will have the form of separate spots which are spread due to different atom velocities in the beam. In the case of the classical behavior of the atom magnetic moment the trace on a plate must have the form of a band with the maximum thickness corresponding to the initial direction of the beam.

This experiment was made for a beam of silver atoms. This beam was split into a nonuniform magnetic field with a force which was directed perpendicular to the beam. As a result, a beam was split in two parts after passage of the region of a nonuniform magnetic field. The beam created two spots on a plate, and the minimum was observed in the initial beam direction. This proved the quantum nature of the atom magnetic moment, but the problem appears to be more complex.

A more detailed analysis showed that the magnetic momentum of silver atoms is not due to the rotation of electrons in the Coulomb field of the nucleus. Thus, along with the evidence of the quantum nature of the atom magnetic momentum, this experiment also testified to the existence of an internal magnetic momentum of the electron—the electron spin. The existence of the electron spin also followed from the analysis of the multiplet structure of atomic spectra and the behavior of atomic spectra in a magnetic field. Electron spin was included in theoretical atomic physics by George Eugene Uhlenbeck and Abraham Goudsmith (Netherlands) in 1925.

In 1925 Wolfgang Pauli (Switzerland) suggested the exclusion principle for atomic electrons. According to this principle, which is usually known as the Pauli exclusion principle, two electrons cannot be found in the same state. The electron state means the electron position in a space and a direction of its spin. In the one-electron model, where interactions in an atom are reduced to an effective interaction of each electron with the atomic core, this principle states that all the atomic electrons are characterized by different quantum numbers. One of these quantum numbers is the electron spin. The Pauli exclusion principle is connected with the statistics of Fermi–Dirac for particles having a semi-integer spin. This requires that the total wave function of electrons must be antisymmetric with respect to the transposition of two electrons (i.e., by exchange of their positions in a space and exchange by spins), so that the wave function changes sign at such a transposition. From this follows the repulsion of two electrons in an atom at their approach. Such an interaction is called the exchange interaction and is of importance for atom properties. The contemporary description of the physics of the atom is based on the Bohr concepts of quantum atomic physics and the Pauli exclusion principle, so that in this book we will use them for a description of the properties of atoms and ions.

The Pauli exclusion principle gave the possibility of analyzing many-electron atoms which were outside the Bohr model of the one-electron atom. This, together with the vector model of summation of electron momenta including the electron spins, allowed one to create the self-consistent strict theory of atomic spectra which can confirm and explain the observational data. Up to the end of 1925, the construction of the atom model was finished. At that time all the principal elements of the atom nature which are the basis of the contemporary physics of atoms and atomic particles were understood.

Since the process of the creation of the atomic theory was based on rich experimental material, atomic physics was of importance for the creation of the quantum theory. In 1926 the formalism of quantum mechanics was formulated in both matrix form or operator mathematics, and in the form of the Schrödinger equation. In any case, the check on the validity of the methods of quantum mechanics was made on the basis of the properties of atomic objects. Agreement of the evidence for atomic properties and the atomic theory allowed one, in the end, to change the Bohr postulates by the logical, self-consistent and strict quantum theory of atomic particles. Thus, the development of the atomic theory was of interest not only for atomic objects. It played a key role in the creation of quantum physics that is one of the important advantages of science of twentieth century.

1.9 The Quantum Mechanics of Atoms

Now let us formulate the contemporary understanding and description of the physics of atomic particles. We introduce the mathematical formalism of quantum mechanics which will be used throughout this book. Its basis is the Schrödinger equation for the electron wave function Ψ which has the form

$$\hat{H}\Psi = E\Psi, \tag{1.17}$$

where $\hat{H}$ is the atom Hamiltonian, Ψ is the atom wave function, and E is the atom energy which is the eigenvalue of the Schrödinger equation. Considering the atom as a system of electrons which are placed in the Coulomb field of the nucleus of infinite mass we have, for the atom Hamiltonian,

$$\hat{H} = \sum_i \left(-\frac{\hbar^2}{2m}\Delta_i\right) - \sum_i \frac{Ze^2}{r_i} + \sum_{i,k} \frac{e^2}{|\mathbf{r}_i - \mathbf{r}_k|}, \tag{1.18}$$

where $\hbar$ is the Plank constant, m, e are the electron mass and charge, Z is the ratio of the nucleus charge to the electron charge, $\mathbf{r}_i$ is the coordinate of the ith electron with the nucleus as the origin, and Δ_i is the Laplacian of the ith electron. The first term of the Hamiltonian (1.18) corresponds to the kinetic energy of electrons, the second term describes the interaction of electrons with the nucleus Coulomb field, and the third term accounts for the interaction between electrons. Equation (1.17) with the Hamiltonian (1.18) gives the total information about the atom.

The Pauli exclusion principle is of importance for atomic properties. It prohibits the location of two electrons in the same state and leads to the following symmetry of the wave function of electrons

$$\begin{aligned}&\Psi(\mathbf{r}_1, \sigma_1; \mathbf{r}_2, \sigma_2; \ldots, \mathbf{r}_i, \sigma_i; \ldots, \mathbf{r}_k, \sigma_k; \ldots, \mathbf{r}_n, \sigma_n)\\ &\quad = -\Psi(\mathbf{r}_1, \sigma_1; \mathbf{r}_2, \sigma_2; \ldots, \mathbf{r}_k, \sigma_k; \ldots, \mathbf{r}_i, \sigma_i; \ldots, \mathbf{r}_n, \sigma_n),\end{aligned} \tag{1.19}$$

where $\mathbf{r}_i$ is the coordinate of ith electron, σ_i is its spin projection onto a given direction, and relation (1.19) accounts for the symmetry properties of a system of Fermi particles such that transposition of two particles changes the sign of the wave function. This form of wave function creates an additional interaction in the atom which is called an exchange interaction and will be the object of subsequent analysis. In particular, from this relation it follows that if two electrons with an identical spin state are located in the same point of space, the wave function is equal to zero. The Schrödinger equation (1.17) with the electron Hamiltonian (1.18), and the symmetry of the wave function (1.19) which follows from the Pauli exclusion principle, is the basis for further analysis of the physics of atoms and ions.

1.10 Radiative Transitions between Discrete States of Atoms

The radiative properties of atoms are a matter of principle for the physics of atoms. Atomic spectra determine the positions of excited levels of atoms and ions with high accuracy. This gives much information concerning atomic particles. Hence, the parameters of atomic radiative transitions play a key role in the analysis of the physics of atoms. Below we obtain an expression for the rate of radiative transition between two discrete atomic states which will be used for subsequent analysis.

A radiative transition between two atom states is a result of the interaction between the related atom and electromagnetic field of radiation. Let us consider this process within the framework of the perturbation theory assuming the interaction between the atom and radiation field as a perturbation. This is valid if a typical time of transition between states of the atomic system is large compared to typical atomic times. This criterion is fulfilled if radiation fields are not strong and do not exceed typical atomic fields. For spontaneous radiation it is valid. In addition, the small parameter $e^2/(\hbar c)$ is of importance for radiative transitions. This means that typical atomic velocities are small compared to the light velocity, and due to this criterion an atom is a nonrelativistic system. The simplest form of the interaction operator between the atom and radiation field is the following:

$$V = -\mathbf{ED}, \tag{1.20}$$

where $\mathbf{E}$ is the strength of the radiation electromagnetic field and $\mathbf{D}$ is the operator of the atom dipole moment. This is the strongest interaction between the atom and the field. We start from the analysis of the absorption process

$$A_j + \hbar\omega \rightarrow A_f, \tag{1.21}$$

where the subscripts mean atom states. Below we determine the rate of this process by standard methods. Take the time dependence of the electric field strength as $\mathbf{E} = \mathbf{E}_\omega \cos\omega t$, where ω is the frequency of the electromagnetic field. Using the standard nonstationary perturbation theory, we present the atom wave function in the form

$$\Psi = \psi_j \exp(-i\varepsilon_j t/\hbar) + c_f \cdot \psi_f \exp(-i\varepsilon_f t/\hbar),$$

where ψ_j and ψ_f are eigenfunctions of the nonperturbed Hamiltonian $\hat{H}_o$ with eigenvalues ε_j and ε_f, and $c_f \ll 1$. Then from the Schrödinger equation,

$$i\hbar\frac{\partial\Psi}{\partial t} = (\hat{H}_o + V)\Psi,$$

there follows the equation for the transition amplitude

$$i\hbar\dot{c}_f = V_{jf}\exp(-i\omega_o t) = \langle j|\mathbf{E}_\omega\mathbf{D}|f\rangle\exp(-i\omega_o t),$$

where $\hbar\omega_o = \varepsilon_f - \varepsilon_j$, and the matrix element is taken between the initial (j) and final (f) atomic states of the process (1.21). Note that in this book we denote matrix elements as $\int \psi_j^*(\mathbf{r})\hat{a}\psi_j(\mathbf{r})\,d\mathbf{r}$ by $(\hat{a})_{jf}$ or $\langle j|\hat{a}|f\rangle$, what is more convenient. The obtained expression yields, for the probability $|c_f|^2$ of the process for large times $t \gg \omega^{-1}$,

$$|c_f|^2 = \frac{1}{\hbar^2} |\langle j| \mathbf{E}_\omega \mathbf{D} |f\rangle|^2 \frac{\sin^2[(\omega - \omega_o)t/2]}{(\omega - \omega_o)^2}$$

We neglect $(\omega + \omega_o)^{-1}$ compared to $(\omega - \omega_o)^{-1}$ because $\omega \approx \omega_o$. In the limit $\omega \to \omega_o$ one can introduce the delta-function $\delta(x)$ which satisfies the following relations

$$\delta(0) = \infty; \qquad \delta(x \neq 0) = 0; \qquad \int_{-\infty}^{\infty} \delta(x)\,dx = 1.$$

Then the probability of the related transition per unit time is equal to

$$w_{jf} = \frac{|c_f|^2}{t} = \frac{\pi}{2\hbar} |\langle i |\mathbf{E}_\omega \mathbf{D}| f\rangle|^2 \delta\left[\hbar\omega - (\varepsilon_f - \varepsilon_i)\right].$$

The radiation energy per unit time is

$$\left\langle \frac{E^2}{8\pi} \right\rangle + \left\langle \frac{H^2}{8\pi} \right\rangle = \left\langle \frac{E^2}{4\pi} \right\rangle = \frac{E_\omega^2}{8\pi},$$

where the angle brackets mean averaging over time. Let us introduce a number of photons in one state n_ω and connect it with the average field energy, assuming the radiation field to be concentrated in an interval $d\omega$. We have

$$\frac{E_\omega^2}{8\pi} = \hbar\omega \cdot n_\omega \cdot \frac{2d\mathbf{k}}{(2\pi)^3} = n_\omega \cdot \frac{\omega^3\, d\hbar\omega}{\pi^2 c^3},$$

where $\mathbf{k}$ is the wave vector of the electromagnetic wave. We account for two polarizations of the electromagnetic wave and the dispersion relations for it $\omega = kc$, where c is the light velocity. On the basis of this relation we obtain, for the probability of photon absorption per unit time after averaging over frequencies,

$$w_{jf} = \frac{4\omega^3}{\hbar c^3} \cdot n_\omega \cdot |\langle i |\mathbf{Ds}| f\rangle|^2, \tag{1.22}$$

where $\mathbf{s}$ is a unit vector which characterizes a photon polarization ($\mathbf{E}_\omega = \mathbf{s}E_\omega$). Averaging over polarizations of the radiation field and summation over the final atom states leads to the following expression for the radiation rate:

$$w_{jf} = \frac{4\omega^3}{3\hbar c^3} \cdot |\langle i |\mathbf{D}| f\rangle|^2 g_f \cdot n_\omega, \tag{1.23}$$

where g_f is the statistical weight of the final state.

Usually the rates of the radiative transitions are expressed through the Einstein coefficients A and B which are introduced by the relations

$$w(j, n_\omega \to f, n_\omega - 1) = A \cdot n_\omega; \qquad w(f, n_\omega \to j, n_\omega + 1) = \frac{1}{\tau_{fj}} + B \cdot n_\omega, \tag{1.24}$$

where τ_{fj} is the radiative lifetime of a state f with respect to the spontaneous radiative transition to a state j. Note that formula (1.23) gives the expression for A. From the condition of the thermodynamic equilibrium of photons with atom states i and f, which reflects the detailed balancing principle for radiative transitions in atoms, it follows that

$$\frac{1}{\tau_{fj}} = B = \frac{g_j}{g_f} A, \tag{1.25}$$

where g_j, g_f are the statistical weights of the lower and upper states of transition. This leads to the following expressions for the Einstein coefficients:

$$A = \frac{4\omega^3}{3\hbar c^3} \cdot |\langle j |\mathbf{D}| f\rangle|^2 g_f; \qquad B = \frac{4\omega^3}{3\hbar c^3} \cdot |\langle j |\mathbf{D}| f\rangle|^2 g_j. \tag{1.26}$$

Along with the value of the rate of radiative transitions, formulas (1.23), (1.26) lead to selection rules, i.e., we choose states between which radiative transitions are possible. The relevant case corresponds to the strongest interaction between radiation and the atomic systems. It chooses the so-called "dipole permitted radiative transitions" or "dipole transitions." The matrix element of the operator of the atom dipole moment is not zero for these transitions, and they are the main part of the observed spectra of atoms and ions. We further consider various cases of dipole radiative transitions for atomic particles.

1.11 Radiative Transitions of Atoms Involving Continuous Spectra

Along with the radiative transitions between discrete states of atomic particles, radiative processes with formation or loss of electrons are of interest. As an example of such processes, we consider below the atom photoionization process which proceeds according to the scheme

$$\hbar\omega + A \to A^+ + e. \tag{1.27}$$

The energy conservation law during this process has the form

$$\hbar\omega = J + \frac{\hbar^2 q^2}{2m}, \tag{1.28}$$

where $\hbar\omega$ is the photon energy, J is the atom ionization potential, the atom ionization potential $\hbar^2 q^2/2m$ is the energy of a released electron so that the electron momentum is $p = \hbar/q$, and q is the electron wave vector. The characteristic of this process is the cross section of the process which is the ratio of the probability of this transition per unit time to the flux of incident photons cN_ω, i.e., the photoionization cross section is

$$d\sigma_{\text{ion}} = A_{of}\, n_\omega/(cN_\omega),$$

where n_ω is a number of photons of a given frequency located in one state, N_ω is the number density of these photons, c is the light velocity, and A_{of} is the probability of the related process per unit time which is given by formula (1.26).

Let us use a standard method in order to tranfer from the discrete spectra of photons and electrons to continuous spectra. Introduce a volume Ω which is large compared to a typical atomic volume, and locate inside this volume the atom and photons. Then the total number of photons in this volume, in an interval of frequencies from ω to $\omega + d\omega$, is equal to

$$\Omega N_\omega = \frac{2\int d\mathbf{r}\, d\mathbf{k}}{(2\pi)^3} n_\omega = \Omega \frac{\omega^2\, d\omega}{\pi^2 c^3} n_\omega,$$

where the factor 2 accounts for two possible photon polarizations, $\Omega = \int d\mathbf{r}$, and $\mathbf{k}$ is the photon wave vector which is connected to the photon frequency by the dispersion relation $\omega = kc$. Thus, we have, for the photoionization cross section,

$$d\sigma_{\text{ion}} = A_{of} \frac{\pi^2 c^2}{\omega^2\, d\omega},$$

where A_{of} is the probability of transition per unit time from the initial atom state o in the group of states of continuous spectra which are considered to be discrete due to the introduction of a finite system volume Ω. Let us summarize the cross section over the final states based on formula (1.26) for the rate of transition. We have

$$d\sigma_{\text{ion}} = \frac{4\pi^2 \omega}{3\hbar c\, d\omega} \sum_f |\langle o\, |\mathbf{D}|\, f\rangle|^2 .$$

Summation over the final states of a free electron has the form

$$\sum_f = \int \frac{d\mathbf{r}\, d\mathbf{q}\, d\Theta_\mathbf{q}}{(2\pi)^3} = \Omega \frac{q^2\, dq\, d\Theta_\mathbf{q}}{8\pi^3},$$

where $d\Theta_\mathbf{q}$ is a solid angle which characterizes the direction of a released electron. The energy conservation law (1.28) gives $q\, dq = m\, d\omega/\hbar$. Next, the wave function of the released electron in the main part of an introduced volume has the form $\Omega^{-1/2}\exp(i\mathbf{qr})$. It is convenient to define the matrix element $\langle o\, |\mathbf{D}|\, \mathbf{q}\rangle$ such that the wave function of the released electron has the form of a plane wave $e^{i\mathbf{qr}}$. Transition to this wave function requires us to multiply the above expression for the cross section by the factor $1/\Omega$. Thus, we obtain the following expression for the photoionization cross section:

$$d\sigma_{\text{ion}} = \frac{mq\omega}{6\pi c\hbar^2} |\langle o\, |\mathbf{D}|\, \mathbf{q}\rangle|^2\, d\Theta_\mathbf{q}. \tag{1.29}$$

Note that the internal states of a formed ion and electron correspond to the internal quantum numbers of the atom initial state.

Problems

Problem 1.1. *Consider an atom within the framework of the one-electron atom model according to which an atom includes one valent electron moving in a spherical symmetric self-consistent field of the atomic core. Find the sums* $\sum_f \omega_{if}^n \left|\mathbf{r}_{if}\right|^2$ *for several values n, where* $\hbar\omega_{if}$ *is the difference of the energies for these two states and* $\mathbf{r}$ *is the electron radius-vector.*

This problem allows us to demonstrate the general methods of quantum mechanics. The transition from classical to quantum mechanics corresponds to the change of a physical parameter by its matrix element. Correspondingly, some relations of classical mechanics are the same in matrix form in quantum mechanics. In particular, let us prove that the relation having the form $dx/dt = p_x/m$ in classical mechanics corresponds to the relation $-i\omega_{if}x_{if} = (\hat{p}_x)_{if}$ in quantum mechanics, where $\hat{p}_x = \hbar/i \cdot \partial/\partial x$ is the operator of the linear electron momentum. Indeed, the Hamiltonian of the electron located in the field of the atomic core has the form

$$\hat{H} = -\frac{\hbar^2}{2m}\left(\frac{\partial^2}{\partial x^2} + \frac{\partial^2}{\partial y^2} + \frac{\partial^2}{\partial z^2}\right) + V(r) = \frac{1}{2m}\left(\hat{p}_x^2 + \hat{p}_y^2 + \hat{p}_z^2\right) + V(r),$$

and the family of eigenfunctions of the Hamiltonian ψ_j satisfies the Schrödinger equation $\hat{H}\psi_j = \varepsilon_j\psi_j$. We introduce ω_{jf} as $(\varepsilon_j - \varepsilon_f)/\hbar$. Hence,

$$\begin{aligned}\hbar\omega_{jf}x_{jf} &= \varepsilon_j x_{if} - \varepsilon_f x_{if} = \langle j|\hat{H}x|f\rangle - \langle j|x\hat{H}|f\rangle \\ &= -\frac{\hbar^2}{2m}\left\langle j\left|x\frac{\partial^2}{\partial x^2} - \frac{\partial^2}{\partial x^2}x\right|f\right\rangle = -\frac{\hbar^2}{m}\left\langle j\left|\frac{\partial}{\partial x}\right|f\right\rangle = -\frac{i\hbar}{m}\left(\hat{p}_x\right)_{jf},\end{aligned}$$

i.e.,

$$\left(\hat{p}_x\right)_{jf} = -im\omega_{jf}x_{jf}.$$

Below we also use the regular summation of the matrix elements, so that $\sum_f a_{jf}b_{fj} = (ab)_{jj}$. On the basis of the above relations we have

$$\sum_f (x_{jf})^2 = (x^2)_{jj} = \tfrac{1}{3}(r^2)_{jj}. \tag{1.30}$$

Next,

$$\sum_f \omega_{jf}(x_{jf})^2 = \frac{i}{m}\sum_f (\hat{p}_x)_{if}x_{jf} = \frac{i}{m}(\hat{p}_x x)_{jj}.$$

In the same way we obtain

$$\sum_f \omega_{jf}(x_{jf})^2 = -\sum_f \omega_{fj}(x_{jf})^2 = -\frac{i}{m}(x\hat{p}_x)_{jj}.$$

Summation of these relations yields

$$\sum_f \omega_{jf}(x_{jf})^2 = \frac{i}{2m}(\hat{p}_x x - x\hat{p}_x)_{jj} = \frac{\hbar}{2m}. \tag{1.31}$$

Take the subsequent relation

$$\sum_f \omega_{jf}^2 (x_{jf})^2 = \frac{1}{m^2} \sum_f (\hat{p}_x)_{jf} (\hat{p}_x)_{fj} = \frac{1}{m^2} (\hat{p}_x^2)_{jj}.$$

Because of the spherical symmetry of this problem $(\hat{p}_x^2)_{jj} = (\hat{p}_y^2)_{jj} = (\hat{p}_z^2)_{jj}$, we have $(\hat{p}_x^2)_{jj} = (2m/3)(\varepsilon_j - V_{jj})$, where $V_{jj} = \langle j\, |V|\, j\rangle$. Thus, we have

$$\sum_f \omega_{jf}^2 (x_{jf})^2 = \frac{2}{3m} (\varepsilon_j - V_{jj}). \tag{1.32}$$

For the next sum $\sum_f \omega_{jf}^3 (x_{jf})^2$ we use the classical relation $m\, d^2x/dt^2 = F = -\partial V/\partial x$, so that the matrix form of this expression is the following:

$$\omega_{jf}^2 x_{jf} = \frac{1}{m} \left(\frac{\partial V}{\partial x} \right)_{jf}.$$

Hence

$$\begin{aligned}
\sum_f \omega_{jf}^3 (x_{jf})^2 &= \frac{i}{m^2} \sum_f (\hat{p}_x)_{jf} \left(\frac{\partial V}{\partial x} \right)_{fj} = \frac{i}{m^2} \left((\hat{p}_x) \frac{\partial V}{\partial x} \right)_{jj} \\
&= -\frac{i}{m^2} \left(\frac{\partial V}{\partial x} \hat{p}_x \right)_{jj} = \frac{i}{m^2} \left((\hat{p}_x) \frac{\partial V}{\partial x} - \frac{\partial V}{\partial x} \hat{p}_x \right)_{jj} \\
&= \frac{\hbar}{2m^2} \left(\frac{\partial^2 V}{\partial x^2} \right)_{jj} = \frac{\hbar}{6m^2} (\Delta V)_{jj}.
\end{aligned}$$

From the Poisson equation it follows that $\Delta V = 4\pi e^2 \rho(\mathbf{r})$, where $\rho(\mathbf{r})$ is the charge density of the atomic core. Thus, we find

$$\sum_f \omega_{jf}^3 (x_{jf})^2 = \frac{2\pi\hbar}{3m^2} \langle j\, |\rho(\mathbf{r})|\, j\rangle.$$

In particular, if the field of the atomic core is determined by the Coulomb field of the nucleus of a charge Z, i.e., $\rho(\mathbf{r}) = Z\delta(\mathbf{r})$, this relation is transformed to

$$\sum_f \omega_{jf}^3 (x_{jf})^2 = \frac{2\pi\hbar}{3m^2} \left| \psi_j(0) \right|^2, \tag{1.33}$$

where $\psi_j(\mathbf{r})$ is the wave function of the related state.

The last evaluated sum is

$$\begin{aligned}
\sum_f \omega_{jf}^4 (x_{jf})^2 &= \frac{1}{m^2} \sum_f \left(\frac{\partial V}{\partial x} \right)_{jf} \left(\frac{\partial V}{\partial x} \right)_{fj} = \frac{1}{m^2} \left\langle j \left| \left(\frac{\partial V}{\partial x} \right)^2 \right| j \right\rangle \\
&= \frac{1}{3m^2} \langle j\, |(\nabla V)^2|\, j\rangle.
\end{aligned} \tag{1.34}$$

Note that near the nucleus $\nabla V = Ze\mathbf{r}/r^3$, where Z is the nucleus charge, so that this matrix element is $\sim r^{-4}$ at small r. Hence, if the wave function of a related state j is const at small r, this sum diverges. Practically, this sum does not diverge if the electron angular momentum is not zero.

Problem 1.2. *In the classical limit find the radiation intensity resulting from the interaction of an atomic particle with the field of another particle.*

Using formulas (1.25), (1.26) we have, for the intensity of radiation as a result of the interaction of an atomic particle with the field of a structureless particle, when the internal state of the radiating particle is not changed ($g_f = 1$)

$$I_j = \sum_{\varepsilon_j > \varepsilon_f} \frac{\hbar\omega}{\tau_{fj}} = \sum_{\varepsilon_j > \varepsilon_f} \frac{4\omega^4}{3c^3} |\langle j |\mathbf{D}| f\rangle|^2 = \frac{4}{3c^3} \sum_{\varepsilon_j > \varepsilon_f} \left| \left\langle j \left| \frac{d^2\mathbf{D}}{dt^2} \right| f \right\rangle \right|^2,$$

where we use the relation for matrix elements $\left\langle j \left| d^2\mathbf{D}/dt^2 \right| f \right\rangle = -\omega^2 \langle j |\mathbf{D}| f\rangle$. Summation is made over states whose energy is less than the energy of the initial state.

We use that, in the classical limit, transitions proceed in close states by energy. Hence, the sum is symmetric with respect to $\varepsilon_j > \varepsilon_f$ and $\varepsilon_j < \varepsilon_f$, so that

$$\sum_{\varepsilon_j > \varepsilon_f} \left| \left\langle j \left| \frac{d^2\mathbf{D}}{dt^2} \right| f \right\rangle \right|^2 = \frac{1}{2} \sum_f \left| \left\langle j \left| \frac{d^2\mathbf{D}}{dt^2} \right| f \right\rangle \right|^2 = \frac{1}{2} \left\langle j \left| \left(\frac{d^2\mathbf{D}}{dt^2} \right)^2 \right| j \right\rangle.$$

Next, in the classical limit the diagonal matrix element of an operator coincides with its physical value, i.e., we have, for the radiation intensity,

$$I = \frac{2}{3c^3} \left(\frac{d^2\mathbf{D}}{dt^2} \right)^2. \tag{1.35}$$

CHAPTER 2

The Hydrogen Atom

2.1 The System of Atomic Units

The goal of this book is to analyze the properties of atomic particles as systems consisting of heavy charged nuclei and electrons. Because atoms and ions are quantum systems, their description is based on the Schrödinger equation for atomic electrons. Let us start from the simplest atomic system—the hydrogen atom. This system consists of one bound electron which is located in the Coulomb field of a charged nucleus. At first, for simplicity, we assume the nuclear mass to be infinite. Then the behavior of one bound electron is described by the following Schrödinger equation:

$$-\frac{\hbar^2}{2m}\Delta\Psi - \frac{e^2}{r}\Psi = \varepsilon\Psi, \tag{2.1}$$

where r is the distance of the electron from the center, $\hbar$ is the Plank constant, e is the electron charge, m is the electron mass, and ε is the electron energy which is the eigenvalue of this equation.

Equation (2.1) contains the three-dimensional parameters: $\hbar = 1.05457 \cdot 10^{-34}\,\mathrm{J \cdot s}$, $e = 1.60218 \cdot 10^{-19}\,\mathrm{C}$, and $m = 9.10939 \cdot 10^{-31}\,\mathrm{kg}$. One can compose only one combination of any dimensionality from these parameters. Values of various dimensionalities constructed from the above parameters form the so-called system of atomic units. Some values of the system of atomic units are given in Table 2.1. This system is convenient for the analysis of atomic particles because their parameters usually are of the order of typical atomic values or expressed through them. Hence, atomic units will be used throughout most of this book. In particular, note that a typical atomic velocity is of the order of $e^2/\hbar = 2 \cdot 10^6\,\mathrm{m/s}$, that is, remarkably smaller that the light velocity $c = 3 \cdot 10^8\,\mathrm{m/s}$. This means the possibility in the first approach of neglecting the relativistic effects for the description

Table 2.1. The system of atomic units.

Parameter	Symbol, expression	Value
Length	$a_o = \hbar^2/(me^2)$	$5.2918 \cdot 10^{-11}$ m
Velocity	$v_o = e^2/\hbar$	$2.1877 \cdot 10^{6}$ m/s
Time	$\tau_o = \hbar^3/(me^4)$	$2.4189 \cdot 10^{-17}$ s
Frequency	$\nu_o = me^4/\hbar^3$	$4.1341 \cdot 10^{16}$ s^{-1}
Energy	$\varepsilon_o = me^4/\hbar^2$	$4.3598 \cdot 10^{-18}$ J
Power	$\varepsilon_o/\tau = m^2e^8/\hbar^5$	0.180 W
Electric potential	$\varphi_o = me^3/\hbar^2$	27.212 V
Electric field strength	$E_o = me^5/\hbar^4$	$5.1422 \cdot 10^{11}$ V/m
Linear momentum	$p_o = me^2/\hbar$	$1.9929 \cdot 10^{-24}$ kg $\cdot$ m/s
Number density	$N_o = a_o^{-3}$	$6.7483 \cdot 10^{30}$ m^{-3}
Volume	$V_o = a_o^3$	$1.4818 \cdot 10^{-31}$ m^3
Cross section	$\sigma_o = a_o^2$	$2.8003 \cdot 10^{-21}$ m^2
Rate constant	$k_o = v_o a_o^2 = \hbar^3/(m^2e^2)$	$6.126 \cdot 10^{-15}$ m^3/s
Electric current	$I = e/\tau = me^5/\hbar^3$	$6.6236 \cdot 10^{-3}$ A
Particle flux	$j_o = N_o v_o = m^3e^8/\hbar^7$	$1.476 \cdot 10^{37}$ m^{-2}s^{-1}
Electric current density	$i_o = eN_o v_o = m^3e^9/\hbar^7$	$2.3653 \cdot 10^{18}$ A/m^2
Energy flux	$J = \varepsilon_o N_o v_o = m^4e^{12}/\hbar^9$	$6.436 \cdot 10^{19}$ W/m^2

of atomic particles until the relevant properties are determined by valent electrons. This corresponds to the usage of a small parameter

$$\alpha = \frac{e^2}{\hbar c} = 0.007295 = (137.036)^{-1}. \tag{2.2}$$

The small parameter (2.2) is named the constant of fine structure. This small parameter allows us to consider the hydrogen atom as a nonrelativistic system.

2.2 Electron States of the Hydrogen Atom

The electron states of the hydrogen atom include both a spatial electron distribution and its spin state. In the nonrelativistic approach the electron spin state does not correlate with its spatial distribution. Therefore spin coordinates of the electron are not included in the Schrödinger equation (2.1). This allows one to describe the electron spin state by a spin projection onto a given direction $\sigma = \pm\frac{1}{2}$. Then the spin state does not depend on a spatial wave function of the electron.

The variables of the Schrödinger equation (2.1) are separated into spherical coordinates. Then the spatial electron wave function has the form

$$\Psi(r, \theta, \varphi) = R(r) Y_{lm}(\theta, \varphi), \tag{2.3}$$

where the origin is located in the nucleus, r is a distance of the electron from the center, θ is the polar angle, and φ is the azimutal angle. The angular wave function satisfies the Schrödinger equation

$$\frac{\partial}{\partial \cos\theta}\left(\sin^2\theta \frac{\partial Y_{lm}}{\partial \cos\theta}\right) + \frac{1}{\sin^2\theta}\frac{\partial^2 Y_{lm}}{\partial \varphi^2} + l(l+1)Y_{lm} = 0 \tag{2.4}$$

and is given by the expression

$$Y_{lm}(\theta, \varphi) = \left[\frac{2l+1}{4\pi}\frac{(l-m)!}{(l+m)!}\right]^{1/2} P_l^m(\cos\theta)\exp(im\varphi). \tag{2.5}$$

Here the electron angular momentum l can be a whole number

$$l = 0, 1, 2, \ldots, \tag{2.6}$$

and the momentum projection m can have the following values:

$$m = -l, -(l-1), \ldots, l-1, l. \tag{2.7}$$

In formula (2.5) $P_l^m(\cos\theta\}$ is the Legendre function and the electron angular wave function is normalized in the following way:

$$\int_{-1}^{+1}\int_0^{2\pi} |Y_{lm}(\theta, \varphi)|^2 \, d\cos\theta d\varphi = 1. \tag{2.8}$$

The angular wave functions are given in Table 2.2 for small values of l and m. Note that the requirement of the whole value for the quantum number m can be obtained from expression (2.5) for the angular wave function and from the physical condition $Y_{lm}(\theta, \varphi) = Y_{lm}(\theta, \varphi + 2\pi)$. Then l is introduced as a possible maximum m in accordance with (2.7), and l can have only positive whole values.

Being expressed in atomic units, l is the orbital electron moment and m is its projection onto a given direction. It is of importance that l and m are quantum numbers of the electron, i.e., they can have only certain values. The radial electron wave function according to equation (2.1) and expansion (2.3) satisfies the Schrödinger equation

$$\frac{1}{r}\frac{d^2}{dr^2}(rR) + \left[2\varepsilon + \frac{2}{r} - \frac{l(l+1)}{r^2}\right]R = 0. \tag{2.9}$$

Note that the usage of atomic units in equation (2.9) simplifies the obtained expression.

The analysis of this equation allows one to determine the energy spectrum of the electron. In bound states the electron motion is restricted by a finite space region. Hence, at large distances from the atom center one can neglect the two last terms of equation (2.9). Then from the solution of the modified equation it follows that

Table 2.2. The angular wave function of the hydrogen atom.

l	m	$Y_{lm}(\theta, \varphi)$
0	0	$\frac{1}{\sqrt{4\pi}}$
1	0	$\sqrt{\frac{3}{4\pi}} \cdot \cos\theta$
1	± 1	$\pm\sqrt{\frac{3}{8\pi}} \cdot \sin\theta \cdot \exp(\pm i\varphi)$
2	0	$\frac{1}{4}\sqrt{\frac{5}{\pi}} \cdot \left(\frac{3}{2}\cos^2\theta - 1\right)$
2	± 1	$\pm\sqrt{\frac{15}{8\pi}} \cdot \sin\theta \cdot \cos\theta \cdot \exp(\pm i\varphi)$
2	± 2	$\frac{1}{2}\sqrt{\frac{15}{8\pi}} \cdot \sin^2\theta \cdot \exp(\pm 2i\varphi)$
3	0	$\frac{1}{4}\sqrt{\frac{7}{\pi}} \cdot (5\cos^3\theta - 3\cos\theta)$
3	± 1	$\pm\frac{1}{8}\sqrt{\frac{21}{\pi}} \cdot \sin\theta \cdot (5\cos^2\theta - 1)\exp(\pm i\varphi)$
3	± 2	$\frac{1}{4}\sqrt{\frac{105}{2\pi}} \cdot \sin^2\theta\cos\theta \cdot \exp(\pm 2i\varphi)$
3	± 3	$\pm\frac{1}{8}\sqrt{\frac{35}{\pi}} \cdot \sin^3\theta \cdot \exp(\pm 3i\varphi)$
4	0	$\sqrt{\frac{9}{4\pi}} \cdot \left(\frac{35}{8}\cos^4\theta - \frac{15}{4}\cos^2\theta + \frac{3}{8}\right)$
4	± 1	$\pm\frac{3}{8}\sqrt{\frac{5}{\pi}} \cdot \sin\theta \cdot (7\cos^3\theta - 3\cos\theta) \cdot \exp(\pm i\varphi)$
4	± 2	$\frac{3}{8}\sqrt{\frac{5}{2\pi}} \cdot \sin^2\theta \cdot (7\cos^2\theta - 1) \cdot \exp(\pm 2i\varphi)$
4	± 3	$\pm\frac{3}{8}\sqrt{\frac{35}{\pi}} \cdot \sin^3\theta \cdot \cos\theta \cdot \exp(\pm 3i\varphi)$
4	± 4	$\frac{3}{16}\sqrt{\frac{35}{2\pi}} \cdot \sin^4\theta \cdot \exp(\pm 4i\varphi)$

the strongest dependence of the wave function on a distance r has the form

$$R \sim \exp[\pm\sqrt{(-2\varepsilon)}r].$$

Because the region of the electron motion is restricted, it is necessary to conserve only the sign minus in this expression at $\varepsilon < 0$ (the electron energy is negative). Based on this, let us represent the radial electron wave function in the form

$$R(r) = \frac{1}{r}\exp\left[-\sqrt{(-2\varepsilon)}r\right] f(r), \tag{2.10}$$

so that the function $f(r)$ satisfies the equation

$$f'' - 2\sqrt{(-2\varepsilon)}f' + \frac{2}{r}f - \frac{l(l+1)}{r^2}f = 0. \tag{2.11}$$

Because the wave function is finite at large distances from the center, we have that the function $f(r)$ can grow at large r not sharper than a power function. Assume

$$f \sim r^n, \qquad r \to \infty. \tag{2.12}$$

Then from equation (2.11) it follows that

$$\varepsilon = -\frac{1}{2n^2}, \tag{2.13}$$

and equation (2.11) takes the form

$$f'' - \frac{2}{n}f' + \frac{2}{r}f - \frac{l(l+1)}{r^2}f = 0. \tag{2.14}$$

Note that the normalization condition of the wave function has the form

$$\int_0^\infty R^2(r)r^2\,dr = 1.$$

From this it follows that function $f(r)$ must be finite at $r \to 0$. This allows one to neglect the second and third terms of equation (2.11) compared to the first term of this equation in the limit $r \to 0$. A general solution of the obtained equation has the form, at small r,

$$f = C_1 r^{l+1} + C_2 r^{-l}.$$

The above condition gives $C_2 = 0$. This leads to an additional requirement for the function $f(r)$ and the parameter n. Indeed, from this it follows that function $f(r)$ has the form of a polynomial, and n must be whole, so that

$$f(r) = \sum_{k=l+1}^{n} a_k r^k,$$

where a_k are numerical coefficients. From this it follows that

$$n \geq l + 1. \tag{2.15}$$

For instance, in the simplest case $l = 0$, equation (2.14) takes the form

$$f'' - \frac{2}{n} f' + \frac{2}{r} f = 0,$$

and its solution for a normalized wave function is

$$f = Cr, \quad n = 1; \qquad f = C(r^2 - 2r), \quad n = 2, \text{ etc.}$$

where C is the normalization constant. In a general case the radial wave function of the electron in the hydrogen atom has the form

$$R_{nl}(r) = \frac{1}{n^{l+2}} \frac{2}{(2l+1)!} \sqrt{\frac{(n+l)!}{(n-l-1)!}} (2r)^l \times \exp\left(-\frac{r}{n}\right) F\left(-n+l+1, 2l+2, \frac{2r}{n}\right), \quad (2.16)$$

where F is the degenerated hypergeometric function. The expressions of the radial wave function of the electron for the hydrogen atom are given in Table 2.3 for the lowest electron states.

Let us sum up the above results. Because n is a whole number, the electron energy takes discrete values according to formula (2.13). Then the lower energetic state of the hydrogen atom corresponds to $n = 1, l = 0, \varepsilon = -1/2$, in accordance with formulas (2.6), (2.13), and (2.15). The lowest atomic state is called the ground atomic state. From this it follows that the electron binding energy for the ground state of the hydrogen atom is equal to $me^4/(2\hbar^2) = 13.6\,\text{eV}$. The spatial state of the electron in the hydrogen atom is characterized by three quantum numbers n, l, m. These numbers, together with the spin projection onto a given direction σ, are called the electron quantum numbers. The number n is called the principal quantum number; the numbers n and l are whole positive numbers and, according to formulas (2.6), (2.7), (2.15), we have $n \geq l + 1$; $l \geq |m|$. Usually the electron state in the hydrogen atom is characterized only by the quantum numbers n and l, because the states of the hydrogen atom are degenerated for the quantum numbers m and σ, i.e., the same electron energy corresponds to different values of m and σ at the same values of n and l. Therefore for notation of the electron states of the hydrogen atom only the quantum numbers n and l are used usually. Then the principal quantum number is given first, and then the notations *s, p, d, f, g, h*, etc., are given for states with $l = 0, 1, 2, 3, 4, 5$. For example, the ground state of the hydrogen atom is denoted as $1s$ ($n = 1, l = 0$), and the notation $5g$ corresponds to $n = 5, l = 4$.

Thus, the electron states of the hydrogen atom are degenerated because different values of m and σ correspond to the same electron energy. The degree of degeneration, i.e., the number of states with the same energy, is equal to

$$2\sum_{l=0}^{n-1}(2l+1) = 2n^2, \quad (2.17)$$

Table 2.3. Radial wave functions R_{nl} of the hydrogen atom.

State	R_{nl}
$1s$	$2\exp(-r)$
$2s$	$\frac{1}{\sqrt{2}}\left(1-\frac{r}{2}\right)\exp(-r/2)$
$2p$	$\frac{1}{\sqrt{24}}r\exp(-r/2)$
$3s$	$\frac{2}{3\sqrt{3}}\left(1-\frac{2r}{3}+\frac{2r^2}{27}\right)\exp(-r/3)$
$3p$	$\frac{2}{27}\sqrt{\frac{2}{3}}\cdot r\left(1-\frac{r}{6}\right)\exp(-r/3)$
$3d$	$\frac{4}{81\sqrt{30}}\cdot r^2\exp(-r/3)$
$4s$	$\frac{1}{4}\left(1-\frac{3r}{4}+\frac{r^2}{8}-\frac{r^3}{192}\right)\exp(-r/4)$
$4p$	$\frac{1}{16}\sqrt{\frac{5}{3}}\cdot r\cdot\left(1-\frac{r}{4}+\frac{r^2}{80}\right)\exp(-r/4)$
$4d$	$\frac{1}{64\sqrt{5}}\cdot r^2\cdot\left(1-\frac{r}{12}\right)\exp(-r/4)$
$4f$	$\frac{1}{768\sqrt{35}}\cdot r^3\cdot\exp(-r/4)$

where $2l+1$ is a number of projections of the orbital momentum onto a given direction and 2 is the number of spin projections onto a given direction. Relativistic interactions can lead to the partial elimination of degeneration. The above wave functions allow one to obtain the average parameters of various states of the hydrogen atom. Table 2.4 contains analytical expressions for some average parameters and Table 2.5 gives numerical values for the parameters of the lowest states of the hydrogen atom.

2.3 Fine Splitting of Levels of the Hydrogen Atom

Let us consider the splitting of levels of the hydrogen atom resulting from the simplest relativistic interaction in the atom—the interaction of the electron spin and orbital motion of the electron. This interaction is called the spin-orbit interaction and the corresponding splitting of levels is named the fine splitting of levels. The

Table 2.4. Average values of degrees from the electron radius-vector in the hydrogen atom $\langle r^n \rangle = \int_0^\infty R_{nl}^2(r) r^{n+2}\, dr$.

Parameter	Expression
$\langle r \rangle$	$\frac{1}{2} \cdot \left[3n^2 - l(l+1)\right]$
$\langle r^2 \rangle$	$\frac{n^2}{2} \cdot \left[5n^2 + 1 - 3l(l+1)\right]$
$\langle r^3 \rangle$	$\frac{n^2}{8} \cdot \left[35n^2(n^2-1) - 30n^2(l+2)(l-1) + 3(l+2)(l+1)l(l-1)\right]$
$\langle r^4 \rangle$	$\frac{n^4}{8} \cdot \left[63n^4 - 35n^2(2l^2+2l-3) + 5l(l+1)(3l^2+3l-10) + 12\right]$
$\langle r^{-1} \rangle$	$\frac{1}{n^2}$
$\langle r^{-2} \rangle$	$\frac{1}{n^3(l+1/2)}$
$\langle r^{-3} \rangle$	$\frac{1}{n^3(l+1) \cdot (l+1/2) \cdot l}$
$\langle r^{-4} \rangle$	$\frac{3n^2 - l\,(l+1)}{2n^5 \cdot (l+3/2) \cdot (l+1) \cdot (l+1/2) \cdot l \cdot (l-1/2)}$

Table 2.5. Average parameters of the hydrogen atom.

State	$\langle r \rangle$	$\langle r^2 \rangle$	$\langle r^3 \rangle$	$\langle r^4 \rangle$	$\langle r^{-1} \rangle$	$\langle r^{-2} \rangle$	$\langle r^{-3} \rangle$	$\langle r^{-4} \rangle$
$1s$	1.5	3	7.5	22.5	1	2	—	—
$2s$	6	42	330	2880	0.25	0.25	—	—
$2p$	5	30	210	1680	0.25	0.0833	0.0417	0.0417
$3s$	13.5	207	3442	$6.136 \cdot 10^4$	0.111	0.0741	—	—
$3p$	12.5	180	2835	$4.420 \cdot 10^4$	0.111	0.0247	0.0123	0.0137
$3d$	10.5	126	1701	$2.552 \cdot 10^4$	0.111	0.0148	0.0247	$5.49 \cdot 10^{-3}$
$4s$	24	648	18720	$5.702 \cdot 10^5$	0.0625	0.0312	—	—
$4p$	23	600	16800	$4.973 \cdot 10^5$	0.0625	0.0104	$5.21 \cdot 10^{-3}$	$5.49 \cdot 10^{-4}$
$4d$	21	504	13100	$3.629 \cdot 10^5$	0.0625	0.00625	$1.04 \cdot 10^{-3}$	$2.60 \cdot 10^{-4}$
$4f$	18	360	7920	$1.901 \cdot 10^5$	0.0625	0.00446	$3.72 \cdot 10^{-4}$	$3.7 \cdot 10^{-5}$

spin-orbit interaction results from the interaction of the spin magnetic moment and a magnetic field which is due to the electron motion in an electrical field. The corresponding interaction potential is

$$\hat{V} = -\frac{e\hbar}{mc}\hat{\mathbf{s}}\hat{\mathbf{H}}, \tag{2.18}$$

where $\hat{\mathbf{s}}$ is the electron spin operator, $e\hbar/2mc$ is the Bohr magneton, so that $2e\hbar\hat{\mathbf{s}}/mc$ is the electron magnetic momentum, and $\mathbf{H}$ is the magnetic field strength.

The magnetic field strength in the frame of axes where the electron is at rest is given by the Lorenz formula

$$\hat{\mathbf{H}} = \frac{1}{c}[\mathbf{E}\hat{\mathbf{v}}] = \frac{1}{c}\left[\frac{e\mathbf{r}}{r^3} \times \frac{\hat{\mathbf{p}}}{m}\right] = \frac{e\hbar}{mc}\frac{\hat{\mathbf{l}}}{r^3},$$

where $\mathbf{E} = e\mathbf{r}/r^3$ is the electric field strength which is created by the central charge on a distance $\mathbf{r}$ from it and $\hat{\mathbf{v}}$, $\hat{\mathbf{p}}$, $\hat{\mathbf{l}}$ are the operators of the electron velocity, momentum, and orbital momentum, respectively. From this it follows that for the interaction operator,

$$\hat{V} = -\left(\frac{e\hbar}{mc}\right)^2 \frac{1}{r^3} \cdot \hat{\mathbf{l}}\hat{\mathbf{s}}.$$

The above consideration is not consistent because we based it on classical laws for the analysis of a quantum system. It leads to an error in the numerical factor of the resultant expression. The consistent, but more cumbersome method for obtaining this expression from the relativistic quantum equation of the electron motion changes the numerical factor in the above expression

$$\hat{V} = -\frac{1}{2}\left(\frac{e\hbar}{mc}\right)^2 \frac{1}{r^3}\hat{\mathbf{l}}\hat{\mathbf{s}}. \tag{2.19}$$

As a matter of fact, formula (2.19) describes the interaction of two magnetic momenta and has the form $\hat{V} = -\hat{\mathbf{m}}_1 \cdot \hat{\mathbf{m}}_2/r^3$, where $\hat{\mathbf{m}}_1 = e\hbar\hat{\mathbf{l}}/(2mc)$ is the operator of the magnetic moment due to the orbital motion of the electron, and $\hat{\mathbf{m}}_2 = e\hbar\hat{\mathbf{s}}/(mc)$ is the operator of the magnetic moment due to the electron spin. Thus the spin-orbit interaction can be represented as an interaction of magnetic momenta. From (2.19) it follows that the ratio of the spin-orbit splitting and a typical electron energy is of the order of the square of the small parameter (2.2), i.e., this ratio is of the order of 10^{-4}. Taking into account the interaction (2.19), we obtain, as a new quantum number the total electron momentum which is a sum of the orbital and spin momenta,

$$\hat{\mathbf{j}} = \hat{\mathbf{l}} + \hat{\mathbf{s}}.$$

Thus, instead of the set of electron quantum numbers $lms\sigma$ ($s = 1/2$ is the electron spin) we obtain the new set of quantum numbers jm_jls, where m_j is the projection of the total electron momentum onto a given direction. Because the maximum projection of the total electron momentum m_j is $l + 1/2$ or $l - 1/2$, depending on the mutual direction of the electron spin and orbital momentum, we have $j = l \pm 1/2$. In the representation of the total electron momentum, the mean value of the operator $\hat{\mathbf{l}}\hat{\mathbf{s}}$ follows from the relation

$$\left\langle \hat{\mathbf{j}}^2 \right\rangle = \left\langle \hat{\mathbf{l}}^2 \right\rangle + 2\left\langle \hat{\mathbf{l}}\hat{\mathbf{s}} \right\rangle + \left\langle \hat{\mathbf{s}}^2 \right\rangle,$$

where the angle brackets mean the averaging over this state. From this we obtain

$$2\left\langle \hat{\mathbf{l}}\hat{\mathbf{s}} \right\rangle = j(j+1) - l(l+1) - s(s+1).$$

Because $j = l \pm 1/2$, the spliting of the level of a state with the orbital electron momentum l is equal

$$\Delta\varepsilon_f = \frac{2l+1}{4}\left(\frac{e\hbar}{mc}\right)^2\left\langle \frac{1}{r^3} \right\rangle.$$

On the basis of the expression $\left\langle 1/r^3 \right\rangle = [n^3(l+1)(l+1/2)l]^{-1}$ for the hydrogen atom we obtain for the fine splitting of levels of the hydrogen atom in atomic units

$$\Delta\varepsilon_f = \left(\frac{e^2}{\hbar c}\right)^2 \frac{1}{2n^3 l(l+1)}. \tag{2.20}$$

Thus, the fine splitting of levels of the hydrogen atom is determined by the interaction of the spin and orbital momenta of the electron. It is absent for states with zero orbital momentum and drops with increase both the principal quantum number and orbital momentum. For the hydrogen atom state with $n = 2, l = 1$ this splitting is maximal and its ratio to the ionization potential of the hydrogen atom in the ground state is equal to $[e^2/(\hbar c)]^2/4 = 1.33 \cdot 10^{-5}$. This proves a weakness of the relativistic interactions for the hydrogen atom.

2.4 The Lamb Shift

The other type of relativistic interaction which causes the splitting of the hydrogen atom levels $l = 0$ and $l = 1$ $(j = 1/2)$ is determined by the interaction of electron motion in an electric field of the central charge with the electromagnetic field of a vacuum. This splitting is called the Lamb shift. The principal significance of the Lamb shift is such that it was one of the first real confirmations of the vacuum nature and allows one to present a vacuum as a set of zero oscillations of electromagnetic waves.

Let us determine the shift of the energy of a bound electron as a result of an electron "trembling" under the action of zero vibrations of the vacuum. Denote by $\delta\mathbf{r}$ a shift of the electron coordinate under the action of a vacuum field. Then the corresponding change of the Hamiltonian $\delta\hat{H}$ describing the electron behavior is

$$\delta\hat{H} = \sum_i \frac{\partial\hat{V}}{\partial x_i}\delta x_i + \frac{1}{2}\sum_{i,k}\frac{\partial^2\hat{V}}{\partial x_i \partial x_k}\delta x_i \delta x_k,$$

where $\hat{V}$ is the interaction potential of the electron and the atomic core, and x_i, x_j are components of the vector $\mathbf{r}$.

Because the electron shift is a random value, its average value is equal to zero: $\langle \delta x_i \rangle = 0$, $\langle \delta x_i \delta x_k \rangle = 0$, if $i \neq k$. The symmetry of the system gives $\langle \delta x_i^2 \rangle = \delta \mathbf{r}^2/3$, and in the first order of the perturbation theory the electron energy shift $\Delta\varepsilon$ is equal to

$$\Delta\varepsilon = \frac{1}{6}\left\langle \sum_i \frac{\partial^2 \hat{V}}{\partial \mathbf{r}^2} \right\rangle \langle \delta \mathbf{r}^2 \rangle .$$

The Poisson equation for the electric potential of the core φ has the form $\Delta\varphi = 4\pi\rho(\mathbf{r})$, where $\rho(\mathbf{r})$ is the nucleus density. Taking $\rho = Z\delta(\mathbf{r})$, where Z is the core charge, and accounting for $\hat{V} = -\varphi$ we obtain

$$\Delta\varepsilon = -\frac{2\pi}{3} Z \left|\psi(0)\right|^2 \langle \delta \mathbf{r}^2 \rangle \tag{2.21}$$

As is seen, the Lamb shift in the related approach of the perturbation theory takes place only for zero orbital momentum of the electron because, in other cases, $\psi(0) = 0$. Hence below we are restricted by the case $l = 0$ and use the expression for the electron wave function at the origin $|\psi(0)|^2 = Z^3/(\pi n^3)$, that gives

$$\frac{\Delta\varepsilon}{J_n} = \frac{4}{3}\frac{Z^2}{n} \langle \delta \mathbf{r}^2 \rangle ,$$

where $J_n = Z^2/(2n^2)$ is the ionization potential for a given state.

An action of the vacuum electromagnetic field is included in the term $\langle \delta \mathbf{r}^2 \rangle$. Because the electron "trembling" due to each frequency is random, we have $\langle \delta \mathbf{r}^2 \rangle = \sum_\omega \langle \delta \mathbf{r}_\omega^2 \rangle$, where $\delta \mathbf{r}_\omega$ is the electron vibration amplitude under the action of electromagnetic waves of frequency ω. Assuming $\omega \gg \omega_o$, where ω_o is a typical atomic frequency, we have the following equation for the electron motion

$$m \frac{d\delta \mathbf{r}_\omega}{dt^2} = \mathbf{E}_\omega e^{i\omega t} ,$$

where $\mathbf{E}_\omega$ is the electric field strength. From this it follows that

$$\left|\delta \mathbf{r}_\omega\right|^2 = \left|\mathbf{E}_\omega\right|^2 / \omega^4 \qquad \text{and} \qquad \langle \delta \mathbf{r}^2 \rangle = \sum_\omega \frac{|\mathbf{E}_\omega|^2}{\omega^4} .$$

A number of states of the vacuum electromagnetic field per unit of phase space is equal to

$$dn = 2\frac{d\mathbf{k}\, d\mathbf{r}}{(2\pi)^3} = \frac{\omega^2\, d\omega\, d\mathbf{r}}{\pi^2 c^3} ,$$

where $k = \omega/c$ is the wave vector of the electromagnetic wave and $d\mathbf{r}$ is the space element. From this we have

$$\langle \delta \mathbf{r}^2 \rangle = \int \frac{|\mathbf{E}_\omega|^2\, d\omega\, d\mathbf{r}}{\pi^2 c^3 \omega^2} .$$

Since $|\mathbf{E}_\omega|^2 \, d\mathbf{r}/(4\pi)$ is the energy density for electromagnetic oscillations of a given frequency and $\hbar\omega/2$ is the energy of zero oscillations, we have $|\mathbf{E}_\omega|^2 \, d\mathbf{r}/(4\pi) = \hbar\omega/2$, and this gives

$$\langle \delta \mathbf{r}^2 \rangle = \frac{2}{\pi c^3} \int \frac{d\omega}{\omega}$$

This integral has a logarithmic divergence. Since we used the motion equation of a free electron for a bound one, it is necessary to use a typical atomic frequency $\omega \sim \omega_o$ as the lower limit; the upper limit we take from the relation $\hbar\omega \sim mc^2$, because the virtual electron transition under the action of such a frequency leads to an increase in the electron mass, i.e., the oscillation amplitude is less than the used one. Thus with the logarithmic accuracy, we have, for the Lamb shift of s-level of the hydrogen atom or hydrogenlike ion (in the usual units),

$$\frac{\Delta\varepsilon}{J_n} = \frac{8Z^2}{3\pi n} \left(\frac{e^2}{\hbar c} \right)^3 \left[\ln \left(\frac{\hbar c}{Z e^2} \right)^2 + C \right].$$

This used method does not allow us to determine the numerical coefficient C which is of the order of unity and is small compared to the logarithmic term. Table 2.6 lists values of C for the excited states obtained on the basis of the accurate method. On the basis of the Table 2.6 data, the above formula for the Lamb shift of the hydrogen atom or hydrogenlike ion can be written in the form

$$\frac{\Delta\varepsilon}{J_n} = \frac{8Z^2}{3\pi n} \left(\frac{e^2}{\hbar c} \right)^3 [7.6 - 2 \ln Z]. \tag{2.22}$$

The accuracy of this formula for the hydrogen atom and helium ion is approximately 1%. Note that the Lamb shift for the hydrogen atom with $n = 2$ is $0.0353 cm^{-1}$ and corresponds to the difference of energies of $2s_{1/2}$ and $2p_{1/2}$-levels, while the spin-orbit splitting respects to the distance $0.366 cm^{-1}$ between the levels $2p_{1/2}$ and $2p_{3/2}$.

2.5 Superfine Splitting and the Isotope Shift of Levels of the Hydrogen Atom

The superfine splitting of atomic levels corresponds to the interaction of the orbital electron momentum and the nucleus spin. As a result of this interaction, the total atom momentum F becomes a quantum atomic number which is a sum of the total electron momentum j and nucleus spin I. The superfine splitting is determined

Table 2.6. Values of the numerical parameter C for excited states of the hydrogen atom.

n	1	2	3	4	∞
A	−2.351	−2.178	−2.134	−2.116	−2.088

by the interaction of the electron magnetic moment μ_e and the nuclear magnetic moment μ_i. This interaction can be estimated as $V \sim \mu_e \mu_i / r^3$. This gives an estimate for the superfine splitting

$$\Delta\varepsilon_{sf} \sim \left\langle \frac{1}{r^3} \right\rangle \frac{e\hbar}{mc} \frac{e\hbar}{Mc}, \tag{2.23}$$

where M is the nuclear mass. As is seen, the superfine splitting contains an additional small parameter m/M and is, by three–four orders of magnitude, lower than the fine splitting of atomic levels. If the electron orbital moment is zero, we have a "contact" interaction of momenta, and the above estimation contains a factor a_o^{-3} instead of $1/r^3$. In addition, the quadrupole nuclear moment can be essential in some cases of electron–nucleus interaction. Usually the superfine splitting of levels is small compared to the fine one, i.e., this interaction conserves the electron quantum numbers of more strong interactions.

Along with magnetic interactions, the isotope shift of levels can be of importance for their positions. The essential part of the isotopic shift is due to a finite nuclear mass. Indeed, the reduced mass of the electron and nucleus, which must be used in the Schrödinger equation (2.1) instead of the electron mass m, is equal to $m(1 - m/M)$. This must be included in all the expressions for electron energies, instead of the electron mass m, that lead to the isotopic shift of levels of the order of $\Delta\varepsilon \sim \varepsilon m/M$, where ε is a typical atomic energy.

The difference of the wavelengths for identical radiative transitions and different nuclei-isotopes is equal to $\lambda_1 - \lambda_2 = \lambda(m/M_1 - m/M_2)$, where λ_1, λ_2 are the transition wavelengths for the related isotopes, M_1, M_2 are the nuclear masses in these cases, and λ is the wavelength of the transition for an infinite nuclear mass. In the case of a proton and deuteron $m/M_1 - m/M_2 = 2.72 \cdot 10^{-4}$. This shift characterizes the relative difference of the wavelengths for identical transitions. The other mechanism of the isotope shift is determined by a nucleus size. But this effect is essential only for heavy atoms. Note that because nuclei-isotopes have different nuclear momenta, this leads to a different superfine splitting of levels. Thus, this leads to a superfine structure of levels.

2.6 Radiative Transitions of the Hydrogen Atom and the Grotrian Diagrams

The radiation spectrum of atomic particles contains essential information about the character of the interaction in the atom because of the high accuracy of the wavelengths of radiative transitions. The probability of a spontaneous radiative transition per unit time is given by formula (1.26):

$$B_{fj} = \frac{1}{\tau_{fj}} = \frac{4\omega^3}{\hbar c^3} \left|(D_x)_{fj}\right|^2 g_j, \tag{2.24}$$

where ω is the transition frequency, D_x is the projection of the operator of the dipole moment onto the direction of radiation polarization, g_j is the statistical weight of the final state, and the matrix element is taken between the initial and final states of the transition. This formula accounts for the first term of expansion of this rate over a small parameter v/c, where v is a typical electron velocity. Hence, restricted by dipole radiative transitions, we extract the strongest radiative transitions for a nonrelavitistic atomic system. Being guided by such systems, we are restricted by the dipole radiative transitions, and we will consider as forbidden other types of interactions in atomic systems which are weak compared to the dipole interactions for nonrelavitistic systems.

Let us analyze the selection rules for the dipole radiative transitions of the hydrogen atom. The operator of the electron dipole moment is $e\mathbf{r}$, where $\mathbf{r}$ is the radius-vector of the electron. Then the selection rules are determined by the properties of the matrix element for the angular wave function (2.5), and give

$$\Delta l = \pm 1, \qquad \Delta m = 0, \pm 1. \tag{2.25}$$

The observational spectrum of radiation and absorption of the hydrogen atom respects to the above selection rules.

The convenient characteristic of a radiative transition is the oscillator strength, which is introduced on the basis of the following formula (in usual units),

$$f_{jf} = \frac{2m\omega}{3\hbar} \left|\langle j\left|\mathbf{D}\right| f\rangle\right|^2 g_f . \tag{2.26}$$

It is of importance that the oscillator strength is a dimensionless parameter which satisfies the sum rule. In the case of the hydrogen atom this rule has the form (1.31)

$$\sum_f f_{jf} = 1, \tag{2.27}$$

where summation is made over both the discrete and continuous states of an electron in the Coulomb field of the nucleus. The oscillator strength is connected with the Einstein coefficient $A_{fi} = 1/\tau_{fi}$ by the relation

$$f_{jf} = 1.499 \cdot 10^{-14} B_{fj} \lambda^2 \frac{g_f}{g_j} . \tag{2.28}$$

Table 2.7 gives quantities of the oscillator strength and radiative lifetimes $\tau_{fj} = 1/B_{fj}$ for transitions between the lowest states of the hydrogen atom.

It is convenient to give the positions of atomic levels together with the radiation and absorption atomic spectrum on the basis of the diagram in Fig. 2.6 for the hydrogen atom. Such a diagram extracts levels of the same symmetry so that the corresponding atomic states are characterized by the same quantum numbers. This diagram is called the Grotrian diagram and is used for the analysis of atomic states. The Grotrian diagram separates atomic states in groups with identical quantum numbers and includes information about both low atomic states and the radiative transition involving them. Therefore, we will use Grotrian diagrams for the analysis of various atoms.

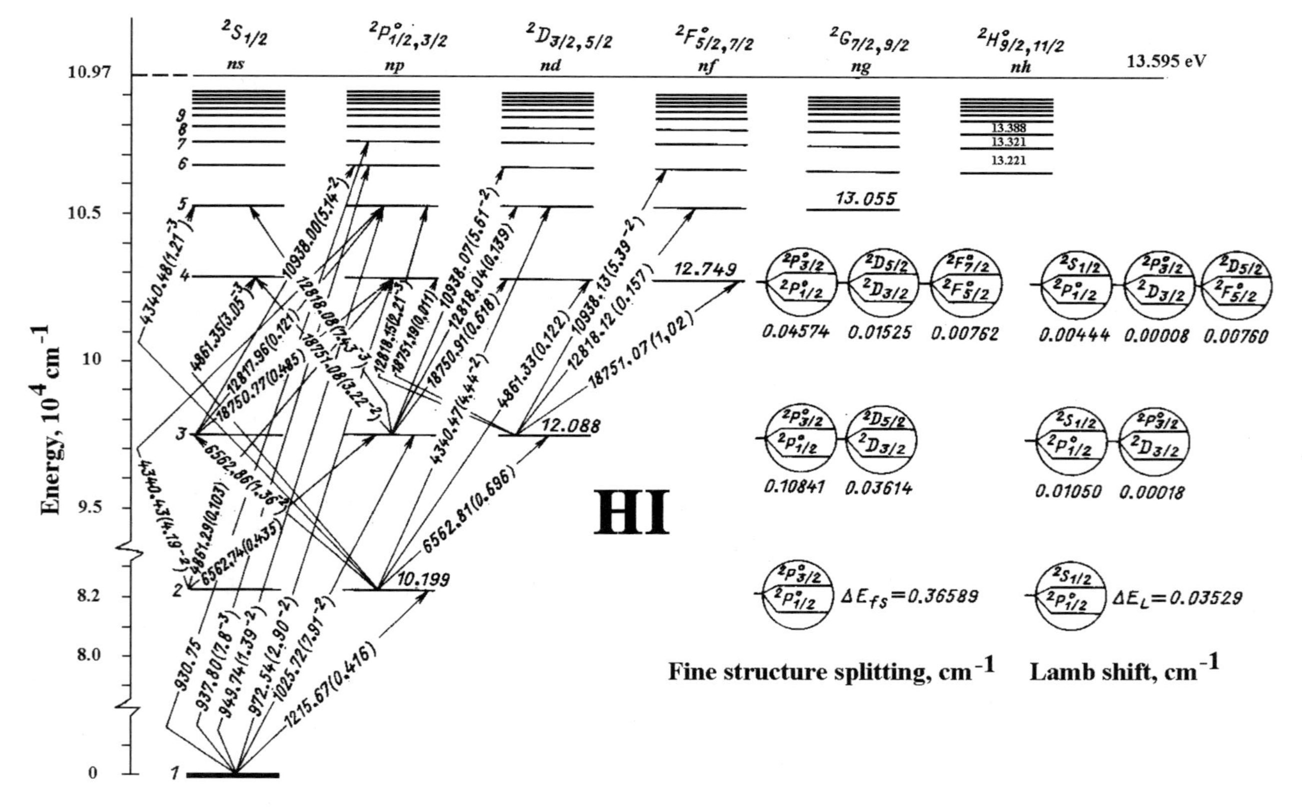

Figure 2.1. The Grotrian diagram for the hydrogen atom.

Table 2.7. Parameters of radiative transitions of the hydrogen atom.

Transition	f_{jf}	τ_{fj}, ns		Transition	f_{jf}	τ_{fj}, ns
1s–2p	0.4162	1.6		3p–4s	0.032	230
1s–3p	0.0791	5.4		3p–4d	0.619	36.5
1s–4p	0.0290	12.4		3p–5s	0.007	360
1s–5p	0.0139	24		3p–5d	0.139	70
2s–3p	0.4349	5.4		3d–4p	0.011	12.4
2s–4p	0.1028	12.4		3d–4f	1.016	73
2s–5p	0.0419	24		3d–5p	0.0022	24
2p–3s	0.014	160		3d–5f	0.156	140
2p–3d	0.696	15.6		4s–5p	0.545	24
2p–4s	0.0031	230		4p–5s	0.053	360
2p–4d	0.122	26.5		4p–5d	0.610	70
2p–5s	0.0012	360		4d–5p	0.028	24
2p–5d	0.044	70		4d–5f	0.890	140
3s–4p	0.484	12.4		4f–5d	0.009	70
3s–5p	0.121	24		4f–5g	1.345	240

2.7 Classical Electron in the Coulomb Field

Let us consider the behavior of an electron in a bound or free state when its energy is small compared to a typical atomic energy. This corresponds to both highly excited hydrogen atom and to a free slow electron which moves in the Coulomb field. In both cases the electron behavior is described by classical laws. The electron wave functions in these cases are used for determination of the cross section of transitions in these states. Below we determine the wave function of the electron in the classical region of its motion. For a highly excited atom this is given by formula (2.16) which has the form, in the limit $n \to \infty$ in the region $r \ll n^2$ (we assume the nucleus charge to be unity),

$$\begin{aligned} R_{nl}(r) &= \frac{1}{n^{l+2}} \frac{2}{(2l+1)!} \sqrt{\frac{(n+l)!}{(n-l-1)!}} (2r)^l \exp\left(-\frac{r}{n}\right) \\ &\quad \times F\left(-n+l+1, 2l+2, \frac{2r}{n}\right) \\ &= \sqrt{\frac{2}{n^3 r}} J_{2l+1}\left(\sqrt{8r}\right). \end{aligned} \tag{2.29}$$

From this we have in the region $(l+1/2)^2 \ll r \ll n^2$, where the electron binding energy is small compared to the Coulomb interaction potential between the electron and nucleus

$$R_{nl}(r) = \frac{2^{1/4}}{\pi^{1/2} n^{3/2} r^{3/4}} \cos\left[\sqrt{8r} - (2l+1)\frac{\pi}{2} - \frac{\pi}{4}\right]. \tag{2.30}$$

The classical region of the electron motion is $r_1 \leq r \leq r_2$, where r_1, r_2 are the turning points. Their positions are determined by the expressions $r_{1,2} = n^2 \pm \sqrt{n^2 - l(l+1)}$. In this region the electron wave function has the form

$$R_{nl}(r) = \frac{2^{1/2}}{\pi^{1/2} n^{3/2} r} \cdot \cos\left[\int_{r_1}^{r} dr' \sqrt{\frac{2}{r'} - \frac{1}{n^2} - \frac{(l+1/2)^2}{r'^2}} - \frac{\pi}{4}\right]$$

$$= \frac{2^{1/2}}{\pi^{1/2} n^{3/2} r} \cdot \cos\Phi, \tag{2.31}$$

$$\Phi = \sqrt{2r - \frac{r^2}{n^2} - \left(l + \frac{1}{2}\right)^2} + n \arcsin\left[\frac{r - n^2}{n\sqrt{n^2 - (l+1/2)^2}}\right]$$

$$- \left(l + \frac{1}{2}\right) \arcsin\left(\frac{n[r - (l + \frac{1}{2})^2\}}{r\sqrt{n^2 - \left(l + \frac{1}{2}\right)^2}}\right) + (n - l - 1)\frac{\pi}{2}.$$

Considering the motion of a free electron, let us expand its wave function $\psi_{\mathbf{q}}$ on spherical angular functions

$$\psi_{\mathbf{q}} = \frac{1}{2q} \sum_{l=0}^{\infty} i^l (2l+1) e^{i\delta_l} R_{ql}(r) P_l(\cos\theta_{\mathbf{qr}}), \tag{2.32}$$

where $\mathbf{q}$ is the electron wave vector, δ_l is the scattering phase, and $\theta_{\mathbf{qr}}$ is the angle between the vectors $\mathbf{q}$ and $\mathbf{r}$. Using the normalization condition for the electron wave function

$$\int \psi_{\mathbf{q}}(\mathbf{r}) \psi_{\mathbf{q}'}^*(\mathbf{r}) = (2\pi)^3 \delta(\mathbf{q} - \mathbf{q}'),$$

we have the following normalization condition for radial wave functions

$$\int_0^{\infty} R_{ql}(r) R_{q'l}(r) r^2 \, dr = 2\pi \delta(q - q').$$

In the absence of an interaction between the electron and nucleus, the expansion of this function has the form of a plane wave

$$\psi_{\mathbf{q}}(\mathbf{r}) = e^{i\mathbf{qr}} = \sqrt{\frac{\pi}{2qr}} \sum_{l=0}^{\infty} i^l (2l+1) J_{l+1/2}(qr) P_l(\cos\theta_{\mathbf{qr}}). \tag{2.33}$$

Because in the limit $x \to \infty$ we have $J_{l+1/2}(x) = \sqrt{2/(\pi x)}\sin(x - \pi l/2)$, the radial wave function has the form, far from the center,

$$R_{ql}(r) = \sqrt{\frac{\pi}{2qr}} J_{l+1/2}(qr) = \frac{2}{r}\sin(qr - \pi l/2). \tag{2.34}$$

In a general case, the radial wave function of a free electron which moves in the Coulomb field of a nucleus center has the form

$$R_{ql}(r) = \sqrt{\frac{8\pi q}{1 - \exp(-2\pi/q)}} \cdot \prod_{s=1}^{l} \sqrt{s^2 + 1/q^2} \cdot \frac{(2qr)^l}{(2l+1)!} \cdot e^{iqr} \cdot F\left(\frac{i}{q} + l + 1, 2l + 2, 2iqr\right). \tag{2.35}$$

Far from the scattered Coulomb center ($qr \gg 1/q, qr \gg l + 1$) this function is transformed into

$$R_{ql}(r) = \frac{2}{r}\cos\left[qr + \frac{1}{q}\ln(2qr) - \frac{\pi}{2}\cdot(l+1) - \frac{\pi}{4} + \delta_l\right], \tag{2.36a}$$

where the scattering phase δ_l for an electron in the Coulomb field is $\delta_l = \arg(l + 1 - i/q)$. Near the Coulomb center $q^2 \ll 1/r$ we have

$$R_{nl}(r) = \sqrt{\frac{4\pi q}{r}} \cdot J_{2l+1}(\sqrt{8r}), \tag{2.36b}$$

and in the limiting case $l + 1/2 \ll r \ll 1/q^2$ this formula gives

$$R_{nl}(r) = \left(\frac{8q^2}{r^3}\right)^{1/4} \cos\left[\sqrt{8r} - (2l+1)\frac{\pi}{2} - \frac{\pi}{4}\right]. \tag{2.37}$$

Comparing the expressions for the radial wave function of the electron in the bound and free states, we obtain the following formula for their ratio in the region where the Coulomb potential is higher than both the binding energy of the bound electron and the energy of the free electron ($1/r \gg q^2, 1/n^2$):

$$R_{ql}(r)/R_{nl}(r) = \sqrt{2\pi q n^3}. \tag{2.38}$$

This formula is useful for the comparison of parameters which characterize transitions with the formation of highly excited and ionized states of the hydrogen atom.

2.8 Photoionization of a One-Electron Atom

The spherical symmetry of the electron wave function in atoms simplifies the analysis of various processes in them and allows us to exclude angular wave functions from consideration. The radial wave function of a free electron determines the matrix elements which are responsible for the radiative transitions between an electron bound and free states. As an example of this let us consider the photoionization process. If one valent electron partakes in this process, the total cross section of this process is equal to, according to formula (1.29),

$$\sigma_{\text{ion}} = \frac{q\omega}{6\pi c} \int |\langle \gamma l \, |\mathbf{r}| \, \mathbf{q} \rangle|^2 \, d\Theta_{\mathbf{q}},$$

where l is the angular momentum of the electron in the initial state, γ includes other quantum numbers of the electron, $\mathbf{q}$ is the wave vector of the electron in the final state, and the wave function of the released electron tends to $e^{i\mathbf{qr}}$ far from the atomic core. Using the expansion (2.31) for the final electron state, we express this cross section through the matrix elements of radial wave functions. The electron wave function of the initial state has the form

$$\psi_l(\mathbf{r}) = R_l(r) \cdot \sqrt{\frac{2l+1}{4\pi}} P_l(\cos\theta_{\mathbf{rs}}),$$

where $\theta_{\mathbf{rs}}$ is the angle between vectors $\mathbf{r}$ and a direction $\mathbf{s}$ onto which the projection of the electron momentum is equal to zero.

Let us fulfill the integration over angles. We have

$$\begin{aligned}
\int |\langle \gamma l \, |\mathbf{r}| \, \mathbf{q} \rangle|^2 \, d\Theta_{\mathbf{q}} &= \int d\mathbf{r} \, d\mathbf{r}' \mathbf{r}\mathbf{r}' \psi_l(\mathbf{r}) \psi_l(\mathbf{r}') \psi_{\mathbf{q}}(\mathbf{r}) \psi_{\mathbf{q}}^*(\mathbf{r}') \, d\Theta_{\mathbf{q}} \\
&= \frac{2l+1}{4\pi} \cdot \int r^3 \, dr (r')^3 \, dr' \cos\Theta_{\mathbf{rr}'} R_l(r) R_l(r') \\
&\quad \cdot P_l(\cos\theta_{\mathbf{rs}}) P_l(\cos\theta_{\mathbf{r}'\mathbf{s}}) \cdot \frac{1}{4q^2} \\
&\quad \times \sum_{p,n} (2p+1)(2n+1) i^{p-n} e^{i(\delta_p - \delta_n)} R_{qp}(r) R_{qn}(r') \\
&\quad \cdot P_p(\cos\theta_{\mathbf{qr}}) P_n(\cos\theta_{\mathbf{qr}'}) \, d\Theta_{\mathbf{q}}.
\end{aligned}$$

Using the theorem of summation of the Legendre polynomials and the condition of their orthogonality, we have

$$\int P_p(\cos\theta_{\mathbf{qr}}) P_n(\cos\theta_{\mathbf{qr}'}) \, d\Theta_{\mathbf{q}} = \frac{4\pi}{(2n+1)} \delta_{np} P_n(\cos\theta_{\mathbf{rr}'}),$$

where δ_{np} is the Kronecker delta symbol, and we obtain the following relation:

$$\begin{aligned}
\int |\langle \gamma l \, |\mathbf{r}| \, \mathbf{q} \rangle|^2 \, d\Theta_{\mathbf{q}} &= \int d\Theta_{\mathbf{r}} \, d\Theta_{\mathbf{r}'} \cdot \sum_n (2l+1)(2n+1) \\
&\quad \cdot |K_{nl}|^2 \cos\theta_{\mathbf{rr}'} P_l(\cos\theta_{\mathbf{rs}}) P_n(\cos\theta_{\mathbf{r}'\mathbf{s}}).
\end{aligned}$$

Here the following notation is used for the matrix element from radial wave functions:

$$K_{nl} = \frac{1}{2q} \int_0^\infty r^3 \, dr \, R_l(r) R_{qn}(r). \tag{2.39}$$

Based on the recurrent relation $(2n+1)x P_n(x) = (n+1) P_{n+1}(x) + n P_{n-1}(x)$, we divide the integral into two parts. Using once more the theorem of summation of the Legendre polynomials and the condition of their orthogonality, we find

$$\int |\langle \gamma l \,|\mathbf{r}|\, \mathbf{q}\rangle|^2 \, d\Theta_{\mathbf{q}} = \frac{(4\pi)^2}{(2l+1)} \left[l K_{l,l-1}^2 + (l+1) K_{l,l+1}^2 \right].$$

From this it follows that for the photoionization cross section

$$\sigma_{\text{ion}} = \frac{8\pi q \omega}{3c(2l+1)} \left[l K_{l,l-1}^2 + (l+1) K_{l,l+1}^2 \right]. \quad (2.40)$$

Thus, we express the photoionization cross section through the matrix elements (2.39) which are taken on the radial wave functions of the valent electron only.

2.9 Hydrogenlike Ions

As a matter of fact, the above analysis for the hydrogen atom is suitable for hydrogenlike ions whose nuclear charge Z differs from 1. Then it is necessary to change the parameter e^2 in equation (2.1) by Ze^2. Making this change in the resultant expressions, one can transfer from the hydrogen atom to the hydrogenlike ions. In particular, the electron energy (2.13) in the hydrogenlike ions is, in accordance with formula (2.13),

$$\varepsilon = -\frac{Z^2 m e^4}{2\hbar^2 n^2}. \quad (2.41)$$

Because we use atomic units, it is possible to transfer in the above expressions to hydrogenlike ions by the change $r \to r/Z$. As for the behavior of a free electron in the Coulomb field of a nucleus, the change of the nucleus charge corresponds to the transformation $q \to qZ$. For example, the electron wave function of the $1s$-state which is given in Table 2.3 for hydrogenlike ions is $R_{10} = 2Z^{3/2}\exp(-Zr)$, and the average size of the electron orbit for $1s$-state is equal to $\langle r \rangle = 3/(2Z)$ according to the data of Table 2.5. In the same way, one can find other atomic parameters of hydrogenlike ions and obtain the dependence of these parameters on the nucleus charge Z. In particular, the fine and superfine splitting of levels according to (2.20), (2.23) is $\Delta\varepsilon \sim Z^4$, the Lamb shift (2.22) is $\sim Z^4$, the radiative lifetime of an excited state due to a one-photon dipole radiative transition is, according to (2.24), $\sim Z^4$, etc.

Note that a different Z-dependence for some parameters can lead to a quality change of atomic properties with an increase in Z. For example, the radiative lifetime of the $2s$-state, which is determined by the two-photon decay of the state, is equal for the hydrogen atom 0.122 s and remarkably exceeds the radiative lifetime of the $2p$-state which is $1.6 \cdot 10^{-9}$ s. Hence the $2s$-state is a metastable one. In the case of multicharged hydrogenlike Fe-ions ($Z = 26$) the lifetime of the $2s$-state is $3.5 \cdot 10^{-10}$ s, while $3.5 \cdot 10^{-15}$ s is the lifetime of the $2p$-state. For the multicharged hydrogenlike uranium ions ($Z = 92$) these lifetimes are respectively equal, $5.1 \cdot 10^{-15}$ s and $2.1 \cdot 10^{-17}$ s. At is seen, now the $2s$-state is not a metastable one.

Problems

Problem 2.1. *Determine the shift of the electron energy of the hydrogenlike ions if the nucleus charge Z is not concentrated in a point, but is distributed uniformly in a ball of radius r_o.*

The electric field strength of a uniform charged ball of radius r_o is equal to, according to the Gauss theorem,

$$\mathbf{E} = -Zer\mathbf{n}/r_o^3, \quad r < r_o; \qquad \mathbf{E} = -Ze\mathbf{n}/r^2, \quad r > r_o.$$

where $\mathbf{n}$ is the unit vector directed along a line joining the electron and nucleus. Then the interaction energy between the electron and positive nucleus charge, which is distributed in a ball of radius r_o, is equal to, at a distance r from the center,

$$V = -\frac{Ze^2}{r_o} \cdot \left(\frac{3}{2} - \frac{r^2}{2r_o^2}\right), \quad r \le r_o; \qquad V = -\frac{Ze^2}{r}, \quad r \ge r_o.$$

Let us calculate the energy shift within the framework of the perturbation theory if the zero-th approximation of the perturbation theory corresponds to location of the nuclear charge at the ball center. Then the perturbation operator in this problem is equal to

$$\delta V = -\frac{3Ze^2}{2r_o} + \frac{Ze^2r^2}{2r_o^3} + \frac{Ze^2}{r}, \quad r \le r_o.$$

From this we obtain the shift of the electron energy for the first order of the perturbation theory

$$\Delta\varepsilon = \int \delta V\, |\psi|^2\, d\mathbf{r} = |\psi(0)|^2 \int \delta V\, d\mathbf{r} = \frac{Ze^2r_o^2}{10}\, |\psi(0)|^2\,,$$

where $\psi(0)$ is the electron wave function in the center. In particular, for the ground state of the hydrogenlike ions we have $\psi(0) = 2(Z/a_o)^{3/2}$, where a_o is the Bohr radius. This gives

$$\Delta\varepsilon = \frac{2}{5}\frac{Z^4e^2}{a_o}\left(\frac{r_o}{a_o}\right)^2 = \frac{4}{5}J\left(\frac{Zr_o}{a_o}\right)^2,$$

where $J = Z^2e^2/2a_o$ is the ionization potential of the ground state of the hydrogenlike ions. The size of a light nuclei is of the order of $r_o \sim 10^{-13}$ cm, and hence the relevant correction is $\Delta\varepsilon/J \sim 10^{-9} \div 10^{-10}$. Therefore, for light nuclei, this correction is small compared to the other effects.

Problem 2.2. *Determine the splitting of levels 2s and 2p of the hydrogen atom in a constant electric field if this splitting remarkably exceeds the fine structure splitting of the levels, and the level splitting is linear with respect to the electric field strength.*

An electric field eliminates the degeneracy which takes place for states with a given principal quantum number by neglecting the relativistic effects. Our goal is to find the electron eigenwave functions for the hydrogen atom, with respect to the Hamiltonian,

$$\hat{H} = \hat{H}_o + V,$$

where $\hat{H}_o$ is the Hamiltonian of the electron in the hydrogen atom and the operator V is responsible for the interaction of the electron with the electric field and is equal to $V = Er\cos\theta$, where E is the electric field strength, and r, θ are spherical electron coordinates. Let us take as a basis the wave functions of the hydrogen atom for $n = 2$ in terms of the quantum numbers lm and calculate the matrix elements for the operator V between these wave functions. Because of the parity with respect to the polar angle θ and the dependence on the azimutal angle φ, only one matrix element between the states $lm = 10$ and 00 is not zero. Thus, the following functions are the eigenfunctions of the Hamiltonian $\hat{H}$:

$$\Psi_1 = \frac{1}{\sqrt{2}}(\psi_{00} + \psi_{10}), \qquad \Psi_2 = \frac{1}{\sqrt{2}}(\psi_{00} - \psi_{10}),$$
$$\Psi_3 = \psi_{11}, \qquad \Psi_4 = \psi_{1,-1}.$$

The matrix element $\langle \psi_{00} | V | \psi_{10} \rangle = -3E$. This gives for the energy of the related levels

$$\varepsilon_1 = -\frac{1}{8} - 3E, \qquad \varepsilon_2 = -\frac{1}{8} + 3E, \qquad \varepsilon_{3,4} = -\frac{1}{8}.$$

Note that this result is valid if the fine splitting of levels is small. This criterion has the following form in the usual units ($\Delta\varepsilon_f = 0.365\,\text{cm}^{-1}$): $E \gg 10^4$ V/cm.

Let us represent the expressions of the wave functions for $m = 0$. We have

$$\Psi_1 = \frac{1}{4\sqrt{\pi}} e^{-r/2}\left(1 - \frac{r-z}{2}\right), \qquad \Psi_2 = \frac{1}{4\sqrt{\pi}} e^{-r/2}\left(1 - \frac{r+z}{2}\right).$$

It is convenient to introduce the parabolic coordinates

$$\xi = r + z, \qquad \eta = r - z, \qquad d\mathbf{r} = (\xi + \eta)\, d\xi\, d\eta\, d\varphi/4.$$

The above wave functions have the following form in these coordinates:

$$\Psi_1 = \frac{1}{4\sqrt{\pi}} \exp\left(-\frac{\xi + \eta}{4}\right)\left(1 - \frac{\eta}{2}\right),$$
$$\Psi_2 = \frac{1}{4\sqrt{\pi}} \exp\left(-\frac{\xi + \eta}{4}\right)\left(1 - \frac{\xi}{2}\right).$$

As is seen, for the considering problem the variables are separated into parabolic coordinates. Introducing

$$\Psi_k = \frac{1}{\sqrt{4\pi}} X_k(\xi) Y_k(\eta),$$

we have, for the related states,

$$X_1 = \frac{1}{\sqrt{2}} e^{-\xi/4}, \qquad Y_1 = \frac{1}{\sqrt{2}} e^{-\eta/4}\left(1 - \frac{\eta}{2}\right),$$

$$X_2 = \frac{1}{\sqrt{2}} e^{-\xi/4}\left(1 - \frac{\xi}{2}\right), \qquad Y_2 = \frac{1}{\sqrt{2}} e^{-\eta/4}.$$

Separation of the variables into parabolic coordinates is the general property of the hydrogen atom located in an external electric field. Then the splitting of the energetic levels is linear with respect to the electric field strength. This is called the linear Stark effect.

Problem 2.3. *The hydrogen atom is located in a weak electric field. Using parabolic coordinates, determine the splitting of levels in the linear approximation with respect to the electric field. One can neglect the relativistic effects.*

The electron Hamiltonian in atomic units has the form

$$\hat{H} = -\frac{1}{2}\Delta - \frac{1}{r} - Ez,$$

where r is the electron distance from the center, z is its projection onto the direction of the electric field, and E is the electric field strength. Introduce the parabolic coordinates

$$\xi = r + z, \qquad \eta = r - z, \qquad d\mathbf{r} = (\xi + \eta)\, d\xi\, d\eta\, d\varphi/4.$$

Since

$$\Delta = \frac{4}{\xi + \eta}\frac{\partial}{\partial \xi}\left(\xi \frac{\partial}{\partial \xi}\right) + \frac{4}{\xi + \eta}\frac{\partial}{\partial \eta}\left(\eta \frac{\partial}{\partial \eta}\right) + \frac{1}{\xi\eta}\frac{\partial^2}{\partial \varphi^2},$$

one can separate the variables into parabolic coordinates. Then write the electron wave function in the form

$$\Psi = \frac{1}{\sqrt{2\pi}} X(\xi) Y(\eta) e^{im\varphi}.$$

Substituting this expression in the Schrödinger equation $\hat{H}\Psi = \varepsilon\Psi$ and separating the variables in this equation, we have

$$\frac{d}{d\xi}\left(\xi \frac{dX}{d\xi}\right) + \left(\frac{1}{2}\varepsilon\xi + \beta_1 - \frac{m^2}{4\xi} + \frac{E}{4}\xi^2\right) X = 0;$$

$$\frac{d}{d\eta}\left(\eta \frac{dY}{d\eta}\right) + \left(\frac{1}{2}\varepsilon\eta + \beta_2 - \frac{m^2}{4\eta} - \frac{E}{4}\eta^2\right) Y = 0.$$

Here β_1, β_2 are the separation constants which are connected by the relationship

$$\beta_1 + \beta_2 = 1.$$

Because $d\mathbf{r} = (\xi + \eta)\, d\xi\, d\eta\, d\varphi/4$, we have from the normalization condition,

$$\frac{1}{4}\int_0^\infty X^2(\xi)\xi\, d\xi \int_0^\infty Y^2(\eta)\, d\eta + \frac{1}{4}\int_0^\infty X^2(\xi)\, d\xi \int_0^\infty Y^2(\eta)\eta\, d\eta = 1.$$

Obtaining the eigenvalues β_1, β_2 for each Schrödinger equation and using the relationship between these values, one can determine the discrete spectrum of these values. First let us consider this problem in the absence of an electric field.

The analysis of the spectrum of equations for $X(\xi)$ and $Y(\eta)$ is analogous to the analysis of the Schrödinger equation (2.9). Introduce a parameter γ by the relation $\varepsilon = -\gamma^2/2$. For the analysis of the Schrödinger equation for $X(\xi)$ let us write the electron wave function in the form

$$X(\xi) = \exp(-\gamma\xi/2)\xi^{m/2}u(\xi),$$

where m is the modulus of the magnetic quantum number. This form of the wave function accounts for the character of its damping at large ξ and uses the requirement of a discrete spectrum of β_1. On the basis of the above expansion we have the following equation for the function $u(\xi)$:

$$\xi\frac{d^2u}{d\xi^2} + (m+1-\xi)\frac{du}{d\xi} + \left(\frac{\beta_1}{\xi} - \frac{m-1}{2}\right)u = 0.$$

The second condition for the wave function $X(\xi)$ follows from the normalization relation according to which at small ξ the wave function $X(\xi)$ cannot grow stronger than $\xi^{-1/2}$. Correspondingly, $u(\xi)$ grows, at small ξ, stronger than $\xi^{-(m+1)/2}$. Let us present this function as

$$u(\xi) = \xi^{\nu}\sum_{k=0}^{n_1} a_k\xi^k.$$

At small ξ we have $u(\xi) = a_o\xi^{\nu}$. Substituting this in the expression for $u(\xi)$ and being restricted by the main term of the expansion on ξ, we obtain

$$\nu(\nu+m) = 0.$$

Solving this equation we find $\nu = 0$ for $m = 0$ and $\nu = 0$, $\nu = -m$ for $m \neq 0$. Since $\nu < (m+1)/2$, the solution $\nu = -m$ does not satisfy the physical requirements for $m > 0$. Thus the only solution of this equation is $\nu = 0$. Thus the function $u(\xi)$ is

$$u(\xi) = \sum_{k=0}^{n_1} a_k\xi^k.$$

From the physical requirements it follows that this series is finite, because in the opposite case this series would diverge at large ξ in the exponential way. Let us consider the case of large ξ. Restricting by the main term $u(\xi) \sim \xi^{n_1}$, we obtain

$$n_1 = \beta_1/\gamma - (m+1)/2.$$

This relation gives the discrete spectrum of the Schrödinger equation for $X(\xi)$. The constant β_1 is the eigenvalue of this equation, and the quantum number n_1 characterizes the state which is described by the Schrödinger equation for ξ in the absence of an electric field. The second Schrödinger equation for $Y(\eta)$ is analogous to the equation for $X(\xi)$. Introducing the quantum number n_2 for the equation for $Y(\eta)$ we have, by analogy with the previous formula,

$$n_1 = \beta_2/\gamma - (m+1)/2.$$

The electron principal quantum number is $n = 1/\gamma$. From the condition $\beta_1 + \beta_2 = 1$, and from expressions for β_1 and β_2 we have

$$n = n_1 + n_2 + m + 1.$$

Let us analyze the results obtained. The variables in the Schrödinger equation for the hydrogen atom are separated in parabolic coordinates. The parabolic quantum numbers n_1, n_2 pass a series of whole numbers from 0 up to $n - m - 1$. The total number of states with a given value n of the principal quantum number is equal to

$$2\sum_{m=1}^{n-1}(n-m) + n = n^2,$$

where m is the modulus of the magnetic quantum number, and n is a number of states for $m = 0$. From comparison with formula (2.17) it follows that the number of states for a quantum number n in the case of parabolic coordinates is the same as in the case of spherical coordinates. But the eigenwave functions for the parabolic coordinates differ from those corresponding to spherical coordinates. They have no spherical symmetry, but can be constructed as a combination of spherical wave functions. An example of such a type is considered in the previous problem for $n = 2$.

The normalized eigenfunctions of the bound electron of the hydrogen atom in parabolic coordinates are given by the following expressions

$$\Psi(\xi, \eta, \varphi) = \frac{e^{im\varphi}}{\sqrt{2\pi}} \frac{\sqrt{2}}{n^2} \exp\left(-\frac{\xi + \eta}{2n}\right) \cdot \left(\frac{\xi\eta}{n^2}\right)^{m/2} u_{n_1 m}\left(\frac{\xi}{n}\right) u_{n_2 m}\left(\frac{\eta}{n}\right),$$

where

$$u_{km}(x) = \frac{(k!)^{1/2}}{[(k+m)!]^{3/2}} \cdot L_{k+m}^m(x) = \frac{1}{m!}\left[\frac{(k+m)!}{k!}\right]^{1/2} F(k, m+1, x),$$

where $L_{k+m}^m(x)$ is the Lagger polynomial, and $F(-k, m+1, x)$ is the hypergeometric function. Note that here m is the modulus of the magnetic quantum number, i.e., $m \geq 0$.

Now let us determine the splitting of levels of the hydrogen atom in a constant electric field. It is convenient to use parabolic coordinates because even in the presence of an electric field the variables are separated into parabolic coordinates. This means that energy levels of the hydrogen atom which are split in an external electric field are characterized by parabolic quantum numbers in contrast to spherical coordinates. In particular, this follows from the analysis of the previous problem where the eigenwave function of the excited hydrogen atom located in an electric field is a combination of spherical wave functions of the hydrogen atom.

The interaction operator of the hydrogen atom with an electric field is equal to $\hat{V} = -Ez = -E(\xi - \eta)/2$. In the first approach of the perturbation theory the energy of states is equal to

$$\varepsilon_{nn_1n_2m} = -\frac{1}{2n^2} - \frac{E}{2}\left\langle nn_1n_2m\left|\xi - \eta\right| nn_1n_2m\right\rangle.$$

Calculating the matrix element, we obtain

$$\varepsilon_{nn_1n_2m} = -\frac{1}{2n^2} - \frac{3}{2}En(n_1 - n_2).$$

In particular, for the case $n = 2$ which was considered in the previous problem we have the following states: $n_1 = 1, n_2 = m = 0$; $n_2 = 1, n_1 = m = 0$; and two states $n_1 = n_2 = 0, m = 1$. Levels of the last two states are not shifted in the linear approach at small electric field strengths, and the shift for the first two states is equal to $\pm 3E$ in accordance with the result of the previous problem.

Problem 2.4. *Determine the rate of decay of an excited hydrogen atom in an external electric field.*

The bound state of an electron of the hydrogen atom in an electric field is not stable. Indeed, there is a region, at large distances from the center in the direction of the electric field, where the electron has a continuous spectrum. Transition of the electron in this region (see Fig. 2.2) leads to decay of the bound state. Thus the energy level of the electron which is located in the field of the Coulomb center has a finite width. This width is determined by the time of the electron transition in the region of continuous spectrum. According to Fig. 2.2 where the cross section of the potential energy in a space is given, this region starts from the distance in the direction of the electric field $z_o = |\varepsilon| / E$, where ε is the electron binding energy and E is the electric field strength. Within the framework of quantum mechanics, this transition is a tunnel transition as follows from Fig. 2.2. Below we determine the time of this transition and hence the width of the electron level.

For this goal we use the Schrödinger equation for the electron wave function in parabolic coordinates which was given in the previous problem. We take another form of expansion of the wave function which is more convenient for this problem

$$\Psi = \frac{e^{im\varphi}}{\sqrt{2\pi}} \cdot \frac{\Phi(\xi)F(\eta)}{\sqrt{\xi\eta}}.$$

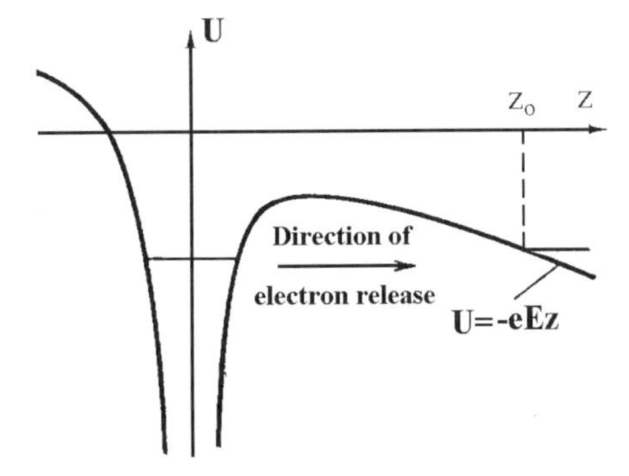

Figure 2.2. The cross section of the electron potential energy for $U(R)$ for the hydrogen atom in a constant electric field.

Correspondingly, the Schrödinger equation for these functions has the form

$$\frac{d^2\Phi}{d\xi^2} + \left(-\frac{\gamma^2}{4} + \frac{\beta_1}{\xi} - \frac{m^2-1}{4\xi^2} + \frac{E}{4}\xi\right)\Phi = 0,$$
$$\frac{d^2F}{d\eta^2} + \left(-\frac{\gamma^2}{4} + \frac{\beta_2}{\eta} - \frac{m^2-1}{4\eta^2} - \frac{E}{4}\eta\right)F = 0,$$

where the electron energy is given by the formula $\varepsilon = -\gamma^2/2$, and the separation constants β_1, β_2 in accordance with the results of the previous problem are equal to

$$\beta_1 = \gamma\left(n_1 + \frac{m+1}{2}\right), \qquad \beta_2 = \gamma\left(n_2 + \frac{m+1}{2}\right),$$

which correspond to the electron energy

$$\varepsilon_{nn_1n_2m} = -\frac{1}{2n^2} - \frac{3}{2}En(n_1 - n_2),$$

i.e., $\gamma = 1/n + \frac{3}{2}En^2(n_1 - n_2)$.

The probability of the electron transition through a barrier per unit time is equal to

$$w = \int_S \mathbf{j}\,d\mathbf{S},$$

where

$$j = \frac{i}{2}\left(\Psi\frac{\partial\Psi^*}{\partial z} - \Psi^*\frac{\partial\Psi}{\partial z}\right)$$

is the electron flux through the barrier, and S is the cross section of the barrier. We use that for small electric field strengths the main contribution to the electron current gives a small region of the barrier cross section near z_o. Use the cylindrical coordinates ρ, z, φ in this region and $\rho \ll z$. Under this condition the connection between the cylindrical and parabolic coordinates has the form

$$\xi = r + z \approx 2z, \qquad \eta = \sqrt{\rho^2 + z^2} \approx \frac{\rho^2}{2z} \approx \rho^2/\xi.$$

The element of the barrier surface is equal to

$$dS = \rho\,d\rho\,d\varphi = \frac{\xi}{2}\,d\eta\,d\varphi.$$

Substituting these formulas into the expression for the rate of decay of this state, we have

$$w = \int_0^\infty \frac{F^2(\eta)d\eta}{\eta}\frac{i}{2}\left(\Phi\frac{d\Phi^*}{d\xi} - \Phi^*\frac{d\Phi}{d\xi}\right).$$

Using expressions for the electron wave functions which are obtained in the previous problem, we get

$$w = \frac{i}{2}\left(\Phi\frac{d\Phi^*}{d\xi} - \Phi^*\frac{d\Phi}{d\xi}\right).$$

In the region near the center, where one can neglect the action of the electric field, the wave function $\Phi(\xi)$ is

$$\Phi = \frac{\sqrt{2}}{n} \cdot \frac{1}{\left[n_1!\,(n_1+m)!\right]^{1/2}} \cdot \left(\frac{\xi}{n}\right)^{n_1+(m+1)/2} \exp\left(-\frac{\xi}{2n}\right).$$

For the analysis of the solution for the Schrödinger equation for the wave function $\Phi(\xi)$, we find the expression for this wave function in the vicinity of the point ξ_o which separates a classically available region from a region where the location of a free classical electron is forbidden. The position of the point ξ_o is given by the equation $p\,(\xi_o) = 0$, where p is the classical electron momentum which, in accordance with the Schrödinger equation for $\Phi\,(\xi)$, is determined by the expression

$$p^2 = -\frac{\gamma^2}{4} + \frac{\beta_1}{\xi} - \frac{m^2-1}{4\xi^2} + \frac{E}{4}\xi.$$

The quasi-classical solutions of the Schrödinger equation $\Phi'' + p^2\Phi = 0$, right and left from the turning point, have the form

$$\Phi = \frac{iC}{\sqrt{p}} \cdot \exp\left(i\int_{\xi_o}^{\xi} p\,d\xi - \frac{i\pi}{4}\right), \quad \xi > \xi_o;$$
$$\Phi = \frac{iC}{\sqrt{p}} \cdot \exp\left(-\int_{\xi_o}^{\xi} |p|\,d\xi\right), \quad \xi < \xi_o.$$

Substituting the first of these expressions into the formula for the decay rate, we obtain

$$w = |C|^2\,.$$

Thus, the probability of the electron tunnel transition per unit time is expressed through the wave function amplitude on the boundary of the region of the free electron motion. In order to determine this value, it is necessary to join the expressions for the wave function near the center and near the boundary of the classical region. In the case of small electric field strengths there is a broad region of ξ where, on one hand, we can neglect the action of the electric field and, on the other hand, the classical expression for the wave function is valid. Joining in this region the expression for the wave function in neglecting the electric field with the classical wave function, we find the value C and, correspondingly, the rate of atom decay

$$w = \frac{\gamma^3\left(4\gamma^3/E\right)^{2n_1+m+1}}{(n_1+m)!n_1!} \cdot \exp\left(-\frac{2\gamma^3}{3E}\right).$$

On one hand, in the region where the solutions are joined we neglect the action of the electric field that corresponds to the condition $\xi \ll \xi_o = \gamma^2/E$ and, on the other hand, in this region we use the quasi-classical wave function that is valid, if $\gamma^2/4 \gg \beta_1/\xi$. This leads to the following criterion of validity of the above expression for the decay rate

$$E \ll \frac{\gamma^4}{\beta_1}.$$

Thus at small values of the electric field strength E the probability of the atom decay per unit time depends exponentially on the value $1/E$. Note that in the previous problem we obtained the expansion of the value γ over a small parameter E. In this case it is necessary to use two expansion terms of γ over E, because the second term of the expansion of γ determines the numerical coefficient for w. In this approximation we have

$$\frac{2\gamma^3}{3E} = \frac{2}{3n^3E} + 3(n_1 - n_2),$$

and the probability of decay of the hydrogen atom per unit time in a weak electric field is equal to:

$$w = \frac{\left(\frac{4}{3n^3E}\right)^{2n_1+m+1}}{n^3\,(n_1+m)!n_1!} \cdot \exp\left(-\frac{2}{3n^3E} + 3n_2 - 3n_1\right).$$

In particular, from this formula it follows for the rate of decay for the hydrogen atom in the ground state,

$$w = \frac{4}{3E} \cdot \exp\left(-\frac{2}{3E}\right).$$

Problem 2.5. *Determine the relative intensities of the radiative transitions from the hydrogen atom state with the quantum numbers nlm in states with quantum numbers $n'l'$ and different electron momentum projections m' onto a given direction.*

According to formula (1.22), the intensities of radiative transitions in the hydrogen atom case are proportional to the value $|\langle nlm|z|n'l'm'\rangle|^2$, where z is the electron coordinate in the direction of the photon polarization. Representing the electron wave function in the form (2.3) and integrating over angles, we find the matrix elements which determine the related radiative transitions

$$\left|\langle nlm\,|z|\,n', l+1, m\rangle\right|^2 = \frac{(l+1)^2 - m^2}{(2l+3)(2l+1)}\left|\langle nl\,|r|\,n', l+1\rangle\right|^2,$$

$$\left|\langle nlm\,|z|\,n', l-1, m\rangle\right|^2 = \frac{l^2 - m^2}{(2l+1)(2l-1)}\left|\langle nl\,|r|\,n', l-1\rangle\right|^2,$$

$$\left|\langle nlm\,|x+iy|\,n', l+1, m+1\rangle\right|^2 = \frac{(l+m+2)(l+m+1)}{(2l+3)(2l+1)}\left|\langle nl\,|r|\,n', l+1\rangle\right|^2,$$

$$\left|\langle nlm\,|x+iy|\,n', l-1, m+1\rangle\right|^2 = \frac{(l-m)(l-m-1)}{(2l+1)(2l-1)}\left|\langle nl\,|r|\,n', l-1\rangle\right|^2,$$

$$\left|\langle nlm\,|x-iy|\,n', l+1, m-1\rangle\right|^2 = \frac{(l-m+2)(l-m+1)}{(2l+3)(2+1)}\left|\langle nl\,|r|\,n', l+1\rangle\right|^2,$$

$$\left|\langle nlm\,|x-iy|\,n',l-1,m-1\rangle\right|^2=\frac{(l+m)(l+m-1)}{(2l+1)(2l-1)}\left|\langle nl\,|r|\,n',l-1\rangle\right|^2,$$

where

$$\langle nl|r|nl'\rangle=\int_0^\infty R_{nl}(r)R_{n'l'}(r)r^3\,dr,$$

and the radial wave functions $R_{nl}(r)$ are given by formula (2.16). These relations allow us to obtain the distribution on polarizations and final states for transitions with a given photon energy. In particular, from this it follows that

$$\sum_{m'}\left|\langle nlm\,|\mathbf{r}|\,n',l+1,m'\rangle\right|^2=\frac{l+1}{2l+1}\left|\langle nl\,|r|\,n',l+1\rangle\right|^2,$$

$$\sum_{m'}\left|\langle nlm\,|\mathbf{r}|\,n',l-1,m'\rangle\right|^2=\frac{l}{2l+1}\left|\langle nl\,|r|\,n',l-1\rangle\right|^2.$$

This proves that the lifetime of an excited state nlm of the hydrogen atom with respect to radiative transition in a lower state does not depend on the momentum projection m' onto a given direction.

Problem 2.6. *Determine the photoionization cross section of a hydrogenlike ion near the threshold.*

The photoionization cross section is given by formula (2.40) and has the following form in the relevant case:

$$\sigma_{\text{ion}}=\frac{8\pi q\omega}{3c}K_{0,1}^2,$$

where $\omega=Z^2/2+q^2/2$ and $q\ll Z$. The radial wave function of the initial state is $R_o=2Z^{3/2}e^{-Zr}$ and the radial wave function of a free electron state is

$$R_{q1}(r)=\frac{2}{3}\sqrt{2\pi qZ}re^{-iqr}F\left(i\frac{Z}{q}+2,4,2iqr\right).$$

In the limit of small q the matrix element is equal to

$$K_{01}(q)=\frac{1}{3}\sqrt{2\pi q}Z^2\int_0^\infty r^4dre^{-Zr}F\left(i\frac{Z}{q}+2,4,2iqr\right)$$

$$=\frac{8\sqrt{2\pi q}}{Z^3}\cdot F\left(i\frac{Z}{q}+2,5,4,\frac{2iq}{Z}\right)=\frac{8i}{Z^2e^2}\sqrt{\frac{\pi}{2q}},$$

where we use the relation for the hypergeometric functions

$$F(\alpha,\beta,\gamma,x)=\frac{(\gamma-\alpha)}{\gamma}\cdot(1-x)^\alpha F\left(\alpha,\gamma-\beta,\gamma+1,\frac{x}{x-1}\right)$$

$$+\frac{\alpha}{\gamma}\cdot(1-x)^{-\alpha}F\left(\alpha+1,\gamma-\beta,\gamma+1,\frac{x}{x-1}\right),$$

and are restricted by the first term of the expansion of the obtained expression over a small parameter q/Z. Substitution of the expression for the matrix element in the formula for the photoionization cross section which now has the form

$$\sigma_{\text{ion}} = \frac{4\pi q Z^2}{3c} K_{0,1}^2,$$

leads to the following value of the threshold cross section:

$$\sigma_{\text{ion}} = \frac{2^9 \pi^2}{3e^4 Z^2 c},$$

where e is the base of the natural logarithm. One can see that this threshold cross section does not depend on the energy of a released electron and is equal to $6.3 \cdot 10^{-18}$ cm^2 for the hydrogen atom.

Problem 2.7. *Determine the cross section of a radiative attachment of a slow electron to a nucleus of charge Z with the formation of a hydrogenlike ion in the ground state.*

The relevant process proceeds according to the scheme

$$e + A^{+Z} \to A^{+(Z-1)} + \hbar\omega,$$

where A^{+Z} is a nucleus of charge Z. This is the inverse process with respect to the photoionization process, so that the rates of both processes are connected by the principle of detailed balance. The condition of equilibrium of these processes leads to the following relation between the cross sections of photoionization σ_{ion} and photorecombination σ_{rec}:

$$c \frac{2\, d\mathbf{k}\, d\mathbf{r}}{(2\pi)^3} g_a \sigma_{\text{ion}} = v \frac{d\mathbf{q}\, d\mathbf{r}}{(2\pi)^3} g_e g_i \sigma_{\text{rec}},$$

where $\mathbf{k}$ is the photon wave vector, $\mathbf{q}$ is the electron wave vector, v is the electron velocity, c is the light velocity, $g_e = 2$, and g_i, g_a are the statistical weights of the electron, ion A^{+Z} and ion $A^{+(Z-1)}$. This relation corresponds to the equilibrium, so that in some volume containing one photon and one atom, the numbers of transitions per unit time with ionization and recombination must be coincided. Then taking into account the energy conservation law (1.27) and the dispersion relation $\omega = kc$ between the photon frequency ω and wave vector k, we obtain

$$\sigma_{\text{rec}} = \frac{g_i}{g_a} \frac{k^2}{q^2} \cdot \sigma_{\text{ion}}.$$

Now we have $g_a = 2$, $g_i = 1$, and on the basis of the threshold value of the photoionization cross section which was obtained in the previous problem we have

$$\sigma_{\text{rec}} = \frac{256\pi^2 Z^2}{3e^4 c^3 q^2},$$

where e is the base of the natural logarithm. Using the numerical values of the parameters of this formula, we obtain (in atomic units)

$$\sigma_{\text{rec}} = 6 \cdot 10^{-6} \frac{Z^2}{q^2}.$$

Problem 2.8. *Determine the average coefficient of the photorecombination of the electrons and nuclei of a charge Z under the assumption that a hydrogenlike ion is formed in the ground state and under the Maxwell distribution of electrons on velocities when the average velocity of electrons is relatively small.*

The photorecombination coefficient of electrons and ions is equal to

$$\alpha = \langle v\sigma_{\text{rec}} \rangle \, ,$$

where v is the electron velocity, σ_{rec} is the photorecombination cross section, and the angle brackets mean averaging over the Maxwell distribution function of electrons on velocities. In the case of slow electrons one can use the result of the previous problem for the recombination cross section, so that the recombination coefficient is equal to

$$\alpha = \frac{256\pi^2 Z^2}{3e^4 c^3} \left\langle \frac{1}{q} \right\rangle .$$

Since for the Maxwell distribution function we have

$$\left\langle \frac{1}{q} \right\rangle = \sqrt{\frac{2}{\pi T_e}} ,$$

where T_e is the electron temperature, the recombination coefficient takes the form

$$\alpha = \frac{256\pi\sqrt{2\pi}\, Z^2}{3e^4 c^3 \sqrt{T_e}} = \alpha_o Z^2 T_e^{-1/2} ,$$

where e is the base of the natural logarithm, so that $\alpha_o = 1.5 \cdot 10^{-13}\, \text{cm}^3/\text{s}$ if T_e is expressed in eV.

CHAPTER 3

Two-Electron Atoms and Ions

3.1 The Pauli Exclusion Principle and Symmetry of the Atomic Wave Function

The atom is a quantum system consisting of a charged nucleus and electrons. Because of a large nuclear mass compared to the electron mass, one can assume the nuclear mass to be infinite. Then an atom is a system consisting of a Coulomb center and electrons, and atom states are determined by the electron behavior in the field of the Coulomb center. The Pauli exclusion principle is of importance for the atom structure. This principle follows from the Fermi–Dirac statistics for electrons and prohibits location of two electrons in the same state. The Pauli exclusion principle leads to an additional interaction in the related electron system. Indeed, if a new electron is added to an atomic core, it cannot be located near an atomic nucleus where other atomic electrons are found. Hence an atom occupies a more larger size than in the case without this principle. The additional interaction between electrons due to the Pauli exclusion principle is called the exchange interaction.

The mathematical formulation of the Pauli exclusion principle consists of the requirement that the electron wave function would be antisymmetric with respect to the transposition of electrons, which is given by formula (1.19):

$$\begin{aligned}&\Psi(\mathbf{r}_1, \sigma_1, \ldots, \mathbf{r}_k, \sigma_k, \ldots, \mathbf{r}_m, \sigma_m, \ldots, \mathbf{r}_n, \sigma_n)\\ &\quad = -\Psi(\mathbf{r}_1, \sigma_1, \ldots, \mathbf{r}_m, \sigma_m, \ldots, \mathbf{r}_k, \sigma_k, \ldots, \mathbf{r}_n, \sigma_n),\end{aligned} \tag{3.1}$$

where $\mathbf{r}_k$ is the coordinate of the k-electron, σ_k is the projection of the electron spin onto a given direction, and the atom consists of n electrons. From relation (3.1) it follows that if states of m and k-electrons are the same (i.e., $\mathbf{r}_k = \mathbf{r}_m$, $\sigma_k = \sigma_m$), then the wave function and, respectively, the electron density, are equal to zero. Thus, the peculiarity of an electron system is such that the interaction potential of an electron with an atomic core depends on the positions of the other electrons, i.e., the exchange interaction of electrons in an atom is of importance. Note that in a system of many electrons collective effects such as plasma waves can occur. But the Pauli exclusion principle states that collective phenomena are weak because of the positions of individual electrons are separated. Therefore, the best model for atomic electrons is a one-electron approach when the wave function of an atom is a sum of products of one-electron wave functions. Then the interaction of an atomic electron with an atomic core includes the exchange interaction of this electron with other electrons due to the Pauli exclusion principle and an electrostatic interaction of this electron with the atomic core. Below we consider this approach for two electron systems—the helium atom and heliumlike ions.

3.2 The Helium Atom

The helium atom or heliumlike ions consist of a heavy nucleus of a positive charge Z and two electrons. We assume one electron to be found in a state described by a wave function Φ, and the state of the other electron is described by a wave function φ. Then in the one-electron approach the electron wave function of this system has the form, accounting for the Pauli exclusion principle,

$$\Psi(1,2) = \frac{1}{\sqrt{2}} \cdot \begin{vmatrix} \Phi(1) & \Phi(2) \\ \varphi(1) & \varphi(2) \end{vmatrix}, \tag{3.2}$$

where the argument shows an electron which is described by this wave function. The wave function of an individual electron is a product of space and spin wave functions. Consider the case when the spatial electron wave functions are identical. Denoting the coordinate wave function by $\psi(\mathbf{r})$ and the spin wave functions by η we have, for the total wave function,

$$\Psi(1,2) = \frac{1}{\sqrt{2}} \cdot \begin{vmatrix} \psi(\mathbf{r}_1)\eta_+(1) & \psi(\mathbf{r}_2)\eta_+(2) \\ \psi(\mathbf{r}_1)\eta_-(1) & \psi(\mathbf{r}_2)\eta_-(2) \end{vmatrix} = \psi(\mathbf{r}_1)\psi(\mathbf{r}_2)S(1,2), \tag{3.3}$$

where

$$S(1,2) = \frac{1}{\sqrt{2}} \cdot \left[\eta_+(1)\eta_-(2) - \eta_-(1)\eta_+(2)\right],$$

where $\mathbf{r}_1$, $\mathbf{r}_2$ are the space coordinates of the electrons, $\eta_+(i)$, $\eta_-(i)$ are the spin wave functions of the ith electron with the spin projection onto a given direction $\frac{1}{2}$ or $-\frac{1}{2}$ respectively, and $S(1,2)$ is the antisymmetric spin wave function of electrons which corresponds to the total spin of electrons $S = 0$. This form of the wave function is called the Slater determinant.

Let us consider the Schrödinger equation for the two-electron atom and formulate the basic mathematical approaches which are useful for the analysis of this and more complicated atoms. If the atom is found in the ground or lower excited states, the variation method is useful for calculation of the atom energy. This is as follows. Let Ψ_k be eigen-wave functions of the atom and let ε_k be the corresponding values of the energy, so that $\hat{H}$ is the Hamiltonian of the electrons. Then the above wave functions satisfy the Schrödinger equation

$$\hat{H}\Psi_k = \varepsilon_k \Psi_k,$$

and the values Ψ_o, ε_o correspond to the ground state of the related atom. Take a wave function Ψ which is close to Ψ_o but differs from it. Let us expand the wave function Ψ over the eigenwave functions of the Hamiltonian $\hat{H}$,

$$\Psi = \sum_k a_k \Psi_k,$$

and the normalization condition gives

$$\sum_k |a_k|^2 = 1.$$

From this, on the basis of the condition of orthogonality of functions Ψ_k, we have

$$\langle \Psi \hat{H} \Psi \rangle = \sum_k |a_k|^2 \varepsilon_k = \varepsilon_o + \sum_k |a_k|^2 (\varepsilon_k - \varepsilon_o).$$

From this it follows that the functional $\langle \Psi \hat{H} \Psi \rangle$ with any function Ψ exceeds the energy of the ground state ε_o. Then among a certain class of wave functions with varying parameters the best wave function is such it that leads to the minimum of $\langle \Psi \hat{H} \Psi \rangle$. This is the basis of the variational principle which allows one to determine the wave function and the energy for the atom ground state with high accuracy. For the analysis of the first excited state it is required that a trial function should be orthogonal to the wave function Ψ_o of the ground state. In this way, one can spread the variational method for the first excited states of the considering atom.

Let us use the variational method for determination of the energy of the ground state of the helium atom or heliumlike ions. Take, as a test wave function, a product of one-electron functions of a hydrogenlike ion with a varying charge

$$\Psi = C \exp[-Z_{\text{ef}}(r_1 + r_2)], \tag{3.4}$$

where $C = Z_{\text{ef}}^6/\pi^2$ is the normalization constant and r_1, r_2 are the distances from the center for the corresponding electron. We use the atomic system of units. Write the Hamiltonian of the heliumlike ion in the form

$$\hat{H} = -\frac{1}{2}\Delta_1 - \frac{1}{2}\Delta_2 - \frac{Z}{r_1} - \frac{Z}{r_2} + \frac{1}{|\mathbf{r}_1 - \mathbf{r}_2|} = \hat{H}_o + V_1 + V_2, \tag{3.5}$$

where

$$\hat{H}_o = -\frac{1}{2}\Delta_1 - \frac{1}{2}\Delta_2 - \frac{Z_{\text{ef}}}{r_1} - \frac{Z_{\text{ef}}}{r_2},$$

$$V_1 = -\frac{(Z - Z_{\text{ef}})}{r_1} - \frac{(Z - Z_{\text{ef}})}{r_2}, \qquad V_2 = \frac{1}{|\mathbf{r}_1 - \mathbf{r}_2|}. \tag{3.6}$$

The test wave function is the eigenfunction with respect to the Hamiltonian $\hat{H}_o$, so that $\hat{H}_o \Psi = -Z_{\text{ef}}^2 \Psi$. According to the virial theorem, or as a result of direct evaluation, one can find the average value $\langle Z_{ef}/r\rangle$ for the hydrogenlike ion with a charge Z_{ef} which is equal to Z_{ef}^2. From this it follows that $\langle\Psi\,|V_1|\,\Psi\rangle = -Z_{ef}(Z - Z_{ef})$. For the calculation of the matrix element from the last term let us expand it over the Legendre polynomials

$$|\mathbf{r}_1 - \mathbf{r}_2|^{-1} = \sum_{n=0}^{\infty} \frac{r_<^n}{r_>^{n+1}} P_n(\cos\theta),$$

where θ is the angle between vectors $\mathbf{r}_1$ and $\mathbf{r}_2$ and $r_>$ and $r_<$ correspond to the larger and smaller values among r_1 and r_2. Because the wave function does not depend on angles, the integral is determined by the first expansion term, and this matrix element is equal to

$$\left\langle \Psi \frac{1}{|\mathbf{r}_1 - \mathbf{r}_2|} \Psi \right\rangle = 2\frac{Z_{ef}^6}{\pi^2} \cdot \int_0^{\infty} \exp(-2Z_{ef}r_1)4\pi r_1^2\, dr_1 \int_{r_1}^{\infty} 4\pi r_2\, dr_2 = -\frac{5}{8}Z_{ef}.$$

The factor 2 is due to the separation of the space into two regions: $r_2 > r_1$ and $r_2 < r_1$, and each of these regions gives the same contribution into the integral because of the symmetry of the problem. Thus we have, for the electron energy,

$$\varepsilon = -Z_{ef}^2 + 2Z_{ef}(Z_{ef} - Z) + \frac{5}{8}Z_{ef} = -Z_{ef} \cdot \left(2Z - \frac{5}{8} - Z_{ef}\right).$$

The condition of the energy minimum gives the optimal values of the parameters. We have

$$Z_{ef} = Z - \frac{5}{16}; \qquad \varepsilon = -Z_{ef}^2 = -Z^2 + \frac{5}{8}Z - \frac{25}{256}. \tag{3.7}$$

Let us determine the ionization potential of a heliumlike ion or helium atom. Note that after the release of one electron the second electron is found in the ground state of the hydrogenlike ion with the nuclear charge Z, so that its binding energy according to formula (2.41) is $Z^2/2$. Thus the first ionization potential J of the heliumlike ion is

$$J = -\varepsilon - \frac{Z^2}{2} = \frac{Z^2}{2} - \frac{5}{8}Z + \frac{25}{256}. \tag{3.8}$$

Table 3.1 gives a comparison of formula (3.8) with the precise values of the ionization potential. As is seen, the variation principle, in spite of its simplicity, allows one to determine the ionization potential with satisfactory accuracy.

Let us estimate the difference between the trial wave function and the precise function on the basis of the above results. For this goal we expand the trial wave function (3.4) over the system of eigenfunctions $\{\Psi_k\}$ of the Hamiltonian of the

Table 3.1. The ionization potentials of the helium atom and heliumlike ions.

Atom, ion	He	Li^+	Be^{++}	C^{4+}	O^{6+}
Accurate J, eV	24.56	75.64	153.9	392.3	739.3
Formula (3.8) for J in eV	23.07	74.08	152.3	390.4	737.3

atom $\hat{H}$, so that we have $\hat{H}\Psi_m = \varepsilon_k\Psi_k$, where ε_k are the precise energies of the atomic states. We have

$$\Psi = a_o\Psi_o + \sum_{k\neq 0} a_k\Psi_k.$$

If the test wave function Ψ is close to the precise wave function of the ground state Ψ_o, the coefficient a_o would be close to 1 and $\sum_{k\neq 0}|a_k|^2 = 1 - |a_o|^2$ would be small. For estimation of these values we use the expression

$$\langle\Psi\hat{H}\Psi\rangle = \sum_k |a_k|^2\varepsilon_k = \varepsilon_o + \sum_k |a_k|^2(\varepsilon_k - \varepsilon_o).$$

For simplicity, we assume that the test wave function, being expanded over the eigenwave function of the Hamiltonian along with the wave function of the ground atomic state, contains an admixture of the first excited state only. This means that only a_o and a_1 are not zero and from the above expression for the energy we obtain

$$a_1^2 = \frac{\Delta E}{\varepsilon_1 - \varepsilon_2},$$

where $\Delta E = \langle\Psi\hat{H}\Psi\rangle - \varepsilon_o$ is an error in the determination of the atom energy. Note that because the wave functions of the relevant state are real, the values a_o and a_1 are also real.

Let us estimate from this, for example, the degree of coincidence of the best wave function (3.4) with the accurate wave function for Be^{++}. As follows from the data of Table 3.1, in this case $\Delta E = 1.7$ eV, and the excitation energy of the lower state (state 2^1S) is equal to 122 eV. This gives $a_1^2 = 0.014$ and $a_1 \approx 0.1$. The total binding energy of two electrons of the beryllium ion is $\varepsilon_o = 322$ eV. As is seen, the variation method in the simplest form allows us to determine the electron energy with an accuracy of 0.5%, and the admixture of the excited states to the wave function is about 10%. This gives an estimation of the accuracy of the used variation method.

3.3 Self-Consistent Electric Field in Two-Electron Atoms

The variation method allows us to choose the best wave function with respect to the real one from a relevant class of wave functions. Symmetric combinations of one-electron wave functions are chosen as a trial wave function in the simplest versions of the method. These one-electron wave functions are eigenfunctions of the Hamiltonian which is a sum of one-electron Hamiltonians. In particular, the wave function (3.4) is the eigenfunction of a Hamiltonian which is a sum of one-electron Hamiltonians. Usage of this function is analogous to a change of the electron–electron interaction potential $1/|\mathbf{r}_1 - \mathbf{r}_2|$ by an effective interaction potential $(Z - Z_{ef})\cdot(1/r_1 + 1/r_2)$, i.e., the interaction of two electrons is changed by an effective shield of the nucleus charge. Thus, the introduction of a test wave function as a product of one-electron wave functions is analogous to the usage of an effective self-consistent interaction potential instead of a real one. The form of this effective interaction potential is determined by the form of the test wave functions.

The introduction of an effective interaction potential can be made in a general form without the use of a certain type of a self-consistent field. Let us make this for the helium atom in the ground state assuming a wave function $\psi(\mathbf{r})$ to be corresponding to each electron. Then the interaction potential between electrons, which acts on the first electron and is averaged over coordinates of the second electron, has the form

$$\int \frac{1}{|\mathbf{r}_1 - \mathbf{r}_2|} |\psi(\mathbf{r}_2)|^2 \, d\mathbf{r}_2.$$

Changing by this expression the interaction potential acted on the first electron, we obtain the following Schrödinger equation:

$$\left[-\frac{1}{2}\Delta + \frac{Z}{r_1} + \int \frac{1}{|\mathbf{r}_1 - \mathbf{r}_2|} |\psi(\mathbf{r}_2)|^2 \, d\mathbf{r}_2\right] \psi(\mathbf{r}_1) = \varepsilon \psi(\mathbf{r}_1), \tag{3.9}$$

where the energy 2ε is equal to the total energy of two electrons.

Equation (3.9) is called the Hartri equation. In this equation the interaction potential between electrons is changed by a self-consistent potential which is obtained by averaging of the interaction potential of electrons over coordinates of the second electron. In this equation we assume the wave function of each electron to be identical. If these functions are different and they are denoted as $\psi(\mathbf{r})$ and $\varphi(\mathbf{r})$, the corresponding Hartri set of equations takes the form

$$\left[-\frac{1}{2}\Delta + \frac{Z}{r} + \int \frac{1}{|\mathbf{r}\_\mathbf{r}'|} \left|\varphi(\mathbf{r}')\right|^2 d\mathbf{r}'\right] \psi(\mathbf{r}) = \varepsilon_1 \psi(\mathbf{r}),$$

$$\left[-\frac{1}{2}\Delta + \frac{Z}{r} + \int \frac{1}{|\mathbf{r} - \mathbf{r}'|} \left|\psi(\mathbf{r}')\right|^2 d\mathbf{r}'\right] \varphi(\mathbf{r}) = \varepsilon_2 \varphi(\mathbf{r}). \tag{3.10}$$

The total electron energy is equal, in this case, to $\varepsilon_1 + \varepsilon_2$.

The method of changing the interaction potential between electrons by a self-consistent interaction potential is called the Hartri approach. This approximation does not account for the symmetry of the wave function. In order to take into account the wave function symmetry, we represent the spatial wave function of two electrons in the form

$$\Psi(\mathbf{r}_1, \mathbf{r}_2) = \frac{1}{\sqrt{2}} \cdot \left[\psi(\mathbf{r}_1)\varphi(\mathbf{r}_2) \pm \psi(\mathbf{r}_2)\varphi(\mathbf{r}_1)\right],$$

and assume the functions $\psi(\mathbf{r})$ and $\varphi(\mathbf{r})$ to be orthogonal, i.e., $\int \psi^*(\mathbf{r})\varphi(\mathbf{r})\,\mathbf{dr} = 0$. In addition, the wave functions $\psi(\mathbf{r})$ and $\varphi(\mathbf{r})$ are normalized

$$\int \psi^*(\mathbf{r})\psi(\mathbf{r})\mathbf{dr} = \int \varphi^*(\mathbf{r})\varphi(\mathbf{r})\mathbf{dr} = 1.$$

For simplicity, below we take the functions $\psi(\mathbf{r})$ and $\varphi(\mathbf{r})$ to be real. Let us take the Hamiltonian of electrons in the form [see formula (3.5)]:

$$\hat{H} = -\frac{1}{2}\Delta_1 - \frac{1}{2}\Delta_2 - \frac{Z}{r_1} - \frac{Z}{r_2} + \frac{1}{|\mathbf{r}_1 - \mathbf{r}_2|} = \hat{h}_1 + \hat{h}_2 + \frac{1}{|\mathbf{r}_1 - \mathbf{r}_2|},$$

where

$$\hat{h}_1 \equiv \hat{h}(\mathbf{r}_1) = -\frac{1}{2}\Delta_1 - \frac{Z}{r_1}, \qquad \hat{h}_2 \equiv \hat{h}(\mathbf{r}_2) = -\frac{1}{2}\Delta_2 - \frac{Z}{r_2}.$$

Substituting the total wave function into the Schrödinger equation $\hat{H}\Psi = \varepsilon\Psi$, multiplying the Schrödinger equation subsequently by $\varphi(\mathbf{r}_2)$ or $\psi(\mathbf{r}_2)$, and integrating over coordinates of the second electron, we obtain the following set of equations:

$$\begin{aligned}
&\hat{h}\psi(\mathbf{r}) + \left[\int \varphi(\mathbf{r}')\hat{h}\varphi(\mathbf{r}')\mathbf{dr} + \int \frac{1}{|\mathbf{r}-\mathbf{r}'|}\varphi^2(\mathbf{r}')^2 d\mathbf{r}'\right]\psi(\mathbf{r}) \\
&\quad \pm \left[\int \varphi(\mathbf{r}')\hat{h}\psi(\mathbf{r}')d\mathbf{r}' + \int \frac{1}{|\mathbf{r}-\mathbf{r}'|}\varphi(\mathbf{r}')\psi(\mathbf{r}')d\mathbf{r}'\right]\varphi(\mathbf{r}) = \varepsilon\psi(\mathbf{r}), \\
&\hat{h}\varphi(\mathbf{r}) + \left[\int \psi(\mathbf{r}')\hat{h}\psi(\mathbf{r}')d\mathbf{r}' + \int \frac{1}{|\mathbf{r}-\mathbf{r}'|}\psi^2(\mathbf{r}')d\mathbf{r}'\right]\varphi(\mathbf{r}) \\
&\quad \pm \left[\int \psi(\mathbf{r}')\hat{h}\varphi(\mathbf{r}')d\mathbf{r} + \int \frac{1}{|\mathbf{r}-\mathbf{r}'|}\psi(\mathbf{r}')\varphi(\mathbf{r}')d\mathbf{r}'\right]\psi(\mathbf{r}) = \varepsilon\varphi(\mathbf{r}). \quad (3.11)
\end{aligned}$$

The set of equations (3.11) introduces a self-consistent atomic field simultaneously with taking into account the symmetry of the wave function. As follows from this set of equations, the different one-electron wave functions correspond to symmetric and antisymmetric wave functions. This set of equations corresponds to the so-called Hartri–Fock approach. It accounts for the exchange interactions in the atom along with the one-electron electrostatic interactions. Therefore, the Hartri–Fock approach describes the real atoms well and is the basis of the numerical calculations for atomic systems.

From the analysis of the set of equations (3.11) it follows that if the electrons of two-electron atoms are found in different states, the atom energy depends on the symmetry of the wave function. Indeed, according to the Pauli exclusion principle, the total wave function of the electrons is antisymmetric with respect to the transposition of electrons. The total wave function is a product of the spin and spatial wave functions. The spin wave function of two electrons is symmetric with respect to the transposition of spins if the total electron spin is equal to one and is antisymmetric if the total spin is zero. Therefore, the spatial wave function is symmetric with respect to the transposition of electron coordinates if the total electron spin is zero, and is antisymmetric if the total electron spin is one. Since the energy of the two-electron atom depends on the symmetry of the spatial wave function of electrons, the energy depends on the total atom spin. Thus, although the Hamiltonian of electrons does not depend on electron spins, the spin state of the electrons influences the total electron energy through the wave function symmetry. The difference of the energies of states for a two-electron atom with the same one-electron states, but different values of the total spin, is called the exchange splitting of electron levels. Note that the exchange interaction potential in light atoms remarkably exceeds the relativistic interactions which depend on the spin states. The reason is that the exchange interaction does not include the small parameter (2.1)—the constant of fine structure.

Let us use the above analysis for the helium atom state with the electron shell $1sns$. There are two such atomic states with total spins $S = 0$ and $S = 1$. Since the states with $S = 1$ electrons have an identical direction of spins, according to the Pauli exclusion principle the second electron is repulsed from the region where the first electron is located, because two these electrons cannot be located at the same space point. In the state with $S = 0$, such a prohibition is absent, and an external electron shields an internal electron in the most degree than in the state with $S = 1$. Hence, the states with $S = 1$ are characterized by more high ionization potentials than those with $S = 0$ (see also Fig. 3.3 where the Grotrian diagram for the helium atom is given). As is seen, the nature of the exchange splitting of levels does not connect with the relativistic effects.

3.4 Self-Consistent Electric Field for Many-Electron Atoms

The above concept of a self-consistent field can be spread for many-electron atoms. Then an effective interaction between electrons also has a one- and two-electron character. Let us consider a many-electron atom on the basis of the self-consistent field concept. For this atom the electron Hamiltonian (1.18) can be divided into parts in the following way:

$$\hat{H} = \sum_{j=1}^{n} \hat{h}_j + \sum_{j=1}^{n} U_j + \sum_{j,k=1}^{n} v_{jk}, \tag{3.12}$$

where

$$\hat{h}_j = -\frac{1}{2}\Delta_j + V(r_j), \qquad U_j = -\frac{Z}{r_j} - V(r_j), \qquad v_{jk} = \frac{1}{|\mathbf{r}_j - \mathbf{r}_k|}.$$

The operator $\hat{h}_j$ includes the kinetic energy of a j-electron and its interaction with a self-consistent field, the last term accounts for the interaction between electrons.

Let us introduce the atom wave functions as a combination of one-electron wave functions. Then, due to the symmetry (1.19) with respect to the transposition of electrons, the atom wave function can be written in the form of the Slater determinant

$$\Psi = \frac{1}{\sqrt{n}} \begin{vmatrix} \psi_1(1)\chi_1(1) & \psi_1(2)\chi_1(2) & \dots & \psi_1(n)\chi_1(n) \\ \psi_2(1)\chi_2(1) & \psi_2(2)\chi_2(2) & \dots & \psi_2(n)\chi_2(n) \\ \dots & \dots & \dots & \dots \\ \psi_n(1)\chi_n(1) & \psi_n(2)\chi_n(2) & \dots & \psi_n(n)\chi_n(n) \end{vmatrix}, \tag{3.13}$$

where ψ is the spatial wave function, χ is the spin wave function, the subscript means the electron state, and the argument indicates the number of an electron to

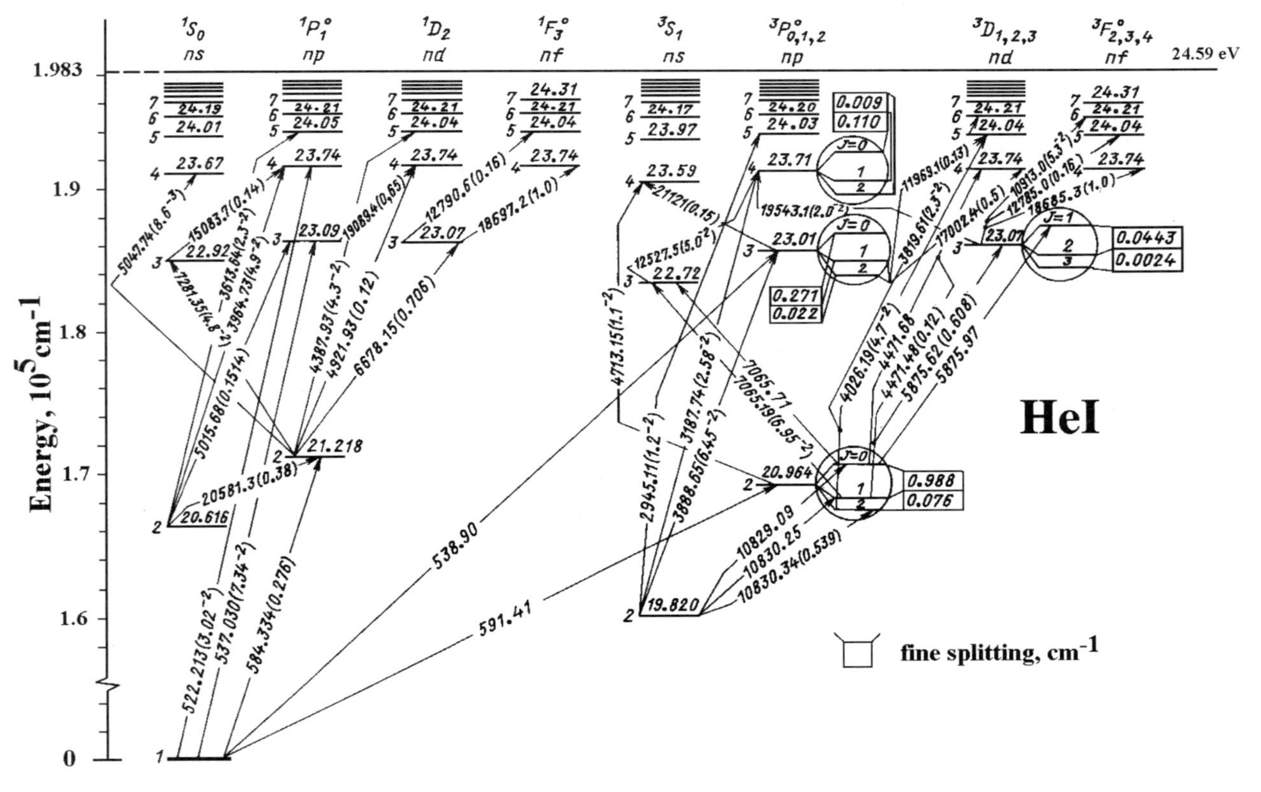

Figure 3.1. The Grotrian diagram for the helium atom.

which the wave function corresponds. The wave function in the form of the Slater determinant changes the sign as a result of the transposition of both electrons and states. Assume that we choose the optimal one-electron wave functions which are eigenfunctions of the Hamiltonians $\hat{h}_j$. Let us determine, on the basis of these functions, the atom energy which has the form

$$E = \langle \Psi | \hat{H} | \Psi \rangle = \sum_{j=1}^{n} \varepsilon_j + \sum_{j=1}^{n} \langle j|U|j\rangle - \sum_{j,k=1}^{n} \langle jk|v|kj\rangle a_{jk}. \tag{3.14}$$

Summation in the last term is made over all j and k; the matrix elements of formula (3.14) have the form

$$\langle j|U|j\rangle = \int \psi_j^*(\mathbf{r})U(\mathbf{r})\psi_j(\mathbf{r})\,d\mathbf{r};$$

$$\langle jk|v|kj\rangle = \int \psi_j^*(\mathbf{r})\psi_k^*(\mathbf{r}')v(\mathbf{r},\mathbf{r}')\psi_k(\mathbf{r})\psi_j(\mathbf{r}')\,d\mathbf{r}\,d\mathbf{r}'.$$

The value a_{jk} is equal to zero, if the spin projections are different for states j and k; this value is one, if the spin projections are coincident for these states. We obtain formula (3.14) on the basis of (3.12) and (3.13) by taking into account the orthogonality of the one-electron wave functions, i.e.,

$$\int \psi_j^*(\mathbf{r})\psi_k(\mathbf{r})\,d\mathbf{r} = \delta_{jk}.$$

The structure of expression (3.14) for the atom energy is such that only one- and two-electron interactions are present in the approach of a self-consistent field for a many-electron system. The first and second terms account for the electron energy if electrons are located in a self-consistent field which is averaged over the positions of other electrons. The last term is called the exchange interaction potential and is a result of the Pauli exclusion principle. This interaction reflects the character of the symmetry of the wave function (1.19) with respect to the transposition of electrons.

Thus, the above formalism allows us to describe a many-electron atomic system on the basis of the self-consistent field concept and the exchange interaction between electrons. Using the one-electron approximation, one can reduce the problem to the two-electron approximation. Let us determine the ionization potential of an atomic particle within the framework of this concept. Assuming that the wave function of electrons of the atomic core does not change as a result of the release of a valent electron we obtain, from formula (3.14) for the atom ionization potential,

$$J = E(n-1) - E(n) = \varepsilon_j + \langle j|U|j\rangle - 2\sum_{k=1}^{n-1} \langle jk|v|kj\rangle, \tag{3.15}$$

where summation in the last term is made over electrons with the same spin projection that has the removed electron j. This form of the atom ionization potential reflects the character of the interaction of atomic electrons.

Table 3.2. The dependence on Z for some parameters of multicharged heliumlike ions.

Parameter	Z-dependence
Ionization potential	Z^2
Exchange interaction	Z
Spin-orbit interaction	Z^4
Rate of one-photon dipole radiation	Z^4
Rate of two-photon dipole radiation	Z^8

3.5 Heliumlike Ions

Heliumlike ions have two electrons located in the Coulomb field of a nucleus charge Z. Above we considered some examples of these ions, now we concentrate on multicharged helium ions in the case of $Z \gg 1$. Table 3.2 gives the dependence of some parameters of these ions on Z. Because of a different Z-dependence for some parameters, the interaction character varies with an increase Z. Indeed, the role of relativistic interactions grows with an increase in Z, whereas the role of the exchange interactions drops. Table 3.3 contains examples which confirm this fact. As is seen from Table 3.3, the relative contribution of the relativistic interactions to the electron energy increases with the growth of Z, while the contribution of exchange interactions decreases.

As for the radiative transitions of multicharged heliumlike ions, the rates of forbidden transitions increase more strongly with an increase in Z than those for the permitted transitions. Some examples for radiative transitions from the lowest excited states to the ground state are given in Table 3.4. Note that if we approximate Z-dependence for the rate of the radiative transitions at large Z as Z^n, then the exponent is equal to $n = 3.6$ for the transition $2^1P \to 1^1S$, $n = 5.8$ for the transition $2^1S \to 1^1S$, $n = 7.3$ for the transition $2^3S \to 1^1S$, and $n = 9.5$ for the transition $2^3P \to 1^1S$. Because of the different Z-dependencies for forbidden and permitted transitions, for large Z the terms "forbidden" and "permitted" lost their meaning. In particular, as follows from Table 3.4, the radiative lifetimes of levels 2^3P and 2^1P have the same order of magnitude at large Z, while they differ by several orders of magnitude at small Z.

Table 3.3. Relative values of the exchange and spin-orbit interactions in heliumlike ions. Here $\varepsilon_{ex}(2^3S)$ is the excitation energy of the metastable state 2^3S, Δ is the difference of excitation energies for levels 2^1S and 2^3S, and $\Delta\varepsilon(2^3P_2 - 2^3P_0)$ is the fine splitting of levels $2^3P_2 - 2^3P_0$. The ratios are given in percent.

Ion	Li^+	Ne^{+8}	Ca^{+18}	Zn^{+28}	Zr^{+38}	Sn^{+48}
Z	3	10	20	30	40	50
$\Delta/\varepsilon_{ex}(2^3S)$, %	3.2	1.1	0.60	0.42	0.38	0.29
$\Delta\varepsilon(2^3P_2 - 2^3P_0)/\varepsilon_{ex}(2^3S)$,%	0.001	0.025	0.14	0.35	0.61	1.1

Table 3.4. Times of radiative transitions for heliumlike ions.

Ion	Li^+	Ne^{+8}	Ca^{+18}	Zn^{+28}	Zr^{+38}	Sn^{+48}
Z	3	10	20	30	40	50
$\tau(2^3S \to 1^1S)$, s	49	$9.2 \cdot 10^{-5}$	$7.0 \cdot 10^{-8}$	$1.1 \cdot 10^{-9}$	$6 \cdot 10^{-11}$	$6 \cdot 10^{-12}$
$\tau(2^1S \to 1^1S)$, s	$5.1 \cdot 10^{-4}$	$1.0 \cdot 10^{-7}$	$1.2 \cdot 10^{-9}$	$1.0 \cdot 10^{-10}$	$2 \cdot 10^{-11}$	$4 \cdot 10^{-12}$
$\tau(2^3P \to 1^1S)$, s	$5.6 \cdot 10^{-5}$	$1.8 \cdot 10^{-10}$	$2.1 \cdot 10^{-13}$	$8.1 \cdot 10^{-15}$	$1.4 \cdot 10^{-15}$	$5 \cdot 10^{-16}$
$\tau(2^1P \to 1^1S)$, s	$3.9 \cdot 10^{-11}$	$1.1 \cdot 10^{-13}$	$6.0 \cdot 10^{-15}$	$1.3 \cdot 10^{-15}$	$5 \cdot 10^{-16}$	$2 \cdot 10^{-16}$

Problems

Problem 3.1. *Determine the ionization potential of the heliumlike ion in the ground state considering the interaction between electrons as a perturbation.*

Use the electron Hamiltonian in the form (3.5):

$$\hat{H} = -\frac{1}{2}\Delta_1 - \frac{1}{2}\Delta_2 - \frac{Z}{r_1} - \frac{Z}{r_2} + \frac{1}{|\mathbf{r}_1 - \mathbf{r}_2|} = \hat{H}_o + V,$$

where

$$V = \frac{1}{|\mathbf{r}_1 - \mathbf{r}_2|}$$

is taken as a perturbation. Then the nonperturbed wave function is a product of the one-electron wave functions of the ground state of the hydrogenlike ion with the nucleus charge Z:

$$\Psi = \psi(r_1)\psi(r_2), \qquad \psi(r) = \frac{Z^{3/2}}{\sqrt{\pi}} \cdot \exp(-Zr).$$

The wave function Ψ is the eigenfunction of the Hamiltonian $\hat{H}_o$, so that $\hat{H}_o\Psi = -Z^2\Psi$. Then in the first order of the perturbation theory we have, for the electron energy,

$$\varepsilon = \langle \Psi \hat{H} \Psi \rangle = -Z^2 + \left\langle \Psi \left| \frac{1}{|\mathbf{r}_1 - \mathbf{r}_2|} \right| \Psi \right\rangle.$$

The average value of the potential of the electron–electron interaction is evaluated above and is equal to

$$\left\langle \Psi \left| \frac{1}{|\mathbf{r}_1 - \mathbf{r}_2|} \right| \Psi \right\rangle = -\frac{5}{8}Z.$$

This gives, for the total electron energy of the heliumlike ion,

$$\varepsilon = -Z^2 + \frac{5}{8}Z.$$

From this it follows that, for the ionization potential of the heliumlike ion in the ground state,

$$J = -\frac{Z^2}{2} - \varepsilon = \frac{Z^2}{2} - \frac{5}{8}Z.$$

In particular, in the case of the helium atom, we have that this formula gives a lower value compared to the result (3.8) of the variation method. The difference of these ionization potentials is $\frac{25}{256} = 2.66\,\text{eV}$. Thus, the perturbation theory leads to a lower value of the ionization potential, and the perturbation theory is worse for the analysis of the ground state of the helium atom than the variation method in the simplest form.

Problem 3.2. *Calculate the energy of heliumlike ions in the ground state within the framework of the perturbation theory where the nonperturbed wave functions correspond to the hydrogenlike ion with an effective charge. Prove that, if the effective charge is obtained on the basis of the variation method, the energy change for the first order of the perturbation theory is zero.*

Take the wave function (3.4),

$$\Psi = C\exp[-Z_{\text{ef}}(r_1 + r_2)], \qquad C = Z_{\text{ef}}^6/\pi^2,$$

as a nonperturbed wave function. This is the eigenfunction of the Hamiltonian

$$\hat{H}_o = -\frac{1}{2}\Delta_1 - \frac{1}{2}\Delta_2 - \frac{Z_{\text{ef}}}{r_1} - \frac{Z_{\text{ef}}}{r_2}.$$

Let us present the Hamiltonian in the form (3.5), (3.6):

$$\hat{H} = \hat{H}_o + V,$$

where

$$V = -\frac{(Z - Z_{\text{ef}})}{r_1} - \frac{(Z - Z_{\text{ef}})}{r_2} + \frac{1}{|\mathbf{r}_1 - \mathbf{r}_2|}.$$

On the basis of the above matrix elements we have, in the first order of the perturbation theory,

$$\langle \Psi V \Psi \rangle = -2Z_{\text{ef}}(Z - Z_{\text{ef}}) + \frac{5}{8}Z_{\text{ef}} = -2Z_{\text{ef}} \cdot \left(Z - Z_{\text{ef}} - \frac{5}{16}\right).$$

From this it follows that if we take the value of Z_{ef} on the basis of the variation method ($Z = Z_{\text{ef}} - 5/16$), this expression is zero. This proves that the wave function with this value of Z_{ef} is the closest to the accurate wave function of the heliumlike ions among the hydrogenlike wave functions.

Problem 3.3. *Determine the ionization potential of the helium atom in the ground state and the heliumlike lithium ions by the variation principle. Take the test wave function in the form* $\Psi = C[\exp(-\alpha r_1 - \beta r_2) - \exp(-\beta r_1 - \alpha r_2)]$, *with varying parameters* α *and* β.

The electron energy

$$\varepsilon = \frac{\langle \Psi \hat{H} \Psi \rangle}{\langle \Psi \Psi \rangle}$$

is expressed through a set of standard integrals. Evaluation of these integrals leads to the electron energy

$$\varepsilon = -Z(\alpha + \beta) + \alpha\beta + \frac{\frac{(\alpha-\beta)^2}{2} + \frac{\alpha\beta}{\alpha+\beta} + \frac{\alpha^2\beta^2}{(\alpha+\beta)^3} + \frac{20\alpha^3\beta^3}{(\alpha+\beta)^5}}{1 + \frac{64\alpha^3\beta^3}{(\alpha+\beta)^6}}.$$

The minimum of this expression for the helium atom gives the values of the parameters $\alpha = 1.189$ and $\beta = 2.183$. This leads to the following value of the ionization potential $J = 23.83\,\text{eV}$. Note that in the case of the use of one varying parameter ($\alpha = \beta = 1.6875$) the ionization potential is equal to 23.06 eV, while its accurate value is 24.56 eV. In the case of the lithium ion we have that the energy minimum corresponds to the values of the parameters $\alpha = 2.08$ and $\beta = 3.29$. This yields, for the ionization potential of the lithium ion, $J = 74.80\,\text{eV}$. The variation of one parameter gives ($\alpha = \beta = 2.6875$) the ionization potential 74.08 eV, and its accurate value is 75.64 eV.

As follows from the above analysis, usage of the second varying parameter leads to a decrease in the error in the ionization potential. In both cases $\beta > Z$, i.e., the parameters α and β are not an effective charge which acts on the corresponding electron. Note that in both cases the value $(\alpha + \beta)/2$ is close to Z_{ef}—an effective charge which is the varying parameter for the wave function with one varying parameter.

Problem 3.4. *Determine the asymptotic expression for the wave function of the valent electron on the basis of the wave function of the previous problems.*

The electron radial wave function at large distances from the center satisfies the Schrödinger equation

$$-\frac{1}{2r} \cdot \frac{d^2}{dr^2}(r\psi) - \frac{1}{r}\psi = -\frac{\gamma^2}{2}, \qquad r\gamma \gg 1, \quad r\gamma^2 \gg 1,$$

where $\gamma^2/2 = J$ is the ionization potential of the atom. The solution of this equation has the form

$$\psi = Ar^{1/\gamma - 1}e^{-r\gamma}, \qquad r\gamma \gg 1, \quad r\gamma^2 \gg 1.$$

This problem consists of the calculation of the parameter A which characterizes the amplitude of electron location outside the atomic core. We evaluate this value on the basis of the following considerations. The test wave function of the variation method is close to the accurate function in a region where electrons are located. The electron energy is determined only by this region. Hence, the test wave function cannot give the accurate asymptotic expression. Nevertheless, there is a region where the electron density is not small, but where the asymptotic expression for the electron density is valid. Joining the test and asymptotic wave functions in this region, one can find the parameter A. The accuracy of this operation is the higher, the more accurate the test function. Correspondingly, the more accurate the test wave function, the wider is the region where the test and asymptotic wave functions are close.

We introduce the electron number density for a two-electron atom as

$$\rho(\mathbf{r}) = \int \Psi^2(\mathbf{r}, \mathbf{r}')(r')^2\, dr',$$

and this is normalized by the condition $\int \rho(\mathbf{r})r^2\, dr = 2$. The asymptotic expression for the electron density is

$$\rho(r) = 2A^2 r^{-0.511} e^{-2.687r}.$$

We first use for electrons the hydrogenlike wave function of Problem 3.2,

$$\Psi(r, r') = C \exp\left[-Z_{\text{ef}}(r + r')\right], \qquad C = Z_{\text{ef}}^6/\pi^2,$$

where r, r' are the distances of the electrons from the nucleus, and $Z_{\text{ef}} = 27/16$ follows from the variation principle for the helium atom in the ground state. This wave function gives, at a large distance r from the center,

$$\rho(r) = 2\int_0^\infty \left|\Psi(r', r)\right|^2 (r')^2\, dr' = 8Z_{\text{ef}}^3 \exp(-2Z_{\text{ef}}r) = 38.4 \exp(-3.375r),$$

and the comparison of this formula with the asymptotic expression of the electron density gives, for the asymptotic coefficient,

$$A_1^2(r) = 19.2 r^{0.511} \exp(-0.688r).$$

This value is denoted by 1 in Fig. 3.2.

On the basis of the expression of the wave function which is used in the previous problem we have

$$\rho(r) = 4 \cdot \left[\frac{1}{\alpha^3\beta^3} + \frac{64}{(\alpha+\beta)^6}\right]^{-1} \cdot \left[\frac{e^{-2\alpha r}}{\beta^3} + \frac{16e^{-(\alpha+\beta)r}}{(\alpha+\beta)^3} + \frac{e^{-2\beta r}}{\alpha^3}\right].$$

Using the numerical values of the previous problem, we obtain

$$\rho(r) = 3.817e^{-2.378r} + 16.57e^{-3.372r} + 23.62e^{-4.366r}.$$

Comparing this expression with the asymptotic one we obtain, for the asymptotic coefficient,

$$A_2^2(r) = r^{0.511}\left(1.908e^{0.309r} + 8.286e^{-0.685r} + 11.81e^{-1.679r}\right)$$

In order to determine the error in the value A, let us use the asymptotic expression of the electron wave function in accounting for two terms of expansion. Then the asymptotic expression for the wave function has the form

$$\psi = Ar^{1/\gamma - 1}e^{-r\gamma} \cdot \left(1 + \frac{a}{r}\right), \qquad \text{where} \quad a = -\frac{1}{2\gamma} \cdot \left(\frac{1}{\gamma} - 1\right).$$

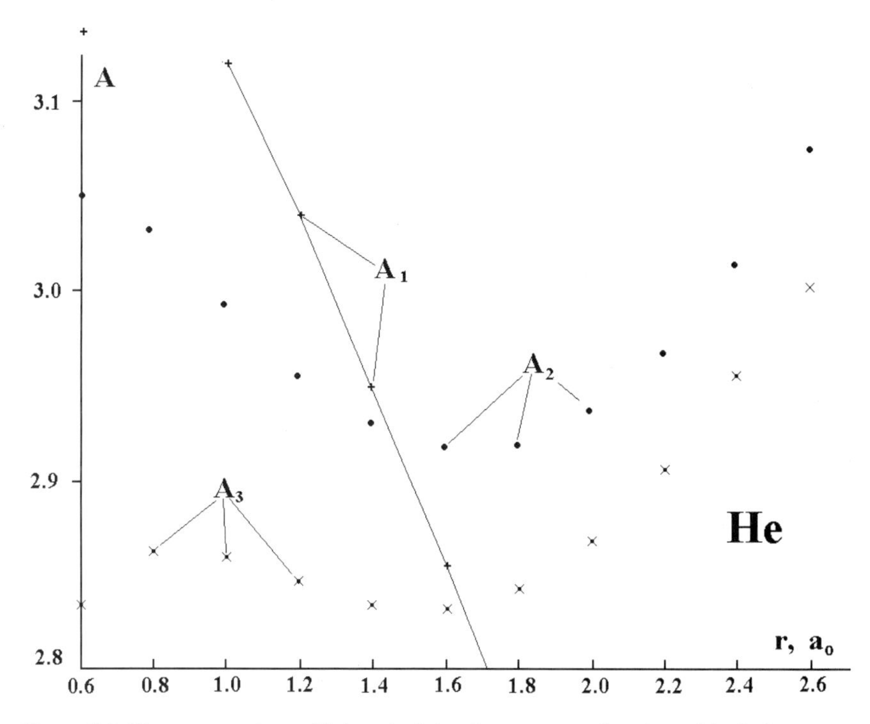

Figure 3.2. The asymptotic coefficient A of the electron wave function of the helium atom according to the results of Problem 3.4.

Then, repeating the above operations, we obtain, for the asymptotic coefficient in the first approximation,

$$A_3^2 = \frac{A_2^2}{1 + 0.095/r},$$

where the value A_2^2 was calculated above. Figure 3.2 contains the values A_1, A_2, A_3 as a function of r. One can see that the functions $A_2(r)$ and $A_3(r)$ vary weakly near the maximum of these functions, and that the function $A_2(r)$ has a maximum at $r = 1.65$, while the function $A_3(r)$ has a maximum at $r = 1.55$ for $A_1(r)$. The function (3.9) has a maximum at $r = 0.75$. Averaging these functions in the distance range $r = 0.4 \div 2.6$, we find that the asymptotic coefficient according to formula (3.10) is $A_2 = 2.98 \pm 0.05$ in this region and according to formula (3.11) it is $A_3 = 2.87 \pm 0.07$, so that on the basis of these data we have

$$A = 2.9 \pm 0.1.$$

The use of the hydrogenlike wave functions of electrons gives $A_1 = 2.8 \pm 0.3$ in this range, i.e., the error in the case of the use of a simple wave function increases several times. This example shows that the identical wave functions of valent electrons lead to a remarkable error for the asymptotic coefficient.

Problem 3.5. *Determine the ionization potential of the helium atom and the lithium ion in the metastable state $1s2s(2^3S)$ on the basis of the variation method with the use of the hydrogen wave functions such that the $1s$-electron is located in the Coulomb field of the nucleus charge, and the effective charge acted on the second electron is varied.*

Because the spin wave function for the relevant state with $S = 1$ is symmetric with respect to transposition of the spins, the spatial wave function is antisymmetric with respect to transposition of the electron coordinates. Therefore, the spatial electron wave function can be represented in the form

$$\Psi = \frac{\psi(r_1)\varphi(r_2) - \psi(r_2)\varphi(r_1)}{\sqrt{2(1 - S^2)}},$$

where ψ is the wave function of the $1s$-electron and φ is the wave function of the $2s$-electron. These wave functions are described by the expressions

$$\psi(r) = \left(\frac{Z^3}{\pi}\right)^{1/2} \exp(-Zr), \qquad \varphi(r) = \left(\frac{\alpha^3}{8\pi}\right)^{1/2} \left(1 - \frac{\alpha r}{2}\right) \exp\left(-\frac{\alpha r}{2}\right),$$

where Z is the nucleus charge and α is the effective charge of the atomic core. The overlapping integral from these wave functions is equal to $S = \langle\psi\varphi\rangle = (2x)^{3/2}(x - 1)/(x + 1/2)^4$, where $x = Z/\alpha \gg 1$.

It is essential that the 2^3S-state is the lowest energetic state among the states with antisymmetric coordinate wave functions. This is orthogonal to the wave function of the ground state due to the symmetry of the spin function. This fact simplifies the procedure, because in the case of the 2^1S-state it is necessary to use such test spatial wave functions which are ortogonal to the wave function of the ground state. Divide the Hamiltonian into parts in the following way:

$$\hat{H} = -\frac{1}{2}\Delta_1 - \frac{1}{2}\Delta_2 - \frac{Z}{r_1} - \frac{Z}{r_2} + \frac{1}{|\mathbf{r}_1 - \mathbf{r}_2|} = \hat{h}_1 + \hat{h}_2 - \frac{Z - \alpha}{r_2} + \frac{1}{|\mathbf{r}_1 - \mathbf{r}_2|},$$

where

$$\hat{h}_1 = -\frac{1}{2}\Delta_1 - \frac{Z}{r_1}, \qquad \hat{h}_2 = -\frac{1}{2}\Delta_2 - \frac{\alpha}{r_2}.$$

On the basis of the relations

$$\left(-\frac{1}{2}\Delta - \frac{Z}{r}\right)\psi(r) = -\frac{Z^2}{2}\psi(r), \qquad \left(-\frac{1}{2}\Delta - \frac{\alpha}{r}\right)\varphi(r) = -\frac{\alpha^2}{8}\varphi(r),$$

we obtain, for the energy of the related state,

$$\varepsilon = \frac{\langle[\psi(r_1)\varphi(r_2) - \psi(r_2)\varphi(r_1)]\hat{H}[\psi(r_1)\varphi(r_2)]\rangle}{1 - S^2}$$
$$= -\frac{Z^2}{2} - \frac{\alpha^2}{8} - \frac{(Z - \alpha)}{1 - S^2}(A - BS) + \frac{K - L}{1 - S^2},$$

where

$$A = \left\langle \varphi \left| \frac{1}{r} \right| \varphi \right\rangle, \qquad B = \left\langle \varphi \left| \frac{1}{r} \right| \psi \right\rangle,$$

$$K = \left\langle \psi(r_1)\varphi(r_2) \left| \frac{1}{|\mathbf{r}_1 - \mathbf{r}_2|} \right| \psi(r_1)\varphi(r_2) \right\rangle,$$

$$L = \left\langle \psi(r_1)\varphi(r_2) \left| \frac{1}{|\mathbf{r}_1 - \mathbf{r}_2|} \right| \psi(r_2)\varphi(r_1) \right\rangle.$$

Calculation of these matrix elements by the standard methods gives

$$A = \frac{\alpha}{4}, \qquad B = \frac{\alpha}{2} \frac{(2x)^{3/2}(x - 1/2)}{(x + 1/2)^3}, \qquad L = \frac{\alpha x^3}{8} \cdot \frac{(20x^3 - 30x + 13)}{(x + 1/2)^7},$$

$$K = \frac{\alpha}{4} \cdot \left[1 - \frac{4x + 1}{4(x + 1/2)^3} + \frac{5x + 1}{4(x + 1/2)^4} - \frac{3(6x + 1)}{32(x + 1/2)^5} \right].$$

In particular, in the case $x = 1$ (i.e., $Z = \alpha$), these integrals are equal to

$$A = \frac{Z}{4}, \qquad B = \frac{4\sqrt{2}}{27} Z = 0.210Z,$$

$$K = \frac{17}{81} Z = 0.210Z, \qquad L = \frac{16}{729} Z = 0.0219Z.$$

The above expressions give, for the electron energy,

$$\varepsilon = -\frac{Z^2}{2\cdot} \cdot \left[1 + \frac{1}{4x^2} + \frac{x - 1}{2x^2} \cdot \frac{1 - 16x^3(x - 1)(x - 1/2)/(x + 1/2)^7}{1 - 8x^3(x - 1)^2/(x + 1/2)^8} \right]$$

$$+ \frac{Z}{4x\left[1 - [8x^3(x - 1)^2/(x + 1/2)^8]\right]}$$

$$\cdot \left[1 - \frac{4x+1}{4(x+1/2)^3} + \frac{5x+1}{4(x+1/2)^4} - \frac{3(6x+1)}{32(x+1/2)^5} - \frac{x^3(20x^2 - 30x + 13)}{2(x+1/2)^7} \right].$$

This expression must be optimized over parameter x. Table 3.5 contains the results which follow from such an optimization. The ionization potential of the atom is equal to $J = -\varepsilon - Z^2/2$. There are, in parentheses, the accurate values of the ionization potential for the helium atom and heliumlike ions.

As is seen, the use of the variation method with a variation of one parameter leads to an error in the energy of electrons of the order of 0.1 eV. This error is lower than in the case of the helium atom or heliumlike ions in the ground state

Table 3.5. Parameters from the optimization of the wave function of the metastable helium atom or heliumlike ions.

Atom, ion	Z	$x_{\min}$	$\alpha_{\min}$	$-\varepsilon$, eV	J, eV
$He(2^3S)$	2	1.289	1.552	58.96	4.53(4.79)
$Li^+(2^3S)$	3	1.167	2.571	138.85	16.40(16.62)
$Be^{++}(2^3S)$	4	1.118	3.578	252.78	35.08(35.30)

where this error is of the order of 1 eV. The reason of this fact is such that in the relevant case the electron orbits are separated, and hence the electron correlation is not so essential for the electron energy. This allows us to find a closer function to the accurate function by the use of one varying parameter. Note that in the relevant case we vary an effective charge of the atomic core in which field the excited electron is located. From the data in Table 3.5 it follows that the shielding of the nucleus field by a $1s$-electron is close for the relevant cases. Indeed, the action of the internal electron on the excited electron is identical to the presence of the negative charge near the center of the value $0.448e$ for the helium atom, $0.429e$ for the lithium ion, and $0.422e$ for the two-charged beryllium ion.

Problem 3.6. *Within the framework of the perturbation theory determine the exchange splitting of the 2^3S- and 2^1S-states for the helium atom and heliumlike ions.*

In the relevant case the electron shell has the configuration $1s2s$, and the coordinate wave function of electrons has the form

$$\Psi = \frac{\psi(r_1)\varphi(r_2) \pm \psi(r_2)\varphi(r_1)}{\sqrt{2(1 \pm S^2)}},$$

where ψ is the wave function of the $1s$-electron, φ is the wave function of $2s$-electron, r_1, r_2 are the coordinates of the electrons, and $S = \langle \varphi \mid \psi \rangle$ is the overlapping integral (see Problem 3.5). The electron Hamiltonian has the form

$$\hat{H} = -\frac{1}{2}\Delta_1 - \frac{1}{2}\Delta_2 - \frac{Z}{r_1} - \frac{Z}{r_2} + \frac{1}{|\mathbf{r}_1 - \mathbf{r}_2|} = \hat{H}_o + \frac{1}{|\mathbf{r}_1 - \mathbf{r}_2|},$$

where the last term is considered as the perturbation. The wave functions ψ and φ are the eigenfunctions of the Hamiltonian

$$\hat{H}_o\Psi = (-Z^2/2 - Z^2/8)\Psi,$$

and the overlapping integral is equal to zero. Thus, the electron energy of the relevant states is

$$\varepsilon = -\frac{5}{8}Z^2 + \left\langle \Psi \left| \frac{1}{|\mathbf{r}_1 - \mathbf{r}_2|} \right| \Psi \right\rangle.$$

This gives

$$\varepsilon = -\frac{5}{8}Z^2 + K \pm L,$$

where

$$K = \left\langle \psi(r_1)\varphi(r_2) \left| \frac{1}{|\mathbf{r}_1 - \mathbf{r}_2|} \right| \psi(r_1)\varphi(r_2) \right\rangle,$$

$$L = \left\langle \psi(r_1)\varphi(r_2) \left| \frac{1}{|\mathbf{r}_1 - \mathbf{r}_2|} \right| \psi(r_2)\varphi(r_1) \right\rangle.$$

The plus sign corresponds to the 2^1S-state and the minus sign refers to the 2^3S-state. Values of the integrals were calculated in the previous problem and now they

Table 3.6. Ionization potentials of the excited helium atom and heliumlike ions.

State	He	Li^+	Be^{++}
2^1S	0.99(3.97)	11.68(14.72)	29.19(32.25)
2^3S	3.38(4.79)	15.27(16.62)	33.96(35.30)

are equal to

$$K = \frac{17}{81}Z, \qquad L = \frac{16}{729}Z.$$

This gives the ionization potential for the first order of the perturbation theory

$$J = -\varepsilon - \frac{Z^2}{2} = \frac{Z^2}{8} - \frac{17}{81}Z \pm \frac{16}{729}Z,$$

where the minus sign corresponds to the 2^1S-state and the plus sign refers to 2^3S-state. Table 3.6 gives values of the ionization potentials calculated on the basis of this formula. The accurate values of the ionization potentials are given in parentheses.

As is seen, the relevant version of the perturbation theory leads to the error for the 2^3S-state approximately 1.3–1.6 eV, while the variation method of the previous problem gives an error of 0.2–0.3 eV. The reason for a high error in the perturbation theory is explained by neglecting the shielding of the external electron by the internal one. This effect is stronger for the 2^1S-state. For this reason the error in the case of the 2^1S-state is higher and is about 3 eV. As is seen, the perturbation theory is not suitable even for a rough estimation of the ionization potential of the 2^1S helium state.

Problem 3.7. *Within the framework of the perturbation theory determine the ionization potential of the helium atom or heliumlike ions in the 2^1S-state. Assume that the nonperturbed state corresponds to the location of the 1s-electron in the Coulomb field of charge Z, and charge $Z-1$ acts on the 2s-electron.*

From the results of the previous problem it follows that the shielding of the external electron by the internal one is of importance. In the previous problem we neglected this effect, and in this problem we assume the total shielding of the external electron by the internal one. Using the expressions of Problem 3.5 we have, for the electron energy of the 2^1S-state,

$$\varepsilon = \frac{\langle[\psi(r_1)\varphi(r_2) + \psi(r_2)\varphi(r_1)]\hat{H}[\psi(r_1)\varphi(r_2)]\rangle}{1+S^2}$$
$$= -\frac{Z^2}{2} - \frac{\alpha^2}{8} - \frac{(Z-\alpha)}{1+S^2}(A+BS) + \frac{K+L}{1+S^2},$$

where the definition of the integrals of this expression is given in Problem 3.5. On the basis of the expressions of these integrals in Problem 3.5 we have, for the electron energy,

$$\varepsilon = -\frac{Z^2}{2} \cdot \left[1 + \frac{1}{4x^2} + \frac{x-1}{2x^2} \cdot \frac{1 + \frac{16x^3(x-1)(x-1/2)}{(x+1/2)^7}}{1 + \frac{8x^3(x-1)^2}{(x+1/2)^8}} \right]$$
$$+ \frac{Z}{4x} \cdot \left[\frac{1 - \frac{4x+1}{4(x+1/2)^3} + \frac{5x+1}{4(x+1/2)^4} - \frac{3(6x+1)}{32(x+1/2)^5} + \frac{x^3(20x^2-30x+13)}{2(x+1/2)^7}}{1 + \frac{8x^3(x-1)^2}{(x+1/2)^8}} \right].$$

The results for the ionization potentials are given in Table 3.7. We use, in Table 3.7, both the total shielding of an external electron ($\alpha = Z - 1$, $x = Z/(Z-1)$) and its partial shielding of the nucleus charge which is taken as the same as in the case of the 2^3S-state of Problem 3.5. As is seen, the real values of the ionization potentials are located between the calculated values. This means that the shielding in the case of the 2^1S-state is more than in the case of the 2^3S-state. The error is of the order of 1 eV and is more than in the case of the variation method in Problem 3.5 for the 2^3S-state. This is clear because the variation method for optimization over the shielding charge is used.

Problem 3.8. *Determine the exchange splitting of levels of the helium atom and heliumlike ions for the electron shell* $1s2s$ *within the framework of the perturbation theory. Ascertain the dependence of this splitting on the nucleus charge in the limit of large* Z.

The exchange splitting of levels in this case is the difference of energies of the 2^1S- and 2^3S-states. According to the formulas of Problem 3.5, this value is equal to

$$\Delta = \varepsilon(2^1S) - \varepsilon(2^3S)$$
$$= -\frac{(Z-\alpha)}{1+S^2}(A+BS) + \frac{K+L}{1+S^2} + \frac{(Z-\alpha)}{1-S^2}(A-BS) - \frac{K-L}{1-S^2},$$

where the definitions of the integrals and their values are given in Problem 3.5. Because the overlapping integral is small $S \ll 1$, this formula takes the form

$$\Delta = 2L - 2(Z-\alpha)(B - AS)S - 2KS^2.$$

Note that $Z - 1 < \alpha < Z$. The lower limit corresponds to the total shielding of the external electron by the internal electron, and the upper limit corresponds to the absence of this shielding. First let us analyze this expression in the limit of large Z.

Table 3.7. The ionization potentials in eV for the 2^1S-states of the helium atom and heliumlike ions.

Atom, ion	Total shielding	Partial shielding (Problem 3.5)	Accurate values
He	4.60	3.00	3.97
Li^+	15.66	13.61	14.72
Be^{++}	33.30	31.08	32.25

Table 3.8. Parameters of the potential of exchange interaction for the electron shell $1s2s$.

Atom, ion	α	Δ, eV	α_{ac}	Δ_{ac}, eV
He	1.55	1.52	1.40	0.82
Li^+	2.57	2.78	2.40	1.90
Be^{++}	3.58	4.00	3.40	3.05

The strongest dependence on Z contains the first term of this formula which varies proportional to Z at large Z; the second term does not depend on Z, and the third term is $\sim 1/Z$. Hence in the limit of large Z this formula gives $\Delta = 32Z/729$. If we express Δ in eV, this formula gives $\Delta = 1.2Z$ (eV). Table 3.8 lists values of the exchange splitting of levels with the use of values for the effective charge α which was obtained in Problem 3.5 for the 2^3S-state by the variation method. Table 3.8 also contains accurate values of the exchange splitting Δ_{ac} and values of the parameter $\alpha(\alpha_{ac})$ which lead to the observational values. As is seen, the value $Z - \alpha_{ac}$ in all cases is equal to 0.6, while for the 2^3S-states it lies between 0.42 and 0.45 (see Table 3.5). This shows that the effective shielding charges are different for the 2^1S- and 2^3S-states.

The above analysis for the energy levels of the helium atom in the ground and lowest excited states gives a general presentation about the validity of the perturbation and variation methods for calculation of the positions of their energy levels. On the basis of this analysis one can conclude that the perturbation theory is not reliable for this goal, while the variation method can provide suitable accuracy for the positions of the energy levels.

Problem 3.9. *Determine the cross section of photoionization of the helium atom and heliumlike ions for large photon energy compared to the atom ionization potential.*

We use formula (2.40) for the cross section of one-electron photoionization by taking into account two facts. First, two electrons can take part in this process independently, so that it is necessary to multiply the cross section (2.41) by two. Second, the state of the combined electrons is changed as a result of the release of other electrons. Hence, an additional factor occurs in formula (2.41) which is the projection of the wave function of the final state of the combined electron onto the wave function of its initial state. As a result, the photoionization cross section (2.41) has the form

$$\sigma_{\text{ion}} = \frac{16\pi q\omega}{3c} K_{01}^2 S^2,$$

where K_{01} is the matrix element for the radial wave functions of the released electrons and the overlapping integral S is equal in the relevant case

$$S = 4Z^{3/2} Z_{\text{ef}}^{3/2} \int_0^\infty r^2\, dr \exp[-(Z + Z_{\text{ef}})r] = \frac{8Z^{3/2} Z_{\text{ef}}^{3/2}}{(Z + Z_{\text{ef}})^3}.$$

According to formula (3.7), below we take $Z_{\rm ef} = Z - \frac{5}{16}$, and the factor S is equal to

$$S = \left(1 - \frac{5}{16Z}\right)^{3/2} \left(1 - \frac{5}{32Z}\right)^{3}.$$

In particular, for the helium atom $S = 0.99$, and this value is taken to be unity for the heliumlike positive ions.

The matrix element for the radial wave functions is equal to

$$K_{01} = \frac{1}{2q} \int_0^\infty r^3\, dr\, R_o(r) R_{q1}(r),$$

where, according to (2.32),

$$R_{q1}(r) = \sqrt{\frac{2\pi q}{r}} J_{3/2}(qr) = \frac{2}{r}\left[\cos(qr) - \frac{\sin(qr)}{qr}\right].$$

Evaluating the matrix element for large q, we take the integral by parts and express it through the parameters of the wave function of the initial state at $r = 0$. Indeed, we have

$$\begin{aligned} K_{01} &= \frac{1}{q} \int_0^\infty r^2\, dr\, R_o(r) \left[\cos(qr) - \frac{\sin(qr)}{qr}\right] \\ &= \frac{1}{q^5} \left[\frac{d^2(rR_o)}{dr^2} - \frac{d^3(r^2 R_o)}{dr^3}\right]_{r=0}. \end{aligned}$$

Using the radial function of the initial state in the form $R_o(r) = 2Z_{\rm ef}^{3/2} \exp(-Z_{\rm ef} r)$, where $Z_{\rm ef} = Z - \frac{5}{16}$, we find $K_{01} = 8Z_{\rm ef}^{5/2}/q^5$. From this we have for the cross section of photoionization, taking into account a high-photon energy ($\omega = q^2/2 \gg Z_{\rm ef}^2$),

$$\sigma_{\rm ion} = \frac{512\pi}{3cq^7} Z_{\rm ef}^5.$$

In particular, for the helium atom this formula has the form

$$\sigma_{\rm ion} = \sigma_o \frac{Z_{\rm ef}^5}{q^7},$$

where $\sigma_o = 1.1 \cdot 10^{-16}\,{\rm cm}^2$, if q is expressed in atomic units, and $q \gg Z_{\rm ef}$.

CHAPTER 4

Light Atoms

4.1 Quantum Numbers of Light Atoms

The atom is a system of bound electrons located in the Coulomb field of a heavy nucleus. The energy levels of this quantum system are determined by the Coulomb interaction of electrons with the center, the Coulomb interaction between electrons, and the exchange interaction between electrons due to the Pauli exclusion principle. Restricted by these interactions, we neglect the relativistic effects. This is valid for atoms with a small charge Z of the nucleus. We call atoms the light ones if the relativistic interactions in them are small compared to the electrostatic and exchange ones. Light atoms are the object of this chapter.

The Hamiltonian of the electrons of a light atom has the form (1.18),

$$\widehat{H} = -\frac{1}{2}\sum_i \Delta_i - \sum_i \frac{Z}{r_i} - \sum_{i,j} \frac{1}{|\mathbf{r}_i - \mathbf{r}_j|}. \tag{4.1}$$

Here and below we use atomic units i, j are numbers of electrons, $\mathbf{r}_i$ is the coordinate of the ith electron, and Z is the nucleus charge. Introduce the eigenfunctions Ψ_n of the Hamiltonian which are solutions of the Schrödinger equation

$$\hat{H}\Psi_n = \varepsilon_n \Psi_n, \tag{4.2}$$

where n is the number of the state, and ε_n are the eigenvalues of this equation. Let us introduce quantum numbers of a light atom such that the operators of the corresponding physical values commute with the Hamiltonian. Such operators are the total spin and total orbital momentum of electrons. We prove this below.

The operator of the total electron spin is equal to

$$\hat{\mathbf{S}} = \sum_i \hat{\mathbf{s}}_i, \tag{4.3}$$

where $\hat{\mathbf{s}}_i$ is the operator of the spin of the ith electron. If the wave function of electrons is antisymmetric with respect to the transposition of electrons, the function $\hat{\mathbf{S}}\Psi$ satisfies this condition also. In addition, the atom Hamiltonian does not depend on electron spins, so that the total spin of the electrons is the quantum number of the light atom, i.e., each atom state is characterized by the total atom spin S and by its projection M_S onto a given direction.

The operator of the total atom orbital momentum is

$$\hat{\mathbf{L}} = \sum_j \hat{\mathbf{l}}_j, \qquad \hat{\mathbf{l}}_j = \left[\hat{\mathbf{r}}_j \times \hat{\mathbf{p}}_j\right], \tag{4.4}$$

where $\hat{\mathbf{l}}_j$ is the operator of the orbital momentum for the ith electron, $\hat{\mathbf{p}}_j = -i(\partial/\partial \mathbf{r}_j)$ is the operator of the linear momentum of the jth electron, and the brackets mean the vector product of vectors. The function $\hat{\mathbf{L}}\Psi = \sum_j \hat{\mathbf{l}}_j \Psi$ satisfies the condition (3.1), if this condition is fulfilled for the Ψ-function due to the symmetry of the operator $\hat{\mathbf{L}} = \sum_j \hat{\mathbf{l}}_j$ with respect to transposition of the electrons. Let us determine the commutator $\{\hat{\mathbf{L}}\hat{H}\}$. If this commutator is zero, the eigenfunction of the Hamiltonian is, simultaneously, the eigenfunction of the operator $\hat{\mathbf{L}}$, i.e., the total orbital momentum of the atom L is its quantum number. The operator $\hat{\mathbf{L}} = \sum_j \hat{\mathbf{l}}_j$ commutes with the first term of the Hamiltonian (4.1). This commutes with the second term of this formula because the orbital momentum conserves in a central force field. Thus

$$\{\hat{\mathbf{L}}\hat{H}\} = \sum_{i,j} \left\{\hat{\mathbf{L}}, \frac{1}{|\mathbf{r}_i - \mathbf{r}_j|}\right\}.$$

Let us take the z-projection of this commutator and extract one term from this formula. We have

$$\left\{\hat{L}_z, \frac{1}{|\mathbf{r}_k - \mathbf{r}_j|}\right\} = \left\{\hat{l}_{zk} + \hat{l}_{zj}, \frac{1}{|\mathbf{r}_k - \mathbf{r}_j|}\right\},$$

because the operator of the momenta of the other electrons does not depend on the coordinates of the kth- and jth-electrons. The operator of the electron orbital moment is $\hat{l}_z = x\hat{p}_y - y\hat{p}_z$, where $\hat{\mathbf{p}} = -i\nabla$, and the distance between electrons is

$$|\mathbf{r}_k - \mathbf{r}_j| = \left[(x_k - x_j)^2 + (y_k - y_j)^2 + (z_k - z_j)^2\right]^{1/2},$$

where x_k, y_k, z_k are the components of the radius-vector of the kth-electron. On the basis of these expressions we have

$$\begin{aligned}
&\left\{\hat{l}_{zk} + \hat{l}_{zj}, \frac{1}{|\mathbf{r}_k - \mathbf{r}_j|}\right\} \\
&\quad = (\hat{l}_{zk} + \hat{l}_{zj})\frac{1}{|\mathbf{r}_k - \mathbf{r}_j|} \\
&\quad = -i\left(x_k\frac{\partial}{\partial y_k} - y_k\frac{\partial}{\partial x_k} + x_j\frac{\partial}{\partial y_j} - y_j\frac{\partial}{\partial x_j}\right)
\end{aligned}$$

$$\times \left[(x_k - x_j)^2 + (y_k - y_j)^2 + (z_k - z_j)^2\right]^{1/2}$$

$$= i \frac{\left[x_k(y_k - y_j) - y_k(x_k - x_j) + x_j(y_j - y_k) - x_j(y_j - y_k)\right]}{\left[(x_k - x_j)^2 + (y_k - y_j)^2 + (z_k - z_j)^2\right]^{3/2}} = 0.$$

Thus $\{\hat{L}_z \hat{H}\} = 0$ and, because the z-direction is random, we have $\{\hat{\mathbf{L}}\hat{H}\} = 0$. From this it follows that the eigenfunction of the Hamiltonian, which satisfies the Schrödinger equation (4.2), is simultaneously the eigenfunction of the operator of the total orbital momentum of the $\hat{\mathbf{L}}$-electrons. Therefore, the eigenstate of the light atom is characterized by a certain orbital momentum L and its projection M_L onto a given direction. Because the atomic field does not include a certain direction, the atomic states are degenerated with respect of the projection of the total orbital momentum of the M_L-electrons.

Thus, neglecting the relativistic interaction, one can characterize an atom as a system of bound electrons in a central Coulomb field by the quantum numbers $LSM_L M_S$, and the states of this system are degenerated with respect to the quantum numbers M_L and M_S, i.e., each state is degenerated $(2M_L + 1)(2M_S + 1)$ times. Therefore, the atomic terms are characterized by the quantum numbers L and S only. In this case the atom states with the orbital momentum $L = 0, 1, 2, 3, 4, 5,$ etc., are denoted by the letters S, P, D, F, G, etc. The atom spin is given in the form of the value $2S + 1$ (a number of spin projections) left above from the letter of the orbital momentum. For example, the notation of the atomic term ^2P means that the atomic orbital momentum is $L = 1$ and its spin is $S = 1/2$.

4.2 The Atom Shell Model

The quantum numbers of atoms $LSM_L M_S$ help us to classify some states of the atom. But the formal description of the problem in the form of the Schrödinger equation (4.2) does not allow us to obtain information about the related atom because the variables in this equation are not separated. For this goal it is convenient to use a simple model which was used in the previous chapter for two-electron atoms. Introducing a self-consistent field, we change the Hamiltonian of electrons (4.1) by a model one which includes a self-consistent field. This model Hamiltonian has the form

$$\hat{H} = \sum_i \left[-\frac{1}{2}\Delta_i + V(r_i)\right], \tag{4.5}$$

where $V(r_i)$ is the potential of the self-consistent field. The Hamiltonian (4.5) allows us to separate variables in the Schrödinger equation (4.2), and the equations for individual electrons take the form

$$\left[-\frac{1}{2}\Delta_i + V(r_i)\right]\psi_i(\mathbf{r}_i) = \varepsilon_i \psi_i(\mathbf{r}_i), \tag{4.6}$$

where $\psi_i(\mathbf{r}_i)$ is the wave function of the ith electron and ε_i is the energy of this electron. The electron state in a central force field $V(r_i)$ is characterized by its orbital momentum l and the momentum projection m onto a given direction. Because of the atom symmetry, degeneration with respect to m takes place.

Let us determine the number states of an individual electron by analogy with the hydrogen atom, taking into account the lower state with a given l, a principal quantum number $n = l + 1$. Then, if one chooses the effective potential of the self-consistent field $V(r)$ in equation (4.6) as the Coulomb one, the electron states with the same n and different l will have the same energy. Because the basis of the self-consistent field potential $V(r)$ is the Coulomb interaction, the introduction of the principal quantum number of electrons in this way provides a correct sequence for the series of the energy levels. In this case the higher n is, the higher lies the energy level. Thus, introducing an effective self-consistent atomic field, we obtain that a state of an individual electron in the atom is characterized by the quantum numbers $nlm\sigma$, where m is the projection of the electron orbital momentum onto a given direction, and $\sigma = \pm\frac{1}{2}$ is the spin projection. The energy of an individual electron is degenerated with respect to the quantum numbers m and σ because of the radial symmetry of the effective self-consistent field.

Now let us construct the state and wave function of atomic electrons on the basis of the above one-electron states. Then place electrons in one-electron states and use the Pauli exclusion principle, so that two electrons cannot be located in the same state. Symmetrizing the total wave function of electrons as a product of one-electron functions, we obtain this wave function. Because the states with the lowest energy are of interest, this operation corresponds to filling some cells with certain quantum numbers. Let us make this procedure. The lowest state by the energy of electrons has the quantum numbers $n = 1, l = m = 0$. There are two such states which differ by the spin projection onto a given direction. These states are denoted by $1s$, and two electrons in these states are denoted as $1s^2$. The helium atom in the ground state has such a structure of the electron shell.

The following two states by energy are characterized by the quantum numbers $n = 2, l = m = 0$, and these states are denoted as $2s^2$. The successive six states are characterized by the quantum numbers of electrons $n = 2, l = 1, m = 0, \pm 1$. Although in the case of the Coulomb interaction the electron states with $n = 2$ and $l = 0, 1$ have the same energy, one can conclude for a real self-consistent field that the state $l = 0$ is characterized by a higher electron binding energy than the state $l = 1$, because an electron is located closer to the nucleus where the shielding by other electrons is weaker. The above operation corresponds to the distribution of electrons over states with different quantum numbers nl. The group of states with the same nl is named an electron shell, and the related atomic model is called the electron shell model. The following notations are used in this case. The principal quantum number of the electron state n is given as a number, the electron orbital momentum is denoted by a letter, so that notations s, p, d, f, g, h correspond to the states $l = 0, 1, 2, 3, 4, 5$, and a number of electrons with the same nl is given as a superscript. For example, the state of the sulphur atom $\mathrm{S}(1s^2 2s^2 2p^6 3s^2 3p^4)$ means that this atom has two electrons on the shell $n = 1, l = 0$, two electrons on the shell $n = 2, l = 0$, six electrons on the shell $n = 2, l = 1$, two electrons on the shell $n = 3, l = 0$, and the unfilled valent shell of this atom contains four electrons with the quantum numbers $n = 3, l = 1$. Often, when interactions or processes involving atoms are determined by valent electrons, it is convenient to take into account only the valent electron shell. Then for the above example we have $S(3p^4)$.

The electron shell of an atom includes only part of the quantum numbers which are used for the description of an atom state. Several different levels of the energy which are described by additional quantum numbers correspond to unfilled atomic shells. The simplest example of such a type is $He(1s2s)$. Two states 2^3S and 2^1S of a different energy relate to this electron shell. They differ by the total spin of electrons. The shell atom model is useful for the classification of atom states and understanding the atom nature. This model can be used for the evaluation of energies of light atoms.

4.3 Parentage Scheme of Atoms

The shell atom model corresponds to a one-electron description of the atom. Then the wave function of atomic electrons is the combination of the products of one-electron wave functions. These combinations can be taken on the basis of the symmetry (3.1) of the electron wave function. In reality the radial symmetry of atomic fields and the character of summation of the momenta of individual electrons in the total atom momenta simplifies this combination. Often the related properties of the atom are determined by valent electrons only. Then it is convenient to extract the wave function of one of the valent electrons from the total wave function of electrons. The operation of the extraction of the wave function of one valent electron from the total electron wave function of the atom is called the parentage scheme of the atom. On the basis of the total symmetry of the atom electron system this operation can be made on the basis of the formula

$$\Psi_{LSM_LM_S}(1, 2, \ldots, n) = \frac{1}{\sqrt{n}} \hat{P} \sum_{L'M'_LS'M'_Sm\sigma} G_{L'S'}^{LS}(l, n),$$

$$\begin{bmatrix} l & L' & L \\ m & M'_L & M_L \end{bmatrix} \begin{bmatrix} \frac{1}{2} & S' & S \\ \sigma & M'_S & M_S \end{bmatrix} \psi_{l\frac{1}{2}m\sigma}(1) \cdot \Psi_{L'S'M'_LM'_S}(2, \ldots, n), \quad (4.7)$$

where n is the number of valent electrons. The operator $\hat{P}$ transposes the positions and spins of a test electron which is described by argument 1 and the atomic electrons, LSM_LM_S are the quantum numbers of the atom, $L'S'M'_LM'_S$ are the quantum numbers of the atomic core, $l\frac{1}{2}m\sigma$ are the quantum numbers of an extracted valent electron, and $G_{L'S'}^{LS}(l, n)$ is the so called fractional parentage coefficient or the Racah coefficient which is responsible for the connection of an electron with the atomic core in the formation of the atom. It is of importance that as a result of the removal of one valent electron from the atom the atomic core can be found in

Table 4.1. Fractional parentage coefficients for valent s- and p-electron shells. The electron shell and the state term for the atom and atomic core are indicated.

Atom	Atomic core	$G_{L'S'}^{LS}$	Atom	Atomic core	$G_{L'S'}^{LS}$
$s(^2S)$	(^1S)	1	$p^3(^2P)$	$p^2(^1S)$	$\sqrt{2/3}$
$s^2(^1S)$	$s(^2S)$	1	$p^4(^3P)$	$p^3(^4S)$	$-1/\sqrt{3}$
$p(^2P)$	(^1S)	1		$p^3(^2D)$	$\sqrt{5/12}$
$p^2(^3P)$	$p(^2P)$	1		$p^3(^2P)$	$-1/2$
$p^2(^1D)$	$p(^2P)$	1	$p^4(^1D)$	$p^3(^4S)$	0
$p^2(^1S)$	$p(^2P)$	1		$p^3(^2D)$	$\sqrt{3/4}$
$p^3(^4S)$	$p^2(^3P)$	1		$p^3(^2P)$	$-1/2$
	$p^2(^1D)$	0	$p^4(^1S)$	$p^3(^4S)$	0
	$p^2(^1S)$	0		$p^3(^2D)$	0
$p^3(^2D)$	$p^2(^3P)$	$1/\sqrt{2}$		$p^3(^2P)$	1
	$p^2(^1D)$	$-1/\sqrt{2}$	$p^5(^2P)$	$p^4(^3P)$	$\sqrt{3/5}$
	$p^2(^1S)$	0		$p^4(^1D)$	$1/\sqrt{3}$
$p^3(^2P)$	$p^2(^3P)$	$-1/\sqrt{2}$		$p^4(^1S)$	$1/\sqrt{15}$
	$p^2(^1D)$	$-\sqrt{5/18}$	$p^6(^1S)$	$p^5(^2P)$	1

a finite number of states. Table 4.1 lists the values of fractional parentage coefficients for s- and p-electron shells. In the case of d- and f-electrons, removal of one valent electron can lead to different states of the atom with the same values, L and S. Then the atom state is described by one more quantum number v—the state seniority.

Fractional parentage coefficients satisfy some relations. In particular, from the condition of normalization of the wave function it follows that

$$\sum_{L'S'v} \left[G_{L'S'}^{LS}(l, n, v)\right]^2 = 1. \tag{4.8}$$

The number of electrons of a filled electron shell equals $4l + 2$. There is some analogy between removal of one vacancy from a shell containing $n + 1$ vacancies and $4l + 3 - n$ electrons, and the case of removal of one electron from the shell containing n electrons. This correspondence is expressed in the form of the formula

$$G_{L'S'}^{LS}(l, n, v) = (-1)^{L+L'+S+S'-l-1/2} \cdot \left[\frac{n(2S'+1)(2L'+1)}{(4l+3-n)(2S+1)(2L+1)}\right]^{1/2} \cdot G_{L'S'}^{LS}(l, 4l+3-n, v). \tag{4.9}$$

The mathematical formalism based on the use of the Clebsh–Gordan coefficients and the fractional parentage coefficients possesses a central place in the atom theory. This formalism allows one to take into account the symmetry of atomic particles for the analysis of their properties. In particular, it gives the selection rule for radiative transition. The rate of radiative transition between two atomic states

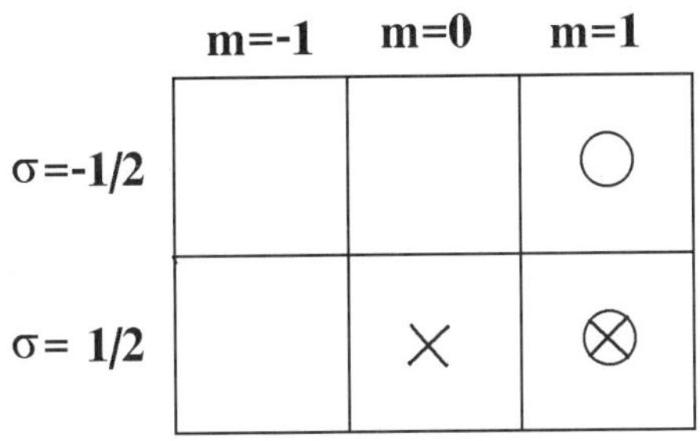

	m=-1	m=0	m=1
σ=-1/2			○
σ= 1/2		×	⊗

Figure 4.1. The distribution in the electron shell of valent p-electrons for atoms of the fourth group of the periodical system of elements for states with maximum spin (circles) and maximum orbital momentum (crosses).

is proportional to the square of the matrix element of the dipole moment operator between states of transition. Within the framework of the scheme of the light atom, this transition is possible between states with $\Delta S = 0$ and $\Delta L = 0, \pm 1$.

The above analysis allows us to construct the Grotrian diagrams in total analogy with the hydrogen atom for various light atoms (see Figs. 4.1–4.7). These diagrams show the positions of low states of atoms and the parameters of intense radiative transitions. Taking into account the radiative transitions for the excited states of atoms, one can divide these states into two groups. Lower excited states of atoms, from which is possible a radiative transition to the ground state are called resonantly excited atomic states. If a radiative transition is forbidden from the lower excited state to the ground one, this is a metastable excited state. Metastable states are characterized by high lifetimes in gaseous and plasma systems. Below (in Chapter 6) we give some examples of these states.

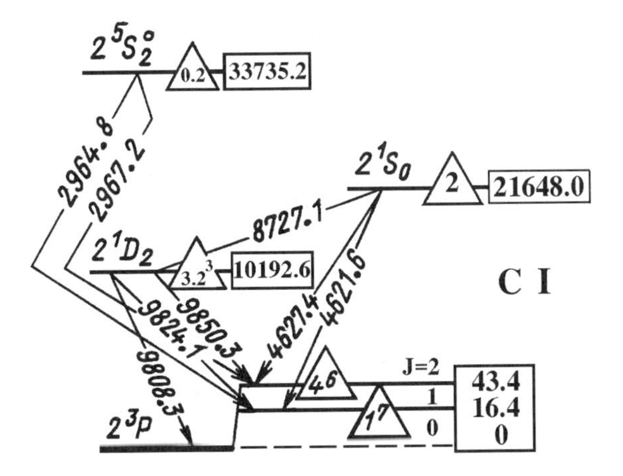

Figure 4.2. Parameters of the metastable states of the carbon atom. These states are characterized as the same electron shell as the ground state.

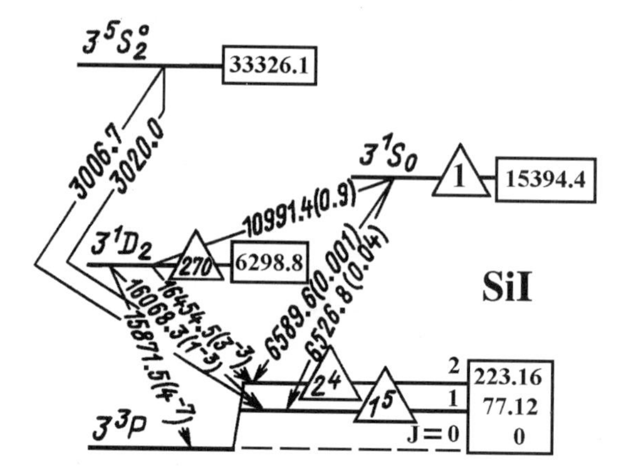

Figure 4.3. Parameters of the metastable states of the silicon atom. These states are characterized by the same electron shell as the ground state.

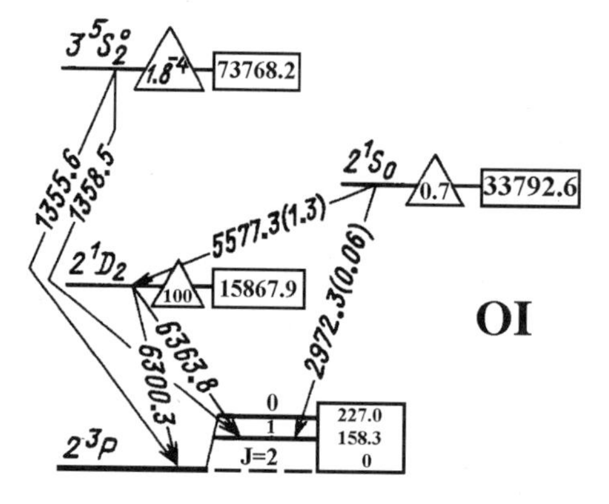

Figure 4.4. Parameters of the metastable states of the oxygen atom. These states are characterized by the same electron shell as the ground state.

4.4 Asymptotic Behavior of Atomic Wave Functions

The parentage scheme allows us to separate one valent electron from other electrons. Now we use it in the case when the distance of one valent electron from the nucleus remarkably exceeds an average atom size. Then the parentage scheme gives a correct description of an atom, while above we considered it as a convenient atom model. Let us present in the related case the wave function of an extracted electron of formula (4.7) in the form

$$\psi_{l\frac{1}{2}m\sigma} = R_l(r)Y_{lm}(\theta, \varphi)\chi_\sigma,$$

where R_l is the radial wave function, $Y_{lm}(\theta, \varphi)$ is the electron angle wave function, and χ_σ is the electron spin function. In the course of the removal of an electron from the atom, only the radial wave function varies. Hence, below we concentrate

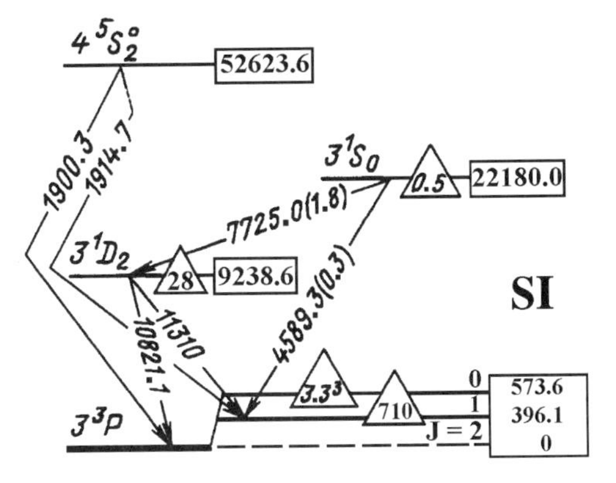

Figure 4.5. Parameters of the metastable states of the sulfur atom. These states are characterized by the same electron shell as the ground state.

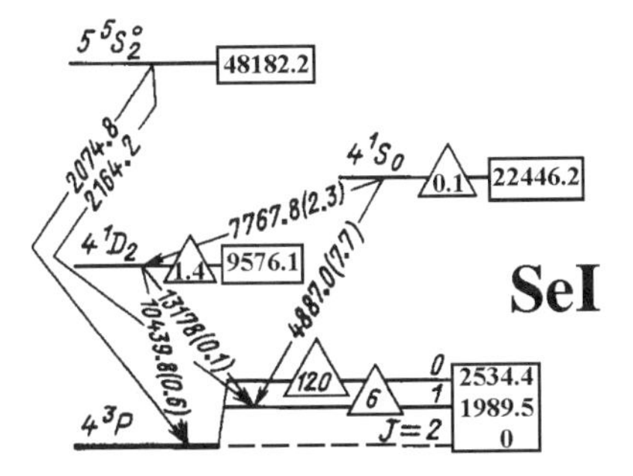

Figure 4.6. Parameters of the metastable states of the selenium atom. These states are characterized by the same electron shell as the ground state.

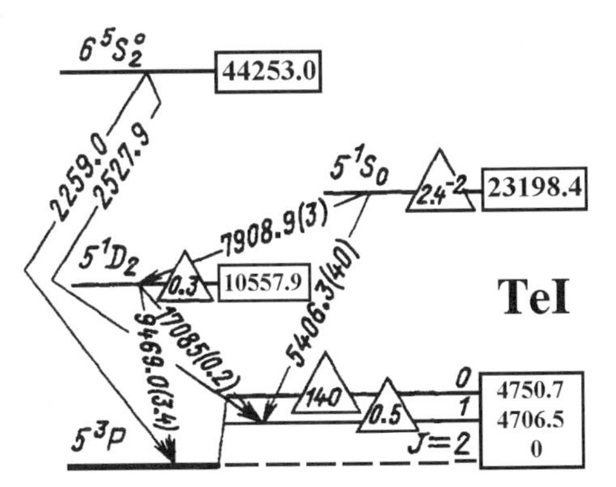

Figure 4.7. Parameters of the metastable states of the tellurium atom. These states are characterized by the same electron shell as the ground state.

our attention on this electron wave function. Let us find the asymptotic expression for the radial wave function. Because at large distances from the atom an exchange interaction is not essential, we neglect this. Next, the self-consistent field potential far from the atom coincides with the Coulomb field of the atomic core. Thus, the Schrödinger equation for the radial function of a test valent electron has the form

$$\frac{1}{r} \cdot \frac{d^2}{dr^2}(rR_l) + \left[\varepsilon - \frac{2}{r} + \frac{l(l+1)}{r^2}\right] R_l(r) = 0. \tag{4.10}$$

Introducing the energy parameter by the relation $\varepsilon = -\gamma^2/2$, we obtain the asymptotic solution of this equation at large r,

$$R_l(r) = Ar^{1/\gamma-1}e^{-r\gamma}, \qquad r\gamma \gg 1; \quad r\gamma^2 \gg 1. \tag{4.11}$$

The coefficient A is determined by the electron behavior in an internal atom region where the related electron is located and formula (4.11) is violated. This coefficient can be obtained by comparison of the asymptotic wave function (4.11) with that at moderate distances from the nucleus, as it is made in Problem 3.4 for the helium atom in the ground state. Indeed, the numerical methods of the solution of the Schrödinger equation allow us to determine the electron wave function in a region where electrons are mostly located. Such a solution leads to an error at large distances from the nucleus, because these positions of the electrons give a small contribution to the atom energy. An increase in the accuracy of the electron wave function makes it correct in a widre region of electron distances from the center. Then there is a region of distances of the test valent electron from the center where, on one hand, the asymptotic expression (4.11) is valid and, on the other hand, a numerical wave function is correct. Comparing these functions, one can find the parameter A. An example of the determination of parameter A by this method for the helium atom is given in Problem 3.4. Table 4.2 contains values of parameter A obtained in this way by using the results of the Hartri–Fock calculations.

Note that if the Coulomb interaction with the atomic core takes place in the basic region of the electron location, i.e., equation (4.10) is valid for all the distances of the electron from the center, we have the following expression for the asymptotic coefficient for the valent s-electron

$$A = \frac{\gamma^{3/2}(2\gamma)^{\frac{1}{\gamma}}}{\Gamma\left(\frac{1}{\gamma}\right)}. \tag{4.12}$$

4.5 Fine Splitting of Levels of Light Atoms

Within the framework of the related scheme, relativistic interactions in the atom are weak. Hence, the state of the atom is characterized, in the first place, by the structure of its electron shell, and also by a certain orbital L and spin S momenta of the atom. For the related scheme of the atom shell model, electron states with the same shell structure and different values of L and S are separated by energy, due to exchange interaction which is determined by the symmetry of the electron wave

Table 4.2. Asymptotic parameters of valent electrons.

Atom, state	Shell	γ	A	Atom, state	Shell	γ	A
He(^1S)	$1s^2$	1.344	2.87 (A)	K(^2S)	$4s$	0.565	0.52 (C)
Li(^2S)	$2s$	0.630	0.82 (B)	Ca(^1S)	$4s^2$	0.670	0.95 (C)
Be(^1S)	$2s^2$	0.828	1.62 (B)	Cu(^2S)	$4s$	0.754	1.29 (A)
B(^2P)	$2p$	0.781	0.88 (C)	Zn(^1S)	$4s^2$	0.831	1.69 (C)
C(^3P)	$2p^2$	0.910	1.30 (C)	Ga(^2P)	$4p$	0.664	0.60 (C)
N(^4S)	$2p^3$	1.034	1.5 (C)	Ge(^3P)	$4p^2$	0.762	1.29 (C)
O(^3P)	$2p^4$	1.000	1.3 (C)	As(^4S)	$4p^3$	0.850	1.58 (C)
F(^2P)	$2p^5$	1.132	1.59 (C)	Se(^3P)	$5p^4$	0.847	1.52 (C)
Ne(^1S)	$2p^6$	1.228	1.75 (C)	Br(^2P)	$4p^5$	0.932	1.83 (B)
Na(^2S)	$3s$	0.615	0.74 (B)	Kr(^1S)	$4p^6$	0.932	2.22 (B)
Mg(^1S)	$3s^2$	0.750	1.32 (B)	Rb(^2S)	$5s$	0.554	0.48 (C)
Al(^2P)	$3p$	0.663	0.61 (C)	Sr(^1S)	$5s^2$	0.647	0.86 (C)
Si(^3P)	$3p^2$	0.774	1.10 (C)	Ag(^2S)	$5s$	0.746	1.18 (C)
P(^4S)	$3p^3$	0.878	1.65 (C)	Cd(^1S)	$5s^2$	0.813	1.6 (C)
S(^3P)	$3p^4$	0.873	1.11 (C)	In(^2P)	$5p$	0.652	0.58 (C)
Cl(^2P)	$3p^5$	0.976	1.78 (C)	Sn(^3P)	$5p^2$	0.735	1.02 (C)
Ar(^1S)	$3p^6$	1.076	2.11 (B)	Sb(^4S)	$5p^3$	0.797	1.67 (C)
Te(^3P)	$5p^4$	0.814	1.65 (C)	Au(^2S)	$6s$	0.823	1.57 (A)
I(^2P)	$5p^5$	0.876	1.94 (C)	Hg(^1S)	$6s^2$	0.876	1.96 (A)
Xe(^1S)	$5p^6$	0.944	2.4 (C)	Tl(^2P)	$6p$	0.670	0.55 (C)
Cs(^2S)	$6s$	0.535	0.42 (C)	Pb(^3P)	$6p^2$	0.738	1.09 (C)
Ba(^1S)	$6s^2$	0.619	0.78 (B)	Bi(^4S)	$6p^3$	0.732	1.43 (C)

function. The group of states of identical L and S is called "the electron term" of the electron system. Until we neglect the relativistic interactions, the energy levels of one electron term are degenerated. Thus, the concept of the light atom requires the weakness of relativistic interactions compared to exchange interactons. Action of the relativistic effects on the light atom can be taken into account on the basis of the perturbation theory within the framework of the above concept. The strongest relativistic interaction is the spin-orbit interaction. The corresponding splitting of the electron term is called fine splitting. Below we estimate the fine splitting of levels for a light atom.

The operator of the spin-orbit interaction has the form

$$\hat{V} = -a\hat{\mathbf{L}}\hat{\mathbf{S}}, \tag{4.13}$$

and for the hydrogen atom the coefficient a is given by formula (2.19). Change the Coulomb field in the hydrogen atom by a self-consistent one, V in other atoms leads to the following expression for this value, if we take into account partaking of all the valent electrons in this interaction:

$$a = \frac{1}{2c}\sum_k \left\langle \frac{1}{r_k}\frac{dV}{dr_k} \right\rangle, \tag{4.14}$$

where k is an electron number.

If we add the interaction operator (4.13) to the Hamiltonian of electrons (4.1), we obtain that the accurate quantum number is now the total electron momentum (see Problem 4.3):

$$\hat{\mathbf{J}} = \hat{\mathbf{L}} + \hat{\mathbf{S}}. \tag{4.15}$$

From this relation it follows that

$$\langle 2\hat{\mathbf{L}}\hat{\mathbf{S}} \rangle = \langle \hat{\mathbf{J}}^2 \rangle - \langle \hat{\mathbf{L}}^2 \rangle - \langle \hat{\mathbf{S}}^2 \rangle = J(J+1) - L(L+1) - S(S+1).$$

The spin-orbit interaction leads to the splitting of an atomic term into several sublevels. The distance between neighboring sublevels is equal to

$$\varepsilon_J - \varepsilon_{J-1} = -\frac{a}{2}J(J+1) + \frac{a}{2}(J-1)J = -aJ, \tag{4.16}$$

i.e., this difference is proportional to J. This relationship for light atoms is known as the Lande rule.

In accordance with the character of the summation of momenta, the quantity J takes the values

$$J = |L-S|, \qquad |L-S|+1, \ldots, L+S.$$

Hence each electron term is split into $2\min(L, S)+1$ levels as a result of spin-orbit interaction. A certain value of the quantum number J corresponds to each sublevel. Hence, the eigenstates of the atom are characterized by the quantum numbers $LSJM_J$, where M_J is the projection of the total momentum of electrons onto a given direction. Each of the states has a $(2J+1)$-multiplied degeneration. Note that the number of degenerated levels in the absence of the relativistic interactions is $\sum_J(2J+1) = (2L+1)(2S+1)$. Notation of the total atom momentum is given as the subscript below the notation of the orbital moment. For example, $^2P_{3/2}$ means that this state of the atom is described by the quantum numbers $S = 1/2$, $L = 1$, $J = 3/2$.

Note that the Lande rule is valid for the case of LS-coupling when the relativistic interactions are small compared to the exchange interactions. Then, in the case of several fine states, a certain relation takes place for the distances between levels which correspond to different total electron momenta. Table 4.3 contains some examples of fine splitting for the ground states of atoms and ions whose ground term is ^4F. Let us denote the difference of energies of the $^4F_{3/2}$- and $^4F_{9/2}$-levels by Δ_f. Then, within the framework of the scheme of LS-coupling, the distance between $^4F_{3/2}$ and $^4F_{5/2}$ is equal to $\frac{5}{21}\Delta_f$, and the distance between the $^4F_{3/2}$- and

Table 4.3. Positions of sublevels of fine structure (in cm^{-1}) for some atoms and ions whose electron term of the ground state is ^4F.

Atom	Shell	ε ($^4F_{3/2}$)	ε ($^4F_{5/2}$)	ε ($^4F_{7/2}$)	ε ($^4F_{9/2}$)
V	$3d^34s^2$	0	137	324	553
			(132)	(316)	
Co	$3d^74s^2$	1809	1407	816	0
			(1378)	(775)	
Rh	$4d^85s$	3473	2598	1530	0
			(2646)	(1488)	
Ti^+	$3d^24s$	0	94	226	393
			(94)	(225)	
Zr^+	$4d^25s$	0	315	763	1323
			(315)	(756)	
Ru^+	$4d^75s^2$	3105	2494	1523	0
			(2366)	(1331)	

$^4F_{7/2}$-levels is $\frac{4}{7}\Delta_f$. Values calculated on the bases of these formulas are given in Table 4.3 in parentheses, and comparison of the calculated and real values of the fine splitting of levels allows us to estimate the validity of the LS-coupling scheme for these cases. Note the competition between s- and d-shells in the related case. Nevertheless, the LS-coupling scheme for the summation of electron momenta describes the observed positions of the levels of fine structure. As is seen from the data of Table 4.3, the Lande rule is better valid, the lighter the atom or ion is.

The fine atom structure is detected in the spectra of absorption and radiation of atoms. Hence it gives some information about atoms. Because of the fine splitting of levels, atomic spectra include series of closed spectral lines which are called multiplets. Let us find the relation between the intensity of individual spectral lines—multiplets. These values are proportional to the value $|\langle SJM_J|\mathbf{D}|S'J'M_{J'}\rangle|^2$. The electron wave function of the state described by the quantum numbers $LSJM_J$ is expressed through wave functions of the states of the quantum numbers LSM_LM_S by the Clebsh–Gordan coefficients, which are responsible for the summation of the orbital and spin atom momentum into the total atom momenta

$$\Psi_{LSJM_J} = \sum_{M_L,M_S} \begin{bmatrix} L & S & J \\ M_L & M_S & M_J \end{bmatrix} \Psi_{LSM_LM_S}.$$

Averaging over the initial states of the transition and summarizing over the final states of the transition we obtain, for the relative probability of a transition between the states of a given fine structure

$$w(\gamma LSJ \to \gamma' L'S'J') = (2L+1)(2J'+1) \begin{Bmatrix} L' & 1 & L \\ J & S & J' \end{Bmatrix}^2 w_o, \qquad (4.17)$$

where γ, γ' are quantum numbers which are not connected with atom momenta, the brackets mean the $6j$-symbol of Wigner, and w_o does not depend on the quantum numbers L', S', J' of the final state. The probability of transition is not zero, if

$$J - J' = 0, \pm 1. \qquad (4.18)$$

This is the selection rule for radiative transitions. The relative probability of transition is normalized by the condition

$$\sum_{J'} w(\gamma LSJ \to \gamma' L'S'J') = w(\gamma LS \to \gamma' L'S'). \qquad (4.19)$$

Let us introduce the oscillator strength as

$$f_{jk} = 2(\varepsilon_k - \varepsilon_j) g_k |\langle j|D_x|k\rangle|^2,$$

where j, k are the initial and final states of transition, and g_k is the statistical weight of the final state. Usually the oscillator strength is averaged over initial states of the transition and is summed up over the final states. Then the relation between the oscillator strengths in the absence of the fine splitting and in its presence has the form

$$f(\gamma LS \to \gamma' L'S') = \sum_{J,J'} \frac{(2J+1)}{(2S+1)(2L+1)} f(\gamma LSJ \to \gamma' L'S'J'),$$

and from the definition of this value it follows that

$$f(\gamma LS \to \gamma' L'S') = -\frac{(2L'+1)}{(2L+1)} f(\gamma' L'S' \to \gamma LS).$$

From relations (4.14), (4.16) it follows that

$$f(\gamma LSJ \to \gamma' L'S'J') = \frac{(2J+1)(2J'+1)}{(2S+1)} \begin{Bmatrix} L' & 1 & L \\ J & S & J' \end{Bmatrix}^2 f(\gamma LS \to \gamma' L'S'). \qquad (4.20)$$

This formula yields the relative intensity for separate lines of a certain multiplet. In particular, Table 4.4 gives the relative intensities of spectral lines for the radiative transition $^2P \to {}^2D$ which is the ratio of the oscillator strength $f(\gamma LSJ \to \gamma' L'S'J')$ to $f(\gamma LS \to \gamma' L'S')$. By definition, the sum of these oscillator strengths is equal to unity.

Thus, the hierarchy of the interactions in light atoms leads to certain quantum numbers of the atom. In the related case the exchange splitting of atom levels is large compared to the spin-orbital interactions. Then, by neglecting the spin-orbital interactions, the atom quantum numbers are $LM_L SM_S$. Taking into account the spin-orbital interactions within the framework of the perturbation theory, we obtain $LSJM_J$ as the atom quantum numbers. This character of interaction in the

Table 4.4. The relative intensity of multiplets in the radiative transition $^2P \to {}^2D$.

Transition	Relative oscillator strength
$^2P_{1/2} \to {}^2D_{3/2}$	1/3
$^2P_{3/2} \to {}^2D_{3/2}$	1/15
$^2P_{3/2} \to {}^2D_{5/2}$	3/5

atom and summation of atom momenta into the total momentum J is called the LS-coupling scheme. Note that if we increase the relativistic interactions such that this scheme of summation of atom momenta does not work, the total number of electron states for a given electron shell is conserved. Hence, if the LS-coupling scheme is used beyond the limits of its validity, it is necessary to consider the LS-numbers as a form of designation of the electron states, but not as quantum numbers.

4.6 Periodic System of Elements and Atoms with Valent s-Electrons

The electrons of unfilled electron shells or the electrons of external shells are called valent electrons. Atoms with an identical orbital momentum of valent electrons and the same number of valent electrons, we call atoms with identical valent shells. One can expect that atoms with identical valent shells have the same properties related to their chemistry and spectrum. This fact is the basis of systematization of elements in the form of the periodic system of elements. Each of the eight groups of the periodic system of elements includes atoms with valent s- and p-electrons. These groups form columns in the periodic system of elements. Along with atoms with valent s- and p-electrons, which form the main groups of elements, atoms with valent d-electrons are included in the periodic system of elements. If such atoms have no analogy with atoms of filling s- and p-shells, they form their own subgroups. Atoms with filling f-shells are taken out of the periodic table and form an additional table of lanthanides and actinides.

Thus, the periodic system of elements reflects the fact that atoms with identical valent shells have identical properties. This confirms the atom shell model and gives the way for systematization of atoms of various elements. Below we analyze the analogy of the atom spectral properties for atoms with valent s- and p-electrons. Atoms of the first group of the periodic system of elements have one valent s-electron. This group includes hydrogen, atoms of alkali metals Li, Na, K, Rb, Cs, and atoms of coin metals Cu, Ag, Au. The ground state of these atoms corresponds to the term $^2S_{1/2}$, the first excited state $^2P_{1/2}$ corresponds to the transition of the valent electrons in the p-state. Hence, the first excited state of atoms of the first group is the resonance one, so that a dipole radiative transition is possible between this and the ground state. Table 4.5 lists the parameters of the first excited state of atoms of the alkali metals. Note that the oscillator strength of absorption has the following form in this case

$$f(s \to p) = \frac{2\Delta\varepsilon}{3} |\langle s|\mathbf{r}|p\rangle|^2.$$

This formula is written in atomic units; $\Delta\varepsilon$ is the difference of energies of these states, $\mathbf{r}$ is the electron radius-vector, and summation over all states of the related p-state is made in this formula. Neglecting the fine splitting of levels and taking into account the expression for the oscillator strength

$$f(s \to p) = f(^2S_{1/2} \to {}^2P_{1/2}) + f(^2S_{1/2} \to {}^2P_{3/2}),$$

we have

$$f(^2S_{1/2} \to {}^2P_{1/2}) = \frac{1}{2} f(^2S_{1/2} \to {}^2P_{3/2}) = \frac{1}{3} f(s \to p).$$

Table 4.5 contains τ—the lifetime of the state $^2P_{1/2}$ which is close to that of $^2P_{3/2}$. Analyzing the spin-orbit splitting of levels of the lowest excited states for atoms of the first group, we take into account that this splitting is determined by one electron. Then let us use for this goal formula (2.20) with an effective charge Z_{ef} of the atomic core by a change the parameter e^2 in formula (2.20) by the value $Z_{ef}e^2$. Comparing the formula with measured values, one can find this effective charge Z_{ef}, which determines the fine splitting. As follows from the data of Table 4.5, the fine splitting of heavy alkali atoms is determined by the location of the valent electron inside the previous shell.

Atoms of the coin metals Cu, Ag, Au, which are located in the first group of the periodic system of elements, have partially the same properties as alkali metal atoms until these properties are determined by the valent s-electron. But the d-electrons of the previous shell have a relatively small excitation energy and give a contribution to the spectrum of these atoms. This fact is demonstrated by Table 4.6 where the energies are given for the lowest excited states of these atoms. As is

Table 4.5. Parameters of the resonantly excited states of atoms of the first group of the periodic system of elements with a valent ns-shell.

Atom	H	Li	Na	K	Rb	Cs
n	1	2	3	4	5	6
$\varepsilon_{ex}(^2P_{1/2})$, eV	10.20	1.85	2.10	1.61	1.56	1.39
$\varepsilon_{ex}(^2P_{3/2})$, eV	10.20	1.85	2.10	1.62	1.59	1.45
$\lambda(^2S_{1/2} \to {}^2P_{1/2})$, nm	1215.7	6707.9	5895.9	7699.0	7947.6	8943.5
$\lambda(^2S_{1/2} \to {}^2P_{3/2})$, nm	1215.7	6707.8	5890.0	7664.9	7800.3	8521.1
$f(p \to s)$	0.416	0.741	0.955	1.05	0.99	1.2
τ, ns	1.6	27	16	26	29	30
$\Delta\varepsilon(^2P_{1/2} - {}^2P_{3/2})$, cm^{-1}	0.365	1.35	17.2	57.6	237.6	554.1
Z_{ef}	1	1.4	2.6	3.5	5.0	6.2

Table 4.6. Parameters of atoms of coin metals (J is the ionization potential, τ is the radiative lifetime of s-state, ε_{ex} is the excitation energy of the corresponding state).

Atom	Cu	Ag	Au
Shell	$3d^{10}4s$	$4d^{10}5s$	$5d^{10}6s$
J, eV	7.726	7.576	9.226
τ, ns	7	7	6
$\varepsilon_{ex}(3d^9 4s^2, {}^2D_{5/2})$, eV	1.389	3.749	1.136
$\varepsilon_{ex}(3d^9 4s^2, {}^2D_{3/2})$, eV	1.642	4.304	2.658
$\varepsilon_{ex}(3d^{10} 4p, {}^2P_{1/2})$, eV	3.786	3.664	4.632
$\varepsilon_{ex}(3d^{10} 4p, {}^2P_{3/2})$, eV	3.816	3.778	5.105

Table 4.7. Parameters of the lowest states for atoms of the second group of the periodic system of elements.

Atom	Be	Mg	Ca	Zn	Sr	Cd	Ba	Hg
Ground state	2^1S_0	3^1S_0	4^1S_0	4^1S_0	5^1S_0	5^1S_0	6^1S_0	6^1S_0
Excited shell	$2s2p$	$3s3p$	$4s4p$	$4s4p$	$5s5p$	$5s5p$	$6s6p$	$6s6p$
$\varepsilon({}^3P_0)$, eV	2.72	2.71	1.88	4.01	1.78	3.73	1.52	4.67
$\varepsilon({}^3P_1)$, eV	2.72	2.71	1.89	4.03	1.80	3.80	1.57	4.89
$\varepsilon({}^3P_2)$, eV	2.72	2.72	1.90	4.08	1.85	3.94	1.68	5.464
$\varepsilon({}^1P_1)$, eV	5.28	5.11	2.93	5.80	2.69	5.42	2.24	6.78
$\lambda({}^1S_0 \to {}^3P_0)$, nm	—	457.11	657.28	307.59	689.26	326.11	791.13	253.65
$\lambda({}^1S_0 \to {}^1P_0)$, nm	234.86	285.21	422.67	213.86	460.73	228.80	553.55	184.95
$\tau({}^1P_0)$, ns	1.9	2.0	4.6	5.0	1.4	1.7	8.4	1.3
$\Delta\varepsilon({}^3P_0 - {}^3P_1)$, cm^{-1}	0.64	20.1	52.2	190	187	542	371	1767

seen, the excitation energy of the d-electron is lower than or is comparable to that of the valent s-electron.

Atoms of the second group of the periodic system of elements have the valent shell s^2, so that in the lowest excited state one electron tranfers to p-state. The helium atom is an exclusion from this, and its lowest excited states relate to the electron shell He($1s2s$). The lowest excited states of the helium atom are metastable states, and the excitation energy for the state He(2^3S) is equal to 19.82 eV, the excitation energy of the metastable state He(2^1S) is equal to 20.62 eV. The following state of the helium atom He(2^3P) can decay as a result of the emitting of a dipole photon with transition in the state He(2^3S), i.e., this state is a resonantly excited one with the radiative lifetime 98 ns. The resonantly excited state He(2^1P), which decays with the emitting of a dipole photon and transition in the ground state, has the excitation energy 21.22 eV and the radiative lifetime 0.56 ns. Table 4.6 contains parameters of the lowest excited states for other atoms of the second group of the periodic system of elements. The absence of a monotonous dependence of excitation energies on the atom number testifies to the influence of internal shells on atom parameters, especially if the atom has the previous d^{10}-shell. In addition, the excitation of a valent electron in a p-state can compete with its excitation in a d-state. For example, the excitation energies of the states $4s4d(^3D_{1,2,3})$ and $4s4d(^1D_2)$ of the calcium atom are lower than that for the resonantly excited state $4s4p(^1P_1)$ which is included in Table 4.7. The lowest excited states of the barium atom Ba$(6s5d, {}^3D_{1,2,3})$ and Ba$(6s5d, {}^1D_2)$ are characterized by the excitation energies 1.12, 1.14, 1.19, and 1.41 eV, respectively. The barium atom states of Table 4.7 have a higher excitation energy.

4.7 Atoms with Valent p-Electrons

Now let us consider atoms of the third group of the periodic system of elements whose valent shell contains one p-electron. The ground state of these atoms is ${}^2P_{1/2}$, the first excited state is ${}^2S_{1/2}$ which corresponds to transition of the valent electron from the np to the $(n+1)s$-electron shell. An exception to this rule is the boron atom whose first excited states corresponds to the transition from the electron shell $2s^22p$ to the shell and states $(2s2p^2)^4P_{1,2,3}$. Table 4.8 gives some parameters for the ground, ${}^2P_{1/2,3/2}$, and resonantly excited, ${}^2S_{1/2}$, states of atoms of the third group of the periodic system of elements. One can see that the fine splitting of the ground state of the related atoms is created by the location of the valent electron in the internal shells.

Atoms with p^2, p^3, and p^4 valent shells have several terms corresponding to the lowest state of the electron shell. At the beginning we considered atoms with a p^2-shell which are characterized by the statistical weight $C_6^2 = 15$. One can obtain all these states by the distribution of electrons over cells with different momenta and spin projections, as is shown in Fig. 4.1. According to the Pauli exclusion principle, only one electron can be placed in one cell. Let us arrange electrons over the states. First we extract the state with the maximum orbital moment (it is

Table 4.8. Parameters of the lowest states for atoms of the third group of the periodic system of elements.

Atom	B	Al	Ga	In	Tl
Ground shell	$2p$	$3p$	$4p$	$5p$	$6p$
Excited shell	$3s$	$4s$	$5s$	$6s$	$7s$
ε_{ex}, eV	4.96	3.14	3.07	3.02	3.28
$\Delta\varepsilon({}^2P_{1/2} - {}^2P_{3/2})$, cm^{-1}	15.2	112	896	2213	7793
Z_{ef}	2.5	4.2	7.0	8.8	12
$\lambda({}^2P_{1/2} - {}^2S_{1/2})$, nm	249.68	394.40	403.30	410.18	377.57
$\tau({}^2S_{1/2})$, ns	3.6	6.8	6.2	7.4	7.6

marked by open circles in Fig. 4.1). Take a direction of the quantum axis such that the projection of the orbital momentum of each electron onto this axis is equal to one. Then the projection of the total atom orbital momentum onto this axis equals 2, and the total spin projection is zero. Thus the term of this state is ^1D, and the statistical weight of this state (the number of projections of the atom orbital momentum) equals 5. Now let us construct the state with the maximum projection of total spin. We obtain that the spin projection is one and that the total orbital momentum projection is also one. Thus this state corresponds to the term ^3P, which includes nine states with different projections of the total orbital momentum and spin. Hence, the extracted terms ^1D and ^3P include 14 $(5+9)$ states from the total number of 15. It is necessary to add one more term ^1S to these terms, i.e., the terms of atoms of the fourth group of the periodic system of elements with the valent shell p^2 are ^3P, ^1D, ^1S.

The energetic sequence of these terms is determined by the Hund rule. According to this rule, the lowest state, by energy, has the maximum spin. The lowest state among the terms with identical spins has the maximum electron momentum. Thus the order of terms for atoms of the fourth group of the periodic system of elements is ^3P, ^1D, ^1S. Table 4.9 gives the energetic parameters of the lowest states for atoms of the fourth group of the periodic system of elements, and Figs. 4.2 and 4.3 contain the spectral parameters of the carbon and silicon atoms.

Atoms of the sixth group of the periodic system of elements have the same order of electron terms as atoms of the fourth group, because the shell in this case can be constructed from two p-holes. Then the order of terms is ^3P, ^1D, ^1S, but the order of states of fine structure is reversible with respect to atoms of the fourth group so that the lowest state is $^3P_2$. Table 4.10 presents energetic parameters of the lowest states for atoms of the sixth group of the periodic system of elements, and Figs. 4.4–4.7 give the spectral parameters of these states. Note that in the case of the tellurium atom, the order of levels of fine structure violates both, due to the strong relativistic interactions and to the interaction of external and internal shells.

Now let us consider atoms of the fifth group of the periodic system of elements with the valent shell p^3. The statistical weight or total number of states for this shell is equal to $C_6^3 = 20$. In order to find these states, let us distribute electrons on shells as was made for atoms of the fourth group of the periodic system of

Table 4.9. Energetic parameters of the lowest states for atoms of the fourth group of the periodic system of elements.

Atom	C	Si	Ge	Sn	Pb
Shell	$2p^2$	$3p^2$	$4p^2$	$5p^2$	$6p^2$
$\varepsilon_{ex}(^1D)$, eV	1.26	0.78	0.88	1.07	2.66
$\varepsilon_{ex}(^1S)$, eV	2.68	1.91	2.03	2.13	3.65
$\varepsilon_{ex}(^3P_1)$, cm^{-1}	16.4	77.1	557	1692	7819
$\varepsilon_{ex}(^3P_2)$, cm^{-1}	43.4	223	1410	3428	10650

Table 4.10. Energetic parameters of the lowest states for atoms of the sixth group of the periodic system of elements.

Atom	O	S	Se	Te
Shell	$2p^4$	$3p^4$	$4p^4$	$5p^4$
$\varepsilon_{ex}(^1D)$, eV	1.97	1.14	1.19	1.31
$\varepsilon_{ex}(^1S)$, eV	4.19	2.75	2.78	2.88
$\varepsilon_{ex}(^3P_1)$, cm^{-1}	158	396	1990	4751
$\varepsilon_{ex}(^3P_2)$, cm^{-1}	227	574	2534	4706

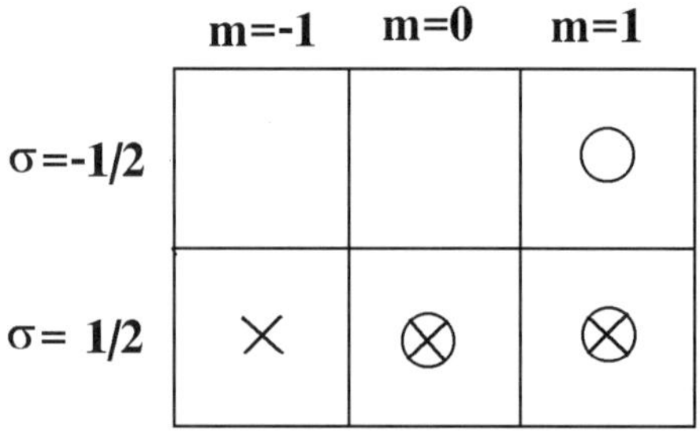

Figure 4.8. The distribution in the electron shell of valent p-electrons for atoms of the fifth group of the periodical system of elements for states with maximum spin (circles) and maximum orbital momentum (crosses).

elements. The state with the maximum projection of the total spin is indicated in Fig. 4.8 by crosses. This state has the spin projection 3/2 and the orbital momentum projection 0 onto a given direction, i.e., it corresponds to the term ^4S. The state with the maximum projection of the orbital momentum is shown in Fig. 4.8 by open circles. This corresponds to the term ^2D. Hence, the extracted terms ^4S, ^2D include 14 (4 + 10) states from 20. One more term is characterized by the spin projection 1/2 and the orbital momentum projection 0 or 1. Evidently, it is the term ^2P which includes six states. Thus the terms of the shell p^3 are ^4S, ^2D, ^2P. These terms are written in the order of their energetic levels in accordance with the Hund rule. Table 4.11 contains energetic parameters of the lowest states for atoms of the fifth group of the periodic system of elements, and Figs. 4.9 and 4.10 contain spectral parameters of the lowest states of the nitrogen and phosphorus atoms.

Atoms of the seventh group of the periodic system of elements (atoms of halogens) have the valent shell p^5 (or one p-hole). Therefore the term of the ground state of the halogen atoms coincides with that of the atoms of the third group of the periodic system of elements, but the sublevels of fine structure have the inverse sequence. The ground state of halogen atoms is $^2P_{3/2}$, the lowest excited state corresponds to the electron shell p^4s. Table 4.12 gives parameters of the lowest states of halogen atoms.

Table 4.11. Energetic parameters of the lowest states for atoms of the fifth group of the periodic system of elements.

Atom	N	P	As	Sb	Bi
Shell	$2p^3$	$3p^3$	$4p^3$	$5p^3$	$6p^3$
$\varepsilon_{ex}(^2D_{5/2})$, eV	2.38	1.41	1.31	1.06	1.42
$\varepsilon_{ex}(^2P_{1/2})$, eV	3.58	2.32	2.25	2.03	2.68
$\Delta\varepsilon(^2D_{5/2} - {}^2D_{3/2})$, cm^{-1}	8.7	15.6	322	1342	4019
$\Delta\varepsilon(^2P_{1/2} - {}^2P_{3/2})$, cm^{-1}	0.39	25.3	461	2069	10927

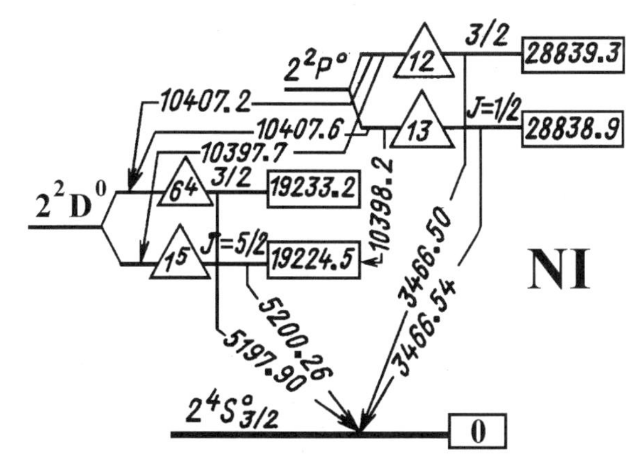

Figure 4.9. Parameters of the metastable states of the nitrogen atom. These states are characterized by the same electron shell as the ground state.

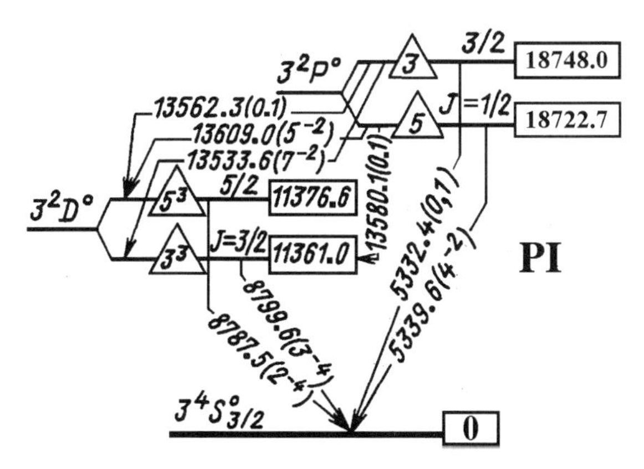

Figure 4.10. Parameters of the metastable states of the phosphine atom. These states are characterized by the same electron shell as the ground state.

Table 4.12. Energetic parameters for halogen atoms.

Atom	F	Cl	Br	I
Shell	$2p^5$	$3p^5$	$4p^5$	$5p^5$
J, eV	17.42	12.97	11.81	10.45
Lowest excited term	$2p^43s, {}^4P_{5/2}$	$3p^44s, {}^4P_{5/2}$	$4p^45s, {}^4P_{5/2}$	$5p^46s, {}^4P_{5/2}$
$\varepsilon_{ex}({}^4P_{5/2})$, eV	12.70	8.92	7.86	6.77
$\Delta\varepsilon({}^2P_{3/2} - {}^2P_{1/2})$, cm^{-1}	404	881	3685	7603

Table 4.13. Parameters of the lowest excited states of inert gas atoms. Here J is the ionization potential, ε_{ex} is the excitation energy from the ground state, λ is the wavelength of the corresponding transition, and τ is the radiative lifetime of the indicated state.

Atom	Ne	Ar	Kr	Xe
J, eV	21.56	15.76	14.00	12.13
Excited shell	$2p^53s$	$3p^54s$	$4p^55s$	$5p^56s$
$\varepsilon_{ex}({}^3P_2)$, eV	16.62	11.55	9.92	8.32
$\varepsilon_{ex}({}^3P_1)$, eV	16.67	11.62	10.03	8.44
$\varepsilon_{ex}({}^3P_0)$, eV	16.72	11.72	10.53	9.45
$\varepsilon_{ex}({}^1P_1)$, eV	16.82	11.83	10.64	9.57
$\lambda({}^1S_0 \to {}^3P_1)$, nm	743.72	1066.6	1235.8	1469.6
$\tau({}^3P_1)$,ns	25	10	4.5	3.6
$\lambda({}^1S_0 \to {}^1P_0)$, nm	735.90	1048.2	1164.8	1295.6
$\tau({}^1P_1)$, ns	1.6	2.0	4.5	3.5

Atoms of the eighth group of the periodic system of elements, or atoms of inert gases, have the closed electron shell p^6, so that the ground state of the atoms of inert gases is ${}^1S_0$. These atoms are characterized by large ionization potentials and excitation energies (see Table 4.13). The first excited states correspond to the valent electron shell p^5s. Although we use for these states the notations corresponding to the scheme of LS-coupling, this scheme is not correct for the excited atoms of inert gases (see also Chapters 5 and 6), so that in this case the values L and S are designations of states only. This will be considered in detail in the next chapter.

4.8 Systematization of the Electron Structures of Atoms

Thus, the shell concept of the atom structure allows us to analyze the general properties of the atoms of various elements and their spectra. Let us sum up this analysis. We are restricted by atoms with valent s- and p-electrons which form the basic groups of the periodic system of elements and include about half of the existing elements. These elements are met more often than other elements in nature and laboratory systems. This justifies our choice.

Some properties of the analyzed atoms are marked in Table 4.14. Note that though the used scheme of LS-coupling is often not grounded, our general conclusions are valid. In other words, if the relativistic interactions are not small compared to the exchange ones, and L, S are not quantum numbers, our general conclusions remain. In particular, the total electron momentum which is the sum of the total orbital and spin momenta continues to be a quantum number. Along with this, the number of levels is independent of the scheme of momentum summation, but the notation of levels by quantum numbers L, S is only the form to denote a state and has no physical sense, in contrast to the case of the validity of LS-coupling. Violation of the used scheme of LS-coupling proceeds when the atom structure becomes complicated. An increase of the orbital momentum of the valent electrons favours to this. Table 4.15 contains statistical parameters of the electron shells which are characteristics of the atom complexity.

If a given electron shell has several electron terms, the choice of the ground state term is made on the basis of the Hund rule. According to the Hund rule, the energies of states with the same electron shell are placed in the order of decrease of the total atom spin, so that the ground state has the maximum total electron spin. States of the same electron spin are placed in the order of decrease of the total orbital momentum, so that the lowest state by energy has the maximum orbital momentum. The Hund rule follows from the character of the exchange interaction within the limits of one electron shell. The order of sublevels of fine structure is the following. If the shell is filled by less than half, the lowest state of fine structure has the minimum total electron momentum J. If the shell is filled by more than half, the lowest sublevel is characterized by the maximum total momentum J. Though this rule is considered as a semiempirical one, it follows from the character of the interaction in the atom.

Table 4.14. General properties of atoms of the periodic system of elements (m denotes a metastable state and r relates to a resonantly excited one).

Number of group	Valent shell	Fine structure	Lowest excited state
1	s	−	r
2	s^2	−	m
3	p	+	m, r
4	p^2	+	m
5	p^3	−	m
6	p^4	+	m
7	p^5	+	m, r
8	p^6	−	m, r

Table 4.15. Electron states of atoms with filling shells.

Configuration of shell	Number of terms	Number of levels	Statistic weight
s	1	1	2
s^2	1	1	1
p, p^5	1	2	6
p^2, p^4	3	5	15
p^3	3	5	20
d, d^9	1	2	10
d^2, d^8	5	9	45
d^3, d^7	8	19	120
d^4, d^6	18	40	210
d^5	16	37	252
f, f^{13}	1	2	14
f^2, f^{12}	7	13	91
f^3, f^{11}	17	41	364
f^4, f^{10}	47	107	1001
f^5, f^9	73	197	2002
f^6, f^8	119	289	3003
f^7	119	327	3432

Problems

Problem 4.1. *The atom valent shell contains two electrons. Express the two-electron wave function of the atom through a one-electron wave function within the framework of the LS-coupling scheme.*

The atom state is characterized by the quantum numbers LSM_LM_S, where L is the orbital momentum of the atom, S is its total spin, and M_L, M_S are the projections of the orbital momentum and spin onto a given direction. Using the character of summation of electron momenta in the total atom momentum we have, for the electron wave function in this case,

$$\Psi_{LSM_LM_S}(1,2) = \sum_{m_1\sigma_1 m_2\sigma_2} \begin{bmatrix} l & l & L \\ m_1 & m_2 & M_L \end{bmatrix} \begin{bmatrix} \frac{1}{2} & \frac{1}{2} & S \\ \sigma_1 & \sigma_2 & M_S \end{bmatrix} \times \psi_{l\frac{1}{2}m_1\sigma_1}(1) \cdot \psi_{l\frac{1}{2}m_2\sigma_2}(2).$$

Here l is the electron orbital momentum, arguments of the wave functions mark electrons to which they correspond, and the subscripts of the wave functions are quantum numbers of the states. Each one-electron wave function is normalized by unity. Because of the property of the Clebsh–Gordan coefficients

$$\sum_{m_1 m_2} \begin{bmatrix} l & l & L \\ m_1 & m_2 & M_L \end{bmatrix}^2 = 1,$$

the two-electron wave function is also normalized by one.

The wave function of the system of two electrons requires the fulfillment of condition (3.1) according to which the transposition of two electrons leads to a change of its sign. For the same electron momenta the electron transposition is analogous to the operation $m_1\sigma_1 \rightleftharpoons m_2\sigma_2$. Using the property of the Clebsh–Gordan coefficients

$$\begin{bmatrix} l & l & L \\ m_1 & m_2 & M_L \end{bmatrix} = (-1)^{2l-L} \begin{bmatrix} l & l & L \\ m_2 & m_1 & M_L \end{bmatrix}.$$

From this it follows that if two electrons located in the same state with the orbital moment l, the value $2l - L + 1 - S$, must be even, as follows from the condition (3.1). Thus, two-electron states of the considering electron shell exist only with even quantity $L + S$. In particular, condition (3.1) prohibits terms ^3D, ^3S, ^1P for the electron shell p^2. Hence only 15 states ^3P, ^1D, and ^1S can exist in this case, while without condition (3.1) 36 states become possible.

Let us construct two-electron wave functions as a product of spatial $\varphi_{lm}(\mathbf{r})$ and spin η_σ one-electron wave functions. We have, for the two-electron wave functions,

$$\Psi_{LSM_LM_S}(1,2) = \Phi_{LM_L}(\mathbf{r}_1, \mathbf{r}_2) \cdot \chi_{SM_S}(1,2),$$

where

$$\Phi_{LM_L}(\mathbf{r}_1, \mathbf{r}_2) = \sum_{m_1, m_2} \begin{bmatrix} l & l & L \\ m_1 & m_2 & M_L \end{bmatrix} \varphi_{lm_1}(\mathbf{r}_1)\varphi_{lm_2}(\mathbf{r}_2),$$

$$\chi_{SM_S}(1,2) = \sum_{\sigma_1, \sigma_2} \begin{bmatrix} \frac{1}{2} & \frac{1}{2} & S \\ \sigma_1 & \sigma_2 & M_S \end{bmatrix} \eta_{\sigma_1}(1)\eta_{\sigma_2}(2).$$

The spin wave functions of two electrons have the form

$$\chi_{00} = \frac{1}{\sqrt{2}}\left[\eta_+(1)\eta_-(2) - \eta_-(1)\eta_+(2)\right]; \qquad \chi_{11} = \eta_+(1)\eta_+(2);$$

$$\chi_{10} = \frac{1}{\sqrt{2}}\left[\eta_+(1)\eta_-(2) + \eta_-(1)\eta_+(2)\right]; \qquad \chi_{1,-1} = \eta_-(1)\eta_-(2).$$

The total orbital moment can change from 0 up to $2l$. In particular, Table 4.16 contains the Clebsh–Gordan coefficients which determine the two-electron wave functions in the case $l = 1$.

Table 4.16 contains values of the Clebsh–Gordan coefficients for the positive values of m_1 and M_L. In the case of negative values of these quantities one can

Table 4.16. The Clebsh–Gordan coefficients $C = \begin{bmatrix} 1 & 1 & L \\ m_1 & m_2 & M \end{bmatrix}$.

L	M	m_1	m_2	C
0	0	0	0	$-\frac{1}{\sqrt{3}}$
0	0	1	-1	$\frac{1}{\sqrt{3}}$
1	0	0	0	0
1	0	1	-1	$\frac{1}{\sqrt{2}}$
1	1	1	0	$\frac{1}{\sqrt{2}}$
2	0	0	0	$\sqrt{\frac{2}{3}}$
2	0	1	-1	$\frac{1}{\sqrt{6}}$
2	1	1	0	$\frac{1}{\sqrt{2}}$
2	2	1	1	1

use the following relations:

$$\begin{bmatrix} l & l & L \\ m_1 & m_2 & M_L \end{bmatrix} = (-1)^{2l-L} \begin{bmatrix} l & l & L \\ -m_1 & -m_2 & -M_L \end{bmatrix} = (-1)^{2l-L} \begin{bmatrix} l & l & L \\ m_2 & m_1 & M_L \end{bmatrix}.$$

Problem 4.2. *Determine the order of the electron terms and relative distance between their levels for an atom with the valent electron shell* p^2 *within the framework of the* LS*-coupling scheme.*

Using the two-electron wave function of the previous problem, we calculate the energy of the corresponding term

$$\varepsilon_{LS} = \langle LSM_L M_S | \hat{H} | LSM_L M_S \rangle.$$

Here the Hamiltonian of the electrons has the form

$$\hat{H} = -\frac{1}{2}\Delta_1 - -\frac{1}{2}\Delta_2 + U(r_1) + U(r_2) + \frac{1}{|\mathbf{r}_1 - \mathbf{r}_2|} = \hat{h}_1 + \hat{h}_2 + \frac{1}{|\mathbf{r}_1 - \mathbf{r}_2|},$$

where $\mathbf{r}_1, \mathbf{r}_2$ are the coordinates of the corresponding electrons and $U(r)$ is the potential of the self-consistent central field. Because the Hamiltonian of the electrons does not depend on spins, we have

$$\varepsilon_{LS} = \langle LM_L | \hat{H} | LM_L \rangle,$$

and the one-electron matrix element is equal to $\langle \varphi_{lm'} | \hat{h} | \varphi_{lm} \rangle = \varepsilon_o \delta_{mm'}$. This gives

$$\varepsilon_{LS} = 2\varepsilon_o + \left\langle \Phi_{LM} \left| \frac{1}{|\mathbf{r}_1 - \mathbf{r}_2|} \right| \Phi_{LM} \right\rangle,$$

and we change the notation M_L by M. Thus the exchange splitting of levels within the limits of one shell is determined by the matrix element from the operator $1/|\mathbf{r}_1 - \mathbf{r}_2|$.

Let us extract the angular wave function from the one-electron wave function $\varphi_{lm}(\mathbf{r}) = R_l(r)Y_{lm}(\theta, \varphi)$, and use the relation

$$\frac{1}{|\mathbf{r}_1 - \mathbf{r}_2|} = \sum_n \frac{r_<^n}{r_>^{n+1}} \cdot \frac{4\pi}{2n+1} \sum_q Y_{nq}(\theta_1, \varphi_1) Y_{nq}^*(\theta_2, \varphi_2),$$

where $r_> = \max(r_1, r_2)$, and $r_< = \min(r_1, r_2)$. Thus, we have

$$\left\langle \Phi_{LM} \left| \frac{1}{|\mathbf{r}_1 - \mathbf{r}_2|} \right| \Phi_{LM} \right\rangle = \sum_n a_n F_n, \; F_n = \int \frac{r_<^n}{r_>^{n+1}} R_l^2(r_1) r_1^2 \, dr_1 R_l^2(r_2) r_2^2 \, dr_2,$$

and

$$a_n = \frac{4\pi}{2n+1} \sum_q \sum_{m_1, m_2, m_1', m_2'} \begin{bmatrix} l & l & L \\ m_1 & m_2 & M \end{bmatrix} \begin{bmatrix} l & l & L \\ m_1' & m_2' & M \end{bmatrix}$$
$$\cdot \left\langle l m_1 | Y_{nq}(\theta_1, \varphi_1) | l m_1' \right\rangle \left\langle l m_2 | Y_{nq}(\theta_1, \varphi_1) | l m_2' \right\rangle .$$

Using the expression for the matrix element

$$\langle lm | Y_{nq}(\theta_1, \varphi_1) | lm' \rangle = \sqrt{\frac{2n+1}{4\pi}} \cdot \begin{bmatrix} l & n & l \\ m & q & m' \end{bmatrix} \cdot \begin{bmatrix} l & n & l \\ 0 & 0 & 0 \end{bmatrix},$$

we obtain

$$a_n = \begin{bmatrix} l & n & l \\ 0 & 0 & 0 \end{bmatrix}^2 \sum_{m_1, m_2, m_1', m_2', q} \begin{bmatrix} l & l & L \\ m_1 & m_2 & M \end{bmatrix}$$
$$\cdot \begin{bmatrix} l & l & L \\ m_1' & m_2' & M \end{bmatrix} \cdot \begin{bmatrix} l & n & l \\ m_1 & q & m_1' \end{bmatrix},$$

$$\begin{bmatrix} l & n & l \\ m_2' & q & m_2 \end{bmatrix} = (-1)^{2l+L+n} \cdot (2l+1) \cdot \begin{bmatrix} l & n & l \\ 0 & 0 & 0 \end{bmatrix}^2 \cdot \begin{Bmatrix} l & l & L \\ l & l & n \end{Bmatrix},$$

where { } means the $6j$-symbol of Wigner. Because l is the whole number, n is an even number, so that we obtain

$$a_n = (-1)^L (2l+1) \cdot \begin{bmatrix} l & n & l \\ 0 & 0 & 0 \end{bmatrix}^2 \cdot \begin{Bmatrix} l & l & L \\ l & l & n \end{Bmatrix}.$$

Since

$$\begin{bmatrix} l & 0 & l \\ 0 & 0 & 0 \end{bmatrix} = 1 \quad \text{and} \quad \begin{Bmatrix} l & l & L \\ l & l & 0 \end{Bmatrix} = (-1)^L \cdot \frac{1}{2l+1},$$

we have $a_0 = 1$.

Table 4.17.

Atom	δ	γ
C	0.47	0.002
O	0.47	0.004
Si	0.40	0.006
S	0.41	0.015
Ge	0.41	0.042
Se	0.38	0.072
Sn	0.45	0.11
Te	0.37	0.16
Pb	0.66	0.26

In the case of the p^2-shell it is necessary to account for two terms in the sum because $\begin{bmatrix} 1 & n & 1 \\ 0 & 0 & 0 \end{bmatrix}$ is not equal to zero only for $n = 0$ and $n = 2$. Therefore, the expression for the atom energy has the form

$$\varepsilon_{LS} = 2\varepsilon_o + F_0 + a_2 F_2.$$

The values of a_2 for $L = 0, 1, 2$ are equal, respectively, to $\frac{2}{5}$, $-\frac{1}{5}$, and $\frac{1}{25}$. From this it follows that the lowest state by energy for the considering electron shell corresponds to the electron term ^3P. The energies of the states of this shell are equal to

$$\varepsilon(^1D) = \varepsilon(^3P) + \frac{6}{25}F_2; \qquad \varepsilon(^1S) = \varepsilon(^3P) + \frac{3}{5}F_2.$$

From this we obtain the following relation for the relative positions of the energy levels of a given electron shell:

$$\delta = \frac{\varepsilon(^1D) - \varepsilon(^3P)}{\varepsilon(^1S) - \varepsilon(^3P)} = 0.4.$$

Table 4.17 yields values of the parameter δ for atoms with p^2- and p^4-electron shells. We use the average energy of excitation of the three lowest levels $\Delta\varepsilon_P$ as the energy of the electron term ^3P, the fourth level is taken as the electron term ^1D, and the fifth level is taken as the electron term ^1S. Alongside this parameter, the ratio of the fine splitting to the exchange interaction

$$\gamma = \frac{\Delta\varepsilon_P}{\varepsilon(^1S) - \varepsilon(^3P)}$$

is given in Table 4.17. The last relation allows us to ascertain the role of the spin-orbit interaction for a given atom.

Problem 4.3. *Determine the quadrupole moment of an atom with a p^n nonfilled shell within the framework of the LS-coupling scheme.*

The atomic quadrupole moment is

$$Q = 2e\sum_i \langle r_i^2 P_2 \cos\theta_i \rangle = 2e\sum_i \frac{l_i(l_i+1) - 3m_i^2}{(2l_i-1)(2l_i+3)}\overline{r_i^2},$$

where e is the electron charge, r_i, θ_i are the spherical coordinates of the ith electron, and l_i, m_i are the momentum of this electron and its projection on the field direction. Since

$$\sum_{l=-m}^{m}\left[l(l+1) - 3m^2\right] = 0,$$

the filled electron shells do not give a contribution to the atom quadrupole moment, and this is determined by the valent electrons.

Using expression (4.7) for the atomic wave function we obtain, for the atomic quadrupole moment after summation over the spin projections,

$$Q(LSM_L) = n\sum_{ls\mu} q_\mu \left|G_{ls}^{LS}(l_e, n)\right|^2 \begin{bmatrix} l_e & l & L \\ \mu & M_L - \mu & M_L \end{bmatrix}^2,$$

where L, M_L, S are the orbital momentum, its projection on the field direction, and the atom spin, respectively, l_e is the momentum of the valent electron, μ is its projection of the field direction, n is the number of identical valent electrons, l and s are the orbital momentum and spin of the atomic core, and the one-electron quadrupole moment is equal to

$$q_\mu = 2e\frac{l_e(l_e+1) - 3\mu^2}{(2l_e-1)(2l_e+3)}\overline{r^2},$$

and the value $\overline{r^2}$ relates to a valent electron. Considering the case of a p^n-electron shell, we have

$$Q(p^n, LSM_L) = \frac{2n}{5}e\overline{r^2}\sum_{ls\mu}(2 - 3\mu^2)\left|G_{ls}^{LS}(p^n)\right|^2 \begin{bmatrix} 1 & l & L \\ \mu & M_L - \mu & M_L \end{bmatrix}^2.$$

From this formula one can find the general properties of the atomic quadrupole moment. The first property

$$Q(p^n, LS, M_L) = Q(p^n, LS, -M_L),$$

which follows from the transformation $\mu,\ M_L \to -\mu,\ -M_L$ in the above expression for the atomic quadrupole moment. The second property of the atomic quadrupole moment uses the analogy between an electron and a hole. This gives

$$Q(p^n, LS, M_L) = -Q(p^{6-n}, LS, M_L).$$

Note that the quadrupole momenta of one electron and one hole with identical quantum numbers have a different sign. This can be explained by a different charge sign for an electron and a hole. The third property relates to summation over

Table 4.18. Values $Q(p^n, LS, M_L)/\overline{r^2}$ for the ground states of atoms with p^n-shells.

State	$p\,^2P$	$p^2\,^3P$	$p^3\,^4S$	$p^4\,^3P$	$p^5\,^2P$
$M_L = 0$	4/5	$-4/5$	0	4/5	$-4/5$
$\|M_L\| = 1$	$-2/5$	2/5	—	$-2/5$	2/5

projections of the atomic momentum which is analogous to an average over the field direction and gives

$$\sum_{M_L} Q(p^n, LS, M_L) = 0.$$

The above formula for the atomic quadrupole momentum allows one to determine this value for atoms containing several electrons with nonzero momentum in a nonfilled shell. In particular, Table 4.18 contains the reduced quadrupole momenta of atoms in the ground states with a nonfilled p-shell.

Problem 4.4. *Prove that if the spin-orbit interaction in the atom corresponds to the interaction of an electron spin with the orbital momentum of the same electron, then the total electron momentum, which is the sum of the total electron orbital momentum and total spin, is conserved.*

The operator of the spin-orbit interaction in the related case has the form

$$\hat{V} = -a \sum_i \hat{\mathbf{l}}_i \hat{\mathbf{s}}_i.$$

Here i is the electron number and the coefficient a is equal to

$$a = \frac{1}{2}\left(\frac{e\hbar}{mc}\right)^2 \left\langle \frac{1}{r} \cdot \frac{dU}{dr} \right\rangle,$$

where $U(r)$ is the potential of a self-consistent field. Let us evaluate the commutator between the above operator of the spin-orbit interaction and the operator of the total electron momentum

$$\hat{\mathbf{J}} = \hat{\mathbf{L}} + \hat{\mathbf{S}} = \sum_i \hat{\mathbf{l}}_i + \sum_i \hat{\mathbf{s}}_i.$$

Because the operator of the total momentum of a given electron commutes with that of another electron, we have

$$\left\{\hat{\mathbf{J}}, V\right\} = -A \sum_i \left\{\hat{\mathbf{l}}_i + \hat{\mathbf{s}}_i, \hat{\mathbf{l}}_i \hat{\mathbf{s}}_i\right\}.$$

Let us take the x-component of one term of this formula. Using the commuting relations for one electron

$$\{\hat{l}_x, \hat{l}_y\} = i\hat{l}_z; \qquad \{\hat{l}_x, \hat{l}_z\} = -i\hat{l}_y; \qquad \{\hat{s}_x, \hat{s}_y\} = i\hat{s}_z; \qquad \{\hat{s}_x, \hat{s}_z\} = i\hat{s}_y,$$

we obtain $\{\hat{l}_x + \hat{s}_x, \hat{\mathbf{l}}\hat{\mathbf{s}}\} = i\hat{l}_z\hat{s}_y - i\hat{l}_y\hat{s}_z + i\hat{l}_y\hat{s}_z - i\hat{l}_z\hat{s}_y = 0$. Thus, for an individual electron, the commutator is $\left\{\hat{\mathbf{l}}_i + \hat{\mathbf{s}}_i, \hat{\mathbf{l}}_i\hat{\mathbf{s}}_i\right\} = 0$. Therefore, this sum is equal to zero, and the total momentum of atom electrons is a quantum number if the spin-orbit interaction occurs independently for each electron.

Problem 4.5. *Express the oscillator strength of an atom with one valent electron through the matrix element of its distance from the nucleus within the framework of the LS-coupling scheme.*

We use the general expression for the oscillator strength, which for a one-electron atom has the form

$$f(\gamma, l \to \gamma', l') = \frac{2\omega}{3} \sum_{m'} |\langle \gamma l m | \mathbf{r} | \gamma' l' m' \rangle|^2,$$

where $\mathbf{r}$ is the electron radius-vector and $\gamma l m$ are the quantum numbers of the valent electron. Note that the averaging over photon polarizations and summation over projections of the electron momentum in the final state correspond to the averaging over projections of the orbital momentum in the initial state. Use the properties of the related matrix element

$$\sum_{m'} |\langle \gamma l m | \mathbf{r} | \gamma' l' m' \rangle|^2 = \frac{l_{\max}}{2l+1} |\langle \gamma l | r | \gamma' l' \rangle|^2,$$

where $l_{\max}$ is the maximum value among l and l'. Then we have, for the oscillator strength,

$$f(\gamma, l \to \gamma', l') = \frac{2\omega l_{\max}}{3(2l+1)} \left|\langle \gamma l | r | \gamma' l' \rangle\right|^2.$$

This yields, for the oscillator strength of transition between states of a given fine structure, as follows from formula (4.14):

$$\begin{aligned} f(\gamma, l, j \to \gamma', l', j') &= (2l+1)(2J'+1) \begin{Bmatrix} l' & 1 & l \\ j & \frac{1}{2} & j' \end{Bmatrix}^2 \cdot f(\gamma, l \to \gamma', l') \\ &= \frac{2\omega}{3} l_{\max}(2j'+1) \cdot \begin{Bmatrix} l' & 1 & l \\ j & \frac{1}{2} & j' \end{Bmatrix}^2 \cdot |\langle \gamma l | r | \gamma' l' \rangle|^2. \end{aligned}$$

Problem 4.6. *Within the framework of the parentage scheme express the cross section of atom photoionization through the one-electron matrix elements.*

The photoionization cross section is given by formula (1.28) and has the form, in atomic units,

$$\sigma_{\text{ion}} = \frac{q\omega}{6\pi c} \int |\langle LS | \mathbf{D} | L'S', \mathbf{q} \rangle|^2 \, d\Theta_{\mathbf{q}},$$

where the parameters L, S characterize the initial atom state, the parameters L', S' correspond to the atom core, and $\mathbf{q}$ is the wave vector of a released electron. Let us use the parentage scheme (4.7) of the atom and express the matrix element through the one-electron matrix elements. The atom dipole moment operator is $\mathbf{D} = \sum_j \mathbf{r}_j$, where the sum is taken over the valent electrons. Take the wave function of the initial state in the form (4.7) in accordance with the parentage scheme. The wave function of the final state we take in the identical form

$$\Psi = \frac{1}{\sqrt{n}} \hat{P} \psi_{\mathbf{q}}(1) \Psi_{L'S'}(2, 3, \ldots, n),$$

and for simplicity we assume that the wave functions of the atomic core are identical for the initial and final states. Then we obtain, for the matrix element,

$$\left\langle LS \left| \mathbf{D} \right| L'S', \mathbf{q} \right\rangle = \sqrt{n} G_{L'S'}^{LS} \begin{bmatrix} l & L' & L \\ m & M' & M \end{bmatrix} \langle lm | \mathbf{r} | \mathbf{q} \rangle,$$

where M is the projection of the atom momentum, M' is the projection of the atomic core momentum, and m is the free electron momentum onto a given direction.

Let us average the square of the matrix element over the initial state and sum up over the final states. Then we have

$$\begin{aligned} \left| \left\langle LS \left| \mathbf{D} \right| L'S', \mathbf{q} \right\rangle \right|^2 &= n (G_{L'S'}^{LS})^2 \cdot \frac{1}{2L+1} \sum_{m,M',M} \begin{bmatrix} l & L' & L \\ m & M' & M \end{bmatrix}^2 |\langle lm | \mathbf{r} | \mathbf{q} \rangle|^2 \\ &= n (G_{L'S'}^{LS})^2 \cdot \frac{1}{2l+1} \sum_m |\langle lm | \mathbf{r} | \mathbf{q} \rangle|^2. \end{aligned}$$

This gives, for the cross section of photoionization,

$$\sigma_{\text{ion}} = \frac{\omega}{6\pi c} \cdot \frac{n}{2l+1} \sum_m q (G_{L'S'}^{LS})^2 \cdot \int \sum_m |\langle lm \, | \mathbf{r} | \, \mathbf{q} \rangle|^2 \, d\Theta_{\mathbf{q}}.$$

Note that the integral inside the sum does not depend on m because the integration over the electron directions of a free electron motion leads to averaging over the momentum projection of a valent electron in the initial electron state. Expressing the integral through the one-electron matrix elements according to formula (2.40) we obtain, for the cross section of a light atom within the framework of the parentage scheme,

$$\sigma_{\text{ion}} = \frac{8\pi\omega}{3c} \cdot \frac{n}{2l+1} \sum_{L'S'} q (G_{L'S'}^{LS})^2 \cdot \left[l K_{l-1} + (l+1) K_{l+1} \right],$$

where summation is made over possible channels of the atom decay, and the matrix element K_l is given by formula (2.39),

$$K_{l\pm1} = \frac{1}{2q} \int_0^\infty r^3 dr \, R_l(r) R_{l\pm1}(q, r),$$

where $R_l(r)$ is the radial wave function of the bound electron and $R_{l\pm1}(q, r)$ is the radial wave function of the released electron whose wave vector is q and the angular momentum is $l \pm 1$.

CHAPTER 5

The Structure of Heavy Atoms

5.1 Atoms with Valent d- and f-Electrons

The shell model of atoms is based on a certain hierarchy of interactions inside atoms. The Pauli exclusion principle and a self-consistent atomic field are the basis of the atom shell structure. This allows us to characterize an atom state by a configuration of its shell. An exchange interaction within a valent electron shell leads to the splitting of levels inside a noncompleted shell. This exchange interaction is determined by a symmetry of the electron wave function and makes the total orbital atom momentum L, and the total atom spin S, to be quantum numbers of the atom. Correlation interactions between electrons, as well as relativistic interactions, are assumed to be weaker than the exchange interactions. They lead to the splitting of levels with given values of L and S. The total electron momentum of the atom J arises due to the relativistic interactions in the atom. Thus, the electron state of a light atom is characterized by quantum numbers which describe the configuration of its electron shell, by the total orbital atom momentum L, the total atom spin S, and the total atom momentum J which is a sum of the orbital and spin atom momenta. In this scheme of summation of electron momenta in the total atom momentum J, the LS-coupling scheme, the orbital momenta of individual electrons are summed up in the total orbital momentum of the atom L, spins of electrons are summed up into the total atom spin S and, further, the total orbital and spin momenta of the atom are summed up in the total electron momentum of the atom J. This scheme of summation of electron momenta takes place for light atoms.

Let us use this scheme for atoms with valent d-electrons. Although this model may be not realistic for these atoms, it allows us to analyze their specifics. Table 5.1 lists the electron terms of atoms with valent d-electrons within the framework of the atom shell model and the LS-coupling scheme. These terms correspond to the lowest configuration of the electron shell. There are C_{10}^{n} (a number of combinations from 10 to n) states of the electron shell d^n. As follows from Table 5.1, the atomic structure in this case is more complex than in the case of the p-shell. In contrast to the s- and p-electron shells, in some cases there are two different terms with the same quantum numbers L and S (then its number is given in parentheses). In order to distinguish these states, an additional quantum number is introduced which is called the seniority and is denoted by v. Let us demonstrate the introduction of this quantum number on a simple example of the state ^1S for the electron shell d^4. We now compose the electron wave function for this atom state as a combination of the products of two-electron wave functions, and the separation of electrons in pairs is made in such a way that the total spin momentum of each electron pair is equal to zero. Then the wave functions for each pair of electrons correspond to the states ^1S, ^1D, ^1G, and these states are the same for both pairs because the total orbital momentum of four electrons is equal to zero. In this way we must refuse from the state ^1S for an electron pair, because the Pauli exclusion principle forbids us to have this state for each electron pair. Thus we have to compose the wave function of the state ^1S of four electrons from the combinations of the two-electron wave functions $d^2(^1D)$, $d^2(^1D)$ and $d^2(^1G)$, $d^2(^1G)$. One can compose two such combinations, i.e., there are two states $d^4(^1S)$, which can be distinguished by the seniority quantum numbers $v = 0$ and $v = 2$. The parity of the quantum number of seniority coincides with the parity of the number of electrons in the shell. The maximum value of the seniority quantum number does not exceed the number of electrons in the shell.

Classifying atoms and distributing them over the periodical system of elements, we used assumptions which are valid for light atoms. These assumptions of the LS-coupling scheme are not well fulfilled for heavy atoms because the energies of electrons placed on the d- and f-shells are close to those of the next s- or p-shells; in addition, the relativistic interactions for heavy atoms cannot be small compared to exchange interactions. Nevertheless, the simple scheme of the atom construction through a sequence of electron shells is convenient in this case, although it requires

Table 5.1. Electron terms of atoms with the filling electron shell d^n.

n	Electron terms	Total number of states	Number of electron terms
0, 10	^1S	1	1
1, 9	^2D	10	2
2, 8	$^1S, {}^3P, {}^1D, {}^3F, {}^1G$	45	9
3, 7	$^2P, {}^4P, {}^2D(2), {}^2F, {}^4F, {}^2G, {}^2H$	120	19
4, 6	$^1S(2), {}^3P(4), {}^1D(2), {}^3D, {}^5D,$		
	$^1F, {}^3F(2), {}^1G(2), {}^3G, {}^3H, {}^1J$	210	40
5	$^2S, {}^6S, {}^2P, {}^4P, {}^2D(3), {}^2F(2),$		
	$^4F, {}^2G(2), {}^4G, {}^2H, {}^2J$	252	37

Table 5.2. The ground state of atoms with an unfilled d-shell.

Unfilled electron shell	Term of the ground state	Atoms with this shell in the ground state
d	$^2D_{3/2}$	Sc(3d),Y(4d),La(5d),Lu(5d),Ac(6d),Lr(6d)
d^2	$^3F_2$	Ti(3d^2), Zr(4d^2), Hf(5d^2), Th(6d^2)
d^3	$^4F_{3/2}$	V(3d^3), Ta(5d^3)
d^4	$^5D_0$	W(5d^4)
d^5	$^6S_{5/2}$	Mn(3d^5),Tc(4d^5),Re(5d^5)
d^6	$^5D_4$	Fe(3d^6), Os(5d^6)
d^7	$^4F_{9/2}$	Co(3d^7), Ir(5d^7)
d^8	$^3F_4$	Ni(3d^8)
d^9	$^2D_{5/2}$	—

Table 5.3. The ground state of atoms with an unfilled f-shell.

Unfilled f-shell	Term of the ground state	Atoms with this shell in the ground state
f	$^2F_{5/2}$	—
f^2	$^3H_4$	—
f^3	$^4I_{9/2}$	Pr(4f^3)
f^4	$^5I_4$	Nd(4f^4)
f^5	$^6H_{5/2}$	Pm(4f^5)
f^6	$^7F_0$	Sm(4f^6), Pu(5f^6)
f^7	$^8S_{7/2}$	Eu(4f^7), Am(5f^7)
f^8	$^7F_6$	—
f^9	$^6H_{15/2}$	Tb(4f^9), Bk(5f^9)
f^{10}	$^5I_8$	Dy(4f^{10}), Cf(5f^{10})
f^{11}	$^4I_{15/2}$	Ho(4f^{11}), Es(5f^{11})
f^{12}	$^3H_6$	Er(4f^{12}), Fm(5f^{12})
f^{13}	$^2F_{5/2}$	Tm(4f^{13}), Md(5f^{13})

some correction. Further we consider atoms with d- and f-valent shells within the framework of LS-coupling (see Tables 5.2 and 5.3). Then the choice of the ground state is based on the Hund rule, as above. Let us demonstrate this in the example of atoms with d^4-electron shells.

For the determination of possible states of the atom with d^4-electron shells it is necessary to compose a scheme as in Fig. 4.8 and place four electrons in free cells. According to the Hund rule, the lower state by energy is characterized by the maximum total spin among the possible spins, and has the maximum orbital momentum among states with maximum spin. Then it is convenient to choose an axis where all the electron spins are directed in the same way, so that the total electron spin and its projection are equal to 2 for the electron shell d^4. The maximum projection of the orbital momentum is 2 $(2+1+0-1)$ as well as the total orbital momentum. Next, if the shell is filled by less than halves, according to the Hund rule, the total atom momentum which is the sum of the spin and orbital momenta, is minimal among the possible total momenta. If the electron shell is filled by more than halves, this momentum is maximal from the possible total value of spin and orbital atom momenta. Thus, for the related electron shell the term of the ground state is $^5D_0$. In the case of the electron shell d^6, according to the Hund rule, the spin and orbital momentum remain the same, but the order of the sequence of the fine structure of levels is changed. Therefore, the term of the ground state for the electron shell d^6 is $^5D_4$. Note that the use of notations of the LS-coupling for atomic states is often a convenient form of writing, but not for notations of the real quantum numbers of the atom.

The analysis of the excited states of atoms with unfilled d- and f-shells is more complex than in the case of the p-shell. This is connected with both an increase of the momentum of valent electrons and also with the competition of different electron shells. The last fact does not allow us to prognosticate the behavior of the excited levels within the framework of a simple scheme. As a demonstration of this fact, we consider the lower excited states of atoms with the electron shell nd, $(n+1)s^2$ (see Table 5.4). In this case, the excitation corresponds to reconstruction of the electron shell because the states with different electron shells are close by energy. In addition, for the yttrium atom, levels of the electron shell $4d^25s$ are close to the levels of the states with the electron shell $5s^25p$. Indeed, the excitation energy of the yttrium atom in the state $5s^25p(^2P_{1/2})$ is equal to 1.30 eV, and the excitation energy of the state $5s^25p(^2P_{3/2})$ is equal to 1.41 eV, i.e., this doublet is overlapped with the multiplet $5d^25s$ (see Table 5.4).

The other example of the closeness of energies for different electron shells is given in Table 5.5 where atoms with the filling $3d$-shell are considered. The previous electron shell of these atoms is $4s^2$. Table 5.5 contains values of the excitation energy ε for states of the lowest electron shell $4s3d^{n+1}$, if the ground state

Table 5.4. States of atoms with the electron shell nd, $(n+1)s^2$, and the excitation energies ε for states with the electron shell nd^2, $(n+1)s$.

Atom	Sc	Y	La
Electron shell of the ground state	$3d4s^2$	$4d5s^2$	$5d6s^2$
Fine splitting of the ground state $(^2D_{3/2}-{}^2D_{5/2}\}$, eV	0.021	0.066	0.13
Electron shell of excited states	$3d^24s$	$4d^25s$	$5d^26s$
$\varepsilon(^4F_{3/2})$, eV	1.43	1.36	0.33
$\varepsilon(^4F_{5/2})$, eV	1.43	1.37	0.37
$\varepsilon(^4F_{7/2})$, eV	1.44	1.40	0.43
$\varepsilon(^4F_{9/2})$, eV	1.45	1.43	0.51

of the atom corresponds to the electron shell $4s^2 3d^n$. The ionization potentials for the ground states of the atoms J are given for comparison. In all cases the excitation energy of a new electron shell is remarkably less than the atom ionization potential. This means competition of the electron shells in the course of atom excitation. Note that the $3d$-shell is the lowest shell by energy among the d- and f-electron shells. Hence, the competition of electron shells in the course of filling of the d- and f-electron shells is significant. For example, the electron shell of the ground state of the zirconium atom is $5s^2 4d^2$. One would expect that the following atom has the electron shell $5s^2 4d^3$, while the electron shell of the ground state of the niobium atom is $5s4d^4$. This effect is stronger for atoms with an f-electron shell. For example, the atoms of europium, gadolinium, and terbium, which are a sequence of atoms with an increase in the number of electrons, have the electron shell of the ground states $4f^7$, $4f^7 5d$, and $4f^9$, respectively.

The above examples confirm the closeness of different electron shells for atoms with filling d- and f-electron shells. This leads to the violation of simple schemes for the construction of the ground states of atoms. For this reason a universal scheme of the sequence of excited states of such atoms does not exist.

5.2 The Thomas–Fermi Atom Model

The closeness of different electron shells complicates the description of the properties of heavy atoms which are determined by valent electrons. On the contrary, the analysis of the properties which are determined by internal electrons is simplified. Indeed, an internal region of the atom, where the electron density is determined by several electron shells, can be analyzed on the basis of statistical methods. Statistical methods are the basis of the Thomas–Fermi atom model which accounts for the self-consistent field of internal electrons in average. This model is correct for

Table 5.5. Parameters of atoms with the filling d-electron shell.

Atom	Shell of ground state	J, eV	Excited shell	ε, eV
Sc	$4s^2 3d$	6.56	$4s3d^2$	1.43
Ti	$4s^2 3d^2$	6.82	$4s3d^3$	0.81
V	$4s^2 3d^3$	6.74	$4s3d^4$	0.26
Cr	$4s^2 3d^4$	6.77	$4s3d^5$	0.96
Mn	$4s^2 3d^5$	7.43	$4s3d^6$	2.11
Fe	$4s^2 3d^6$	7.90	$4s3d^7$	0.86
Co	$4s^2 3d^7$	7.86	$4s3d^8$	0.43
Ni	$4s^2 3d^8$	7.64	$4s3d^9$	0.025
Cu	$4s3d^{10}$	7.73	$4s^2 3d^9$	1.39

the internal region of a heavy atom where the electron density is large. We describe this atom model below.

The phase volume of electrons in a given space point is equal to

$$2\frac{d\mathbf{p}\,d\mathbf{r}}{(2\pi\hbar)^3} = \frac{p^2 dp\,d\mathbf{r}}{\pi^2\hbar^3},$$

where the factor 2 accounts for two projections of the electron spin, $\mathbf{p}$ is the electron momentum, and $d\mathbf{r}$ is the space element. From this it follows that, for a degenerated electron gas, the electron number density N is connected with its maximum momentum p_o by the relation

$$N = \int \frac{p^2\,dp}{\pi^2\hbar^2} = \frac{p_o^3}{3\pi^2\hbar^2}. \tag{5.1}$$

Then the implied assumption has the form (we use atomic units below):

$$p_o r_o \gg 1, \tag{5.2}$$

where r_o is a typical size on which the number density of electrons and other atomic parameters vary remarkably.

The maximum electron momentum p_o and the potential of the self-consistent field φ are connected by the relation

$$\frac{p_o^2}{2} = e\varphi. \tag{5.3}$$

The electric field potential satisfies the boundary condition $\varphi = 0$ far from the nucleus $r \to \infty$, where the electron density tends to zero, and hence $p_o \to 0$. The potential of the self-consistent electron field of an atom satisfies the Poisson equation

$$\Delta\varphi = 4\pi N. \tag{5.4}$$

For convenience we change the sign of this equation compared to that used in electrostatics. Expressing the right-hand side of equation (5.4) in accord with formulas (5.1), (5.2), we obtain the equation for the potential φ of the self-consistent field

$$\Delta\varphi = \frac{8\sqrt{2}}{3\pi}\varphi^{3/2}. \tag{5.5}$$

This equation is called the Thomas–Fermi equation. It is convenient to use the reduced variables

$$x = 2\left(\frac{4}{3\pi}\right)^{2/3} Z^{1/3} r = 1.13 Z^{1/3} r; \qquad \varphi = \frac{Z}{r}\chi(x). \tag{5.6}$$

Because the potential of the self-consistent field does not depend on angular variables, so that $\Delta\varphi = (1/r)(d^2/dr^2)(r\varphi)$, and the Thomas–Fermi equation (5.5) can be transformed to the form

$$x^{1/2}\frac{d^2\chi}{dx^2} = \chi^{3/2}. \tag{5.7}$$

Since the potential of the self-consistent field coincides with the Coulomb field of the nucleus charge near the nucleus $\varphi = Z/r$, the boundary condition near the center has the form

$$\chi(0) = 1. \tag{5.8a}$$

The other boundary condition corresponds to the absence of the electric charge far from the center

$$\chi(\infty) = 0. \tag{5.8b}$$

The numerical solution of equation (5.7) with the boundary conditions (5.8) is given in Table 5.6. In particular, $\chi'(0) = -1.588$, i.e., the electric potential of the self-consistent field near the center has the form

$$\varphi(r) = Z/r - 1.794Z^{4/3}. \tag{5.9}$$

The second term of this expression is the electric potential which is created by atomic electrons in the atom center.

The Thomas–Fermi atom model is valid in the atom region where the electron density is high enough in accordance with the criterion (5.2). Since an atom size is of the order of the Bohr radius, and the electron number density in the region of location of the valent electrons is of the order of unity, the Thomas–Fermi model is not valid in this region. Hence, this model can be used only in an internal region of the atom. A typical size of this model is equal to $Z^{-1/3}$ according to formula (5.6). Thus, a small parameter of the Thomas–Fermi model is

$$Z^{-1/3} \ll 1. \tag{5.10}$$

Fulfillment of this criterion provides the validity of the Thomas–Fermi model for internal electrons.

The Thomas–Fermi model allows us to determine the parameters of a heavy atom which are given by internal electrons. In particular, let us find the dependence of the total electron energy of an atom on the atom charge Z within the framework of the Thomas–Fermi model. The total electron energy is equal to

$$\varepsilon = T + U, \tag{5.11}$$

where T is the total kinetic energy of electrons and U is the potential electron energy which is the sum of the interaction potentials of electrons with the nucleus and between electrons

$$U = U_1 + U_2 = -\int \frac{Z}{r} N(\mathbf{r})\, d\mathbf{r} + \frac{1}{2} \int \frac{N(\mathbf{r})}{|\mathbf{r} - \mathbf{r}'|}\, d\mathbf{r}\, d\mathbf{r}'. \tag{5.12}$$

Let us determine the dependence of each of these terms on Z by taking into account that each integral is determined by an atom region $x \sim 1 (r \sim Z^{-1/3})$. In this region a typical electron number density is $N \sim Z^2$, a typical electron momentum is $p \sim p_o \sim N^{-1/3} \sim Z^{2/3}$, and a typical volume of this region is

Table 5.6. The potential of the self-consistent electric field for the Thomas–Fermi atom model.

x	$\chi(x)$	$-\chi'(x)$		x	$\chi(x)$	$-\chi'(x)$
0	1.000	1.588		2.4	0.202	0.0900
0.02	0.972	1.309		2.6	0.185	0.0793
0.04	0.947	1.199		2.8	0.170	0.0702
0.06	0.924	1.118		3.0	0.157	0.0625
0.08	0.902	1.052		3.2	0.145	0.0558
0.10	0.882	0.995		3.4	0.134	0.0501
0.2	0.793	0.794		3.6	0.125	0.0451
0.3	0.721	0.662		3.8	0.116	0.0408
0.4	0.660	0.565		4.0	0.108	0.0369
0.5	0.607	0.489		4.5	0.0919	0.0293
0.6	0.561	0.429		5.0	0.0788	0.0236
0.7	0.521	0.380		5.5	0.0682	0.0192
0.8	0.485	0.339		6.0	0.0594	0.0159
0.9	0.453	0.304		6.5	0.0522	0.0132
1.0	0.424	0.274		7.0	0.0461	0.0111
1.2	0.374	0.226		7.5	0.0410	0.095
1.4	0.333	0.189		8.0	0.0366	0.081
1.6	0.298	0.160		8.5	0.0328	0.070
1.8	0.268	0.137		9.0	0.0296	0.060
2.0	0.243	0.118		9.5	0.0268	0.053
2.2	0.221	0.103		10	0.0243	0.048

$\int d\mathbf{r} \sim 1/Z$. From this it follows that the electron kinetic energy is

$$T \sim \int \frac{p^2}{m} N \, d\mathbf{r} \sim Z^{7/3}. \tag{5.13}$$

The energy of the interaction of electrons with the nucleus is

$$U_1 \sim \int \frac{Ze^2}{r} N \, d\mathbf{r} \sim Z^{7/3}.$$

The energy of the interaction between electrons is equal to

$$U_2 = \frac{1}{2} \int \frac{e^2 N(\mathbf{r})}{|\mathbf{r} - \mathbf{r}'|} d\mathbf{r} \, d\mathbf{r}' \sim Z^{7/3}.$$

As follows from this, the total binding energy of electrons varies with the nucleus charge as $Z^{7/3}$, i.e., the binding energy per one electron is of the order of $Z^{4/3}$.

Note that this value is of the order of Z^2 for electrons which are located near the nucleus, and does not depend on Z for valent electrons.

5.3 Exchange Interaction in Heavy Atoms

Considering the self-consistent field within the framework of the Thomas–Fermi atom model, we neglect the exchange interaction of electrons, assuming it to be small. Now we calculate the exchange interaction of electrons for the Thomas–Fermi atom. The exchange interaction potential which acts on the j-electron is given by formula (3.15):

$$V_{\text{ex}}(j) = -2\sum_k \left\langle jk \left| \frac{1}{|\mathbf{r}-\mathbf{r}_k|} \right| kj \right\rangle,$$

where the summation is made over atomic electrons whose spin projection coincides with that of the test electron j, $\mathbf{r}$ is the coordinate of the test electron, and $\mathbf{r}_k$ is the coordinate of other electrons with the same spin projection.

The wave functions for electrons for the Thomas–Fermi atom are plane waves, and their wave vector lies in the interval $0 < q_k < q_F$, where q_F is the Fermi wave vector which in atomic units coincides with the Fermi momentum p_o—the boundary momentum of electrons in a degenerated electron gas. Let us place the electrons in unit volume because the final result does not depend on a fictitious volume in which electrons are located. Then the electron wave function has the form $\psi = \exp(i\mathbf{qr})$, so that the exchange interaction potential is equal to

$$V_{\text{ex}}(j) = -2 \cdot \int \exp\left[i(\mathbf{q}_j - \mathbf{q}_k)(\mathbf{r}-\mathbf{r}')\right] \cdot \left|\mathbf{r}-\mathbf{r}'\right|^{-1} \cdot \frac{d\mathbf{q}_k}{(2\pi)^3}\, d\mathbf{r}',$$

where $\mathbf{q}_j$ is the wave vector of the test electron, and we take into account only atomic electrons with the same spin projection as the test electron. The average exchange potential is

$$\langle V_{\text{ex}}(j)\rangle = \int_0^{p_o} V_{\text{ex}}(j)\frac{3q_j^2 dq_j}{p_o^3} = -\frac{3p_o}{2\pi}. \tag{5.14}$$

An estimate of this integral can be obtained in a simple way, because from its expression it follows that it is proportional to p_o. If we include the exchange interaction potential in the Thomas–Fermi equation, we obtain so-called Thomas–Fermi–Dirac equation which partially takes into account the exchange interaction of electrons. But the correction due to this exchange potential is small in the basic atom region where the Thomas–Fermi model is valid. Indeed, in this region $\chi \sim 1$, i.e., $r \sim Z^{-1/3}$. Then the kinetic energy per electron (and, correspondingly, the self-consistent field potential) is of the order of $p_o^2 \sim N^{-2/3} \sim Z^{4/3}$. The average exchange interaction potential per one electron according to formula (5.14) is of the order of $p_o \sim N^{-1/3} \sim Z^{2/3}$. Thus, in the region of validity of the Thomas–Fermi atom model, the exchange interaction potential gives a small contribution to the self-consistent field potential.

5.4 Schemes of Summation of Electron Momenta in Atoms

Analyzing the properties of a light atom, we neglected the relativistic effects in the zeroth approach of the perturbation theory, and this was taken into account as a perturbation in the first approximation. In addition, we assumed the exchange interaction of electrons for the same electron shell to be small compared to the electrostatic interaction. This was the basis of the shell model of the atom. Then at the first stage we take the parameters of the electron shell of the atom as its quantum numbers. Next, we include the exchange interactions inside the valent electron shell that allow us to characterize the atomic states by the total orbital momentum L and spin momentum S as quantum numbers of the atom. At the following stage of consideration, we include the relativistic interactions which are small compared to the exchange interactions. Then along with the quantum numbers L, S, we obtain the total atom momentum J as its quantum number. This relation between the related types of interaction in the case of two electrons corresponds to the summation of two orbital momenta of electrons into the total orbital momentum of the atom L and the summation of two electron spins into the total spin of the atom S. This is the LS-coupling scheme of the summation of electron momenta in atoms.

If relativistic interaction is compared to an electrostatic splitting of levels, the values L and S are not quantum numbers. But the total atom momentum is the quantum number both in the case of weak and strong relativistic interactions. Therefore, depending on the relation between the electrostatic and relativistic interactions, different methods are possible for summation of electron momenta into the total atom momentum. Below we consider the possible schemes of summation of electron momenta in the total atomic moment. They can be as follows in the case of two electrons

$$\begin{aligned}
&\mathbf{l}_1 + \mathbf{l}_2 =: \mathbf{L}, \mathbf{s}_1 + \mathbf{s}_2 = \mathbf{S}, && S + L = J \quad (LS\text{-coupling}),\\
&\mathbf{l}_1 + \mathbf{s}_1 \ = \mathbf{j}_1, \mathbf{l}_2 + \mathbf{s}_2 = \mathbf{j}_2, && j_1 + \mathbf{j}_2 = \mathbf{J} \quad (jj\text{-coupling}),\\
&\mathbf{l}_1 + \mathbf{l}_2 =: \mathbf{L}, \mathbf{L} + \mathbf{s}_1 = \mathbf{K}, && \mathbf{K} + \mathbf{s}_2 = \mathbf{J} \quad (LK\text{-coupling}),\\
&\mathbf{l}_1 + \mathbf{s}_1 \ = \mathbf{j}_1, \mathbf{j}_1 + \mathbf{l}_2 = \mathbf{K}, && \mathbf{K} + \mathbf{s}_2 = \mathbf{J} \quad (JK\text{-coupling}).
\end{aligned}$$

Each of these summation schemes is valid in the case that if the interaction potential, which is responsible for summation of momenta in the third sum, is small compared to the other potentials of interaction, so that one can choose the intermediate momenta as quantum numbers. In particular, in the case of LS-coupling, the spin-orbital interaction is small compared to the electrostatic and exchange interactions. Hence L and S are quantum numbers in this case. In reality, the basic relativistic interactions in atoms are interactions of the orbital and spin of the same electron, or the interaction of two orbital momenta as well as two spins of different electrons. This means that the combinations $\hat{\mathbf{l}}_1\hat{\mathbf{s}}_1$, $\hat{\mathbf{l}}_2\hat{\mathbf{s}}_2$, $\hat{\mathbf{l}}_1\hat{\mathbf{l}}_2$, $\hat{\mathbf{s}}_1\hat{\mathbf{s}}_2$ are the strongest ones in the Hamiltonian of electrons. This condition chooses two of four possible schemes of summation: LS- or jj-coupling. If relativistic interactions are compared to electrostatic or exchange splitting of levels, an intermediate case between LS- and jj-coupling takes place.

5.5 Filling of Electron Shells for jj-Coupling

Let us consider an atom in which the spin-orbit relativistic interaction, which is determined by the term $a\hat{\mathbf{l}}\hat{\mathbf{s}}$ in the electron Hamiltonian, exceeds both the electrostatic interaction of levels which is given by the terms $\hat{\mathbf{l}}^2/2r^2$ in the atom Hamiltonian and the exchange interactions which are characterized by a certain symmetry of the wave function of valent electrons. In this case the total atomic momentum can be composed as a sum of the total momenta of individual electrons, and the total momentum of each electron is the sum of its orbital momentum and spin. In the case of jj-coupling the scheme of construction of the atom electron shell differs from that for LS-coupling. Below we demonstrate this scheme for the filling of a p-shell of atoms.

Table 5.7 lists the results of the construction of terms in the case of the filling of p-electron shells within the framework of LS- and jj-schemes. Let us analyze the peculiarities of the jj-coupling scheme. p-Electrons have total momentum 1/2 or 3/2, and the electron terms of atoms are characterized by the same total electron momentum for both LS- and jj-coupling. The lowest electron states of the atom have the maximum number of electrons with the total momentum 1/2. Note that from the Pauli exclusion principle it follows that the maximum number of electrons with total momentum 1/2 is equal to 2, and the maximum number of electrons with total electron momentum 3/2 is equal to 4 in accordance with the total number of momentum projections onto a given direction. Compose the wave function of two electrons with the same total momentum j within the framework of the jj-coupling scheme. This has the form

$$\Psi_{JM}(1,2) = \hat{P} \sum_{m,m'} \begin{bmatrix} j & j & J \\ m & m' & M \end{bmatrix} \psi_{jm}(1)\psi_{jm'}(2), \tag{5.15}$$

where $\hat{P}$ is the operator of the transposition of electrons, J, M are the total momentum of electrons and its projection onto a given direction, and m, m' are the projections of momenta of individual electrons onto this direction. Analyzing the wave function (5.15), note that the transposition of two electrons leads to the change of the wave function sign, and the transposition of the two first columns of the Clebsh–Gordan coefficient requires us to multiply this expression by $(-1)^{2j-J}$. From this it follows that the value $2j - J$ must be odd because, in the other case, the transposition of two electrons and the transposition of columns of the Clebsh–Gordan coefficient yields $\Psi \to -\Psi$, i.e., $\Psi = 0$. Thus, we obtain that the total momentum of the shell $[1/2]^2$ is zero and cannot be one, and the total momentum of the shell $[3/2]^2$ may be 0 or 2, and cannot be 1 or 3. The prohibition of the state with the total momentum $J = 2j$ follows directly from the Pauli exclusion principle because, in the opposite case for a certain direction of the axis, we obtain that

Table 5.7. Electron shells of atoms with p-valent electrons.

LS-shell	LS-term	J	jj-shell	J
p	^2P	1/2	$[1/2]^1$	1/2
	^2P	3/2	$[3/2]^1$	3/2
p^2	^3P	0	$[1/2]^2$	0
	^3P	1	$[1/2]^1[3/2]^1$	1
	^3P	2	$[1/2]^1[3/2]^1$	2
	^1D	2	$[3/2]^2$	2
	^1S	0	$[3/2]^2$	0
p^3	^4S	3/2	$[1/2]^2[3/2]^1$	3/2
	^2D	3/2	$[1/2]^1[3/2]^2$	3/2
	^2D	5/2	$[1/2]^1[3/2]^2$	5/2
	^2P	1/2	$[1/2]^1[3/2]^2$	1/2
	^2P	3/2	$[3/2]^3$	3/2
p^4	^3P	2	$[1/2]^1[3/2]^3$	2
	^3P	0	$[1/2]^2[3/2]^2$	0
	^3P	1	$[1/2]^1[3/2]^3$	1
	^1D	2	$[1/2]^1[3/2]^3$	2
	^1S	0	$[3/2]^4$	0
p^5	^2P	3/2	$[1/2]^2[3/2]^3$	3/2
	^2P	1/2	$[1/2]^1[3/2]^4$	1/2
p^6	^1S	0	$[1/2]^2[3/2]^4$	0

two electrons have the same projection j of the total momentum. The character of summation of momenta chooses certain states for the total momentum. If electrons have total momenta j and j', their total momentum ranges from $\left|j - j'\right|$ to $j + j'$. Thus, the symmetry of the total wave function of electrons and the character of summation of momenta chooses certain states of the total system of electrons and leads to the prohibition of some states of the system. Note once more that the expression for the two-electron atom wave function of a certain total electron momentum as a product of one-electron wave functions has the identical form within the framework of both LS- and jj-coupling schemes (see also Table 5.7).

The real character of the coupling of electron momenta in atoms follows from the analysis of the positions of the electron levels. We demonstrate this in the example of the electron shell p^2. Both LS- and jj-coupling schemes give the same sequence for the total momentum of electrons, which is $J = 0, 1, 2, 2, 0$. If the LS-scheme is valid and the exchange interaction dominates in the atom, the three lowest levels are close and are separated from the other two levels by $J = 2$

and $J = 0$. In particular, in the carbon case, the excitation energies for the above states are 0, 0.002, 0.005, 1.264 and 2.654, eV, respectively, which corresponds to the LS-coupling scheme.

If the jj-scheme is valid, the lowest atom states are grouped in a singlet $\left(\left[\frac{1}{2}\right]^2\right)$ and two doublets $\left(\left[\frac{1}{2}\right]\left[\frac{3}{2}\right]\right.$ and $\left.\left[\frac{3}{2}\right]^2\right)$. For example, in the case of the Pb atom with the electron shell $6p^2$ the energies of the lowest levels are 0, 0.969, 1.320, 2.660, and 3.653 eV. As is seen, the LS-coupling scheme is not valid in this case, and the jj-coupling scheme is suitable for the lowest doublet. Thus, the LS-coupling scheme is working for light atoms and an intermediate coupling scheme describes heavy atoms. Accounting for the total electron momentum is the atom quantum number for any relationship between the exchange and spin-orbit interactions, we shall use further the notations of the LS-coupling scheme and keep in mind that although this scheme is not valid for heavy atoms, the sequence of levels according to the total electron momentum becomes correct.

5.6 Excited States of the Atoms of Inert Gases

Let us consider the peculiarities of summation of electron momenta into the total atomic momentum in the example of excited inert gas atoms. First, consider the lowest group of excited atoms of inert gases which have the electron shell $np^5(n+1)s$. An inert gas ion has the electron shell np^5 and ground state $^2P_{3/2}$. Denote by Δ_f the fine splitting of ion levels, i.e., the distance between levels of the ion states $^2P_{3/2}$ and $^2P_{1/2}$, and this will be used as a typical energy of the spin-orbital interaction for excited atoms. There are four different energy levels for the lowest atomic states. If we use the scheme of LS-coupling, these states are $^3P_2$, $^3P_1$, $^3P_0$, $^1P_1$. In accordance with the Hund rule, we set these states in turn of atom excitation. Within the framework of the jj-coupling scheme, this sequence of states is the following: $s\left[\frac{3}{2}\right]_2$, $s\left[\frac{3}{2}\right]_1$, $s'\left[\frac{1}{2}\right]_0$, $s'\left[\frac{1}{2}\right]_1$. These notations are close to the usual notations of summation of momenta for the case of jj-coupling which usually has the form $\left[j_1, j_2\right]_J$ and we denote that summation of momenta j_1 and j_2 yields the total momentum J. The so-called Pashen notations are used often for the excited atoms of inert gases due to their simplicity. Then the relevant states in turn of the excitation of the atom are denoted as $1s_5$, $1s_4$, $1s_3$, and $1s_2$. Below we take the level $1s_5$ as zero and denote the difference of the excitation energies of this and the other state as ε_4, ε_3, and ε_2, i.e., ε_2 is the difference of energies for levels $1s_2$ and $1s_5$ (see Fig. 5.1). Table 5.8 contains some parameters of the lower excited states for atoms of inert gases. We briefly analyze these data below.

One can extract two terms in the Hamiltonian $\hat{H} = -a\hat{\mathbf{l}}\hat{\mathbf{s}}_1 - b\hat{\mathbf{s}}_1\hat{\mathbf{s}}_2$ for this modeling of the relevant levels (see Problem 5.5). Here $\hat{\mathbf{l}}$ is the operator of the orbital momentum of the hole, $\hat{\mathbf{s}}_1$ is the spin operator of the hole, and $\hat{\mathbf{s}}_2$ is the spin operator of the excited valent s-electron. If $a \gg b$, jj-coupling is valid. Then the total momentum of the hole is the quantum number. In addition, in this case, $\varepsilon_2 - \varepsilon_3 \ll \Delta_f$, $\varepsilon_4 \ll \Delta_f$, and $\varepsilon_2, \varepsilon_3 \cong \Delta_f$. As follows from the data of Table 5.8,

jj - notations	LS - notations	Pashen notations	Number of states
$(n+1)s'\ [1/2]^0_1$	$^1P_1$	$1s_2$	3
$(n+1)s'\ [1/2]^0_0$	$^3P_0$	$1s_3$	1
$(n+1)s\ [3/2]^0_1$	$^3P_1$	$1s_4$	3
$(n+1)s\ [3/2]^0_2$	$^3P_2$	$1s_5$	5

Figure 5.1. Notations of the first excited levels of the atoms of inert gases.

Table 5.8. Energies of the Lower Excited States of Atoms of Inert Gases

Atom	Δ_f, cm^{-1}	ε_3/Δ_f	ε_4/Δ_f	$(\varepsilon_2-\varepsilon_3)/\Delta_f$	$\Delta\varepsilon_{2p}/\Delta_f$
Ne	780.3	0.996	0.54	0.09	6.0
Ar	1432.0	0.984	0.43	0.10	3.2
Kr	5370.1	0.972	0.18	0.13	1.4
Xe	10537	0.866	0.11	0.11	1.2

these relations are roughly fulfilled, i.e., jj-coupling is valid in this case. Note that the states, of this group are separated from the following excited states, which were denoted by Pashen as $2p$-levels, and their electron shell is $np^5(n+1)p$.

The excited state $(n+1)s$ of the valent electron is a resonantly excited one because the radiative transition $np \to (n+1)s$ is permitted in the dipole approach. But only two of the four states of this group are resonantly excited states. For the extraction of the resonantly excited states, let us represent the wave function of electrons for the excited atom in the form

$$\Psi_{JM} = \sum_{m,\sigma_1,\sigma_2} \begin{bmatrix} \frac{1}{2} & 1 & j \\ \sigma_1 & m-\sigma_1 & m \end{bmatrix} \begin{bmatrix} \frac{1}{2} & j & J \\ \sigma_2 & m & M \end{bmatrix} \psi_m \chi_{\sigma_1} \eta_{\sigma_2}, \tag{5.16}$$

Where σ_1, σ_2, $m-\sigma_1$ are projections of the spin of the atom core, the spin of the valent electron and the orbital momentum of the atomic core onto a given direction, respectively, j is the total momentum of the atomic core, and J is the total atom momentum. Although this expression is written within the framework of jj-coupling, it is valid in a general case because the total atom momentum J is a quantum number. Separating the atom into the atomic core and valent electron, present the wave function of the ground atom state as the product of their spin and spatial wave functions. In a general case this is not correct, but because the electron shell is filled for the ground atom state, spins of the valent electron and atomic core have opposite directions, i.e., their total spin is zero. This allows one to present the total wave function for the ground atom state as a product of the spatial and spin wave functions of electrons, that is, the atom wave function for the ground state has the form

$$\Phi = \varphi_o \cdot \frac{1}{\sqrt{2}}(\chi_+\eta_- - \chi_-\eta_+), \tag{5.17}$$

where χ_+, χ_-, η_+, η_- are the spin wave functions of the atomic core and valent electron with the spin projection $\pm 1/2$ onto a given direction, respectively and φ_o is the spatial wave function for the ground state of the system of the atomic core and valence electron. We use in formula (5.17) that the total spin of the atom in the ground state is equal to zero. The probability of the radiative transition per unit time is proportional to the square of the matrix element of the dipole moment operator which does not depend on electron spins. Hence, it is convenient to project the wave function of the excited state (5.16) upon the spin wave function of the ground state (5.17). This gives

$$\langle \Psi_{JM} \mid \Phi \rangle = \frac{1}{\sqrt{2}} \left(\begin{bmatrix} \frac{1}{2} & 1 & j \\ \frac{1}{2} & M & M+\frac{1}{2} \end{bmatrix} \begin{bmatrix} \frac{1}{2} & j & J \\ -\frac{1}{2} & M+\frac{1}{2} & M \end{bmatrix} - \begin{bmatrix} \frac{1}{2} & 1 & j \\ -\frac{1}{2} & M & M-\frac{1}{2} \end{bmatrix} \begin{bmatrix} \frac{1}{2} & j & J \\ \frac{1}{2} & M-\frac{1}{2} & M \end{bmatrix} \right). \tag{5.18}$$

Below we show that this value is equal to zero for the states $1s_3$ and $1s_5$. Let us take, for the lowest excited states $1s_5$ or $[\frac{3}{2}]_2$, such an axis for which $M = 2$. Then all the Clebsh–Gordan coefficients except the last one are equal to zero because their momentum projections exceed their values. Then the matrix element is zero. This results from the symmetry of the wave function. Indeed, the wave function (5.16) for $M = 2$ has the form $\Psi_{22} = \psi_{+1}\chi_+\eta_+$, i.e., the total spin function of the valent electron and atomic core corresponds to their total spin $S = 1$, while it is equal to zero for the ground atom state. For the states $1s_5$ or $[\frac{1}{2}]_0$ formula (5.18) has the form

$$\langle \Psi_{00} \mid \Phi \rangle = \frac{1}{\sqrt{2}} \left(\begin{bmatrix} \frac{1}{2} & 1 & \frac{1}{2} \\ \frac{1}{2} & 0 & \frac{1}{2} \end{bmatrix} \begin{bmatrix} \frac{1}{2} & \frac{1}{2} & 0 \\ -\frac{1}{2} & \frac{1}{2} & 0 \end{bmatrix} - \begin{bmatrix} \frac{1}{2} & 1 & \frac{1}{2} \\ -\frac{1}{2} & 0 & -\frac{1}{2} \end{bmatrix} \begin{bmatrix} \frac{1}{2} & \frac{1}{2} & 0 \\ \frac{1}{2} & -\frac{1}{2} & 0 \end{bmatrix} \right). \tag{5.19}$$

As is seen, the second term of this expression is equal to the first one and eliminates it. Thus, the states $1s_5$ and $1s_3$ are metastable states and the radiative transition from these states in the ground state is prohibited (see also Problem 5.6).

Let us consider the next group of excited states for the atoms of inert gases with the electron shell $np^5(n+1)p$. The total number of different levels for this group of states is 10. Within the framework of LS-coupling these states are

$$^3D_3, {}^3D_2, {}^3D_1, {}^3P_2, {}^3P_1, {}^3P_0, {}^3S_1, {}^1D_2, {}^1P_1, \quad {}^1S_0.$$

If we use the notation of jj-coupling, these states are

$$[3/2, 3/2]_3, [3/2, 3/2]_2, [1/2, 3/2]_2, [3/2, 1/2]_2, [3/2, 3/2]_1,$$
$$[1/2, 3/2]_1, [3/2, 1/2]_1, [1/2, 1/2]_1, [3/2, 3/2]_0, [1/2, 1/2]_0.$$

Thus, we have, in this case, one state with the total electron momentum 3, three states with the total momentum 2, four states with the total momentum 2, and two states with the total electron momentum 0. The value $\Delta\varepsilon_{2p}$ of Table 5.8 is the difference of the energies of extreme levels of this group. This shows that for Ne, Ar, and Kr atoms the electrostatic and exchange interactions are compared to the fine splitting of the ion levels (see also Problem 5.5). In these cases both the LS-coupling and jj-coupling schemes cannot describe these states, i.e., they are suitable only for a rough description. Hence, experimental information for the positions of levels and other parameters of these states can be a basis for different realistic models.

Figures 5.2–5.5 contain parameters of the relevant excited states and radiative transitions with their participation. Pashen notations and designations of LS- and jj-couplings are used. The radiative lifetimes are given in these figures inside triangles, radiative transitions are shown by arrows, in discontinuity of these arrows the waves lengths of these transitions are given in Å, and the frequencies of transitions are given in parentheses and are expressed in 10^6s^{-1}.

5.7 Grotrian Diagrams for Atoms with Filling d- and f-Electron Shells

The above analysis shows the complexity of the spectra of atoms with unfilled d- and f-shells. First, such electron momenta lead to a large number of electron terms for this configuration of the electron shell. Second, this effect is reinforced due to the competition of different electron shells. The closeness of the excitation energies for different electron shells does not allow one to describe excited electron terms on the basis of a simple model due to the excitation of certain valent electrons. Third, the complexity of spectra is due to the disturbance of simple schemes of summation of electron momenta into the total momentum of the atom. All this hampers the extraction of multiplet levels corresponding to certain electron terms. These facts are responsible for a complex spectrum of these atoms in contrast

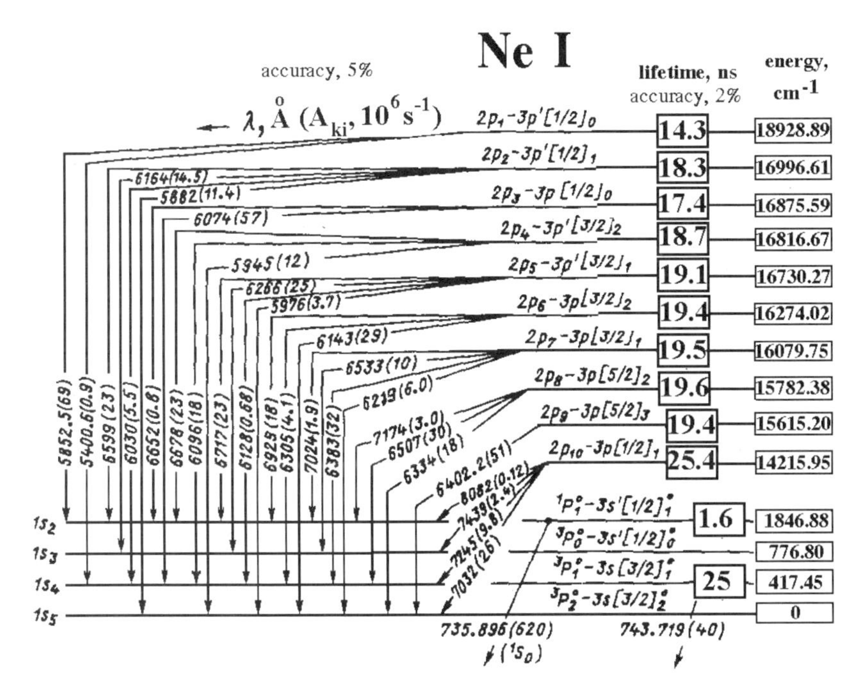

Figure 5.2. Parameters of the excited states of the neon atom.

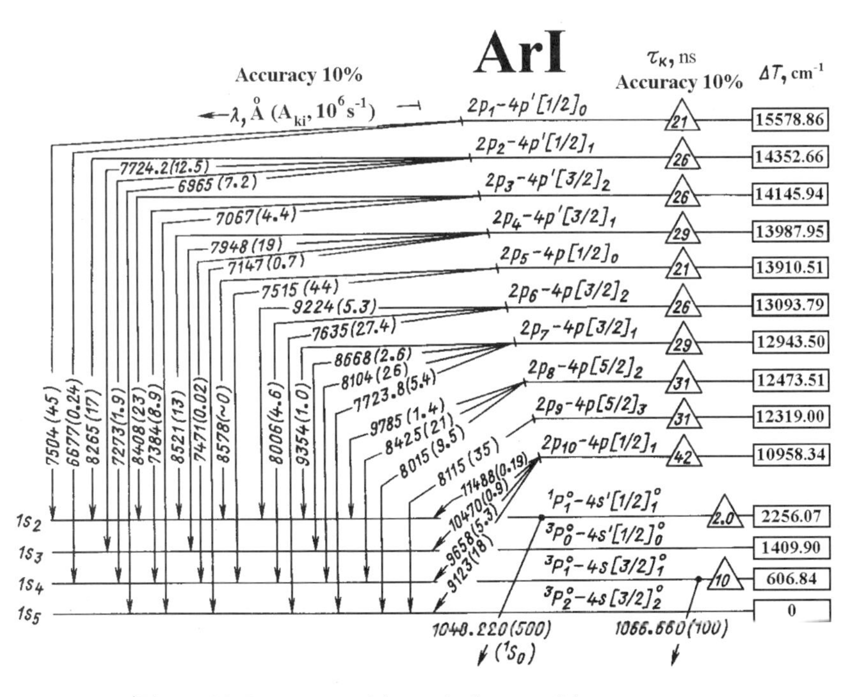

Figure 5.3. Parameters of the excited states of the argon atom.

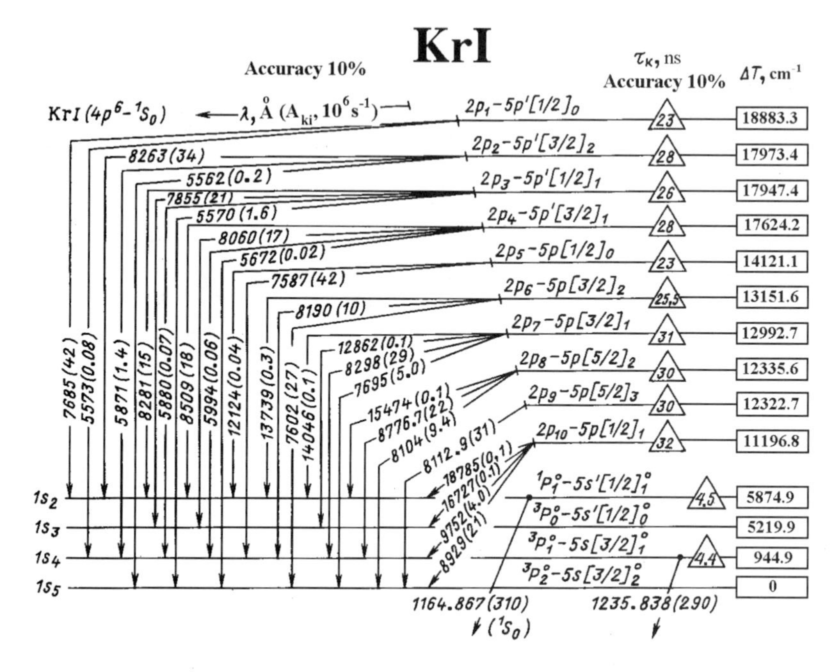

Figure 5.4. Parameters of the excited states of the krypton atom.

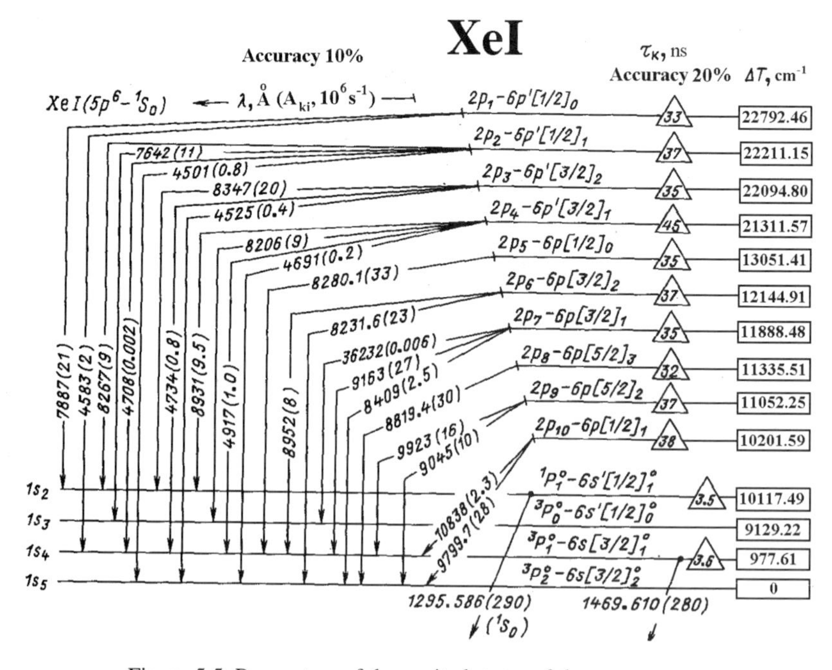

Figure 5.5. Parameters of the excited states of the xenon atom.

to atoms with s- and p-electron shells. In addition, they hamper decoding of the spectra of atoms and ions as well as their interpretation.

It is convenient to represent the spectra of atoms by filling d- and f-electron shells on the basis of the Grotrian diagrams. Then it is necessary to extract all the quantum numbers of the atom. One of them is the parity of the state. This is characterized by the parity of the electron wave function for the reflection of electrons with respect to the plane which is perpendicular to the quantum axis and passes through the atom center. In the case of LS-coupling, the atom parity is determined by these quantum numbers, and introduction of the atom parity does not give new information. In particular, the parity of one electron is $(-1)^l$, i.e., s-, d-, g-electron states are even and p-, f-states are odd. For complex atoms the parity of states remains a quantum number. It is essential that dipole radiative transitions are permitted only between states with different parities, because the operator of the dipole moment of the atom is odd with respect to this operation. This makes it convenient to use the parity of states for classifications of atomic states. Thus, one can divide atomic states into even and odd states, and the radiative transitions inside one group are forbidden. Usually, the levels of complex atoms are denoted within the framework of the scheme of LS-coupling, and numbers the L, S are used only as a form of notation. Then states with the same numbers L and S are denoted for even states by the letters a, b, c, d, etc., as they are excited, and by the letters x, y, z, v, w for odd states.

5.8 Correlation and Collective Effects in Atoms

Analyzing an atom as a physical system, consisting of a heavy Coulomb center and bound electrons, we use a one-electron model for its description. Within the framework of this model, we change the action of other electrons and the Coulomb center on a test electron by a self-consistent field, which does not depend on the coordinates of other electrons and also takes into account the exchange interaction between electrons. This model distributes electrons over electron shells so that because, for a spherical self-consistent field, each shell is characterized by the principal quantum number and angular electron momentum and, according to the Pauli exclusion principle, only one electron can be found in one state of each electron shell, so that this state is also described by the projection of the angular momentum and spin onto a given direction. In this way, we neglect the relativistic interactions, correlation effects, and collective effects in atoms. Let us consider errors due to such approximations.

As for the relativistic effects, they are mostly connected with spin-orbit interaction. If we restrict the spin-orbit interaction in atoms, one can construct the strict atom theory for different relations between the exchange and spin-orbit interactions. Above we considered some examples of this. Briefly, we have a different way of obtaining the total momentum of electrons depending on these interactions. When the exchange interaction dominates, we get the LS-coupling scheme of the atom, where the angular electron momenta are summed into the total atom angular momentum, spins of individual electrons are summed into the atom total spin, and then the total atom angular momentum and spin are summed into the total atom momentum. The LS-coupling scheme is typical for light atoms. On the contrary, for heavy atoms the jj-scheme is preferable for summation of electron momenta. In this case the angular momentum and spin of an individual electron are summed into the total momentum of an individual electron, and then the total momenta of electrons are summed into the total atom momentum. This scheme is more suitable for heavy atoms. Note two peculiarities of these schemes. First, when an electron shell is completed, i.e., its total momentum is zero, both schemes are equivalent. Second, when we have a filling shell, a number of electron states for this shell is identical for both schemes of summation of momenta. These peculiarities allow us to use the electron shell model outside the limits of the validity of the LS-coupling scheme. Thus, these relativistic interactions can be taken into account in the atom model without complication.

Neglecting the correlation effects, we assume the atom wave function to be a product of one-electron wave functions with their transposition property due to the Pauli exclusion principle. Owing to correlation effects the electron shell model is violated because the electron wave function cannot be constructed from one-electron functions, and below we estimate the accuracy of the one-electron approximation for atoms. We take this error on the basis of the data of Table 3.1 and Problem 3.3 for the helium atom. In this case, taking the best atom wave function consisting of one-electron exponential wave functions with the same degrees of exponents, we have the error 1.5 eV for the ionization potential. When the exponents of the one-electron functions are different (Problem 3.3), this error is 0.7 eV. Evidently, one can expect an error to be of this order of magnitude, if we take the best combination of one-electron wave functions, that is, these functions are solutions of the Hartri–Fock equation. Taking the error in the ionization potential in this case to be ~ 0.3 eV and a typical excitation energy of the helium atom to be ~ 30 eV for states which are responsible for the correlation effect, we obtain the admixture of excited states into the wave function of the ground state of the helium atom to be $\sim 10\%$. This means that the admixture of two-electron wave functions in the total wave function of the ground state is $\sim 10\%$.

Note that the correlation effect is stronger when interacting electrons have close space distributions, and the relative role of these effects is enforced as electrons are excited. Nevertheless, from the above estimations one can take roughly an admixture of a two-electron wave function into the wave function of two electrons to be $\sim 10\%$. From this it follows that one can neglect the correlation effects when we construct the atom scheme. But these effects are of importance for accurate evaluation of the atom energy and atom wave function. Especially, the correlation interaction is significant for the evaluation of parameters which are determined by two-electron interactions. An example of this is the width of the autoionization level of the system $A(nl + nl) \rightarrow A(n'l') + e$. As a result of this process, one bound electron transfers into the lower state in the field of an atomic core, and the second electron releases.

As for the collective effects, they are important for a classical system of electrons in a plasma. The spread collective phenomenon consists of the excitation of plasma waves, i.e., oscillations of the electron component as a whole. In an atom, such excitations are forbidden because of the Pauli exclusion principle which does not allow for electrons to transfer into occupied states. This is valid for an atom in the ground state. Under strong excitations, plasma collective excitations can exist in the atom, in principle. But this requires a strong excitation of the atom with the participation of many electrons, when the excitation energy exceeds the atom ionization potential by several orders of magnitude.

Problems

Problem 5.1. *Determine the total electron energy of a heavy atom, the total kinetic energy of electrons, the interaction energy of electrons with the nucleus and energy of interaction between electrons within the framework of the Thomas–Fermi model.*

The total electron energy is given by formula (5.11): $\varepsilon = T + U$ and, according to the virial theorem for a system of particles with the Coulomb interaction, we have $2T + U = 0$. This gives

$$\varepsilon = -T = \frac{U}{2}.$$

The average kinetic energy of electrons at a given point is equal to

$$\left\langle \frac{p^2}{2m} \right\rangle = \int_0^{p_o} \frac{p^2}{2}\, d\mathbf{p} = \frac{3}{5} \cdot \frac{p_o^2}{2}.$$

We use that the electron distribution in the momentum space corresponds to the internal part of the Fermi sphere $p \le p_o$. From this, on the basis of formulas (5.3), (5.4), (5.6), we have

$$T = \int \frac{3}{5} \cdot \frac{p_o^2}{2} N(\mathbf{r})\, d\mathbf{r} = \frac{3}{5} \cdot \int \varphi \frac{\Delta\varphi}{4\pi}\, d\mathbf{r} = \frac{3}{5} \cdot \frac{Z^{7/3}}{b} \int \chi\chi''\, dx,$$

$$\text{where} \quad b = \frac{1}{2} \cdot \left(\frac{3\pi}{4}\right)^{2/3} = 0.885.$$

The integral is equal to

$$\int_0^\infty \chi''\chi\, dx = -\chi(0)\chi'(0) - \int_0^\infty \left[\chi'\right]^2 dx = -\chi'(0) - \int_0^\infty \left[\chi'\right]^2 dx.$$

Thus we obtain, for the total kinetic energy of electrons within the framework of the Thomas–Fermi model,

$$T = \frac{3}{5} \cdot \frac{Z^{7/3}}{b} \cdot \left[1.588 - - \int_0^\infty \left[\chi'\right]^2 dx\right] = 0.769 Z^{7/3}.$$

The potential energy of electrons U is the sum of the energy of interaction of electrons with the nucleus

$$U_1 = -\int \frac{Z}{r} N \, d\mathbf{r},$$

and the energy of interaction between electrons

$$U_2 = \int \frac{1}{|\mathbf{r} - \mathbf{r}'|} N(\mathbf{r}) N(\mathbf{r}') \, d\mathbf{r} \, d\mathbf{r}'.$$

As is seen, the value

$$\int \frac{1}{|\mathbf{r} - \mathbf{r}'|} N(\mathbf{r}') \, d\mathbf{r}'$$

is the electric potential which is created by electrons in a point $\mathbf{r}$, i.e.,

$$\int \frac{1}{|\mathbf{r} - \mathbf{r}'|} N(\mathbf{r}') \, d\mathbf{r}' = \frac{Z}{r} - \varphi.$$

This gives

$$U_2 = -\frac{1}{2} \varphi N(\mathbf{r}) \, d\mathbf{r} + \frac{1}{2} \int \frac{Z}{r} N(\mathbf{r}) \, d\mathbf{r} = -\frac{1}{2} \int \varphi N(\mathbf{r}) \, d\mathbf{r} - \frac{U_1}{2}.$$

Because the expression for the total kinetic energy has the form

$$T = \int \frac{3}{5} \cdot \frac{p_o^2}{2} N(\mathbf{r}) \, d\mathbf{r} = \frac{3}{5} \cdot \int \varphi N(\mathbf{r}) \, d\mathbf{r},$$

we have $U_2 = -5T/6 - U_1/2$. According to the virial theorem we obtain $U = U_1 + U_2 = -2T$. Thus we have the following relations: $U_1 = -7T/3$, $U_2 = T/3$. As is seen, the energy of interaction between electrons is seven times less than the interaction energy of electrons with the nucleus.

Let us calculate the energy of interaction of electrons with the nucleus. We have

$$U_1 = -\int \frac{Z}{r} N(\mathbf{r}) \, d\mathbf{r} = -\int \frac{Z}{r} \frac{\Delta\varphi}{4\pi} d\mathbf{r} = Z \frac{d}{dr} (r\varphi)|_{r=0}$$
$$= \frac{Z^{7/3}}{b} \chi'(0) = -1.794 Z^{7/3}.$$

From this we have, for the other energetic parameters of the Thomas–Fermi atom,

$$T = -\frac{3}{7} U_1 = 0.769 Z^{7/3}, \qquad \varepsilon = -T = -0.769 Z^{7/3},$$
$$U_2 = -\frac{U_1}{7} = 0.256 Z^{7/3}, \qquad U = -2T = -1.538 Z^{7/3}.$$

In addition, we have the following relation for the Thomas–Fermi reduced function

$$\int_0^\infty \left[\chi'\right]^2 dx = -\frac{2}{7} \chi'(0) = 0.454.$$

Problem 5.2. *The Tietz approximation for the Thomas–Fermi function has the form $\chi(x) = (1 + ax)^{-2}$. Determine the parameter a from the condition that this function leads to the correct total number of atomic electrons.*

We use the integral relation $\int N(\mathbf{r})\,d\mathbf{r} = Z$. This gives the following relation for the Thomas–Fermi function:

$$\int_0^\infty \sqrt{x}\chi^{3/2}(x)\,dx = 1.$$

Then the Tietz approximation gives $a = 0.536$.

Note that the Tietz approximation is working in the region where the electrons are mainly located. In the region near the nucleus this approximation leads to remarkable errors. For instance, in this approximation $\chi'(0) = -1.072$, while its accurate value is -1.588. On the contrary, the value of the integral $\int_0^\infty [\chi']^2\,dx = 0.429$, which is close to its accurate value of 0.454 (see Problem 5.1). For the total kinetic energy of electrons which is proportional to the value $-\chi'(0) - \int_0^\infty \chi']^2\,dx$, the Tietz approximation takes the result which is in 1.76 less than the accurate value.

Problem 5.3. *Determine the shift of the energy of a K-electron as a result of the shielding of a K-electron by atomic electrons within the framework of the Thomas–Fermi model.*

The first order of the perturbation theory gives, for the shift of the energy of K-electron,

$$\Delta\varepsilon = \int |\psi(\mathbf{r})|^2 \varphi\,d\mathbf{r} = V(0),$$

where $\mathbf{r}$ is the coordinate of the K-electron, $\psi(\mathbf{r})$ is its wave function, the φ is the electric potential which is created by other electrons. We consider φ as a perturbation and assume that it varies weakly on distances of the order of the size of the orbit of the K-electron. Indeed, the size of the K-orbit is of the order of $1/Z$, and a typical size of the Thomas–Fermi model is of the order of $Z^{-1/3}$, i.e., the small parameter of the using perturbation theory is $Z^{-2/3} \ll 1$.

On the basis of the results of Problem 5.1, we have

$$V(0) = -\int \varphi N\,d\mathbf{r} = 1.794Z^{4/3},$$

so that the energy of the K-electron, by accounting for the related shift, has the form

$$\varepsilon = -\frac{Z^2}{2} + 1.794Z^{4/3}.$$

As is seen, the correction due to the shielding of electrons is relatively small and corresponds to expansion over a small parameter $Z^{-2/3}$. Note that this effect is stronger than the influence of the second K-electron. Indeed, accounting for the action of the second K-electron within the framework of the variation method yields, according to formula (3.8) for the ionization potential of the K-electron,

$$J = \frac{Z^2}{2} - \frac{5Z}{8} + \frac{25}{256},$$

i.e., shielding due to the second K-electron leads to the shift of its energy by the value $5Z/8$. This gives us that the correction to the electron energy due to external electrons is $0.35/Z^{1/3}$, i.e., the action of external electrons is more remarkable than that from the second K-electron.

Summing up the above results, we obtain the following expression for the ionization potential of the K-electron:

$$J = \frac{1}{2}(Z - \sigma)^2,$$

where the shielding charge created under the action of other atomic electrons is equal to

$$\sigma = 1.794Z^{1/3} + 5/8.$$

Problem 5.4. *Determine minimum values of the nucleus charge Z at which electrons with an orbital moment l occur in the electron shell of the ground atom state. Use the Thomas–Fermi atom model.*

The equation for the radial wave function of a valent electron $R(r)$ by analogy with equation (2.9) has the form

$$\frac{1}{r} \cdot \frac{d^2(rR)}{dr^2} + \left[2\varepsilon - 2\varphi(r) - \frac{l(l+1)}{r^2}\right] R = 0,$$

where r is an electron distance from the center, ε is the electron energy, and $\varphi(r)$ is the potential of a self-consistent field. An electron state is bound if, in some distance region,

$$2\varepsilon - 2\varphi(r) - \frac{l(l+1)}{r^2} > 0.$$

We have for the bound state $\varepsilon < 0$, i.e., $-2r^2\varphi(r) > l(l+1)$. Taking the Thomas–Fermi self-consistent field and using the variables of formulas (5.6) we have, from this relation,

$$1.77Z^{2/3}x\chi(x) > l(l+1).$$

Taking the maximum of the function $x\chi(x)$, which is equal to 0.486 at $x = 2.1$ (see Table 5.6), we obtain the criterion

$$0.86Z^{2/3} > l(l+1).$$

According to this formula, d-electrons occur in the shell of the ground state of elements starting from $Z = 18$, f-electrons can occur at $Z = 52$, and g-electrons occur at $Z = 112$. In reality, the first d-electron is observed in the electron shell of the ground atom state at $Z = 21$ (Sc), and their first f-electron occurs at $Z = 58$ (Ce). As is seen, the above simple analysis gives correct estimate.

Problem 5.5. *Take the Hamiltonian for the first excited states of inert gas atoms in the form* $\hat{H} = -a\hat{\mathbf{l}}\hat{\mathbf{s}}_1 - b\hat{\mathbf{s}}_1\hat{\mathbf{s}}_2$, *where* $\hat{\mathbf{l}}$ *is the atomic core orbital momentum operator,* $\hat{\mathbf{s}}_1$ *is the spin operator of the atomic core, and* $\hat{\mathbf{s}}_2$ *is the spin operator of an excited electron. Determine the relative positions of the lowest excited states if one electron transfers from a p-shell to the lowest excited s-shell.*

The first term of this Hamiltonian accounts for the spin-orbit interaction for the atomic core, and the sign of this term allows us to further consider a p-hole as a p-electron. The second term corresponds to the exchange interaction between internal electrons and the excited electron. First let us consider the limiting cases. Neglecting the exchange interaction ($b = 0$), we obtain two levels of fine structure with the total momenta of the atomic core $j = 1/2$, 3/2, and the distance between them, $\Delta_f = \frac{3}{2}a$, corresponds to the spin-orbit splitting of the atomic core. The lowest state relates to $j = 3/2$ in accordance with the chosen sign of this term. Neglecting the spin-orbit interaction ($a = 0$), we obtain two levels with the total electron spin $S = 0$, 1.The lowest state corresponds to $S = 1$, and the distance between levels is $\Delta_{\text{ex}} = b$.

Taking into account both terms of the Hamiltonian, we have that the total momentum $\hat{\mathbf{J}} = \widehat{\mathbf{l}} + \hat{\mathbf{s}}_1 + \hat{\mathbf{s}}_2$ ($\widehat{\mathbf{l}}$ is the operator of the core angular momentum) is the quantum number because this operator commutes with the Hamiltonian. The total number of states is the product of the number of projections of the atomic core orbital momentum (3), atomic core spin (2), and excited electron spin (2), i.e., the total number of states is equal to 12. These states relate to the total momentum $J = 0$, 1, 2, and there are two different levels for $J = 1$. As is seen, the total number of projections of the total momentum, i.e., the total number of states, is equal to 12.

In order to find the positions of the related four energy levels, it is necessary to construct the wave functions of these states from wave functions of the orbital momenta and spins. Denote by ψ_m the wave function of the atomic core with a projection m of the orbital momentum onto a given direction. Correspondingly, the wave functions χ_+, χ_- correspond to the spin projections 1/2 and $-1/2$ of the atomic core onto a given direction, and the wave functions η_+, η_- describe the spin states of the excited electron with the spin projections 1/2 and $-1/2$. The atom wave function is a product of the above wave functions, and we have 12 different combinations of such products. Our task is to find eigenfunctions of the Hamiltonian that require us to fulfill some operations with operators of the angular momentum and spins. We have the following relations for the orbital momentum operator:

$$\hat{l}_z\psi_m = m\psi_m, \qquad \hat{l}_+\psi_{-1} = \sqrt{2}\psi_0; \qquad \hat{l}_+\psi_0 = \sqrt{2}\psi_1; \qquad \hat{l}_+\psi_1 = 0;$$
$$\hat{l}_-\psi_{-1} = 0; \qquad \hat{l}_-\psi_0 = \sqrt{2}\psi_{-1}; \qquad \hat{l}_-\psi_1 = \sqrt{2}\psi_0,$$

where $\hat{l}_+ = \hat{l}_x + i\hat{l}_y$, $\hat{l}_- = \hat{l}_x - i\hat{l}_y$, and we use $l = 1$. The identical relations we have for spin operators

$$\hat{s}_{1z}\chi_+ = \frac{1}{2}\chi_+; \qquad \hat{s}_{1z}\chi_- = -\frac{1}{2}\chi_-; \qquad \hat{s}_{1+}\chi_+ = 0;$$

$$\hat{s}_{1-}\chi_+ = \chi_-; \qquad \hat{s}_{1+}\chi_- = \chi_+; \qquad \hat{s}_{1-}\chi_- = 0,$$

and the same relations we have for the operator $\hat{\mathbf{s}}_2$. Here $\hat{s}_+ = \hat{s}_x + i\hat{s}_y$; $\hat{s}_- = \hat{s}_x - i\hat{s}_y$.

Let us denote the wave function of an atom state by the total momentum J and its projection M by Ψ_{JM}. Take the state with $J = 2$, $M = 2$ whose wave function is $\Psi_{22} = \psi_1\chi_+\eta_+$ and calculate the energy of this state. We have

$$\hat{H}\Psi_{22} = -a\hat{l}_z\hat{s}_{1z}\Psi_{22} - b\hat{s}_{1z}\hat{s}_{2z}\Psi_{22} = \varepsilon_5\Psi_{22},$$

where the energy of this state ε_5 (we denote this in Pashen notation as $1s_5$) is equal to

$$\varepsilon_5 = -\frac{a}{2} - \frac{b}{4}.$$

This energy corresponds to states with $J = 2$ and any momentum projection onto a given direction.

Next, we construct the wave function of the state with $J = 0$, as a result of summation of the orbital moment $\hat{\mathbf{l}}$ and the total spin $\hat{\mathbf{s}}_1 + \hat{\mathbf{s}}_2$, which has the form

$$\Psi_{00} = \frac{1}{\sqrt{3}}\psi_1\chi_-\eta_- + \frac{1}{\sqrt{3}}\psi_{-1}\chi_+\eta_+ - \frac{1}{\sqrt{6}}\psi_0\chi_+\eta_- - \frac{1}{\sqrt{6}}\psi_0\chi_-\eta_+.$$

We have

$$\hat{\mathbf{l}}\hat{\mathbf{s}}_1\Psi_{00} = -\Psi_{00}; \qquad \hat{\mathbf{s}}_1\hat{\mathbf{s}}_2\Psi_{00} = \frac{1}{4}\Psi_{00}, \quad \text{and} \quad \hat{H}\Psi_{00} = \varepsilon_3\Psi_{00}.$$

The energy of this state as the eigenvalue of the Hamiltonian is equal to

$$\varepsilon_3 = a - \frac{b}{4}.$$

In order to determine the energies of states with $J = 1$, let us consider states with the moment projection $M - 1$. The wave functions of these states can be constructed from $\varphi_1 = \psi_0\chi_+\eta_+$, $\varphi_2 = \psi_1\chi_-\eta_+$, $\varphi_3 = \psi_1\chi_+\eta_-$. Extract from these functions the wave function for the state $J = 2$, $M = 1$, which has the form

$$\Phi_1 = \Psi_{21} = \frac{1}{\sqrt{2}}\varphi_1 + \frac{1}{2}\varphi_2 + \frac{1}{2}\varphi_3.$$

We obtain

$$\hat{\mathbf{l}}\hat{\mathbf{s}}_1\Psi_{21} = \frac{1}{2}\Psi_{21}; \qquad \hat{\mathbf{s}}_1\hat{\mathbf{s}}_2\Psi_{21} = \frac{1}{4}\Psi_{21}, \quad \text{and} \quad \hat{H}\Psi_{21} = \varepsilon_5\Psi_{21}.$$

The energy ε_5 is the eigenvalue for this wave function and is given by the above formula. Taking two other wave functions to be normalized to unity, orthogonal to this function, and orthogonal each to other, we get

$$\Phi_2 = -\frac{1}{\sqrt{2}}\varphi_1 + \frac{1}{2}\varphi_2 + \frac{1}{2}\varphi_3; \qquad \Phi_3 = \frac{1}{\sqrt{2}}\varphi_2 - \frac{1}{\sqrt{2}}\varphi_3.$$

Next, we have

$$\hat{\mathbf{l}}\hat{\mathbf{s}}_1\Phi_2 = -\frac{1}{2}\Phi_2 - \frac{1}{\sqrt{2}}\Phi_3; \qquad \hat{\mathbf{l}}\hat{\mathbf{s}}_1\Phi_3 = -\frac{1}{\sqrt{2}}\Phi_3,$$

$$\text{and} \quad \hat{\mathbf{s}}_1\hat{\mathbf{s}}_2\Phi_2 = \frac{1}{4}\Phi_2, \qquad \hat{\mathbf{s}}_1\hat{\mathbf{s}}_2\Phi_3 = -\frac{3}{4}\Phi_3.$$

Calculating the matrix elements of the Hamiltonian on the basis of these relations, we obtain the following secular equation for the eigenvalues of the Hamiltonian:

$$\begin{vmatrix} \frac{a}{2} - \frac{b}{4} - \varepsilon & \frac{a}{\sqrt{2}} \\ \frac{a}{\sqrt{2}} & \frac{3}{4}b - \varepsilon \end{vmatrix} = 0.$$

The solution of this equation gives the state energies

$$\varepsilon_{2,4} = \frac{1}{4}(a+b) \pm \frac{1}{4}\sqrt{9a^2 - 4ab + 4b^2}.$$

In the limiting case $b = 0$, when one can neglect the exchange interaction, we have $\varepsilon_4 = -a/2$, $\varepsilon_2 = a$ (i.e., $\varepsilon_4 = \varepsilon_5$ and $\varepsilon_2 = \varepsilon_3$). In the other limiting case $a = 0$, when one can neglect the spin-orbit interaction, we have $\varepsilon_4 = -b/4$, $\varepsilon_2 = 3b/4$, i.e., $\varepsilon_4 = \varepsilon_5 = \varepsilon_3$ and the exchange splitting is equal to b.

Let us take the position of the lowest excited level of an inert gas atom as zero ($\varepsilon_5 = 0$). Then from the obtained formulas it follows for positions of the other energy levels that

$$\varepsilon_{2,4} = \frac{3}{4}a + \frac{1}{2}b \pm \frac{1}{4}\sqrt{9a^2 - 4ab + 4b^2}; \qquad \varepsilon_3 = \frac{3}{2}a.$$

Table 5.9 contains the results which follow from comparison of this formula with the positions of energy levels ε_2, ε_3, ε_4 for real inert gas atoms. In this Table Δ_f is the fine splitting of levels for the corresponding free ion. As is seen, this value is close to ε_3 (see also Table 5.8). The exchange interaction splitting b according to the above formulas is equal to $b = \varepsilon_4 + \varepsilon_2 - \varepsilon_3$. As follows from the data of Table 5.9, the exchange splitting depends slowly on the atom type. Table 5.9 also gives the value $\sqrt{9a^2/4 + b^2 - ab}$ which, according to the obtained formulas, is equal to the difference $\varepsilon_2 - \varepsilon_4$. Comparison of these values and also of Δ_f with $\varepsilon_3 = 3a/2$ for real atoms shows that the above Hamiltonian includes the main part of the interaction in the lower excited atoms of inert gases.

Note that along with the positions of the $1s_2$ and $1s_4$ energy levels, the above operations allow us to find the expressions for the wave functions of these states.

Table 5.9. Parameters of the first excited states of inert gas atoms (all the parameters are expressed in cm^{-1}).

Atom	Δ_f	ε_3	b	$\varepsilon_4 - \varepsilon_2$	$\sqrt{9a^2/4 + b^2 - ab}$	x	c_3^2
Ne	780	777	1487	1430	1430	−1.67	0.071
Ar	1432	1410	1453	1649	1653	−0.72	0.207
Kr	5370	5220	1600	4930	4923	0.038	0.481
Xe	10537	9129	1967	9140	8674	0.16	0.423

Indeed, representing the wave functions, Ψ_2 and Ψ_4, of these states in the form

$$\Psi_2 = c_2\Phi_2 + c_3\Phi_3, \quad \text{and} \quad \Psi_4 = -c_3\Phi_2 + c_2\Phi_3,$$

and from the Schrödinger equation $\widehat{H}\Psi_{2,4} = \varepsilon\Psi_{2,4}$ we have the set of equations for the coefficients of the wave function expansion

$$\left(\langle\Phi_2\widehat{H}\Phi_2\rangle - \varepsilon\right)c_2 + \langle\Phi_2\widehat{H}\Phi_3\rangle c_3 = 0,$$
$$\langle\Phi_3\widehat{H}\Phi_2\rangle c_2 + \left(\langle\Phi_3\widehat{H}\Phi_3\rangle - \varepsilon\right)c_3 = 0.$$

Here we account for the wave functions Φ_i and the coefficients c_i are real values. This is the secular set of equations for the determination of energy levels. Simultaneously, this allows us to find the expansion coefficients. Indeed, introducing

$$x = \frac{\langle\Phi_2\widehat{H}\Phi_2\rangle - \langle\Phi_3\widehat{H}\Phi_3\rangle}{\langle\Phi_3\widehat{H}\Phi_2\rangle} = \frac{1}{2\sqrt{2}}\left(1 - \frac{2b}{a}\right) = \frac{1}{2\sqrt{2}}\left(1 - \frac{3b}{\Delta_f}\right),$$

and accounting for the normalization condition $c_2^2 + c_3^2 = 1$, we get

$$c_{2,4}^2 = \frac{\sqrt{1+x^2} \pm x}{2\sqrt{1+x^2}}.$$

Table 5.9 contains values of these parameters for excited atoms under consideration.

Problem 5.6. *Determine the relative lifetimes of the four lowest states of inert gas atoms if the radiative transitions proceed between the valence electron shells np^6 and $np^5(n+1)s$. Use the results of the previous problem and neglect the energy difference between states of the excited states in the analysis of radiative transitions.*

Above is shown (formulas (5.18), (5.19)) that only two states from four of the lowest excited atom shells are connected with the ground state by the dipole radiative transitions. Now we obtain this result in a direct way from the expression for the intensity of the dipole radiative transitions. Use formula (1.25), (1.26) according to which the radiative lifetime of an excited state is proportional to the square of the matrix element of the dipole moment operator

$$\frac{1}{\tau_f} \sim |\langle 0|\mathbf{D}|f\rangle|^2,$$

where the indices 0, f refer to the ground and related excited states. In accordance with the notation of the previous problem and formula (5.17), we present the wave function of the ground atom state in the form

$$\Psi_0 = \varphi_0 \cdot \frac{1}{\sqrt{2}}(\chi_+\eta_- - \chi_-\eta_+),$$

where χ, η are the spin wave functions of the core and test valent electron, and φ_0 is the total spatial wave function of the core and test electron for the ground atom state. The wave function of the lowest excited state with the projection of the total momentum 2 onto a given direction has the form

$$\Psi_5 = \psi_1 \chi_+ \eta_+,$$

where ψ_1 is the spatial wave function of the core and test electron that has the projection one of the angular momentum onto a given direction. As is seen, the matrix element $\langle \Psi_0 | \mathbf{D} | \Psi_5 \rangle = 0$, because of the orthogonality of the spin wave functions (compare this with formula (5.18)).

Taking the expression for the wave function of the state $1s_3$ from the previous problem as

$$\Psi_2 = \frac{1}{\sqrt{3}} \psi_1 \chi_- \eta_- + \frac{1}{\sqrt{3}} \psi_{-1} \chi_+ \eta_+ - \frac{1}{\sqrt{6}} \psi_0 \chi_+ \eta_- - \frac{1}{\sqrt{6}} \psi_0 \chi_- \eta_+,$$

one can see that the matrix element $\langle \Psi_0 | \mathbf{D} | \Psi_3 \rangle = 0$, because of the orthogonality of the spin wave functions of the core and test electron (compare thus with formula (5.19)). Thus the states $1s_5$, $1s_3$ are metastable states and are characterized by an infinite radiative lifetime with respect to dipole radiation.

For the analysis of the radiative lifetimes of the states $1s_4$ and $1s_2$, let us use expressions for the wave functions of the states $1s_5$, $1s_4$, $1s_2$ with the momentum projection $M = 1$. These have the form

$$\Phi_1 = \Psi_{21} = \frac{1}{\sqrt{2}} \varphi_1 + \frac{1}{2} \varphi_2 + \frac{1}{2} \varphi_3;$$

$$\Phi_2 = -\frac{1}{\sqrt{2}} \varphi_1 + \frac{1}{2} \varphi_2 + \frac{1}{2} \varphi_3; \qquad \Phi_3 = \frac{1}{\sqrt{2}} \varphi_2 - \frac{1}{\sqrt{2}} \varphi_3.$$

Write the dipole moment operator in the form

$$\mathbf{D} = \sum_m \left(\mathbf{i} \sin\theta_m \cos\varphi_m + \mathbf{j} \sin\theta_m \sin\varphi_m + \mathbf{k} \cos\theta_m \right) r_m,$$

where **i**, **j**, **k** are the unit vectors directed along the x-, y-, z-axes respectively, and the subscript m corresponds to mth electron whose spherical coordinates are r_m, θ_m, φ_m. Using the expressions of previous Problem for the basis wave functions

$$\varphi_1 = \psi_0 \chi_+ \eta_+, \qquad \varphi_2 = \psi_1 \chi_- \eta_+, \qquad \varphi_3 = \psi_1 \chi_+ \eta_-,$$

we obtain, for the matrix elements,

$$\langle \Psi_0 | \mathbf{D} | \varphi_1 \rangle = 0; \qquad \langle \Psi_0 | \mathbf{D} | \varphi_2 \rangle = C\left(-\mathbf{i} + i\mathbf{j} \right); \qquad \langle \Psi_0 | \mathbf{D} | \varphi_3 \rangle = C\left(\mathbf{i} - i\mathbf{j} \right).$$

Thus, the rate of the radiative transition is proportional to the amplitude of the wave function Φ_3 into the wave function of the excited state. From this we get, for the ratio of the radiative lifetimes of the resonantly excited states,

$$\frac{\tau(1s_2)}{\tau(1s_4)} = \frac{c_2^2}{c_3^2} = \frac{\sqrt{1+x^2} - |x|}{\sqrt{1+x^2} + |x|}.$$

Table 5.10 contains the radiative lifetimes, $\tau(1s_2)$ and $\tau(1s_4)$, of the resonantly excited states of inert gas atoms, the ratio of these quantities, and that according to

Table 5.10. Radiative lifetimes for the lower resonantly excited states of inert gas atoms.

Atom	Ne	Ar	Kr	Xe
$\tau(1s_2)$, ns	1.6	2.0	3.2	3.5
$\tau(1s_4)$, ns	25	10	3.5	3.6
$\tau(1s_2)/\tau(1s_4)$	16	5	1.1	1.0
c_2^2/c_3^2	13	3.8	1.1	1.4

formulas of the previous problem when the excitation state is described by the exchange and spin-orbital interaction potentials. One can see that related interactions are responsible for the properties of the lower excited states of inert gas atoms.

CHAPTER 6

Excited Atoms

6.1 Metastable and Resonantly Excited Atoms

The lowest excited states of atoms can be divided into two groups: metastable and resonantly excited states. Resonantly excited atoms can emit photons and transfer to the ground state as a result of the dipole interaction with radiation fields. For metastable states this radiation transition is prohibited, and the radiative lifetime for metastable states is several orders of magnitude more than that for resonantly excited states. Hence, metastable atoms can be accumulated in gases, while resonantly excited atoms decay fast. Quenching of resonantly excited atoms generates resonant photons which have a relatively small free path length in a gas consisting of atoms of this sort, because such photons are absorbed effectively by other atoms of this sort in the ground state. Thus, metastable and resonantly excited atoms differ strongly in their behavior in gaseous and plasma systems. With an increase in the excitation energy this difference between the excited states disappears because an excited atom can transfer in different atomic states with a lower energy as a result the dipole radiation. Hence, the separation of excited atomic states in metastable and resonantly excited states corresponds to only the first excited states. Note that the strongly excited states of atoms form a special group of highly excited or Rydberg states of atoms whose radiative lifetime is intermediate between metastable and resonantly excited states. Next, the prohibition of the radiative transition from an excited state to the ground state can be abolished in the following orders of the perturbation theory, constructed on the basis of a weak interaction between an atom and radiation field. Therefore the lowest excited states of some atoms can have an intermediate nature between metastable and resonantly excited states. For these states the radiation dipole transition in lower states is prohibited, but the radiative lifetime is not so large. It especially relates to heavy atoms for which relativistic interactions are remarkable and the competition

Table 6.1. Parameters of the metastable states of atoms; ε_{ex} is the excitation energy of the state, λ is the wavelength of radiative transition to the ground state, and τ is the radiative lifetime of the metastable state.

Atom, state	ε_{ex}, eV	λ, nm	τ, s	Atom, state	ε_{ex}, eV	λ, nm	τ, s
$H(2^2S_{1/2})$	10.20	121.6	0.12	$P(3^2D_{5/2})$	1.41	878.6	$5 \cdot 10^3$
$He(2^3S_1)$	19.82	62.56	7900	$P(3^2P_{1/2,3/2})$	2.32	1360	4
$He(2^1S_0)$	20.62	60.14	0.02	$S(3^1D_2)$	1.15	1106	28
$C(2^1D_2)$	1.26	983.7	3200	$S(3^1S_0)$	2.75	772.4	0.5
$C(2^1S_0)$	2.68	462.4	2	$Cl(3^2P_{1/2})$	0.109	11100	80
$N(2^2D_{5/2})$	2.38	520.03	$1.4 \cdot 10^5$	$Se(4^1D_2)$	1.19	1160	1.4
$N(2^2D_{3/2})$	2.38	519.8	$6.1 \cdot 10^4$	$Se(4^1S_0)$	2.78	464.0	0.1
$N(2^2P_{1/2,3/2})$	3.58	1040	12	$Br(4^1P_{1/2})$	0.46	2713	0.9
$O(2^1D_2)$	1.97	633.1	100	$Te(5^1D_2)$	1.31	1180	0.28
$O(2^1S_0)$	4.19	557.7	0.76	$Te(5^1S_0)$	2.88	474.0	0.025
$F(2^2P_{1/2})$	0.050	24700	660	$I(5^2P_{1/2})$	0.94	1315	0.14
$P(3^2D_{3/2})$	1.41	880.0	$3 \cdot 10^3$	$Hg(6^3P_0)$	4.67	265.6	1.4

of different electron shells partially abolishes the prohibition for the radiative transition in the ground state. Below we leave such atom states aside.

6.2 Generation and Detection of Metastable Atoms

Table 6.1 contains the parameters of some of the metastable states of atoms. Metastable atoms are formed in a gas or weakly ionized plasma under the action of electric fields, UV-radiation, electron beams, and other ways of the excitation of gaseous systems. As a result of the relaxation of such a system, some excited atoms transfer to metastable states, and metastable atoms are accumulated in the gas due to a high lifetime. For example, under optimal conditions of operation of an He–Ne laser, the number density of metastable helium atoms $He(2^3S_1)$ in a gas-discharge plasma is of the order of 10^{12} cm^{-3}, when the number density of helium atoms is of the order of 10^{17} cm^{-3}, and the electron number density is of the order of 10^{11} cm^{-3}. Thus, the number density of metastable atoms in this gas-discharge plasma is higher than the electron number density. A higher value of metastable helium atoms is observed in a cryogenic gaseous discharge ($\sim 10^{13}$–$10^{14} cm^{-3}$) where relaxation processes, which lead to the decay of metastable states, are weaker due to a low temperature.

An effective method of generation of some metastable atoms is based on the photodissociation of certain molecules. In spite of small efficiency, this method is characterized by high selectivity. As an example, consider the photodissociation process of ozone

$$O_3 + \hbar\omega \to O(^1D) + O_2(^1\Delta_g). \tag{6.1}$$

According to measurements, in the region of wavelengths $\lambda = 254$–$310\,\mathrm{nm}$, the quantum yield of this channel is equal to unity. The reason is such that in this range of wavelengths the exciting term of the ozone molecule is the repulsive one which corresponds to the channel (6.1) of dissociation of the ozone molecule. Process (6.1) determines the presence of the metastable atoms $O(^1D)$ in the Earth's stratosphere at altitudes of 20–30 km where the number density of these metastable oxygen atoms is of the order of $10^2\,\mathrm{cm}^{-3}$. For the generation of the metastable oxygen atoms $O(^1S)$ on the basis of this process, UV-radiation with wavelengths of 105–130 nm is used. Along with the ozone molecule, the photodissociation of the molecules CO_2 and NO_2 is used for the generation of the metastable oxygen atoms $O(^1S)$, where the quantum yield of the metastable oxygen atoms $O(^1S)$ can reach unity. The photodissociation method is used for the generation of different metastable atoms. Metastable bromine atoms $Br(4^2P_{1/2})$ are formed as a result of the photodissociation of the molecules CF_3Br, for formation of the metastable iodine atoms $I(5^2P_{1/2})$ the photodissociation of molecules CF_3I, C_2F_5I, C_3F_7I is used. These processes of generation of metastable iodine atoms are the basis of the iodine laser which operates due to the transition

$$I(5^2P_{1/2}) \to I(5^2P_{3/2}) + \hbar\omega. \tag{6.2}$$

In this laser the energy of a pulse UV-lamp is transformed into the energy of laser radiation, and this process is characterized by high yield parameters.

Various methods are used for the detection of metastable atoms. If these atoms are located in an atomic beam, a convenient method of measurement of their flux

Table 6.2. Radiative transitions for detection of the metastable atoms of inert gases (the notations of excited states for the scheme of LS-coupling are used for the helium atom, for other inert gas atoms the Pashen notations are used; λ is the wavelength of the transition, and A_{fi} is the Einstein coefficient for this transition).

Atom,state	Transition	λ, nm	A_{fi}, $10^7\,\mathrm{s}^{-1}$
$He(2^3S)$	$2^3S_1 \to 2^3P_1$	1083	1.02
$He(2^3S)$	$2^3S_1 \to 3^3P_1$	388.9	0.95
$He(2^3S)$	$2^3S_1 \to 4^3P_1$	318.8	0.56
$He(2^1S)$	$2^1S_0 \to 2^1P_1$	2058.2	0.20
$He(2^1S)$	$2^1S_0 \to 3^1P_1$	501.6	1.34
$He(2^1S)$	$2^1S_0 \to 4^1P_1$	396.5	0.69
$Ne(1s_5)$	$1s_5 \to 2p_9$	640.2	5.1
$Ar(1s_5)$	$1s_5 \to 2p_9$	811.5	3.5
$Kr(1s_5)$	$1s_5 \to 2p_9$	811.3	3.1
$Xe(1s_5)$	$1s_5 \to 2p_9$	904.5	1.0

is the determination of a current of electron emission as a result of the interaction of this beam with a metallic surface. If the excitation energy of metastable atoms exceeds the work function of metals, the quenching of metastable atoms due to their interaction with a metallic surface leads to the release of electrons. If metastable atoms are found in a gas, they can be quenched by admixtures located into the gas. The atoms or molecules of admixtures are excited in collisions with metastable atoms, and a part of the excitation energy of metastable atoms is transferred to admixture particles during quenching. The intensity of emission of these excited particles characterizes the number density of the metastable atoms. This method is convenient for determination of the relative number density of metastable atoms in a gas when this value varies in time.

For determination of the absolute values of the number densities of metastable atoms in a gas, the method of absorption is used for a resonant radiation with respect to metastable atoms. This radiation transfers metastable atoms into more excited states. Hence, measurement of the absorption coefficient for this resonance radiation allows one to determine the number density of metastable atoms. With the use of contemporary experimental techniques, this method is convenient if the number density of metastable atoms exceeds 10^7–$10^8\,\mathrm{cm}^{-3}$. Table 6.2 represents the parameters of radiative transitions which are used for determination of the number density of the metastable atoms of inert gases.

6.3 Metastable Atoms in Gas Discharge and Gas Lasers

Let us consider the basic processes involving metastable atoms which take place in gaseous and plasma systems. The Penning process is of principle for the generation of electrons in some gas discharge plasmas. An example of this process is

$$\mathrm{Ne}(1s_5) + \mathrm{Ar} \rightarrow \mathrm{Ne} + \mathrm{Ar}^+ + e. \tag{6.3}$$

The excitation energy of the metastable state exceeds the ionization potential of an admixture atom, so that the excitation energy is consumed in this process for the ionization of an admixture atom. Hence, if a discharge is burnt in a gas with a high ionization potential of atoms, an admixture with a small ionization potential of atoms can change the parameters of the discharge because now, instead of the atom ionization, it is enough to transfer it into the metastable state. In this case the change of parameters of the gas discharge is observed at the concentrations of admixture atoms of the order of 0.01–0.1%.

The presence of metastable atoms can influence the spectroscopy of an excited atom if quenching of the metastable atoms proceeds through formation of the excited atoms of an admixture, and further these atoms emit radiation. This process is the basis of the operation of the He–Ne laser which uses the process of excitation transfer from metastable helium atoms to neon atoms

$$\begin{aligned}
&\mathrm{He}\left(2^3S\right) + \mathrm{Ne}\left(2p^6\right) \rightarrow \mathrm{He}\left(1^1S_0\right) + \mathrm{Ne}\left(2p^5 4s\right); \\
&\mathrm{He}\left(2^1S\right) + \mathrm{Ne}\left(2p^6\right) \rightarrow \mathrm{He}\left(1^1S_0\right) + \mathrm{Ne}\left(2p^5 5s\right).
\end{aligned} \tag{6.4}$$

Here the electron terms of the helium atom and electron shells of the neon atom are shown for the relevant processes. Because the final state of the electron shell of the neon atom can correspond to several electron terms, the He–Ne laser generates radiation at various wavelengths. The highest intensities correspond to transitions with wavelengths of 3.39μ and 0.633μ from the electron shell $2p^54s$ and 1.15μ from the electron shell $2p^55s$.

There is a group of lasers which are similar to the He–Ne laser, and the first stage of processses in these lasers is the formations of heiium metastable atoms. Let us consider such lasers whose first stage is the formation of helium metastable atoms. One can extract from among these lasers which are used helium with an admixture of metallic vapor (Cd, Hg, Au, Zn, etc.). These lasers are operated on transitions between the excited states of ions. For example, a Cd laser can emit radiation due to several transitions, and the most intense transitions are the following:

$$\mathrm{Cd}^+(4d^{10}4f, {}^2F_{5/2} \rightarrow 4d^{10}5d, {}^2D_{3/2}), \qquad \lambda = 533.7\,\mathrm{nm}, \qquad (6.5a)$$

$$\mathrm{Cd}^+(4d^95s^2, {}^2D_{5/2} \rightarrow 4d^95p^2, {}^2P_{3/2}), \qquad \lambda = 441.6\,\mathrm{nm}. \qquad (6.5b)$$

The initial excited state of the cadmium ion is formed as a result of the excitation transfer process

$$\mathrm{He}(2^3S) + \mathrm{Cd} \rightarrow \mathrm{He} + (\mathrm{Cd}^+)^* + e. \qquad (6.6)$$

Metastable atoms are characterized by a large radiation lifetime. Hence in gaseous and plasma systems their lifetime is determined by the collision and transport processes. These atoms are accumulated in gases in relatively high amounts, and their quenching allows one to transfer the excitation energy to other degrees of freedom. This fact is used in the above gaseous lasers. In this case, metastable states are not the upper states of a laser transition because of a high radiative lifetime. One can decrease this value as a result of the formation of a bound state with other atomic particles. Then the prohibition of radiation is partially abolished due to the interaction between bound atomic particles. For example, the radiative lifetime of the oxygen metastable atom $\mathrm{O}(^1S)$ is equal to 0.8 s, while the radiative lifetime of the molecule $\mathrm{ArO}(^1S)$ is equal to $3.8 \cdot 10^{-3}$ s, and the radiative life time of the molecule $\mathrm{XeO}(^1S)$ is equal to $3 \cdot 10^{-4}$ s. This fact is used in lasers which use the molecules ArO, KrO, XeO with a small dissociation energy for the emission of laser radiation in the vicinity of the transition $\mathrm{O}(^1S_0) \rightarrow \mathrm{O}(^1D_2) + \hbar\omega$, and also in lasers with the molecules KrSe, XeSe which are working in the vicinity of the transitions

$$\mathrm{Se}(^1S_0) \rightarrow \mathrm{Se}(^1D_2) + \hbar\omega \quad \text{and} \quad \mathrm{Se}(^1S_0) \rightarrow \mathrm{Se}(^3P_1) + \hbar\omega.$$

6.4 Resonantly Excited Atoms in Gases

The resonantly excited states of atoms are characterized by small radiative lifetimes (see Table 6.3), because the dipole radiative transition is permitted from these states to the ground state. These states are effectively excited in a weakly ionized plasma by electron impact. Further resonantly excited states decay as a result of quenching in collisions with atomic particles or by means of the emission of radiation. The radiation of these atoms is used in the gas-discharge sources of light. This principle is used, in particular, in mercury lamps.

Processes involving resonantly excited atoms are used in various devices including lasers. In such lasers it is used that the resonantly excited states of atoms are formed in collisions with electrons more effectively than in other states. This leads to the creation of an inverse population of levels. But since the lifetime of the resonantly excited states is relatively small, such lasers operate in a pulse regime. As an example, we consider the copper laser. The ground state of this atom is characterized by the electron shell $3d^{10}4s$, and the resonantly excited state corresponds to the transition of the valent electron from the $4s$- to $4p$-state. The excitation energy of the electron state $4^2P_{1/2}$ is equal to 3.79 eV, and for the electron state $4^2P_{3/2}$ the excitation energy is 3.82 eV. Metastable states of the copper atom with the electron shell $3d^94s^2$ have a lower excitation energy. The excitation energy of state $3^2D_{5/2}$ is equal to 1.39 eV, and for state $3^2D_{3/2}$ it is equal to 1.64 eV. The copper laser is operated from both a gaseous discharge and electron beam. Then the resonantly excited states are formed by electron impact more effectively than metastable states and are used as the upper levels of laser transitions. This laser, as other lasers of this type (the nitrogen laser, the laser on the vapor of lead and thallium, and ion lasers such as argon and krypton lasers), operates in the pulse regime.

In accordance with the nature of the resonantly excited states, a convenient method of their excitation consists of the irradiation of a gas or vapor by resonant radiation. This method is known as optical pumping and, it gains new applications with the creation of lasers. This method is also used for the generation of the so-called photoresonant plasma. In this case, a small volume of gas or vapor is irradiated from a pulse light source such that the wavelength of the generated photons coincides with that of the spectral lines of this gas. Then the photons are effectively absorbed by gaseous atoms, and further collisions of the excited atoms lead to atom ionization and the formation of free electrons. The photon source can be tuned not only to transitions from the ground state of gaseous atoms, but also to transitions between excited states or states of its ion. Then an initial formation of these excited atoms or ion states is required. This process leads to the transformation of the energy of incident radiation into the plasma energy. As a result, a photoresonant plasma with specific properties is formed in some space element, and the character of evolution of this plasma is used. The other application of optical pumping consists of the creation of new lasers with larger wavelengths than that of incident radiation.

New phenomena occur if a polarized light source is used for optical pumping. Let us consider the process of the irradiation of an atomic gas when atoms are excited by a polarized resonance radiation, and subsequently excited atoms return to the ground state by emitting photons. Since the probability of the radiation of each excited atom does not depend on a photon direction and its polarization, the process of the emission of photons does not change the average atom momentum, while the absorption process influences this value. As a result, the average momentum of gaseous atoms varies. This phenomenon is known as the atom alignment process or optical alignment. As an example of optical alignment, consider this process in a cell with rubidium vapor ^{87}Rb. The nuclear momentum of this isotope is equal to 3/2, the atom spin in the ground state is 1/2, and the atom orbital momentum is equal to zero in the ground state. Hence, the atom ground state level is split into two superfine levels, so that the total atom momentum, which is a sum of the electron spin and nucleus momenta, is equal to $F = 1$ for the lower superfine state and to $F = 2$ for the upper superfine state. As a result of the exposure of atoms by resonant circularly polarized radiation, the projection of the total atom momentum onto the photon direction increases by one after the absorption of each photon (at the corresponding choice of the polarization direction). Hence, if we neglect the relaxation processes which lead to the randomizing of the momentum direction, in the end we can obtain the vapor with an identical direction of atom momenta. When this condition is reached, the absorption coefficient for the polarized light decreases. This change of the absorption coefficient can be evidence of the optical alignment of atoms.

The optical alignment of atoms is used in various methods and devices. One of these is the rubidium maser which operates between states of the superfine structure of ^{87}Rb atoms. According to the above consideration, optical pumping increases the atom momentum, i.e., as a result of this process a number of atoms with the moment $F = 2$ becomes more than that in the equilibrium case. Because the level with $F = 2$ is the upper one, this process leads to creation of the inverse population of superfine levels. This provides the operation of a quantum generator. The rubidium maser is the main element of the corresponding quantum standard of frequency.

The phenomenon of the optical alignment of atoms is the basis of operation of magnetometers. As an example, we consider a magnetometer whose working gas is helium ^{4}He (the abundance of this isotope is practically 100%, the nuclear momentum is equal to zero). Gas discharge is burnt in a cell filled with helium so that a remarkable amount of metastable atoms He(2^3S) are formed in this cell (the number density of metastable atoms is of order of $10^{10}\,\mathrm{cm}^{-3}$, while the number density of atoms in the ground state is of the order of $10^{16}\,\mathrm{cm}^{-3}$). These metastable atoms are irradiated by resonant polarized light which transfers them to the state 2^3P. Further atoms He(2^3P) emit photons and return in the metastable state He(2^3S). As a result, metastable helium atoms obtain a primary direction of the electron spin which is determined by the direction of incident radiation.

If this gas discharge cell is located in a magnetic field, the metastable level is split into three levels depending on their spin projection onto the magnetic field direction, and the difference of the level energies is proportional to the magnetic field strength. Transitions between these levels can result from radiofrequency electromagnetic waves which are in resonance with magnetic sublevels. The process of absorption of the radiowave leads to the establishment of equilibrium distribution over magnetic sublevels, while optical pumping violates the equilibrium distribution. Therefore, the absorption of a radiofrequency signal causes an increase in

Table 6.3. Parameters of the lowest resonantly excited states of some atoms; ε_{ex} is the excitation energy of the state, λ is the wavelength of radiative transition to the ground state, and τ is the radiative lifetime of the lowest resonantly excited state.

Atom, state	ε_{ex}, eV	λ, nm	τ, *ns*	Atom, state	ε_{ex}, eV	λ, nm	τ, ns
$H(2^1P_1)$	10.20	121.567	1.60	$Zn(4^1P_1)$	5.796	228.8	1.4
$He(2^1P_1)$	21.22	58.433	0.555	$Ga(5^2S_{1/2})$	3.073	403.41	7.1
$He(3^1P_1)$	23.09	53.703	1.72	$Kr(1s_4)$	10.03	123.58	3.5
$Li(2^2P_{1/2})$	1.85	670.791	27.3	$Kr(1s_2)$	10.64	116.49	3.2
$Li(2^2P_{3/2})$	1.85	670.776	27.3	$Rb(5^2P_{1/2})$	1.560	794.76	29
$Ne(1s_4)$	16.671	74.372	25	$Rb(5^2P_{3/2})$	1.589	780.03	27
$Ne(1s_2)$	16.848	73.590	1.6	$Sr(5^3P_1)$	1.798	689.26	$2.1 \cdot 10^4$
$Na(3^2P_{1/2})$	2.102	589.59	16.4	$Sr(5^1P_1)$	2.690	460.73	6.2
$Na(3^2P_{3/2})$	2.104	589.00	16.3	$Cd(5^3P_1)$	3.801	326.1	2400
$Mg(3^1P_1)$	4.346	285.21	2.1	$Cd(5^1P_1)$	5.417	228.80	1.7
$Al(4^2S_{1/2})$	3.143	394.51	6.8	$In(6^2S_{1/2})$	3.022	410.29	7.6
$Ar(1s_4)$	11.62	106.67	10	$Xe(1s_4)$	8.44	146.96	3.6
$Ar(1s_2)$	11.83	104.82	2.0	$Xe(1s_2)$	9.57	129.56	3.5
$K(4^2P_{1/2})$	1.610	769.90	27	$Cs(6^2P_{1/2})$	1.386	894.35	31
$K(4^2P_{3/2})$	1.617	766.49	27	$Cs(6^2P_{3/2})$	1.455	852.11	31
$Ca(4^3P_1)$	1.886	657.28	$4 \cdot 10^5$	$Ba(6^3P_1)$	1.57	791.13	300
$Ca(4^1P_1)$	2.933	422.67	5.2	$Ba(6^1P_1)$	2.24	553.55	8.5
$Cu(4^2P_{1/2})$	3.786	327.40	7.0	$Hg(6^3P_1)$	4.887	253.7	118
$Cu(4^2P_{3/2})$	3.817	324.75	7.2	$Hg(6^1P_1)$	6.704	184.9	1.6
$Zn(4^3P_1)$	4.006	307.6	$2.5 \cdot 10^4$	$Tl(7^2S_{1/2})$	3.283	37.57	7.6

the absorption coefficient for optical pumping. The determination of a resonance frequency in this system gives the distance between magnetic sublevels, and correspondingly yields the value of the magnetic field. This method allows one to measure magnetic fields of very small strength, up to 10^{-7}Oe.

For the above version of the optical magnetometer, the resonant frequency of an electromagnetic field is determined on the basis of the dependence on the frequency of the radiowave for the absorption coefficient of optical pumping which causes transitions the between excited atom states. Optical pumping violates a random distribution for magnetic sublevels of the ground atom state, and this causes variation in the absorption coefficient of an optical signal. Another version of this method uses resonant transitions from the ground atom state. This takes place in the method of the Double Radio-optical Resonance which was first realized in 1952 by Brossel and Bitter (France) for mercury atoms. A cell contains a mercury vapor including even isotopes so that their nuclear momentum is zero. This vapor is excited by circularly polarized light with wavelength 253.7 nm which transfers mercury atoms from the ground state 6^1S_0 into the resonantly excited state 6^3P_1. Since the atom orbital momentum in the ground state is zero, the circular polarized radiation excites atoms in the state with momentum projection $M = 1$. Hence, the decay of excited atoms creates a polarized radiation in each direction. One can vary the distribution of atoms over momentum projections applying a constant magnetic field which splits the magnetic sublevels, or by variation of an alternating (radiofrequency) magnetic field which causes transitions between the magnetic sublevels. Measurement of the resonant frequency for a magnetic field signal, which causes the strongest transitions between magnetic sublevels, allows one to determine the magnetic field strength in this case.

The interaction of polarized resonant radiation with atoms leads to various phenomena which, along with the above devices, give information about atomic parameters, their processes of their interaction and collision. The use of polarized radiation is the basis of various methods which allows one to measure the atomic characteristics.

6.5 Detection of Individual Atoms

Due to the small width of spectral lines resulting from the radiation of excited atoms in gaseous or plasma systems, the spectral lines related to atoms of different elements can be separated. This is the basis of a strong method of the identification of excited atoms on the basis of their spectra. The measurement of the radiation spectrum of a hot gas, vapor, or plasma allows one to determine the content of various elements in this system. One can trace the history of this method from 1860–1861, when Bunsen and Kirchhoff discovered cesium and rubidium on the basis of the analysis of their spectra. Add to this another historical example when in 1871 Lokier (France) discovered one more element—helium—due to its spectral line in the radiation of the Sun. The method of spectral emission was further developed and became the main diagnostic method in metallurgy. Within the framework of this method, a small amount of a metallic object is introduced into an arc discharge in the form of dust or solution. Then the atoms of iron and other metals are evaporated and excited in an arc plasma. This analysis is based on comparison of the intensities for certain spectral lines of this plasma, and the ratio of these intensities yields the relative concentrations of different admixtures to iron in the sample analyzed. This method allows one to determine reliably the content of admixtures up to their concentrations of 10^{-5}–10^{-3} with respect to iron.

The optohalvanic method of determination of the admixture concentration is more precise. It uses the measurement of electric current through a plasma which is under the action of tuned monochromatic radiation whose frequency varies continuously. If this radiation is found in resonance with atoms in the ground or excited states, absorption of this radiation influences the electric current through the plasma. Then the electric current, as a function of the frequency of laser radiation, contains some narrow resonances and downfalls corresponding to certain transitions between the atomic states. The positions of these peculiarities in the electric current and their intensities give information about the content of admixtures in the plasma. This method is used both for a gas discharge plasma and flames, and allows one to determine the concentration of admixtures of the concentration up to 10^{-11}. The difficulty of this method is in the establishment of the numerical connection between current variations and the concentration of the corresponding elements, because this connection depends on many factors.

Below we present the two-step ionization method for the detection of individual atoms. Due to the transformation of atoms of a certain sort into ions, this method is characterized by high selectivity and sensitivity. The first stage of the method is the excitation of a certain atom state by resonance radiation, and the second stage is the photoionization of formed excited atoms. Because of the small photon energy, the radiation does not act on atoms of other types in a gas. As a demonstration of the possibilities of this method, we describe in detail the first version of this method which was developed by Hurst, Naifi, and Yang from the Oak Ridge National Laboratory in 1977. Then cesium atoms were detected in gaseous systems. The excitation of cesium atoms in the state $7^2P_{3/2}$ was made by laser radiation with the wavelength 455.5 nm. Because the cross section of this transition exceeds the photoionization cross section of the excited atom by several orders of magnitude, for the second stage a lamp of a wide spectrum was used. The ionization potential of the excited cesium state $7^2P_{3/2}$ is equal to 1.17 eV. Cesium ions formed in the above two-step process are detected by the ion detector. This method allowed one to detect one cesium atom among 10^{19} other atoms or molecules. This sensitivity is higher by several orders of magnitude, than if we had tried to remove one person from among all the Earth's inhabitants. Note that the method of two-step ionization is suitable for the detection of elements with a low-ionization potential.

6.6 Properties of Highly Excited Atoms

Highly excited states are a specific group of atom states. They are sometimes called Rydberg states. A valent electron of a highly excited atom has a weak bond with the atomic core and can be ionized as a result of the collision with atomic particles or underaction of an external field. The radiative lifetime of highly excited atoms is remarkably large due to the weak interaction of the valent electron with the atomic core, and this parameter for highly excited atoms is intermediate between the resonantly excited and metastable atoms.

In order to describe a highly excited atom, take the interaction potential between a valent electron and an atomic core in the form

$$U(r) = -\frac{e^2}{r} + V(r), \tag{6.7}$$

where r is the distance of the electron from the center, and $V(r)$ is the potential of a short-range interaction of the electron and atomic core. This interaction is essential only in the region where electrons of the atomic core are located. If we neglect the short-range interaction, the ionization potential of a highly excited atom coincides with that of an excited hydrogen atom which is equal to, according to formula (2.13),

$$J_n = \frac{Ry}{n^2},$$

where Ry is the Rydberg constant—the ionization potential of the hydrogen atom in the ground state, and n is the principal quantum number of the electron. Taking into account the short-range interaction within the framework of the perturbation theory, we have

$$J_n = \frac{Ry}{n^2} + \langle V(r) \rangle.$$

The last term is equal to the order of magnitude $\langle V(r) \rangle \sim V(a) \cdot w$ where $V(a) \sim Ry$ is a short-range interaction at distances of the order of atomic sizes, and $w \sim n^{-3}$ is the probability of location of the valent electron in the region of the atomic core. Accounting for $n \gg 1$, represent this formula in the form

$$J_n = \frac{Ry}{(n - \delta_l)^2} = \frac{Ry}{n_*^2}, \qquad n_* = n - \delta_l. \tag{6.8}$$

The parameter δ_l, which characterizes a shift of the level under the action of a short-range interaction of the excited electron and atomic core, is called the quantum defect. The value n_* is called the effective principal quantum number.

The quantum defect decreases strongly with an increase in the electron orbital momentum l, because the higher the electron orbital momentum, the less the probability for the electron to locate near the atomic core. Next, the quantum defect increases with an increase in the nucleus charge because an increase in the atomic number leads, on one hand, to an increase in the short-range interaction and, on the other hand, shifts the electron quantum number for the ground atom state. For example, the ionization potential of the cesium atom is equal to 3.89 eV which corresponds to $n_* = 1.87$ according to formula (6.8) for the ground atom state where $n = 6$. Hence the quantum defect of the ground state is equal to $\delta_l = 4.13$. This varies weakly with excitation of the atom, so that one can expect that the quantum defect of the highly excited cesium atom is about 4 for the s-electron. This is confirmed the data of Table 6.4 where values of the quantum defects for the alkali metal atoms are represented.

Analyzing the radiative properties of the highly excited atoms, note that radiative transitions in closed excited states by energy give a small contribution to the radiative lifetime of this state, i.e., the radiative lifetime of a highly excited state is determined mainly by transitions to low lying levels. Because the probability per unit time for the radiative transition w_{rad} to a low lying level is proportional to the probability of the location of a highly excited electron in a region of the order of an atom size, we have $w_{\text{rad}} \sim n_*^{-3}$. Thus we have, for the radiative lifetime of a

Table 6.4. Quantum defect of alkali metal atoms.

Electron state	Li	Na	K	Rb	Cs
s	0.40	1.35	2.16	3.13	4.05
p	0.047	0.855	1.72	2.64	3.56
d	0.001	0.015	0.24	1.35	2.47
f	0	0	0.009	0.016	0.032

Table 6.5. The parameter of the radiative lifetime $\tau_l(ns)$ for highly excited atoms of alkali metals.

Electron moment	Li	Na	K	Rb	Cs
s	0.84	1.36	1.21	1.18	1.3
p	3.4	2.7	3.9	2.9	3.4
d	0.47	0.93	2.6	1.4	0.7
f	1.1	1.0	0.76	0.66	0.67

highly excited atom,

$$\tau_{nl} = \frac{\tau_l}{n_*^3}, \tag{6.9}$$

where τ_l does not depend on n_*. Table 6.5 contains values of the parameter τ_l for alkali metal atoms.

6.7 Generation and Detection of Highly Excited Atoms

The first stage of the investigations of highly excited atom methods of their generation included processes of the charge exchange of ions on atoms or molecules, the process of the excitation of atoms by electron impact, and the process of molecule dissociation accompanied by excitation of its fragments. In all these processes, the relative probability of the formation of highly excited atoms is proportional to the factor n^{-3}, that is, the probability of location of an excited electron in the region of the atomic core. Hence the probability of the formation of a highly excited state is small in these processes, and a mixture of excited atoms with different n can be formed as a result of these processes. The contemporary method of the generation of highly excited atoms uses a dye-laser which allows one to obtain excited atoms in a given state only. As a result, we have a mixture of atoms where the main part of the atoms is found in the ground state, and a small part of the atoms is found in a given excited state. Atoms of other highly excited states are practically absent in this mixture. One- and two-photon absorption processes are used for the generation of highly excited atoms. One can extract from these processes two-photon processes when absorbed photons move in opposite directions. Then the Doppler shift of one absorbed photon is compensated for by the Doppler shift due to the absorption of the other photon. This leads to the identical absorption of atoms moving with different velocities, i.e., the Doppler shift of the spectral lines is eliminated and the absorption spectral line becomes narrow. This method, called the two-photon spectroscopy without Doppler broadening, is spread for the generation and detection of highly excited atoms.

The detection of highly excited atoms can be made from the analysis of their radiation spectrum. But an increase in the principal quantum number of an excited electron leads to a strong decrease in the radiation intensity, so that this method is applied for small n (usually $n \leq 20$). The other method of detection of a highly excited atom is based on ionization of the atom and the registration of a formed ion. Different collision methods of the ionization of highly excited atoms are possible for their detection, such as collisions with electrons in a thermoionic detector and collisions with atoms and interaction with a metallic surface. But because these methods are not selective with respect to atomic states, they are not widely practiced. The selective method of the detection of highly excited atoms uses the ionization of these atoms in a constant electric field of moderate strength. In particular, in a typical scheme of the study of highly excited atoms, an atomic beam with an admixture of highly excited atoms passes between the plates of a capacitor of a certain electric field strength in this region.

Let us analyze the ionization process of highly excited atoms in a constant electric field. Then the potential acting on a weakly bound electron in the main region of its location is equal to

$$U = -\frac{e^2}{r} - eEz, \tag{6.10}$$

where E is the electric field strength, and z is the projection of the electron radius-vector $\mathbf{r}$ on the electric field direction. The first term corresponds to the interaction of the electron with the Coulomb field of the atomic core and the second term corresponds to the interaction with the electric field. Figure 6.1 gives the cross section of this interaction potential. As is seen, this interaction potential creates a barrier which locks the electron into the atom region. The maximum decrease in the barrier takes place in the plane of the nucleus and field ($x = y = 0$), and is $U_{\max} = -2e^{3/2}E^{1/2}$. Let us introduce the critical electric field strength E_{cr}, such that the barrier disappears at the plane $x = y = 0$. Then we obtain, on the basis of formula (6.8),

$$E_{\mathrm{cr}} = \frac{E_o}{16n_*^4}, \tag{6.11}$$

where $E_o = 5.1 \cdot 10^9$ V/cm is the atomic value of the electric field strength (see Table 2.1). One can expect that at fields of the order of E_{cr} the decay time of a highly excited atom in the electric field is of the order of a typical atomic time for an excited electron, which is $\tau_n \sim \tau_o/n^3$ where $\tau_o = 2.4 \cdot 10^{-17}$ s is the atomic time.

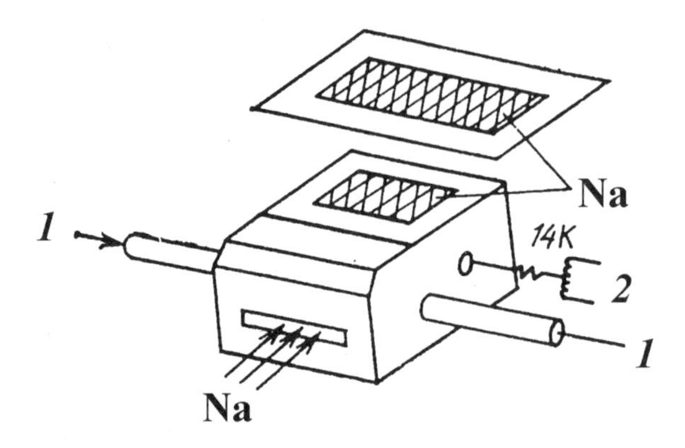

Figure 6.1. The scheme of the detector of infrared radiation. A beam of sodium atoms passes through a gap in the camera with the temperature of the walls at 14 K. Then the atom beam is excited by two laser beams and (1) it moves in the region of action of an external source of infrared radiation (2). Next, this beam moves between plates of the capacitor in an external electric field which has such strength that the atoms Na(23P) are ionized, while the atoms Na(22D) are not. Forming ions are collected by a collector.

Let us make numerical estimations for $n = 20$–30. Formula (6.11) gives $E_{\rm cr} \approx 10^4$ V/cm, i.e., these fields can be reached under laboratory conditions. Next, $\tau_n \approx (2\text{–}6) \cdot 10^{-13}$ s. A typical time of location of a highly excited atom in the field region, if it moves with an atomic beam between the capacitor plates, is $\tau_{dr} \sim l/v \sim 10^{-6}$ s, if we take the beam velocity $v \sim 10^5$ cm/s and a field size $l \sim 1mm$. Because of $\tau_{dr} \gg \tau_n$, this corresponds to the used electric field strengths $E < E_{\rm cr}$.

Let us determine a typical time of the atom decay, if the electric field strength is close to the critical one $E_{\rm cr} - E \ll E_{\rm cr}$. Then the time of the tunnel electron transition is equal to

$$\tau \sim \tau_o n^3 \exp\left(-\frac{C}{\hbar} \cdot \sqrt{m\Delta U} \cdot \Delta z\right),$$

where C is a numerical factor which is determined by a barrier shape, ΔU is the minimum difference of the potential energy and the level energy, and Δz is the width of a barrier region where the potential energy exceeds the energy of the level. We have

$$\Delta U = 2e^{3/2}E_{\rm cr}^{1/2} - 2e^{3/2}E^{1/2} = e^{3/2} \cdot \frac{E_{\rm cr} - E}{E_{\rm cr}^{1/2}} \quad \text{and} \quad \Delta z = e^{1/2}\frac{E_{\rm cr} - E}{E_{\rm cr}}.$$

Then, using formula (6.11), we obtain

$$\tau \sim \tau_o n_*^3 \exp\left[\frac{an_*(E_{\rm cr} - E)}{E_{\rm cr}}\right], \tag{6.12}$$

where the numerical factor in this formula is $a \sim 10$. The probability of the decay of a highly excited atom after passage through the field region is equal to

$$W = 1 - \exp\left(-\frac{\tau_{dr}}{\tau}\right).$$

Let us make estimates for the parameters $n = 30$, $\tau_{dr} = 10^{-6}$ s, $a = 10$. Determine the interval of the electric field values at which the decay probability varies from 10% to 90%. We obtain that this leads to variation of $(E_{\rm cr} - E)/E_{\rm cr}$ from 4% to 5%. In this example, variation of the decay probability of the highly excited atom from 20% to 80% corresponds to variation of the electric field strength by 0.4%, i.e., the value $(E_{\rm cr} - E)/E_{\rm cr}$ varies in the limits from 4.5% to 4.9%. Note that the transition from the state with $n = 30$ to the state with $n = 31$ corresponds to a decrease in $E_{\rm cr}$ by 12%. Thus, because of a strong dependence of the decay probability on the electric field strength, the excited atom decay proceeds in a narrow range of the electric field strengths.

This is the result of the principle which is the basis for the identification of highly excited atom states. In the related example, the relative variation of the electric field strength, which corresponds to the transition from an almost total conservation of excited atoms of a given state up to their almost total decay, is approximately 1%. An electric current resulting from the ionization of atoms in a given state occurs in this range of variation of the electric field. The electric field corresponding to the decay of the state with the nearest value of n differs from this one by approximately 10%. Thus two neighboring excited states with the same electron momentum can be selected reliably. Moreover, this method allows one to separate states with the same n, but different l for $l = 0, 1, 2$. Thus the ionization of highly excited atoms in a constant electric field is the reliable method for the identification of states of highly excited atoms.

6.8 Radiation of a Classical Electron in the Coulomb Field of a Nucleus

If a classical electron moves in the Coulomb field of a nucleus, the action of this field creates an acceleration of the electron and causes the emission of radiation. If this electron is found in a bound state, i.e., we have a highly excited atom, this effect leads to the radiative transition of the atom in less excited states. For a free electron this process can lead to capture in bound states, i.e., to recombination. Let us determine the cross section of the related process for a free electron. This cross section can be represented in the form

$$\frac{d\sigma}{d\omega} = \int_0^\infty 2\pi\rho\, d\rho B_\omega, \tag{6.13}$$

where $d\sigma$ corresponds to an interval of frequencies of emitted photons from ω to $\omega + d\omega$, ρ is the impact parameter of collisions, and B_ω is the radiation power per unit frequency. We use the general expression for the probability of radiation per unit time (1.26) and change the matrix elements by their Furie components in the classical limit. Then we have

$$B_\omega \sim \frac{\hbar\omega}{\omega} w_{fi} \sim \frac{\omega^3}{c^3} r_\omega^2, \tag{6.14}$$

where r_ω is the Furie component from the electron coordinate.

Note that a strong interaction of the electron with the Coulomb center takes place at small impact parameters. The law of energy conservation for the electron energy has the form

$$\varepsilon = \frac{v_\infty^2}{2} = \frac{v^2}{2} - \frac{Z}{r} + \frac{l^2}{r^2},$$

where ε is the electron energy, v_∞ is its velocity far from the Coulomb center, v is the electron velocity at a distance r from the center, $l = \rho v_\infty$ is the rotation momentum of the electron which is large compared to that in the classical limit. Neglecting the electron energy far from the center compared to the interaction energy in the region of radiation, we find, from this relation for the distance of closest approach r_o at a given impact parameter ρ,

$$r_o \sim l^2/Z \sim \rho^2 \varepsilon/Z.$$

The size of an interaction region which is responsible for the emission of photons of a frequency ω is given by the estimate $r_o \sim v/\omega$, and the electron velocity in this region is determined by interaction with the Coulomb center, so that $v \sim \sqrt{Z/r_o}$. From this it follows that

$$\omega \sim Z^{1/2} r_o^{-3/2}, \quad \text{i.e., } r_o \sim Z^{1/3}\omega^{2/3}.$$

This leads to the following estimate for the impact parameters of collisions which give the main contribution to the radiation of photons of a frequency ω,

$$\rho^2 \sim \frac{Zr_o}{\varepsilon} \sim \frac{Z^{4/3}}{\varepsilon\omega^{2/3}}.$$

Next,

$$r_\omega \sim \frac{r_o}{\omega} \sim \frac{Z^{1/3}}{\omega^{5/3}};$$

from this we find, for the cross section of the emission of photons of a given frequency ω,

$$\frac{d\sigma}{d\omega} \sim \rho^2 B_\omega \sim \frac{\rho^2\omega^2 r_\omega^2}{c^3} \sim \frac{Z^2}{c^3\varepsilon\omega}. \tag{6.15}$$

This formula gives the dependence of the radiation cross section on the parameters of the problem. A stricter, but cumbrous analysis of this problem allows one to determine a numerical coefficient for this formula. With this numerical coefficient the above formula has the following form, in usual units,

$$\frac{d\sigma}{d\omega} = \frac{8\pi Z^2 e^6}{3\sqrt{3}mc^3\varepsilon\hbar\omega}. \tag{6.16}$$

This formula describes the radiation of a classical electron which is free at the start. At the end of the process, the electron can be both free and in a bound state. Hence, formula (6.16) corresponds to both bremsstrahlung and to recombination radiation.

Let us analyze the criterion of the validity of this formula. We assume the related effect to be classical, so that in the radiation region the interaction potential is small compared to the quantum value. This gives $Z/r_o \ll Z^2$, i.e., $r_o \gg Z^{-1}$. Since $r_o \sim \frac{Z^{1/3}}{\omega^{2/3}}$, from this it follows that

$$\omega \ll Z^2. \tag{6.17}$$

The other criterion requires large electron momenta of collisions for the classical character of electron motion. This gives $l \sim \rho\sqrt{\varepsilon} \gg 1$, and since $\rho \sim Z^{2/3}\omega^{-1/3}\varepsilon^{-1/2}$, from this we again obtain the criterion (6.17). Along with these assumptions, we used a close approach to the center during the electron motion. This gives $r_o \ll \rho$, and leads to the following criterion:

$$\omega \gg \varepsilon^{3/2}/Z. \tag{6.18}$$

Thus, formula (6.16) is valid for small electron velocities compared to the typical atomic velocities and corresponds to large frequencies of emitting radiation.

Let us use formula (6.5) for the deduction of the cross section of capture of a slow electron on a highly excited level as a result of photon emitting. The energy conservation law in this process has the form

$$\omega = \varepsilon + \frac{Z^2}{2n^2},$$

where ω is the photon energy and ε is the initial electron energy. From this it follows that $d\omega = |Z^2 dn/n^3|$. Let us define the photorecombination cross section with the formation of an atom in a state with the principal quantum number n as

$$\sigma_{\text{rec}}(n) = \frac{d\sigma}{dn}.$$

Then formula (6.16) gives

$$\sigma_{\text{rec}}(n) = \frac{8\pi Z^4}{3\sqrt{3}c^3 \varepsilon\omega n^3}. \tag{6.19}$$

Using the connection between the cross sections of photoionization and photorecombination as reversible processes (see Chapter 2), we get from this, for the cross section of photoionization of a highly excited state with a principal quantum number n, if this process proceeds under the action of a photon of a frequency ω,

$$\sigma_{\text{ion}}(n) = \frac{8\pi Z^4}{3\sqrt{3}c\omega^3 n^5}. \tag{6.20}$$

Formulas (6.19), (6.20) are called the Kramers formulas. In order to estimate their accuracy, we use formula (6.19) in the case of photorecombination with the participation of a slow electron and the formation of a hydrogenlike ion in the ground state. Then this formula yields ($n = 1$, $\varepsilon - q^2/2$, $\omega - Z^2/2$):

$$\sigma_{\text{rec}}(n) = \frac{32\pi Z^2}{3\sqrt{3}c^3 q^2}, \tag{6.21}$$

where q is the wave vector of the incident electron. Although this case is found beyond the limits of the validity of the Kramers formulas, its result differs from the accurate result by 25%.

6.9 Application of Highly Excited Atoms

Various precise devices can be constructed on the basis of highly excited atoms. Below we consider in detail one type of detector of submillimeter radiation. It was constructed in Germany in the 1980s and gives a representation of the possibilities of measuring techniques by using highly excited atoms. The concept of this device is based on transitions between highly excited atom states under the action of thermal radiation. Then, on the basis of measurement of the rates of transitions between highly excited states, one can determine the intensity of thermal radiation. In this version of the device, which will be described below, a sodium atomic beam is excited by two dye lasers. One of these lasers with lengthwave $\lambda = 589\,\text{nm}$ creates resonantly excited atoms as a result of the transition $3^2S_{1/2} \to 3^2P_{1/2}$. The second laser with wavelength about 415 nm transfers excited atoms in the state 22D. After this a beam of atoms with an admixture of highly excited atoms passes through a region where it interacts with a thermal radiation, and then is directed to the region of an electric field where highly excited atoms are detected. The atom beam is surrounded by walls with a low temperature (14 K) in order to escape the excitation of highly excited atoms by background thermal radiation. The scheme of this device and levels of transition are shown in Fig. 6.2.

Let us give some parameters of the device. Approximately 0.1% of the atoms are transferred into the state $22D$, and the initial flux of highly excited atoms in the beam is of the order of $10^{11}\,\text{s}^{-1}$. The radiative lifetime of Na($22D$) is about 10^{-5} s, so that at used drift time through the device approximately 5% of the excited atoms reach the region of the electric field. The transition $22D \to 23P$ (95 GHz) was used for the analysis of the action of thermal radiation. Then the

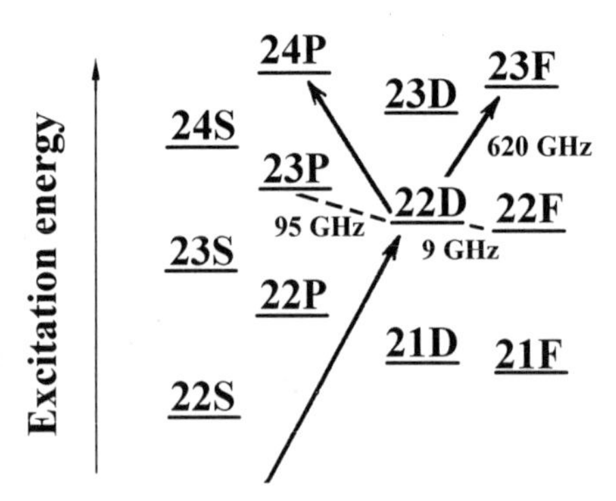

Figure 6.2. The scheme of the excited levels of the sodium atom which are used for operation of the detector of infrared radiation.

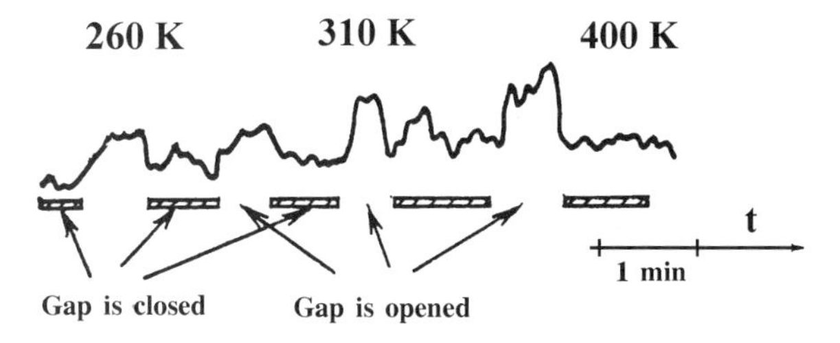

Figure 6.3. The dependence of the yield signal of the detector of infrared radiation depending on the temperature of the radiation source.

electric field was taken in such a way that it must decompose the highly excited atoms $23P$ and cannot act on the highly excited atoms $22D$. The rate of transition $22D \to 23P$ under the action of thermal radiation at room temperature is about $1300\,\mathrm{s}^{-1}$. For an external source with temperature 300 K this corresponds to the flux of atoms in the state $23P$ at approximately $10^5\,\mathrm{s}^{-1}$. The quantum efficiency of this device is $3 \cdot 10^{-3}$, i.e., approximately 300 excited atoms per second are detected if the temperature of the thermal source is 300 K. This detailed description of the device allows one to understand the complexity and possibilities of contemporary measuring techniques on the basis of highly excited atoms (see also Fig. 6.3).

6.10 The Cooling of Atoms

Optical pumping allows one to govern the behavior of atoms under the action of resonant radiation. Above we described some schemes of devices which use excited atoms through optical pumping. Below we concentrate on the problem of the cooling of atoms where superlow temperatures are reached on the basis of optical pumping combined with other methods with atom participation. We start from the Doppler cooling of atoms within the framework of a simple scheme, where free atoms are irradiated by a monochromatic laser radiation from six sides, so that for each axis of the rectangular coordinate system two laser beams move in direct and opposite directions. For a description of the real character of the interaction of these laser beams with individual atoms we assume the width of the spectral line of laser radiation to be zero in the scales under consideration, and these lines are shifted into the red side in comparison with the resonant line.

In this process the atom obtains the momentum $\Delta p = \hbar k = \hbar\omega/c$ as a result of the absorption of one photon, where k is the photon wave vector, c is the speed of light, and $\omega = kc$ is the radiation frequency. When the laser line is shifted to the red side with respect to the atom resonant line, the radiation is mostly absorbed by the laser beam moving toward the atom and, because the direction of the emitting photon is random, the atom is decelerated as a result of the absorption and radiation of one laser photon. The ratio of the probabilities to absorb the photon from two laser beams moving in opposite directions is

$$\frac{(\Delta\omega - kv_x)^2 + \Gamma^2/4}{(\Delta\omega + kv_x)^2 + \Gamma^2/4}, \tag{6.22}$$

where $\Delta\omega$ is the difference between the frequencies of the resonant atom transition and the laser spectral line, v_x is the projection of the atom velocity onto the direction of the laser beams, and Γ is the width for the spectral line of the resonant transition. Evidently, the minimum atom velocity corresponds to the minimal spectral line width which is the radiative width of the spectral line $\Gamma = 1/\tau$, where τ is the radiative lifetime. From this it follows that the minimum temperature of cooling due to this method is $T_{\mathrm{Dop}} \sim \Gamma$. This accurate analysis gives the Doppler limit $T_{\mathrm{Dop}} = \Gamma/2$. In particular, in the sodium case this limit is $T_{\mathrm{Dop}} = 240\,\mu\mathrm{K}$, which corresponds to the root mean square velocity of the sodium atoms $\Delta v_D = \sqrt{\overline{v^2}} = 30\,\mathrm{cm/s}$.

Note that this limit for alkali metal atoms is approximately two orders of magnitude higher than the limit due to the recoil of the absorbed atom. Indeed, the variation of the atom velocity as a result of photon absorption is $\Delta v_R = \hbar k/M = \hbar\omega/(Mc)$, where M is the atom mass. The energy which gives the restless atom as a result of photon absorption is

$$E_R = \frac{M\Delta v_R^2}{2} = \frac{(\hbar k)^2}{2M}. \tag{6.23}$$

Usually the temperature $T_R = 2E_R$ is called the recoil temperature limit. In the sodium case we have $\Delta v_R = 3\,\mathrm{cm/s}$, $E_R = 2.4\,\mu\mathrm{K}$, and the temperature which is limited by the atom recoil is $T_R = 2E_R = 5\,\mu\mathrm{K}$. Because the thermal velocity of sodium atoms at room temperature is $5.0 \cdot 10^4\,\mathrm{cm/s}$, the number of absorption acts to stop atoms must exceed 10^4.

In the above consideration of the Doppler cooling, we assume the intensity of the laser beams to be relatively small, so that, depending on the tuning, the radiation acts on atoms in some range of velocities. As a result, the radiation decelerates atoms in this velocity range and compresses the velocity distribution function, transforming it into a narrow peak. Far from this velocity range the velocity distribution function of atoms does not vary because these atoms do not interact with laser radiation. Hence, in the case of the use of weak laser intensities, when the radiation acts on the individual atoms independently, only a portion of the atoms interacts with the laser radiation and, finally, the velocity distribution function obtains a resonance in the range of a strong interaction. Under these conditions, the atoms are not stopped in this scheme.

In real schemes, other mechanisms occur for the interaction of laser beams with atoms which, in combination with the Doppler cooling, allow us to reach the Doppler limit of the atom temperature. Usually atoms are irradiated by laser beams from six sides. The absorption and reabsorption of resonant radiation strongly influences the atom's behavior in the irradiation region. For example, the maximum absorption cross section of the sodium atom under optimal conditions, when the broadening is determined by the radiative lifetime of the excited state, is equal to

$\sigma_{max} = 3\lambda^2/(2\pi) = 1.6 \cdot 10^{-9}\,\text{cm}^2$, where $\lambda = 589\,\text{nm}$ is the wavelength of the sodium resonant radiation and 3 is the ratio of the statistical weights for the upper and lower states of the radiative transition. Under the typical number density of atoms $\sim 10^7\,\text{cm}^{-3}$ the mean free path of resonant photons is of the order of 100 cm, i.e., the atom gas is transparent for photons for the typical size of an irradiated zone ($\sim$ 1 cm) where six laser beams are crossed. On the contrary, the mean free path of atoms is relatively small even at low fluxes of resonant photons. For example, a low laser intensity 1 W/cm^2, which corresponds to the flux of the resonant photons $3 \cdot 10^{18}\,\text{cm}^{-2}\,\text{s}^{-1}$ for sodium, and the absorption act under optimal conditions is characterized by the rate $5 \cdot 10^9\,\text{s}^{-1}$. Because the thermal velocity of sodium atoms at room temperature is $5 \cdot 10^4$ cm/s, this leads to the mean free path of atoms of the order of $0.1\,\mu$m. Of course, this is the lower estimate for the mean free path of atoms under the action of the resonant photons of a given intensity, but from this it follows that under typical conditions atoms are closed in the region where laser beams are crossed. Thus, the laser resonant radiation creates, for atoms, a viscous medium which is called the optical molasses. A strong interaction between atoms through this viscous medium allows one to reach the Doppler limit for the atom temperature.

The disadvantage of this method is such that the absorption rate varies significantly in the course of the stopping of atoms from thermal velocities. In order to improve the laser action on atoms, one can use "chirping" by variation of the laser frequency in time or space. The last effect is partially fulfilled in magneto-optical traps, where the shift of the resonant frequency varies in space due to the spatial variation of the magnetic field.

One more cooling mechanism within the framework of the above scheme takes place for a polarized laser beam. If the alkali metal atom in the ground state is located in the field of the linearly polarized radiation of high intensity, then the electron level is split into two levels depending on the spin projection onto the direction of the electric field of this electromagnetic wave. For the linear polarization of two beams moving in opposite directions, the polarization of the total field of the electromagnetic waves varies in space and, simultaneously, the splitting of energy levels varies in space as it shown is Fig. 6.4. The rate of excitation correlates with the spin projection, and this correlation depends on the tuning of laser signals with respect to the resonant transition. When the laser radiation frequency is shifted to the red side with respect to the resonant frequency, the dominant processes for absorption of the atom in the ground electron state, and for its emission from the resonantly excited state, are shown in Fig. 6.4. These processes lead to an additional cooling of atoms. The cooling due to polarization gradients is called the Sisuphus effect. This effect allows one to reach the temperature of the recoil limit.

The other problem is to keep cool atoms in some space region because even gravitational forces influence their motion. In particular, the restless sodium atoms reach the recoil velocity Δv_R through time $t = g/\Delta v_R \sim 0.003$ s (g is the free-fall acceleration). In order to keep the atoms in some space region, various traps are used. Usually they are based on some spatial configurations of magnetic fields and are called magneto-optic traps. Because of the slowness of captured atoms, the depth of the potential well for these atoms is correspondingly small.

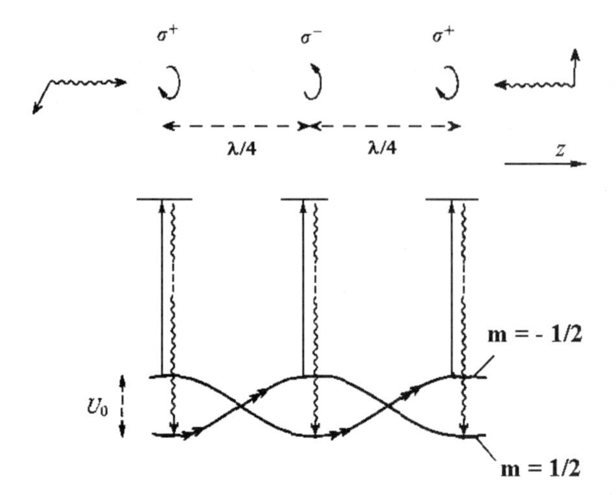

Figure 6.4. The Sisyphus effect of cooling as a result of the interaction of two counter-propagating laser beams with ortogonal linear polarization. (a). The resultant polarization is spatially modulated with a period $\lambda/2$, so that on a distance $\lambda/4$ it changes from σ^+ to σ^- and vice versa. (b). The ground state of atoms with the total electron momentum $1/2$ is split under the action of the radiation field, and the spliting is spatially modulated. The correlation causes the emission process near the potential hills, while the absorption proceeds near the potential wells. As a result, the atom is cooled because of the optical pumping.

The recoil limit is not of principle, though the above cooling schemes do not achieve temperatures below the recoil limit since the atom cooling results from the absorption of individual photons. In order to provide the so-called "subrecoil cooling," it is necessary to use the simultaneous absorption of two photons having close energies and moving in opposite directions. Two methods are now used for this goal. The first one, "Velocity Selective Coherent Population Trapping" uses the interference of two laser signals of close energies. As a result, the atom in the ground state transfers between the magnetic sublevels or levels of superfine structure. Due to the tuning of laser signals, the interaction of laser signals with atoms decreases with a decrease in their velocity, so that the atoms of zero velocity do not partake in these processes. The second method of subrecoil cooling, Raman cooling, is based on the character of stimulated Raman transitions. These methods are the second stage of cooling, and its first stage leads to the recoil limit by means of the above methods. As a result of two-photon transitions, with a small variation of the atom electron energy, the recoil is small which decreases the limiting atom temperature. These methods use fine aspects of the interaction of the interference of photons during their interaction with an individual atom, so that we leave them outside this consideration. The temperatures which are reached by the methods of subrecoil cooling, as a result two-photon transitions, are of the order of several nK.

Cold atoms may be used in various fine devices. As an example, below we consider atomic clocks. In standard schemes, the atomic clocks contain two microwave resonators which are excited by the same generator. An atomic beam crosses these resonators subsequently, and the frequencies of signals are compared due to the microwave transitions between the superfine atomic states. The smaller the atomic velocity, the narrower the resonance line which determines the stability of the atomic clocks.

The main scheme of the atomic clocks, based on slow atoms from the laser cooling, uses the so-called "Zacharias atomic fountain." In this scheme slow atoms are located in a magneto-optic cavity and are thrown up under the action of a laser pulse. When the atoms move up and down, they pass through the resonator cavity twice and interact with the resonator at these times. The comparison of the transition frequencies at these times allows us to determine them with high accuracy. This method provides a higher stability of atomic clocks in comparison with the previous method using a thermal beam of atoms, because time between two signals is now longer by two orders of magnitude than in the case of a thermal beam. As a result, this method provides a high stability of atomic clocks (10^{-15}–10^{-16}).

Of course, the above description does not pretend to give the state of the problem of the stopping and cooling of atoms by resonant radiation. This problem includes many other ideas, concepts, and applications. Our goal is to give a general representation of this problem.

Problems

Problem 6.1. *Determine the wave function of a highly excited electron in the main region of its location and the average size of the highly excited atom for the case $l \ll n$.*

The radial wave function of the electron satisfies equation (2.9):

$$\frac{d^2}{dr^2}[rR(r)] + \left[2\varepsilon + \frac{2}{r} - \frac{l(l+1)}{r^2}\right][rR(r)] = 0,$$

where $\varepsilon = 1/(2n_*^2)$ is the electron energy and n_* is the effective principal quantum number of the electron. Writing this equation in the form

$$\frac{d^2}{dr^2}[rR(r)] + p^2(r)\,[rR(r)] = 0,$$

we have that the classical electron momentum is positive at $r_1 < r < r_2$, where

$$r_{1,2} = n_*^2 \pm n_* \cdot \sqrt{n_*^2 - l(l+1)}.$$

Thus the region of classical motion of the electron is $r_1 < r < r_2$, and beyond this region the electron wave function decreases in an exponential way. Then we assume that the electron is located only in the region $r_1 < r < r_2$. The quasi classical solution of the Schrodinger equation in this region has the form

$$f = rR(r) = \frac{C}{\sqrt{p}} \cdot \sin\left(\int_{r_1}^{r} p\,dr + \frac{\pi}{4}\right), \qquad p^2(r) = -\frac{1}{n_*^2} + \frac{2}{r} - \frac{l(l+1)}{r^2},$$

where we account for $f(r_1) = 0$. One can determine the normalization constant C by taking into account that the wave function has a large number of knots ($\sim n$). Then changing $\sin^2 \varphi$ in the normalization integral by 1/2 we obtain, in accordance with formula (2.31),

$$C = \frac{1}{n_*^{3/2}} \cdot \sqrt{\frac{2}{\pi}}.$$

In the case $l \ll n$ we have $r_1 = 0, r_2 = 2n^2$, so that introducing the new variable $t = \sqrt{r}/(n_* \sqrt{2})$ with the range of its variation $0 < t < 1$, we obtain

$$rR(r) = \frac{Ct^{1/2}}{(1 - t^2)^{1/4}} \sin \varphi.$$

Here

$$\varphi = \int_0^r \sqrt{2/r - n_*^{-2}}\, dr - \pi/4 = 2n_* \left(t\sqrt{1 - t^2} + \arcsin(t + \delta) \right),$$

where the phase δ is of the order of one. Thus the quasi-classical wave function of an excited electron can be taken in the form

$$rR(r) = \sqrt{\frac{2}{\pi}} \cdot \frac{t^{1/2}}{n_*(1 - t^2)^{1/4}} \cdot \sin \left[2n_* \left(t\sqrt{1 - t^2} + \arcsin t \right) \right], \qquad 0 < t < 1.$$

The above quasi-classical expression for the electron wave function allows one to determine the mean parameters of the excited electron. In particular, from this it follows that

$$\langle r \rangle = \langle n|r|n \rangle = 3n_*^2/2; \qquad \left\langle r^2 \right\rangle = \left\langle n \left| r^2 \right| n \right\rangle = 5n_*^4/2,$$

which coincides with the data of Table 2.5 for the excited hydrogen atom.

The obtained expression for the electron wave function of a highly excited atom in a classical region of electron motion also allows one to obtain the quasi-classical expression in the classical forbidden region of electron coordinates. We have, for $n \gg l$,

$$R(r) = \frac{C}{2r\sqrt{|p|}} \cdot \exp\left(- \int_{r_1}^r |p|\, dr \right), \qquad C = \sqrt{\frac{2}{\pi}} \frac{1}{n_*^{3/2}},$$

$$|p(r)| = \sqrt{\frac{1}{n_*^2} - \frac{2}{r}}, \qquad r_1 = 2n_*^2.$$

Problem 6.2. *Obtain the quasi-classical asymptotic expressions for the electron angular wave function for $l \gg 1$.*

Let us first consider the case $m = 0$. Then, according to formula (2.5), the angular wave function has the form

$$Y_{lo} = \sqrt{\frac{2l + 1}{4\pi}} P_l(\cos \theta),$$

and our goal is to obtain asymptotic expressions for the Legendre polynomials $P_l(\cos\theta)$. This function satisfies the Schrödinger equation

$$\frac{d}{d\cos\theta}(1-\cos^2\theta)\frac{d}{d\cos\theta}P_l(\cos\theta)+l(l+1)P_l(\cos\theta)=0.$$

It is convenient to represent this equation in the form

$$\frac{d^2}{d\theta^2}P_l+ctg\theta\frac{d}{d\theta}P_l+l(l+1)P_l=0.$$

By the introduction of the new variable $f(\theta)=\sqrt{\sin\theta}\,P_l(\cos\theta)$, we transform the equation to the form

$$\frac{d^2f}{d\theta^2}+\left[\left(l+\frac{1}{2}\right)^2+\frac{1}{4\sin^2\theta}\right]f=0.$$

This can be represented in a quasi-classical form

$$\frac{d^2f}{d\theta^2}+p^2f=0,$$

where the value

$$p=\sqrt{\left[\left(l+\frac{1}{2}\right)^2+\frac{1}{4\sin^2\theta}\right]}$$

is the effective electron momentum. The condition of the validity of the quasi-classical approach $d(1/p)/d\theta\ll 1$ corresponds to the conditions

$$l\theta\gg 1,\qquad (\pi-\theta)l\gg 1.$$

In this range of the argument values the quasi-classical solution of the above equation is

$$P_l(\cos\theta)=C\frac{\sin\left[(l+1/2)\theta+\delta\right]}{\sqrt{\sin\theta}}.$$

In order to determine constants in this solution, let us consider the accurate equation for small angles

$$\frac{d^2}{d\theta^2}P_l+\frac{1}{\theta}\frac{d}{d\theta}P_l+(l+1/2)^2P_l=0.$$

Here we change $l(l+1)$ to $(l+1/2)^2$. The solution of this equation, accounting for $P_l(0)=1$, has the form

$$P_l(\cos\theta)=J_o\left[(l+1/2)\theta\right],$$

where J_o is the Bessel function. At small values of the argument this expression has the following asymptotic form:

$$P_l(\cos\theta)=\sqrt{\frac{2}{\pi l}}\cdot\frac{\sin\left[(l+1/2)\theta+\pi/4\right]}{\sqrt{\theta}}.$$

This solution must coincide with the quasi-classical solution in the region of the validity of both solutions $1/l \ll \theta \ll 1$. This comparison allows us to determine the coefficients in the quasi-classical expression for the Legendre polynomial

$$P_l(\cos\theta) = \sqrt{\frac{2}{\pi l}} \cdot \frac{\sin\left[(l+1/2)\theta + \pi/4\right]}{\sqrt{\sin\theta}}.$$

Now let us consider the cases $m \neq 0$ and $m \ll l$. Then the angular wave function is given by formula (2.5) and has the form

$$Y_{lm} = \sqrt{\frac{2l+1}{4\pi}} P_l^m(\cos\theta) e^{im\varphi},$$

and the asymptotic expression for $P_l^m(\cos\theta)$ can be found by the above method and has the form

$$P_l^m(\cos\theta) = \sqrt{\frac{2}{\pi l}} \cdot \frac{\sin[(l+1/2)\theta + m\pi/2 + \pi/4)]}{\sqrt{\sin\theta}}, \qquad l \gg m, \quad l\theta \gg 1.$$

Problem 6.3. *Determine the decay rate for an excited atom in a constant electric field.*

This problem is identical to Problem 2.5 for the decay of an excited hydrogen atom in a constant electric field. But in this case an excited electron is located in a spherical self-consistent field of the atomic core. Hence, the electron state is characterized by the quantum numbers nlm, and the electric field does not mix states of different l. The wave function of the excited electron satisfies the following Schrödinger equation:

$$\left[-\frac{1}{2}\Delta + U(r) - Ez\right]\Psi = \varepsilon\Psi,$$

where E is the electric field strength, and the electron energy is $\varepsilon = -\gamma^2/2$. At distances r near the atomic core one can neglect the electric field, so that the electron wave function has the form $\Psi = R_l(r)Y_{lm}(\theta, \varphi)$ where r, θ, φ are spherical coordinates, and the radial wave function satisfies the following equation:

$$\frac{d^2}{dr^2}[rR_l(r)] - \left[\gamma^2 - 2U(r) + \frac{l(l+1)}{r^2}\right][rR_l(r)] = 0.$$

Far from the atomic core the self-consistent field is transformed into the Coulomb field $[U(r) \to -1/r]$, and the radial wave function is

$$R_l(r) = Ar^{1/\gamma - 1}e^{-r\gamma},$$

where A is the asymptotic coefficient (see Chapter 4).

It is convenient to introduce parabolic coordinates in the region where the electron penetrates the barrier. By analogy with Problem 2.5, we have

$$\xi = r + z, \quad \eta = r - z; \qquad \Psi = \frac{e^{im\varphi}}{\sqrt{2\pi}} \cdot \frac{\Phi(\xi)F(\eta)}{\sqrt{\xi\eta}},$$

and the Schrödinger equation is transformed to the form

$$\Phi'' + \left(-\frac{\gamma^2}{4} + \frac{\beta_1}{\xi} - \frac{m^2-1}{4\xi^2} + \frac{E}{4}\xi\right)\Phi = 0;$$
$$F'' + \left(-\frac{\gamma^2}{4} + \frac{\beta_2}{\eta} - \frac{m^2-1}{4\eta^2} - \frac{E}{4}\eta\right)F = 0.$$

These equations follow from the initial Schrödinger equation if we change the self-consistent field potential $U(r)$ by the Coulomb potential $(-1/r)$. The separation constants are connected by the relation $\beta_1 + \beta_2 = 1$.

Let us connect the parabolic and spherical coordinates near the axis which is directed along the electric field and passes through the atomic center. We have

$$r = (\xi + \eta)/2 \approx \xi; \qquad \eta = \sqrt{\rho^2 + z^2} - z \approx \rho^2/(2z) = r\theta^2/2 = \xi\theta^2/4.$$

The use of the expression for the Legendre function at small angles

$$P_l^m(\cos\theta) \approx \frac{(l+m)!}{2^m m!(l-m)!} \cdot \sin^m\theta, \qquad \theta \ll 1,$$

leads to the following expression for the angular wave function

$$Y_{lm}(\theta,\varphi) = \left[\frac{(2l+1)(l+m)!}{2(l-m)!}\right]^{1/2} \cdot \frac{\theta^m}{2^m m!} \cdot \frac{e^{im\varphi}}{\sqrt{2\pi}}.$$

This gives, for the asymptotic expression of the wave function in the region $r\gamma^2 \gg 1, \theta \ll 1$:

$$\Psi = A\cdot\frac{e^{im\varphi}}{\sqrt{2\pi}}\cdot\left(\frac{\xi}{2}\right)^{1/\gamma-1-m/2}\cdot e^{-\gamma\xi/2}\cdot\left(\frac{\eta}{2}\right)^{m/2}\cdot e^{-\gamma\eta/2}\cdot\frac{1}{m!}\cdot\left[\frac{(2l+1)(l+m)!}{2(l-m)!}\right]^{1/2}.$$

Using this wave function in the Schrödinger equation for $\Phi(\xi)$ and $F(\eta)$, we find the separation constants for the related region

$$\beta_1 = 1 - \frac{\gamma}{2}(m+1), \qquad \beta_2 = \frac{\gamma}{2}(m+1).$$

The probability of the electron transition through the barrier is equal to

$$w = \int_S \mathbf{j}\,d\mathbf{s}, \qquad \mathbf{j} = \frac{i}{2}\cdot(\Psi\nabla\Psi^* - \Psi^*\nabla\Psi),$$

where j is the electron flux through the surface S which is perpendicular to the electric field direction. We have $ds = \rho\,d\rho\,d\varphi$, and because $\eta = \sqrt{\rho^2+z^2} - z \approx \rho^2/(2z)$, this gives $ds = \frac{\xi}{2}\,d\eta\,d\varphi$. We use the fact that the wave function $F(\eta)$ is the same both before and after the barrier. In addition, the integral w is determined by the spatial region near the axis. From this it follows that

$$w = \int j\frac{\xi}{2}\,d\eta\,d\varphi = \frac{m!}{\gamma^m}\cdot\frac{i}{2}\left(\Phi\frac{d\Phi^*}{d\xi} - \Phi^*\frac{d\Phi}{d\xi}\right).$$

In the spatial region, where one can neglect the action of the electric field, the wave function $\Phi(\xi)$ has the form

$$\Phi(\xi) = A \cdot \left(\frac{\xi}{2}\right)^{1/\gamma - 1 - (m+1)/2} \cdot e^{-\gamma\xi/2} \cdot \frac{1}{m!} \cdot \left[\frac{(2l+1)(l+m)!}{2(l-m)!}\right]^{1/2}.$$

Let us consider quasi-classical solutions near the boundary of the classical region $\xi_o = \gamma^2/E$. In the classical region of the electron motion $\xi > \xi_o$ the wave function $\Phi(\xi)$ has the form

$$\Phi(\xi) = i\frac{C}{\sqrt{p}} \cdot \exp\left(i\int_{\xi_o}^{\xi} p\, d\xi - i\frac{\pi}{4}\right), \qquad \xi > \xi_o.$$

In the classical forbidden region of the electron motion this wave function is

$$\Phi(\xi) = \frac{C}{\sqrt{|p|}} \cdot \exp\left(-\int_{\xi_o}^{\xi} |p|\, d\xi\right), \qquad \xi < \xi_o,$$

where

$$p^2 = -\frac{\gamma^2}{4} + \frac{\beta_1}{\xi} - \frac{m^2 - 1}{4\xi^2} + \frac{E}{4}\xi.$$

Using the expression for the wave function in the classical region of the electron motion we have, for the frequency of the atom decay,

$$w = \frac{m!}{\gamma^{m+1}} \cdot |C|^2.$$

Thus, our goal is to determine the coefficient C. Then let us compare the above expressions for the wave function $\Phi(\xi)$ in the region $1/\gamma^2 \ll \xi \ll \xi_o$, where, on one hand, one can neglect the electric field and, on the other hand, the asymptotic expression for the wave function is valid. We have

$$|p|^2 = \frac{\gamma^2}{4} - \frac{\beta_1}{\xi} - \frac{E}{4}\xi = \frac{E}{4} \cdot (\xi - \xi_o) \cdot \left(1 - \frac{4\beta_1}{E\xi\xi_o}\right), \qquad \xi_o = \frac{\gamma^2}{E} - \frac{4\beta_1}{\gamma^2},$$

and the second term in the expression for ξ_o is small compared to the first one. Thus we have, for the integral,

$$\int_{\xi_o}^{\xi} |p|\, d\xi = \frac{\sqrt{E}}{2} \cdot \int_{\xi_o}^{\xi} \sqrt{\xi_o - \xi}\left(1 - \frac{2\beta_1}{E\xi\xi_o}\right) d\xi = \frac{1}{3}\frac{\gamma^3}{E} - \frac{1}{2}\gamma\xi - \frac{\beta_1}{\gamma}\ln\left(\frac{4\gamma^2}{E\xi}\right).$$

Using in this expression $\beta_1/\gamma \ll \xi \ll \xi_o$, we obtain the following form for the quas-iclassical wave function before the barrier:

$$\Phi(\xi) = C \cdot \sqrt{\frac{\gamma}{2}} \cdot \left(\frac{E\xi}{4\gamma^2}\right)^{\beta_1/\gamma} \exp\left(\frac{\gamma^3}{3E} - \frac{\gamma\xi}{2}\right).$$

Comparing this with the asymptotic expression of the wave function in the region of a small electric field, we find the coefficient C,

$$C = A \cdot \sqrt{\frac{\gamma}{2}} \cdot \left(\frac{2\gamma^2}{E}\right)^{\beta_1/\gamma} \exp\left(-\frac{\gamma^3}{3E}\right) \cdot \frac{1}{m!} \cdot \left[\frac{(2l+1)(l+m)!}{2^m(l-m)!}\right]^{1/2}.$$

From this we obtain the following expression for the decay probability per unit time

$$w = \frac{A^2}{2} \cdot \left(\frac{2\gamma^2}{E}\right)^{2/\gamma - m - 1} \exp\left(-\frac{2\gamma^3}{3E}\right) \cdot \frac{(2l+1)(l+m)!}{(2\gamma)^m (l-m)!m!}. \tag{6.24}$$

Note that this expression is obtained under other conditions compared with Problem 2.5. In this case, an excited atom state is characterized by the spherical quantum numbers nlm, while under the conditions of Problem 2.5 parabolic quantum numbers are the atom quantum numbers. Hence, the results of these two problems coincide only in the case of one state, when the degeneration of levels is absent. In particular, for the hydrogen atom in the ground state ($\gamma = 1, l = m = 0$, $A = 2$), this formula gives

$$w = \frac{4}{3E} \cdot \exp\left(-\frac{2}{3E}\right), \tag{6.25}$$

which coincides with the result of Problem 2.4. The other common case corresponds to the states $n_1 = n_2 = 0, m = n - 1$ for parabolic quantum numbers and to $l = m = n - 1$ for spherical quantum numbers. These quantum numbers correspond to the same state. If the electron is located in the Coulomb field, i.e., for the hydrogen atom, the radial wave function of the relevant state is

$$R_l(r) = \left(\frac{2}{n}\right)^{3/2} \cdot \frac{1}{\sqrt{2n!}} \cdot \left(\frac{2r}{n}\right)^{n-1} \cdot e^{-r/n},$$

so that

$$C = \frac{1}{\sqrt{2n!}} \cdot \left(\frac{2}{n}\right)^{n+1/2}.$$

From this we obtain, for the decay rate the same formula as in Problem 2.5,

$$w = \frac{1}{n^2 n!} \cdot \left(\frac{4}{n^3 E}\right)^n \exp\left(-\frac{2}{3n^3 E}\right). \tag{6.26}$$

Note that the result of this problem corresponds to relatively small fields when the level splitting due to the electric field is small compared to distances between these levels. The result of Problem 2.4 corresponds to the opposite criterion. This result is suitable for states with $l = 0, 1$.

Problem 6.4. *Find the electric field strengths of the disappearance of the spectral lines of sodium of* 474.8 nm *and* 475.2 nm *which correspond to the transitions* $7^2S_{1/2} \rightarrow 3^2P_{1/2,3/2}$. *The radiative lifetime of the state* $\mathrm{Na}(7^2S_{1/2})$ *is equal to* $\tau_{\mathrm{rad}} = 2.7 \cdot 10^{-7}$ s *and the ionization potential of this state is* 0.43 eV.

The disappearance of these spectral lines is due to decay of the excited atoms in the electric field. The electric field strength of the disappearance of spectral lines we found from the relation $w\tau_{\mathrm{rad}} \sim 1$, where w is the decay rate for the excited state and τ_{rad} is the radiative lifetime of the excited state. Because of the strong dependence of the decay rate on the electric field strength, below we use this relation in the form $w\tau_{\text{rad}} = 1$. According to the expression of the previous problem, the rate of ionization of the s-electron is equal to

$$w = \frac{A^2}{2} \cdot \left(\frac{2\gamma^2}{E}\right)^{2/\gamma - 1} \exp\left(-\frac{2\gamma^3}{3E}\right).$$

The asymptotic coefficient A for the bound s-electron can be found from the assumption that in the main region of the electron's location its interaction with the atomic core is the Coulomb interaction. This leads to formula (4.12) which has the form

$$A^2 = \frac{\gamma(2\gamma)^{2/\gamma}}{\Gamma^2(1/\gamma + 1)}.$$

In the case under consideration we have the following parameters: $\gamma = 0.178$; $\tau_{\text{rad}} = 1.12 \cdot 10^{10}$; and $A^2 = 1.25 \cdot 10^{-12}$. The equation $w\tau_{\text{rad}} = 1$ takes the form $2.6 \cdot 10^{11} \cdot x^{10.25} e^{-x} = 1$ where $x = 2\gamma^3/E$. The solution of this equation $x = 70$ corresponds to the electric field strength $E = 28$ kV/cm. Note that we used the small parameter E/E_{cr}, where E_{cr} is the critical electric field strength (formula (6.11)). Under the related conditions this parameter is equal to 0.8, i.e., they are found near the boundary of validity of the used approach.

Problem 6.5. *Determine the critical electric field strength for an excited electron in the Coulomb field with parabolic quantum number $n_2 = 0$ on the basis of the condition that the barrier disappears at this electric field.*

The electron wave function in parabolic coordinates has the form, in accordance with the formulas of Problem 2.5,

$$\Psi = \frac{e^{im\varphi}}{\sqrt{2\pi}} \cdot \frac{\Phi(\xi)F(\eta)}{\sqrt{\xi\eta}},$$

and the Schrödinger equations can be written in the form

$$\frac{d^2\Phi}{d\xi^x} + p_\xi^2 = 0, \qquad \frac{d^2F}{d\eta^2} + p_\eta^2 = 0,$$

where

$$p_\xi^2 = \frac{\varepsilon}{4} + \frac{\beta_1}{\xi} - \frac{m^2 - 1}{4\xi^2} + \frac{E}{4}\xi; \qquad p_\eta^2 = \frac{\varepsilon}{4} + \frac{\beta_2}{\eta} - \frac{m^2 - 1}{4\eta^2} - \frac{E}{4}\eta,$$

where ε is the electron energy. The separation constants are connected by the relation $\beta_1 + \beta_2 = 1$. Because the electron is found in an excited state, one can use the quasi-classical conditions of the Bohr quantization

$$\int_{\xi_1}^{\xi_2} p_\xi \, d\xi = \pi\left(n_1 + \frac{1}{2}\right); \qquad \int_{\eta_1}^{\eta_2} p_\eta \, d\eta = \pi\left(n_2 + \frac{1}{2}\right),$$

where $\xi_1, \xi_2, \eta_1, \eta_2$ are the electron turning points, i.e., $p_\xi(\xi_{1,2}) = 0$ and $p_\eta(\eta_{1,2}) = 0$, and n_1, n_2 are the parabolic quantum numbers of the weakly bound electron which characterize a number of knots for the corresponding wave function. The above quasi-classical conditions of quantization are valid for $n_1 \gg 1, n_2 \gg 1$.

Note that the condition of the barrier disappearance has the form

$$\frac{dp_\xi}{d\xi}(\xi_2) = 0.$$

This relation, together with the Bohr equations of quantization, establishes the connection between the critical electric field E_{cr}, the separation constants β_1, β_2, and the electron energy ε. Below we consider this problem for the simplest case $m = 0, n_2 \ll n_1$. This corresponds to $n_2 = 0, \beta_2 = 0, \beta_1 = 1$. Because in this case the turning point ξ_2 relates, simultaneously, to a maximum of p_ξ, we have

$$\xi_2 = \frac{2}{\sqrt{E_{\mathrm{cr}}}}, \qquad |\varepsilon| = 2\sqrt{E_{\mathrm{cr}}}, \quad \text{and} \quad p_\xi^2 = \frac{E_{\mathrm{cr}}}{4\xi} \cdot (\xi_2 - \xi)^2.$$

Since in this case $\xi_1 = 0$, we obtain

$$\int_0^{\xi_2} |p_\xi|\, d\xi = \frac{2^{5/2}}{3E_{\mathrm{cr}}^{1/4}} = \pi n_1.$$

Because $n_1 = n$, we have

$$E_{\mathrm{cr}} = \frac{2^{10}}{(3\pi n)^4} = \frac{0.130}{n^4}; \qquad |\varepsilon| = \frac{64}{9\pi^2 n^2} = \frac{0.720}{n^2}. \tag{6.27}$$

Thus, action of the electric field on a highly excited atom in the relevant state lowers the energy level. The critical electric field strength according to this formula is almost twice as high according to formula (6.11). Note that the relevant state decays more effectively than other states with the same n, because the electron distribution is stretched in the direction of the electric field. For this distribution the electron is concentrated in the region where its interaction with the field corresponds to attraction. Hence, interaction with the field leads to lowering of the level. Interaction with the electric field for other levels of a given n can lead to a rise in the electron energy. But the critical electric field strength must be higher than in the relevant case which is preferable with respect to the ionization of a weakly bound electron.

Note that our consideration corresponds to diabatic switching of the electric field. This means that until the electric field strength varies from zero up to its final value, transitions between the atomic levels are absent. In the opposite case of the adiabatic evolution of this system, the level energy varies weakly in the course of switching on the electric field. Then at the end of switching at the field, the weakly bound electron will be located on another level, although the distribution inside the atom is not changed. Let us analyze switching of the electric field by the adiabatic way in the considered case $n_2 \approx n$. Then we have, for electron quantum numbers after the establishment of the electric field $n_1' \approx n'$, and from the condition of conservation of the electron energy, that it follows that

$$\frac{1}{2n^2} = \frac{64}{9\pi^2 (n')^2}.$$

The critical electric field strength is equal, in this case, to

$$E_{\rm cr} = \frac{2^{10}}{(3\pi n')^4} = \frac{1}{16n^4},$$

i.e., we obtain formula (6.11) for the critical electric field strength. The coincidence of these expressions is explained by conservation of the electron energy resulting from switching of the electric field. This fact shows the convenience of formula (6.11) for real estimate.

Problem 6.6. *Determine the accuracy of the Kramers formula for the cross section of photoionization of the hydrogenlike ion in the ground state.*

Taking formula (6.20) for the photoionization cross section $\omega = Z^2/2$, $n = 1$, we obtain the Kramers formula for photoionization of the hydrogenlike ion in the ground state

$$\sigma_{\rm ion}^{\rm Kr} = \frac{64\pi}{3\sqrt{3}cZ^2}.$$

Comparing this with the accurate cross section for photoionization of the hydrogenlike ion in the ground state $\sigma_{\rm ion}^{\rm ac}$, which is obtained in Problem 2.6, we have

$$\frac{\sigma_{\rm ion}^{\rm Kr}}{\sigma_{\rm ion}^{\rm ac}} = \frac{e^4}{8\pi\sqrt{3}} = 1.25.$$

Although this case is outside the limits of validity of the Kramers formula, the error is not high.

Problem 6.7. *Determine the rate of the cooling of atoms as a result of the absorption and reabsorption of resonant photons in the one-dimensional case when they are irradiated from two sides by laser beams of identical frequency. This frequency is shifted in the red side with regard to the atomic resonant transition. Neglect the broadening of the spectral line under the action of the laser field.*

We use the expression for the rate of photon absorption in the form

$$w = w_o\left[\frac{1}{(x-y)^2+1} + \frac{1}{(x+y)^2+1}\right], \qquad x = \frac{\Delta\omega}{2\Gamma}, \qquad y = \frac{2kv_x}{\Gamma},$$

where w_o is the maximum rate of photon absorption when the radiation frequency corresponds to the center of the spectral line, $\Delta\omega$ is the shift of the laser line with regard to the resonance, Γ is the width of the spectral line for the absorption process, $k = \omega/c$ is the photon wave vector, and v_x is the atom velocity. Each act of the photon absorption leads to a variation of the atom velocity by the recoil velocity $\pm\Delta v_R$, and the sign is determined by the photon direction, v_x is the atom velocity.

The recoil velocity Δv_R is small in comparison with the typical atom velocity, so that we consider the atom velocity as a continuous variable, and describe the atom evolution in the space of variable y on the basis of the Fokker–Planck equation

$$\frac{\partial f}{\partial t} + u\frac{\partial f}{\partial y} + D\frac{\partial^2 f}{\partial y^2} = 0,$$

where $f(y)$ is the velocity distribution function of atoms, and the drift velocity u and diffusion coefficient D of atoms in the space of variable y are equal to

$$u = u_o\left[\frac{1}{(x-y)^2+1} + \frac{1}{(x+y)^2+1}\right], \qquad u_o = w_o\frac{2k\Delta v_R}{\Gamma},$$

$$D = D_o\left[\frac{1}{(x-y)^2+1} + \frac{1}{(x+y)^2+1}\right], \qquad D_o = 2w_o\left(\frac{k\Delta v_R}{\Gamma}\right)^2.$$

Let us consider the limiting case $y \gg x \gg 1$ so that, in this case,

$$u = u_o\frac{4x}{y^3}, \qquad D = D_o\frac{2}{y^2},$$

where the parameters u_o and D_o correspond to the evolution of an atom in the velocity space under the action of one laser beam which is tuned to the center of the atom absorption line. In the stationary case we obtain the equation

$$u\frac{df}{dy} + D\frac{d^2 f}{dy^2} = 0.$$

Its solution has the form

$$f(y) = C\left(\frac{k\Delta v_R}{2\Delta\omega}\right)^y,$$

where C is a constant. Usually the optimal tuning is of the order of the line width or exceeds it. The parameter $k\Delta v_R/\Gamma$ is usually large, in particular, in the sodium case it is 100. Hence, the distribution function drops strongly at large y when the equilibrium is established. This means that atoms are stoped through some time, and the typical width of the velocity distribution function is of the order of the Doppler limit when $x \sim 1$.

Estimating a typical time of the equilibrium establishment, we note that the obtained nonstationary Fokker–Plank equation has the automodel form, so that at large time the combination y^4/t does not vary in time. Hence, a typical relaxation time, when the initial distribution function has a large width in comparison with the tuning ($y \gg x$) and the width of the absorption spectral line ($y \gg 1$), may be estimated as

$$t \sim \frac{y}{u}, \qquad t \sim \frac{y^2}{D},$$

and both estimates for $x \sim 1$ ($\Delta\omega \sim \Gamma$) give the identical result which has the form

$$t \sim \frac{y}{u} \sim \frac{1}{w_o}\frac{\Gamma}{k\Delta v_R}\frac{\Gamma}{\Delta\omega}\left(\frac{kv}{\Gamma}\right)^4 \sim \frac{1}{w_o}\frac{\Gamma}{\Delta\omega}\frac{v^4}{\Delta v_R \Delta v_D^3}.$$

Note that the value $w_o t$ is the number of absorption acts which lead to the equilibrium establishment if the absorption process proceeds under optimal conditions. Let us consider the sodium case when $\Delta\omega \sim \Gamma$. If we start from thermal velocities at room temperature, we have $v \sim 5 \cdot 10^4$ cm/s, $\Delta v_R = 3$ cm/s, $\Gamma/k \sim 4 \cdot 10^3$ cm/s, and this estimation gives that the equilibrium is established as a result of $\sim 10^8$ absorption acts, while under optimal conditions the decelerating stopping of the atom results from

$$\frac{v}{\Delta v_R} \sim 10^4$$

acts of photon absorption.

CHAPTER 7

Positive and Negative Ions

7.1 Peculiarities of Positive Ions

A positive ion is a system consisting of a Coulomb center of a charge Z and electrons whose total number is less than Z. As the difference between Z and the total number of electrons increases, some parameters of this system vary monotonically. In particular, this follows from Table 7.1 where the ionization potentials are given for atoms and the three first ions. Below we demonstrate the variation of atomic parameters with an increase in the ion charge on the basis of other atomic characteristics.

Table 7.2 gives the parameters of the metastable state of the carbon atom and the same state of the nitrogen ion. The difference of the energies between this and the ground state level is due to an exchange interaction inside the same electron shell. Because of the compression of the electron shells of ions compared to identical atoms, the exchange interaction potential increases for ions and the lifetime decreases correspondingly. Table 7.3 demonstrates the same appropriateness for the fine splitting of levels of the ground state of the iodine atom and xenon ion.

Thus the character of the interaction in ions is the same as in atoms, but the self-consistent field potential in ions is stronger than in atoms with the identical electron shell. Hence, the properties of ions with a small charge are close to those of atoms. The parameters of ions vary monotonically as the ion charge increases. This leads to principal changes of the parameters of valent electrons for ions of large charges.

Table 7.1. Ionization potentials of atoms and ions. The values J_0, J_1, J_2, J_3 are, correspondingly, the ionization potentials of the atom, and the ground states of its one-, two-, and three-charged ions. The ionization potentials are given in eV.

Z	Atom	J_0, eV	J_1, eV	J_2, eV	J_3, eV
1	H	13.598	—	—	—
2	He	24.588	54.418	—	—
3	Li	5.392	75.641	122.45	—
4	Be	9.323	18.211	153.90	217.72
5	B	8.298	25.155	37.931	259.38
6	C	11.260	24.384	47.89	64.49
7	N	14.534	29.602	47.45	77.47
8	O	13.618	35.118	54.936	77.414
9	F	17.423	34.971	62.71	87.14
10	Ne	21.565	40.963	63.46	97.12
11	Na	5.139	47.287	71.620	98.92
12	Mg	7.646	15.035	80.144	109.27
13	Al	5.986	18.829	28.448	119.99
14	Si	8.152	16.346	33.493	45.142
15	P	10.487	19.770	30.203	51.444
16	S	10.360	23.338	34.83	47.305
17	Cl	12.968	23.814	39.61	53.47
18	Ar	15.760	27.630	40.911	59.81
19	K	4.341	31.63	45.81	60.91
20	Ca	6.113	11.872	50.913	67.3
21	Sc	6.562	12.800	24.757	73.49
22	Ti	6.82	13.58	27.49	43.27
23	V	6.74	14.66	29.31	46.71
24	Cr	6.766	16.50	31.0	49.2
25	Mn	7.434	15.640	33.67	51.2
26	Fe	7.902	16.188	30.65	54.8
27	Co	7.86	17.084	33.5	51.3
28	Ni	7.637	18.169	35.3	54.9
29	Cu	7.726	20.293	36.84	57.4
30	Zn	9.394	17.964	39.72	59.57
31	Ga	5.999	20.51	30.7	64.2
32	Ge	7.900	15.935	34.2	45.7
33	As	9.789	18.59	28.4	50.1

Table 7.1. (*cont.*)

Z	Atom	J_0, eV	J_1, eV	J_2, eV	J_3, eV
34	Se	9.752	21.16	30.82	42.95
35	Br	11.814	21.81	35.90	47.3
36	Kr	14.000	24.360	36.95	52.5
37	Rb	4.177	27.290	39.2	52.6
38	Sr	5.694	11.030	42.88	56.28
39	Y	6.217	12.24	20.525	60.61
40	Zr	6.837	13.13	23.1	34.419
41	Nb	6.88	14.32	25.0	37.7
42	Mo	7.099	16.16	27.2	46.4
43	Tc	7.28	15.26	29.5	—
44	Ru	7.366	16.76	28.5	—
45	Rh	7.46	18.08	31.1	—
46	Pd	8.336	19.43	32.9	—
47	Ag	7.576	21.49	34.8	—
48	Cd	8.994	16.908	37.47	—
49	In	5.786	18.87	28.0	57.0
50	Sn	7.344	14.632	30.50	40.74
51	Sb	8.609	16.53	25.32	44.16
52	Te	9.010	18.6	27.96	37.42
53	I	10.451	19.131	33.0	—
54	Xe	12.130	20.98	31.0	45
55	Cs	3.894	23.15	33.4	46
56	Ba	5.212	10.004	35.8	47
57	La	5.577	11.1	19.18	49.9
58	Ce	5.539	10.8	20.20	39.76
59	Pr	5.47	10.6	21.62	38.98
60	Nd	5.525	10.7	22.1	40.4
61	Pm	5.58	10.9	22.3	41.0
62	Sm	5.644	11.1	23.4	41.4
63	Eu	5.670	11.24	24.9	42.7
64	Gd	6.150	12.1	20.6	44.0
65	Tb	5.864	11.5	21.9	39.4
66	Dy	5.939	11.7	22.8	41.4
67	Ho	6.022	11.8	22.8	42.5
68	Er	6.108	11.9	22.7	42.7

Table 7.1. *(cont.)*

Z	Atom	J_0, eV	J_1, eV	J_2, eV	J_3, eV
69	Tm	6.184	12.1	23.7	42.7
70	Yb	6.254	12.18	25.05	43.6
71	Lu	5.426	13.9	20.96	45.25
72	Hf	6.8	14.9	23.3	33.4
73	Ta	7.89	—	—	—
74	W	7.98	—	—	—
75	Re	7.88	—	—	—
76	Os	8.73	—	—	—
77	Ir	9.05	—	—	—
78	Pt	8.96	18.56	—	—
79	Au	9.226	20.5	34	43
80	Hg	10.438	18.76	34.2	46
81	Tl	6.108	20.43	29.85	—
82	Pb	7.417	15.033	31.94	42.33
83	Bi	7.286	16.7	25.56	45.3
84	Po	8.417	—	—	—
85	At	9.0	—	—	—
86	Rn	10.75	—	—	—
87	Fr	4.0	—	—	—
88	Ra	5.278	10.15	—	—
89	Ac	5.2	11.75	20	—
90	Th	6.1	11.9	18.3	28.7
91	Pa	6.0	—	—	—
92	U	6.194	11.9	20	37
93	Np	6.266	—	—	—
94	Pu	6.06	—	—	—
95	Am	6.0	—	—	—
96	Cm	6.02	—	—	—
97	Bk	6.23	—	—	—
98	Cf	6.30	—	—	—
99	Es	6.42	—	—	—
100	Fm	6.5	—	—	—
101	Md	6.6	—	—	—

Table 7.2. Parameters of the metastable state ${}^1S_0$ of the atom and ion with the electron shell $1s^2 2s^2 2p^2$.

Atom, ion	Excitation energy, eV	Radiative lifetime, s
$C(^1S)$	2.684	2.0
$N^+(^1S)$	4.053	0.92

Table 7.3. Parameters of the upper state of the fine structure of the iodine atom and xenon ion with the ground state p^5 of the electron shell.

Atom, ion	Excitation energy, eV	λ, μ	τ, s
$I(^2P_{1/2})$	0.94	1.315	0.13
$Xe^+(^2P_{1/2})$	1.31	0.949	0.05

7.2 Multicharged Ions

Multicharged ions are characterized by a large difference between the nucleus charge Z and the total number of electrons. This simplifies the description of the system. Let us represent the Hamiltonian of electrons in the form

$$\hat{H} = \sum_i \left(-\frac{1}{2}\Delta_i - \frac{Z}{r_i} \right) + V_{ee} + V_{\rm rel}, \qquad (7.1)$$

where the two terms in parentheses include the kinetic energy of electrons and their interaction with the Coulomb field of the nucleus, the next term, $V_{ee} = \sum_{i,j} \left|\mathbf{r}_i - \mathbf{r}_j\right|^{-1}$, is the interaction potential between electrons, and the last term $V_{\rm rel}$ includes all the relativistic interactions. The analysis of the multicharged heliumlike ions (Chapter 3) shows that the electron binding energy varies as Z^2 at large nucleus charges Z, the exchange interaction potential which corresponds to the operator V_{ee} in formula (7.1) varies as Z, and the spin-orbit interactions, which are characterized by the operator $V_{\rm rel}$, vary as Z^4. Hence, the relative role of the exchange and relativistic interactions can change with an increase in Z. Nevertheless, one can assume the terms V_{ee} and $V_{\rm rel}$ of formula (7.1) to be small compared to the main term of this formula. This is the basis of the perturbation theory which is valid for multicharged ions.

Let us construct the perturbation theory for multicharged ions. In the zeroth approximation we neglect the terms V_{ee} and $V_{\rm rel}$, and the Pauli exclusion principle must be taken into account. Thus the obtained form of the Hamiltonian leads to one-electron approximation, i.e., the eigenfunction of the Hamiltonian in the zeroth approximation is a product of one-electron wave functions which correspond to the location of each electron in the Coulomb field of the charge Z. The Pauli exclusion principle requires the distribution of electrons over the states of the hydrogenlike ion, so that we obtain the shell atom scheme. Exchange and relativistic interactions lead to a relatively small shift of electron levels. Note that in contrast to light atoms, where the electron shell concept is a convenient model for the description of atoms, in the case of multicharged ions this concept corresponds to the correct solution of the problem on the basis of the perturbation theory.

In the first order of the perturbation theory we introduce a real self-consistent field which accounts for the exchange and relativistic interactions. Let us divide this additional interaction into long- and short-range parts. A long-range self-consistent field creates a shielding charge which partially screens the Coulomb field of a nucleus charge. A short-range interaction leads to a shift of electron levels. As a result, we have in the first order of the perturbation theory that a multicharged ion can be considered as a sum of independent electrons which occupy positions in the corresponding shells of the hydrogenlike ion. Then the total electron energy is $E = \sum_j \varepsilon_j$, and the energy of each electron corresponds to its energy in the hydrogenlike ion with the above corrections:

$$\varepsilon_j = -\frac{(Z - \sigma_{nj})^2}{2(n - \delta_{nj})^2}, \tag{7.2}$$

where σ_{nj} is the shielding charge due to internal electrons, and δ_{nj} is the quantum defect of the electron state which takes into account a short-range interaction involving this electron. In this representation we take as electron quantum numbers the principal quantum number of the electron n and the total electron momenta j. This means that relativistic interactions are stronger than exchange interactions. This is valid for internal electrons at large Z, while for intermediate Z the electron quantum numbers are n and l—the orbital electron momentum.

The parameters σ_{nj} and δ_{nj} depend on the filling of subsequent shells. Let us demonstrate this in the example of a $1s$-electron. Then if a multicharged ion contains one electron we have, according to formula (2.41), $\sigma = 0$. For the heliumlike ion formula (3.8) gives $\sigma = 0.75$. From the results of Problem 5.3 it follows that if subsequent electron shells are filled, this parameter is equal to $\sigma = 0.75 + 1.794Z^{1/3}$. We collect, in Table 7.4, values of the parameters of formula (7.2) in the case when the relevant valent electron is the last one, so that the electron energy in formula (7.2) corresponds to the ionization potential of the relevant multicharged ion which is

$$J = \frac{(Z - \sigma)^2}{2(n - \delta)^2}. \tag{7.3}$$

The parameters of Table 7.4 are evaluated on the basis of the ionization potentials of the corresponding multicharged ions. As follows from the data of Table 7.4, the parameter δ is small which allows us to use the hydrogenlike positions of levels; the parameter $\sigma < n$, where n is the number of electrons of the multicharged ion.

Let us study the character of change of the ionization potential of a multicharged ion as we remove its valent electrons. If we remove electrons of the same shell, the ionization potential of the multicharged ion varies slowly. Transition to a new shell of valent electrons leads to a strong jump of the ionization potential. This reflects the nature of the electron shell model which is based on a strong exchange interaction between electrons resulting from the prohibition to locate two electrons

Table 7.4. Parameters of formula (7.3) for the ionization potential of multicharged ions.

Shell	Atom of this shell	σ	δ
$1s$	H	0	0
$1s^2$	He	0.75	0.01
$2s$	Li	1.76	0.03
$2s^2$	Be	2.20	0.02
$2p$	B	3.23	0.02
$2p^2$	C	2.87	0.01
$2p^3$	N	4.52	0.01
$2p^4$	O	5.46	0.01
$2p^5$	F	6.16	0.01
$2p^6$	Ne	6.97	0.02
$3s$	Na	8.50	0.08
$3s^2$	Mg	9.15	0.08
$3p$	Al	10.36	0.08
$3p^2$	Si	10.96	0.08
$3p^3$	P	11.63	0.08
$3p^4$	S	12.56	0.09
$3p^5$	Cl	13.32	0.11
$3p^6$	Ar	14.23	0.16

in a state with the same quantum numbers. Below we demonstrate the fact of a jump of the ionization potential for the transition of multicharged ions from the Ne-electron shell to the Na-electron shell. Taking the value of the quantum defect to be zero for the related states, we obtain the expression for the jump of the ionization potential of a multicharged ion as a result of the transition from $3s$- to $2p$-shells:

$$\frac{J_{2p}}{J_{3s}} = \frac{9}{4}\frac{(Z-\sigma_{2p})^2}{(Z-\sigma_{3s})^2} = \frac{9}{4}\cdot\left[1+\frac{2(\sigma_{3s}-\sigma_{2p})}{(Z-\sigma_{3s})}\right] = \frac{9}{4}\left(1+\frac{3.7}{Z-8.5}\right). \quad (7.4)$$

Here we use formula (7.3) for the ionization potential, J_{2p}, J_{3s} are the ionization potentials of multicharged ions if these have the valent electron shells $2p^6$ and $3s$, respectively, and the values of the shielding factors σ_{2p}, σ_{3s} are taken from Table 7.4. Formula (7.4) gives the value of the ionization potential jump 2.94 for the Ar ion and 2.44 for the Fe ion.

7.3 Negative Ions

Negative ions correspond to a bound state of the electrons and atoms. Because of the short-range character of this interaction, only several stable states of a negative ion exist. Since this bond is determined by the exchange interaction of electrons and atoms, atoms with closed electron shells usually do not have stable negative ions. In addition, the two-charged negative ions of atoms do not exist in the stable state. Figure 7.3 contains the contemporary values of the electron binding energy in negative ions. This value is called the electron affinity of atoms and is denoted by EA. The maximum binding energy corresponds to the negative ions with a filled valent electron shell. The EA of the halogen atoms exceeds 3 eV, but the maximum EA of the atoms which correspond to the chlorine atom, and is equal to 3.61 eV, is smaller than the lowest ionization atom potential which corresponds to the cesium atom and is equal to 3.89 eV.

The existence of the excited states of negative ions is of interest. Because of the short-range interaction between the electron and atom in the negative ions, a finite number of their bound states is possible. Evidently, there are preferable conditions for the excited states of negative ions exist for elements of the four groups of the periodical table of elements. For example, from the data of Fig. 7.3 it follows that carbon has one excited state of the negative ion and silicon has two excited states. In addition, aluminum, scandium, and ytterbium have excited states of negative ions.

The electron affinity of atoms is determined by various methods and during almost a century of study of the negative ions these methods varied. Now the best method for this goal is based on the photodetachment of negative ions by laser radiation. The accuracy of this method exceeds, by one–three orders of magnitude, the accuracy of other methods and, therefore, almost all the data of Fig. 7.3 are obtained by this method. This method is used in different modifications. In the first modification the photon energy varies and the threshold of the photodetachment process is determined. This corresponds to the electron binding energy in a negative ion. In another modification of this method, the photon energy is fixed and the spectrum of the released electrons is measured, so that the electron affinity of an atom is the difference between the photon energy and the energy of the released electron. In both versions, negative ions are located in an atom beam, and the main problem of this method is to approximate precisely the threshold dependencies of the process that determines the accuracy of the obtained data.

7.4 The Electron Wave Function of Negative Ion

Because the electron binding energy in a negative ion is remarkably lower than that in the corresponding atom, the size of a negative ion is large compared to the atom size. But the bond is determined by the exchange interaction of electrons, i.e., by an electron region where valent atomic electrons are located. Hence, in

PERIODICAL SYSTEM OF ELEMENTS - NEGATIVE IONS

Period \ Group	I	II	III	IV	V	VI	VII	VIII		
1	$_1$H Hydrogen $1s^2\,^1S_0$ 0.75420							$_2$He Helium $2s\,^2S_{1/2}$ absent		
2	$_3$Li Lithium $2s^2\,^1S_0$ 0.6180	$_4$Be Berillium $2p\,^2P_{1/2}$ absent	$_5$B Boron $2p^2\,^3P_0$ 0.27972	$_6$C Carbon $2p^3$ $^4S_{3/2}$ 1.2621 $^2D_{3/2}$ 0.033	$_7$N Nitrogen $2p^4\,^3P_0$ absent	$_8$O Oxygen $2p^5\,^2P_{3/2}$ 1.4611103	$_9$F Fluorine $2p^6\,^1S_0$ 3.401190	$_{10}$Ne Neon $2p^63s\,^2S_{1/2}$ absent		
3	$_{11}$Na Sodium $3s^2\,^1S_0$ 0.54793	$_{12}$Mg Magnesium $3p\,^2P_{1/2}$ absent	$_{13}$Al Aluminium $3p^2$ $^3P_0$ 0.4328 $^1D_2$ 0.11	$_{14}$Si Silicon $3p^3$ $^4S_{3/2}$ 1.385 $^2D_{3/2}$ 0.5272 $^2D_{5/2}$ 0.5255 $^2P_{1/2}$ 0.029	$_{15}$P Phosphorus $3p^4\,^3P_0$ 0.7465	$_{16}$S Sulfur $3p^5\,^2P_{3/2}$ 2.077104	$_{17}$Cl Chlorine $3p^6\,^1S_0$ 3.61269	$_{18}$Ar Argon $3p^64s\,^2S_{1/2}$ absent		
4	$_{19}$K Potassium $4s^2\,^1S_0$ 0.5015	$_{20}$Ca Calsium $4p\,^2P_{1/2}$ 0.0245	$_{21}$Sc Scandium $3d4s^24p$ $^3F_2$ 0.19 $^3D_1$ 0.04	$_{22}$Ti Titanium $3d^34s^2\,^4F_{3/2}$ 0.08	$_{23}$V Vanadium $4d^44s^2\,^5D_0$ 0.52	$_{24}$Cr Chromium $3d^54s^2\,^6S_{5/2}$ 0.66	$_{25}$Mn Manganese $3d^64s^2\,^5D_4$ absent	$_{26}$Fe Iron $3d^74s^2\,^4F_{9/2}$ 0.151	$_{27}$Co Cobalt $3d^84s^2\,^3F_4$ 0.662	$_{28}$Ni Nickel $3d^94s^2\,^2D_{5/2}$ 1.16
	$_{29}$Cu Copper $3d^{10}4s^2\,^1S_0$ 1.235	$_{30}$Zn Zinc $4p\,^2P_{1/2}$ absent	$_{31}$Ga Gallium $4p^2\,^3P_0$ 0.43	$_{32}$Ge Germanium $4p^3$ $^4S_{3/2}$ 1.2327 $^2D_{3/2}$ 0.4014 $^2D_{5/2}$ 0.3773	$_{33}$As Arsenic $4p^4\,^3P_0$ 0.81	$_{34}$Se Selenium $4p^5\,^2P_{3/2}$ 2.02067	$_{37}$Br Bromine $4p^6\,^1S_0$ 3.363590	$_{36}$Kr Krypton $4p^65s\,^2S_{1/2}$ absent		
5	$_{37}$Rb Rubidium $5s^2\,^1S_0$ 0.48592	$_{38}$Sr Strontium $5p\,^2P_{1/2}$ 0.048	$_{39}$Y Yttrium $4d5s^25p$ $^3F_2$ 0.31 $^3D_1$ 0.16	$_{40}$Zr Zirconium $4d^35s^2\,^4F_{3/2}$ 0.43 52	$_{41}$Nb Niobium $5d^45s^2\,^5D_0$ 0.89	$_{42}$Mo Molibdenum $4d^55s^2\,^6S_{5/2}$ 0.748	$_{43}$Tc Technetium $4d^65s^2\,^5D_4$ 0.6	$_{44}$Ru Ruthenium $4d^75s^2\,^4F_{9/2}$ 1.0	$_{45}$Rh Rhodium $4d^85s^2\,^3F_4$ 1.137	$_{46}$Pd Palladium $4d^95s^2\,^2D_{5/2}$ 0.562
	$_{47}$Ag Silver $4d^{10}5s^2\,^1S_0$ 1.302	$_{48}$Cd Cadmium $5p\,^2P_{1/2}$ absent	$_{49}$In Indium $5p^2\,^3P_0$ 0.40	$_{50}$Sn Tin $5p^3$ $^4S_{3/2}$ 1.1121 $^2D_{3/2}$ 0.3976 $^2D_{5/2}$ 0.3046	$_{51}$Sb Antimony $5p^4$ $^3P_2$ 1.047 $^3P_1$ 0.714 $^3P_0$ 0.700 $^1D_2$ 0.131	$_{52}$Te Tellurium $5p^5\,^2P_{3/2}$ 1.9708	$_{53}$I Iodine $5p^6\,^1S_0$ 3.05904	$_{54}$Xe Xenon $5p^66s\,^2S_{1/2}$ absent		
6	$_{55}$Cs Cesium $6s^2\,^1S_0$ 0.47163	$_{56}$Ba Barium $6p\,^2P_{1/2}$ 0.15	$_{57}$La Lanthanum $5d^26s^2\,^3F_2$ 0.5	$_{58}$Hf Hafnium $5d^36s^2\,^4F_{3/2}$ absent	$_{73}$Ta Tantalum $5d^46s^2\,^5D_0$ 0.32	$_{74}$W Tungsten $5d^56s^2\,^6S_{5/2}$ 0.815	$_{75}$Re Rhenium $5d^66s^2\,^5D_4$ 0.2	$_{76}$Os Osmium $5d^76s^2\,^4F_{9/2}$ 1.1	$_{77}$Ir Iridium $5d^86s^2\,^3F_4$ 1.5638	$_{78}$Pt Platinum $5d^96s^2\,^2D_{5/2}$ 2.128
	$_{79}$Au Gold $5d^{10}6s^2\,^1S_0$ 2.30863	$_{80}$Hg Mercury $6p\,^2P_{1/2}$ absent	$_{81}$Tl Thallium $6p^2\,^3P_0$ 0.4	$_{82}$Pb Lead $6p^3\,^4S_{3/2}$ 0.364	$_{83}$Bi Bismuth $6p^4\,^3P_0$ 0.95	$_{84}$Po Polonium $6p^5\,^2P_{3/2}$ 1.9	$_{85}$At Astatine $6p^6\,^1S_0$ 2.8	$_{86}$Rn Radon $6p^67s\,^2S_{1/2}$ absent		
7	$_{87}$Fr Francium $7s^2\,^1S_0$ 0.46									

Shell of valent electrons
Symbol
Atomic number
Element
Electron term
Electron binding energy
$3d^74s^2\,^4F_{9/2}$
26 Fe
Iron
0.151

Figure 7.1. The electron affinities of atoms for some elements of the periodical system. An error may be in the last letter.

the main region of location of a weakly bound electron of the negative ion one can neglect its interaction with an atomic field. This gives a simple solution for the electron wave function because, in the main region of electron location, the interaction potential between the electron and atom is equal to zero. Hence the radial wave function of the weakly bound electron $R(r)$ satisfies the Schrödinger equation

$$\frac{d^2}{dr^2}[rR(r)] - \left[\gamma^2 + \frac{l(l+1)}{r^2}\right] \cdot [rR(r)] = 0, \tag{7.5}$$

where the electron binding energy in the negative ion is equal to $EA = \hbar^2\gamma^2/(2m)$ and l is the electron orbital moment. The solution of equation (7.5) is expressed through the Macdonald function

$$R_l(r) = \frac{C}{\sqrt{r}} K_{l+1/2}(\gamma r).$$

In particular, for valent s- and p-electrons this formula has the form

$$R_s(r) = \frac{A}{r} e^{-\gamma r}; \qquad R_p(r) = \frac{A}{r} e^{-\gamma r} \cdot \left(1 + \frac{1}{\gamma r}\right). \tag{7.6}$$

Note that because the electron binding energy is determined by an internal electron region, the parameter γ corresponds to the boundary condition and is not the eigenvalue of equation (7.5).

The wave function (7.5) is convenient for the analysis of such parameters of a negative ion which are determined by its external region. Using the wave function in such a form, we extract a weakly bound electron from the valent electrons. Its wave function differs from that of other valent electrons, but the position of a weakly bound electron can occupy any valent electron. This is taken into account by the symmetry condition (3.1). Let us demonstrate the character of the extraction of a weakly bound electron in the example of a negative ion with the valent electron shell ns^2. In the related representation the wave function has the form

$$\Psi(r_1, r_2) = \varphi(r_<)\psi(r_>),$$

where $r_< = \min(r_1, r_2)$, $\dot{r}_> = \max(r_1, r_2)$, i.e., the weakly bound electron is assumed to be the valent electron which has the largest distance from the nucleus. The normalization condition for a weakly bound electron has the form

$$\int \psi^2(r)\, d\mathbf{r} = 1.$$

Note that in spite of the smallness of an atom region compared to the negative ion size, this introduces a contribution into the wave function normalization. One can demonstrate this in the example of the hydrogen negative ion. Let us consider a model of the negative hydrogen ion in which the wave function of a weakly bound electron is equal to zero for $r < r_o$, and is given by expression (7.6) for $r \geq r_o$. Then, from the normalization condition, we have

$$A = \sqrt{2\gamma} \cdot e^{\gamma r_o}. \tag{7.7}$$

Table 7.5. Parameters of a weakly bound electron in negative ions.

Ion, state	nl	γ	B
$H^-(^1S)$	10	0.236	1.16
$Li^-(^1S)$	20	0.212	1.9
$C^-(^4S)$	21	0.305	0.74
$O^-(^2P)$	21	0.328	0.52
$F^-(^1S)$	21	0.500	0.70
$Na^-(^1S)$	30	0.201	1.9
$Al^-(^3P)$	31	0.192	0.8
$Si^-(^4S)$	31	0.320	0.9
$P^-(^3P)$	31	0.266	0.75
$S^-(^2P)$	31	0.391	1.05
$Cl^-(^1S)$	31	0.516	1.2
$K^-(^1S)$	40	0.192	2.0
$Cu^-(^1S)$	40	0.301	2.4
$Br^-(^1S)$	41	0.498	1.4
$Rb^-(^1S)$	50	0.189	2.5
$Ag^-(^1S)$	50	0.309	2.5
$I^-(^1S)$	51	0.475	2.8

According to the above consideration, the parameter γr_o must be small, and for the hydrogen negative ion we have $\gamma r_o = 0.14$. But the factor is the expression of the electron density $B^2 = e^{2\gamma r_o} = 1.3$, and we obtain a remarkable error if we take $\gamma r_o = 0$. Thus, the region inside the atom introduces a contribution into the wave function normalization. Based on this analysis, one can use the following simple and realistic model for a weakly bounded electron of the negative ion. The electron wave function is taken in the asymptotic form

$$\psi(r) = \frac{A}{r\sqrt{4\pi}} \cdot e^{-\gamma r} = \frac{B}{r}\sqrt{\frac{\gamma}{2\pi}} \cdot e^{-\gamma r}, \qquad \gamma r \gg 1, \tag{7.8}$$

and the coefficient A is determined by the electron behavior in the region occupied by other valent electrons. If we take the dependence (7.8) at all r we obtain, from the normalization condition, $B = 1$. Table 7.5 lists the parameters of a weakly bound electron for some negative ions.

Let us find the asymptotic coefficient by a standard method (see Problem 3.4), comparing the wave function of a valent electron of the negative ion with its asymptotic expression (7.8). We use the Chandrasekhar wave function of electrons as the wave function of the hydrogen negative ion that has the form

$$\Psi(r_1, r_2) = C \cdot \left[\exp(-\alpha r_1 - \beta r_2) - \exp(-\beta r_1 - \alpha r_2)\right] (1 + c|\mathbf{r}_1 - \mathbf{r}_2|), \tag{7.9}$$

and the variation principle gives the following parameters of this wave function: $\alpha = 1.039$, $\beta = 0.283$, $c = 0$, for the two-parameter form of the wave function, and $\alpha = 1.075$, $\beta = 0.478$, $c = 0.312$, if we use three parameters. The electron affinity of the hydrogen atom is 0.367 eV in the first case and 0.705 eV in the second case instead of the accurate value 0.754 eV.

Repeating the operations of Problem 3.4 we obtain, for the asymptotic coefficient square,

$$A_1^2(r) = r^2(0.0695e^{-0.096r} + 0.540e^{-0.852r} + 3.44e^{-1.608r}) \quad (7.10a)$$

in the first case, and

$$\begin{aligned} A_2^2(r) = r^2 &\left(0.102 + 0.0411r + 0.0064r^2\right) e^{-0.486r} \\ &+ \left(0.660 + 0.218r + 0.0341r^2\right) e^{-1.063r} \\ &+ \left(2.20 + 0.468r + 0.0729r^2\right) e^{-1.68r} \end{aligned} \quad (7.10b)$$

in the second case. Figure 7.2 contains the asymptotic coefficients obtained on the basis of these formulas. As a result, we have $A_1 = 1.07 \pm 0.03$ on the basis of formula (7.10a) and $A_2 = 1.19 \pm 0.01$ if we use formula (7.10b), and r ranges in both cases from 2 to 5. Here only the statistical error is indicated, and the real

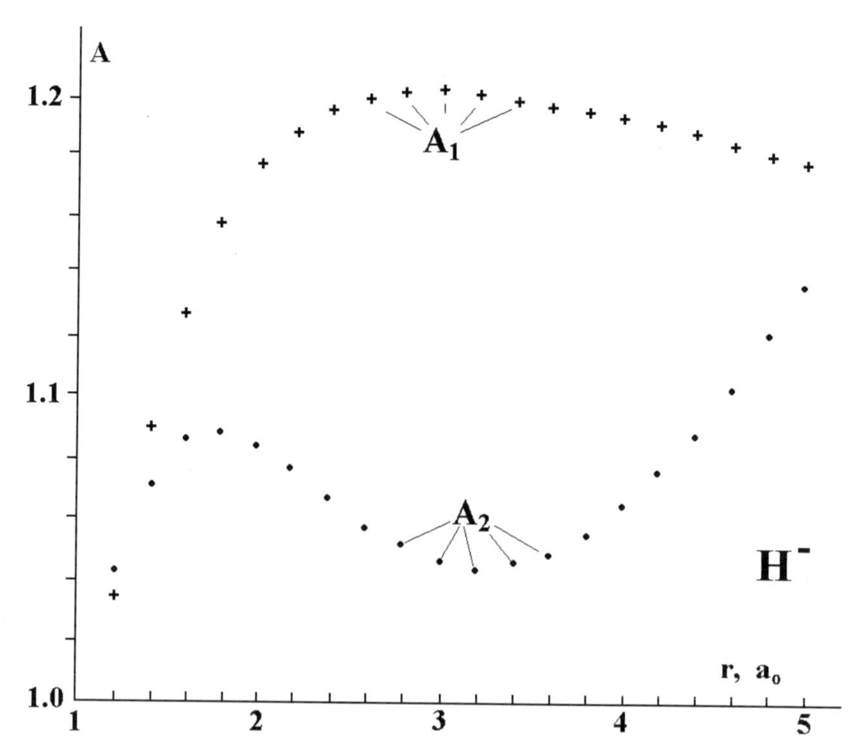

Figure 7.2. Determination of the asymptotic coefficient A of the electron wave function for the hydrogen negative ion on the basis of formulas (7.10).

accuracy of this asymptotic coefficient is worse. Nevertheless, the accuracy of the asymptotic coefficient for a negative ion is usually better than that for an atom, because of the absence of the interaction of a weakly bound electron with its core outside the atom. Both formulas (7.10) give, for the asymptotic coefficient, $A = 1.13 \pm 0.06$ in the range $r = 2$–5. This corresponds to $B = A/\sqrt{2\gamma} = 1.64 \pm 0.08$ or, if we reduce this value to one valent electron, we obtain $B = 1.16 \pm 0.06$. Within the framework of the model of a finite radius (7.7) we have $r_o = 2.1 \pm 0.2$.

7.5 Photodetachment of Negative Ions and Photoattachment of Electrons to Atoms

Let us consider the process of photodetachment of the negative ion which proceeds according to the scheme

$$A^- + \hbar\omega \to A + e. \tag{7.11}$$

The characteristic of this elementary process is the cross section whose expression is obtained in Problem 7.1 and has the form

$$\sigma_{\text{det}} = \frac{q\omega}{6\pi c} \cdot \int |\langle 0|\mathbf{D}|\mathbf{q}\rangle|^2 d\Theta_{\mathbf{q}}, \tag{7.12}$$

where $d\Theta_{\mathbf{q}}$ is the element of the solid angle in which the released electron is moving, $\mathbf{D}$ is the dipole moment operator so that the wave function of the initial state is denoted by 0, and the wave function of the released electron has the form of the plane wave $e^{i\mathbf{qr}}$ far from the atom.

Formula (7.12) gives a certain selection rule for this process. In addition, the symmetry of the wave function of the valent electrons leads to the relation between the intensities of the process with the formation of the atom in different states. These values are given in Tables 7.6 and 7.7 for the valent p-shell and for the case when the photon energy remarkably exceeds the energy difference between the related atom levels. The normalization of the process intensity is such that the total intensity is equal to the number of valent electrons. As is seen, the intensities of the separate multiplets depend on the coupling scheme. Hence, the measurement of the relative intensities of multiplets in the photodetachment of negative ions allows one to ascertain the scheme of coupling of valent electrons. This statement is also valid for the process of atom photoionization.

Let us consider the process of the photoattachment of an electron to an atom which is the reversible process with respect to the process of the photodetachment of negative ions (7.11) and proceeds according to the scheme

$$e + A \to A^- + \hbar\omega. \tag{7.13}$$

The cross sections of processes (7.11) and (7.13) are connected by the principle of detailed balance. Below we obtain the relation between these cross sections by using the basis of the principle of detailed balance. Let us take some spatial volume where the electrons, photons, atoms, and negative ions are found in equilibrium, so

Table 7.6. The relative intensity of the final states of the formed atom as a result of negative ion photodetachment within the framework of LS-coupling.

Initial shell and term	Final state	Intensity
$p^2, {}^3P_0$	${}^2P_{1/2}$	1.333
$p^2, {}^3P_0$	${}^2P_{3/2}$	0.667
$p^3, {}^4S_{3/2}$	${}^3P_0$	0.333
$p^3, {}^4S_{3/2}$	${}^3P_1$	1.000
$p^3, {}^4S_{3/2}$	${}^3P_2$	1.667
$p^4, {}^3P_2$	${}^4S_{3/2}$	1.333
$p^4, {}^3P_2$	${}^2D_{3/2}$	0.167
$p^4, {}^3P_2$	${}^2D_{5/2}$	1.500
$p^4, {}^3P_2$	${}^2P_{1/2}$	0.167
$p^4, {}^3P_2$	${}^2P_{3/2}$	0.833
$p^5, {}^2P_{3/2}$	${}^3P_0$	0.167
$p^5, {}^2P_{3/2}$	${}^3P_1$	0.750
$p^5, {}^2P_{3/2}$	${}^3P_2$	2.083
$p^5, {}^2P_{3/2}$	${}^1D_2$	1.667
$p^5, {}^2P_{3/2}$	${}^1S_0$	0.333

Table 7.7. The relative intensity of the final states of the formed atom as a result of negative ion photodetachment within the framework of jj-coupling.

Initial shell	J_i	Final shell	J_f	Intensity
$(1/2)^2$	0	$(1/2)^1$	1/2	2.000
$(1/2)^2(3/2)^1$	3/2	$(1/2)^2$	0	1.00
$(1/2)^2(3/2)^1$	3/2	$(1/2)^1(3/2)^1$	1	0.750
$(1/2)^2(3/2)^1$	3/2	$(1/2)^1(3/2)^1$	2	0.250
$(1/2)^2(3/2)^2$	2	$(1/2)^2(3/2)^1$	3/2	2.00
$(1/2)^2(3/2)^2$	2	$(1/2)^1(3/2)^2$	3/2	0.800
$(1/2)^2(3/2)^2$	2	$(1/2)^1(3/2)^2$	5/2	0.200
$(1/2)^2(3/2)^3$	3/2	$(1/2)^2(3/2)^2$	0	0.500
$(1/2)^2(3/2)^3$	3/2	$(1/2)^1(3/2)^3$	2	2.500
$(1/2)^2(3/2)^3$	3/2	$(1/2)^1(3/2)^3$	1	0.750
$(1/2)^2(3/2)^3$	3/2	$(1/2)^1(3/2)^3$	2	1.250
$(1/2)^2(3/2)^4$	0	$(1/2)^2(3/2)^3$	3/2	4.000
$(1/2)^2(3/2)^4$	0	$(1/2)^1(3/2)^4$	1/2	2.000

that the number of decays and formations of negative ions is identical. If we take, in this volume, one atom and one electron at the beginning, this relation obtains the form

$$c\frac{2\,d\mathbf{k}\,d\mathbf{r}}{(2\pi)^3}g_i\sigma_{\text{det}} = v\frac{2\,d\mathbf{q}\,d\mathbf{r}}{(2\pi)^3}g_a\sigma_{\text{at}}, \tag{7.14}$$

where σ_{det} is the cross section of process (7.11), σ_{at} is the cross section of process (7.13), $\mathbf{k}$, $\mathbf{q}$ are the wave vectors of the photon and electron, respectively, v is the electron velocity, g_i, g_a are the statistical weights of the ion and atom, $d\mathbf{r}$ is the value of an extracted volume, and the factor 2 in the left-hand side of the relation accounts for two photon polarizations.

Using the dispersion relation for photons $\omega = ck$ and the law of energy conservation $\hbar\omega = EA + \hbar^2q^2/(2m)$ (we return to the usual units), we have the following relation from formula (7.14),

$$\sigma_{\text{at}} = \frac{g_i}{g_a}\cdot\frac{k^2}{q^2}\cdot\sigma_{\text{det}}. \tag{7.15}$$

This formula connects the cross sections of the direct and reversible processes. Make estimates on the basis of this formula. Taking the electron energy of the order of EA, we obtain $k^2/q^2 \sim \hbar\omega/(mc^2)$. Because $mc^2 = 500\,\text{keV}$ and $\hbar\omega \sim \text{EA} \sim 1\,\text{eV}$, the cross section of the photoattachment of the electron to atom is five-six orders of magnitude smaller than the cross section of the photodetachment of negative ions.

Let us consider the case when the valent shell of the negative ion is s^2. Then, as follows from Problem 7.2, the cross section of the photodetachment of the negative ion is

$$\sigma_{\text{det}} = \frac{8\pi B^2\gamma v_e^3}{3\omega^3 c}\cdot\frac{e^2}{\hbar},$$

and the cross section of the electron attachment to the atom is equal to ($g_a = 1$, $g_i = 1$):

$$\sigma_{\text{det}} = \frac{4\pi B^2}{3c^3}\cdot\frac{e^2\hbar}{m^2}\cdot\frac{\gamma v}{\omega}. \tag{7.16}$$

This cross section has the maximum at $q = \gamma$, so that $\gamma v/\omega = 1$ at the maximum, and the maximum attachment cross section is

$$\sigma_{\text{det}} = \frac{4\pi B^2}{3c^3}\cdot\left(\frac{e^2}{\hbar c}\right)^3\cdot a_o^2. \tag{7.17}$$

In particular, for electron attachment to the hydrogen atom the maximum cross section is equal to $6\cdot 10^{-23}\,\text{cm}^2$, that is, six orders of magnitude lower than the maximum cross section of the photodetachment of the hydrogen negative ion.

7.6 Bremsstrahlung from the Scattering of Electrons and Atoms

The interaction between an electron and atom, which is responsible for the formation of a negative ion, also leads to radiation resulting from electron–atom scattering. If the electron remains free after radiation, the radiation process is bremsstrahlung and proceeds according to the scheme

$$e(\mathbf{q}) + A \longrightarrow e(\mathbf{q}') + A + \hbar\omega, \tag{7.18}$$

where $\mathbf{q}$, $\mathbf{q}'$ are the electron wave vectors before and after scattering. The energy conservation law gives for this process (in the usual units)

$$\frac{\hbar^2 q^2}{2m} = \frac{\hbar^2 (q')^2}{2m} + \hbar\omega. \tag{7.19}$$

Our goal is to determine the cross section of process (7.18).

For simplification of operation with the wave functions of a continuous spectrum, we use the standard method (see Chapter 1) by placing the system of colliding electrons and atoms into a cell of large volume Ω. Then the radiation rate is equal to, according to formulas (1.26),

$$dw_{\rm br} = \frac{4\omega^3}{3\hbar c^3} \cdot |\langle \mathbf{q}|\mathbf{D}|\mathbf{q}'\rangle|^2 \frac{\Omega\, d\mathbf{q}'}{(2\pi)^3}. \tag{7.20}$$

We assume the atom to be structureless, so that the electron spin state is conserved as a result of scattering, and the electron statistical weight is unity. Note that the electron wave function far from the atom is $e^{i\mathbf{q}\mathbf{r}}/\sqrt{\Omega}$, because it is normalized to unity, and the electron flux is $\hbar\mathbf{q}/(\Omega m)$. Hence, the cross section of bremsstrahlung is

$$\frac{d\sigma_{br}}{d\omega} = \frac{dw_{br}}{d\omega} \cdot \frac{m\Omega}{\hbar q} = \frac{m^2 q' \Omega^2}{6\pi^3 \hbar^3 c^3 q} \cdot \int |\langle \mathbf{q}|\mathbf{D}|\mathbf{q}'\rangle|^2 d\Theta_{\mathbf{q}'}, \tag{7.21}$$

where $d\Theta_{\mathbf{q}'}$ is the element of the solid angle in which the electron is scattered and we use the law of energy conservation $q'\,dq' = -m d\omega/\hbar$.

Now let us transfer to the electron wave functions which are plane waves far from the atom, i.e.,

$$\psi_{\mathbf{q}} \longrightarrow e^{i\mathbf{q}\mathbf{r}}, \quad \text{if } r \longrightarrow \infty.$$

Transition to such wave functions leads to multiplication of the matrix element $\langle \mathbf{q}|\mathbf{D}|\mathbf{q}'\rangle$ by $1/\Omega$. The dipole moment operator is $\mathbf{D} = e\mathbf{r}$, where $\mathbf{r}$ is the electron radius-vector. Thus, we have, for the bremsstrahlung cross section,

$$\frac{d\sigma_{\rm br}}{d\omega} = \frac{m^2 e^2 q'}{6\pi^3 \hbar^3 c^3 q} \cdot \int \left|\mathbf{r}_{\mathbf{q}\mathbf{q}'}\right|^2 d\Theta_{\mathbf{q}'}, \quad \text{where} \quad \mathbf{r}_{\mathbf{q}\mathbf{q}'} = \int \psi_{\mathbf{q}}(\mathbf{r})\mathbf{r}\psi_{\mathbf{q}'}(\mathbf{r})\, d\mathbf{r}. \tag{7.22}$$

In the case of the spherical symmetry of a scattered atom, it is convenient to expand the integral (7.22) over spherical harmonics. Repeating operations for the deduction of formula (2.40) we get, in this case,

$$\frac{d\sigma_{\rm br}}{d\omega} = \frac{32m\omega^3 q'}{3\hbar c^3 a_o q} \cdot \sum_l (l+1)\left(K_{l+1,l}^2 + K_{l,l+1}^2\right),$$

$$\text{where} \quad K_{ll'} = \int u_l(q,r)\, u_{l'}(q',r) r\, dr, \tag{7.23}$$

and the spherical harmonics $u_l(q,r)$ of the electron wave functions have the following asymptotic form:

$$u_l(q,r) = \frac{1}{q}\sin(qr - \pi l/2 + \delta_l), \quad \text{if} \quad r \longrightarrow \infty.$$

Here $\delta_l(q)$ is the scattering phase for electron–atom collisions.

In particular, for a slow electron we have

$$\delta_o = -Lq,$$

where L is the atom scattering length, and the scattering phases with $l \geq 1$ are relatively small and we neglect them. The cross section of the elastic scattering of a slow electron on this atom is

$$\sigma_{\rm ea} = 4\pi L^2.$$

In this limiting case one can restrict, by the matrix elements K_{01} and K_{10} in the sum of formula (7.23), and they are equal to

$$K_{01} = K_{10} = \frac{\sin\delta_0}{2q^2(q-q')^2},$$

when $|q - q'| \ll q$. In this case, we have

$$\frac{d\sigma_{br}}{d\omega} = \frac{8}{3\pi}\frac{\sigma_{ea}}{\omega}\frac{e^2}{\hbar c}\frac{\varepsilon}{mc^2}, \tag{7.24}$$

where $\varepsilon = \hbar^2 q^2/(2m)$ is the incident electron energy. Thus, the bremsstrahlung cross section due to electron–atom scattering is small compared to the cross section of elastic scattering and has a logarithmic divergency at small and large frequencies of the emitting radiation.

Problems

Problem 7.1. *Express the cross section of the photodetachment of a negative ion through the matrix element of the valent electron.*

Use the expression for the probability of a radiative transition per unit time

$$B_{oj} = \frac{4\omega^3}{3c^3} \cdot \sum_j |\langle 0|\mathbf{D}|j\rangle|^2 n_\omega,$$

where n_ω is the number of photons in one state with a frequency ω in the flux of incident radiation, ω is the frequency of an absorbed photon, c is the velocity of light, and the wave function of the negative ion is normalized by unity. There is no averaging in this formula over the initial states 0. A final state of the transition is a free electron, so that the electron wave function far from the nucleus has the form of a plane wave: $\psi(\mathbf{r}) \to e^{i\mathbf{qr}}$, $r \to \infty$, where $\mathbf{q}$ is the wave vector of a free electron. Within this framework of formal operations, we introduce a volume Ω in which an electron is located that allows us to transfer from the continuous to discrete spectra of electrons in the final state. Because an introduced volume remarkably exceeds an atom volume, in the main part of this volume the interaction between the electron and atom is small. Thus the normalized wave function far from the atom has the form $\Omega^{-1/2}e^{i\mathbf{qr}}$. As a result, we change the matrix element $\langle 0|\mathbf{D}|j\rangle$ to the expression $\langle 0|\mathbf{D}|\mathbf{q}\rangle\Omega^{-1/2}$, where the plane wave $e^{i\mathbf{qr}}$ is used as the wave function of the state $\mathbf{q}$.

Let us sum up the expression for the probability of the photon absorption over the final states. This corresponds to multiplying by the statistical weight of the final state which is equal to

$$g_f = \int \frac{d\mathbf{q}\,d\mathbf{r}}{(2\pi)^3} = \Omega \frac{q^2\,dq\,d\Theta_{\mathbf{q}}}{8\pi^3},$$

where $d\Theta_{\mathbf{q}}$ is the element of a solid angle of motion of the released electron. Note that the electron state of the formed atom and the spin direction of a free electron are determined through the selection rules. Therefore, we take the statistical weight of the atom and electron to be unity. The energy conservation law gives

$$\omega = EA + \frac{q^2}{2},$$

where EA is the electron binding energy in the negative ion. From this it follows that $d\omega = q\,dq$, and the expression for the probability of the photodetachment per unit time takes the form

$$dB_{oj} = \frac{4\omega^3}{3c^3} \cdot |\langle 0\,|\mathbf{D}|\,\mathbf{q}\rangle|^2\, n_\omega q\, d\omega \cdot \frac{d\Theta_{\mathbf{q}}}{8\pi^3},$$

where the indices 0, $\mathbf{q}$ correspond to the initial and final states of the system under consideration.

Dividing this expression by the photon flux cN_ω, we obtain the cross section of the photodetachment. Here N_ω is the photon number density. Taking into account that n_ω is the number of photons located in one state of a frequency we have ω, for the number density of photons,

$$dN_\omega = \frac{2d\mathbf{k}}{(2\pi)^3} \cdot n_\omega = \frac{8\pi k^2\,dk}{8\pi^3} \cdot n_\omega = \frac{\omega^2\,d\omega}{\pi^2 c^3} \cdot n_\omega,$$

where the factor 2 accounts for two polarizations of photons and $\mathbf{k}$ is the photon wave vector which is connected by the dispersion relation with the photon frequency $\omega = kc$. Thus we obtain, for the cross section of the photodetachment,

$$d\sigma_{\text{det}} = \frac{dB_{0j}}{cdN_\omega} = \frac{q\omega}{6\pi c} \cdot |\langle 0 |\mathbf{D}| \mathbf{q}\rangle|^2 \, d\Theta_{\mathbf{q}}. \tag{7.25}$$

Problem 7.2. *Determine the cross section of the photodetachment of a negative ion with the valent electron shell* s^2.

Let us express the matrix element from the dipole moment operator through the radius-vector of the valent electron $\mathbf{r}$. We have

$$|\langle 0 |\mathbf{D}| \mathbf{q}\rangle|^2 = 2 \cdot |\langle 0 |\mathbf{r}| \mathbf{q}\rangle|^2 ,$$

and we account for the location of two electrons in the valent electron shell. Evaluating the matrix element on the basis of the wave function of the bound electron (7.8) and the plane wave as the wave function of the free electron, we obtain

$$|\langle 0 |\mathbf{r}| \mathbf{q}\rangle| = 4B \cdot \frac{\sqrt{2\pi q}}{\gamma^2 + q^2} \cdot \mathbf{n},$$

where $\mathbf{n}$ is the unit vector directed along $\mathbf{q}$. Using the formula of the previous problem for the cross section of the photodetachment and integrating the cross section over the angle of motion of a free electron, we obtain

$$\sigma_{\text{det}} = \frac{8\pi B^2 \gamma v_e^3}{3\omega^3 c} \cdot \frac{e^2}{\hbar}. \tag{7.26}$$

Now, for convenience, we return to the usual units from the atomic units, so that the conservation energy law is $\hbar\omega = \hbar^2(q^2 + \gamma^2)/(2m)$ and the velocity of the free electron is $v_e = \hbar q/m$.

According to the structure of the cross section, this has the maximum at $q = \gamma$ which is equal to

$$\sigma_{\text{det}} = \frac{8\pi B^2}{3\gamma^2} \cdot \frac{e^2}{\hbar c}. \tag{7.27}$$

In particular, for the negative hydrogen ion the maximum photodetachment cross section is equal to $1.4a_o^2 = 0.40\,\text{Å}^2$.

Note that the assumption used is valid in the range of the cross section maximum because the corresponding matrix element is determined by the region of the valent electron where the interaction potential of the electrons and atoms is small. The obtained formula for the cross section of the photodetachment is not valid at large photon energies which are compared to the atom ionization potential. Then the assumption of a small interaction of the valent electrons and atoms is violated.

Problem 7.3. *Find the connection between the electron binding energy in a negative ion with an s-valent electron and the scattering length of a slow s-electron on the atom which is the basis of the related negative ion. Neglect the polarization interaction between the electron and atom.*

The wave function of a slow s-electron outside the action of an atomic field has the following form, according to formula (2.35),

$$R_{qo}(r) = \frac{1}{r}\sin(qr - \delta_o),$$

where the scattering phase is equal to, at a small electron wave vector, $\delta_o = -Lq$. This relation is valid for a slow electron $q \ll 1$ and is the definition of the scattering length L. From this it follows that the wave function of a slow electron in the atom vicinity has the form $R_{qo}(r) = \text{const}\,(1 - L/r)$. Expanding the wave function (7.7) of the bound s-electron in the vicinity of the atom we have $R_s(r) = \text{const}\,(1 - \gamma r)$. Comparing these expressions, which must be identical near the atom, we obtain

$$L\gamma = 1. \tag{7.28a}$$

Let us deduce the same relation in the case of the finite radius model, when the wave function of a test electron is given by formula (7.6) for $r \geq r_o$ and is zero at lower distances. For determination of the parameter r_o one can use an identical behavior of a free and bound electron outside the atom near its boundary. Indeed, the wave function of a slow free s-electron outside the action of an atomic field has the form

$$R_q(r) = \frac{1}{r}\sin(qr - \delta_o),$$

where q is the electron wave vector, r is the electron distance from the atom, and the scattering phase of the s-electron δ_o is equal, at a small electron wave vector, to $\delta_o = -Lq$ where L is the scattering length for a slow electron on the atom. This relation is valid for a slow electron $q << 1$ and is the definition of the scattering length L. As follows from this, the logarithm derivative of this wave function on the atom surface is

$$\left.\frac{d\ln[rR_q(r)]}{dr}\right|_{r=r_o} = \frac{1}{r_o - L},$$

and the logarithm derivative of a bound electron, according to formula (7.6), is equal, on the atom surface, to

$$\left.\frac{d\ln[rR_q(r)]}{dr}\right|_{r=r_o} = -\gamma.$$

Because, inside an atom and in the vicinity of its boundary, the behavior of the free and bound electrons is identical, the logarithmic derivatives of their wave functions coincide on the atom boundary, which gives

$$r_o = L - 1/\gamma. \tag{7.28b}$$

In particular, in the case of the hydrogen negative, ion we have $L = 5.8, \gamma = 0.235$, so that $r_o = 1.55$, whereas for the asymptotic coefficient which follows from formulas (7.10) we have $r_o = 2.1 \pm 0.2$.

Problem 7.4. *Consider the previous problem by taking into account the polarization interaction between the electron and the atom.*

The Schrödinger equation for the wave function of the bound s-electron $R_s(r)$ in the negative ion, by taking into account the atom polarizability, has the following form instead of (7.5):

$$\frac{d^2}{dr^2}\left[rR_s(r)\right]+\left(\frac{\alpha}{r^4}-\gamma^2\right)\left[rR_s(r)\right]=0,$$

where α is the atom polarizability. This wave function satisfies the boundary condition

$$\frac{d\ln\left[rR_s(r)\right]}{dr}=-\frac{1}{L}. \tag{7.29}$$

Using this boundary condition and expression (7.10) for the wave function far from the atom, one can connect these expressions by solving the Schrödinger equation in the middle region of distances between the electron and the atom. Introducing a reduced variable $x=r(\gamma^2/\alpha)^{1/4}$ and the small parameter $\beta=(\alpha\gamma^2)^{1/4}$ of the perturbation theory, we obtain the above Schrödinger equation in the form

$$\frac{d^2\varphi}{dr^2}+\beta^2\left(\frac{1}{x^4}-1\right)\varphi=0,$$

where $\varphi = rR_s(r)$. Below we analyze this problem for small values of the parameter β.

Use the perturbation operator of the perturbation theory in the form

$$V=-\beta^2,\quad x\leq 1;\qquad V=\frac{\beta^2}{x^4},\quad x\geq 1.$$

Then the Shrödinger equation in the zero approximation has the form

$$\frac{d^2\varphi}{dr^2}+\frac{\beta^2}{x^4}\varphi=0,\quad x\leq 1;\qquad \frac{d^2\varphi}{dr^2}-\beta^2\varphi=0,\quad x\geq 1.$$

This equation has the following solution

$$\varphi=Ax\sin\left(\frac{\beta}{x}+\delta\right),\quad x\leq 1;\qquad \varphi=B\sqrt{2\beta}e^{-\beta x},\quad x\geq 1,$$

where for $x\geq 1$ we use the solution in the form of (7.10). In order to join these solutions, it is necessary to equalize the logarithmic derivatives of the wave functions at $x=1$. The logarithmic derivative of the wave function for $x\leq 1$ is

$$\frac{\varphi'}{\varphi}=\frac{1}{x}-\beta\cot\left(\frac{\beta}{x}+\delta\right).$$

We compare this logarithmic derivative of the electron wave function near the atom with that of a slow free electron when the wave function is $\varphi=\text{const}\,(1-L/r)$ where L is the electron–atom scattering length. Note that the electron wave function has a finite number of knots in an internal atom region, so that at small x the argument $\beta/x+\delta$ is small. Then expanding $\cot(\beta/x+\delta)$ we find, at small x,

$$\frac{\varphi'}{\varphi}=\frac{\delta}{\beta}.$$

Comparing this logarithmic derivative with that of a slow free electron, we have $\delta = -\beta^2/(\gamma L)$. Next, equalizing the logarithmic derivatives of the above expressions for the electron wave function at $x = 1$, we find the following relation between the parameters of the problem

$$\tan(\beta + \delta) = \frac{\beta}{1+\beta}. \tag{7.30}$$

In particular, for the hydrogen negative ion ($\gamma^2 = 1/18$, $\alpha = 9/2$) we obtain formula from (7.30) for the electron–atom scattering length $L = 6.7$ in the case when the total electron–atom spin is zero, while the accurate value is $L = 5.8$. Note that the used small parameter for this case $\beta = 1/\sqrt{2}$, i.e., the used approximation is not well for this case.

Problem 7.5. *Determine the normalization coefficient in expression* (7.10) *for the wave function of a valence s-electron by taking into account a weak polarization interaction between the electron and atom.*

If the electron wave function has the form (7.10) at all distances r from the atom, we have $B = 1$. Below we determine the next term of the expansion of B over the small parameter $\beta = (\alpha\gamma^2)^{1/4} \ll 1$, where α is the atom polarizability, $\gamma^2/2$ is the electron affinity of the atom. For this goal we use the solution of the Schrödinger equation which corresponds to the zeroth order of expansion of the electron wave function over this small parameter and was obtained in the previous problem. This has the form

$$\varphi = Ax\sin\left(\frac{\beta}{x} + \delta\right), \quad x \le 1; \qquad \varphi = B \cdot \sqrt{2\beta} \cdot e^{-\beta x}, \quad x \ge 1.$$

Combining these wave functions at $x = 1$ and normalizing the total electron wave function, we have

$$B^2 e^{-2\beta}\left[1 + \frac{2\beta(\beta^2 + \beta\delta + \delta^2/3)}{(\beta+\delta)^2}\right] = 1.$$

Restricted by the first expansion terms over the small parameter β and accounting for $\delta \sim \beta^2$, we find

$$B^2 = \frac{1}{1 - 2\beta^2 - 2\delta}. \tag{7.31}$$

In particular, for the hydrogen negative ion ($\gamma^2 = 1/18$, $\alpha = 9/2$, $L = 5.8$, $\delta = -0.37$) we obtain $B^2 = 1.4$, which corresponds to the accurate value $B^2 = 1.3 \pm 0.2$. Note that the small parameter of the theory is not small for this case $\beta = 1/\sqrt{2}$.

Problem 7.6. *Determine the photorecombination coefficient for the collisional process of a slow electron and the bare nucleus of charge Z if the forming hydrogenlike ion is found in a highly excited state with the principal quantum number n, and the electrons have the Maxwell distribution function on velocities.*

On the basis of the Kramers formula (6.19) we have, for the photorecombination coefficient,

$$\alpha_{\text{rec}} = \langle v\sigma_{\text{rec}} \rangle = \frac{16\pi Z^4}{3\sqrt{3}c^3n^3} \left\langle \frac{1}{v} \frac{1}{\varepsilon + Z^2/(2n^2)} \right\rangle,$$

where v is the electron velocity, $\varepsilon = v^2/2$ is the electron energy, Z is the nucleus charge, and the angle brackets mean an average over the electron velocities with the Maxwell distribution function. After integration over the electron velocities, we obtain

$$\alpha_{\text{rec}} = -\frac{16\sqrt{2\pi}}{3\sqrt{3}} \frac{Z^2}{c^3n^3T_e^{3/2}} \exp\left(\frac{Z^2}{2n^2T_e}\right) \text{Ei}\left(-\frac{Z^2}{2n^2T_e}\right), \tag{7.32}$$

where T_e is the electron temperature. The limiting expressions for the photorecombination coefficient are the following:

$$\alpha_{\text{rec}} = -\frac{32\sqrt{2\pi}}{3\sqrt{3}} \frac{Z^2}{c^3nT_e^{1/2}}, \quad T_e \gg Z^2/n^2$$

and

$$\alpha_{\text{rec}} = -\frac{16\sqrt{2\pi}}{3\sqrt{3}} \frac{Z^2}{c^3n^3T_e^{3/2}} \ln\left(\frac{2n^2T_e}{Z^2\gamma}\right), \quad T_e \ll Z^2/n^2.$$

Here $\gamma = \exp C = 1.78$, where $C = 0.577$ is the Euler constant.

Problem 7.7. *Express the cross section of the bremsstrahlung of an electron in the field of a multicharged ion through the electron Furie components of the electron coordinates on the basis of formula* (7.23).

We first evaluate the integral

$$I = \int \hbar\omega\, d\sigma_{br},$$

which is proportional to the radiation power due to electron scattering on multicharged ions. In this formula $d\sigma_{\text{br}}$ is the cross section of the bremsstrahlung which is determined by formula (7.23), so that

$$I = \int \frac{32m\omega^4\, d\omega q'}{3c^3a_oq} \cdot \sum_l (l+1)\left(K_{l+1,l}^2 + K_{l,l+1}^2\right) = \int \frac{64\omega^4}{3c^3q} \left(q'\right)^2 dq' \int K_{ll}^2 l\, dl.$$

Above we use the connection between the final electron wave vector q' and the radiation frequency $d\omega = \hbar q'\, dq'/m$ and the classical character of scattering that allows us to replace the sum by an integral and neglect the difference between l and $l+1$. Introducing the matrix element

$$J_{ll} = -\omega^2 K_{ll} = \int u_l(q,r)\, u_l(q',r) \frac{d^2r}{dt^2} dr,$$

we have

$$I = \int \frac{64}{3c^3 q} \left(q'\right)^2 dq' \int J_{\bar{l}l}^2 l\, dl.$$

From the condition of normalization of the electron radial wave functions

$$\int u_l(q', r) u_l(q', r) \left(q'\right)^2 dq' = 2\pi \delta(r - r')$$

we have

$$\int J_{\bar{l}l}^2 \left(q'\right)^2 dq' = \int \left(q'\right)^2 dq' \int u_l(q, r) u_l(q', r) \frac{d^2 r}{dt^2} dr \int u_l(q, r') u_l(q', r') \frac{d^2 r'}{dt^2} dr'$$
$$= 2\pi \int u_l^2(q, r) \left(\frac{d^2 r}{dt^2}\right)^2 dr = 2\pi \left\langle \left(\frac{d^2 r}{dt^2}\right)_{ll}^2 \right\rangle.$$

Here the angle brackets mean an average over the electron positions because $u_l^2(q, r)\, dr$ is the probability of locating the electron at distances from r to $r + dr$ from the ion. In this way, considering the electron motion to be classical, we replace the electron momentum l by the impact parameter of collision $\rho = l/q$, and take a certain trajectory of the electron motion which corresponds to a definite dependence $r(t)$. It is convenient to transfer to Furie components because the matrix elements are transformed into Furie components in the classical case. We have

$$\int J_{\bar{l}l}^2 l\, dl = \frac{1}{q^2} \int \rho\, d\rho \int_{-\infty}^{\infty} e^{-i\omega t} \left(\frac{d^2 r}{dt^2}\right)_\omega dt \int_{-\infty}^{\infty} e^{i\omega' t'} \left(\frac{d^2 r}{dt^2}\right)_{\omega'} dt',$$

where the Furie components are

$$\left(\frac{d^2 r}{dt^2}\right)_\omega = \frac{1}{2\pi} \int_{-\infty}^{\infty} e^{-i\omega t} \left(\frac{d^2 r}{dt^2}\right) dt, \qquad \left(\frac{d^2 r}{dt^2}\right) = \int_{-\infty}^{\infty} e^{-i\omega t} \left(\frac{d^2 r}{dt^2}\right)_\omega d\omega,$$

and we use the relation

$$\int_{-\infty}^{\infty} e^{i(\omega - \omega')t}\, dt = 2\pi \delta(\omega - \omega').$$

In addition, we use the relation between the Furie components

$$\left(\frac{d^2 r}{dt^2}\right)_\omega = -\omega^2 r_\omega.$$

Thus, we finally obtain

$$\frac{d\sigma_{\text{br}}}{d\omega} = \frac{8\pi e^2 \omega^3}{3\hbar c^3} \cdot \int_0^\infty r_\omega^2 2\pi \rho\, d\rho. \tag{7.33}$$

Problem 7.8. *Determine the cross section of the bremsstrahlung for an electron moving in the field of a multicharged ion of charge Z using the quasi-classical character of this process.*

This problem was considered in Chapter 6, and we now evaluate the bremsstrahlung cross section in another way. Assuming the criterion (6.17) to be fulfilled, we consider the electron to move along a classical trajectory. Then on the basis of formula (1.34) we have, for the total energy of photons ΔE emitted as a result of the movement of the electron along a given trajectory,

$$\Delta E = \int_{-\infty}^{\infty} \frac{2e^2}{3c^3} \left(\frac{d^2\mathbf{r}}{dt^2} \right)^2 dt,$$

where $\mathbf{r}(t)$ is the electron radius-vector in the coordinate frame whose origin coincides with the multicharged ion, and we take into account that the varying part of the dipole moment of this system is $e\mathbf{r}$. Let us introduce the Furie component

$$\left(\frac{d^2\mathbf{r}}{dt^2} \right)_\omega = \frac{1}{2\pi} \int_{-\infty}^{\infty} e^{-i\omega t} \left(\frac{d^2\mathbf{r}}{dt^2} \right) dt,$$

and use the relation between the Furie components

$$\left(\frac{d^2\mathbf{r}}{dt^2} \right)_\omega = i\omega \left(\frac{d\mathbf{r}}{dt} \right)_\omega = -\omega^2 \mathbf{r}_\omega .$$

Formula (7.33) has the form

$$\Delta E = \int_0^{\infty} S_\omega d\omega, \; S_\omega = \frac{8e^2\omega^4}{3c^3} \left| \mathbf{r}_\omega \right|^2 . \tag{7.34}$$

Here S_ω is the spectral power of radiation for this trajectory of the electron. Note that transition to the classical case leads to the replacement of the matrix elements in the quantum case by the Furie components of the corresponding values in the classical case. Next, under strict consideration, the lower and upper limits in the last formula are determined by the criteria (6.18) and (6.17), respectively. But for frequencies which give the main contribution to ΔE this change is not essential.

Integrating the spectral power over the possible trajectories of the electron and dividing by the photon energy $\hbar\omega$, we transfer to the cross section of the bremsstrahlung which is given by

$$\frac{d\sigma_{br}}{d\omega} = \int_0^{\infty} \frac{S_\omega}{\hbar\omega} \cdot 2\pi\rho \, d\rho, \tag{7.35a}$$

where ρ is the impact parameter of the collision of the electron and multicharged ion that is modeled by a restless Coulomb center. We present this formula in the following form:

$$\begin{aligned} \frac{d\sigma_{\rm br}}{d\omega} &= \frac{16\pi e^2\omega^3}{3\hbar c^3} \int_0^{\infty} \left(|x_\omega|^2 + |y_\omega|^2 \right) \rho \, d\rho \\ &= \frac{16\pi e^2\omega}{3\hbar c^3} \int_0^{\infty} \left(\left| \left(\frac{dx}{dt} \right)_\omega \right|^2 + \left| \left(\frac{dy}{dt} \right)_\omega \right|^2 \right) \rho \, d\rho, \end{aligned} \tag{7.35b}$$

where x, y are the electron coordinates in the plane of motion. This formula coincides with formula (7.33) that is obtained in another way.

We now use the law of electron motion in the field of the multicharged ion of charge Z and which it is convenient to represent in the parametric form

$$x = \sqrt{\rho^2 + \rho_o^2} - \rho_o ch\xi, \, y = \rho \, sh\xi, t = \frac{1}{v}\left(\sqrt{\rho^2 + \rho_o^2} sh\xi - \rho_o \xi\right),$$

where ξ is the parameter, and $\rho_o = Ze^2/(mv^2)$, so that m is the electron mass, and v is the electron velocity far from the Coulomb center of charge Z.

First we make estimates as was done in Chapter 6. According to criterion (6.18), we have a small parameter

$$\gamma = \frac{mv^3}{Ze^2\omega} \ll 1, \tag{7.36}$$

and we have estimates

$$\rho \sim \rho_o \gamma^{1/3}, \qquad r_{\min} \sim \rho_o \gamma^{2/3},$$

where ρ is the typical impact parameter of collisions which determines the emitting of photons of frequency ω, and $r_{\min}$ is the distance of closest approach which corresponds to such impact parameters. From this we determine the Furie components, in particular,

$$\left|\left(\frac{dx}{dt}\right)_\omega\right| \sim \left|\left(\frac{dy}{dt}\right)_\omega\right| \sim \frac{v_{\min}}{\omega} \sim \frac{v}{\omega}\gamma^{-1/3}.$$

Here $v_{\min}$ is the typical electron velocity at the distance of closest approach whose value follows from the relation

$$mv_{\min}^2 \sim Ze^2/r_{\min}.$$

From this we have for the bremsstrahlung cross section

$$\frac{d\sigma_{\rm br}}{d\omega} \sim \frac{Z^2e^6}{3\hbar\omega c^3 m^2 v^2}.$$

Of course, this estimate corresponds to formula (6.16).

We now evaluate the Furie components. We have for the exponent

$$\exp(-i\omega t) = \exp\left[-\frac{i}{\gamma}\left(\sqrt{1 + \frac{\rho^2}{\rho_o^2}} \operatorname{sh}\xi - \xi\right)\right]$$

and because $\gamma \ll 1$, small ξ give the main contribution to the Furie components. Accounting for $\rho \sim \rho_o\gamma^{-1/3} \ll \rho_o$ and for $\gamma \ll 1$, and expanding over small ξ, we obtain for the exponent

$$\exp(-i\omega t) = \exp\left[\frac{i}{\gamma}\left(-\frac{\rho^2}{2\rho_o^2}\xi + \frac{\xi^3}{6}\right)\right].$$

From this it follows that $\xi \sim \gamma^{1/3}$ gives the main contribution to the integral. This leads to the above estimates for the Furie components and the bremsstrahlung cross section. Next, from this we have

$$\left|\left(\frac{dx}{dt}\right)_\omega\right| = \frac{\rho_o}{2\pi}\left|\int_{-\infty}^{\infty}\exp\left[\frac{i}{\gamma}\left(-\frac{\rho^2}{2\rho_o^2}\xi+\frac{\xi^3}{6}\right)\right]\operatorname{sh}\xi\, d\xi\right|$$

$$= \frac{\rho^2}{\pi\sqrt{3}\rho_o}K_{1/3}\left(\frac{\rho^3}{3\rho_o^3\gamma}\right),$$

$$\left|\left(\frac{dy}{dt}\right)_\omega\right| = \frac{\rho}{2\pi}\left|\int_{-\infty}^{\infty}\exp\left[\frac{i}{\gamma}\left(-\frac{\rho^2}{2\rho_o^2}\xi+\frac{\xi^3}{6}\right)\right]\operatorname{ch}\xi\, d\xi\right|$$

$$= \frac{\rho^2}{\pi\sqrt{3}\rho_o}K_{2/3}\left(\frac{\rho^3}{3\rho_o^3\gamma}\right).$$

Substituting these expressions for the Furie components in formula (7.35b), we get, using the variable $z = \rho^3/(3\rho_o^3\gamma) = \rho^3 m^2 v^3\omega/(3Z^2e^4)$,

$$\frac{d\sigma_{\text{br}}}{d\omega} = \frac{16Z^2e^6}{3\hbar c^3\omega m^2v^2}\int_0^{\infty}\left(K_{1/3}^2(z)+K_{2/3}^2(z)\right)z\,dz = \frac{16\pi Z^2e^6}{3\sqrt{3}\hbar c^3\omega m^2v^2}, \quad (7.37)$$

which coincides with formula (6.16).

CHAPTER 8

The Autoionization States of Atoms and Ions

8.1 The Auger Process and Autoionization States

In the course of the study of the ionization of atoms under the action of X-rays in the Wilson camera, Auger (France) revealed in 1925 tracks of V-form which was evidence of the release of two electrons from one atom. The first electron was formed as a result of absorption of the X-ray photon. The analysis showed that release of the second electron was connected with the new phenomenon that was called the Auger process. Indeed, absorption of the X-ray photon leads to tearing out an electron from an internal shell and to the formation of a vacancy there. This vacancy is subsequently filled by an electron from an external shell. The energy excess is transferred to another electron of an external shell which leads to its release.

According to the contemporary representations, the Auger process is an example of the decay of an autoionization state. The excitation energy of an autoionization state exceeds the atom ionization potential. Therefore, the autoionization state is a bound state of an atom whose discrete level lies above the boundary of continuous spectrum. In particular, these states result from the excitation of two atomic electrons whose total excitation energy exceeds the ionization potential of the atom. Until we neglect the interaction between electrons, this state corresponds to a discrete level. But interaction between electrons leads to the decay of this state so that one electron transfers into a lower state, and the second electron takes an energy excess that causes its release. Since the lifetime of an atom in an autoionization state is large compared to typical atomic times, one can consider the autoionization level as a quasi-discrete level. Because of the possibility of its decay, the autoionization level is characterized by a width Γ, which is small compared to the excitation energy of this level ε. This criterion $\Gamma \ll \varepsilon$ allows us to consider the autoionization level as a quasi-discrete level.

8.2 Types of Autoionization States

Let us consider various kinds of autoionization states. The first state corresponds to the above character of the excitation of two electrons, when the energies of the excitation of each electron are of the same order of magnitude, and the total excitation energy exceeds the atom ionization potential. As an example of this type, the autoionization state is an excited state of the helium atom He($2s^2$, ^1S) or heliumlike ion. Taking, for simplicity, the nuclear charge $Z \gg 1$, we obtain the excitation energy $3Z^2/4$, which exceeds the ionization potential $Z^2/2$ of the heliumlike ion, and this state is the autoionization state.

The other type of autoionization state is realized when the atom excitation energy, which is the sum of the excitation energy of a valent electron and the atom core, exceeds the atom ionization potential. For example, the ground state of the krypton ion is $Kr^+(4p^5, {}^2P_{3/2})$, and the lowest excited state of this ion is characterized by a change of the fine structure of the state; the excitation energy of this state $Kr^+(4p^5, {}^2P_{1/2})$ is 0.666eV. If an excited state of a krypton atom is formed such that its atomic core is found in the upper state $^2P_{1/2}$, of fine structure, and the potential ionization of an excited electron is lower than 0.666eV, this state is the autoionization state. This autoionization state can decay with transition of the ion in the state $^2P_{3/2}$ which leads to the release of the excited electron. This takes place for the krypton atom if the effective principal quantum number of the excited electron is $n^* > 5$.

The third type of autoionization state is a bound state whose level is found above the boundary of continuous spectra, but decay of the autoionization state is prohibited by a conservation of quantum numbers. Examples of such states are $He^-(1s2s2p, {}^4P)$, $Li(1s2s2p, {}^4P)$ whose levels are located lower than the metastable states $He(1s2s, {}^3S)$ and $Li^+(1s2s, {}^3S)$. Hence, the only channel of decay of these states corresponds to formation of the helium atom or lithium ion in the ground state, i.e., the decay process proceeds according to the scheme

$$He^-(1s2s2p, {}^4P) \to He(1s^2, {}^1S) + e. \tag{8.1}$$

As it is seen, the total spin in the initial channel of the process is equal to 3/2, while in the final channel it is 1/2. From the spin conservation law it follows that this transition proceeds only due to a weak relativistic interaction which violates the prohibition on this process. Hence, the lifetime of this autoionization state is larger by several orders of magnitude than that of the autoionization states of other types.

The fourth type of autoionization state corresponds to excitation of the internal electron and formation of a vacancy in an internal electron shell which is further filled as a result of the transition of an electron from an external electron shell. The excess of energy is consumed on tearing off one or several electrons from the external shells. Filling of the vacancy can result from the radiative transition of an electron from an external shell that leads to the generation of an X-ray photon. The process of the filling of an internal vacancy can be a stepwise process. Then several shells can partake in this process, and several electrons can be released.

This type of autoionization state was the beginning of the investigations of the atomic autoionization states.

8.3 Decay of Internal Vacancies

An internal atom vacancy occurs as a result of the atom excitation by X-ray photons, or by electrons or ion impact which lead to removal of an internal electron. Then a vacancy is formed in an internal electron shell. The notation of these vacancies is given in Table 8.1. Note that if an electron is removed from an internal shell, its quantum numbers coincide with those of the vacancy formed, since the corresponding electron shell is filled at the beginning. The internal vacancy formed can be filled in two ways. In the first case, an electron transfers to this shell from an external shell, and the released energy is consumed on tearing off another external electron. The other way of filling to the vacancy corresponds to a radiative transition of an electron this shell from an external one, and an X-ray photon is generated. The first channel of the filling of the vacancy corresponds to the Auger process, and the scheme of this process can be written, for example, in this way: $K \to L_1 L_2$. This means that a vacancy in a K-shell is filled by one electron from the L_1 or L_2-shells, and the second electron is released from these shells. Note that the filling of a vacancy by electrons from the same shell is possible. Then electrons must have a larger moment compared to the vacancy. This is known as the Coster–Kronig process.

Let us consider the energetics of this process in the simplest case when a vacancy is formed on a K-shell. At a large nucleus charge $Z \gg 1$ the energy which is consumed on the ionization of a K-electron is equal to

$$\varepsilon_K = Z^2/2. \tag{8.2a}$$

Table 8.1. Notations of the vacancies in atoms; the subscript of the notation of the orbital electron momentum indicates the total vacancy momentum; the electron with these quantum numbers is removed for formation of the vacancy.

n	$s_{1/2}$	$p_{1/2}$	$p_{3/2}$	$d_{3/2}$	$d_{5/2}$	$f_{5/2}$	$f_{7/2}$
1	K						
2	L_1	L_2	L_3				
3	M_1	M_2	M_3	M_4	M_5		
4	N_1	N_2	N_3	N_4	N_5	N_6	N_7
5	O_1	O_2	O_3	O_4	O_5	O_6	O_7
6	P_1	P_2	P_3	P_4	P_5		
7	Q_1						

Let this vacancy be filled by electrons from an L-shell. The energy of formation of the vacancy in an L-shell at large Z is

$$\varepsilon_L = Z^2/8. \tag{8.2b}$$

In the case of the radiative channel of the vacancy filling, a $2p$-electron of L_2- or L_3-shells transfers to a K-shell, so that the photon energy at large Z is equal to

$$\hbar\omega = \varepsilon_K - \varepsilon_L = 3Z^2/8. \tag{8.3}$$

If the K-vacancy is filled due to the Auger process, two electrons of the L-shell partake in this process, and the energy of a released electron is equal to

$$\hbar\omega = \varepsilon_K - 2\varepsilon_L = Z^2/4. \tag{8.4}$$

Formulas (8.3), (8.4) give the approximate energetic parameters of the process of the filling of the K-vacancy if electrons of the L-vacancies participate in this process. We used the excessive values (8.2) for the energies of the vacancy formation since we do not take into account the shielding of the nucleus charge by electrons. Some parameters of the process under consideration are given in Table 8.2 where they are compared with those according to formulas (8.2)–(8.4).

Let us analyze the competition of the above processes. Introduce the quantum yields of the X-ray photon w_r as

$$w_r = \frac{\Gamma_r}{\Gamma_r + \Gamma_a}, \tag{8.5}$$

Table 8.2. Parameters of decay of K-vacancies in atoms of inert gases. In this table $\hbar\omega$ is the photon energy resulting from the transition $L_2 \to K$, λ is the wavelength of the photon, and ε is the electron energy released as a result of the Auger process.

Atom	Ne	Ar	Kr	Xe
Z	10	18	36	54
ε_K, keV	0.870	3.206	14.33	34.56
ε_{L_1}, keV	0.048	0.326	1.92	5.45
ε_{L_2}, keV	0.022	0.251	1.73	5.11
$2\varepsilon_K/Z^2$	0.64	0.73	0.82	0.87
$\hbar\omega$, keV	0.848	2.955	12.60	29.46
λ, nm	1.462	0.4196	0.0984	0.0421
$8\hbar\omega/(3Z^2)$	0.83	0.89	0.95	0.99
ε, keV	0.748	2.509	10.39	23.52
$4\varepsilon/Z^2$	1.10	1.14	1.18	1.19

where Γ_r is the width of the energy level of the autoionization state due to photon emitting and Γ_a is the width due to the Auger process. Because the values Γ_r and Γ_a, expressed in atomic units, are the probabilities of the corresponding transitions per unit time, one can determine the Z-dependence of these values in the same way. Indeed, within the framework of the perturbation theory, the probability of vacancy filling per unit time is

$$\Gamma = 2\pi |\langle \Psi_i | V | \Psi_f \rangle|^2 \rho_f, \tag{8.6}$$

where V is the operator of interaction which causes this transition, Ψ_i, Ψ_f are the wave functions of the initial and final states of the transition, and ρ_f is the density of the final states per unit energy interval. In the case of the Auger process this transition is determined by the interaction between electrons so that $V \sim e^2/r$, and in the equation for the wave function we must replace the interaction parameter e^2 by Ze^2. This gives the dependence Z^{-2} for the matrix element, and since the width Γ_a is expressed in energy units, it contains an additional factor Z^2. As a result, we obtain from this consideration that Γ_a does not depend on Z. Figure 8.1 gives the Z-dependence for the width of the autoionization level when the K-vacancy is filled as a result of the Auger process with the participation of L-electrons. The rise of the curve at large Z is due to the relativistic effects. This dependence characterizes the accuracy of the above consideration. Analyzing the radiation rate Γ_r, on the basis of formula (1.24), we find its dependence on the problem parameters with Z. We have $\Gamma_r \sim \omega^3 D^2$, where $\omega \sim Z^2$ is the transition frequency and $D \sim 1/Z$ is the matrix element of the operator of the dipole moment for this transition. From this it follows that $\Gamma_r \sim Z^4$, so that the larger the nucleus charge, the larger is the quantum yield (8.5) of the radiative decay of the K-vacancy. Figure 8.2 contains the Z-dependence for the quantum yield of radiation of the X-photon as a result of the decay of the K-vacancy. The data of Fig. 8.2 confirm the obtained conclusion.

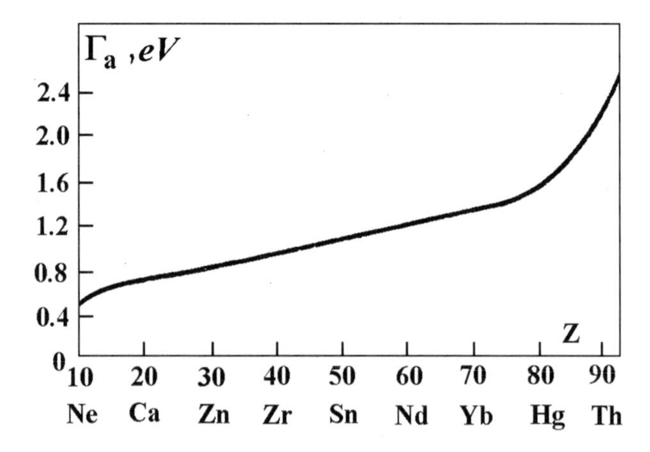

Figure 8.1. The width of autoionization levels of the K-vacancy as a function of Z.

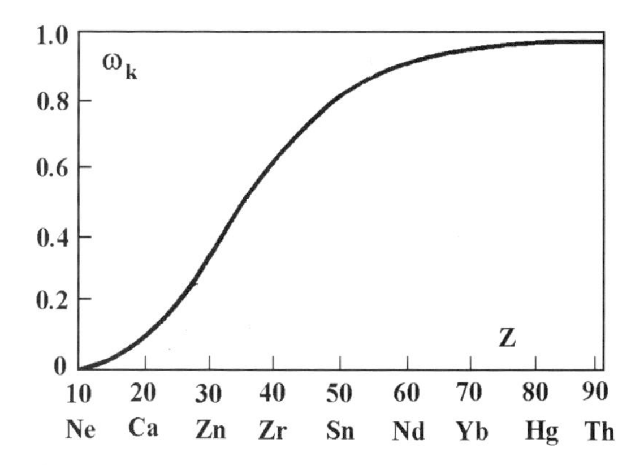

Figure 8.2. The quantum yields of X-ray photons as a result of decay of the K-vacancy.

8.4 Dielectron Recombination

A specific mechanism of the recombination of electrons and multicharged ions occurs due to formation of the autoionization states. This process proceeds according to the scheme

$$e + A^{+Z} \longleftrightarrow \left[A^{+(Z-1)}\right]^{**}; \qquad \left[A^{+(Z-1)}\right]^{**} \to A^{+(Z-1)} + \hbar\omega, \tag{8.7}$$

where $\left[A^{+(Z-1)}\right]^{**}$ means the autoionization state of a multicharged ion. This process, which is called "dielectron recombination," leads to the formation of a multicharged ion of a lower charge. Let us find the expression for the coefficient of dielectron recombination. The balance equation for the number density of multicharged ions in the autoionization state N_a has the form

$$\frac{dN_a}{dt} = N_e N_i k - N_a \Gamma_a - \frac{N_a}{\tau}, \tag{8.8}$$

where N_e is the number density of electrons, N_i is the number density of recombining ions, k is the rate constant of formation of the autoionization state, and $1/\tau$ is the frequency of a radiative transition for the autoionization state. Because the typical time of recombination is small compared to the time of establishment of the equilibrium for autoionization states, one can take the derivation to be zero. This leads to the following expressions for the number density of multicharged ions in the autoionization state N_a and for the intensity of recombination $J_a = N_a/\tau$,

$$N_a = \frac{N_e N_i k}{\Gamma_a + 1/\tau}; \; J_a = \frac{N_a}{\tau} = \frac{N_e N_i k}{\Gamma\tau + 1}. \tag{8.9}$$

In order to determine the rate constant k of the formation of the autoionization state, let us consider the case $\Gamma_a \tau \gg 1$, so that the radiative process does not influence the equilibrium number density of ions in the autoionization state. Then the equilibrium between the autoionization state and the states of continuous spectrum which are determined by the first process (8.7), leads to the Saha formula

$$\frac{N_e N_i}{N_a} = \frac{g_e g_i}{g_a} \left(\frac{T_e}{2\pi} \right)^{3/2} e^{\varepsilon_a / T_e}, \tag{8.10}$$

where T_e is the electron temperature, ε_a is the excitation energy of the autoionization state, and g_e, g_i, g_a are the statistical weights of the electron, ion and autoionization state, respectively. Comparing formulas (8.9) and (8.10) under the condition $\Gamma_a \tau \gg 1$ we obtain, for the rate constant of electron capture on the autoionization state,

$$k = \frac{g_a}{g_e g_i} \cdot \left(\frac{2\pi}{T_e} \right)^{3/2} \exp\left(-\frac{\varepsilon_a}{T_e} \right) \cdot \Gamma_a. \tag{8.11}$$

Note that this expression does not depend on the number density of ions in the autoionization state, but that the character of the distribution of electrons on energies can influence it. This formula uses the Maxwell distribution function of electrons on energies since the equilibrium between electrons was used for its deduction. Next, because the rate constant (8.11) does not depend on the kinetics of processes, it does not depend on the parameter $\Gamma_a \tau$. Hence, it can be used at any value of this parameter. From this it follows that, for the coefficient of dielectron recombination,

$$\alpha_d = \frac{J_a}{N_e N_i} = \frac{\Gamma_a}{\Gamma_a \tau + 1} \cdot \frac{g_a}{g_e g_i} \cdot \left(\frac{2\pi}{T_e} \right)^{3/2} \exp\left(-\frac{\varepsilon_a}{T_e} \right). \tag{8.12}$$

This expression corresponds only to the related autoionization state. If several autoionization states give a contribution to the rate constant of recombination, summation over the autoionization states must be made in formula (8.12). The process of dielectron recombination is of importance for a rare plasma with multicharged ions, mostly for astrophysical plasmas.

8.5 Dielectron Recombination and Photorecombination

The recombination of electrons and multicharged ions in a rare plasma can proceed through both autoionization states and by a direct photoprocess. Below we compare the rates of these processes. We use the results of Problem 7.6, for the coefficient of the photorecombination of electrons and multicharged hydrogenlike ions, which are the following:

$$\alpha_{\text{ph}} = \frac{64\sqrt{\pi}}{3\sqrt{3}} \cdot \frac{Z}{c^3} \cdot F\left(\frac{Z^2}{2n^2 T_e} \right), \qquad F(x) = -x^{3/2} e^x \, \text{Ei}(-x), \tag{8.13}$$

i.e.,

$$F(x) = x^{1/2}, \quad x \gg 1, \quad \text{and} \quad F(x) = x^{3/2} \ln \frac{1}{\gamma x}, \quad x \ll 1; \quad \gamma = e^C = 1.78,$$

where Z is the nucleus charge, n is the principal quantum number, c is the velocity of light, and $c = 137$ in atomic units. We compare this with the coefficient of dielectron recombination which is given by formula (8.12),

$$\alpha_d = gw \left(\frac{4\pi n^2}{Z^2} \right)^{3/2} x^{3/2} \exp\left(-\frac{\varepsilon_a}{T_e} \right), \tag{8.14}$$

where

$$w = \frac{\Gamma_a}{\Gamma_a \tau + 1}, \qquad g = \frac{g_a}{g_e g_i}.$$

Below, for definiteness, we take $n = 2$, and the autoionization state is such that both electrons are found in this state. Then $\varepsilon_a = Z^2/4$, $\varepsilon_a/T_e = Z^2/(4T_e) = 2x$, and the ratio of the above rate constants is

$$\frac{\alpha_d}{\alpha_{\mathrm{ph}}} = \frac{3\pi\sqrt{3}c^3 gw}{Z^4} \cdot \left[-\frac{e^{-3x}}{\mathrm{Ei}(-x)} \right]. \tag{8.15}$$

This ratio, as a function of the electron temperature has a maximum at $x = 0.25$ where the dielectron recombination process gives the maximum contribution to the total recombination coefficient, is equal to

$$\frac{\alpha_d}{\alpha_{\mathrm{ph}}} = \frac{\tau_o \cdot gw}{Z^4}, \tag{8.16}$$

where $\tau_o = 1.5 \cdot 10^{-10}$ s.

Let us analyze the contribution from dielectron recombination into the total recombination rate as a function of the nucleus charge. The rate of dielectron recombination is determined by the slower of two processes, radiative and non-radiative decay of the autoionization state, because $1/w = 1/\Gamma_a + \tau$. At small values of the nucleus charge Z, this process is the radiative transition $w = 1/\tau$, and $\tau \sim Z^{-4}$. Hence at small Z the ratio (8.16) does not depend on Z. At large Z this value $w = \Gamma_a$, and Γ_a does not depend on Z, so that the ratio (8.16) varies as Z^{-4}. From this it follows that dielectron recombination is of importance at moderate Z. Table 8.3 gives values of this ratio in the case of recombination with participation of the hydrogenlike ion of iron, where the main channels of dielectron recombination are taken into account. As is seen, dielectron recombination gives a small contribution to the total recombination coefficient compared to photorecombination. The collisional recombination of electrons and ions is significant at high electron number densities and is not essential for rare plasmas.

A small contribution of dielectron recombination in the above case can be explained by a large excitation energy of the autoionization states for hydrogenlike and heliumlike multicharged ions. If the K-shell of multicharged ions is filled, the excitation energy of the autoionization states can be small. This corresponds to lower electron temperatures where the relative contribution of dielectron recombination to the total one is higher. Let us demonstrate this in the example of the dielectron recombination of a lithiumlike multicharged ion if it develops according

Table 8.3. Main channels of dielectron recombination involving multicharged hydrogenlike ions of iron.

State	Notation	g	Γ_a, 10^{13}s^{-1}	$\frac{1}{\tau}$, 10^{13}s^{-1}	w, 10^{13}s^{-1}	$\alpha_d/\alpha_{\text{ph}}$,%
$(2s^2)^1S$	A	0.25	32	10	8.2	0.65
$(2p^2)^3P_2$	E	1.25	12	25	7.9	3.2
$(2p^2)^1D_2$	G	1.25	25	28	13	5.3
$(2p^2)^1D_2$	O	1.25	25	15	9.5	3.8
$(2p^2)^1D_2$	R	1.25	25	14	9.0	3.6
$(2s2p)^1P_1$	S	0.75	20	28	12	2.7

to the scheme

$$A^{+Z}(1s^2 2s) + e \to A^{+(Z-1)}(1s^2 2pnl) \to A^{+(Z-1)}(1s^2 2snl) + \hbar\omega, \qquad (8.17)$$

where nl are the quantum numbers of a captured electron. The excitation energy of the autoionization state is equal, in this case, to

$$\varepsilon_a = \frac{Z^2}{2(2-\delta_{2s})^2} - \frac{Z^2}{2(2-\delta_{2p})^2} - \frac{(Z-1)^2}{2n^2}, \qquad (8.18)$$

where δ_{2s}, δ_{2p} are the quantum defects of the corresponding states. From (8.18) it follows that there are many autoionization states with different n and l which can partake in dielectron recombination. Because the radiative transition $2p \to 2s$ is enough strong, these autoionization states are characterized by relatively high-level widths and are excited effectively. Next, the excitation energy of some autoionization states is not high and they can be excited at small electron energies. Therefore, in this case, dielectron recombination can introduce the main contribution into the rate of recombination of multicharged ions and electrons. The total coefficient of dielectron recombination in this case can remarkably exceed the photorecombination coefficient.

8.6 Satellites of the Resonant Spectral Lines of Multicharged Ions

The process of dielectron recombination is of interest, not only because its contribution to the recombination coefficient of electrons and multicharged ions. As a result of this process, new spectral lines appear in a rare plasma involving electrons and multicharged ions. Hence this process gives additional information about plasma parameters on the basis of the positions and intensities of spectral lines. These lines accompany the corresponding resonance spectral lines of multicharged ions and hence are called satellites of resonance spectral lines or, simply satellite spectral lines. First these lines were observed by Edlen and Tirren in 1939 for the transitions of $1s2p \to 1s^2$ of heliumlike ions during investigations of the spec-

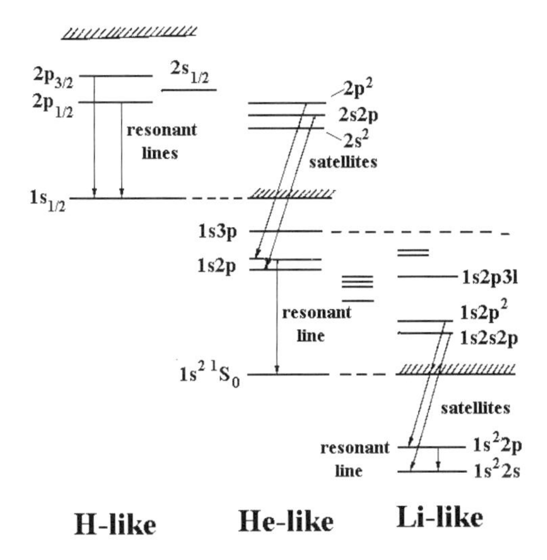

Figure 8.3. Scheme of levels of hydrogenlike, heliumlike, and lithiumlike ions. Resonant and satellite radiative transitions in the radiation spectrum of a plasma.

tra of light elements in a vacuum spark. Figure 8.3 indicates the character of the radiative transitions from the autoionization states of heliumlike and lithiumlike multicharged ions which lead to the generation of satellite photons with respect to resonant photons emitted by the resonantly excited hydrogenlike and heliumlike multicharged ions.

In order to understand the nature of satellites, let us consider the simplest case when satellites correspond to the resonance transitions $2p \to 2s$ of a hydrogenlike ion. Then the two resonant lines L_{α_1}, and L_{α_2} would be observed in the plasma spectrum which respect to the Laiman transitions $2p_{1/2} \to 2s_{1/2}$ and $2p_{3/2} \to 2s_{1/2}$. Along with this, nearby energy radiative transitions correspond to radiative quenching of the autoionization states $2s^2 \to 1s2p$, $2p^2 \to 1s2p$, and $2s2p \to 1s2s$. The photon energy for these transitions is close to that of the transition $2p \to 1s$, which is equal to $3Z^2/8$, where Z is the nucleus charge. This value is evaluated in neglecting the interaction between electrons. Accounting for this interaction leads to a small shift of the photon energy. Partial shielding of the nucleus field for one electron by the other one gives the shift of satellite spectral lines in the "red side" with respect to the resonance spectral lines. This is confirmed by the data of Tables 8.4 and 8.5, where satellite lines of iron are given with respect to the resonance lines of hydrogenlike and heliumlike iron ions. Figure 8.4 gives examples of the spectra of iron multicharged ions which are formed in plasmas of various types. These spectra we include satellite of resonant spectral lines.

The data of Tables 8.4 and 8.5 confirm an abundance of satellite spectral lines in the spectra of multicharged ions. As is seen, in all cases, the satellite spectral lines are longer than the corresponding resonant spectral lines. The relative intensity of the satellite spectral lines depends on the parameters of the corresponding autoionization state, in particular, it is determined by the width of an autoioniza-

Table 8.4. Parameters of transitions in the vicinity of the resonant $2p \to 1s$ transition of the hydrogenlike multicharged iron ions.

Initial shell	Final shell	Transition	Notation	λ, 10^{-3} nm
		Resonant transitions		
$2p$	$1s$	$^2P_{3/2} \to 1^2S_{1/2}$	L_{α_1}	177.70
$2p$	$1s$	$^2P_{1/2} \to 1^2S_{1/2}$	L_{α_2}	178.24
		Satellite transitions		178.24
$2p^2$	$1s2p$	$^1S_0 \to {}^1P_1$	A	181.04
$2p^2$	$1s2p$	$^1S_0 \to {}^3P_1$	B	180.19
$2p^2$	$1s2p$	$^3P_0 \to {}^1P_1$	C	180.09
$2p^2$	$1s2p$	$^3P_1 \to {}^1P_1$	D	179.83
$2p^2$	$1s2p$	$^3P_2 \to {}^1P_1$	E	179.66
$2p^2$	$1s2p$	$^1S_0 \to {}^1P_1$	F	178.24
$2p^2$	$1s2p$	$^1D_2 \to {}^1P_1$	G	179.13
$2p^2$	$1s2p$	$^3P_1 \to {}^3P_0$	H	179.83
$2p^2$	$1s2p$	$^3P_0 \to {}^3P_1$	I	179.24
$2p^2$	$1s2p$	$^3P_1 \to {}^3P_1$	K	178.99
$2p^2$	$1s2p$	$^3P_2 \to {}^3P_1$	M	178.82
$2p^2$	$1s2p$	$^1S_0 \to {}^3P_1$	N	177.41
$2p^2$	$1s2p$	$^1D_2 \to {}^3P_1$	O	178.30
$2p^2$	$1s2p$	$^3P_1 \to {}^3P_2$	P	179.36
$2p^2$	$1s2p$	$^3P_2 \to {}^3P_2$	Q	179.19
$2s2p$	$1s2s$	$^1D_2 \to {}^3P_2$	R	178.67
$2s2p$	$1s2s$	$^1P_1 \to {}^1S_0$	S	178.66
$2s2p$	$1s2s$	$^3P_1 \to {}^1S_0$	T	179.98
$2s2p$	$1s2s$	$^1P_1 \to {}^3S_1$	U	177.85
$2s2p$	$1s2s$	$^3P_0 \to {}^3S_1$	V	179.25
$2s2p$	$1s2s$	$^3P_1 \to {}^3S_1$	X	179.16
$2s2p$	$1s2s$	$^3P_2 \to {}^3S_1$	Y	178.75

The data of Tables 8.4, 8.5 are taken from R.K. Janev, L.P. Schevelko, and L.P. Presnyakov. *Physics of Highly Charged Ions*. (Springer, Berlin, 1985)

tion level. Thus, the positions and intensities of satellite spectral lines give much information about a related plasma consisting of electrons and multicharged ions. Reliable methods of diagnostics of such plasmas are developed on the basis of this information. In reality, the neighboring satellite spectral lines are overlapping. This is demonstrated in Fig. 8.4 where the spectra are represented of various plasmas containing multicharged iron ions. The treatment of such spectra allows one to

Table 8.5. The parameters of radiative transitions for the heliumlike multicharged iron ion and their satellites.

Initial shell	Final shell	Transition	Notation	λ, 10^{-3} nm
		Resonant transitions		
$1s2p$	$1s^2$	$^1P_1 \to {}^1S_0$	w	185.00
$1s2p$	$1s^2$	$^3P_2 \to {}^1S_0$	x	185.51
$1s2p$	$1s^2$	$^3P_1 \to {}^1S_0$	y	185.91
$1s2s$	$1s^2$	$^3S_1 \to {}^1S_0$	z	186.77
		Satellite transitions		
$1s2p^2$	$1s^22p$	$^2P_{3/2} \to {}^2P_{3/2}$	a	186.22
$1s2p^2$	$1s^22p$	$^2P_{3/2} \to {}^2P_{1/2}$	b	185.78
$1s2p^2$	$1s^22p$	$^2P_{1/2} \to {}^2P_{3/2}$	c	186.72
$1s2p^2$	$1s^22p$	$^2P_{1/2} \to {}^2P_{1/2}$	d	186.28
$1s2p^2$	$1s^22p$	$^4P_{5/2} \to {}^2P_{3/2}$	e	187.27
$1s2p^2$	$1s^22p$	$^4P_{3/2} \to {}^2P_{3/2}$	f	187.43
$1s2p^2$	$1s^22p$	$^4P_{3/2} \to {}^2P_{1/2}$	g	186.99
$1s2p^2$	$1s^22p$	$^4P_{1/2} \to {}^2P_{3/2}$	h	187.67
$1s2p^2$	$1s^22p$	$^4P_{1/2} \to {}^2P_{1/2}$	i	187.06
$1s2p^2$	$1s^22p$	$^2D_{5/2} \to {}^2P_{3/2}$	j	186.59
$1s2p^2$	$1s^22p$	$^2D_{3/2} \to {}^2P_{1/2}$	k	186.31
$1s2p^2$	$1s^22p$	$^2D_{3/2} \to {}^2P_{3/2}$	l	186.75
$1s2p^2$	$1s^22p$	$^2S_{1/2} \to {}^2P_{3/2}$	m	185.67
$1s2p^2$	$1s^22p$	$^2S_{1/2} \to {}^2P_{1/2}$	n	185.24
$1s2s^2$	$1s2s2p$	$^2S_{1/2} \to {}^2P_{3/2}$	o	189.68
$1s2s^2$	$1s2s2p$	$^2S_{1/2} \to {}^2P_{1/2}$	p	189.23
$1s2s2p$	$1s^22s$	$(^1P)^2P_{3/2} \to {}^2S_{1/2}$	q	186.10
$1s2s2p$	$1s^22s$	$(^1P)^2P_{1/2} \to {}^2S_{1/2}$	r	186.36
$1s2s2p$	$1s^22s$	$(^3P)^2P_{3/2} \to {}^2S_{1/2}$	s	185.63
$1s2s2p$	$1s^22s$	$(^3P)^2P_{1/2} \to {}^2S_{1/2}$	t	185.71
$1s2s2p$	$1s^22s$	$^4P_{3/2} \to {}^2S_{1/2}$	u	187.38
$1s2s2p$	$1s^22s$	$^4P_{1/2} \to {}^2S_{1/2}$	v	187.48

determine the electron temperature T_e, the temperature T_Z, which characterized the ratio of the number density of ions of different charge, and other parameters, especially if this plasma is nonequilibrium. This method is the main method of the diagnostics of astrophysical, laser and hot plasmas.

Usually the resonant transitions in a plasma containing multicharged ions are stronger than satellite ones. As follows from a comparison of the coefficients of

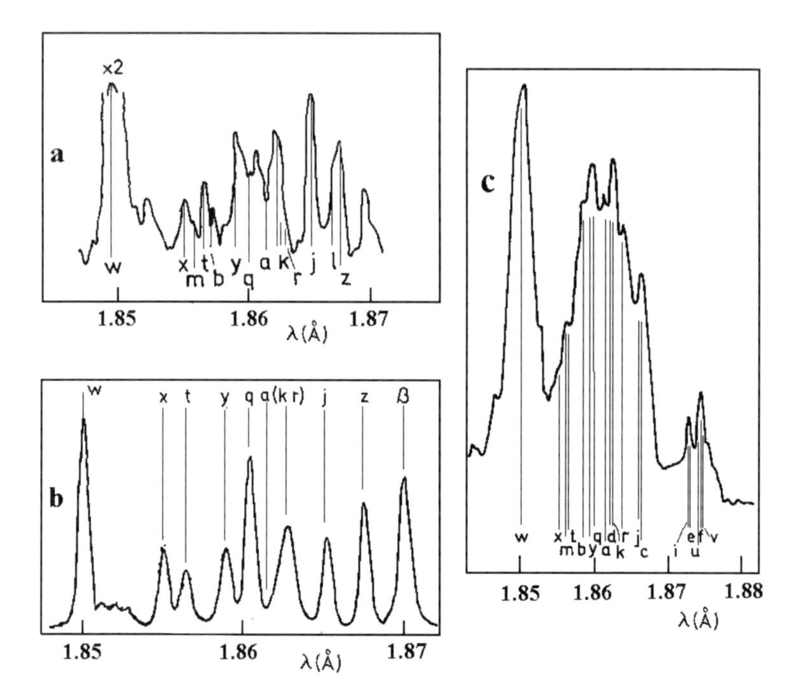

Figure 8.4. Radiative transitions between states of iron multicharged ions including the dielectron recombination with participation of heliumlike multicharged ions which leads to the generation of the sattelite spectral lines. (a) Plasma of Sun corona; (b) plasma of vacuum spark; (c) Tokomak plasma.

dielectron recombination and photorecombination, this takes place for the transitions of hydrogenlike and heliumlike multicharged ions, because the energy of excitation of the autoionization states in these cases is high enough. This is not so for the autoionization levels of lithiumlike multicharged ions, so that in this case satellite spectral lines can be more intense than the resonant ones.

8.7 Photoionization of Atoms Through Autoionization States

Autoionization states influence the absorption spectrum of atoms in the continuous region. The excitation of atoms in autoionization states results from the absorption of photons accompanied by decay of these states and reflects the character of the absorption process. If the radiative transitions in autoionization states are the dipole permitted ones, this creates the resonance structure of the photoionization cross section. This is demonstrated by the data of Fig. 8.5 where the photoionization cross section of inert gases, as a function of the wavelength, is given.

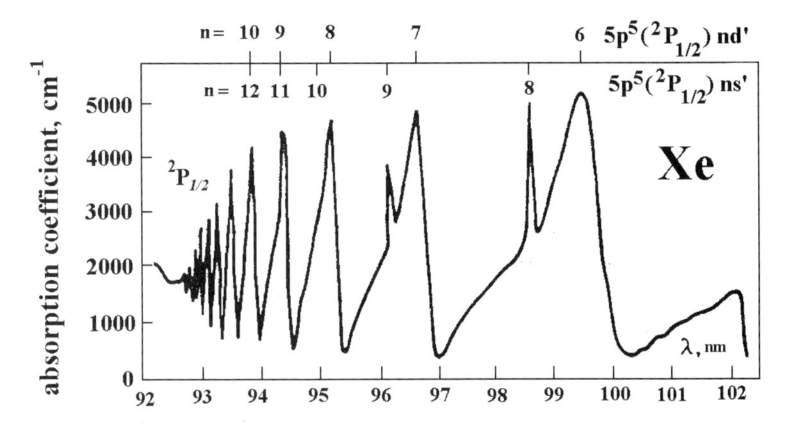

Figure 8.5. Photoabsorption spectrum of krypton involving formation of the autoionization states of krypton atoms.

Below we analyze the peculiarities of the photoionization cross section resulting from the interaction between the autoionization states and the states of continuous spectra. Let us take in the zeroth order of the perturbation theory the absence of the autoionization states. Then this wave function ψ_ε is normalized in the standard way

$$\langle\psi_\varepsilon|\hat{H}|\psi_{\varepsilon'}\rangle = \varepsilon\delta(\varepsilon - \varepsilon'), \tag{8.18a}$$

where $\hat{H}$ is the Hamiltonian, and ε is the energy of a given state of the continuous spectra. The autoionization state is described by the wave function ψ_a and can be considered in the zeroth order of the perturbation theory as a discrete state which is characterized by the energy ε_a,

$$\langle\psi_a|\hat{H}|\psi_a\rangle = \varepsilon_a. \tag{8.18b}$$

Interaction of the autoionization state and the states of the continuous spectra are determined by the matrix element

$$\langle\psi_\varepsilon|\hat{H}|\psi_a\rangle = V_\varepsilon, \tag{8.19}$$

and the width of the autoionization state, in the first order of the perturbation theory, is $\Gamma = 2\pi\,|V_\varepsilon|^2$. Taking into account this interaction between the autoionization state and the states of the continuous spectra we obtain, for the wave functions of these states which correspond to an energy E in the first order of the perturbation theory,

$$\Phi_a = \psi_a + \int \frac{dE\psi_E V_E}{\varepsilon - E}; \qquad \varepsilon_a(E) = \varepsilon_a + \int \frac{|V_E|^2\, dE}{\varepsilon - E}. \tag{8.20}$$

The main value of the integral is taken in these formulas. Accounting for the interaction between the autoionization state and the states of continuous spectra we have, for the electron wave function of the continuous spectra in the two-level approximation,

$$\Psi_\varepsilon = \frac{V_\varepsilon^* \Phi_a + [\varepsilon - \varepsilon_a(\varepsilon)]\, \psi_\varepsilon}{\left[(\varepsilon - \varepsilon_a)^2 + \pi^2 \left|V_\varepsilon\right|^4\right]^{1/2}}. \tag{8.21}$$

The wave function Ψ_ε includes the interference of states of the continuous spectra and this autoionization state. This wave function allows us to compare the photoionization cross section in the absence and presence of an autoionization state. For this goal let us introduce the Fano parameter

$$q = \frac{\langle \psi_o \left|\mathbf{D}\right| \Phi_a \rangle}{\pi V_\varepsilon \langle \psi_o \left|\mathbf{D}\right| \Psi_\varepsilon \rangle}, \tag{8.22}$$

where ψ_o is the wave function of the initial state at the photoionization process, and $\mathbf{D}$ is the operator of the atom dipole moment. Using this parameter, we obtain the ratio of the photoionization cross section σ in the considering case to this value σ_o, when the autoionization state is absent:

$$\frac{\sigma}{\sigma_o} = \frac{\left|\langle \psi_o \left|\mathbf{D}\right| \Psi_\varepsilon \rangle\right|^2}{\left|\langle \psi_o \left|\mathbf{D}\right| \psi_o \rangle\right|^2} = \frac{(q + \xi)^2}{1 + \xi^2}, \tag{8.23}$$

where

$$\xi = \frac{\varepsilon - \varepsilon_a(\varepsilon)}{\pi \left|V_\varepsilon\right|^2} = 2\frac{\varepsilon - \varepsilon_a(\varepsilon)}{\Gamma}.$$

Thus we have for the photoionization cross section, by taking into account the autoionization state,

$$\sigma = \sigma_o \cdot \left[1 + \frac{q^2 - 1 + 2q\xi}{1 + \xi^2}\right]. \tag{8.24}$$

Above we assume that the related autoionization state and states of continuous spectra have the same symmetry, and hence they interact. In this case, if a part of these states corresponds to a different symmetry and hence does not interact with the autoionization state, the photoionization cross section, in the absence of the autoionization state, has the form

$$\sigma_c = \sigma_o + \sigma_1,$$

where σ_1 corresponds to the transition in the noninteracting part of the states of continuous spectra. Taking into account the autoionization state, and replacing the cross section σ_o with that of formula (8.24), we obtain the Fano formula

$$\sigma = \sigma_o \cdot \left[1 + \rho^2 \cdot \frac{q^2 - 1 + 2q\xi}{1 + \xi^2}\right], \qquad \text{where} \quad \rho^2 = \frac{\sigma_o}{\sigma_c} = \frac{\sigma_o}{\sigma_o + \sigma_1}. \tag{8.25}$$

Formula (8.25) leads to a special form of the cross section which includes both a resonance and a dip of the cross section that is demonstrated by the data of Fig. 8.6. The relative depth of the dip and the height of the resonance are determined by the Fano parameter q which can have different values. As for the parameter ρ, this is close to unity, if the atomic core state is the same for both channels of the

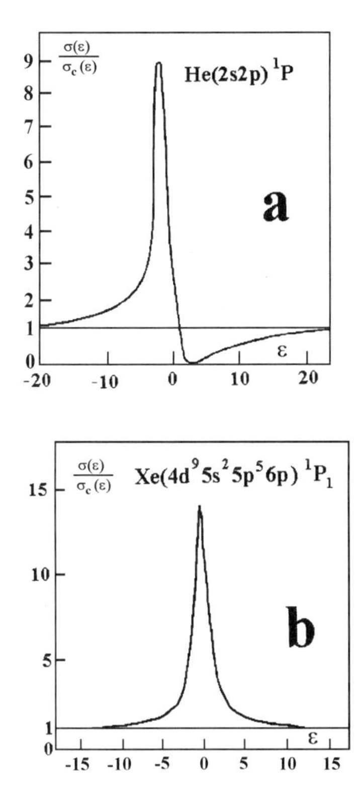

Figure 8.6. Fano profiles of spectral lines near the autoionization resonance of inert gas atoms.

photoionization process, i.e., for the process of direct photoionization and through an autoionization state. For instance, let us consider the photoionization of the helium atom in the vicinity of the formation of the autoionization level $2s2p$. This direct photoionization leads to this formation of $\mathrm{He}^{+}(1s)$ and a free p-electron, and the decay of the autoionization state $\mathrm{He}(2s2p)$ gives the same final state of the related system. Hence, in this case, we have $\sigma_1 = 0$ and $\rho^2 = 1$.

Another case takes place for the autoionization states whose decay changes the internal state of a formed ion. For example, consider the photoionization process of the xenon atom which corresponds to the excitation of an internal $4d$-electron, i.e., the process proceeds according to the scheme

$$\mathrm{Xe}(4d^{10}5s^25p^6;{}^1S_0) + \hbar\omega \rightarrow \mathrm{Xe}(4d^95s^25p^66p;{}^1P_0) \rightarrow \mathrm{Xe}^+ + e.$$

The formed vacancy is filled by an electron from the $5s$- or $5p$-shells which are close to the vacancy and interact with each other effectively. As a result of this direct photoionization, an electron is released from the $5s$ or $5p$-shells. Thus, the final states of the process are different for the related channels. Indeed, the formed xenon ion resulting from this direct photoionization process has the lowest electron shell by an energy of $\mathrm{Xe}^+(4d^{10}5s^25p^5)$, while if the process proceeds through the autoionization state, it leads to the formation of an excited xenon ion with the electron shells $Xe^+(4d^{10}5s^25p^46p)$ or $Xe^+(4d^{10}5s5p^56p)$. As a result, the parameter ρ is small in this case. In particular, as follows from Fig. 8.6, this parameter is equal to $\rho^2 = 3 \cdot 10^{-4}$ for xenon, while the Fano parameter is equal to $q = 200$, i.e., $\rho^2 q^2 = 13$. Then the resonance in the photoionization cross section is without a dip, and the maximum cross section in $1 + \rho^2 q^2 = 14$ exceeds the cross section far from the resonance.

Problems

Problem 8.1. *Determine the shielding charge σ for the autoionization states when two electrons are found in the identical states $(nl)^2$ in the field of a core of charge Z.*

These autoionization states are called the Wannier–Rydberg states by analogy with the character of the ionization of an atom near the threshold which was analyzed by G. Wannier. The most probable behavior of electrons, in the process of the ionization of an atom by electron impact near the threshold, is such that both electrons move from the core in opposite directions and are found at almost identical distances from the core. Using this analogy, for the bound states of two highly excited electrons located in the Coulomb field of an atomic core, we assume that the probable distribution of electrons corresponds to their location at identical distances from the opposite sides.

The interaction potential for two electrons located in the Coulomb field of charge Z is

$$U(\mathbf{r}_1, \mathbf{r}_2) = -\frac{Z}{r_1} - \frac{Z}{r_2} + \frac{1}{|\mathbf{r}_1 - \mathbf{r}_2|},$$

where $\mathbf{r}_1$, $\mathbf{r}_2$ are the coordinates of electrons and the origin is found in the core. Reducing the system under consideration to hydrogenlike ions, we introduce the shielding charge σ and replace the above interaction potential by the effective potential

$$U_{\text{ef}}(\mathbf{r}_1, \mathbf{r}_2) = -\frac{Z - \sigma}{r_1} - \frac{Z - \sigma}{r_2}.$$

Evidently, for the electron distribution which corresponds to the above character of the removal of two electrons in the ionization process $\mathbf{r}_1 = \mathbf{r}_2$, this replacment corresponds to $\sigma = 1/4$. Thus, the Wannier character of the behavior of identical highly excited bound electrons in the Coulomb field of the core gives the energy of the two electrons, by analogy with (7.3), in the form

$$E = -\frac{(Z - 1/4)^2}{(n - \delta)^2},$$

where δ is the quantum defect.

Problem 8.2. *Determine the relation between the width of the autoionization level* Γ *and the cross section of excitation of the atomic core by electron impact if an atom in the autoionization state consists of a weakly excited atomic core and a highly excited valent electron. Consider the interaction between electrons as a perturbation.*

Decay of the relevant autoionization state proceeds according to the scheme

$$b, nl \to a, \varepsilon l',$$

where a, b are the ground and weakly excited states of the atomic core, nl are the electron quantum numbers in a highly excited state, and ε, l' are the energy and orbital moment of the released electron. Though we divide excitations in an excitation of the core and that of a highly excited electron, the decay of the autoionization state proceeds due to the interaction between these degrees of freedom. Denoting the operator of this interaction by V we have, for the rate of decay of the autoionization state an expression similar to formula (8.6),

$$\Gamma = 2\pi |\langle b, nl|V|a, \varepsilon l'\rangle|^2 \rho_f,$$

and the interaction operator is $V = \sum_j \left|\mathbf{r} - \mathbf{r}_j\right|^{-1}$, so that $\mathbf{r}$ is the coordinate of the valent electron and $\mathbf{r}_j$ are the coordinates of the core electrons. Evidently, if the excitation energy of the atomic core is small compared to the excitation energy of the valent electron, in the spatial region which is responsible for the transition, one can neglect both by the energy of the excited electron and by the energy of the ion excitation. Then the rate of decay of the autoionization state, and the rate of excitation of the ion state in collisions with a slow electron, must be in a simple relation since they correspond to the same character of electron–ion interaction.

Let us write the expression, for the quenching cross section, for the related ion state by electron impact, which proceeds according to the scheme

$$b, ql \to a, q'l',$$

where q, q' are the wave vectors of the electron before and after the process. In order to transfer to the limit $q \to 0$, let us use the wave functions of the electron in continuous spectrum. Far from the center they correspond to a plane wave so that the electron flux is equal to q. Considering the cross section of the process as the ratio of the transition per unit time to the flux of incident particles, we have

$$\sigma(b, ql \to a, q'l') = \frac{2\pi}{q} \left|\langle b, ql\,|V|\,a, q'l'\rangle\right|^2 \rho_f.$$

Using formulas for the width of the autoionization state and this cross section we have

$$\frac{\Gamma}{\sigma(b, ql \to u, q'l')} = q \left|\frac{\psi_{nl}}{\psi_{ql}}\right|^2,$$

where ψ_{ql} is the wave function of a free electron which is determined by the formulas of Chapter 2. We assume the binding energy of a weakly bound electron

$1/(2n^2)$ to be small compared to the ion excitation energy $\Delta\varepsilon$, which allows us to consider the wave functions of the continuous spectrum to be the same for both processes. Besides, we suppose that the same angle parameters of electrons in these processes correspond to the same character of summation of electron and ion momenta in these cases.

Comparing the formulas for the wave functions of weakly bounded and free electrons in the Coulomb field (see Chapter 2), we obtain

$$\left|\frac{\psi_{nl}}{\psi_{ql}}\right|^2 = \frac{Z^2 q}{2\pi^2 n^3 (2l+1))}.$$

From the principle of detailed balance it follows that the cross section of direct and opposite processes are connected by the relation

$$g_b q^2 \sigma(b, ql \to a, q'l') = g_a (q')^2 \sigma(a, q'l' \to b, ql),$$

where g_b, g_a are the statistical weights of the ion in these states. In the limit $q \to 0$, when $(q')^2 = 2\Delta\varepsilon$, we obtain

$$\frac{\Gamma(b, ql \to a, \varepsilon l')}{\sigma(a, \Delta\varepsilon, l \to a, \varepsilon = 0, l')} = \frac{g_a}{g_b} \cdot \frac{Z^2 \Delta\varepsilon}{\pi^2 n^3 (2l+1))}.$$

It is essential in this formula that the cross section of the ion excitation $\sigma(a, \Delta\varepsilon, l \to a, \varepsilon, l')$ tends to a constant near the threshold of this process.

Part II

Atoms in Interactions and Collisions

CHAPTER 9

Atoms in External Fields

9.1 Atoms in Magnetic Fields

The rotation motion of the valent electrons of atoms creates an electric current and, hence, is responsible for the atom's magnetical moment which is proportional to the mechanical momentum of electrons. It is important that the proportionality coefficient between the magnetic and mechanical momenta is different for the spin and orbital momenta. Indeed, the magnetic moment $\boldsymbol{\mu}_{\text{orb}}$ which is determined by the orbital electron momentum $\mathbf{l}$ is equal to

$$\boldsymbol{\mu}_{\text{orb}} = -\mu_B \mathbf{l},$$

and the magnetic moment $\boldsymbol{\mu}_{\text{sp}}$ which is determined by the electron spin $\mathbf{s}$ is equal to

$$\boldsymbol{\mu}_{\text{sp}} = -2\mu_B \mathbf{s},$$

where $\mu_B = e\hbar/2mc = 9.274 \cdot 10^{-28}$ J/G is the so-called Bohr magneton. From these relations we obtain, for the atom magnetic moment,

$$\boldsymbol{\mu} = -\mu_B(\mathbf{L} + 2\mathbf{S}), \tag{9.1}$$

where $\mathbf{L}$, $\mathbf{S}$ are the spin and angular momenta of the atom. The interaction potential of the atom magnetic moment in the linear approach over the magnetic field strength $\mathbf{H}$ has the form

$$V = -\mu_B \mathbf{H}(\hat{\mathbf{L}} + 2\hat{\mathbf{S}}). \tag{9.2}$$

The average atom magnetic moment depends on the atom's quantum numbers and the magnetic field strength H. If the magnetic field is high enough so that the magnetic field splitting is large as compared to the fine structure splitting, the atomic quantum numbers in the case of the LS-coupling scheme are LSM_LM_S, where M_L and M_S are the projections of the angular and spin momenta onto the direction of the magnetic field. In this case, the level shift is given by

$$\Delta\varepsilon = -\mu_B H(M_L + 2M_S). \tag{9.3}$$

This formula is also valid at any magnetic field strength if $S = 0$ or $L = 0$. This case is known as the normal Zeeman effect, or the Pashen–Back effect.

Let us consider the other limiting case of a weak magnetic field, so that within the framework of the LS-coupling scheme the atom state is described by the quantum numbers $LSJM_J$, where M_J is the total electron momentum projection onto the direction of the magnetic field. Let us take the interaction operator in this case as

$$V = -\mu_B H(\hat{J}_z + \hat{S}_z), \tag{9.4}$$

where the z-axis directed along the magnetic field. This operator gives, for the level shift in the first approach of the perturbation theory,

$$\Delta\varepsilon = -g\mu_B M_J H, \tag{9.5}$$

where the parameter g is the Lande factor which is equal to

$$g = 1 + \frac{\langle LSJM_J\,|S_z|\,LSJM_J\rangle}{\langle LSJM_J\,|J_z|\,LSJM_J\rangle}. \tag{9.6}$$

We can use that the action of the operator $\hat{S}_z$ on the wave function does not change the spin projection and, hence, the total momentum projection onto the direction of the magnetic field does not vary. Hence all the nondiagonal matrix elements of this operator are zero, and only the diagonal matrix elements for the same M_J are not zero. In order to determine the Lande factor let us use the relation

$$\langle LSJ|\hat{\mathbf{S}}_z|LSJ\rangle = a\langle LSJ|\hat{\mathbf{J}}_z|LSJ\rangle, \tag{9.7}$$

which means that the average value of the atom spin directs along the average total electron momentum of the atom because it is the only vector in this problem. In particular, from this relation it follows that $g = 1 + a$, and below we evaluate the value a. Note that the absence of averaging over M_J in the above formula means that the quantization axis can have any direction, that is, the vector $\mathbf{J} = \langle LSJM_J|\hat{\mathbf{J}}|LSJM_J\rangle$ does not depend on the direction of this axis. If we multiply both sides of formula (9.7) by $\mathbf{J}$, since this vector does not depend on the direction of the quantization axis, one can introduce this vector inside the averaging. This gives

$$\langle LSJ|\hat{\mathbf{J}}\hat{\mathbf{S}}|LSJ\rangle = a\langle LSJ|\hat{\mathbf{J}}^2|LSJ\rangle.$$

These matrix elements do not depend on the value of the momentum projection M_J onto the magnetic field. Calculating the matrix element on the basis of the relation $(\hat{\mathbf{J}} - \hat{\mathbf{S}})^2 = \hat{\mathbf{L}}^2$, we obtain

$$g = 1 + \frac{J(J+1) + S(S+1) - L(L+1)}{2J(J+1)}. \tag{9.8}$$

The case (9.5) of the splitting of atom levels under the action of a magnetic field is known as the anomal Zeeman effect.

9.2 Atoms in Electric Fields

Energy levels are shifted and split under the action of an electric field. Let us consider this effect assuming the electric field strength to be small compared to the typical atomic values. Then one can evaluate the action of the electric field on the basis of the perturbation theory where the electric field strength is a small parameter. Then the result of the first order of the perturbation theory is not zero if degeneration of the states on the angular momentum takes place. This is valid for the hydrogen atom (see Problem 2.2), for a highly excited atom in the case that if the electric field strength is relatively high, so that the corresponding splitting of levels remarkably exceeds the distance between the neighboring energy levels corresponding to the different orbital momenta. This case is called the linear Stark effect, and then the average dipole moment of the atom is not zero.

If the average atom dipole moment is zero, the shift and splitting of the energy levels corresponds to the second order of the perturbation theory and is called the quadratic Stark effect. Representing the perturbation operator in the form $V = -\mathbf{DE}$, where $\mathbf{E}$ is the electric field strength directed along the z-axis and $\mathbf{D}$ is the operator of the atom dipole moment, we obtain for the shift of an energy level

$$\Delta\varepsilon = -\frac{\alpha E^2}{2},$$

$$\alpha = 2\sum_f \frac{|(D_z)_{of}|^2}{\varepsilon_f - \varepsilon_o}. \tag{9.9}$$

Here $\Delta\varepsilon$ is the energy shift under the action of the electric field, α is the atom's polarizability, so that D_x is the projection of the operator of the atom's dipole moment onto the electric field direction, the subscript o refers to the atom's state under consideration, the subscript f corresponds to its other states, and ε_f is the energy of these states. Formula (9.9) states that the polarizability of the ground state is positive ($\varepsilon_o < \varepsilon_f$), whereas it can have any sign for excited states. In addition, the polarizability is conserved as a result of the transformation $x \to -x$, i.e., it does not depend on the sign of the atom momentum projection onto the electric field direction

$$\alpha(\gamma, LSM_LM_S) = \alpha(\gamma, LS, -M_L, -M_S);$$

$$\alpha(\gamma, LSJM_J) = \alpha(\gamma, LSJ, -M_J), \tag{9.10}$$

where the atom quantum numbers are given in parentheses. Table 9.1 contains values of the atom's polarizabilities α.

If an atom is found in a nonuniform constant electric field, an additional interaction due to the field nonuniformity is added to the interaction operator, and is equal to

$$V = \frac{1}{6}\left(\frac{\partial E}{\partial x}Q_{zx} + \frac{\partial E}{\partial y}Q_{zy} + \frac{\partial E}{\partial z}Q_{zz}\right), \tag{9.11}$$

Table 9.1. Polarizabilities of some atoms in the ground states.

Atom	α, a.u.	Atom	α, a.u.	Atom	α,a.u.
H	4.5	Al	46	Se	25
He	1.383	Si	36	Br	21
Li	164	P	24	Kr	16.77
Be	38	S	20	Rb	320
B	20	Cl	14.7	Sr	190
C	12	Ar	11.08	In	69
N	7.4	K	290	Xe	27.3
O	5.4	Ca	160	Cs	400
F	3.8	Cu	45	Ba	270
Ne	2.670	Ga	55	Hg	36
Na	159	Ge	41	Tl	51
Mg	72	As	29		

where the electric field strength is directed along the z-axis. The tensor Q_{ik} is called the quadrupole moment tensor and is determined by the formula

$$Q_{ik} = \sum_j \langle (q_j)_{ik} \rangle, \qquad \langle (q_j)_{ik} \rangle = \langle 3(r_j)_i (r_j)_k - r_j^2 \delta_{ik} \rangle, \tag{9.12}$$

where j is the electron number, the subscripts i, k refer to the corresponding coordinates (x, y, z), $(q_j)_{ik}$ is the quadrupole moment tensor of the jth electron, δ_{ik} is the Kronecker delta symbol, and the average is made over the electron distribution in the atom.

9.3 Atom Decay in Electric Fields

A free electron in a constant electric field has a continuous energy spectrum, so that it includes any negative values of energy. Hence, if an atom is located in an electric field, an atomic electron can transfer from the atomic region to the region of the continuous spectrum (see Fig. 2.2). Thus, atomic levels become quasi-discrete levels if an electric field acts on the atom. The level width $\Gamma = 1/\tau$, where τ is the atom lifetime in the electric field. The atom decay corresponds to the tunnel transition of a valent electron from the atom region to the region of free electron motion.

Let us estimate the principal dependence of the level width on the electric field strength for weak fields. Evidently, the level width or the rate of atom decay is proportional to the electron flux through the barrier, i.e.,

$$\Gamma \sim |\psi(z_o)|^2 , \tag{9.13}$$

where ψ is the electron wave function and z_o is the boundary of the barrier (see Fig. 2.2). The Schrödinger equation for the electron under consideration has the form

$$-\frac{1}{2}\Delta\psi + V\psi - Ez\psi = -\frac{\gamma^2}{2}\psi, \qquad (9.14)$$

where V is the effective potential of the atomic core which acts on the related electron, $\gamma^2/2$ is the ionization potential for a given atom state, and E is the electric field strength. From this it follows that for the barrier boundary $z_o = \gamma^2/(2E)$, and for a weak electric field $V(z_o) \ll \gamma^2$, i.e., one can neglect the atomic field for consideration of a tunnel transition of the electron. Next, since $z_o \gg 1$, one can use the quasi-classical solution of the Schrödinger equation. Taking $\psi = \exp(-S)$ and assuming $S \gg 1$, i.e., $S'' \ll (S')^2$ we obtain, from the Schrödinger equation,

$$\frac{1}{2}\left(\frac{dS}{dz}\right)^2 = \frac{\gamma^2}{2} - Ez = E(z_o - z).$$

Since $S(0) \sim 1$, we have from this, for the level width,

$$\Gamma \sim \psi^2(z_o) \sim \exp\left[-2\int_0^{z_o}\sqrt{2E(z_o - z)}\,dz\right] \sim \exp\left(-\frac{2\gamma^3}{3E}\right). \qquad (9.15)$$

This is the principal dependence of the level width on the electric field strength which is valid for atoms, negative ions, and excited atoms at low E, i.e., when the factor $2\gamma^3/(3E)$ is large.

Problems

Problem 9.1. *Find the eigenstates and positions of the magnetic levels of the hydrogen atom in the ground state which is located in a constant magnetic field. An atom nucleus is proton with the nuclear spin $i = 1/2$ and the superfine splitting in the hydrogen atom is $\delta\varepsilon = 1420.4$ MHz.*

The interaction which causes the superfine splitting of levels of the hydrogen atom in a magnetic field consists of two parts. The first part corresponds to the interaction of the electron **s** and nuclear **i** spins, so that the corresponding interaction operator has the form $\hat{V}_1 = A\hat{\mathbf{s}}\hat{\mathbf{i}}$. This interaction leads to the superfine splitting of levels so that in the absence of a magnetic field the total atom momentum $\hat{\mathbf{F}} = \hat{\mathbf{s}} + \hat{\mathbf{i}}$ characterizes the atomic eigenstate. Since

$$2\langle\hat{\mathbf{s}}\hat{\mathbf{i}}\rangle = \langle\hat{\mathbf{F}}^2\rangle - \langle\hat{\mathbf{s}}^2\rangle - \langle\hat{\mathbf{i}}^2\rangle = F(F+1) - s(s+1) - i(i+1),$$

where the angle brackets mean the average over states, and the difference of the energies of the states of superfine structure is

$$\delta\varepsilon = \frac{A}{2}F(F+1)\Big|_0^1 = A.$$

The other part of the interaction operator is responsible for the interaction of the spin momenta with the magnetic field. Because the electron magnetic moment exceeds the proton magnetic moment by more than three orders of magnitude, one can neglect the interaction of the proton spin with the magnetic field. This gives the interaction operator $\hat{V}_2 = -2\mu\hat{\mathbf{s}}\mathbf{H}$, where $\mu = e\hbar/(2mc) = \mu_B$ is the electron magnetic moment, and $\mathbf{H}$ is the magnetic field strength. Thus, the Hamiltonian of the system of spins under consideration in a magnetic field has the form

$$\hat{H} = A\hat{\mathbf{s}}\hat{\mathbf{i}} - 2\mu\hat{\mathbf{s}}\mathbf{H}.$$

This system has four states (2×2). Let us find the eigenstates of this system and the corresponding energies of these states. For this goal we take a suitable basis of wave functions $\{\psi_i\}$ and calculate the matrix elements $\langle\psi_i \hat{H} \psi_k\rangle$. The condition that the determinant of the matrix $\varepsilon\delta_{ik} - \langle\psi_i \hat{H} \psi_k\rangle$ is equal to zero allows us to determine the positions of the energy levels for this system.

Let us take, as a basis, the eigenstates for the interaction potential $\hat{V}_1 = A\hat{\mathbf{s}}\hat{\mathbf{i}}$. The perturbation potential V_2 conserves the momentum projection onto the magnetic field direction. For this reason two states with total spin 1 and spin projections onto the direction of the magnetic fields ± 1 remain the eigenstates of the Hamiltonian. The energies of these states are equal to

$$\varepsilon_{1,2} = A \pm \mu H,$$

if the position of the singlet level in the absence of the magnetic field is taken as zero. The wave functions of two other states are

$$\psi_{3,4} = \frac{1}{\sqrt{2}}\left(\chi_+\eta_- \pm \chi_-\eta_+\right),$$

where χ, η are the spin wave functions of the electron and proton respectively, the subscript $+$ corresponds to the projection $+1/2$ onto the direction of the magnetic field, and the subscript $-$ corresponds to the projection $-1/2$. From this we obtain the following energy matrix:

$\varepsilon - A$	$-\mu H$
$-\mu H$	ε

.

Above we use the relations $\hat{s}_z\psi_3 = \psi_4/2$, $\hat{s}_z\psi_4 = \psi_3/2$. From the energy matrix we have the following expressions for the energies of these levels:

$$\varepsilon_{3,4} = \frac{A}{2} \pm \left(\frac{A^2}{4} + \mu^2 H^2\right)^{1/2}. \tag{9.16}$$

Let us introduce the reduced variables $\varepsilon_i' = \varepsilon_i/A$, $x = H/H_o$, where $H_o = A/\mu = 1015\,\mathrm{G}$. In these variables the positions of the energy levels are given by the expressions (see also Fig. 9.1):

$$\varepsilon_{1,2}' = 1 \pm x, \qquad \varepsilon_{3,4}' = 1/2 \pm \sqrt{1/4 + x^2}.$$

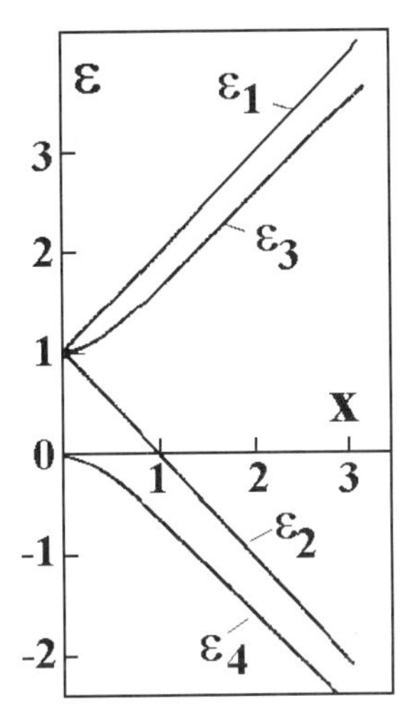

Figure 9.1. Decay in an electric field.

Problem 9.2. *Within the framework of the LS-coupling scheme determine the positions of the levels for an atom with angular momentum L and spin $S = 1/2$ located in a constant magnetic field.*

Based on a standard method of diagonalization of the energy matrix, we include in the Hamiltonian a spin-orbit interaction and the interaction potential (9.2) between the atom and the magnetic field, so that the Hamiltonian of the system under consideration has the form

$$\hat{H} = A\hat{\mathbf{L}}\hat{\mathbf{S}} - \mu_B \mathbf{H}(\hat{\mathbf{L}} + 2\hat{\mathbf{S}}).$$

Let us construct the energy matrix by taking the basis in the form of wave functions which correspond to a certain projection of the atom angular momentum M and a certain spin projection σ onto the magnetic field direction. The total number of states of the system under consideration is equal to $2(2L + 1)$. One can separate two states from the other states which correspond to the atom's total momentum $J = L + 1/2$, and its projection onto the magnetic field direction is $\pm(L + 1/2)$. For these states we have

$$\langle M, 1/2|\hat{H}|M, 1/2\rangle = \frac{AL}{2} - \mu_B H(L + 1);$$
$$\langle -M, -1/2|\hat{H}| - M, -1/2\rangle = \frac{AL}{2} + \mu_B H(L + 1).$$

The other secular equation for the energy matrix $|\varepsilon - H_{ik}| = 0$ is divided into the $2L$ pair equations because the first term of the Hamiltonian mixes states with M, 1/2 and $M + 1, -1/2$, and the second term is diagonal on this basis. The energy matrix for one of the $2L$ elements is presented in Table 9.2.

Table 9.2.

	$M, 1/2$	$M+1, -1/2$
$M, 1/2$	$\frac{A}{2}M - \mu_B H(M+1)$	$\frac{A}{2}\sqrt{(L+M+1)(L-M)}$
$M+1, -1/2$	$\frac{A}{2}\sqrt{(L+M+1)(L-M)}$	$-\frac{A}{2}(M+1)M - \mu_B HM$

From the solution of the secular equation we obtain, for the eigenvalues of the energy,

$$\varepsilon = -\frac{A}{4} - \mu_B H\left(M+\frac{1}{2}\right) \pm \sqrt{\frac{A^2}{4}\left(L+\frac{1}{2}\right)^2 + \frac{1}{4}\mu_B^2 H^2 - \frac{A}{2}\mu_B H\left(M+\frac{1}{2}\right)}. \tag{9.17}$$

In particular, from this one can find the positions of levels in the limiting cases. In the case of a weak magnetic field ($\mu_B H \ll A$) we have $\varepsilon_1 = -(A/2)L$, $\varepsilon_2 = (A/2)\left(L+\frac{1}{2}\right)$. In a strong magnetic field ($\mu_B H \gg A$) the positions of these levels are $\varepsilon_1 = -\mu_B H(M+1)$, $\varepsilon_2 = -\mu_B HM$. Thus, in a weak field this atomic term includes two energy levels with the atom's total momentum $J = L \pm \frac{1}{2}$. In a strong magnetic field this term includes the $2L+2$ energy levels (the energy levels for the quantum numbers M, 1/2 and $M+1, -1/2$ are coincident), and at moderate magnetic fields this term contains the $4L+2$ energy levels.

Problem 9.3. *Calculate the polarizability for the negative ion with a valent s^2-electron shell. Assume that the typical size of the negative ion remarkably exceeds that of an atom which is the basis of the negative ion.*

The wave function ψ_o of a valent electron, in the basic region of its location, satisfies the Schrödinger equation

$$-\frac{1}{2}\Delta\psi_o = -\frac{\gamma^2}{2}\psi_o,$$

where $\gamma^2/2$ is the binding energy of the valent s-electron in the negative ion. The solution of this equation is given by formula (7.8),

$$\psi_o = \frac{B}{r}\sqrt{\frac{\gamma}{2\pi}}e^{-\gamma r},$$

where r is the distance between the electron and nucleus. The polarizability of the negative ion is given by formula (9.9),

$$\alpha = 2\sum_f \frac{|(D_z)_{of}|^2}{\varepsilon_f - \varepsilon_o} = 4\sum_f \frac{z_{of}^2}{\varepsilon_f - \varepsilon_o}.$$

We account for two identical valent electrons in the negative ion, and z is the projection of the electron radius-vector onto the electric field direction. Let us introduce the operator $\hat{g}$ which satisfies the relation $z_{of} = (\varepsilon_k - \varepsilon_o)g_{of}$, so that $\hat{z} = i(d\hat{g}/dt)$ and the operators $\hat{z}$ and $\hat{g}$ are connected by the relationship

$$\hat{H}\hat{g}\psi_o - \hat{g}\hat{H}\psi_o = z\psi_o$$

with the boundary conditions $\psi_k r\hat{g}\psi_o \to 0$ at $r \to 0$ for any resonantly excited state k. Let us take the function $g(r)$ in the form $g = \varphi(r)\cos\theta$, where θ is the angle between the electron radius-vector and the z-axis, i.e., $\cos\theta = z/r$. In the basic region of the electron distribution, where one can neglect the interaction between the valent electron and the atomic core, this equation has the form

$$-\frac{1}{2}\varphi'' + \gamma\varphi' + \frac{1}{r^2}\varphi = r.$$

The general solution of this equation is

$$\varphi = \frac{r^2}{2\gamma} + C_1\frac{r\gamma + 1}{r\gamma} + C_2\frac{r\gamma - 1}{r\gamma}\exp(r\gamma),$$

where C_1, C_2 are the integration constants. From the boundary conditions for $r \to 0$ and $r \to \infty$ it follows that $C_1 = C_2 = 0$ and $g = (r^2/2\gamma)\cos\theta$. Hence the polarizability of the negative ion is

$$\alpha = 4\sum_k \frac{z_{ok}^2}{\varepsilon_k - \varepsilon_o} = 4\sum_k z_{ok}g_{ko} = 4\left\langle\frac{r^3}{2\gamma}\cos^2\theta\right\rangle = \frac{2\langle r^3\rangle}{3\gamma}.$$

Using the asymptotic wave function for a valent s-electron of a negative ion (7.8) which is

$$\psi_o = B\sqrt{\frac{\gamma}{2\pi}}e^{-\gamma r}/r,$$

we have

$$\langle r^3\rangle = \frac{3B^2}{4\gamma^2},$$

and the polarizability of the negative ion is

$$\alpha = \frac{B^2}{2\gamma^4}. \tag{9.18}$$

In particular, in the case of the hydrogen negative ion we use the parameters of Table 7.6 ($\gamma = 0.236$, $B = 1.15$), and have for the polarizability $\alpha = 213$. This is a relatively high value in comparison with the polarizabilities of atoms (Table 9.1), because of the large size of the negative ion.

Problem 9.4. *Evaluate the polarizability of the hydrogen atom in the ground state.*

Using, for the evaluation of the polarizability, the method of the previous problem, we have the following expression for the polarizability of the hydrogen atom:

$$\alpha = 2\langle zg\rangle = \frac{2}{3}\langle r\varphi(r)\rangle.$$

Here the average is made on the basis of the wave function of the electron wave function for the hydrogen atom in the ground state, and the function $\varphi(r)$ satisfies the equation

$$-\frac{1}{2}\varphi'' + \varphi' - \frac{1}{r}\varphi' + \frac{1}{r^2}\varphi = r.$$

Here r, θ are the electron spherical coordinates and the following boundary conditions are fulfilled for the function $\varphi(r)$:

$$r^2\varphi(r) \to 0 \quad \text{at } r \to 0, \quad \text{and} \quad \exp(-r)\varphi(r) \to 0 \quad \text{at } r \to \infty.$$

The solution of the above equation, which satisfies the second condition, has the form

$$\varphi(r) = \frac{r^2}{2} + r + C\left(2 + \frac{2}{r} + \frac{1}{r^2}\right).$$

The second condition gives $C = 0$, so that

$$\varphi(r) = \frac{r^2}{2} + r.$$

From this it follows that for, the polarizability of the hydrogen atom in the ground state,

$$\alpha = \frac{2}{3}\langle r\varphi(r)\rangle = \frac{1}{3}\langle r^3 + 2r^2\rangle = \frac{9}{2}.$$

Problem 9.5. *Determine the polarizability of the negative ion with a valent s^2-shell on the basis of a direct method.*

We use formula (9.9) for the negative ion polarizability and take into account that the negative ion has only one bound state. Then, locating the negative ion in a volume Ω, we obtain the following expression, for the polarizability of the negative ion on the basis of formula (9.9),

$$\alpha = 2\sum_k \frac{|(D_z)_{ok}|^2}{\varepsilon_k - \varepsilon_o} = 4\sum_k \frac{z_{ok}^2}{\varepsilon_k - \varepsilon_o} = 4\int \frac{\Omega\, d\mathbf{q}}{(2\pi)^3}\frac{|z_{o\mathbf{q}}|^2}{(EA + q^2/2)}.$$

We account for two identical valent electrons in the negative ion; in this formula z is the projection of the electron radius-vector onto the electric field direction and $EA = \gamma^2/2$ is the electron binding energy in the negative ion. Thus our task is to calculate the matrix element $z_{o\mathbf{q}}$ and the integral for the polarizability of the negative ion. Neglecting the interaction of the valent electron and atomic core in the principal region of electron location we have, for the wave function of the bound electron, according to formula (7.8),

$$\psi_o = B\sqrt{\frac{\gamma}{2\pi}}e^{-\gamma r}/r,$$

where r is the distance between the electron and nucleus. In this case, the wave function of a free electron is given in the form of a plane wave $\psi_{\mathbf{q}} = \Omega^{-1/2}\exp(i\mathbf{q}\mathbf{r})$.

From this we have, for the matrix element,

$$z_{o\mathbf{q}} = 4iB\sqrt{\frac{2\pi\gamma}{\Omega}}\frac{q\cos\vartheta}{(q^2+\gamma^2)^2},$$

where ϑ is the angle between the z-axis and the $\mathbf{q}$-direction. This yields, for the polarizability of the related negative ion,

$$\alpha = \int \frac{2\Omega q^2\,dq\,d\cos\vartheta}{\pi^2(q^2+\gamma^2)}\left|z_{o\mathbf{q}}\right|^2 = \frac{B^2}{2\gamma^4}.$$

As is seen, this formula coincides with formula (9.18) of Problem 9.3. This deduction allows us to estimate the criterion of the validity of this result. The polarizability is determined by the electron wave vectors $q \sim \gamma$. For example, the maximum of the function under the integral takes place at $q = 0.8\gamma$. This corresponds to large distances of the electron from the nucleus. In particular, for $q = \gamma$ the maximum under the integral for the matrix element $z_{o\mathbf{q}}$ corresponds to $rq = 2.04$. Because large distances of the electrons from the nucleus r give the basic contribution to the integral, this justifies the use of asymptotic expressions for the wave functions of the bound and free electrons for calculation of the polarizability of the negative ion.

Problem 9.6. *Determine the upper and lower limits for the polarizability of a light atom in the ground state. Compare this with the values for the hydrogen and helium atoms in the ground state.*

All the terms of formula (9.9) for the atom's polarizability are positive if the atom is found in the ground state. Hence, replacing the energy difference $\varepsilon_f - \varepsilon_o$ in this formula, by the excitation energy in the lowest resonantly excited state $\Delta\varepsilon$, we obtain the upper limit for the polarizability

$$\alpha < \frac{2\sum_f\left|(D_z)_{of}\right|^2}{\Delta\varepsilon} = \frac{2\langle o|D_z^2|o\rangle}{\Delta\varepsilon}.$$

The other limit for the polarizability follows from the sum rule for the oscillator strengths

$$\sum_i f_{oi} = n,$$

where n is the number of valent electrons. Multiplying this relation by formula (9.9) for the polarizability, we obtain

$$\begin{aligned}\alpha n &= 4\sum_{i,f}\left|(D_z)_{oi}\right|^2\left|(D_z)_{of}\right|^2\frac{\varepsilon_i-\varepsilon_o}{\varepsilon_f-\varepsilon_o} \\ &= 2\sum_{i,f}\left|(D_z)_{oi}\right|^2\left|(D_z)_{of}\right|^2\left[\frac{\varepsilon_i-\varepsilon_o}{\varepsilon_f-\varepsilon_o}+\frac{\varepsilon_f-\varepsilon_o}{\varepsilon_i-\varepsilon_o}\right] \\ &= 2\sum_{i,f}\left|(D_z)_{oi}\right|^2\left|(D_z)_{of}\right|^2\left[2-\frac{(\varepsilon_i-\varepsilon_f)^2}{(\varepsilon_f-\varepsilon_o)(\varepsilon_i-\varepsilon_o)}\right].\end{aligned}$$

Because the second term in parentheses is positive we obtain, as a result of neglecting this term,

$$\alpha n > 4\left(\left\langle o\left|D_z^2\right|o\right\rangle\right)^2.$$

Summarizing the above results, we have

$$\frac{4(\langle o|D_z^2|o\rangle)^2}{n} < \alpha < \frac{2\langle o|D_z^2|o\rangle}{\Delta\varepsilon}. \tag{9.19}$$

This relation gives the range of the polarizability values. In particular, for the hydrogen atom in the ground state ($n = 1$, $\Delta\varepsilon = 3/8$) we have $\left(\left\langle o\left|D_z^2\right|o\right\rangle\right) = \left\langle o\left|r^2\right|o\right\rangle/3 = 1$, where $r$is the electron radius-vector for the hydrogen atom. Then the obtained relation yields

$$4 < \alpha < 5.3,$$

while the real value of the polarizability of the hydrogen atom is equal to 4.5.

In the case of the helium atom in the ground state we use the hydrogenlike wave function (3.4). Then we have

$$\langle o|D_z^2|o\rangle = \frac{2}{3}\langle o|r^2|o\rangle = \frac{2}{Z_{ef}^2},$$

where the factor 2 accounts for two electrons in the helium atom. Then the above relation for the polarizability takes the form ($n = 2$):

$$\frac{8}{Z_{\rm ef}^4} < \alpha < \frac{4}{Z_{\rm ef}^2\Delta\varepsilon}. \tag{9.20}$$

Using the parameters of the helium atom $Z_{\rm ef} = 1.69$, $\Delta\varepsilon = 0.780$, we obtain the above relation in the form

$$0.98 < \alpha < 1.80,$$

while the real value of the helium polarizability is equal to 1.38. As is seen, for both cases the atom polarizability lies close to the middle between the obtained limits, so that the formula for the atom polarizability

$$\alpha = \frac{\langle o|D_x^2|o\rangle}{\Delta\varepsilon} + \frac{2\left(\langle o|D_x^2|o\rangle\right)^2}{n} \tag{9.21}$$

is valid with an accuracy of several percent.

Problem 9.7. *Find the dependence of the polarizability of a highly excited atom on the principal quantum number.*

The polarizability of a highly excited atom is determined by the formula

$$\alpha = 2\sum_{n'l'm'} \frac{\left|\langle nlm\,|z|\,n'l'm'\rangle\right|^2}{\varepsilon_{n'l'm'} - \varepsilon_{nlm}},$$

where nlm are the quantum numbers of a given highly excited state, and the selection rule gives $l' = l \pm 1$. This follows from the expression for the angular wave function and leads to the expression for the matrix element

$$\langle nlm \,|r \cos\theta|\, n'l'm'\rangle = \langle nl\,|r|\,n'l'\rangle \left[\sqrt{\frac{(l+1)^2-m^2}{(2l+3)(2l+1)}}\delta_{l',l+1}\delta_{mm'} + \sqrt{\frac{l^2-m^2}{(2L+1)(2l-1)}}\delta_{l',l-1}\delta_{mm'}\right].$$

From this we obtain the following formula for the polarizability of a highly excited atom

$$\alpha = 2\frac{(l+1)^2-m^2}{(2l+3)(2l+1)}\sum_{n'}\frac{|\langle nl\,|r|\,n',l+1\rangle|^2}{\varepsilon_{n',l-1}-\varepsilon_{nl}} + 2\frac{l^2-m^2}{(2L+1)(2l-1)}\sum_{n'}\frac{|\langle nl\,|r|\,n',l-1\rangle|^2}{\varepsilon_{n'l-1}-\varepsilon_{nl}}.$$

If the atom is found in the s state, this formula does not contain the second term.

The values of n' which are close to n give the main contribution to the result. Since $\varepsilon_{n',l\pm 1} - \varepsilon_{nl} \sim n^{-3}$, and because $\langle nl|r|n'l \pm 1\rangle \sim n^2$, we have

$$\alpha \sim n^7. \tag{9.22}$$

Note that the sign of the polarizability depends of the positions of the levels.

Problem 9.8. *Determine the polarizability of a highly excited atom for states with a small value of the quantum defect.*

We consider a highly excited state with the principal quantum numbers $n \gg l$ and $\delta_{l-1} \ll 1$. Since $\delta_{l-1} > \delta_l > \delta_{l+1}$, we have that the quantum defects are small, so that the main contribution to the polarizability gives $n' = n$. Then using the formula of the previous problem for the polarizability of a highly excited atom and restricting, in this formula, the terms with $n' = n$, we have

$$\alpha = \frac{9n^7}{2(\delta_{l-1}-\delta_l)}\cdot\frac{l^2-m^2}{(2l-1)(2l+1)} - \frac{9n^7}{2(\delta_l-\delta_{l+1})}\cdot\frac{(l+1)^2-m^2}{(2l+1)(2l+3)}. \tag{9.23}$$

Since δ_l depends sharply on l, the second term is large compared to the first one.

Let us give the criterion that the electric field strength is small and does not mix states with different values of the electron angular momenta. Then the level shift due to the electric field $\alpha E^2/2$ is small compared to the distance between levels with the same n and neighboring values l which is equal to $(\delta_{l-1} - \delta_l)/n^3$. This criterion has the form

$$E \ll \frac{\delta_{l-1}-\delta_l}{n^5}. \tag{9.24}$$

If this criterion is fulfilled, nlm remain the quantum numbers of a highly excited atom.

Problem 9.9. *Determine the quadrupole moment of a light atom with valent p-electrons.*

We use formula (9.12) for the tensor of the atom quadrupole moment which has the form

$$Q_{ik} = \sum_j \langle (q_j)_{ik} \rangle, \qquad \langle (q_j)_{ik} \rangle = \langle 3(r_j)_i (r_j)_k - r_j^2 \delta_{ik} \rangle,$$

where j is the electron number, indices i, k refer to the corresponding coordinates (x, y, z), and q_{ik} is the quadrupole moment of an individual electron. Expressing the atom wave function of a light atom according to the parentage scheme (4.7), one can conclude that the atom core does not influence the atom quadrupole moment. Next, the nondiagonal elements of this tensor are equal to zero, so that the mean values of the quadrupole tensor moment are equal to

$$Q_{zz} = \sum_j (r_j)^2 \langle P_2(\cos\theta_{zj}) \rangle,$$

where r_j is the radius-vector of the j-electron, $P_2(\cos\theta_{zj}) = 3\cos^2\theta_{zj} - 1$, and θ_{zj} is the angle between the direction of the radius-vector of the jth electron and the z-axis. From the relation for Q_{zz} because

$$\cos^2\theta_{xj} + \cos^2\theta_{yj} + \cos^2\theta_{zj} = 1,$$

we have

$$Q_{xx} + Q_{yy} + Q_{zz} = 0.$$

In addition, the filled shell does not give a contribution to the atom quadrupole moment.

Let us consider an atom with a noncompleted electron shell. Using the formula

$$\langle P_2(\cos\theta) \rangle = \frac{l(l+1) - 3m^2}{(2l-1)(2l+1)},$$

where l is the orbital momentum of a valent electron, m is its projection onto the z-axis, and we have, from the formula for the atom's quadrupole moment,

$$Q_{zz} = \langle r^2 \rangle \cdot \sum_j \frac{l(l+1) - 3m_j^2}{(2l-1)(2l+1)}, \tag{9.25}$$

where m_j is the momentum projection of the jth electron onto the z-axis. This sum for a filled shell is equal to zero because

$$\sum_{m=-l}^{l} m^2 = \frac{(2l+1)((l+1)l}{3}.$$

Let us use the parentage scheme for summation of the parameters of an atom core and the electron in the atoms quantum numbers LM_LSM_S on the basis of formula (4.7). Although the atomic core does not make a contribution to the atom's quadrupole moment, it can influence the quadrupole moment through the character of summation of momenta of the core and valence electron into the atom's total momenta. Thus, within the framework of the parentage scheme of the momenta summation (4.7), the expression for the atom quadrupole moment has the form

$$Q_{zz} = \langle r^2 \rangle \cdot \sum_{L'S'M'_L m} \left(G^{LS}_{L'S'}\right)^2 \begin{bmatrix} l & L' & L \\ m & M'_L & M_L \end{bmatrix}^2 \langle lm \, |q_{zz}| \, lm \rangle \,, \tag{9.26}$$

where lm are the angular momenta and their projection onto the z-axis for a valent electron, LM_L are these quantum numbers for the atom, $L'M'_L$ are the same quantum numbers for the atom core, and q_{zz} is the quadrupole moment of a valent electron with quantum numbers lm. In particular, for the quadrupole moment of a p-electron, we have

$$\langle 10 \, |q_{zz}| \, 10 \rangle = \frac{4}{5} \langle r^2 \rangle, \qquad \langle 1, \pm 1 \, |q_{zz}| \, 1, \pm 1 \rangle = -\frac{2}{5} \langle r^2 \rangle .$$

Thus, finally we have, for the atom quadrupole moment,

$$Q_{zz} = \langle r^2 \rangle \cdot \sum_{LS'} \left(G^{LS}_{L'S'}\right)^2 \left\{ \frac{4}{5} \begin{bmatrix} 1 & L' & L \\ 0 & M_L & M_L \end{bmatrix}^2 - \frac{2}{5} \begin{bmatrix} 1 & L' & L \\ 1 & M_L - 1 & M_L \end{bmatrix}^2 - \frac{2}{5} \begin{bmatrix} 1 & L' & L \\ -1 & M_L + 1 & M_L \end{bmatrix}^2 \right\} .$$

Problem 9.10. *Find a shift of the 3^2P levels of the sodium atom under the action of a constant electric field. Consider any relation between this shift and the fine structure splitting.*

The splitting and shift of this state level both proceed as a result of the spin-orbit interaction and under the action of the electric field. Let us take into account both effects within the framework of the standard methods when the Hamiltonian matrix is constructed on a certain basis of the atomic wave functions, and the eigenvalues of the state energies follow from the diagonalization of this matrix. Then we include the spin-orbit interaction in the Hamiltonian by the term $-A\hat{\mathbf{l}}\hat{\mathbf{s}}$ where $\hat{\mathbf{l}}$ is the operator of the angular moment and $\hat{\mathbf{s}}$ is the spin operator of the valent electron. Note the symmetry of the problem with respect to the change of sign of the momentum projection onto the electric field direction. Hence, we take only states with positive values of the projection of the total electron momentum onto the electric field direction. Table 9.3 gives the obtained energy matrix and U_o, U_1 in this table are shifts of the energy levels under the action of the electric field.

The obtained matrix energy leads to the secular equation $|\varepsilon - H_{ik}| = 0$, where $|\,|$ is the determinant of this secular equation which yields

$$[(\varepsilon - U_o)(\varepsilon - U_1 + A/2) - A^2/2](\varepsilon - U_1 + A/2) = 0.$$

Table 9.3.

	$m=0, \sigma=1/2$	$m=1, \sigma=-1/2$	$m=1, \sigma=1/2$
$m=0, \sigma=1/2$	U_o	$A\sqrt{2}$	0
$m=1, \sigma=-1/2$	$A\sqrt{2}$	$U_1-A/2$	0
$m=1, \sigma=1/2$	0	0	$U_1+A/2$

From this it follows that for the positions of the energy levels

$$\varepsilon_1 = U_1 - A/2;$$

$$\varepsilon_{2,3} = (U_o + U_1)/2 + A/4 \pm \sqrt{\frac{(U_o - U_1)^2}{4} + \frac{A}{4}(3U_o + U_1) + \frac{9}{16}A^2}. \quad (9.27)$$

The first value relates to the state with projection of the total momentum 3/2, the second and third values relate to the value of this projection 1/2. In the limit of small electric fields, when one can neglect the terms U_o and U_1, the positions of the energy levels correspond to states with total electron momentum 1/2 and 3/2; these are equal to $\varepsilon_1 = \varepsilon_2 = -A/2$, $\varepsilon_3 = A$. In the opposite limiting case we have $\varepsilon_1 = \varepsilon_2 = U_1$, $\varepsilon_3 = U_o$.

The second order of the perturbation theory for expansion over the electric field strength gives $U_o = -\alpha_o E^2$, $U_1 = -\alpha_1 E^2$, where α_o, α_1 are the atom's polarizabilities for the projections of the angular momentum onto the electric field directions 0 and 1, respectively. For determination of these values we use the relations for the matrix elements of the operator $z = r\cos\theta$,

$$\langle \gamma l m \, |r\cos\theta| \, \gamma', l+1, m\rangle = \sqrt{\frac{(l+1)^2 - m^2}{(2l+3)(2l+1)}} \cdot \langle \gamma l \, |r\cos\theta| \, \gamma', l+1\rangle;$$

$$\langle \gamma l m \, |r\cos\theta| \, \gamma', l-1, m\rangle = \sqrt{\frac{l^2 - m^2}{(2l+1)(2l-1)}} \cdot \langle \gamma l \, |r\cos\theta| \, \gamma', l-1, m\rangle,$$

where $\gamma l m$ is a set of atom quantum numbers. This gives, for the atom polarizabilities in the related states,

$$\alpha_o = \frac{2}{3}\sum_{\gamma} \frac{(\langle p\,|r|\,\gamma s\rangle)^2}{\varepsilon_{\gamma s} - \varepsilon_p} + \frac{8}{15}\sum_{\gamma} \frac{(\langle p\,|r|\,\gamma d\rangle)^2}{\varepsilon_{\gamma d} - \varepsilon_p}; \qquad \alpha_1 = \frac{2}{5}\sum_{\gamma} \frac{(\langle p\,|r|\,\gamma d\rangle)^2}{\varepsilon_{\gamma d} - \varepsilon_p}, \quad (9.28)$$

where $\gamma s, \gamma d$ are the states with angular electron momenta 0 and 2 which determine the atom polarizability. Note that on introducing the polarizabilities α_o and α_1, we neglect the fine level spliting compared to the energy differences between the levels of different n. The main contribution to the above polarizabilities gives transitions in states with nearby energies to the related state. On the basis of this we obtain, accounting for several terms in the above sums,

$$\alpha_o = 530; \qquad \alpha_1 = 280.$$

Thus the above formulas allow us to evaluate the positions of levels at any electric field strength. In particular, let us find the electric field strength at which the distance between the levels of the fine structure is doubled compared to that in the absence of the electric field. This leads to the value of the electric field strength $E = 6 \cdot 10^6$ V/cm which is close to the critical electric field strength (6.11), $E_{cr} = 7 \cdot 10^6$ V/cm.

Problem 9.11. *Determine the rate of decay of an excited hydrogen atom in an external electric field (this is similar to Problem* 2.4).

The bound state of an electron in the hydrogen atom in an electric field is not stable. Indeed, there is a region at large distances from the center in the direction of the electric field where the electron has continuous spectrum (see Fig. 2.2). The transition of the electron in this region leads to decay of the bound state. Thus the energy level of the electron which is located in the field of the Coulomb center has a finite width. This width is determined by the time of the electron transition in the region of continuous spectrum. According to Fig. 2.2, where the cross section of the potential energy in a space is given, this region starts from a distance in the direction of the electric field $z_o = |\varepsilon|/E$, where ε is the electron binding energy and E is the electric field strength. Below we determine the time of this tunnel transition and hence the width of the electron level.

For this goal we use the Schrödinger equation for the electron wave function in parabolic coordinates. We take the following form of expansion of the wave function which is more convenient for this problem

$$\Psi = \frac{e^{im\varphi}}{\sqrt{2\pi}} \cdot \frac{\Phi(\xi)F(\eta)}{\sqrt{\xi\eta}}.$$

Correspondingly, the Schrödinger equations for these wave functions have the form

$$\frac{d^2\Phi}{d\xi^2} + \left(-\frac{\gamma^2}{4} + \frac{\beta_1}{\xi} - \frac{m^2-1}{4\xi^2} + \frac{E}{4}\xi\right)\Phi = 0,$$
$$\frac{d^2F}{d\eta^2} + \left(-\frac{\gamma^2}{4} + \frac{\beta_2}{\eta} - \frac{m^2-1}{4\eta^2} - \frac{E}{4}\eta\right)F = 0,$$

where the electron energy is given by formula $\varepsilon = -\gamma^2/2$, and the separation constants β_1, β_2 in accordance with the results of Problem 2.3 are equal to

$$\beta_1 = \gamma\left(n_1 + \frac{m+1}{2}\right), \qquad \beta_2 = \gamma\left(n_2 + \frac{m+1}{2}\right),$$

which corresponds to the electron energy

$$\varepsilon_{nn_1n_2m} = -\frac{1}{2n^2} - \frac{3}{2}En(n_1 - n_2),$$

i.e.,

$$\gamma = \frac{1}{n} + \frac{3}{2}En^2(n_1 - n_2).$$

The probability of the electron transition through a barrier per unit time is equal to

$$w = \int_S \mathbf{j}\, d\mathbf{S},$$

where

$$j = \frac{i}{2}\left(\Psi\frac{\partial\Psi^*}{\partial z} - \Psi^*\frac{\partial\Psi}{\partial z}\right)$$

is the electron flux through the barrier, and S is the cross section of the barrier. We use that for small electric field strengths the main contribution to the electron current gives a small region of the barrier cross section near z_o. Use the cylindrical coordinates ρ, z, φ in this region and the condition $\rho \ll z$, when the connection between the cylindrical and parabolic coordinates has the form

$$\xi = r + z \approx 2z, \qquad \eta = \sqrt{\rho^2 + z^2} \approx \frac{\rho^2}{2z} \approx \rho^2/\xi.$$

The element of the barrier surface is equal to $dS = \rho\, d\rho\, d\varphi = (\xi/2)\, d\eta\, d\varphi$. Substituting these formulas into the expression for the rate of decay of this state, we have

$$w = \int_0^\infty \frac{F^2(\eta)d\eta}{\eta}\frac{i}{2}\left(\Phi\frac{d\Phi^*}{d\xi} - \Phi^*\frac{d\Phi}{d\xi}\right).$$

Using the expressions for the electron wave functions which are obtained in Problem 2.3, we obtain

$$w = \frac{i}{2}\left(\Phi\frac{d\Phi^*}{d\xi} - \Phi^*\frac{d\Phi}{d\xi}\right).$$

In the region near the center, where one can neglect the action of the electric field, the wave function $\Phi(\xi)$ is

$$\Phi = \frac{\sqrt{2}}{n}\cdot\frac{1}{[n_1!\,(n_1+m)!]^{1/2}}\cdot\left(\frac{\xi}{n}\right)^{n_1+(m+1)/2}\exp\left(-\frac{\xi}{2n}\right).$$

Let us analyze the solution of the Schrödinger equation for the wave function $\Phi(\xi)$. Let us find the expression for this wave function in the vicinity of the point ξ_o which separates the classically available region from the region where location of the classical electron is forbidden. The position of the point ξ_o is given by equation $p(\xi_o) = 0$, where p is the classical electron momentum which, in accordance with the Schrödinger equation for $\Phi(\xi)$, is given by the expression

$$p^2 = -\frac{\gamma^2}{4} + \frac{\beta_1}{\xi} - \frac{m^2-1}{4\xi^2} + \frac{E}{4}\xi.$$

The quasi-classical solutions of the Schrödinger equation $\Phi'' + p^2\Phi = 0$, right and left of the turning point, have the form

$$\Phi = \frac{iC}{\sqrt{p}} \cdot \exp\left(i\int_{\xi_o}^{\xi} p\,d\xi - \frac{i\pi}{4}\right), \quad \xi > \xi_o;$$

$$\Phi = \frac{iC}{\sqrt{p}} \cdot \exp\left(-\int_{\xi_o}^{\xi} |p|\,d\xi\right), \quad \xi < \xi_o.$$

Substituting the first of these expressions into the formula for the decay rate, we obtain

$$w = |C|^2.$$

Thus the probability of the electron tunnel transition per unit time is expressed through the wave function amplitude on the boundary of the classical region for the electron. In order to determine this value, it is necessary to connect the expressions for the wave function near the center and boundary of the classical region. In the case of small electric field strengths there is a broad region of ξ where, on one hand, one can neglect the action of the electric field and, on the other hand, the classical expression for the wave function is valid. Connecting in this region the expression for the wave function by neglecting the electric field with the classical wave function, we find the value C and, correspondingly, the rate of atom decay

$$w = \frac{\gamma^3 \left(4\gamma^3/E\right)^{2n_1+m+1}}{(n_1 + m)!n_1!} \cdot \exp\left(-\frac{2\gamma^3}{3E}\right). \tag{9.29}$$

The region where both solutions are valid corresponds, on one hand, to neglecting the action of the electric field that is valid under the relation $\xi \ll \xi_o = \gamma^2/E$ and, on the other hand, in this region the quasi-classical wave function is valid, i.e., $\gamma^2/4 \gg \beta_1/\xi$. This leads to the following criterion of the validity of the above expression for the decay rate

$$E \ll \frac{\gamma^4}{\beta_1}.$$

Thus, at small values of the electric field strength E the probability of the atom decay per unit time depends exponentially on the value $1/E$. Note that in Problem 2.3 we obtain the expansion of the value γ over a small parameter which is proportional to the electric field strength E. In this case, it is necessary to use two expansion terms of γ on E, because the second term of the expansion of γ gives a term of the order of one in the exponent of w. In this approximation, we have

$$\frac{2\gamma^3}{3E} = \frac{2}{3n^3E} + 3(n_1 - n_2),$$

and the probability of decay of the hydrogen atom per unit time in a weak electric field is equal to

$$w = \frac{\left(\frac{4}{3n^3E}\right)^{2n_1+m+1}}{n^3\,(n_1+m)!n_1!} \cdot \exp\left(-\frac{2}{3n^3E} + 3n_2 - 3n_1\right). \tag{9.30}$$

In particular, from this formula it follows the formula of Problem 2.4 for the probability of decay per unit time for the hydrogen atom in the ground state is

$$w = \frac{4}{3E} \cdot \exp\left(-\frac{2}{3E}\right).$$

CHAPTER 10

Interactions Involving Atoms and Ions

10.1 Long-Range Interactions of Atoms

The interaction potential of two atoms or an atom and ion at moderate and large distances between them consists of two parts—long-range and exchange interactions. The long-range interaction of two atomic particles is determined by the interaction between their momenta induced by fields of interacting particles. The exchange part of the interaction potential is due to overlapping of the wave functions of electrons whose centers are located on different nuclei. Since these types of interaction are created by different regions of the electron distribution in a space, they can be summed independently into the total interaction potential of atomic particles.

Let us consider, in consecutive order, various types of the long-range interaction of atomic particles. A general scheme for the determination of the long-range interaction potential consists of the expansion of the electron energy of interacting particles over a small parameter r_j/R, where r_j is the distance of the jth electron from the nucleus in whose field the electron is located and R is the distance between nuclei of the interacting particles. Take the interaction operator in the form (in atomic units)

$$V = \frac{Z_1 Z_2}{R} - \sum_j \frac{Z_1}{|\mathbf{R} - \mathbf{r}_j|} - \sum_k \frac{Z_2}{|\mathbf{R} + \mathbf{r}_i|} + \sum_j \frac{1}{|\mathbf{R} - \mathbf{r_j} + \mathbf{r}_i|}, \qquad (10.1)$$

where Z_1, Z_2 are the charges of nuclei, $\mathbf{r}_j$ are the coordinates of electrons of the first atomic particle, and $\mathbf{r}_i$ are the coordinates of the electrons of the second atomic particle. Using the operator (10.1) within the framework of the theory of the perturbation and expanding of this interaction potential over a small parameter r/R, one can analyze various cases of the long-range interaction of atomic particles.

First we consider the long-range ion–atom interaction. Then take $r_i = 0$ in the interaction operator V (10.1), so that it can be presented in the form (the ion charge is assumed to be one and the charge of the atom nuclei is Z)

$$V = \frac{Z}{R} - \sum_j \frac{1}{|\mathbf{R} - \mathbf{r}_j|}. \tag{10.2}$$

The first order of the perturbation theory in expansion over a small parameter r/R gives, for the ion–atom interaction energy,

$$U(R) = \left\langle \frac{1}{R^3} \sum_j r_j^2 P_2(\cos\theta_j) \right\rangle = \frac{Q_{zz}}{R^3}. \tag{10.3}$$

As a matter of fact, this term corresponds to the interaction of the nonuniform electric field of a charged particle with the quadrupole moment of the other atomic particle. Indeed, use formula (9.11) for this part of the interaction potential which has the form

$$U = \frac{1}{6}\frac{\partial E_z}{\partial z} Q_{zz},$$

where the z-axis joins the nuclei and the electric field of a charged particle is $E = R^{-2}$, so that $\partial E_z/\partial z = 2R^{-3}$ (we omit the sign of the interaction potential and assume below that the ion charge is one). From this formula (10.3) follows. The next term of expansion of the interaction potential over the small parameter corresponds to the second order of the perturbation theory and is given by formula (9.9), $U = -\alpha E^2/2$, where the electric field strength of the charge particle at the point of location of another particle is equal to $E = R^{-2}$, i.e., the interaction potential is

$$U(R) = -\frac{\alpha}{2R^4}, \qquad \alpha = 2\sum_k \sum_f \frac{\left|(z_k)_{of}\right|^2}{\varepsilon_f - \varepsilon_o}, \tag{10.4}$$

where the state under consideration is denoted by the subscript o, the subscript f refers to other states of the atomic particle, and k is the electron number.

Now let us consider the long-range interaction of two neutral atomic particles. A general method consists of the expansion of the operator (10.1) over a small parameter r/R, and in the extraction the strongest interaction for each case. This expansion has the following general form

$$V = \sum_{i,k} \sum_{l=1}^{\infty} R^{-l-1} \left[|\mathbf{r}_k - \mathbf{r}_i|^l \, P_l(\mathbf{n}_k\mathbf{n} - \mathbf{n}_i\mathbf{n}) - r_k^l P_l(\mathbf{n}_k\mathbf{n}) - r_i^l P_l(\mathbf{n}_i\mathbf{n}) \right], \tag{10.5}$$

where $\mathbf{n}_k$, $\mathbf{n}_i$, $\mathbf{n}$ are the unit vectors directed along $\mathbf{r}_k$, $\mathbf{r}_i$, and $\mathbf{R}$, respectively, and $P_l(x)$ is the Legendre polynomial. The first two terms of this expansion are

$$V = \frac{\mathbf{D}_1\mathbf{D}_2 - 3(\mathbf{D}_1\mathbf{n})(\mathbf{D}_2\mathbf{n})}{R^3} + \frac{3}{2R^4}\left[(\mathbf{D}_1\mathbf{n})[3(\mathbf{D}_2\mathbf{n})^2 - 1] + (\mathbf{D}_2\mathbf{n})[3(\mathbf{D}_1\mathbf{n})^2 - 1]\right], \tag{10.6}$$

where $\mathbf{D}_1 = \sum_k \mathbf{r}_k$, $\mathbf{D}_2 = \sum_i \mathbf{r}_i$ are operators of the dipole moment of the corresponding atomic particle.

Let us determine the interaction potential of two atomic particles on the basis of the interaction operator (10.6). Since the average dipole moment of an atom is equal to zero for any state, the first order of the perturbation theory, accounting for the first term of (10.6), differs from zero for the interaction of two dipole molecules. Then we have, for the interaction potential,

$$U(R) = \frac{\mathbf{D}_1\mathbf{D}_2 - 3(\mathbf{D}_1\mathbf{n})(\mathbf{D}_2\mathbf{n})}{R^3}, \tag{10.7}$$

where $\mathbf{D}_1$, $\mathbf{D}_2$ are the dipole moments of the corresponding atoms. The first order of the perturbation theory leads to a nonzero result in the case of the interaction of two identical atom states if their states o and f are such that the matrix element $(\mathbf{D})_{of}$ is not zero. Let us denote the wave function of an excited atom by φ and the wave function of a nonexcited atom by ψ, so that the eigenwave functions of this system of interacting atoms are $\psi_s = (1/\sqrt{2})[\psi(a)\varphi(b) + \varphi(a)\psi(b)]$ and $\psi_a = (1/\sqrt{2})[\psi(a)\varphi(b) - \varphi(a)\psi(b)]$. Here the arguments a, b indicate an atom which relates to a given wave function. From this we find the interaction potential of atoms in the related states

$$U_{s,a}(R) = \pm\frac{1}{R^3}\left[\left|(D_x)_{of}\right|^2 + \left|(D_y)_{of}\right|^2 - 2\left|(D_z)_{of}\right|^2\right], \tag{10.8}$$

where the matrix element is $(D_x)_{of} = \langle\psi|D_x|\varphi\rangle$, the + sign corresponds to the symmetric state, the – sign relates to the antisymmetric state, and the z-axis joins the nuclei. Thus, in this case, the interaction of the dipole moments leads to the splitting of the quasi-molecule energy levels.

The next term in the expansion of the interaction potential of two atomic particles with a distance between them accounts for the interaction of a dipole moment of one particle (molecule) with a quadrupole moment of the other atomic particle. This corresponds to taking into consideration the second term of the operator (10.6) in the first order of the perturbation theory, and yields

$$U(R) = \frac{3}{R^4}[(\mathbf{D}_1\mathbf{n})Q_2 + (\mathbf{D}_2\mathbf{n})Q_1], \tag{10.9}$$

where $\mathbf{D}_i$ is the average dipole moment of the corresponding atomic particle, $\mathbf{n}$ is the unit vector directed along the z-axis, and Q is the zz-component of the quadrupole moment tensor for an indicated atomic particle. The next term of the first order of the perturbation theory corresponds to the interaction of two quadrupole momenta of atoms and yields, for the interaction potential,

$$U(R) = \frac{6Q_1Q_2}{R^5}, \tag{10.10}$$

where Q_1, Q_2 are the zz-components of the quadrupole moment tensor for the indicated atoms. This interaction describes the long-range interaction of two atoms with unfilled shells. Averaging over the direction of moments of any atom gives zero for the mean interaction potential.

The second order of the perturbation theory for a long-range interaction of two atoms starts from the term which is proportional to R^{-6}. In this case, on the basis of the operator of the dipole–dipole interaction of atoms, we have

$$U(R) = -\frac{C}{R^6},$$

$$C = \sum_{f,m} \frac{4\left|(D_{1z})_{of}\right|^2 |(D_{2z})_{om}|^2 + \left|(D_{1x})_{of}\right|^2 |(D_{2x})_{om}|^2 + \left|(D_{1y})_{of}\right|^2 \left|(D_{2y})_{om}\right|^2}{E_f - E_o + \varepsilon_m - \varepsilon_o}, \tag{10.11}$$

where the subscript o corresponds to the related states of the interacting atoms, the subscripts f and m refer to possible states of the first and second interacting atom respectively, the $\mathbf{D}_1$, $\mathbf{D}_2$ are the dipole moment operators for the corresponding atoms. E_f indicates the energy of fth state of the first atomic particle, and ε_m is that for mth state of the second particle. The parameter C is called the van der Waals constant. The interaction potential (10.11) corresponds to the interaction of remote atoms with filled electron shells or is the average long-range interaction potential for any atoms.

Let us analyze the expression for the van der Waals constant in the case of the interaction of two atoms with zero angular momenta. Then $\left|(D_x)_{of}\right|^2 = \left|(D_y)_{of}\right|^2 = \left|(D_z)_{of}\right|^2$ for each atom and expression (10.11) for the van der Waals constant has the form

$$C = 6\sum_{f,m} \frac{\left|(D_{1z})_{of}\right|^2 |(D_{2z})_{om}|^2}{E_f - E_o + \varepsilon_m - \varepsilon_o}. \tag{10.12}$$

Considering the case of two identical atoms, when $E_m - E_o = \varepsilon_m - \varepsilon_o$, we have

$$\frac{3}{2}\alpha(D_z^2)_{oo} - C = \sum_{f,m} \frac{3\left|(D_{1z})_{of}\right|^2 |(D_{2z})_{om}|^2}{2(E_f - E_o)} + \sum_{f,m} \frac{3\left|(D_{1z})_{of}\right|^2 |(D_{2z})_{om}|^2}{2(E_m - E_o)}$$

$$- \sum_{f,m} \frac{6\left|(D_{1z})_{of}\right|^2 |(D_{2z})_{om}|^2}{E_f - E_o + \varepsilon_m - \varepsilon_o}$$

$$= \frac{3}{2}\sum_{f,m} \frac{\left|(D_{1z})_{of}\right|^2 |(D_{2z})_{om}|^2 (E_m - E_f)^2}{(E_m - E_o)(E_f - E_o)(E_f + E_m - 2E_o)}.$$

For the ground state of atoms this combination is positive, so that

$$\frac{3}{2}\alpha(D_z^2)_{oo} \geq C.$$

In the case of different interacting atoms this relation takes the form

$$\frac{3}{4}\alpha_1(D_{2z}^2)_{oo} + \frac{3}{4}\alpha_2(D_{1z}^2)_{oo} \geq C, \tag{10.13}$$

where the subscripts 1, 2 refer to the respective atoms. This relation can be used as the basis for approximating formulas for the van der Waals constant. Use the relation for the atom polarizability which was obtained in Problem 9.6 and has the form:

$$\frac{4(\langle o\,|D_z^2|\,o\rangle)^2}{n} < \alpha < \frac{2\langle o\,|D_z^2|\,o\rangle}{\Delta\varepsilon}.$$

Let us replace the matrix element $(D_z^2)_{oo}$ in expression (10.13) on the basis of the right-hand side of this inequality $(D_{2z}^2)_{oo} = 2\alpha\Delta\varepsilon$, where $\Delta\varepsilon$ is the excitation energy for the first resonance level. Then we obtain the following expression for the van der Waals constant, which is called the London formula,

$$C = \frac{3\alpha_1\alpha_2\Delta\varepsilon_1\Delta\varepsilon_2}{2(\Delta\varepsilon_1 + \Delta\varepsilon_2)}, \tag{10.14}$$

where α_1, α_2 is the polarizability and $\Delta\varepsilon_1$, $\Delta\varepsilon_2$ is the excitation energy of the first resonant level for the corresponding atom. Using the approximating relation $(D_{2z}^2)_{oo} = \sqrt{\alpha n}/2$, we obtain the following expression for the van der Waals constant which is called the Kirkwood–Slater formula,

$$C = \frac{3\alpha_1\alpha_2}{2(\sqrt{\alpha_1/n_1} + \sqrt{\alpha_2/n_2})}, \tag{10.15}$$

where n_1, n_2 are the numbers of the valent electrons for the related atoms. Formulas (10.14) and (10.15) are used as the approximations for calculation of the van der Waals constants for the interaction of atoms with filled electron shells or for the van der Waals constants of atoms with unfilled electron shells if the interaction potential is averaged over the momentum directions of atoms. Table 10.1 gives a comparison of formulas (10.14) and (10.15) with accurate values for the van der Waals constants. As is seen, the Kirkwood–Slater formula works better than the London formula.

Table 10.1. Van der Waals constants (in atomic units) for the interaction of identical atoms.

Atoms	London formula	Kirkwood–Slater formula	Accurate value
H–H	5.7	7.2	6.2
He–He	1.1	1.7	1.5
He–H	2.4	3.2	2.8
He–Ne	2.0	3.8	3.1
Ne–Ne	4.5	8.4	6.6
Ar–Ar	40	66	68
Kr–Kr	80	130	130
Xe–Xe	190	260	270
Na–Na	1600	1600	1600
Rb–Rb	3800	3800	3800
Cs–Cs	5100	5200	5200

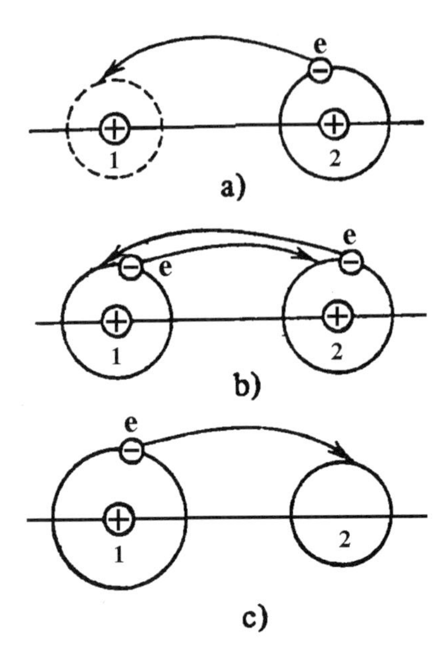

Figure 10.1. Three types of exchange interaction. (a) Electron transferred from the field of one ion or atom to the field of another atomic particle; (b) exchange of two electrons between two atomic particles; (c) the wave function tail of the valence electron reaches another atomic particle and thereby interacts with it.

10.2 Ion–Atom Exchange Interaction at Large Separations

The exchange interaction of atomic particles is determined by the overlapping of the electron wave functions which belong to different atomic centers. One can divide the exchange interaction between atomic particles into three groups (see Fig. 10.1). The first group corresponds to the transition of a valent electron from the field of one ion or atomic particle to the field of another. The second type of exchange interaction corresponds to a simultaneous exchange by two electrons which belong to different atomic centers. The third type is the result of the interaction of a valent electron with the field of a neighboring atomic particle. We now concentrate on the first type of exchange interaction which corresponds to the interaction between an ion and an atom. We begin with the case when a one-electron atom interacts with the parent ion that has the filled electron shell. Then the interaction is determined by the transition of the valent electron from the field of one ion to the field of another ion (see Fig. 10.2a). First we consider the case when the valent electron is found in an s-state so that the related system has two states which can be composed from states with the location of the electron in the field of the first and second ions.

Let us denote by ψ_1 the electron wave function which is centered on the first nucleus and by ψ_2 the wave function centered on the second nucleus. The electron

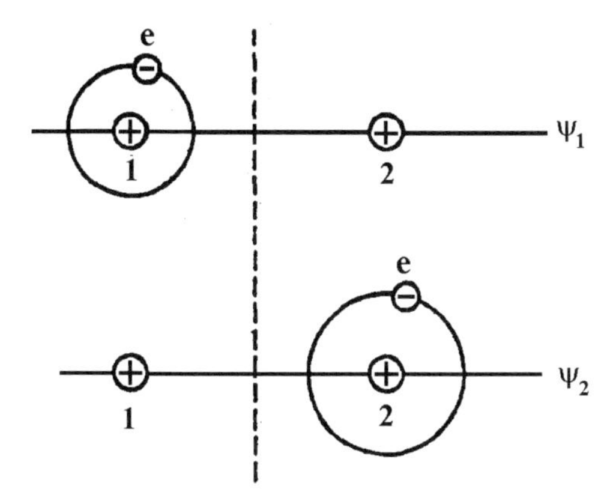

Figure 10.2. Electron in the field of two identical centers. Reflection in the symmetry plane corresponds to the transformation $\psi_1 \to \psi_2$, $\psi_2 \to \psi_1$. This yields the eigenfunctions of the system: the even electron wave function which retains its sign under reflection is equal to $\psi_g = (1/\sqrt{2})(\psi_1 + \psi_2)$, and the odd wave function is equal to $\psi_u = (1/\sqrt{2})(\psi_1 - \psi_2)$.

Hamiltonian has the form

$$\hat{H} = -\frac{1}{2}\Delta + V(r_1) + V(r_2) + \frac{1}{R}, \tag{10.16}$$

where R is the distance between the atomic cores, r_1, r_2 are the distances of the electron from the related nucleus, $V(r)$ is the interaction potential of the electron with ions, and far from the ion this potential is the Coulomb potential $V(r) = -1/r$. Use the symmetry of the problem under consideration, so that the symmetry plane is perpendicular to the line joining the nuclei and bisects it, and the electron reflection with respect to this plane conserves the electron Hamiltonian. Hence, the electron eigenstates can be divided into even and odd states, depending on the property of their wave functions to conserve or change their sign as a result of reflection with respect to the symmetry plane. Evidently, these wave functions are the following compositions of ψ_1 and ψ_2:

$$\psi_g = \frac{1}{\sqrt{2}}(\psi_1 + \psi_2), \qquad \psi_u = \frac{1}{\sqrt{2}}(\psi_1 - \psi_2). \tag{10.17}$$

These wave functions satisfy the Schrödinger equations

$$\hat{H}\psi_g = \varepsilon_g \psi_g, \qquad \hat{H}\psi_u = \varepsilon_u \psi_u, \tag{10.18}$$

where $\varepsilon_g(R)$, $\varepsilon_u(R)$ are the eigenvalues of the energies of these states. Let us define the exchange interaction potential in this case as

$$\Delta(R) = \varepsilon_g(R) - \varepsilon_u(R). \tag{10.19}$$

In order to determine this value at large distances between atoms we use the following method. Let us multiply the first equation by ψ_u^*, the second equation by ψ_g^*, take the difference of the obtained equations, and integrate the result over the volume which is a half-space restricted by the symmetry plane. Since the separation between the nuclei is large, the wave function ψ_2 is zero inside this volume and the wave function ψ_1 is zero outside this volume. Hence $\int_V \psi_u^* \psi_g d\mathbf{r} = 1/2$, and the relation obtained has the form

$$\frac{\varepsilon_g(R) - \varepsilon_u(R)}{2} = \frac{1}{2}\int_V (\psi_u \Delta\psi_g - \psi_g \Delta\psi_u)\, d\mathbf{r} = \frac{1}{2}\int_S \left(\psi_2 \frac{\partial}{\partial z}\psi_1 - \psi_1 \frac{\partial}{\partial z}\psi_2\right) d\mathbf{s},$$

where S is the symmetry plane which limits the integration range; we use relations (10.17) with real wave functions, and the z-axis joins the nuclei. Take the origin of the coordinate system in the center of the line joining the nuclei. Since the electron is found in the s-state in the field of each atomic core, its wave functions in this coordinate system can be represented in the form

$$\psi_1 = \psi\left(\sqrt{(z + R/2)^2 + \rho^2}\right), \qquad \psi_2 = \psi\left(\sqrt{(z - R/2)^2 + \rho^2}\right),$$

where ρ is the distance from the axis in the perpendicular direction to it. Since $d\mathbf{s} = 2\pi\rho\, d\rho$ we have, from the above relation,

$$\begin{aligned}\varepsilon_g(R) - \varepsilon_u(R) &= \int_0^\infty 2\pi\rho\, d\rho \\ &\quad \times \left[\psi\left(\sqrt{(z - R/2)^2 + \rho^2}\right)\frac{\partial}{\partial z}\psi\left(\sqrt{(z + R/2)^2 + \rho^2}\right)\right. \\ &\quad \left. - \psi\left(\sqrt{(z + R/2)^2 + \rho^2}\right)\frac{\partial}{\partial z}\psi\left(\sqrt{(z - R/2)^2 + \rho^2}\right)\right]_{z=0} \\ &= R\int_0^\infty d\rho^2 \frac{\partial}{\partial \rho^2}\psi^2\left(\sqrt{\frac{R^2}{4} + \rho^2}\right) = R\psi^2\left(\frac{R}{2}\right).\end{aligned}$$

In the course of the deduction of this formula we use the obvious relation

$$\frac{\partial}{\partial z}\left[\psi\left(\sqrt{(z + R/2)^2 + \rho^2}\right)\right]_{z=0} = R\frac{\partial}{\partial \rho^2}\psi\left(\sqrt{\frac{R^2}{4} + \rho^2}\right).$$

Now let us connect the molecular wave function $\psi(r)$ of the s-electron with the atomic function ψ_{at} whose behavior at large distances from the atomic core is determined by the Schrödinger equation

$$-\frac{1}{2}\frac{\partial^2}{\partial r^2}(r\psi_{\text{at}}) - \frac{1}{r}\psi_{\text{at}} = -\frac{\gamma^2}{2}\psi_{\text{at}},$$

where $\gamma^2/2$ is the electron binding energy. The solution of this equation is given by formula (4.11),

$$\psi_{\text{at}}(r) = Ar^{1/\gamma - 1}e^{-r\gamma}.$$

Take the molecular wave function in the form $\psi(r) = \chi(\mathbf{r})\psi_{\text{at}}(r)$ and compare the Schrödinger equations for the molecular and atomic wave functions near the axis between the nuclei where one can use the asymptotic form of the interaction potential $V(r) = -1/r$ in formula (10.16) for the electron Hamiltonian. So, we have, from the Schrödinger equation for ψ, neglecting the second derivative of χ near the axis,

$$\gamma \frac{\partial \chi}{\partial r_1} + \left(\frac{1}{R} - \frac{1}{r_2} \right) \chi = 0.$$

Solving this equation, we connect the molecular wave function of the s-electron near the axis with the atomic wave function that allows us to express the exchange interaction potential through the asymptotic parameters of the valent s-electron in the atom

$$\Delta_o = A^2 R^{2/\gamma - 1} e^{-R\gamma - 1/\gamma}. \tag{10.20}$$

In particular, this formula yields, for the exchange interaction of the proton and hydrogen atom in the ground state,

$$\Delta_o = \frac{4}{e} R e^{-R}. \tag{10.21}$$

Formula (10.20) is the asymptotic expression for the exchange interaction potential of a one-electron atom with a valent s-electron and its atomic core. The criterion of validity of this formula has the form

$$R\gamma \gg 1, \, R\gamma^2 \gg 1. \tag{10.22}$$

Formula (10.20) admits a generalization. Consider the interaction of a one-electron atom with the parent ion for an electron angular momentum l and its projection μ onto the molecular axis. Then the electron wave function is $\psi(\mathbf{r}) = Y_{l\mu}(\theta, \varphi)\Phi(r)$, where r, θ, φ are the spherical electron coordinates if its center coincides with a corresponding nuclei, and the z-axis directs along the molecular axis. For determination of the exchange interaction potential we based on deduction of formula (10.20) for the s-electron and make changes in integration over $d\rho$. Then we have

$$\Delta \sim \int_0^\infty \left| Y_{l\mu}(\theta, \varphi) \right|^2 \Phi^2(r) \rho d\rho,$$

where $r = \sqrt{R^2/4 + \rho^2}$ is the distance from each nucleus for the electron located on the symmetry plane. Since $\Phi(r) \sim e^{-\gamma r}$, the integral converges at small $\rho (\rho \sim \sqrt{R/\gamma} \ll R)$. Then $\Phi(r) = \Phi(R/2) e^{-\gamma\rho^2/R}$. This corresponds to the small angles $\theta = 2\rho/r$ and $Y_{l\mu}(\theta, \varphi) \sim \theta^\mu$ for $\theta \ll 1$. Thus we have

$$\Delta_{l\mu} = \Delta_o \int_0^\infty e^{-2\gamma\rho^2/R} \frac{\left| Y_{l\mu}(\theta, 0) \right|^2}{\theta^{2\mu}} \left| \frac{2\rho}{R} \right|^{2\mu} \cdot 4\gamma \frac{\rho d\rho}{R},$$

where the exchange interaction potential Δ_o is given by formula (10.20) and relates to an $s-$transferring electron with the same asymptotic parameters in the atom. Since the exchange interaction potential does not depend on the sign of μ, we take the momentum projection to be positive. Thus, we find, for the exchange interaction potential of a one-electron atom with the parent ion,

$$\Delta_{l\mu} = A^2 R^{2/\gamma-1-\mu} e^{-R\gamma-1/\gamma} \cdot \frac{(2l+1)(l+\mu)!}{(l-\mu)!\mu!(2\gamma)^\mu} = \Delta_o \frac{(2l+1)(l+\mu)!}{(l-\mu)!\mu!(2R\gamma)^\mu}. \tag{10.23}$$

In particular, for p-electrons it follows from this that

$$\frac{\Delta_{11}}{\Delta_{10}} = \frac{3}{R\gamma}.$$

In a general case, when the atom has several valent electrons, we use the LS-coupling scheme for the atom and the parentage scheme of summation of the electron and atomic core momenta into the atom and molecular ion momenta. We consider the case when the energy splitting, due to the orbital momentum projection on the molecular axis for the atom or ion remarkably exceeds the fine splitting of the energy levels. Then the quantum numbers of the system are the atom quantum numbers LSM_LM_S and the quantum numbers of the ion are $L'sM_{L'}m_s$. We sum up the electron momenta l, $\frac{1}{2}$ and the momenta of the atomic core $L's$ into the atom momenta LS, and then the atom spin S and the spin of another atom core s are summed into the total spin I of the molecular ion. Using formula (4.7) for the atom wave function within the framework of the parentage scheme and substituting it into the expression for the exchange interaction potential, we obtain

$$\begin{aligned}
\Delta &= n(G_{L's}^{LS})^2 \sum_{\mu, M'_{L'}, M'_L} \sum_{\sigma,\sigma',m_s,m'_s} \sum_{M_S, M'_S} \\
&\times \begin{bmatrix} l & L' & L \\ \mu & M_{L'} & M'_L \end{bmatrix} \begin{bmatrix} l & L' & L \\ \mu & M'_{L'} & M_L \end{bmatrix} \begin{bmatrix} \frac{1}{2} & s & S \\ \sigma & m_s & M_S \end{bmatrix} \\
&\times \begin{bmatrix} \frac{1}{2} & s & S \\ \sigma' & m'_s & M'_S \end{bmatrix} \begin{bmatrix} s & S & I \\ m_s & M_S & M \end{bmatrix} \begin{bmatrix} s & S & I \\ m'_s & M'_S & M \end{bmatrix} \Delta_{l\mu} \\
&= n(G_{L's}^{LS})^2 \sum_\mu \begin{bmatrix} l & L' & L \\ \mu & M_{L'} & M_L \end{bmatrix}^2 (2S+1) \begin{Bmatrix} s & \frac{1}{2} & S \\ s & I & S \end{Bmatrix} \Delta_{l\mu} \\
&= n \frac{I+\frac{1}{2}}{2s+1} (G_{L's}^{LS})^2 \sum_{\mu, M'_{L'}, M'_L} \begin{bmatrix} l & L' & L \\ \mu & M_{L'} & M'_L \end{bmatrix} \begin{bmatrix} l & L' & L \\ \mu & M'_{L'} & M_L \end{bmatrix} \Delta_{l\mu}
\end{aligned} \tag{10.24}$$

Here M is the projection of the total spin I onto the molecular axis; the result does not depend on this value because the influence of the spin on the exchange interaction is determined by the Pauli exclusion principle only, and the fine structure splitting is assumed to be small. Note that summation of the Clebsh–Gorgan coefficients leads to the $6j$ Wigner symbol which is denoted by braces,and its value is used. Formula (10.24) gives the asymptotic expression for the exchange interaction potential of an atom with its atomic core within the framework of the

LS-coupling scheme. The criterion for the validity of this expression is given by formula (10.22).

Note that the dependence of the ion–atom exchange interaction potential on the total spin is determined by the exchange of electrons which belong to different cores. Hence, at large separations R, the level of splitting due to different total spins I is weaker than $\sim \exp(-2\gamma R)$ and is small compared with the exchange interaction potential which is $\sim \exp(-\gamma R)$. This allows us to use, in formula (10.24), the average molecular spin $\overline{I}$. Next, since the exchange interaction potential decreases with increasing μ as $R^{-\mu}$, one can restrict by one term in the sum (10.24) with a minimal value of μ, and formula (10.24) takes the form

$$\Delta(l\mu, L'M_{L'}s, LM_LS) = \frac{\overline{I} + \frac{1}{2}}{2s+1}\left(G_{L's}^{LS}\right)^2 \begin{bmatrix} l & L' & L \\ \mu & M_{L'} & M_{L'}+\mu \end{bmatrix} \begin{bmatrix} l & L' & L \\ \mu & M_L-\mu & M_L \end{bmatrix} \Delta_{l\mu}, \tag{10.25}$$

where the brackets mean an average over the molecular total spin, μ is the possible minimal projection of the transferring electron momentum onto the molecular axis, and the value $\Delta_{l\mu}$ is given by formula (10.23).

The method used allows us to express this interaction potential through the asymptotic parameters of the valent atomic electron. Formulas (10.20), (10.21), (10.23), (10.24), and (10.25) give the first term of the asymptotic expansion of the ion–atom exchange interaction potential at large distances between the nuclei. The following terms of this expansion over a small parameter of the theory are of the order $1/R$ from the first one.

10.3 The LCAO Method

One can construct the asymptotic theory in another way. Let us take the wave functions ψ_1, ψ_2, and on their bases construct the eigenfunctions ψ_g, ψ_u of the Hamiltonian (10.16) according to formula (10.17). Then the eigenvalues for the energies of the levels follow from equations (10.18), and the exchange interaction potential (10.19) is equal to

$$\Delta = \varepsilon_g - \varepsilon_u = \frac{\langle\psi_g|\hat{H}|\psi_g\rangle}{\langle\psi_g|1|\psi_g\rangle} - \frac{\langle\psi_u|\hat{H}|\psi_u\rangle}{\langle\psi_u|1|\psi_u\rangle} = 2\langle\psi_1|\hat{H}|\psi_2\rangle - 2\varepsilon_o S, \tag{10.26}$$

where $\varepsilon_o = \langle\psi_1|\hat{H}|\psi_1\rangle$, and $S = \int \psi_1^*\psi_2\, d\mathbf{r}$ is the overlapping integral. If we take the wave functions ψ_1, ψ_2 such that they coincide with the atomic functions near the parent nuclei, and account for the action of the other atomic cores in the region between the nuclei, we obtain the above asymptotic expressions for the exchange interaction potential. Hence, formula (10.26) is another form of presentation of the exchange interaction potential. This is valid if the overlapping integral is small, i.e., at large distances between the nuclei.

Let us use formula (10.26) in order to estimate the dependence of the exchange interaction potential at a distance R between the nuclei. Take the electron Hamiltonian as $\hat{H} = \hat{h}_2 - 1/r_1 + 1/R$, where $\hat{h}_2$ is the Hamiltonian of a valent electron in the second atom. Because of the Schrödinger equation for the atomic electron $\hat{h}_2\psi_2 = \varepsilon_o\psi_2$, where $\varepsilon_o = -\gamma^2/2$ is the electron energy for the isolated atom we have, on the basis of formula (10.26), the following expression for the ion–atom exchange interaction potential:

$$\Delta = 2\left\langle \psi_1 \left| \frac{1}{r_1} - \frac{1}{R} \right| \psi_2 \right\rangle,$$

where r_1 is the electron distance from the first nuclei. Taking into account that the volume which determines the overlapping integral is of the order of R^2/γ we obtain, on the basis of the asymptotic expression (4.11) for the wave functions, the following estimate for the exchange interaction potential $\Delta \sim A^2 R^{2/\gamma-1} e^{-R\gamma}$. Of course, this estimate coincides with the dependence of formula (10.20). From this it also follows that the region of the first atom gives a contribution in the exchange interaction potential of the order of $\psi(R) \sim R^{1/\gamma-1} e^{-R\gamma}$. Hence, one can express the exchange interaction potential through the asymptotic parameters of the valent electron with the accuracy $\sim 1/R$. For determination of the subsequent expansion terms of the exchange interaction potential over a small parameter, information is required about the electron behavior inside the atom.

One can use a simple model for the calculation of the exchange interaction potential by replacing the molecular wave functions in formula (10.26) by the atomic functions. This approach is called the LCAO-method (linear combination atomic orbits) and is based on simplified wave functions. The LCAO-method is widely practiced due to its simplicity. Below we calculate, on the basis of this method, the exchange potential for the ion–atom interaction of a one-electron atom with a valent s-electron when the asymptotic expression is given by formula (10.20). Present the electron Hamiltonian (10.16) in the form $\hat{H} = \hat{h}_1 - 1/r_2 + 1/R$, where $\hat{h}_1$ is the electron Hamiltonian in the absence of the second ion. Since ψ_1 is the atomic wave function, we have $\hat{h}_1\psi_1 = \varepsilon_a\psi_1$, where $\varepsilon_a = -\gamma^2/2$ is the electron energy in an isolated atom. Thus the exchange interaction potential in this case is equal to

$$\Delta = 2\left\langle \psi_1 \left| \frac{1}{r_2} \right| \psi_2 \right\rangle - \frac{2}{R} S.$$

Now let us take the asymptotic expression (4.11) for the basis wave function

$$\psi(r) = \frac{1}{\sqrt{4\pi}} A r^{1/\gamma-1} e^{-r\gamma}$$

and use the ellipticity coordinates ξ, η, φ, so that $r_{1,2} = R(\xi \pm \eta)/2$. The ellipticity coordinates are located in the regions $-1 < \eta < 1$, $1 < \xi < \infty$, and the volume element is $d\mathbf{r} = \frac{\pi R^3}{4}(\xi^2 - \eta^2)\, d\xi\, d\eta$. At the start we calculate the overlapping

Table 10.2. The ratio of the ion-atom exchange interaction potential evaluated on the basis of the LCAO-method to the accurate one.

γ	$\frac{1}{3}$	$\frac{1}{2}$	1	2
f	$\frac{e^3}{35}=0.57$	$\frac{e^2}{10}=0.75$	$\frac{e}{3}=0.91$	$\frac{3\pi\sqrt{e}}{16}=0.97$

integral $S=\int \psi_1^*\psi_2\,d\mathbf{r}$, which is equal to

$$S=\frac{A^2}{4}\left(\frac{R}{2}\right)^{\frac{2}{\gamma}+1}\int_{-1}^{1}d\eta\int_{1}^{\infty}d\xi e^{-R\gamma\xi}(\xi^2-\eta^2)^{1/\gamma}$$
$$=\frac{A^2}{4}\left(\frac{R}{2}\right)^{\frac{2}{\gamma}+1}\int_{1}^{\infty}d\xi e^{-R\gamma\xi}\int_{-1}^{1}d\eta(1-\eta^2)^{1/\gamma}$$
$$=\frac{A^2}{4\gamma}\left(\frac{R}{2}\right)^{2/\gamma}e^{-R\gamma}\cdot\frac{\sqrt{\pi}\Gamma(1/\gamma+1)}{\Gamma(1/\gamma+3/2)}.$$

Here we use that the integral converges in the region $\xi-1\sim 1/(R\gamma)$ and, hence, we replace ξ by 1 everywhere except the exponent. This method leads to the following expression for the exchange interaction potential:

$$\Delta_{\rm LCAO}=\frac{A^2}{4}\left(\frac{R}{2}\right)^{2/\gamma-1}e^{-R\gamma}\frac{\sqrt{\pi}\Gamma(1/\gamma+2)}{\Gamma(1/\gamma+3/2)}. \tag{10.27}$$

Although the basis wave functions of the LCAO-method are not correct, this method gives the accurate dependence on separation R for the exchange interaction potential. It is of interest to compare the ratio of formulas (10.27) and (10.20) which have the form

$$f=\frac{\Delta_{\rm LCAO}}{\Delta_{\rm ac}}=\frac{\sqrt{\pi}}{2}\left(\frac{e}{4}\right)^{1/\gamma}\frac{\Gamma(1/\gamma+2)}{\Gamma(1/\gamma+3/2)},$$

where $\Delta_{\rm ac}$ is the asymptotic exchange interaction potential which is given by formula (10.20) and has the accurate dependence on separations at large R. The difference between the numerical factors increases with a decrease in γ. Table 10.2 contains the ratios of the numerical factors for these interaction potentials at some γ. From these data the validity follows the LCAO-method if the electron binding energy is not small.

10.4 Exchange Interactions of Atoms at Large Separations

We now consider the exchange interaction of two atoms for the case of two identical one-electron atoms with valent s-electrons (an example of this is the system of two hydrogen atoms). One can construct two combinations of the atomic wave functions when electrons are located near the parent cores as

$$\Psi_1 = \psi(1a)\psi(2b), \qquad \Psi_2 = \psi(1b)\psi(2a).$$

Here the arguments 1, 2 enumerate electrons and the arguments a, b enumerate nuclei, so that, for example, the wave function $\psi(1a)$ means that the first electron is located in the field of the atomic core a. There are two states of this electron system which are described by the following wave functions:

$$\Psi_s = \frac{1}{\sqrt{2}}(\Psi_1 + \Psi_2), \qquad \Psi_a = \frac{1}{\sqrt{2}}(\Psi_1 - \Psi_2),$$

and the wave function of the symmetric state Ψ_s is conserved as a result of the transposition of electrons while, as a result of this operation, the wave function of the antisymmetric state Ψ_a changes sign. For determination of the exchange interaction potential of two atoms we use a similar method to that used for the ion–atom exchange interaction potential. The eigen electron wave functions satisfy the Schrödinger equations

$$\hat{H}\Psi_s = \varepsilon_s\Psi_s, \qquad \hat{H}\Psi_a = \varepsilon_a\Psi_a,$$

where $\varepsilon_s(R)$, $\varepsilon_a(R)$ are the eigenvalues of the energies of these states. The exchange interaction potential is defined as

$$\Delta(R) = \varepsilon_a(R) - \varepsilon_s(R), \tag{10.28}$$

and the electron Hamiltonian has the form

$$\hat{H} = -\frac{1}{2}\Delta_1 - \frac{1}{2}\Delta_2 - \frac{1}{r_{1a}} - \frac{1}{r_{1b}} - \frac{1}{r_{2a}} - \frac{1}{r_{2b}} + \frac{1}{|\mathbf{r}_1 - \mathbf{r}_2|} + \frac{1}{R}, \tag{10.29}$$

where Δ_1, Δ_2 are the Laplacians of the corresponding electrons, $\mathbf{r}_1$, $\mathbf{r}_2$ are the electron coordinates, r_{1a} is the distance for the first electron from the nucleus a and, in the same manner, the other distances between electrons and nuclei are denoted. Now let us multiply the Schrödinger equation for Ψ_s by Ψ_a, for Ψ_a by Ψ_s, and taking the difference of the obtained equation, we integrate the result over the hyperspace $z_1 \le z_2$ (the z-axis is directed along the line joining the nuclei and the origin is taken in its center; in addition, we also take into account that the wave functions of the s-electrons are real). Thus we obtain

$$\Delta \int_V (\Psi_1^2 - \Psi_2^2)\, d\mathbf{r}_1\, d\mathbf{r}_2$$
$$= \int_V \left[\left(\Psi_1 \frac{\partial^2}{\partial z_1^2}\Psi_2 - \Psi_2 \frac{\partial^2}{\partial z_1^2}\Psi_1\right) + \left(\Psi_2 \frac{\partial^2}{\partial z_2^2}\Psi_1 - \Psi_1 \frac{\partial^2}{\partial z_2^2}\Psi_2\right)\right] d\mathbf{r}_1\, d\mathbf{r}_2.$$

Inside the volume V the wave function Ψ_2 is exponentially small, while the function Ψ_1 is exponentially small outside this volume. Hence $\int_V (\Psi_1^2 - \Psi_2^2)\, d\mathbf{r}_1\, d\mathbf{r}_2 = 1$. Let us transform the integral in the right-hand side of the above integral to an integral over the hypersurface $z_1 = z_2$. Since the transformation $\mathbf{r}_1 \to \mathbf{r}_2$, $\mathbf{r}_2 \to \mathbf{r}_1$ gives $\Psi_1 \to \Psi_2$, $\Psi_2 \to \Psi_1$, the second term of the above equation coincides with the

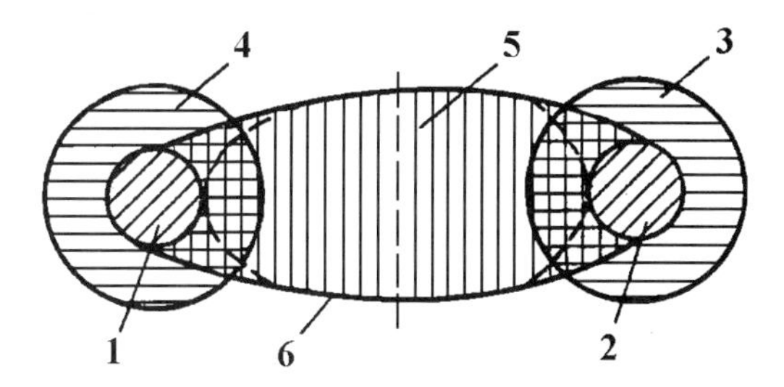

Figure 10.3. Electron regions which determine the exchange interaction potential for two identical atoms at large distances between the nuclei: 1, 2, internal regions of atoms where the electrons are located; 3, 4, regions where the asymptotic expressions for the atomic wave functions are valid; 5, in this region the quasi-classical approach is valid for valence electrons (this is restricted by a dotted line); 6, this region gives the main contribution into the exchange interaction potential of these atoms. Since the volume of region 6 is of the order of R^2, where R is the distance between the nuclei, and regions 1 and 2 possess a volume of the order of an atomic value, on the basis of asymptotic data for atomic wave functions, one can evaluate the exchange interaction potential with an accuracy of the order of $1/R^2$.

first one, and we find

$$\Delta = \int_S \left(\Psi_1 \frac{\partial}{\partial z_1} \Psi_2 - \Psi_2 \frac{\partial}{\partial z_1} \Psi_1 \right) dz\, dx_1\, dy_1\, dx_2\, dy_2, \qquad (10.30)$$

where the hypersurface S is described by the equation $z_1 = z_2 = z$.

We now evaluate this integral on the basis of the standard method which we used for the determination of the ion–atom exchange interaction potential. Let us present the wave functions of the atom system in the form

$$\Psi_1 = \psi(1a)\psi(2b)\chi_1, \qquad \Psi_2 = \psi(1b)\psi(2a)\chi_2.$$

Taking into account that the strongest dependence of the wave functions Ψ_1, Ψ_2 on the electron coordinates is determined by the atomic wave functions and that a region of electron coordinates near the molecule axis gives the main contribution to the integral (see Fig. 10.3), we have

$$\frac{\partial}{\partial z_1} \Psi_2 = \Psi_2 \frac{\partial}{\partial z_1} \ln \psi(1b) = \gamma \Psi_2,$$

and the integral (10.30) is equal to

$$\Delta = 2\gamma \int_S \Psi_1 \Psi_2\, dz\, dx_1\, dy_1\, dx_2\, dy_2$$
$$= 2\gamma \int_S \psi(1a)\psi(2b)\psi(1b)\psi(2a)\chi_1\chi_2\, dz\, dx_1\, dy_1\, dx_2\, dy_2.$$

For determination of the functions $\chi_1(\chi_2)$ we use the Schrödinger equation $\hat{H}\Psi_1 = E\Psi_1(\hat{H}\Psi_2 = E\Psi_2)$ and the expression (10.29) for the Hamiltonian, so that the electron energy in this approach is $E = -\gamma^2$. Taking the origin in the center of the molecular axis and introducing $a = R/2$, we have the equation for χ,

$$\left(\gamma\frac{\partial}{\partial z_1} - \gamma\frac{\partial}{\partial z_2} - \frac{1}{a - z_1} - \frac{1}{a + z_2} + \frac{1}{2a} + \frac{1}{|\mathbf{r}_1 - \mathbf{r}_2|}\right)\chi_1 = 0.$$

Here we use the asymptotic expressions (4.11) for the atomic wave functions and, based on the quasi-classical approach for the wave function, we neglect the second derivative of χ_1. Introducing the new variables $\xi = z_1 + z_2$, $\eta = z_1 - z_2$, we obtain this equation in the form

$$2\gamma\frac{\partial\chi_1}{\partial\eta} + \left(\frac{1}{2a} + \frac{1}{\sqrt{\eta^2 + \rho_{12}^2}} - \frac{1}{a - \xi - \eta} - \frac{1}{a + \xi - \eta}\right)\chi_1 = 0,$$

where $\rho_{12}^2 = (x_1 - x_2)^2 + (y_1 - y_2)^2$, and the variables ρ_{12}, ξ are included in the equation as parameters. Because of the linearity of this equation, one can solve it separately for each of the four terms in parentheses. Then let us present χ_1 in the form $\chi_1 = \varphi_1\varphi_2\varphi_3\varphi_4$ and demonstrate the analysis for the first term which satisfies the equation

$$\gamma\frac{\partial\varphi_1}{\partial\eta} = -\frac{1}{2a}\varphi_1.$$

Its solution is $\varphi_1 = C(\xi)\exp(-\eta/4a\gamma)$ with the boundary condition $\varphi_1 \to 1$, if $z_1 + z_2 > 0$, and $z_2 \to \infty$, i.e., $\varphi_1(2\xi - 2a) = 1$. This gives

$$\varphi_1 = \exp\left(-\frac{1}{2\gamma} - \frac{\eta - 2\xi}{4\gamma a}\right) \quad \text{for } z_1 + z_2 > 0.$$

In a similar way one can find the expression for φ_1 in the region $z_1 + z_2 < 0$ on the basis of the boundary condition $\varphi_1 = 1$, if $z_1 = -a$, i.e., at $\eta = -2a - 2\xi$. This yields

$$\varphi_1 = \exp\left(-\frac{1}{2\gamma} - \frac{\eta + 2\xi}{4\gamma a}\right) \quad \text{for } z_1 + z_2 < 0.$$

Fulfilling the same operations for φ_2, φ_3, φ_4 and then for χ_2, we calculate the integral (10.30) for the exchange interaction potential (10.28) and extract from this the dependence on the distance between the nuclei. The result has the following form:

$$\Delta(R) = \varepsilon_a(R) - \varepsilon_s(R) = A^4 F(\gamma) R^{\frac{7}{2\gamma}-1} e^{-2R\gamma}, \tag{10.31}$$

where A is the asymptotic coefficient for the atomic wave function in formula (4.11), and function $F(\gamma)$ is the following:

$$F(\gamma) = 2^{-1/\gamma-1}\gamma^{-2-1/2\gamma}\Gamma\left(\frac{1}{2\gamma}\right)\int_0^1 e^{u-1/\gamma}(1-u)^{3/2\gamma}(1+u)^{1/2\gamma}\,du. \tag{10.32}$$

Table 10.3 contains values of $F(\gamma)$ for some values of the argument, and Table 10.4 lists the values of the numerical factor in expression (10.31) for the interaction of hydrogen atoms and the atoms of alkali metals.

Table 10.3. Values of the function F on the basis of formula (10.32).

γ	0.80	0.84	0.88	0.92	0.96	1	1.04
F	0.0918	0.0944	0.0968	0.0990	0.1011	0.1031	0.1050
γ	1.08	1.12	1.16	1.20	1.24	1.28	1.32
F	0.1067	0.1083	0.1098	0.1112	0.1124	0.1136	0.1148

Table 10.4. Parameters of the exchange interaction potential for identical atoms of the first group of the periodical system of elements.

Atoms	H	Li	Na	K	Rb	Cs
γ	1	0.630	0.626	0.567	0.557	0.536
A^2	4	0.58	0.56	0.28	0.24	0.17
$100FA^4$	165	2.7	2.5	0.60	0.42	0.21

The criterion of the validity of the asymptotic expression (10.31) for the exchange interaction potential of atoms is (10.22), as well as the case of the ion–atom exchange interaction potential. The other analogy corresponds to the next term of expansion in the asymptotic theory which is of the order of $1/\sqrt{R}$ with respect to the main term (10.31). In the same way, one can find the accuracy with which the exchange interaction potential of two identical atoms can be expressed through the asymptotic parameters of the atomic valence electrons. Indeed, the contribution of regions which determine the normalization of the atomic wave functions into the exchange interaction potential of atoms is of the order of $\psi^2(R) \sim R^{2/\gamma-2}e^{-2R\gamma}$, so that this accuracy is of the order of $R^{-3/2\gamma-1}$.

The principal difference, compared to the ion–atom exchange interaction potential, is a different R-dependence of the exchange interaction potential of atoms obtained on the basis of the asymptotic theory and within the framework of the LCAO-model. In contrast to the ion–atom exchange interaction, in this case the functions χ_1, χ_2 are equal to zero in the middle of the space region which determines the exchange interaction potential. Hence, the molecular wave functions of the system under consideration differ, in principle, from the product of the atomic wave functions. From this it follows that the LCAO-method, with the atomic wave functions as a basis, gives at large R another power dependence on R for the exchange interaction potential of atoms other than that for the accurate asymptotic result.

10.5 Repulsion Interactions of Atoms with Filled Electron Shells

The long-range interaction potential of atomic particles and the exchange interaction potential are determined by different regions of electrons. Hence, the total interaction potential can be constituted as a sum of these. This is not strictly from a mathematical standpoint because the exchange and long-range interaction potentials have different dependence on the distance between nuclei, so that they cannot be terms of the same asymptotic series. But it is convenient, in practice, at separations where these interaction potentials have the same order of magnitude. In particular, the interaction potential of two atoms with valent s-electrons is equal to

$$U_{s,a}(R) = U_l(R) \pm \frac{1}{2}\Delta(R), \tag{10.33}$$

where $U_l(R)$ is the long-range interaction potential of atoms, Δ is the exchange interaction potential of atoms, the $+$ sign corresponds to the symmetric state, and the $-$ sign refers to the antisymmetric state. A long-range interaction potential because of its nature only acts at large distances between atoms; at moderate distances it disappears by transformations in interaction of Coulomb centers whose field is partially shielded by valent electrons. The exchange interaction potential of atoms, which is determined by the overlapping of electron orbits, is conserved at moderate distances between the nuclei. This accounts for partially the shielding of Coulomb charges of the nuclei. Thus one can use the exchange interaction potential for a description of the atom interaction at moderate distances between them.

This model can be successful under certain conditions. If the exchange interaction corresponds to attraction, it competes with the Coulomb repulsion of atoms. Because both potentials contain a remarkable error, in this case such a description cannot be valid. Hence, this model is suitable for the repulsive exchange interaction potential. This potential must be small compared to the typical electron energies, i.e., the distances between atoms are not small. This is the second requirement for this model. These conditions are valid for the interaction of atoms with filled electron shells when the interaction potential is not large. Below we consider this interaction.

In the case of the interaction of atoms of the second group, four valent s-electrons partake in the exchange interaction. The nature of the exchange interaction connects an exchange by electrons which belong to different atoms. But in each exchange only electrons with the same spin state can take part. Hence, in this case, the exchange interaction is put into effect by two pairs of electrons. Thus, the exchange interaction potential is $\Delta(R)$, and this repulsion interaction potential is given by formulas (10.31) and (10.32). Table 10.5 contains the parameters of this interaction potential.

Let us consider the exchange interaction of atoms with valent p-electrons. If the projection of the angular momenta of the interacting electrons onto the molecular axis is $m = 1$, the result for valent s-electrons with the same radial wave function must be multiplied by a factor of the order of θ^4, where θ is a typical angle between the radius-vector of the valence electron and the molecular axis. According to this estimate, electrons with $m = 1$ give the contribution $\sim R^{-2}$ to the exchange interaction potential compared to electrons with zero momentum projection ($m = 0$) onto the molecular axis.

Table 10.5. Parameters of the exchange interaction potential for identical atoms of the second group of the periodic system of elements.

Atoms	γ	A	FA^4
He	1.345	2.8	7.0
Be	0.829	1.9	1.2
Mg	0.756	1.5	0.44
Ca	0.678	0.96	0.069
Sr	0.652	0.87	0.046
Ba	0.620	0.76	0.026
Zn	0.830	1.7	0.78
Cd	0.813	1.6	0.60
Hg	0.878	1.7	0.86

In the case of the interaction of the identical atoms of inert gases we have that the exchange interaction potential is determined by electrons with the momentum projection $m = 0$ onto the molecular axis. One can use expression (10.31) for this interaction potential with the following corrections. First, in this interaction four electrons partake which have a zero momentum projection onto the molecular axis. Second, comparing this case with the above case when each of the interacting atoms has one s-electron, it is necessary to take into account that the angle wave function of the p-electron on the axis is $\sqrt{3}$ times more than that in the case of the valent s-electron. We finally obtain that the exchange interaction potential of two atoms of inert gases with valent p-electrons is equal to $9\Delta(R)$, and $\Delta(R)$ is given by formulas (10.31) and (10.32), where the parameters of the radial wave functions for the valent electrons are used in this formula. Let us present the exchange interaction potential of two identical atoms of inert gases in the form

$$U(R) = DR^{7/2\gamma-1}e^{-2R\gamma}. \tag{10.34}$$

Table 10.6 gives the parameters of this formula for the interaction of the identical atoms of inert gases.

It is convenient to use the logarithm derivative for a description of the repulsion interaction potential of two atoms with filled electron shells in some range of

Table 10.6. Parameters of the exchange interaction potential for the identical atoms of inert gases.

Atoms	Ne	Ar	Kr	Xe
γ	1.26	1.08	1.03	0.944
A	1.9	2.7	2.8	2.0
D	15	51	54	14

Table 10.7. The logarithmic derivation n for the repulsion interaction potential of two atoms of inert gases at the interaction energy 0.3 eV.

Atoms	He	Ne	Ar	Kr	Xe
He	5.9	5.6	5.2	5.5	5.2
Ne	—	7.6	6.6	7.6	6.8
Ar	—	—	6.1	7.0	5.9
Kr	—	—	—	7.7	7.1
Xe	—	—	—	—	6.4

separations which is given by

$$n = -\frac{d \ln U(R)}{d \ln R}. \tag{10.35}$$

Correspondingly, the interaction potential of two atoms in this region of distances has the following form:

$$U(r) = U_o \left(\frac{R_o}{R}\right)^n, \tag{10.36}$$

where r_o, U_o are the parameters. The parameter n characterizes the sharpness of the interaction potential. Table 10.7 contains values of this parameter for the interaction of atoms of inert gases at the distance where the interaction potential equals 0.3 eV. As is seen, this parameter is high enough to provide the validity of the hard sphere model for collisions of these atoms.

If the exchange interaction potential is responsible for the interaction of atoms at moderate distances between them, we have the simple relation for the parameter n as follows from the comparison of formulas (10.34) and (10.36):

$$n = 2R\gamma - \frac{7}{2\gamma} + 1. \tag{10.37}$$

Table 10.8 compares the real value of the parameter n with that obtained from formula (10.37) for the interaction of two identical atoms of inert gases at a distance which corresponds to the repulsive interaction potential 0.3 eV. In this table r_o is the corresponding distance between atoms for the real interaction potential and R_o corresponds to the interaction potential in the form (10.36). From these examples it follows that the asymptotic expression for the exchange interaction potential of two identical atoms with filled electron shells is suitable for a rough description of the interaction at moderate distances between them.

10.6 Exchange Interactions of Different Atoms

The expressions obtained for the exchange interaction potentials of an ion with the parent atom or for two identical atoms can be generalized for the interaction of

Table 10.8. Parameters of the repulsion of two identical atoms of inert gases at the interaction energy 0.3 eV.

	r_o, Å	n	R_o	$2r\gamma - 3.5/\gamma + 1$
He	1.58.	5.9	1.62	6.4
Ne	2.07	7.6	2.01	8.1
Ar	2.85	6.1	3.03	9.4
Kr	2.99	7.7	3.33	9.2
Xe	3.18	6.4	3.43	8.6

different atomic particles with close binding energies of electrons. This requires an additional criterion that the main contribution to this value is determined by the regions of electron coordinates where the electrons are found between the nuclei and near the molecular axis. Let us begin from the ion–atom exchange interaction by taking the electron binding energies of each atomic core to be $\gamma^2/2$ and $\beta^2/2$, so that the asymptotic expressions for the radial electron atomic wave functions (4.11) in the field of the first and second atomic cores are the following:

$$\psi_{\rm at}(r) = A_1 r^{1/\gamma - 1} e^{-r\gamma}, \qquad \varphi_{\rm at}(r) = A_2 r^{1/\beta - 1} e^{-\beta r}.$$

Repeating the operations for the deduction of the asymptotic expression (10.20) in the case of identical atomic particles we have, for the exchange interaction potential of an atom and ion of different sorts with valent s-electrons,

$$\Delta = R\psi\left(\frac{R}{2}\right)\varphi\left(\frac{R}{2}\right),$$

where $\psi(r)$, $\varphi(r)$ are the molecular wave functions which are transformed into the corresponding atomic wave functions $\psi_{\rm at}(r)$, $\varphi_{\rm at}(r)$ when the field of another ion is small enough. Expressing the molecular wave functions through the atomic functions by the same method as above $\psi(r) = \psi_{\rm at}(r)\chi_1$, $\varphi(r) = \varphi_{\rm at}(r)\chi_2$, we will find the previous quasi-classical expressions for the functions χ_1, χ_2. Then we have the expression for the ion–atom exchange interaction potential instead of formula (10.20) in the case of the s-state of the valent electrons when they are located in the field of the atomic cores

$$\Delta = A_1 A_2 R^{1/\beta + 1/\gamma - 1} \exp\left[-\frac{R}{2}(\beta + \gamma) - \frac{1}{2\beta} - \frac{1}{2\gamma}\right]. \qquad (10.38)$$

This formula is a generalization of formula (10.20) and is transformed into (10.20) in the case of the identical atoms.

Let us analyze the criterion of validity of this expression. Along the asymptotic conditions (10.22), this requires that the interaction potential is determined by the region of electron coordinates between the nuclei. In reality, this leads to the criterion

$$R\,|\beta - \gamma| \ll 1. \qquad (10.39)$$

Generalization of expression (10.38), to the case of nonzero orbital electron momenta in atoms, has a similar form to the case of the interaction of an ion with the parent atom because the angle dependence for the electron wave functions is similar in both cases.

In the same way one can generalize the asymptotic expression (10.31), for the exchange interaction potential of two identical atoms with valent s-electrons, to the case of the interaction of different s-atoms with close binding energies of electrons. Repeating the operations which we use for identical atoms we obtain, for the exchange interaction potential of two different one-electron atoms with valent s-electrons,

$$\Delta(R) = A_1^2 A_2^2 F(\beta, \gamma) R^{2/\beta+2/\gamma-1/(\beta+\gamma)-1} \exp\left[-R(\beta+\gamma)\right]. \tag{10.40}$$

Note that the numerical coefficient $F(\gamma)$ in formula (10.32) depends on γ not being strong and the parameters β and γ are close within the framework of the validity of this version of the asymptotic theory according to the criterion (10.39). This allows us to use a simple relation

$$F(\beta, \gamma) = F\left(\frac{\beta+\gamma}{2}\right),$$

where the function $F(x)$ is given by formula (10.32) and its values are given in Table 10.3.

If the criterion (10.39) violates strongly, the exchange interaction potential is determined by the electron region near an atom with a larger ionization potential. Below we consider, as an example of this, the interaction potential of the excited and nonexcited atoms. Since this is determined by an electron region near a nonexcited atom, it is convenient to take the surface of a nonexcited atom as the integration surface S in formula (10.30). One can deduct this formula on the basis of the Schrödinger equations for the wave functions of an excited electron $\Phi(\mathbf{r})$, and $\Psi(\mathbf{r})$ in the cases of the absence and presence of a nonexcited atom

$$-\frac{1}{2}\Delta\Phi + V_o\Phi = E_o\Phi, \qquad -\frac{1}{2}\Delta\Psi + V_o\Psi + V_1\Psi = E_1\Psi,$$

where V_o is the electron interaction potential with its own atomic residue, V_1 is the interaction potential with a nonexcited atom, and E_o, E_1 are the electron energies in the absence and presence of a nonexcited atom, so that the interaction potential of the related atoms is $U(R) = E_1 - E_o$. Making standard operations by multiplying the first equations by Ψ, multiplying the second equations by Φ (for simplicity we assume Ψ and Φ to be real), and integrating the result over a nonexcited atom volume, we then obtain

$$U(R) = \int_S (\Psi\nabla\Phi - \Phi\nabla\Psi)\, d\mathbf{s}. \tag{10.41}$$

We choose a surface S to be found far from the nonexcited atom, so that $V_1 = 0$ on this surface. The wave function of an electron with zero energy near an atom at distance r from its nucleus has the form $C(1 - L/r)$ where L is the electron–atom scattering length. The requirement of the coincidence of functions $\Psi(\mathbf{r})$ and

$\Phi(\mathbf{r})$ far from the perturbation atom gives $\Psi(\mathbf{r}) = \Phi(\mathbf{r})(1 - L/r)$. On the basis of this relation we obtain, for the exchange interaction potential of an excited and nonexcited atoms,

$$U(R) = 2\pi L \Phi^2(\mathbf{r}). \tag{10.42}$$

This formula, which is sometimes called the Fermi formula, determines a shift of the spectral lines of photons emitted as a result of the radiative transitions of excited atoms in the presence of a gas of neutral atoms.

In the same way let us consider the exchange interaction of positive and negative ions in the case when the binding energies of the electron in the negative ion and excited atom $\gamma^2/2$ are coincident. Then taking as the surface S in formula (10.41) the boundary of the atom which is a base of the negative ion we obtain, for the exchange interaction potential,

$$U(R) = \int_S (\psi \nabla \varphi - \varphi \nabla \psi)\, d\mathbf{s}, \tag{10.43}$$

where the wave functions ψ and φ transfer into the wave functions of the excited atom and negative ion when one can neglect the field of the other atomic cores. We use expression (7.8) for the electron wave function in the negative ion

$$\varphi(r) = B\sqrt{\frac{\gamma}{2\pi}} e^{-\gamma r}/r,$$

and take it near the atom to be independent of the field of the positive ion. Here r is an electron distance from the negative ion nucleus, and $\gamma^2/2$ is the binding energy in the negative ion. Since the main dependence of the electron wave function of the excited atom at distance r_2 from its center is $e^{-\gamma r_2}$, the electron wave function on the surface S is equal to

$$\psi(\mathbf{r}) = \psi(\mathbf{R}) \exp(\gamma R - \gamma |\mathbf{R} - \mathbf{r}|) = \psi(\mathbf{R}) \exp(\gamma r \cos\theta),$$

where θ is the angle between the vectors $\mathbf{r}$ and $\mathbf{R}$. Integrating over the atom surface S and using the above relations, we finally find the following expression for the exchange interaction potential of the negative and positive ions

$$\Delta(R) = 2B\sqrt{2\pi\gamma}\,\psi(\mathbf{R}). \tag{10.44}$$

Let us use this formula in the case of the exchange interaction of a negative ion with the parent atom. Then, on the basis of formula for the wave function of the valent s-electron (7.8) in the negative ion, and formula (10.43) we obtain

$$\Delta(R) = 2B\gamma e^{-\gamma R}/R. \tag{10.45}$$

A similar formula follows from the general expression $\Delta = R\psi^2(R/2)$ for the valent s-electron case, where $\psi(r)$ is the molecular wave function of the electron which now coincides with the atomic wave function because of the absence of a Coulomb interaction and which is connected to the radial wave function $\varphi(r)$ by the following formula $\psi(r) = \sqrt{4\pi}\varphi(r)$.

Let us consider one more case of the exchange interaction of atomic particles of different sizes. If an atom interacts with an excited atom of the same sort, an additional symmetry arises. This corresponds to the reflection of electrons in the symmetry plane which is perpendicular to the line joining the nuclei and bisects it. In order to extract this symmetry, let us take one electron to be excited and the other one to be nonexcited, so that we obtain two combinations of the basis wave functions $\Psi_1 = \psi(1a)\varphi(2b)$, $\Psi_1 = \psi(1b)\varphi(2a)$, where we number the valent electrons by the ciphers 1, 2 and the nuclei by the letters a, b; ψ, φ are the wave functions of the nonexcited and excited electrons. Let us take the Hamiltonian of electrons in the form (10.29) and divide it into two parts

$$\hat{H} = -\frac{1}{2}\Delta_1 - \frac{1}{2}\Delta_2 - \frac{1}{r_{1a}} - \frac{1}{r_{1b}} - \frac{1}{r_{2a}} - \frac{1}{r_{2b}} + \frac{1}{|\mathbf{r}_1 - \mathbf{r}_2|} + \frac{1}{R} = \hat{h}_1 + \hat{h}_2,$$

so that

$$\hat{h}_1 = -\frac{1}{2}\Delta_1 - \frac{1}{r_{1a}} - \frac{1}{r_{1b}} + \frac{1}{R}, \qquad \hat{h}_2 = -\frac{1}{2}\Delta_2 - \frac{1}{r_{2a}} - \frac{1}{r_{2b}} + \frac{1}{|\mathbf{r}_1 - \mathbf{r}_2|}.$$

If we take the first electron to be located near the first Coulomb center, and the second electron to be located near the second Coulomb center, and assume the size of the electron orbit of the first electron to be small in comparison with that of the second electron, formula (10.25) leads to the clear and simple expression for the exchange interaction potential

$$\Delta = \Delta_{ia}(1)S_2, \tag{10.46}$$

where $\Delta_{ia}(1)$ is the exchange interaction potential for the interaction of a nonexcited atom with its ion, S_2 is the overlapping integral for the excited electron, so that in the case of the nonexcited s-electrons $\Delta_{ia}(1)$ is determined by formula (10.20), and the overlapping integral is given by the formula $S_2 = \int \varphi(1a)\varphi(2a)\,d\mathbf{r}$, so that $\varphi(1a)$, $\varphi(2a)$ are the atomic wave functions in the field of the corresponding center.

Note that although the above examples do not include all the possible cases of the exchange interaction of atomic particles, they reflect the nature of this interaction and give a representation about its character.

Problems

Problem 10.1. *Determine the potential of the long-range interaction of highly excited and nonexcited atoms at large separations.*

Take the interaction operator according to formula (10.9), $\hat{V}(R) = 3D_z Q_{zz}/R^4$, where D_z is the projection of the dipole moment operator of a nonexcited atom onto the molecule axis and Q_{zz} is the component of the quadrupole moment of the excited atom. The interaction potential of atoms corresponds to the first order of the perturbation theory with respect to the quadrupole moment of the excited atom and to the second order of the perturbation theory with respect to the nonexcited atom. Then we obtain, for the interaction potential,

$$U(R) = \frac{3\alpha Q}{R^6}, \tag{10.47}$$

where α is the polarizability of the nonexcited atom, and Q is the quadrupole moment of an excited electron in its atom.

Problem 10.2. *Determine the interaction potential of two atoms of the same kind if one is found in the ground S-state and the other one is found in the resonantly excited P-state.*

On the basis of formula (10.8) we have, for the interaction potentials in the indicated states of the system of interacting atoms,

$$U_{s\sigma} = -\frac{2d^2}{R^3}, \qquad U_{a\pi} = -\frac{d^2}{R^3}, \qquad U_{s\pi} = \frac{d^2}{R^3}, \qquad U_{a\sigma} = \frac{2d^2}{R^3}, \tag{10.48}$$

where R is the distance between atoms, the subscripts s and a characterize the symmetry of the wave functions of the state (symmetric and antisymmetric, respectively) for the reflection of electrons in the symmetry plane which is perpendicular to the axis joining the nuclei and bisects it, and the subscripts σ, π correspond to the projection of the momentum of the P-atom onto the molecular axis which is equal to zero and one, respectively. Next, the matrix element d is introduced as $\mathbf{n}d = \langle s|\mathbf{D}|p\sigma\rangle$, where $\mathbf{n}$ is the unit vector which is directed along the axis.

Let us transfer from the symmetric and antisymmetric states of the quasi-molecule to the even and odd states which are usually used. The parity of states of such a symmetric system corresponds to the inversion of electrons with respect to the origin which is located in the middle of the molecular axis. In this case the inversion transformation $1a \to 1b$, $1b \to 1a$, $2a \to 2b$, $2b \to 2a$ (1, 2 are the numbers of electrons, a, b are the numbers of nuclei) causes the simultaneous transformation of the electron coordinates $x \to -x$, $y \to -y$, $z \to -z$ near the center where the electron is located. The total transformation of the spatial wave function of electrons depends on the parity of the wave function with respect to the transposition of electrons which is determined by the total electron spin of the system. Note that the inversion of the electron coordinates $x \to -x$, $y \to -y$, $z \to -z$ near the center where the p-electron is located leads to a change of the sign of the electron wave function. From this we establish the correspondence between the quantum numbers of the quasi-molecule. Indeed, we find that the states $^1\Sigma_u$, $^3\Sigma_g$ correspond to state σs, the states $^3\Pi_g$, $^1\Pi_u$ correspond to state πs, $^1\Sigma_g$, $^3\Sigma_u$ correspond to state σa, and the states $^1\Pi_g$, $^3\Pi_u$ correspond to state πa.

Problem 10.3. *Evaluate the possibility of the existence of the bound state for the molecular hydrogen ion* H_2^+ *in the odd* $^2\Sigma_u$*-state.*

The interaction potential in this state is given by formula (10.33) with the use of the polarization interaction potential as a long-range potential, and formula (10.21) for the exchange interaction potential

$$U(R) = -\frac{\alpha}{2R^4} + 2Re^{-R-1}, \tag{10.49}$$

where the polarizability of the hydrogen atom is $\alpha = 9/2$. This potential has the minimum $U(R_o) = -5.8 \cdot 10^{-5}$ at $R_o = 12.8$. This function is equal to zero at $R = 10.8$. This character of the interaction potential allows us to model it by the cut-off polarization potential

$$U(R) = \infty, \quad R < R_1; \qquad U(R) = -\alpha/(2R^4), \quad R > R_1.$$

The parameter R_1 in this potential lies between 11 and 13, and the wave function of nuclei Ψ satisfies the Schrödinger equation

$$-\frac{1}{2\mu}\frac{1}{R}\frac{d^2}{dR^2}(R\Psi) - \frac{\alpha}{2R^4}\Psi = -\frac{\gamma^2}{2\mu}\Psi \quad \text{and} \quad \Psi(R_1) = 0$$

where μ is the reduced mass of nuclei and $\gamma^2/2\mu$ is the binding energy of the nuclei. For a small binding energy we assume the parameters to be $\alpha\mu\gamma^2 \ll 1$. Then one can divide the region of R into two parts, so that at small distances one can neglect the right-hand side term and at large distances one can neglect the second term of the left-hand side of this equation. Combining these solutions in an intermediate region where both approaches are valid, we determine a number of bound states of this system. Thus we have

$$Psi = C_1 \sin\left(\frac{\sqrt{\alpha\mu}}{R} - \delta\right), \qquad R_1 < R \ll \frac{1}{\gamma},$$

and from the condition $\Psi(R_1) = 0$ we have $\delta = \sqrt{\alpha\mu}/R_1$. In the other region $R \gg \sqrt{\alpha\mu}$, we find

$$\Psi = C_2 \exp(-\gamma R)/R.$$

Combining these solutions in a region $\sqrt{\alpha\mu} \ll R \ll 1/\gamma$, we obtain $\tan\delta = \gamma\sqrt{\alpha\mu}$. From this it follows that if $\delta > \pi n$, where n is a whole number, there are n bound states. In the case under consideration the number of bound states of the nuclei corresponds to a whole part of the value $\sqrt{\alpha\mu}/(\pi R_1) = 21/R_1$. Since $11 < R_1 < 13$, one can conclude that there is one bound state of the hydrogen molecular ion in the odd electron state $^2\Sigma_u$.

Problem 10.4. *Determine the exchange interaction potential for the hydrogen atom in the ground state and multicharged ion at large separations when the fields of the proton and multicharged ion are separated by a barrier.*

Let us use formula (10.43) for the exchange interaction potential which has the form

$$\Delta(R) = \int_S (\psi_H \nabla \psi_i - \psi_i \nabla \psi_H)\, d\mathbf{s},$$

where ψ_H is the electron wave function which is centered on the proton, and a multicharged ion is the center of the wave function ψ_i. We choose, as a surface

S in the above formula, the plane that passes through the top of a barrier which separates the proton and multicharged ion fields.

This plane crosses the molecular axis at a distance $R_1 = R(1+\sqrt{Z})^{-1}$ from the proton and at a distance $R_2 = R\sqrt{Z}(1+\sqrt{Z})^{-1}$ from the multicharged ion. Take into account that the main distance dependencies, of the wave functions for the electron located in the field of the proton and multicharged ion, are equal to $\psi_H(r) \sim \exp(-\gamma_H r)$, and $\psi_i(r) \sim \exp(-\gamma_i r)$. Then, evaluating the integral over a surface S, we get

$$\Delta(R) = \frac{2\pi(\gamma_H + \gamma_i)}{\gamma_H/R_1 + \gamma_i/R_2}\psi_H(R_1)\psi_i(R_2) = \frac{4\pi R\sqrt{Z}}{(1+\sqrt{Z})^2}\psi_H(R_1)\psi_i(R_2). \quad (10.50)$$

We use here that $\gamma_H = \gamma_i = 1$. As a matter of fact, this formula is the generalization of formula (10.19) for the case of different Coulomb centers. The condition that the barrier separates regions of the action of the fields of different Coulomb centers has the following form in this case:

$$R \gg 4\sqrt{Z}.$$

The wave functions $\psi_H(r)$, $\psi_i(r)$ in the expression for the exchange interaction potential are the molecular wave functions, i.e., they include the action of the field of another particle. Let us express them through the electron wave functions $\varphi_H(r)$, $\varphi_i(r)$ of the isolated hydrogen atom and hydrogenlike ion on the basis of the standard method. Then we present the connection of these functions in the form $\psi_H(r) = \chi_H \varphi_H(r)$, $\psi_i(r) = \chi_i \psi_i(r)$, and determine functions χ_H, χ_i as a result of quasi-classical solution of the Schrödinger equation in neglecting of the second derivatives from χ_H, χ_i. This yields

$$\chi_H = \left(\frac{R}{R_2}\right)^Z \exp\left(-Z\frac{R_1}{R}\right), \qquad \chi_i = \frac{R}{R_1}\exp\left(-\frac{R}{R_2}\right).$$

From this we have for the exchange interaction potential, using the expression for the electron wave function in the hydrogen atom $\psi_H(r) = e^{-r}/\sqrt{\pi}$:

$$\Delta(R) = 4R\sqrt{\pi}\left(1+\frac{1}{\sqrt{Z}}\right)^{Z-1}\psi_i\left(\frac{R\sqrt{Z}}{1+\sqrt{Z}}\right)\exp\left(\sqrt{Z} - \frac{R\sqrt{Z}}{1+\sqrt{Z}}\right). \quad (10.51)$$

In particular, in the case of the interaction of the hydrogen atom and proton ($Z=1$, $\psi_i = \exp(-R/2)/\sqrt{\pi}$) we obtain from this formula $\Delta(R) = 4Re^{-R-1}$, that is, formula (10.21).

Problem 10.5. *Obtain the expression for the exchange interaction potential of the hydrogen atom in the ground state and a multicharged ion of charge $Z \gg 1$. The electron angular momentum l in the field of a multicharged ion is zero.*

Along with the expression of the previous problem for the exchange interaction potential in the system under consideration, we use the quasi-classical expression for the wave function of a highly excited electron in the field of the multicharged ion which is found in Problem 6.1 and has the form, in this case,

$$R_{n0}(r) = \frac{1}{r} \cdot \frac{C}{2\sqrt{|p|}} \exp\left(-\int_{r_o}^{r} |p|\,dr\right),$$

where

$$|p| = \sqrt{\gamma^2 - \frac{2Z}{r}}, \qquad \gamma = Z/n, \qquad r_o = 2n^2/Z, \qquad C = \sqrt{2}/(n^{3/2}\sqrt{\pi Z}).$$

From this we obtain, for the exchange interaction potential,

$$\Delta_{n0} = \frac{4}{n^{3/2}} \cdot \sqrt{\frac{2}{\pi}} \cdot \frac{b}{1-b} \cdot \left(\frac{1-\sqrt{b}}{1+\sqrt{b}}\right)^{\sqrt{b}} \exp[-Rf(b)], \tag{10.52}$$

where

$$b = \left(1 + \frac{2Z}{R}\right)^{-1}, \qquad f(b) = 1 - \frac{1-b}{2\sqrt{b}} \ln\frac{1+\sqrt{b}}{1-\sqrt{b}}.$$

In the limiting cases this formula gives

$$\Delta_{n0} = \frac{2}{n^{3/2}} \cdot \sqrt{\frac{2}{\pi}} \cdot \exp[-\frac{R^2}{3Z}], \qquad \sqrt{Z} \ll R \ll Z; \tag{10.53}$$

$$\Delta_{n0} = \sqrt{\frac{2}{\pi Z}} \cdot \left(\frac{2R}{Z}\right)^{Z} \cdot e^{-R}, \qquad R \gg Z.$$

Let us analyze the criterion of validity of the above formula for the exchange interaction potential. First, we assume that this interaction is determined by tails of the electron wave functions for the ground state of the hydrogen atom and a highly excited state of the electron in the Coulomb field of charge Z. This leads to the condition

$$Rf(b) \gg 1. \tag{10.54}$$

The other condition requires a closeness of the electron energies for bound states of the electron in the fields of the proton and multicharged ion. The difference of these energies is equal to

$$\Delta\varepsilon = \frac{1}{2} + \frac{Z-1}{R} - \frac{Z^2}{2n^2}.$$

Since this value is small, it chooses certain values of the principal quantum number of the electron in the multicharged ion field. It is necessary to compare the energy difference with the height of the barrier which separates fields of the proton and multicharged ion. The barrier height at a given distance between charges is equal to

$$U_{\max}(R) = \frac{1}{2} - \frac{2\sqrt{Z}+1}{R}.$$

The criterion of validity of the above asymptotic formulas for the exchange interaction potential is

$$\Delta\varepsilon \ll U_{\max}.$$

One more condition requires that the exchange interaction potential is small compared to the barrier height. This criterion is fulfilled if the previous ones are valid.

Problem 10.6. *Connect the exchange interaction potentials of the hydrogen atom and multicharged ion for cases when the electron orbital momentum l in the field of the multicharged ion satisfies the condition $l \ll Z$, where Z is the nuclear charge.*

We are working with formula (10.50) of Problem 10.4 for the exchange interaction potential of the hydrogen atom and multicharged ion. Let us give the criterion when this formula is valid for electron states with an orbital momentum l. Obtaining this formula, we assume the electron wave function of the multicharged ion to be independent of the angle θ between the electron radius-vector and molecular axis. This is valid for $l\theta \ll 1$ and, since $\theta \sim \rho/R$, where ρ is an electron distance from the axis, and the main contribution into the integral gives $\rho^2 \sim R\sqrt{Z}$, this requires fulfillment of the condition

$$l^2 \ll R\sqrt{Z}. \tag{10.55}$$

The electron state in the field of the multicharged ion is described by the quantum numbers nl. Let us connect the electron wave functions with the quantum numbers nl and $n0$. We have, from the quasi-classical expression of the radial electron wave function $R_{nl}(r)$, for the electron quantum number nl,

$$R_{nl}(r) = R_{n0}(r)\exp\left[-\int_{r_o}^{r} dr\sqrt{\gamma^2 - \frac{2Z}{r}} - \int_{r_l}^{r} dr\sqrt{\gamma^2 - \frac{2Z}{r} + \frac{l(l+1)}{r^2}}\right],$$

where $\gamma^2/2$ is the electron binding energy and r_0, r_l are the turning points for the classical electron motion at the indicated orbital momenta; at these distances from the center the corresponding root is zero. From this it follows that, at large r,

$$R_{nl}(r) = R_{n0}(r)\exp\left[-\frac{l(l+1)\gamma}{2Z}\right], \qquad r \gg Z/\gamma^2.$$

We take into account that the resonance with the hydrogen atom corresponds to $\gamma = 1$, and the angular wave function is equal to, at the molecular axis, $Y_{l0}(\theta,\varphi) = \sqrt{(2l+1)/(4\pi)}P_l(\cos\theta)$. From this we have the following connection between the exchange interaction potential and the electron orbital momentum

$$\Delta_{nl} = \sqrt{2l+1}e^{-l(l+1)/2Z}\Delta_{n0}. \tag{10.56}$$

Problem 10.7. *Construct the matrix of the exchange interaction potential of an ion and atom with valent p-electrons. Determine the exchange interaction potential for this system if the quantization axis, on which the momentum projection is zero, forms an angle θ with the molecular axis.*

Our goal is to construct the ion–atom exchange interaction potentials on the basis of formula (10.25) if the atomic particles have p-electron shells and are found in the ground states. For the atoms of Group 3 (one valent p-electron) and the atoms of Group 8 (one valent p-hole) of the periodic table of elements, when the ground states of the atom and ion are ^1S and ^2P, the exchange interaction potential of the interacting atom and ion, according to formula (10.25), gives,

$\Delta(M_L) =$

$M_L = -1$	$M_L = 0$	$M_L = +1$
Δ_{11}	Δ_{10}	Δ_{11}

, (10.57a)

where M_L is the orbital momentum projection for the atom (the elements of Group 3) or ion (the elements of Group 8). For the elements of Groups 4 and 7 of the periodic table, when the ground electron states of the atom and ion are ^3P and ^2P, the matrix of the exchange interaction potential according to formula (10.25) is

$\Delta(M_{L'}, M_L) = \frac{5}{3} \cdot$

	$M_L = -1$	$M_L = 0$	$M_L = +1$
$M_{L'} = -1$	Δ_{10}	Δ_{11}	Δ_{10}
$M_{L'} = 0$	Δ_{11}	$2\Delta_{11}$	Δ_{11}
$M_{L'} = 1$	Δ_{10}	Δ_{11}	Δ_{10}

, (10.57b)

where $M_{L'}$, M_L are the projections of the orbital ion and atom momenta. For the elements of Groups 5 and 6 of the periodic table, with the ground states of the atom and ion ^4S and ^3P, the matrix of the exchange interaction potential has the form

$\Delta(M_{L'}) = \frac{7}{3} \cdot$

$M_{L'} = -1$	$M_{L'} = 0$	$M_{L'} = 1$
Δ_{11}	Δ_{10}	Δ_{11}

. (10.57c)

We take as a quantization axis the direction in which the projection of the electron momentum is zero, and denote by θ the angle between the quantization and molecular axes. By definition, the exchange interaction potential $\Delta(\theta)$ of an atom and its ion, with valent p-electrons, is equal to

$$\Delta(\theta) = \frac{1}{3} \sum_M |d^1_{M0}(\theta)|^2 \Delta_{1M} = \frac{4\pi}{3} \sum_M |Y_{1M}(\theta, \varphi)|^2 \Delta_{1M},$$

where $d^1_{M0}(\theta)$ is the Wigner function of rotation and $Y_{1M}(\theta)$ is the spherical function, so that $4\pi|Y_{1M}(\theta)|^2$ is the probability of finding a state of a momentum projection M at angles θ, φ with respect to the molecular axis. The spherical function satisfies the normalization condition

$$\int_{-1}^{1} d\cos\theta \left|Y_{1M}(\theta)\right|^2 = \frac{1}{4\pi},$$

and $-1 \le \cos\theta \le 1$. From this we have, for the exchange interaction potential of an atom and parent ion in the case of Groups 3 and 8 of the periodic table of elements,

$$\Delta(\theta) = \Delta_{10} \cos^2\theta + \Delta_{11} \sin^2\theta. \tag{10.58a}$$

The matrix (10.57b) gives the ion–atom exchange interaction potential as a function of the angles between the quantization and molecular axes for elements of Groups 4 and 7 of the periodic system

$$\Delta(\theta) = \frac{5}{3}\left[\Delta_{10}\sin^2\theta_1\sin^2\theta_2 + \Delta_{11}(\cos^2\theta_1 + \cos^2\theta_2)\right], \tag{10.58b}$$

where θ_1, θ_2 are the angles between the molecular axis and the quantization axes for the atom and ion, respectively so that the electron momentum projection onto the quantization axis is zero. In the case of Groups 4 and 7 of the periodic system the exchange interaction potential is similar to that for atoms of Groups 3 and 8 and has the form

$$\Delta(\theta) = \frac{7}{3}\cdot\left(\Delta_{10}\cos^2\theta + \Delta_{11}\sin^2\theta\right) \tag{10.58c}$$

Although we are restricted by the ground states of the ion and parent atom, this is a general scheme of construction of the ion–atom exchange interaction potential. Being averaged over the total quasi-molecule spin I, the exchange interaction potential depends on the ion m and atom M_L angular momentum projections onto the molecular axis. This corresponds to the LS-coupling scheme for atoms and ions, i.e., we neglect the spin-orbital interaction. Hence, the above expressions correspond to the following hierarchy of the interaction potentials

$$V_{ex} \gg U(R),\, \Delta(R), \tag{10.59}$$

where V_{ex} is a typical exchange interaction potential for valent electrons inside the atom or ion, $U(R)$ is the long-range interaction potential between the atom and ion at large separations R, and $\Delta(R)$ is the exchange interaction potential between the atom and ion. Within the framework of the LS-coupling scheme for atoms and ions, we assume the excitation energies inside the electron shell to be relatively large, and this criterion is fulfilled for light atoms and ions. In the same manner one can construct the exchange interaction potential matrix for excited states within a given electron shell.

Problem 10.8. *Formulate the selection rules for the ion–atom interaction potential with the transition of one electron.*

Because the exchange interaction potential is determined by the transition of one electron from a valent electron shell and a transferring electron carries a certain momentum and spin, additional selection rules occur for one-electron interaction. In particular, in the case of a transition of a p-electron, the selection rules have the form

$$|L - l| \leq 1, \qquad |S - s| \leq 1/2. \tag{10.60}$$

These selection rules follow the properties of the Clebsh–Gordan coefficients in formula (6). If these conditions are violated, the ion–atom exchange interaction potential is zero in a scale of one-electron interaction potentials. Table 10.9 lists the states of atoms and their ions with valent p-electrons for which the ion–atom one-electron exchange interaction potential is zero.

Table 10.9. The electron configuration and state of an ion and the parent atom with valent p-electrons, so that a one-electron transition is forbidden between these states, and the exchange interaction potential of the ion and the parent atom is zero.

Ion state	Atom state
$p^2(^1D)$	$p^3(^4S)$
$p^2(^1S)$	$p^3(^4S)$
$p^2(^1S)$	$p^3(^2D)$
$p^3(^4S)$	$p^4(^1D)$
$p^3(^4S)$	$p^4(^1S)$
$p^3(^2D)$	$p^4(^1S)$

Problem 10.9. *Within the framework of the delta-function model for a negative ion, analyze the behavior of the electron terms of the quasi-molecule consisting of a negative ion and atom with a valent s-electron.*

The delta-function model for the negative ion assumes a radius of the action of the atom field in a negative ion to be small compared to the size of the negative ion. Hence, one can consider the atomic field as having the delta-function form. Then it is convenient to change the action of the atomic field on a valent electron by a boundary condition for the electron wave function. Let us consider the behavior of an s-electron in the field of two atomic centers on the basis of this model. Outside the atoms the electron wave function satisfies the Schrödinger equation

$$-\frac{1}{2}\Delta\Psi = -\frac{1}{2}\alpha^2\Psi,$$

where $\frac{1}{2}\alpha^2$ is the electron binding energy. The solution of this equation has the form

$$\Psi = Ae^{-\alpha r_1}/r_1 + Be^{-\alpha r_2}/r_2,$$

where r_1, r_2 are the electron distances from the corresponding nucleus. The boundary conditions for the electron wave function near the corresponding atom have the form

$$\frac{d\ln(r_1\Psi)}{dr_1}(r_1 = 0) = -\kappa_1, \qquad \frac{d\ln(r_2\Psi)}{dr_2}(r_2 = 0) = -\kappa_2, \tag{10.61}$$

where $1/\kappa$ is the electron scattering length for the atom or $\kappa^2/2$ is the electron binding energy in the negative ion formed on the basis of this atom.

We have, from the above boundary conditions,

$$-\kappa_1 = -\alpha + \frac{B}{A}e^{-\alpha R}/R; \qquad -\kappa_2 = -\alpha + \frac{A}{B}e^{-\alpha R}/R,$$

where R is the distance between atoms. Excluding the parameters A and B from this equation, we obtain

$$(\alpha - \kappa_1)(\alpha - \kappa_2) - e^{-2\alpha R}/R^2 = 0. \tag{10.62}$$

The solution of this equation allows us to find the dependence of the electron binding energy $\frac{1}{2}\alpha^2$ on the distance between atoms. In particular, the related electron term can intersect the boundary of a continuous spectrum ($\alpha = 0$). The electron bond is broken at $R_c = \sqrt{\alpha_1\alpha_2} = \sqrt{L_1L_2}$, where L_1, L_2 are the electron scattering lengths for the corresponding atom.

Problem 10.10. *An electron is located in the field of two identical atoms with spin 1/2. One can construct the negative ion with zero spin on the basis of the atom. Determine a distance between the negative ion and parent atom when the electron term intersects the boundary of the continuous spectrum within the framework of the model of the delta-function for the electron–atom interaction. Express the distance of the intersection of terms through the electron scattering lengths L_- and L_+ for the singlet and triplet states of the electron–atom system.*

We use the method of the previous problem taking into account a different character of the electron–atom interaction depending on their total spin. Accounting for the total spin of the interacting atom and negative ion to be 1/2, compose the electron wave function in a space between the atoms in the form

$$\Psi = \Phi_1(\mathbf{r})S_a + \Phi_2(\mathbf{r})T_a^- + \Phi_3(\mathbf{r})T_a^+ = \Psi_1(\mathbf{r})S_b + \Psi_2(\mathbf{r})T_b^- + \Psi_3(\mathbf{r})T_b^+,$$

where the nuclei are denoted by a and b, the spin wave functions S, T correspond to the atom spin zero and one, the subscript of the spin wave function indicates a nucleus where the negative ion is formed, and the subscript of the spin wave function gives the spin projection of a free atom. These spin functions have the form

$$S_a = \eta_-(b) \cdot \frac{1}{\sqrt{2}}[\eta_+(a)\eta_- - \eta_-(a)\eta_+],$$

$$T_a^- = \eta_-(b) \cdot \frac{1}{\sqrt{2}}[\eta_+(a)\eta_- + \eta_-(a)\eta_+], \qquad T_a^+ = \eta_+(b)\eta_-(a)\eta_+,$$

and similar notations correspond to the spin functions S_b, T_b^-, T_b^+, when the electron is located in a field of a b-atom. Here the subscript of the spin function η indicates the spin projection onto a given direction ($\pm\frac{1}{2}$), the argument gives an atom to which correspond these spin wave functions, and the electron spin function does not contain an argument. Using the connection between the spin wave functions S_a, T_a^-, T_a^+ and S_b, T_b^-, T_b^+, one can obtain the following relation between the above coordinate wave functions

$$\Psi_1 = \frac{\Phi_1 - \Phi_2}{2} + \frac{\Phi_3}{\sqrt{2}}, \qquad \Psi_2 = -\frac{\Phi_1 - \Phi_2}{2} + \frac{\Phi_3}{\sqrt{2}}, \qquad \Psi_3 = \frac{\Phi_1 + \Phi_2}{\sqrt{2}}.$$

In the region between the atoms the coordinate wave functions satisfy the Schrödinger equation

$$-\frac{1}{2}\Delta\Psi = -\frac{\alpha^2}{2}\Psi,$$

where $\alpha^2/2$ is the electron binding energy. Therefore, we present the electron coordinate wave functions in the form

$$\Phi_i = A_i e^{-\alpha r_a}/r_a + B_i e^{-\alpha r_b}/r_b, \qquad \Psi_i = C_i e^{-\alpha r_a}/r_a + D_i e^{-\alpha r_b}/r_b,$$

where $i = 1, 2, 3$, and r_a, r_b is the electron distance from the corresponding nucleus. Using the boundary conditions (10.61) and expressing the asymptotic behavior of the electron wave functions near atoms through the scattering lengths L_0 and L_1 for the total spin of the atom and electrons 0 and 1, respectively, we have, for the behavior of the above wave functions near the corresponding atoms,

$$\Phi_1 = \text{const}\left(\frac{1}{r_a} - \frac{1}{L_0}\right), \qquad \Phi_{2,3} = \text{const}\left(\frac{1}{r_a} - \frac{1}{L_1}\right), \qquad \text{if } r_a \to 0,$$

$$\Psi_1 = \text{const}\left(\frac{1}{r_b} - \frac{1}{L_0}\right), \qquad \Psi_{2,3} = \text{const}\left(\frac{1}{r_b} - \frac{1}{L_1}\right), \qquad \text{if } r_b \to 0.$$

From this we obtain the following set of equations:

$$\left(\alpha - \frac{1}{L_0}\right)A_1 - \frac{e^{-\alpha R}}{R}B_1 = 0, \qquad \left(\alpha - \frac{1}{L_1}\right)A_{2,3} - \frac{e^{-\alpha R}}{R}B_{2,3} = 0,$$

$$\left(\alpha - \frac{1}{L_0}\right)C_1 - \frac{e^{-\alpha R}}{R}D_1 = 0, \qquad \left(\alpha - \frac{1}{L_1}\right)C_{2,3} - \frac{e^{-\alpha R}}{R}D_{2,3} = 0.$$

This gives the equation for the energy of the electron terms

$$\left[\left(\alpha - \frac{1}{L_0}\right)^2 - \frac{1}{R^2}e^{-2\alpha R}\right]\left\{\frac{1}{R^4}e^{-4\alpha R} - \frac{1}{4R^2}e^{-2\alpha R}\right.$$
$$\times\left[\left(\alpha - \frac{1}{L_0}\right)^2 + 6\left(\alpha - \frac{1}{L_0}\right)\left(\alpha - \frac{1}{L_1}\right) + \left(\alpha - \frac{1}{L_0}\right)^2\right]$$
$$\left. + \left(\alpha - \frac{1}{L_0}\right)^2\left(\alpha - \frac{1}{L_0}\right)^2\right\} = 0.$$

This equation is the product of two factors so that the first factor corresponds to the total spin 3/2, and the second factor corresponds to the total spin of the system 1/2. Since we consider the interaction involving the negative ion whose spin is zero, only the second factor is of interest. This factor is divided into two factors:

$$\left[\frac{1}{R^2}e^{-2\alpha R} - \left(\alpha - \frac{1}{L_0}\right)\left(\alpha - \frac{1}{L_1}\right) + \frac{1}{2R}e^{-\alpha R}\left(\frac{1}{L_0} - \frac{1}{L_1}\right)\right],$$
$$\left[\frac{1}{R^2}e^{-2\alpha R} - \left(\alpha - \frac{1}{L_0}\right)\left(\alpha - \frac{1}{L_1}\right) - \frac{1}{2R}e^{-\alpha R}\left(\frac{1}{L_0} - \frac{1}{L_1}\right)\right] = 0.$$

The first factor corresponds to the odd state of the system and the second factor corresponds to the even state. This symmetry relates to the reflection of the electron with respect to the plane which is perpendicular to the molecular axis and bisects it. The intersection of the electron term with a boundary of continuous spectrum only takes place in the odd state. Since this corresponds to the equation $\alpha = 0$, we obtain the following relation for the distance of the intersection R_c:

$$R_c = 4\left[\sqrt{\frac{1}{L_0^2} + \frac{1}{L_1^2} + \frac{14}{L_0 L_1}} + \frac{1}{L_0} - \frac{1}{L_1}\right]^{-1}. \tag{10.63}$$

In particular, for the case of the interaction of the hydrogen atom with its negative ion ($L_0 = 5.8$, $L_1 = 1.8$) this distance is $R_c = 4.4$.

Problem 10.11. *Determine the asymptotic dependence on a distance between atoms for the exchange interaction potential of the two-charged ion with the parent atom at large separations.*

Use expression (10.26) for the exchange interaction potential

$$\Delta = 2\langle\psi_1|\hat{H}|\psi_2\rangle - 2\varepsilon_o\langle\psi_1|1|\psi_2\rangle,$$

where ε_o is the energy of the isolated ion and atom and ψ_1, ψ_2 are the wave functions of the interacting particles, so that in the first case the second particle is the ion and in the second case the electron wave function is centered on the second nucleus. Let us analyze the general structure of this integral. If we remove one electron from an atom, the second electron is located in the field of a two-charged ion. Hence the asymptotic behavior of the electron wave function makes valent electrons nonequivalent. Thus, take the electron wave function in the form $\psi_1 = \hat{P}\psi(r_{1a})\varphi(r_{2a})$, where the operator $\hat{P}$ transposes the electrons, r_{1a}, r_{2a} is the distance of the corresponding electron from the center a, and the one-electron wave functions $\psi(r)$, $\varphi(r)$ have a different asymptotic form. This chooses the form of the overlapping integrals, and the exchange interaction potential has the following structure:

$$\Delta \sim \langle\psi(r_{1a})\varphi(r_{2a})|\hat{H}|\psi(r_{2b})\varphi(r_{1b})\rangle.$$

We take into account that the wave functions $\psi(r)$, $\varphi(r)$ have a different dependence on the distance from its nucleus. The wave function of a valent electron with a smaller binding energy has the following asymptotic form in the field of a foreign atomic core, as follows from the quasi-classical solution of the Schrödinger equation in this region

$$\psi(r_{1a}) \sim r_{1a}^{1/\gamma - 1} e^{-\gamma r_{1a}} \left(\frac{R}{r_{1b}}\right)^{2/\gamma}.$$

As follows from the structure of the exchange interaction potential, it is determined by overlapping integrals when a weakly bound electron penetrates the region of a foreign atomic particle and overlaps the wave functions corresponding to a stronger bound electron. Then the exchange interaction potential is estimated as $\Delta \sim \psi^2(r_a \to R, r_b \sim 1)$. This leads to the following dependence:

$$\Delta \sim R^{6/\gamma - 2} e^{-2R\gamma}. \tag{10.64}$$

In particular, in the case of the interaction of an α-particle with a helium atom this dependence is

$$\Delta \sim R^{2.45} e^{-2.69R}.$$

CHAPTER 11

Diatomic Molecules

11.1 Quantum Numbers and Types of Bonds in Diatomic Molecules

Diatomic molecules are the simplest systems of bound atoms. The analysis of a diatomic molecule is based on a small parameter m/M, where m is the electron mass and M is a typical nuclear mass. In the first approximation of expansion over this small parameter, one can consider the nuclei to have infinite mass, so that the molecular parameters depend only on the distance R between the nuclei. The dependencies of the energy $E_n(R)$ of certain states of a diatomic molecule on distance R between the nuclei are called the electron terms of the molecule.

In particular, Fig. 11.1 contains the lowest electron terms of the CN-molecule which are typical for stable molecular states. Below we consider the nuclear states of the molecule at weak excitations. Then the electron energy can be expanded near the bottom of the potential well and has the form

$$E_n(R) = E_{no} - k(R - R_e)^2/2,$$

where the distance R_e corresponds to the energy minimum and $k = \partial^2 E_n/\partial R_e^2$. The Schrödinger equation for the nuclear wave function Ψ is given by

$$-\frac{\hbar^2}{2\mu}\Delta_R \Psi + \frac{k(R - R_e)^2}{2}\Psi = \varepsilon_n \Psi,$$

where μ is the reduced mass of the nuclei, and ε_n is the energy of the nuclear degrees of freedom for a given electron term. The solution of the Schrödinger equation allows us to divide the nuclear degrees of freedom in the molecule into vibrational and rotational degrees of freedom, so that the energy of the nuclei in

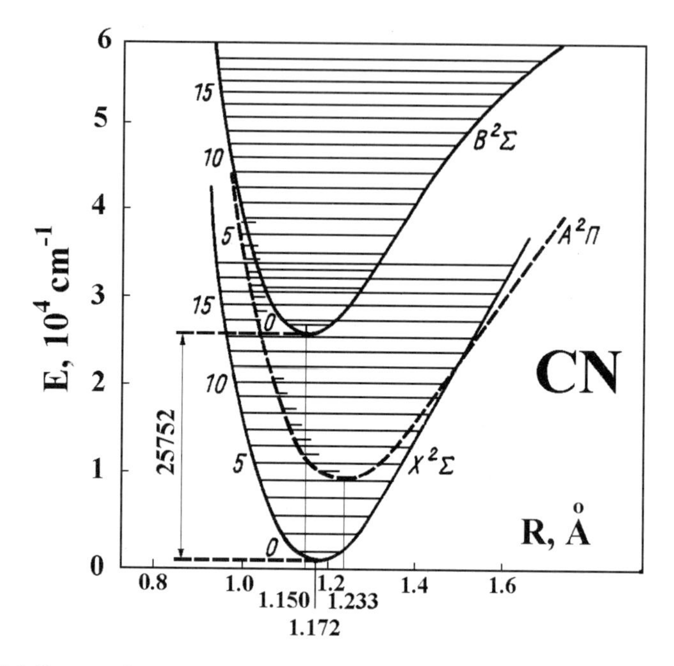

Figure 11.1. Lowest electron terms of a CN-molecule as typical electron terms of stable molecular states. Near the bottom these potential curves have a parabolic form, and a certain number of vibrations is found as an attractive part of these electron terms.

the first approximation takes the form

$$\varepsilon_n = \hbar\omega(v + 1/2) + \frac{\hbar^2}{2\mu R_e^2} J(J+1), \tag{11.1}$$

where v, J are the vibration and rotation quantum numbers which begin from zero and which are whole positive numbers, $\hbar\omega$ is the energy of excitation of the vibration level, and the value

$$B = \frac{\hbar^2}{2\mu R_e^2} \tag{11.2}$$

is called the rotational constant of a molecule. As follows from the Schrödinger equation, $\omega = \sqrt{k/\mu}$. Hence, comparing the values of the typical vibrational $\hbar\omega$ and the rotational B energies with the typical electron energy $\varepsilon_e \sim me^4/\hbar^2$, one can find $\hbar\omega \sim \varepsilon_e\sqrt{m/\mu}$ and $B \sim \varepsilon_e m/\mu$.

The parameters $\hbar\omega$, B are called the spectroscopic parameters of the molecule. Different orders of magnitude for electron, vibration, and rotation energies allow us to separate these degrees of freedom and present the total molecule energy as the sum of these energies. Usually for this relation the following notations are used

$$T = T_e + G(v) + F_v(J), \tag{11.3}$$

where T is the total excitation energy of the molecule, T is the excitation energy of an electron state, $G(v)$ is the excitation energy of a vibrational state, and $F_v(J)$ is the excitation energy of a rotational state. Formula (11.1) includes the main part of the vibrational and rotational energies for a weakly excited molecule. As the molecule is excited in the limits of a given electron term, it is necessary to introduce corrections to formula (11.1) in order to take into account a divergence of the electron term from the parabolic form in the region of excitation. Then the vibrational and rotational energies of a molecule, accounting for the first corrections, take the form

$$G(v) = \hbar\omega_e(v + 1/2) - \hbar\omega_e x(v + 1/2)^2,$$

$$F_v(J) = B_v J(J + 1), \qquad B_v = B_e - \alpha_e(v + 1/2), \tag{11.4}$$

and B_e is given by formula (11.2).

The parameters $\hbar\omega_e$, $\hbar\omega_e x$, B_e, and α_e are the spectroscopic parameters of the diatomic molecule. The subscript e indicates that these parameters correspond to the equilibrium distances between the nuclei for the electron term under consideration. The quantum numbers of the electron terms of the diatomic molecules are determined by the character of the interaction inside the molecule. Three types of interaction are of importance for a diatomic molecule. The first type is described by the interaction potential V_e that corresponds to the so-called interaction between the orbital angular momentum of electrons and the molecular axis. In the absence of other interactions, due to this interaction, an electron term is characterized by the projection of the molecular orbital angular momentum onto the molecular axis. The second type of interaction, whose potential is denoted by V_m, corresponds to the spin-orbit interaction. The third type of interaction in a diatomic molecule is denoted by V_r and accounts for the interaction between the orbital and spin electron momenta with rotation of the molecular axis. This interaction often is called the Coriolis interaction. Depending on the ratio between the potentials of these interactions, a certain character of formation of quantum numbers of the molecule takes place. Possible types of this ratio between the above interaction potentials lead to various limiting cases which are known as the Hund coupling rules. The various types of cases of the Hund coupling are summarized in Table 11.1 together with the quantum numbers which describe an electron term in this case.

Table 11.1 presents the standard classification of the Hund coupling as it was given by R.S. Mulliken in 1930. Sometimes, for the completeness of this scheme, one more case "e′" is added to the above cases with the relation between the interaction potentials $V_m \gg V_r \gg V_e$. But this case does not differ in principle from case "e". Indeed, in this case, the molecule is characterized by the same quantum numbers J and J_N, although the method of obtaining these quantum numbers is different. In case "e" first the momenta **L** and **S** are quantized onto the axis **N**, and then their momentum projections are summed into J_N; in case "e′" first the momenta **L** and **S** are summed in the total electron momentum **J**, and then the total momentum is quantized onto the direction **N** yielding the quantum number J_N.

Table 11.1. The cases of Hund coupling. In the used notations, **L** is the electron angular momentum, **S** is the total electron spin, **J** is the total electron momentum, **n** is the unit vector along the molecular axis, **K** is the rotation momentum of nuclei, Λ is the projection of the angular momentum of electrons onto the molecular axis, Ω is the projection of the total electron momentum **J** onto the molecular axis, S_n is the projection of the electron spin onto the molecular axis, and L_N, S_N, J_N are projections of these momenta onto the direction of the nuclei rotation momentum **N**.

Hund case	Relation	Quantum numbers
a	$V_e \gg V_m \gg V_r$	Λ, S, S_n
b	$V_e \gg V_r \gg V_m$	Λ, S, S_N
c	$V_m \gg V_e \gg V_r$	Ω
d	$V_r \gg V_e \gg V_m$	L, S, L_N, S_N
e	$V_r \gg V_m \gg V_e$	J, J_N

The character of the coupling is changed with excitation of the molecule because the role of the interaction between the axes and molecule rotation grows as the molecule is excited. In addition, rotation of the quasi-molecule axis in the course of the atom collisions is, in principle, for the transitions between multiplets or degenerated atomic states. Nevertheless, it is convenient to use the moleculer states, in the absence of rotation, as the basis states for the study of excited states and atomic transitions. Therefore let us consider case "a" of the Hund coupling and determine the quantum numbers of the diatomic molecule in this case. Neglecting the relativistic interactions, one can present the electron Hamiltonian as a sum of the kinetic energies of the electrons, the Coulomb potential of the interaction of electrons with nuclei, and the Coulomb potential of the interaction between electrons. This Hamiltonian is invariant with respect to the rotation of the molecule around its axis. Hence, the projection of the orbital angular electron momentum onto the molecular axis Λ is a quantum number of the diatomic molecule. The states with a certain value of Λ are denoted by Greek capital letters which are identical to the notation of the orbital angular momentum L of an atom. In particular, molecular states with $\Lambda = 0, 1, 2, 3$ are denoted by the Σ, Π, Δ, Φ quantum numbers.

One more quantum number of the diatomic molecule corresponds to its symmetry for reflection of the electrons in the plane which passes through the molecular axis. If the momentum projection onto the molecular axis is not zero, this operation changes the sign of the momentum projection. Hence, molecular states with $\Lambda \neq 0$ are twice degenerated, because two states with different signs of Λ are characterized by the same energy. For the state with $\Lambda = 0$ the above operation of electron reflection conserves the molecular state, so that the molecular wave function is multiplied by a constant as a result of such a reflection. Because two reflections return the system to the initial state, this constant is equal to ± 1. Thus, the Σ-terms are of two kinds, so that the wave function of a state Σ^+ conserves its sign as a result of the reflection of electrons with respect to a plane passed through the molecular axis, and the wave function of the state Σ^- changes sign under this operation.

Along with the above quantum numbers, the total electron spin S of the molecule, and its projection onto the molecular axis, are quantum numbers of the molecule. Since neither the electron energy nor the space wave function of electrons depend on the spin projection (they depend on the total electron spin through the Pauli exclusion principle), this quantum number does not partake in the state notation. Thus, in the notation of the electron terms of the diatomic molecules are included, the projection of the orbital angular momentum of electrons onto the molecular axis, the multiplicity $2S+1$ with respect to the electron spin which is indicated as a superscript left from the notation of the momentum projection, and for the Σ-states is also given the parity for the reflection of electrons in the plane passed through the molecular axis. For example, the notation $^3\Sigma^-$ relates to the electron term with the electron spin $S = 1$, with zero projection of the orbital momentum onto the molecular axis, and this state is odd for reflection of the molecular electrons with respect to a plane passing through the molecular axis; the notation $^2\Pi$ corresponds to an electron term with the spin $S = 1/2$ and the orbital momentum projection onto the molecular axis $\Lambda = 1$.

Dimers are molecules or molecular ions with identical nuclei. They have the additional symmetry plane which is perpendicular to the line joining the nuclei and bisects it. The reflection of electrons with respect to this plane conserves the Hamiltonian. Because two such reflections return the molecule in the initial state, the wave function of the electrons can conserve or change its sign as a result of one reflection. Usually the parity of the dimer states is characterized with respect to inversion of the electrons. This operation corresponds to a change of signs of the three electron coordinates if the origin of the coordinate frame is taken as the middle of the molecular axis. Since the inversion operation can proceed, as a result of reflection in the plane passing through the molecular axis and the reflection in the symmetry plane which is perpendicular to the molecular axis, both methods of introduction of the dimer parity are identical. The even molecular state is denoted by the subscript g (even) right from the notation of the orbital momentum projection and the odd state, with respect to electron inversion, is denoted by the subscript u (odd). For example, the ground state of the oxygen molecule is denoted as $^3\Sigma_g^-$, so that this state is even for inversion of the electrons. Note that symmetry of the dimers is determined by the identity of the atomic fields and does not depend on the isotope state of the nuclei, while for vibrational and rotational states the isotope state of the nuclei is of importance.

Let us take into account the spin-orbit interaction within the framework of case "a" of the Hund coupling when we want to sum the orbital angular momentum and the spin of an atom into a total electron momentum. In this case it is necessary to use the projection of the total electron momentum Ω onto the molecular axis, instead of the projection of the orbital momentum. In the same way, the parity for reflection of the electrons with respect to a plane passing through the molecular axis corresponds to the total electron wave function instead of the spatial electron wave function in the absence of a spin orbit interaction. The notation of the projection of the total electron momentum Ω is given by a cipher. For example, the electron term $^3\Sigma_g^-$ is divided into the terms 0_g^- and 1_g as a result of accounting for the spin-orbit interaction. Note that the operation of the reflection of electrons in symmetry planes can act on the spatial and total electron wave functions in a different way. If we assume, as usual, that the spin wave function is even with respect to these reflections, we obtain the correspondence between the operations with the spatial and total wave functions of the electrons.

The description of the molecular states within the framework of cases "a" or "c" of the Hund coupling is a convenient model for electron terms. The corresponding notation of electron terms are used even if the criteria of these cases are violated. Usually, the sequence of electron terms is denoted by the capital letters X, A, B, C, and so on, in the order of electron excitation. The notation of the electron term follows from these letters in accord with the notation of cases "a" or "c" of the Hund coupling. For example, the sequence of the electron terms of the molecule NO is $X^2\Pi$, $A^2\Sigma^+$, $B^2\Pi$, and so on. For some molecules, along with the capital letters, lowercase letters are used in order to distinguish the states of the different symmetry. For example, singlet states of the hydrogen molecule and other similar dimers are denoted by capital letters, while for the triplet states lowercase letters are used. The sequence of the electron terms of a certain molecule reflect the history of the study of this molecule. Since some excited states of the molecule were discovered when notation of the neighboring electron terms were introduced, they are denoted by a letter with a prime or two primes. In addition, the sequence of letters may not correspond to the excitation of the molecule because of errors in the positions of terms in the course of the study of this molecule. Briefly, there are only general regulations for the notation of the electron terms of molecules, and the adopted notation cannot be in correspondence with a universal scheme.

Thus, we concentrate our attention on cases "a" and "c" of the Hund coupling as a convenient model or as a basis for the accurate description of both a molecule and a quasimolecule which is formed in the course of the collisions of atoms. Let us determine the corrections in the molecule energy due to the interaction between electron and nuclear momenta, which allows us to estimate the reality of this approximation. In formulas (11.1), (11.3) above we neglect this interaction. We now obtain the expression for the rotational energy of the molecule taking this interaction into account. We introduce the total molecular momentum $\mathbf{J} = \mathbf{L} + \mathbf{S} + \mathbf{K} = \mathbf{j} + \mathbf{K}$, where $\mathbf{L}$ is the orbital angular momentum of the electrons, $\mathbf{S}$ is the total electron spin, $\mathbf{j} = \mathbf{L} + \mathbf{S}$ is the total electron momentum, and $\mathbf{K}$ is the nuclear rotation momentum. We now replace these values by their operators and take into account that the total molecular momentum is the quantum number of the molecule because its operator commutes with the Hamiltonian of the molecule. We denote the eigenvalues of this operator by J, and this is the quantum number for any case of the Hund coupling.

Returning to the deduction of formula (11.1) we have, for the nuclear rotational energy which is connected with the angular wave function for the relative motion of the nuclei,

$$\varepsilon_{\mathrm{rot}} = \langle B(R)\hat{\mathbf{K}}^2 \rangle = B(R)\langle (\hat{\mathbf{J}} - \hat{\mathbf{j}})^2 \rangle, \tag{11.5}$$

where $B(R) = \hbar^2/(2\mu R^2)$ and the angle brackets mean an average over the electron and nuclear states of the molecule. Since the total momentum of the molecule is the quantum number, one can extract it by taking into account the relation $\langle \hat{\mathbf{J}}^2 \rangle = J(J+1)$. This gives

$$\varepsilon_{\text{rot}} = B(R)\left[J(J+1) - 2\langle \hat{\mathbf{J}} \rangle \langle \hat{\mathbf{j}} \rangle + \langle \hat{\mathbf{j}}^2 \rangle\right]$$

Let us introduce the unit vector along the molecular axis $\mathbf{n}$ and use the relation $\langle \hat{\mathbf{K}}\mathbf{n} \rangle = 0$ which follows from the definition of the nuclear rotation momentum for a linear molecule. Since, in the related case, we have only one explicit vector $\mathbf{n}$, all the average values of the vectors are directed along $\mathbf{n}$. Hence we have $\hat{\mathbf{j}} = \hat{j}_n \mathbf{n}$, and we find

$$\varepsilon_{\text{rot}} = B(R)\left[J(J+1) - 2\langle (\hat{\mathbf{J}} - \hat{\mathbf{K}})\mathbf{n}\hat{j}_n \rangle + \langle \hat{\mathbf{j}}^2 \rangle\right].$$

Since $\langle \hat{j}_n \rangle = \Omega$, where Ω is the projection of the total electron momentum onto the molecular axis, we obtain:

$$\varepsilon_{\text{rot}} = B(R)\left[J(J+1) - 2\Omega^2 + \left\langle \hat{\mathbf{j}}^2 \right\rangle\right]. \tag{11.6}$$

The last term gives the contribution to the molecular energy, but this does not depend on the quantum numbers J and Ω. Thus the interaction between the electron and nuclear momenta leads to a shift in the total rotation energy of the molecule. This effect does not reflect on the difference between the energies of the rotational levels if the nuclear rotation momentum in formulas (11.1) and (11.5) is replaced by the total rotation momentum of the molecule, and the energy shift is small at large nuclear rotational momenta. Thus, in reality, formula (11.5) is correct, although its deduction was made by neglecting the interaction between the molecule rotation and the electron motion.

11.2 Correlation Diagrams for the Correspondence Between the Molecular and Atomic States

There is a certain connection between the electron terms of diatomic molecules and the quantum numbers of atoms from which the molecule is constructed. The scheme which establishes this conformity is called the correlation diagram. Below we consider some examples of this conformity. We first analyze the electron terms of a molecule consisting of different atoms within the framework of case "a" of the Hund coupling and assume that the LS-coupling scheme (see Chapter 4) is valid for each atom. Then one can produce summation of the orbital and spin momenta of atoms in molecular momenta separately. Based on this, we calculate the number of molecular states with a certain value Λ of the orbital momentum projection onto the molecular axis. Let the molecule consist of atoms with orbital angular momenta L_1 and L_2, respectively, and $L_1 > L_2$. The total number of states, i.e., the number of different projections of atom momenta, is equal to $(2L_1+1)(2L_2+1)$. Using the result of Problem 11.1, we obtain that the number of electron terms with $\Lambda = 0$ is equal to $(2L_2 + 1)$. Because the electron terms with $\Lambda \neq 0$ are degenerated twice, we find the total number of electron terms $L_1(2L_2 + 1)$ which have a nonzero projection of the electron orbital momentum onto the molecular axis. One can calculate the number of electron terms having each momentum projection. There are $(2L_2 + 1)$ electron terms with $\Lambda = 1, 2, \ldots, L_1 - L_2$, i.e., the number of electron terms of the molecule is the same for Λ in the regions from 1 to $L_1 - L_2$; then the number of electron terms decreases by one with an increase in Λ by one and is equal to one at $\Lambda = L_1 + L_2$.

If the molecule consists of identical atoms, a new symmetry occurs which corresponds to the electron inversion with respect to the center of the molecular axis. This leads to some changes in the number of states. If the atoms are found in different states, the number of the corresponding states is doubled, because the excitation can transfer to the other atom. If the atoms are found in the same state, the total number of states of a given Λ is the same as in the case of different atoms. But since the total molecular spin acts on the state parity, the parity of states in each case depends on the molecule spin. Table 11.2 lists the symmetry of the possible molecular states if the molecule is formed from identical atoms in the same states for case "a" of the Hund coupling.

The character of summation of the atomic momenta determines the possible states of the molecule consisting of these atoms. Table 11.3 gives a list of the molecular states for the j–j-coupling scheme inside each atom. Then the molecular momentum is summed from the total (orbital and spin) atomic momenta. The summation of the atomic momenta into the molecular momentum projection corresponds to both cases "a" and "c" of the Hund coupling.

Let us consider the character of the transition from atomic states to molecular states on the basis of some examples. We first consider the correlation diagram for the NH molecule (see Fig. 11.2). In this case one can neglect the spin-orbit interaction, so that the behavior of the molecular terms is determined by the electrostatic and exchange interactions of the electrons in the molecule. We concentrate upon the lowest states of the molecule which are usually determined by the state $2p^3$ of the electron shell of the nitrogen atom. Then the three lowest electron states of the nitrogen atom ^4S, ^2D, ^2P, are separated by energy as a result of the exchange interaction in the atom due to the Pauli exclusion principle. Interaction with the hydrogen atoms leads to the splitting of these levels, and the molecular electron terms include all the possible projections of the orbital momentum and the total electron spin onto the molecular axis. A part of these terms (approximately one-half) corresponds to the attraction of atoms, the other part corresponds to their repulsion.

In this analysis the parity of the states is of importance for the reflection of electrons with respect to a plane passed through the molecular axis. Then the parity of the molecular state is the sum of the parities of the atomic states, and this value is conserved with the variation of separations between the nuclei. The parity of atomic states is denoted by the subscript g for the even state and by the subscript u for the odd state. The total parity of the interacting atoms determines the

Table 11.2. Correlation between the states of two identical atoms and the dimer consisting of these atoms in neglecting the spin-orbit interaction. The number of dimer states of a given symmetry is given in parentheses, otherwise it is 1.

Atomic states	Dimer states
^1S	$^1\Sigma_g^+$
^2S	$^1\Sigma_g^+, {}^3\Sigma_u^+$
^3S	$^1\Sigma_g^+, {}^3\Sigma_u^+, {}^5\Sigma_g^+$
^4S	$^1\Sigma_g^+, {}^3\Sigma_u^+, {}^5\Sigma_g^+, {}^7\Sigma_u^+$
^1P	$^1\Sigma_g^+(2), {}^1\Sigma_u^-, {}^1\Pi_g, {}^1\Pi_u, {}^1\Delta_g$
^2P	$^1\Sigma_g^+(2), {}^1\Sigma_g^-, {}^1\Pi_g, {}^1\Pi_u, {}^1\Delta_g, {}^3\Sigma_u^+(2), {}^3\Sigma_g^-, {}^3\Pi_g, {}^3\Pi_u, {}^3\Delta_u$
^3P	$^1\Sigma_g^+(2), {}^1\Sigma_u^-, {}^1\Pi_g, {}^1\Pi_u, {}^1\Delta_g, {}^3\Sigma_u^+(2), {}^3\Sigma_g^-, {}^3\Pi_g, {}^3\Pi_u, {}^3\Delta_u,$
	$^5\Sigma_g^+(2), {}^5\Sigma_u^-, {}^5\Pi_g, {}^5\Pi_u, {}^5\Delta_g$
^4P	$^1\Sigma_g^+(2), {}^1\Sigma_g^-, {}^1\Pi_g, {}^1\Pi_u, {}^1\Delta_g, {}^3\Sigma_u^+(2), {}^3\Sigma_g^-, {}^3\Pi_g, {}^3\Pi_u, {}^3\Delta_u$
	$^5\Sigma_g^+(2), {}^5\Sigma_u^-, {}^5\Pi_g, {}^5\Pi_u, {}^5\Delta_g, {}^7\Sigma_u^+(2), {}^7\Sigma_g^-, {}^7\Pi_g, {}^7\Pi_u, {}^7\Delta_g$
^1D	$^1\Sigma_g^+(3), {}^1\Sigma_u^-(2), {}^1\Pi_g(2), {}^1\Pi_u(2), {}^1\Delta_g(2), {}^1\Delta_u, {}^1\Phi_g, {}^1\Phi_u, {}^1\Gamma_g$
^2D	$^1\Sigma_g^+(3), {}^1\Sigma_u^-(2), {}^1\Pi_g(2), {}^1\Pi_u(2), {}^1\Delta_g(2), {}^1\Delta_u, {}^1\Phi_g, {}^1\Phi_u, {}^1\Gamma_g,$
	$^3\Sigma_u^+(3), {}^3\Sigma_g^-(2), {}^3\Pi_g(2), {}^3\Pi_u(2), {}^3\Delta_g, {}^3\Delta_u(2), {}^3\Phi_g, {}^3\Phi_u, {}^3\Gamma_u$
^3D	$^1\Sigma_g^+(3), {}^1\Sigma_u^-(2), {}^1\Pi_g(2), {}^1\Pi_u(2), {}^1\Delta_g(2), {}^1\Delta_u, {}^1\Phi_g, {}^1\Phi_u, {}^1\Gamma_g,$
	$^3\Sigma_u^+(3), {}^3\Sigma_g^-(2), {}^3\Pi_g(2), {}^3\Pi_u(2), {}^3\Delta_g, {}^3\Delta_u(2), {}^3\Phi_g, {}^3\Phi_u, {}^3\Gamma_u,$
	$^5\Sigma_g^+(3), {}^5\Sigma_u^-(2), {}^5\Pi_g(2), {}^5\Pi_u(2), {}^5\Delta_g(2), {}^5\Delta_u, {}^5\Phi_g, {}^5\Phi_u, {}^5\Gamma_g$

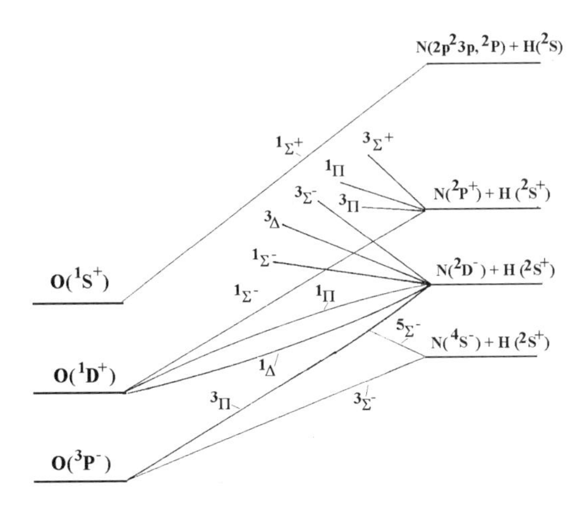

Figure 11.2. The correlation diagram for the electron terms of an NH-molecule. The subscript right above indicates the parity of states for the reflection of electrons in a plane passing through the molecular axis ($y \to -y$). Until the spin-orbit interaction potential is small compared to the exchange one, i.e., fulfilled for this molecule, the total parity of a molecular state is conserved in the course of a change of distances between the nuclei.

Table 11.3. Correlation between the states of two identical atoms and the dimer molecule consisting of these atoms in case "c" of the Hund coupling, if the atom state is characterized by the total electron momentum J. The number of dimer states of a given symmetry is indicated in parentheses, otherwise it is 1.

Atom states	Dimer states
$J = 0$	0_g^+
$J = 1/2$	$1_u, 0_g^+, 0_u^-$
$J = 1$	$2_g, 1_u, 1_g, 0_g^+(2), 0_u^-$
$J = 3/2$	$3_u, 2_g, 2_u, 1_g, 1_u(2), 0_g^+(2), 0_u^-(2)$
$J = 2$	$4_g, 3_g, 3_u, 2_g(2), 2_u, 1_g(2), 1_u(2), 0_g^+(3), 0_u^-(2)$

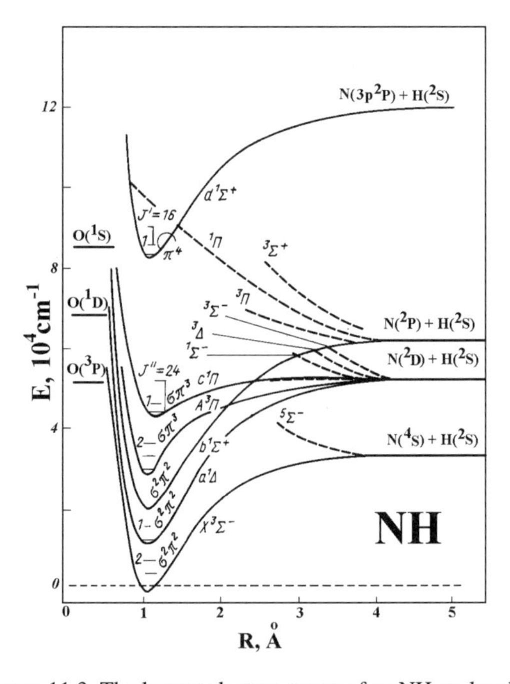

Figure 11.3. The lowest electron terms of an NH-molecule.

parity of the Σ-terms. At small distances between the atoms the electron terms of the NH molecule are transformed in terms of the oxygen atom. Evidently, stable molecular states correspond to such terms which are transformed to the lowest states of the oxygen atom. The positions of the lowest electron states of the NH molecule, which are presented in Fig. 11.3, confirm this assertion.

The case under consideration corresponds to case "a" of the Hund coupling where one can neglect the spin-orbit and the other magnetic interactions in the molecule. The next example corresponds to the interaction of two alkali metal atoms when one of these is found in the ^2S ground state, and the state of the other

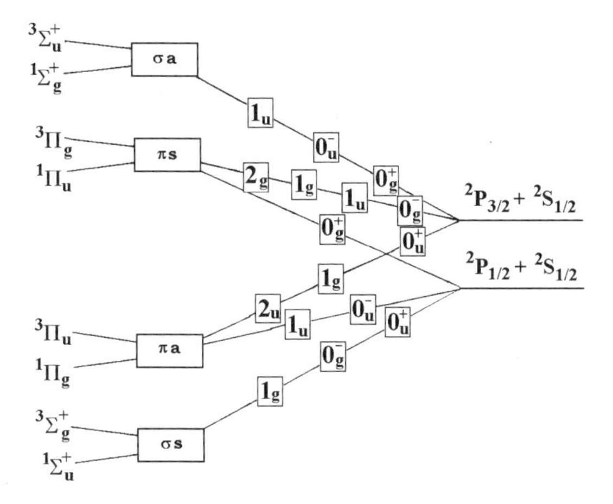

Figure 11.4. The correlation diagram for the electron terms of a quasi-molecule consisting of two identical alkali metal atoms so that one of these is found in the ground ^2S-state and the other is found in the resonance excited ^2P-state.

atom corresponds to the lowest resonance ^2P excited state. At large separations we have two levels of fine structure $^2S +^2 P_{1/2}$ and $^2S +^2 P_{3/2}$. The approach of atoms leads to the splitting of these levels due to the dipole–dipole interaction of atoms. This splitting is usually determined by the parity of the state for the inversion of electrons with respect to the molecule center and depends on the projection of the orbital momentum onto the molecular axis. At large distances between the atoms the fine splitting exceeds that of the dipole–dipole interaction, and at moderate distances between the atoms the spin-orbit interaction introduces a small contribution to the energetic position of the term. These conclusions follow from the correlation diagram for this case and is given in Fig. 11.4.

Thus, at large distances between atoms, case "c" of the Hund coupling is valid for a description of this quasi-molecule. Then the state is characterized by the projection Ω of the total electron moment of the molecule onto the molecular axis and by the symmetry (g, u) of the electron system for inversion with respect of the molecule center. If $\Omega \neq 0$, the electron terms are degenerated on the sign of the momentum projection. An additional symmetry occurs for terms with $\Omega = 0$, which corresponds to the parity of a state for the reflection of electrons with respect to a plane passing through the molecular axis. Note that the reflection symmetry is different for cases "a" and "c" of the Hund coupling. In case "a" the spatial and spin coordinates are separated, and the reflection operation corresponds to the spatial electron coordinates only. As a result of this transformation an individual electron changes the projection of the orbital momentum m on the molecular axis, i.e., this transformation gives $m \to -m$. In case "c" of the Hund coupling, the reflection operation results in the simultaneous change of the orbital and spin momentum projection, i.e., $m \to -m, \sigma \to -\sigma$, where $\sigma = \pm\frac{1}{2}$ is the spin projection of an individual electron onto the molecular axis. Then in the case "c" of Hund coupling, the parity of the molecular state, as a result of the reflection operation, leads to multiplication of the molecular wave function by the factor $(-1)^j$, where j is the total electron momentum of the molecule.

Returning to the interaction of two identical alkali metal atoms in the ground states ^2S and the resonantly excited ^2P states, we have at large separations that two levels of the fine structure of the ^2P term give the origin to several electron terms of the quasi-molecule. The states with the total electron momentum $j = 0, 1$ start from the lower level of this system and form the electron terms $1_g, 1_u, 0_g^-, 0_u^-$ from the states with $j = 1$ and from the electron terms $0_g^+, 0_u^+$ from the states with $j = 0$. The upper electron terms of this system correspond to $j = 1, 2$, and are 2_g, $2_u, 1_g, 1_u, 0_g^+, 0_u^+$ for $j = 2$ and $1_g, 1_u, 0_g^-, 0_u^-$ for $j = 1$. These electron terms are joined in groups with practically identical positions. In particular, if we account for a long-range dipole–dipole interaction only, these terms form four groups, and the positions of these terms in case "a" of the Hund coupling are determined in Problem 10.2. Note that the transition from case "c" to case "a" proceeds at large distances between the atoms because the dipole–dipole interaction is strong. In particular, Table 11.4 gives the values of a typical distance of the transition R_* which is determined by the relation $d^2/R_*^3 = \Delta_f$, where Δ_f is the difference of the $^2P_{3/2} - ^2P_{1/2}$ levels of the excited atom, and the splitting of levels $^1\Sigma_g^+ - ^1\Sigma_u^+$ at distance R_* between the atoms is equal to $4\Delta_f$, as follows from the solution of Problem 10.2. As is seen, the distance R_* is large enough, so that the other types of interaction are small at this separation. This means that the positions of only four electron terms are different at this and larger separations. At smaller distances between atoms, when the exchange interaction of atoms gives a contribution to the interaction potential, the electron terms obtain eight different positions. The correlation diagram in Fig. 11.4 reflects such behavior of the electron terms.

Thus, the behavior of the electron terms is determined by the competition of some interactions, and the variation of the distances between the nuclei can lead to a change in the quantum numbers which characterize the molecular states. We now consider this in the simple example of the interaction of an inert gas ion with the parent atom at large distances between the atoms. Then we take into consideration the fine splitting of the ion level and the long-range ion–atom interaction which depends on the projection of ion orbital momentum m onto the molecular axis and the exchange ion-atom interaction which also depends on m. It is of importance that the long-range and exchange interaction potentials do not depend on the fine structure state of the ion. This allows us to separate these interactions.

Figure 11.5 gives the correlation diagram for this case. Below we find positions of the levels by the standard method. Let us denote by Δ_f the fine splitting of the

Table 11.4. Parameters of the interaction of two atoms of alkali metals in the ground and resonantly excited states.

	Li	Na	K	Rb	Cs
d^2	5.4	6.2	8.8	8.6	11.5
R_*	150	43	32	20	17

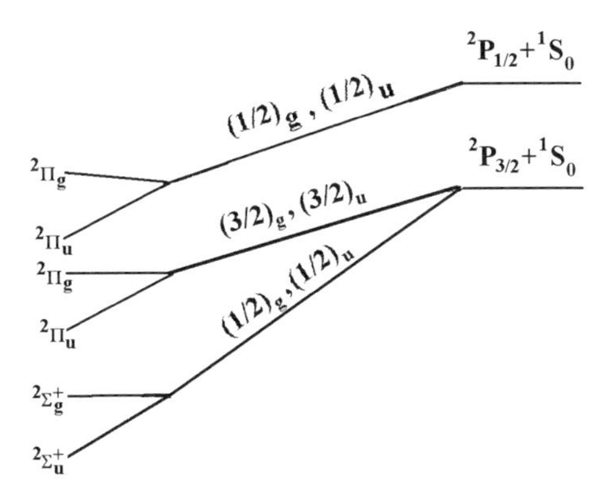

Figure 11.5. The correlation diagram for electron terms of a quasi-molecule which consists of an inert gas ion and the parent atom.

ion levels, U_0, U_1 are the long-range interaction potentials between the ion and atom for the projection of the ion orbital momentum onto the molecular axes 0 and 1, respectively, and we denote the ion–atom exchange interaction potential for the orbital momentum projection onto the molecular axes 0 and 1 by Δ_0 and Δ_1, respectively. The total number of states of the molecule is 12, and the simultaneous change of sign of the orbital momentum projection m and the spin projection σ onto the molecular axis does not change the Hamiltonian. Therefore, we must analyze six states from which three are even and three are odd for the operation of electron reflection, with respect to the symmetry plane which is perpendicular to the molecular axis and passes through its middle. Then using as a basis the states with a corresponding projection of the orbital and spin momenta of the ion onto the molecular axis, we obtain the Hamiltonian matrix H_{if} at these bases wave functions in the form

$H_{if} =$

	$m = 0, \sigma = \frac{1}{2}$	$m = 1, \sigma = -\frac{1}{2}$	$m = 1, \sigma = \frac{1}{2}$
$m = 0, \sigma = \frac{1}{2}$	$U_0 \pm \frac{1}{2}\Delta_0$	$-\frac{\sqrt{2}}{3}\Delta_f$	0
$m = 1, \sigma = -\frac{1}{2}$	$-\frac{\sqrt{2}}{3}\Delta_f$	$U_1 - \frac{1}{3}\Delta_f \pm \frac{1}{2}\Delta_1$	0
$m = 1, \sigma = \frac{1}{2}$	0	0	$U_1 - \frac{1}{3}\Delta_f \pm \frac{1}{2}\Delta_1$

(11.7)

Here the + and − signs relate to the corresponding parity of the state. We define the even and odd parity of a state depending on the conservation or change of sign of the electron wave function as a result of the inversion with respect to the molecule center. In this case the inversion of the electrons of the ion with respect to their centers leads to the change of sign of the electron wave function, while in the atom case the sign of the electron wave function does not change as a result of this operation. From this it follows that the lowest state of the molecular ion is $^2\Sigma_u^+$ and the upper sign in the matrix (11.6) corresponds to the odd state. Note that if we define the state parity with respect to reflection of the electrons in the symmetry plane, which is perpendicular to the molecular axis and passes through the molecule center, the lowest state of the molecular ion attains the symmetry $^2\Sigma_g^+$.

The matrix (11.6) allows us to determine the positions of the energy levels by a standard method on the basis of the secular equation $|\varepsilon - H_{if}| = 0$. This yields, for the energy of the electron states if we take the energy of the lowest level of the system at $R \to \infty$ to be zero

$$\begin{aligned}
\varepsilon_{1,2} &= \frac{U_0 + U_1}{2} \pm \frac{\Delta_0 + \Delta_1}{2} + \frac{\Delta_f}{2} \\
&\quad + \frac{1}{2}\sqrt{[U_0 - U_1 \pm (\Delta_0 - \Delta_1)]^2 - \frac{2}{3}[U_0 - U_1 \pm (\Delta_0 - \Delta_1)]\Delta_f + \Delta_f^2}, \\
\varepsilon_{3,4} &= \frac{U_0 + U_1}{2} \pm \frac{\Delta_0 + \Delta_1}{2} + \frac{\Delta_f}{2} \\
&\quad - \frac{1}{2}\sqrt{[U_0 - U_1 \pm (\Delta_0 - \Delta_1)]^2 - \frac{2}{3}[U_0 - U_1 \pm (\Delta_0 - \Delta_1)]\Delta_f + \Delta_f^2}, \\
\varepsilon_{5,6} &= U_1 - \frac{1}{3}\Delta_f \pm \frac{1}{2}\Delta_1.
\end{aligned} \tag{11.8}$$

Formula (11.8) gives the positions of all six electron terms of the system under consideration in increasing order by energy. Note that at large distances between the nuclei we have $U_0 \gg U_1$, $\Delta_0 \gg \Delta_1$. Since the long-range interaction potential U_0 has an inverse power dependence on R and since Δ_0 varies at large R exponentially, one can assume $U_0 \gg \Delta_0$ at large distances. Then the transition from case "c" to case "a" of the Hund coupling is determined by the relation between the values U_0 and Δ_f. Let us assume that U_0 corresponds to the attraction between the ion and atom (for example, this is the polarization interaction potential between the ion and atom). Then we obtain that only one lowest term corresponds to a strong attraction which varies as $\frac{2}{3}U_0$ at $\Delta_f \gg U_0$ and as U_0 at $\Delta_f \ll U_0$. At smaller distances between the nuclei, when the exchange ion–atom interaction becomes remarkable, this term is split, and the difference between the forming terms is of the order of Δ_0. Thus, the above analysis shows that a number of interacting terms and their behavior with a variation of distances between the atomic particles depends on the competition of different kinds of interaction in this system.

Let us analyze one more aspect of the interaction of electron terms when they are intersected. In reality we have the pseudocrossing of two levels if they belong to terms of the same symmetry. Let us analyze the behavior of two terms near the intersection point, dividing the Hamiltonian of the molecular electrons into two parts $\hat{H} = \hat{H}_o + V_m$, so that the corresponding terms are intersected for the Hamiltonian $\hat{H}_o$. For example, we include in V_m magnetic interactions in the molecule, so that the projection of the orbital momentum onto the molecular axis is the quantum number of the molecule in neglecting V_m, and the electron terms corresponding to different values of this quantum number are intersected. Then the magnetic interaction which mixes the electron terms of different projections of the orbital momentum leads to splitting of the electron terms. Let us study the behavior of the terms near the intersection point, so that the wave function of the first molecular state is $\psi_1(\mathbf{r}, R)$ and for the second state this function is equal to $\psi_2(\mathbf{r}, R)$, where $\mathbf{r}$ includes all the electron coordinates, and R is the distance between the nuclei. Let R_o be the intersection point for the Hamiltonian $\hat{H}_o$, so that we have

$$\left\langle \psi_1(\mathbf{r}, R_o) \left| \hat{H}_o \right| \psi_1(\mathbf{r}, R_o) \right\rangle = \left\langle \psi_2(\mathbf{r}, R_o) \left| \hat{H}_o \right| \psi_2(\mathbf{r}, R_o) \right\rangle.$$

Expanding these matrix elements near the intersection point, we introduce the difference of the inclines F of the electron terms on the basis of the relation

$$\left\langle \psi_1(\mathbf{r}, R) \left| \hat{H}_o \right| \psi_1(\mathbf{r}, R) \right\rangle - \left\langle \psi_2(\mathbf{r}, R) \left| \hat{H}_o \right| \psi_2(\mathbf{r}, R) \right\rangle = F(R - R_o).$$

Denoting the nondiagonal matrix element from the perturbation operator by $V = \langle \psi_1(\mathbf{r}, R) | V_m | \psi_2(\mathbf{r}, R) \rangle$ we neglect, for simplicity, the diagonal matrix elements from the perturbation operator because taking into account these elements only leads to a small shift of the intersection point. Then one can express the parameters of the given electron terms through the parameters F and V in the two-level approximation. Indeed, let us take the eigenfunction of the total Hamiltonian $\hat{H} = \hat{H}_o + V_m$ in the form

$$\Psi = a\psi_1(\mathbf{r}, R) + b\psi_2(\mathbf{r}, R),$$

so that this wave function satisfies the Schrödinger equation $\hat{H}\Psi = E\Psi$. From the orthogonality of the wave functions $\psi_1(\mathbf{r}, R)$ and $\psi_2(\mathbf{r}, R)$ we find the parameters of the above formulas, which are

$$a = \sqrt{\frac{1}{2} + \frac{V}{\sqrt{V^2 + F^2(R - R_o)^2}}}, \qquad b = \sqrt{\frac{1}{2} - \frac{V}{\sqrt{V^2 + F^2(R - R_o)^2}}},$$
$$E_1 - E_2 = \sqrt{V^2 + F^2(R - R_o)^2}. \tag{11.9}$$

These formulas characterize the behavior of the electron terms near the intersection point (see also Fig. 11.5). Thus, due to an additional interaction the electron terms, which are intersected within the framework of a simplified Hamiltonian of electrons, are mixed and the splitting between the related levels is determined by the nondiagonal matrix elements from the interaction potential which mixes these terms. Such a behavior of the electron terms is of importance for the collisional transitions between some atomic states when a certain interaction can cause these transitions or lead to the mixing of states. Then this effect determines the probability of the collisional transition.

11.3 The Parameters of Dimers and Their Ions

Below we consider dimers, i.e., molecules consisting of identical atoms. Because of the additional symmetry, some properties of these molecules can be analyzed at great length. Table 11.5 gives the spectroscopic parameters of these molecules which are determined by formula (11.3). Table 11.6 contains the spectroscopic parameters of the positive ions of dimers. These systems are similar to dimer molecules and have an additional symmetry for the inversion of electron with respect to the middle of the molecular axis. We can obtain the simple relation for the energetic parameters of positive molecular ions if we destroy a dimer in an atom, atomic ion, and electron in two different ways

$$A_2 \to 2A \to A + A^+ + e; \qquad A_2 \to A_2^+ + e \to A + A^+ + e. \tag{11.10}$$

This relation between the energetic parameters of the atomic particles has the form

$$D(A_2^+) = D(A_2) + J(A) - J(A_2), \tag{11.11}$$

where $D(XY)$ is the dissociation energy of a molecular particle XY and $J(X)$ is the ionization potential of an indicated particle.

A similar relation for negative ions has the form

$$D(A_2^-) = D(A_2) + EA(A) - EA(A_2), \tag{11.12}$$

where $EA(X)$ is the electron affinity of a particle X. Table 11.7 contains the spectroscopic parameters (11.4) of the dimer negative ions which are found in the ground electron state.

11.4 The Method of Molecular Orbits

For the analysis of diatomic molecules one can use a model which is similar to the shell atom model. This model is called the method of molecular orbits and is convenient in the case of dimers. Let us introduce a self-consistent potential of the molecular field which acts on electrons and present the molecule as a result of the distribution of electrons by states of this self-consistent field. Evidently, this field has axial symmetry, so that the states of individual electrons are characterized by the projection of the orbital momentum onto the molecular axis. In addition, the parity of the corresponding molecular orbit is the quantum number of electrons of the dimers or their ions. The accepted notation for molecular orbits is σ, π, δ for the molecular orbits of electrons with projections 0, 1, 2 of the electron orbital momentum onto the molecular axis, respectively. The method of molecular orbits possesses a central place in quantum chemistry, allowing one to evaluate the structures of various molecules and to calculate the energies of the corresponding molecular states. In our consideration of the atom interaction this method helps in the analysis of some of the properties of quasi-molecules as systems of interacting atoms. We demonstrate this method below by some examples.

We first return to the case of the NH molecule whose electron terms are given in Fig. 11.3. The valent electrons of this molecule are three p-electrons of the nitrogen atom and one s-electron of the hydrogen atom. These electrons can occupy 3σ- or 3π-orbitals. It is clear that the higher binding energy corresponds to the 3σ-orbital because of the smaller distances of the electron from the core. Therefore, the lowest states by energy correspond to the electron shell $3\sigma^2 3\pi^2$, and this fact is confirmed

Table 11.5. Spectroscopic parameters of some dimers in the ground state. The reduced mass μ accounts for the isotope content of an element and is expressed in atomic units of mass ($1.66056 \cdot 10^{-24}$ g), the energy of the vibrational quantum $\hbar\omega_e$ and the anharmonicity parameter $\hbar\omega_e x$ are given in cm^{-1}, the equilibrium distance between nuclei R_e is taken in Å, the rotational constant B_e is expressed in cm^{-1}, the parameter α_e for the vibration–rotation interaction is taken in 10^{-3} cm^{-1}, and the dimer dissociation energy D is given in eV.

Dimer	Term	μ	$\hbar\omega_e$	$\hbar\omega_e x$	R_e	B_e	α_e	D
Ag_2	$X^1\Sigma_g^+$	53.934	135.8	0.50	2.53	0.049	0.195	1.67
Al_2	$X^3\Pi_u$	13.491	284.2	2.02	2.47	0.205	0.8	0.46
Ar_2	$X^1\Sigma_g^+$	19.974	30.68	2.42	3.76	0.0596	3.64	0.0122
As_2	$X^1\Sigma_g^+$	37.461	429.6	1.12	2.103	0.102	0.333	3.96
Au_2	$X^1\Sigma_g^+$	98.483	190.9	0.42	2.47	0.028	0.66	2.31
B_2	$X^1\Sigma_g^+$	5.405	1059	15.66	1.60	1.216	14.0	2.8
Be_2	$X^1\Sigma_g^+$	4.5061	275.8	12.5	2.45	0.615	81	0.098
Bi_2	$X^1\Sigma_g^+$	104.49	173.1	0.376	2.66	0.082	0.053	2.08
Br_2	$X^1\Sigma_g^+$	39.952	325	1.08	2.28	0.82	0.318	2.05
C_2	$X^1\Sigma_g^+$	6.0055	1855	13.27	1.24	1.899	17.8	5.36
Ca_2	$X^1\Sigma_g^+$	20.04	64.9	1.087	4.28	0.047	0.7	0.13
Cl_2	$X^1\Sigma_g^+$	17.726	559.7	2.68	1.99	0.244	2.1	2.576
Cr_2	$X^1\Sigma_g^+$	25.998	470	14.1	1.68	0.23	3.8	1.66
Cs_2	$X^1\Sigma_g^+$	66.453	42.02	0.082	4.65	0.013	0.026	0.452
Cu_2	$X^1\Sigma_g^+$	31.773	266.4	1.03	2.21	0.109	0.062	1.99
F_2	$X^1\Sigma_g^+$	9.4992	916.6	11.24	1.41	0.89	14.1	1.66
H_2	$X^1\Sigma_g^+$	0.5040	4401	121.3	0.741	60.85	3062	4.478
D_2	$X^1\Sigma_g^+$	1.0070	3116	61.82	0.741	30.44	1079	4.556
T_2	$X^1\Sigma_g^+$	1.5082	2546	41.23	0.741	20.34	.589	4.591
I_2	$X^1\Sigma_g^+$	63.452	214.5	0.615	2.67	0.037	0.124	1.542
K_2	$X^1\Sigma_g^+$	19.549	92.09	0.283	3.92	0.057	0.165	0.551
Kr_2	$X^1\Sigma_g^+$	41.90	24.1	1.34	4.02	0.024	1.0	0.018
Li_2	$X^1\Sigma_g^+$	3.571	351.4	2.59	2.67	0.672	7.04	1.05
Mg_2	$X^1\Sigma_g^+$	12.152	51.08	1.623	3.89	0.093	3.78	0.053
Mo_2	$X^1\Sigma_g^+$	47.97	477	1.51	2.2	0.072	0.48	4.1
N_2	$X^1\Sigma_g^+$	7.0034	2359	14.95	1.098	1.998	17.1	9.579
Na_2	$X^1\Sigma_g^+$	11.495	159.1	0.725	3.08	0.155	0.874	0.731
Ne_2	$X^1\Sigma_g^+$	10.090	31.3	6.48	2.91	0.17	60	0.0037
O_2	$X^3\Sigma_g^-$	7.9997	1580	11.98	1.207	1.445	15.93	5.12
Pb_2	$X^3\Sigma_g^-$	103.6	110.2	0.327	2.93	0.019	0.057	0.083
Rb_2	$X^1\Sigma_g^+$	42.734	57.78	0.139	4.17	0.023	0.047	0.495
S_2	$X^3\Sigma_g^-$	16.03	725.6	2.84	1.89	0.295	1.58	4.37
Se_2	$X^3\Sigma_g^-$	39.48	385.3	0.963	2.16	0.89	0.28	2.9
Si_2	$X^3\Sigma_g^-$	14.043	510.9	2.02	2.24	0.239	1.35	3.24
Sn_2	$X^3\Sigma_g^-$	59.345	186.2	0.261	2.75	0.038	0.1	2.0
Sr_2	$X^1\Sigma_g^+$	43.81	39.6	0.45	4.45	0.019	0.2	0.13
Te_2	$X^3\Sigma_g^-$	63.80	249.1	0.537	2.56	0.040	0.109	2.7
V_2	$X^3\Sigma_g^-$	25.471	537.5	3.34	1.78	0.209	1.4	2.62
Xe_2	$X^1\Sigma_g^+$	65.645	21.12	0.65	4.36	0.013	0.3	0.024

Table 11.6. Spectroscopic parameters of some dimer ions in the ground state. The vibration energy $\hbar\omega_e$ the anharmonicity parameter $\hbar\omega_e x$, and the rotational constant B_e are given in cm^{-1}, the equilibrium distance between the nuclei R_e is taken in Å, and the dissociation energy D of the molecular ion and the ionization potential J of the corresponding dimer are expressed in eV.

Ion	Term	$\hbar\omega_e$	$\hbar\omega_e x_e$	R_e	B_e	D	J
Ag_2^+	$X^2\Sigma_g^+$	118	0.05	2.8	0.040	1.69	7.66
Al_2^+	$X^2\Sigma_g^+$	178	2.0	3.2	0.122	1.4	4.84
Ar_2^+	$X^2\Sigma_u^+$	308.9	1.66	2.43	0.143	1.23	14.5
Be_2^+	$X^2\Sigma_u^+$	502	4.2	2.23	0.752	1.9	7.45
Br_2^+	$X^2\Pi_g$	376	1.13	2.3	0.088	2.96	10.52
C_2^+	$X^4\Sigma_g^-$	1351	12.1	1.41	1.41	5.3	12.15
Cl_2^+	$X^2\Pi_g$	645.6	3.02	1.88	0.265	3.95	11.50
Cs_2^+	$X^2\Sigma_g^+$	32.4	0.051	4.44	0.013	0.61	3.76
Cu_2^+	$X^2\Sigma_g^+$	188	0.75	2.35	0.096	1.8	7.90
F_2^+	$X^2\Pi_g$	1073	9.13	1.32	1.015	3.34	15.47
H_2^+	$X^2\Sigma_g^+$	2323	67.5	1.06	30.21	2.650	15.426
He_2^+	$X^2\Sigma_g^+$	1698	35.3	1.08	7.21	2.47	22.22
Hg_2^+	$X^2\Sigma_u^+$	91.6	0.301	2.8	0.021	0.96	9.4
I_2^+	$X^2\Pi_g$	243	—	2.58	0.040	1.92	9.3
K_2^+	$X^2\Sigma_g^+$	73.4	0.2	4.6	0.042	0.81	4.06
Kr_2^+	$X^2\Sigma_u^+$	178	0.82	2.8	0.051	1.15	12.97
Li_2^+	$X^2\Sigma_g^+$	263.1	1.61	3.12	0.49	1.28	5.14
N_2^+	$X^2\Sigma_g^+$	2207	16.2	1.12	1.932	8.713	15.581
Na_2^+	$X^2\Sigma_g^+$	120.8	0.46	3.54	0.113	0.98	4.80
Ne_2^+	$X^2\Sigma_u^+$	586	5.4	1.75	0.554	1.2	20.4
O_2^+	$X^2\Pi_g$	1905	16.3	1.12	1.689	6.66	12.07
P_2^+	$X^2\Pi_u$	672	2.74	1.98	0.276	5.0	10.56
Rb_2^+	$X^2\Sigma_g^+$	44.5	—	4.8	0.017	0.75	3.9
S_2^+	$X^2\Pi_u$	806	3.33	1.82	0.318	5.4	9.4
Sr_2^+	$X^2\Sigma_g^+$	86	0.54	3.9	0.025	1.1	4.74
Xe_2^+	$X^2\Sigma_g^+$	123	0.63	3.25	0.026	1.03	11.85

Table 11.7. Spectroscopic parameters of some negative ions of dimers in the ground state. The vibration energy $\hbar\omega_e$, the anharmonicity parameter $\hbar\omega_e x_e$, and the rotational constant B_e are expressed in cm^{-1}, the equilibrium distance between the nuclei R_e is taken in Å, and the dissociation energy D of the molecular ion and the electron affinity EA of the corresponding dimer are expressed in eV.

Ion	Term	$\hbar\omega_e$	$\hbar\omega_e x_e$	R_e	B_e	D	EA
Ag_2^-	$X^2\Sigma_u^+$	145	0.9	2.6	0.046	1.37	1.03
Al_2^-	$X^4\Sigma_g^-$	335	—	2.65	0.178	2.4	1.1
Bi_2^-	$X^2\Pi_g$	152	0.53	2.83	0.020	2.8	1.27
Br_2^-	$X^2\Sigma_u^+$	178	0.88	2.81	0.054	1.2	2.55
C_2^-	$X^2\Sigma_g^+$	1781	11.7	1.27	1.746	3.3	3.27
Cl_2^-	$X^2\Sigma_u^+$	277	1.8			1.26	2.38
Cs_2^-	$X^2\Sigma_u^+$	28.4	0.042	4.8	0.011	0.45	0.47
Cu_2^-	$X^2\Sigma_u^+$	196	0.7	2.34	0.097	1.57	0.84
F_2^-	$X^2\Pi_u$	475	5.1	1.92	0.47	1.3	3.08
Li_2^-	$X^2\Sigma_u^+$	233.1	1.92	2.8	0.516	0.88	0.7
O_2^-	$X^2\Pi_g$	1090	10	1.35	1.12	4.16	0.45
P_2^-	$X^2\Pi_u$	640	—	1.98	0.277	4.08	0.59
Pb_2^-	$X^2\Pi_g$	129	0.2	2.81	0.021	1.37	1.66
S_2^-	$X^2\Pi_u$	601	2.16	1.8	0.32	3.95	1.67

by the positions of the potential curves which are given in Fig. 11.3. Next, the radiative dipole transitions in this system correspond to the transitions $\sigma \to \pi$ with conservation of the total molecule spin. Therefore, for this molecule, the radiative dipole transitions are permitted between the electron terms $X^3\Sigma^- \to A^3\Pi$ and $a^1\Delta, b^1\Sigma^+ \to C^1\Pi$.

Another example of the application of the method of molecular orbits corresponds to the interaction of two helium atoms. Based on this method, one can show that the electron term, corresponding to the ground states of atoms at infinite distances between the atoms, intersects the boundary of the continuous spectrum at small separations. This means that if such distances are reached under the collision of two helium atoms, the electron release from one colliding atom is possible.

Let us construct the correlation diagram of the quasi-molecule consisting of two helium atoms. Two atoms $2\mathrm{He}(1s^2)$ form the quasi-molecule $\mathrm{He}_2(1\sigma_g^2 1\sigma_u^2)$ which is transformed to the beryllium atom state $\mathrm{Be}(1s^2 2p^2, {}^1S)$ when the helium nuclei are joined. The excitation energy of the beryllium atom from the ground state $\mathrm{Be}(1s^2 2s^2)$ into the excited state $\mathrm{Be}(1s^2 2s2p)$ is 5.3 eV. Since the excited beryllium atom $\mathrm{Be}(1s^2 2s2p)$ $2s$-electron has a higher binding energy than the $2s$-electron in the ground state of the atom, the excitation energy from the ground state of the beryllium atom $\mathrm{Be}(1s^2 2s^2)$ into the excited state $\mathrm{Be}(1s^2 2p^2)$ exceeds 10.6 eV. The ionization potential of the beryllium atom is 9.32 eV. Thus the electron term of the state which corresponds to two helium atoms in the ground state, at infinite separations, intersects the boundary of the continuous spectrum and is transformed into the autoionization state at small distances between the nuclei, if we consider the pseudointersections with other electron terms as intersections.

One can prove that the above electron term for two helium atoms in the ground state intersects the electron terms corresponding to two excited helium atoms. This means that the collision of two helium atoms in the ground state can lead to the excitation of both atoms. Let us construct a correlation diagram for $\mathrm{He}(1s2s)$ and $\mathrm{He}^+(1s)$ when this quasi-molecule is found in the even state with respect to reflection of the electrons in the plane which is perpendicular to the molecule axis and bisects it. Then the state of the quasi-molecule at intermediate distances between the atoms is $\mathrm{He}_2^+(1\sigma_g^2 2\sigma_g)$. This state corresponds to the ground state of the beryllium ion $\mathrm{Be}^+(1s^2 2s)$ when the nuclei are joined. Therefore, using the above result for the intersection of the electron term $\mathrm{He}_2(1\sigma_g^2 1\sigma_u^2)$ with the boundary of the continuous spectrum, one can find that the electron term of the state $2\mathrm{He}(1s^2)$ intersects any electron term $\mathrm{He}(1s2s) + \mathrm{He}(1snl)$ if the states of the first three electrons are even for the reflection of electrons with respect to the symmetry plane. These examples demonstrate the possibilities of the method of molecular orbits.

11.5 Radiative Transitions in Diatomic Molecules

Let us analyze briefly the specifics of the radiative transitions in diatomic molecules. We will be guided by dipole radiation when the radiation rates are determined by formulas (1.22)–(1.26). Then the peculiarities of the radiative transitions are connected with the properties of the matrix element for the dipole moment operator. For diatomic molecules the dipole moment vector is directed along the molecular axis. This leads to certain selection rules for the quantum numbers of transition.

We first consider the radiative transitions between the electron states of a diatomic molecule. Then, because of the structure of the dipole moment operator, the selection rules permit transitions with conservation of the electron momentum projection onto the molecular axis or change it by one. This momentum projection corresponds to the electron angular momentum projection Λ in case "a" of the Hund coupling or to the projection of the total electron momentum onto the molecular axis Ω in case "c" of the Hund coupling. Simultaneously, a radiative transition in diatomic molecules requires certain selection rules for the rotation quantum numbers. If we separate the rotation degrees of freedom, and take into account the different orders of magnitude for the electron and rotational energies, the dependence on the rotational quantum numbers in formula (1.23) would be the factor of $\langle JM|\mathbf{n}|J'M'\rangle$, where $\mathbf{n}$ is the unit vector directed along the molecular axis, J, M are the rotational quantum numbers for the initial states, and J', M' are the rotational quantum numbers for the final state. This matrix element is reduced to the Clebsh–Gordan coefficients and leads to the following selection rules

$$J - J' = 0, \pm 1; \qquad M - M' = 0, \pm 1. \tag{11.13}$$

An additional selection rule applies for dimer molecules. Because the dipole moment operator is antisymmetric with respect to the inversion of electrons, radiative transitions between the states of a different symmetry ($g \longleftrightarrow u$) only are permitted.

Radiative transitions between the vibration and rotation states are weaker than those between the electron states mainly due to the difference of the photon energies. Radiative transitions between the rotational states are possible only for dipole molecules and are absent for dimer molecules consisting of identical isotopes because their electric center and center of mass are coincident. These conclusions follow from a simple analysis of formula (1.23). The radiative transitions between vibrational states are determined by the dependence of the molecular dipole moment on the distance R between the nuclei. Take this dependence in the form

$$\mathbf{D} = \mathbf{n}\left[D_o + \frac{dD}{dR|_{R_e}}(R - R_e) + \frac{1}{2}\frac{d^2D}{dR^2|_{R_e}}(R - R_e)^2\right], \tag{11.14}$$

where R_e is the equilibrium distance between the nuclei in the molecule and $\mathbf{n}$ is the unit vector directed along the molecular axis. The matrix element for radiative transitions between the vibrational states is determined by the second term in the expansion (11.14). This leads to the selection rule $v' = v - 1$ and to the matrix element $\langle v|R - R_e|v - 1\rangle \sim \sqrt{v/\mu\omega}$, where μ is the reduced mass of the nuclei ω is the transition frequency. Because $\omega \sim \mu^{-1/2}$, one can estimate from this that the ratio of the intensities for vibration and electron radiative transitions is order of μ^{-2}.

Along with the radiative transitions $v \to v-1$, one can expect weaker transitions in other vibration states. In particular, the radiative transitions $v \to v - 2$ are determined by both the second part of formula (11.14) and by the anharmonicity of vibrations. Although the corresponding matrix elements are relatively small, they grow with excitation of the molecule. Hence, for a strong excitation of the molecule such transitions may be significant. For example, in the case of the HF molecule the ratio of the rates of the radiative transitions $w(v \to v - 2)/w(v \to v - 1)$ are equal to 0.12 for $v = 2$ and to 5.6 for $v = 10$; for the molecule DF this ratio is 0.09 and 1.3, respectively, for these vibrational numbers.

Problems

Problem 11.1. *Determine the parity for reflection in a plane which passes through the atom center for the nitrogen atom in the ground state N(^4S) and for the lowest states of the carbon and oxygen atoms.*

This parity can be introduced in the case when the spin and spatial coordinates are separated, and which takes place for the case under consideration. This parity corresponds to the states with zero projection of the orbital momentum onto the axis which is located on the reflection plane. Let us take the coordinate frame such that the momentum projection is taken onto the z-axis and reflection is produced in the xz-plane. Then this reflection leads to the transformation $y \to -y$ and in the spherical or cylindrical coordinates the reflection gives $\varphi \to -\varphi$. This corresponds to a change of sign of the momentum projection for both an individual electron and for the total atom. The operation of electron reflection, whose operator we denote by $\hat{\sigma}$, commutes with the electron Hamiltonian. Hence, the eigenfunctions of the Hamiltonian Ψ are simultaneously eigenfunctions of the reflection operator, i.e., $\hat{\sigma}\Psi = \sigma\Psi$ (below we consider Ψ as the spatial wave functions of the electrons). Our goal is to find the σ eigenvalues of the reflection operator for given states of atoms.

In the case of the nitrogen atom in the ground state ^4S the spin wave functions of all three atoms of the valence electron shell are identical. Therefore the spin wave function of the electron is the product of three one-electron spin functions, and the total electron wave function is the product of the spin and coordinate wave functions. Because the total wave function of the electrons is antisymmetric with respect to the transposition of two the electrons, and the spin wave function is symmetric for this operation, the coordinate wave function is antisymmetric with respect to the transposition of coordinates of any two electrons. Thus the electron spatial wave function can be represented in the form of the Slater determinant

$$\Psi(\mathbf{r}_1, \mathbf{r}_2, \mathbf{r}_3) = \begin{vmatrix} \psi_1(\mathbf{r}_1) & \psi_1(\mathbf{r}_2) & \psi_1(\mathbf{r}_3) \\ \psi_{-1}(\mathbf{r}_1) & \psi_{-1}(\mathbf{r}_2) & \psi_{-1}(\mathbf{r}_3) \\ \psi_0(\mathbf{r}_1) & \psi_0(\mathbf{r}_2) & \psi_0(\mathbf{r}_3) \end{vmatrix},$$

where $\psi_m(\mathbf{r}_i)$ is the space wave function of the ith electron which has the momentum projection m onto the z-axis. As is seen, the reflection operation $\hat{\sigma}$ for electrons is similar to the operation of the transposition of two electrons and, taking into account the symmetry of the spatial wave function with respect to the electron transposition, we find $\sigma = -1$ for the ground state of the nitrogen atom $N(p^3, {}^4S)$.

In the case of the carbon atom the electron wave function is the product of the spin and coordinate wave functions and, since the projection of the orbital momentum of two electrons is zero, the spatial wave function of the electrons has the following structure:

$$\Psi = \sum_m C_{m,-m}\psi_m(\mathbf{r}_1)\psi_{-m}(\mathbf{r}_2),$$

where $C_{m,-m}$ is the Clebsh–Gordan coefficient and m is the momentum projection onto the axis for the first electron. We have

$$\hat{\sigma}\Psi = \hat{\sigma}\sum_m C_{m,-m}\psi_m(\mathbf{r}_1)\psi_{-m}(\mathbf{r}_2) = \sum_m C_{m,-m}\psi_m(\mathbf{r}_2)\psi_{-m}(\mathbf{r}_1).$$

As is seen, the electron reflection leads, in this case, to the same result as the electron transposition. Hence the symmetry of this wave function coincides with the symmetry of the spatial electron wave function with respect to the transposition of the two electrons. Therefore, the state $C(^3P)$ is odd, and the states $C(^1D)$, $C(^1S)$ are even. Considering the oxygen electron shell to consist of two p-holes, we obtain the same result for the oxygen atom as in the carbon case, that is, state $O(^3P)$ is odd, and states $O(^1D)$, $O(^1S)$ are even.

Problem 11.2. *Determine the number of Σ^+- and Σ^--terms of a molecule consisting of atoms with electron orbital momenta L_1 and L_2 ($L_1 > L_2$) within the framework of case "a" of the Hund coupling.*

Since case "a" of the Hund coupling scheme admits the summation of the orbital and spin momenta of atoms independently, the number of Σ-terms of a different parity is the same for each molecular spin. Hence, molecular spin is out of the question. Below, for simplicity, we consider weakly interacting atoms which are located far from each other, although the conclusion for the momentum projection of a formed molecule does not depend on this interaction. We obtain zero projection of the molecule momentum if the momentum projection is equal to M for the first atom and to $-M$ for the second atom. Because of $L_1 > L_2$, there is one such state for each projection momentum of the second atom, i.e., a number of Σ-terms is $2L_2 + 1$ at a given spin state. In this case, the molecular wave function can be constructed as a product of atomic wave functions, so that the molecule wave function for the Σ-terms is given by the formula

$$\Psi_0 = \frac{1}{\sqrt{2}}(\psi_M \varphi_{-M} \pm \psi_{-M}\varphi_M),$$

where the wave function of the first atom with momentum projection M onto the molecular axis is denoted ψ_M and the corresponding wave function of the second atom is denoted by φ_M, the $+$ sign corresponds to the Σ^+-terms and the $-$ sign corresponds to the Σ^--terms. As is seen, the above expression is valid for $M \neq 0$. Thus, in this way, we obtain L_2 electron terms of the symmetry Σ^+ and L_2 electron terms of the symmetry Σ^-.

One more term is described by the wave function $\Psi_0 = \psi_0\varphi_0$. Let us determine its parity. Evidently, the parity of this term is the product of the parities of the atoms forming the molecule $\sigma_1\sigma_2$. Hence, this term has Σ^+-symmetry if $\sigma_1\sigma_2 = 1$, and this term has Σ^--symmetry if $\sigma_1\sigma_2 = -1$. Thus, finally, we obtain that if $\sigma_1\sigma_2 = 1$, there are (L_2+1) terms of Σ^+-symmetry and L_2 terms of Σ^--symmetry. If $\sigma_1\sigma_2 = -1$, we have L_2 terms of Σ^+-symmetry and $(L_2 + 1)$ terms of Σ^--symmetry.

Problem 11.3. *Find the ion–atom exchange interaction potential if the p-electron is located in the field of structureless cores and the fine splitting of levels significantly exceeds the electrostatic splitting.*

This relates to elements of Groups 3 and 8 of the periodical table and the jj-scheme of momentum coupling is valid, so that the electron state is characterized by the quantum numbers jm_j—the total electron momentum and its projection onto a quantization axis. According to the character of summation of the orbital electron momentum and its spin into the total electron momentum, the electron wave function Ψ_{jm_j} is given by

$$\Psi_{jm_j} = \sum_{\mu,\sigma} \begin{bmatrix} \frac{1}{2} & 1 & j \\ \sigma & \mu & m_j \end{bmatrix} \psi_{1\mu}\chi_\sigma,$$

where $\psi_{1\mu}$ is the spatial wave function of the p-electron with a momentum projection μ onto the quantization axis, and χ_σ is the spin function. From this we have the following relation between the exchange interaction potential Δ_{jm_j} within the framework of the jj-coupling scheme for atoms and ions, and the exchange interaction potentials Δ_{1m} for the LS-coupling scheme

$$\Delta_{jm_j} = \sum_{\mu} \begin{bmatrix} \frac{1}{2} & 1 & j \\ \sigma & \mu & m_j \end{bmatrix}^2 \Delta_{1\mu},$$

where $\Delta_{1\mu}$ is given by formula (10.23) and, according to the properties of the Clebsh–Gordan coefficients, $m_j = \sigma + \mu$. This formula for ap-electron can be presented in the form

$$\Delta_{1/2,1/2} = \frac{1}{3}\Delta_{10} + \frac{2}{3}\Delta_{11}, \qquad \Delta_{3/2,1/2} = \frac{2}{3}\Delta_{10} + \frac{1}{3}\Delta_{11}, \qquad \Delta_{3/2,3/2} = \Delta_{11}. \tag{11.15}$$

Introducing an angle θ between the molecular axis and a quantization axis onto which the angular momentum projection is zero we have, for the exchange interaction potentials at a given electron momentum,

$$\Delta_{1/2} = \frac{1}{3}\Delta_{10}+\frac{2}{3}\Delta_{11}, \quad \Delta_{3/2}(\theta) = \left(\frac{1}{6} + \frac{1}{2}\cos^2\theta\right)\Delta_{10}+\left(\frac{1}{3} + \frac{1}{2}\sin^2\theta\right)\Delta_{11}. \tag{11.16}$$

Problem 11.4. *Determine the exchange interaction potential of an ion and its parent atom in the ground states if the atomic particles have p-electron shells and if a jj-scheme of momentum coupling is valid.*

We compose the atom and ion electron shells from two subshells with electron momenta $j = 1/2$ and $j = 3/2$. The results for the ion and atom ground states are given in Table 11.8, where the correspondence is shown between the LS- and jj-coupling schemes. As follows from the data of Table 11.8, the ion–atom exchange interaction potential is simpler in the presence of relativistic interactions because of a lower symmetry of atomic particles in this case.

Problem 11.5. *Find the structure of the exchange interaction potential for the ions and parent atoms with p^2- and p^3-electron shells, i.e., for the elements of Group 5 of the periodic table. Establish the correspondence between the LS- and jj-schemes of momentum coupling.*

The results are given in Table 11.9 where the correspondence between the LS- and jj-coupling schemes are given for individual terms. The values are presented for the jj-coupling scheme, and this is indicated in parentheses for the LS-coupling

Table 11.8. The ground states of atoms with p-electron shells within the framework of the LS- and jj-coupling schemes, and the ion–atom exchange interaction potential (Δ) for cases "c" and "e" of the Hund coupling.

Shell	J	LS-term	jj-shell	Δ
p	1/2	$^2P_{1/2}$	$[1/2]^1$	$\Delta_{1/2}$
p^2	0	$^3P_0$	$[1/2]^2$	$\Delta_{1/2}$
p^3	3/2	$^4S_{3/2}$	$[1/2]^2[3/2]^1$	$\Delta_{3/2}$
p^4	2	$^3P_2$	$[1/2]^1[3/2]^3$	0
p^5	3/2	$^2P_{3/2}$	$[1/2]^2[3/2]^3$	$\Delta_{1/2}$
p^6	0	$^1S_0$	$[1/2]^2[3/2]^4$	$\Delta_{3/2}$

Table 11.9. The exchange interaction potential for atoms of Group 5 of the periodic table system of elements whose atomic electron shell is p^3 with their ions having the electron shell p^2.

LS		$^4S_{3/2}$	$^2D_{3/2}$	$^2D_{5/2}$	$^2P_{1/2}$	$^2P_{3/2}$
	$j-j$	$\left[\left(\frac{1}{2}\right)^2\left(\frac{3}{2}\right)\right]_{3/2}$	$\left[\left(\frac{1}{2}\right)\left(\frac{3}{2}\right)^2\right]_{3/2}$	$\left[\left(\frac{1}{2}\right)\left(\frac{3}{2}\right)^2\right]_{5/2}$	$\left[\left(\frac{1}{2}\right)\left(\frac{3}{2}\right)^2\right]_{1/2}$	$\left[\left(\frac{3}{2}\right)^3\right]_{3/2}$
$^3P_0$	$\left[\left(\frac{1}{2}\right)^2\right]_0$	$\Delta_{3/2}(+)$	$0(+)$	$0(+)$	$0(+)$	$0(+)$
$^3P_1$	$\left[\left(\frac{1}{2}\right)\left(\frac{3}{2}\right)\right]_1$	$\Delta_{1/2}(+)$	$\Delta_{3/2}(+)$	$\Delta_{3/2}(+)$	$\Delta_{3/2}(+)$	$0(+)$
$^3P_2$	$\left[\left(\frac{1}{2}\right)\left(\frac{3}{2}\right)\right]_2$	$\Delta_{1/2}(+)$	$\Delta_{3/2}(+)$	$\Delta_{3/2}(+)$	$\Delta_{3/2}(+)$	$0(+)$
$^1D_2$	$\left[\left(\frac{3}{2}\right)^2\right]_2$	$0(0)$	$\Delta_{1/2}(+)$	$\Delta_{1/2}(+)$	$\Delta_{1/2}(+)$	$\Delta_{3/2}(+)$
$^1S_0$	$\left[\left(\frac{3}{2}\right)^2\right]_0$	$0(0)$	$\Delta_{1/2}(0)$	$\Delta_{1/2}(0)$	$\Delta_{1/2}(+)$	$\Delta_{3/2}(+)$

scheme, that the exchange interaction potential is zero (0) or is not zero (+) for LS-coupling. In particular, for the ground atom and ion states, the exchange interaction potential occupies one cell in Table 11.9, while within the framework of the LS-coupling scheme it is given by the matrix of formula (10.57c).

CHAPTER 12

Atom Interaction in Systems of Many Bound Atoms

12.1 Exchange Interactions of Three Hydrogen Atoms

The interaction of many atoms leads to new properties of systems of many bound atoms. Below we give some examples of this. We first consider the interaction of three atoms and show that this interaction can lead to the intersection of electron terms that is of importance for the collisional processes involving these particles. Let us take three hydrogen atoms in the ground state and analyze the behavior of the electron states which correspond to the atoms depending on the distances between atoms. One can learn from this analysis that the electron terms of the two lowest states of the system are intersected when the hydrogen atoms form a regular triangle.

Assuming the distance between hydrogen atoms to be large, construct the wave functions of the system under consideration, taking the electron wave functions of the individual hydrogen atoms as a basis. The eigenfunctions of these states, corresponding to the projection of the total electron spin 1/2 onto a given direction, are combinations of the following basis wave functions:

$$\Phi_1 = \begin{vmatrix} \psi_a(1)\eta_-(1) & \psi_a(2)\eta_-(2) & \psi_a(3)\eta_-(3) \\ \psi_b(1)\eta_+(1) & \psi_b(2)\eta_+(2) & \psi_b(3)\eta_+(3) \\ \psi_c(1)\eta_+(1) & \psi_c(2)\eta_+(2) & \psi_c(3)\eta_+(3) \end{vmatrix}, \qquad (12.1a)$$

$$\Phi_2 = \begin{vmatrix} \psi_a(1)\eta_+(1) & \psi_a(2)\eta_+(2) & \psi_a(3)\eta_+(3) \\ \psi_b(1)\eta_-(1) & \psi_b(2)\eta_-(2) & \psi_b(3)\eta_-(3) \\ \psi_c(1)\eta_+(1) & \psi_c(2)\eta_+(2) & \psi_c(3)\eta_+(3) \end{vmatrix}, \qquad (12.1b)$$

$$\Phi_3 = \begin{vmatrix} \psi_a(1)\eta_+(1) & \psi_a(2)\eta_+(2) & \psi_a(3)\eta_+(3) \\ \psi_b(1)\eta_+(1) & \psi_b(2)\eta_+(2) & \psi_b(3)\eta_+(3) \\ \psi_c(1)\eta_-(1) & \psi_c(2)\eta_-(2) & \psi_c(3)\eta_-(3) \end{vmatrix}. \qquad (12.1c)$$

Here $\psi_a(i)$, $\psi_b(i)$, $\psi_c(i)$ are the coordinate wave functions of the ith electron which is located in the field of the a, b, and c nuclei, respectively, $\eta_+(i)$, $\eta_-(i)$ are the spin wave functions of ith electron if the spin projection is $1/2$ or $-1/2$ onto a given direction. Let us consider the case when the nuclei form an isosceles triangle. Then the plane which is perpendicular to the triangle plane, and passes through its height, is the symmetry plane of this system. Because of the s-state of electrons of the hydrogen atom, reflection with respect to this symmetry plane gives $\psi_c \to \psi_c$, $\psi_a \to \psi_b$, $\psi_b \to \psi_a$, where ab and ac are equivalent sides of the triangle. Hence this operation of reflection yields $\Phi_1 \to -\Phi_2$, $\Phi_2 \to -\Phi_1$, $\Phi_3 \to -\Phi_3$. Thus, the eigenfunctions of the system of three hydrogen atoms, when these atoms form an isosceles triangle, are an even wave function $C_1(\Phi_1 - \Phi_2)$ that conserves the sign as a result of the reflection of electrons with respect to the symmetry plane, and of two odd functions $C_2(\Phi_1 + \Phi_2) + C_3\Phi_3$, $C_3(\Phi_1 + \Phi_2) - C_2\Phi_3$ that change sign under this operation. If we remove an atom c far from the other two atoms, the even state corresponds to the hydrogen molecule, whereas the other two odd states relate to the triplet state of the hydrogen molecule. In odd states electrons are distributed in a wider region than in an even state, i.e., the binding energy of electrons in the odd states is smaller than in the even state.

If the nuclei form an equilateral triangle, a new symmetry of the system occurs. Let us introduce the operator $\hat{\alpha}$ of the rotation of electrons by an angle $2\pi/3$ around the axis which is perpendicular to the triangle plane and passes through the triangle center. Because this operator commutes with the electron Hamiltonian, the eigenfunctions of the Hamiltonian are eigenfunctions of the operator $\hat{\alpha}$. This operation of rotation of the system corresponds to the transformations $\psi_a \to \psi_c$, $\psi_c \to \psi_b$, $\psi_b \to \psi_a$, so that this operation leads to the following transformations of the wave functions of the system $\Phi_1 \to \Phi_3$, $\Phi_2 \to \Phi_1$, $\Phi_3 \to \Phi_2$. Using this for the determination of the eigenfunctions of the system of three hydrogen atoms, we note that the three rotations return the system to the initial state. Hence the eigenvalues of the operator $\hat{\alpha}$ satisfy the relation $\hat{\alpha}^3 = 1$, i.e., $\alpha = 1, e^{i2\pi/3}, e^{-i2\pi/3}$. The corresponding expressions for the eigenfunctions $\Psi = a_1\Phi_1 + a_2\Phi_2 + a_3\Phi_3$ follow from the relation $\hat{\alpha}\Psi = \alpha(a_1\Phi_1 + a_2\Phi_2 + a_3\Phi_3) = a_1\Phi_3 + a_2\Phi_1 + a_3\Phi_2$. In this way we determine the eigenfunctions of the system that have the form

$$\Psi_{\rm I} = C_1(\Phi_1 + \Phi_2 + \Phi_3);$$
$$\Psi_{\rm II} = C_2(\Phi_1 + \Phi_2 e^{i2\pi/3} + \Phi_3 e^{-i2\pi/3});$$
$$\Psi_{\rm III} = C_3(\Phi_1 + \Phi_2 e^{-i2\pi/3} + \Phi_3 e^{i2\pi/3}).$$

Since $\Psi_{\rm II} = \Psi_{\rm III}^*$, $E_{\rm II} = \langle \Psi_{\rm II}^* | \hat{H} | \Psi_{\rm II} \rangle = \langle \Psi_{\rm II}^* | \hat{H} | \Psi_{\rm II} \rangle^* = E_{\rm III}$, i.e., the levels of states II and III are coincident when the atoms form a regular triangle. Next,

$$\begin{aligned}
\Psi_{\rm II} = \Psi_{\rm III}^* &= C_2(\Phi_1 + \Phi_2 e^{i2\pi/3} + \Phi_3 e^{-i2\pi/3}) \\
&= C_2\left[\frac{1}{2}(\Phi_1 - \Phi_2)(1 + e^{i2\pi/3}) + \frac{1}{2}(\Phi_1 + \Phi_2)(1 - e^{i2\pi/3}) + \Phi_3 e^{-i2\pi/3}\right] \\
&= C_2\left[\Phi_1 + (\Phi_2 + \Phi_3)\cos\frac{2\pi}{3} + i(\Phi_2 - \Phi_3)\sin\frac{2\pi}{3}\right]
\end{aligned}$$

$$= C_2\left[\frac{1}{2}(\Phi_1 - \Phi_3)(1 + e^{-i2\pi/3}) + \frac{1}{2}(\Phi_1 + \Phi_3)(1 - e^{-i2\pi/3}) + \Phi_2 e^{i2\pi/3}\right]$$

i.e., the wave functions Ψ_{II}, Ψ_{III} include combinations of both even and odd functions with respect to reflection in the symmetry planes, while the wave function Ψ_I consists of only odd functions with respect to this operation. Therefore we obtain $E_I > E_{II} = E_{III}$, i.e., the intersection of the electron term of the ground state with the electron term of the first excited electron state takes place in the system of three hydrogen atoms when they form an equilateral triangle.

Let us calculate the energy of states of the systems of three hydrogen atoms in the ground state. Take the above wave functions Φ_1, Φ_2, Φ_3 as the basis wave functions. The normalization wave function Φ_1 has the form

$$\Phi_1 = \frac{1}{\sqrt{6(1+S_{bc})}}\begin{vmatrix} \psi_a(1)\eta_-(1) & \psi_a(2)\eta_-(2) & \psi_a(3)\eta_-(3) \\ \psi_b(1)\eta_+(1) & \psi_b(2)\eta_+(2) & \psi_b(3)\eta_+(3) \\ \psi_c(1)\eta_+(1) & \psi_c(2)\eta_+(2) & \psi_c(3)\eta_+(3) \end{vmatrix}, \qquad (12.2)$$

where $S_{bc} = \langle\Psi_{bc}(1,2)|\hat{H}|\Psi_{bc}(2,1)\rangle$ is the overlapping integral and $\Psi_{bc}(1,2) = \psi_a(1)\psi_b(2)$ in the region where the electrons are found far from the other nuclei. An accurate two-electron wave function, similar to that of formula (10.30), is used for small distances between related electrons. The expressions for the wave functions Φ_2, Φ_3 have a similar form.

For calculation of the matrix elements $H_{ik} = \langle\Phi_i|\hat{H}|\Phi_k\rangle$ we divide the electron Hamiltonian into three parts:

$$\hat{H} = \hat{h}_{ab}(1,2) + \hat{h}_c(3) + V,$$

where

$$\hat{h}_{ab}(1,2) = -\frac{1}{2}\Delta_1 - \frac{1}{2}\Delta_2 - \frac{1}{r_{1a}} - \frac{1}{r_{1b}} - \frac{1}{r_{2a}} - \frac{1}{r_{2b}} + \frac{1}{|\mathbf{r}_1 - \mathbf{r}_2|} + \frac{1}{R_{ab}},$$

$$\hat{h}_c(3) = -\frac{1}{2}\Delta_3 - \frac{1}{r_{3c}},$$

$$V = -\frac{1}{r_{3a}} - \frac{1}{r_{3b}} - \frac{1}{r_{1c}} - \frac{1}{r_{2c}} + \frac{1}{|\mathbf{r}_1 - \mathbf{r}_3|} + \frac{1}{|\mathbf{r}_2 - \mathbf{r}_3|} + \frac{1}{R_{ac}} + \frac{1}{R_{bc}}.$$

Here $\mathbf{r}_i$ is the coordinate of the ith electron, r_{ia} is the distance of the ith electron from the nucleus a, R_{ab} is the distance between the nuclei a and b, the Hamiltonian $\hat{h}_{ab}(1,2)$ describes the system of two hydrogen atoms with nuclei a and b, the Hamiltonian $\hat{h}_c(3)$ corresponds to the hydrogen atom consisting of the third electron and nucleus c, and the operator V accounts for the interaction between these systems. This form is convenient for calculation of the matrix element $\langle\Psi_{ab}(1,2)\psi_c(3)|\hat{H}|\Psi_{ab}(2,1)\psi_c(3)\rangle$. In this case the matrix element from the operator V at large distances R between atoms contains an additional small parameter $1/R$ compared with the overlapping integral S between these wave functions.

Separating the Hamiltonian of two electrons into two- and one-electron Hamiltonians and neglecting the interaction between these systems in the region of the electron coordinates which determine the integrals under consideration, we reduce the three-election problem to a two-electron problem. The matrix elements of the Hamiltonian in the basis Φ_1, Φ_2, Φ_3 are equal to

$$H_{11} = E_o - \frac{1}{2}\Delta_{bc}, \qquad H_{22} = E_o - \frac{1}{2}\Delta_{ac}, \qquad H_{33} = E_o - \frac{1}{2}\Delta_{ab},$$
$$H_{12} = -\frac{1}{2}\Delta_{ab}, \qquad H_{13} = -\frac{1}{2}\Delta_{ac}, \qquad H_{23} = -\frac{1}{2}\Delta_{bc},$$

where Δ_{ab} is the exchange interaction potential for two hydrogen atoms at a distance R_{ab} and E_o is the energy of three hydrogen atoms accounting for a long-range interaction between them. Presenting the energy of the system of three hydrogen atoms in the form $E = E_o + \varepsilon$, we obtain the following secular equation for the electron energy ε:

$$\begin{vmatrix} \varepsilon + \Delta_{bc} & \Delta_{ab} & \Delta_{ac} \\ \Delta_{ab} & \varepsilon + \Delta_{ac} & \Delta_{bc} \\ \Delta_{ac} & \Delta_{bc} & \varepsilon + \Delta_{bc} \end{vmatrix} = 0.$$

This equation is

$$\begin{aligned} &\varepsilon^3 + (\Delta_{ab} + \Delta_{ac} + \Delta_{bc})\varepsilon^2 \\ &\quad + (\Delta_{ab}\Delta_{ac} + \Delta_{ab}\Delta_{bc} + \Delta_{ac}\Delta_{bc} - \Delta_{ab}^2 - \Delta_{ac}^2 - \Delta_{bc}^2)\varepsilon \\ &\quad\quad - \Delta_{ab}^3 - \Delta_{ac}^3 - \Delta_{bc}^3 + 3\Delta_{ab}\Delta_{ac}\Delta_{bc} = 0. \end{aligned}$$

The solution of this equation gives, for the energies of three electron states of the system of three hydrogen atoms when they are found in the ground states at large separations,

$$E_{\mathrm{I}} = E_o - \frac{1}{2}\Delta, \qquad E_{\mathrm{II}} = E_o + \frac{1}{2}\Delta, \qquad E_{\mathrm{III}} = E_o + \frac{1}{2}(\Delta_{ab} + \Delta_{ac} + \Delta_{bc}), \tag{12.3}$$

where

$$\Delta = \sqrt{\frac{1}{2}(\Delta_{ab} - \Delta_{ac})^2 + \frac{1}{2}(\Delta_{ab} - \Delta_{bc})^2 + \frac{1}{2}(\Delta_{ac} - \Delta_{bc})^2}. \tag{12.4}$$

From this solution it follows that if the hydrogen atoms form an equilateral triangle ($\Delta_{ab} = \Delta_{ac} = \Delta_{bc}$), we obtain $\Delta = 0$, and two lowest levels of the system are coincident. In the case where one atom (atom c) is removed to infinity ($\Delta_{ac}, \Delta_{bc} \to 0$), the first state corresponds to the combined singlet state of the hydrogen molecule $E_{\mathrm{I}} = E_o - \frac{1}{2}\Delta_{ab}$, whereas the other two states correspond to the triplet repulsive state of the hydrogen molecule $E_{\mathrm{II}} = E_{\mathrm{III}} = E_o + \frac{1}{2}\Delta_{ab}$.

Thus the analysis of the interaction of three hydrogen atoms shows a specific structure of the electron terms. In particular, the intersections of the surfaces of the electron potential energy for different electron terms may be responsible for collisional transitions. In this case collisions of the hydrogen atom or its isotope with the hydrogen molecule can lead to an exchange between the free and bound atoms or to a dissociation of the molecule if configurations of the atoms are reached which are close to a regular triangle. This example shows that complex systems of atoms can have additional properties.

12.2 The Character of the Interactions of Atoms in Bulk Inert Gases

The attraction between atoms is responsible for the formation of many-atom systems. In the case of the pair character of an atom interaction in a system of many bound atoms, the parameters of this system can be expressed through the parameters of the pair interaction potential of atoms. The analysis of this connection allows one to work out realistic models for the description of the bulk systems of bound atoms. Below we make this analysis for the bulk condensed systems of inert gases where additional simplicity is due to a weak interaction between atoms in the bulk system.

The parameters of the interaction potentials of two inert gas atoms in the repulsion and attraction regions of interaction follow from the analysis of various properties and processes, such as the basis of measurement of a high resolution vacuum ultraviolet absorption spectrum for the ground electronic state which yields the positions of the lowest vibrational and rotational levels of dimers, and from the total and differential collision cross sections resulting from measurements in atomic beams. Use of the measurements of the second virial coefficients of inert gases, transport data, mainly the viscosity and thermal conductivity coefficients of gases and also data for the binding energy of crystals of inert gases at 0 K, gives independent information about the interaction potential of two atoms. Summation of these data leads to enough accurate values of the interactions potential in the range of distances between atoms which determine these data. Table 12.1 contains some parameters of the interaction potentials of two identical atoms of the inert gases. Let us consider their peculiarities. Since an inert gas atom has a filled electron shell, the exchange interaction of two atoms, which is determined by the overlapping of the electron wave functions of different atoms, corresponds to repulsion. This means that attraction in atomic interactions is due to a weak long-range interaction and, hence, the dissociation energy of the dimer molecules of inert gases is small compared to a typical electron energy. From this it follows that three-body interactions in a system of many atoms of inert gases are relatively small and, hence, one can be restricted only by a pair interaction of atoms in such a system, i.e., the total interaction potential of atoms in this system has the form

$$U = \sum_{j,k} U(r_{jk}), \tag{12.5}$$

where $U(R)$ is the pair potential of the atom interaction of two atoms at a distance R between them and r_{jk} is the distance between the atoms j and k. Note the character of atom motion in a bulk system of inert gases. In particular, the dimer molecules of neon and argon are quantum systems because only two vibrational

Table 12.1. The parameters of the interaction between two identical atoms of inert gases and the parameters of bulk inert gases. Here R_e is the equilibrium distance between the nuclei of the dimer, D is the depth of the well in the pair interaction potential of atoms, D_o is the dimer dissociation energy for the ground vibrational state, R_l is the lattice constant at 0 K, so that the distance between nearest neighbors in the crystal is equal to $R_l/\sqrt{2}$, and ε_{sub} is the sublimation energy of the crystal per atom, so that $E = \varepsilon_{\text{sub}} n$ is the binding energy of atoms for the crystal consisting of n atoms.

Parameter	Ne	Ar	Kr	Xe
R_e, Å	3.091	3.756	4.011	4.366
D, K	42.2	143	201	283
D_o, K	24	121	184	267
R_l, Å	4.493	5.311	5.646	6.132
$\varepsilon_{\text{sub}} = E/n$	232	929	1343	1903
$R_e\sqrt{2}/R_l$	0.973	1.000	1.005	1.007
$\varepsilon_{\text{sub}}/6D$	0.92	1.08	1.11	1.12
C_6	6.6	68	130	270
$2R_e^6 D$	10.7	116	242	566

levels exist for the neon dimer, and 13–14 vibration levels are observed for the argon dimer molecule. Nevertheless, we will consider bulk systems of these inert gases as classical systems because they have many degrees of freedom.

From the data of Table 12.1 follows the short-range character of the interaction between the atoms of inert gases in their crystals, i.e., interaction between nearest neighbors determines the parameters of these crystals. Indeed, if we neglect the interaction of nonnearest neighbors, we obtain that the distance between nearest neighbors $a = R_l/\sqrt{2}$ becomes identical to the equilibrium distance for the diatomic molecule R_e, and the binding energy per atom for the classical bulk system in the absence of atom vibrations is $6D$, where D is the binding energy per bond, because each internal atom has 12 nearest neighbors.

Now let us consider a bulk system of atoms with a short-range interaction where only the interaction between the nearest neighbors takes place. In this system atoms can be modeled by balls whose radius is determined by the equilibrium distance between atoms in the corresponding diatomic. Then the profitable configuration of this system is the structure of close packing which corresponds to maximum density of these balls in a volume occupied them. This structure provides the maximum total binding energy of the atoms in the system. Let us transfer to the limit of zero temperature. Then the total binding energy of the system is equal to

$$E = \frac{D}{2}\sum_k k n_k, \tag{12.6}$$

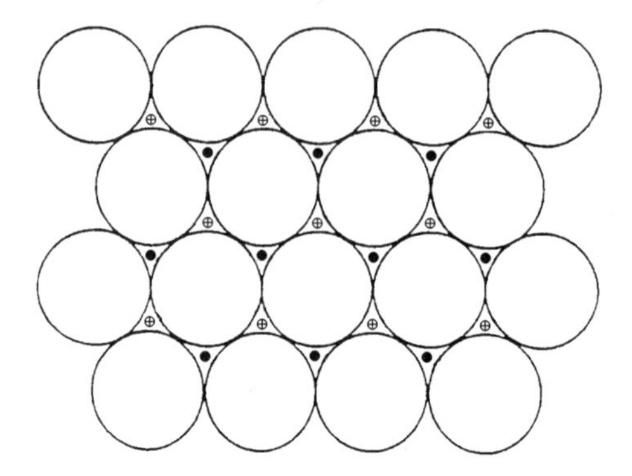

Figure 12.1. Assembling of a lattice of a close-packed structure consisting of atoms which are modeled by hard spheres of a radius a located in the plane. Crosses indicate the projections of centers of atoms of the lower plane onto this one. The distance between these planes is $a\sqrt{2/3}$. The projections of the centers of atoms of the upper plane onto this one are marked by open circles for the hexagonal structure of the lattice and by black circles for the face-centered cubic (fcc) structure of the lattice.

where D is the binding energy per bond, k is the number of bonds between the nearest neighbors, and n_k is the number of atoms with this number of bonds, the factor 1/2 takes into account that each bond includes two atoms.

Let us construct a close-packed structure containing an infinite number of atoms. Denoting the distance between the nearest neighbors by a, we construct an infinite crystal of this structure in the following way. First we arrange the atoms-balls along straight lines, so that the neighboring straight lines are located at the distance $a\sqrt{3}/2$, and each atom of a given line has two nearest neighbors located on the neighboring line. Repeating this operation, such that all the atom centers are located on the same plane, we obtain the plane of the atoms-balls as is shown in Fig. 12.1 for the plane $\{111\}$ where each atom has six nearest neighbors among the atoms of this plane. In constructing the next plane, we place each atom in hollows between three atoms of the previous plane. Thus each atom of this structure, with planes of direction $\{111\}$, has three nearest neighbors among the atoms of the previous and following planes, so that the total number of nearest neighbors for the atoms of the structure of close packing is equal to 12. Then, according to formula (12.6), the total binding energy of the atoms in the case of a short-range interaction potential equals $E = 6nD$, where $n \gg 1$ is the number of atoms of this system.

Note that the structure of close packing can be of two types. There are two possibilities of placing atom balls of the subsequent plane in hollows of this plane with respect to the atoms of the previous plane (see Fig. 12.1). If the projections of the atoms of the previous and subsequent planes onto this plane are coincident, then these atoms form the hexagonal lattice. If these projections of atoms are different, atoms form the face-centered cubic lattice. In the course of the construction of the lattice of atoms, by addition of the new planes, one can change the positions of these planes which leads to transition from the face-centered cubic lattice to the hexagonal lattice or vice versa. This character of the change of symmetry is

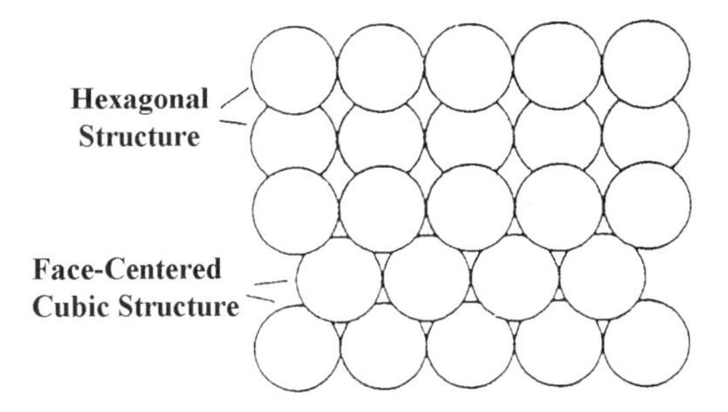

Figure 12.2. Transition between fcc and hexagonal structures of the close-packed lattice (twinning). The lattice is assembled such that the atoms of the next layer are located in hollows between atoms of the previous one. There are two possibilities for the positions of atoms of a new layer (see Fig. 12.1). A change in the order of these layers leads to a change of the lattice structure.

called "twinning" (see Fig. 12.2) and is observed in crystals with the close-packed structure of atoms.

We now analyze the properties of crystals with a pair interaction of atoms and a simple form of the pair interaction potential. We first consider the popular Lennard–Jones interaction potential which has the form

$$U(R) = D \cdot \left[\left(\frac{R_e}{R} \right)^{12} - 2 \left(\frac{R_e}{R} \right)^6 \right], \tag{12.7}$$

where R is the distance between atoms and D is the binding energy of two classical atoms. This interaction potential has a minimum $U = -D$ at$R = R_e$, i.e., R_e is the equilibrium distance between atoms in the diatomic molecule. An attractive feature of the Lennard–Jones interaction potential is that it contains a correct long-range dependence on the distances between atoms. But because of its simplicity, this interaction potential includes a certain connection between a long-range part of the interaction and the parameters of the bond. Indeed, as follows from formula (12.7), at large distances between the atoms, this potential has the form $U(R) = -2DR_e^6/R^6$, which corresponds to the dependence (10.11), $U(R) = -C_6/R^6$. The data of Table 12.1 show that the values of the parameters of a long-range interaction $2DR_e^6$ and C_6 differ by a factor 2. This means that the Lennard–Jones form of the interaction potential cannot satisfactorily describe the interaction of inert gas atoms at large distances between the atoms.

In analyzing the parameters of the Lennard–Jones crystal, which consists of atoms with the Lennard–Jones potential of interaction between them, we use the face-centered cubic structure of this crystal which is observed for real inert gas crystals. Let us determine the average binding energy per atom E/n which is given by

$$\frac{E}{n} = \frac{D}{2} \cdot \left[2C_1 \left(\frac{a}{R_e} \right)^6 - C_2 \left(\frac{a}{R_e} \right)^{12} \right], \tag{12.8}$$

where a is the distance between nearest neighbors in the lattice and the parameters of this formula are equal to (see Problem 12.2), for the face-centered cubic lattice,

$$C_1 = \sum_k \frac{n_k}{k^6} = 14.356, \qquad C_2 = \sum_k \frac{n_k}{k^{12}} = 12.131,$$

where n_k is the number of atoms at a distance ka from the atom under consideration and a is the distance between the nearest neighbors in this formula. Optimization of this formula over the distance a yields, for this parameter and the maximal binding energy,

$$a = \left(\frac{C_2}{C_1}\right)^{1/6} R_e = 0.971 R_e, \qquad \frac{E}{n} = \frac{D}{2} \cdot \frac{C_1^2}{C_2} = 8.61 D. \tag{12.9}$$

In the case of a short-range interaction potential these parameters are equal to

$$a = R_e, \qquad \frac{E}{n} = 6D. \tag{12.10}$$

Table 12.2 contains the average parameters of inert gases in reduced units. Comparison of these data with those of model crystals allows us to ascertain which of the above two forms of the pair interaction potentials better describes the real inert gas crystals. One can add to this comparison that the Lennard–Jones crystal has the hexagonal lattice and, in the case of the crystal with a short-range interaction of atoms, that the face-centered cubic lattice and hexagonal lattice are equivalent. The real crystals of all inert gases (Ne, Ar, Kr, Xe) at low temperatures have the face-centered cubic lattice. Thus a short-range interaction potential is more profitable, for the modeling of systems of many bound atoms of inert gases, than the Lennard–Jones one.

One more peculiarity of the short-range interaction between atoms follows from the analysis of the liquid state of bulk inert gases. The transition from the solid to liquid state is accompanied by a change in the bulk density so that a number of nearest neighbors varies at this transition. In the course of the heating of an inert gas solid from 0 K up to melting point, the distance between the nearest neighbors increases due to excitation of the phonons. We assume that as a result of melting, the distance between the nearest neighbors does not change while the number of nearest neighbors varies. Then one can connect the average number of nearest neighbors q in the liquid state at the melting point from the change bulk density. In a volume V there are $\rho_s V$ atoms in the solid state and $\rho_l V$ atoms in the solid state,

Table 12.2. Properties of model and real crystals with pair interaction atoms at zero temperature.

Parameter	Lennard–Jones crystal	Short-range interaction crystal	Average for crystals of Ne, Ar, Kr, Xe
$R_l/(R_e\sqrt{2})$	0.97	1	1.00 ± 0.02
$\varepsilon_{\text{sub}}/D$	8.61	6	6.4 ± 0.6
Structure	hexag.	fcc or hexag.	fcc

where ρ_s, ρ_l are the density of the solid and liquid inert gases at the melting point. Correspondingly, there are $(\rho_s - \rho_l)V$ vacancies (or voids) in this volume. The formation of each vacancy leads to a loss of 12 bonds, and the number of bonds is equal to $6\rho_s V$ for the solid at the melting point. Hence in the given volume there are $6\rho_s V - 12(\rho_s - \rho_l)V = 6(2\rho_l - \rho_s)V$ bonds for the liquid state. Because, in the liquid state, $6\rho_l V$ atoms are found in a given volume, the average number of nearest neighbors is equal to

$$q = 24 - 12\frac{\rho_s}{\rho_l}. \tag{12.11}$$

We now calculate the average number of nearest neighbors with accounting for the energetics of the system within the framework of the short-range character of the atom interaction in liquid inert gases. The relation

$$\frac{1}{12}H_s(T_m) = \frac{1}{q}H_l(T_m)$$

yields

$$q = \frac{12}{1 + \Delta H/H_l}, \tag{12.12}$$

where H_s, H_l are the enthalphies of atomization per atom for solid and liquid inert gases and $\Delta H = H_s - H_l$ is the variation of this value as a result of melting. Table 12.3 contains values of the average nearest neighbors resulting from formulas (12.11), (12.12). The coincidence of the results of different models testifies to the validity of the short-range character of the interaction of atoms for the bulk inert gas systems.

Note one more peculiarity of the properties of the bulk systems of inert gases. Although the pair interaction potential can be approximated by the use of several numerical parameters, only two of them are of importance, the equilibrium distance between the dimer atoms and the depth of the attraction well. From these parameters and the atom mass m one can compose only one combination of a given dimensionality. This leads to a certain scaling law for various parameters of the

Table 12.3. The reduced parameters of liquid inert gases. Here T_m is the melting point, ρ_s, ρ_l are the densities of condensed inert gases at the melting point in the solid and liquid states, $\rho_o = m\sqrt{2}/R_e^3$ is the typical density (m is the atom mass), ΔH is the fusion energy per atom, H_l is the sublimation energy in the liquid state per atom, and q is the number of nearest neighbors in the liquid state.

	Ne	Ar	Kr	Xe	Average
T_m/D	0.583	0.585	0.576	0.570	0.578 ± 0.006
$(\rho_s - \rho_l)/\rho_o$	0.123	0.116	0.126	0.127	0.123 ± 0.005
$\Delta H/H_l$	0.211	0.182	0.175	0.173	0.185 ± 0.015
q, formula (12.11)	10.10	10.27	10.11	10.19	10.17 ± 0.08
q, formula (12.12)	10.07	10.15	10.14	10.19	10.15 ± 0.04

Table 12.4. Reduced critical parameters of inert gases. T_{cr} is the critical temperature, p_{cr} is the critical pressure and ρ_{cr} is the critical density.

System	Ne	Ar	Kr	Xe	Average
T_{cr}/D	1.052	1.052	1.041	1.025	1.04 ± 0.01
$p_{cr}R_e^3/D$	0.132	0.129	0.132	0.132	0.131 ± 0.001
$\rho_{cr}R_e^3/m$	0.301	0.303	0.301	0.298	0.301 ± 0.002

bulk systems consisting of the interacting atoms of inert gases. For some examples of this scaling, Table 12.4 contains the reduced critical parameters of inert gases, and their coincidence for different inert gases confirms the above statement.

12.3 Short-Range Interactions in Many-Atom Systems

The above analysis shows the short-range character of atom interaction in the bulk systems of inert gases. We now analyze the peculiarities of this interaction in bulk systems and clusters, systems consisting of a finite number of bound atoms. The name for a short-range interaction of atoms is taken from nuclear physics where this term has another meaning. This means that the interaction between two nuclear particles takes place only in a restricted region of the order of a radius of the action of nuclear forces. The potential of a short-range interaction of two nuclear particles has one sign, i.e., it corresponds to attraction or repulsion only, while the interaction potential of two atomic particles varies from their repulsion at small distances between them to attraction at large distances. Then the term "a short-range interaction" in atomic and molecular physics means that the interaction between atomic particles is absent starting from some distance between them. For example, this takes place if the interaction potential $U(R)$ at large distances R between the interacting atomic particles varies as $\sim \exp(-\gamma R)$, where γ is a constant.

In the case of the short-range interaction of atoms, the total interaction potential in the system of many atoms is the sum of the interaction potentials between nearest neighbors only, and the interaction potential of two atoms does not depend on the interaction of these atoms with other atoms. These facts simplify the analysis of such systems and allows one to describe their properties in a simple way. Below we evaluate the energy of clusters with a short-range interaction at zero temperature. Note that large clusters differ from bulk systems because of the quantum character of variation of the cluster energy with an increase in the number of cluster atoms. This is determined by the cluster structure and is of importance up to large cluster sizes.

The binding energy of a large cluster consisting of n atoms at zero temperature, tends, in the limit (12.10), to $E = 6Dn$ at large n, where D is the energy of one bond or the well depth in the pair interaction potential of atoms. Below we express the energetic parameters of clusters in units of D, so that the total binding energy of atoms is equal to the number of bonds between the nearest neighbors. Formula (12.10) corresponds to the first term of expansion of the binding energy of atoms over a small parameter $n^{-1/3}$ and takes into account the fact that the distance between the nearest neighbors in clusters is equal to the equilibrium distance in the diatomic molecule in the case of a short-range pair interaction potential of atoms. Accounting for the second term of expansion of the atom binding energy over the small parameter, we introduce the cluster surface energy $E_{\rm sur}$ from the relation

$$E = 6n - E_{\rm sur}. \tag{12.13}$$

From this, on the basis of formula (12.6) we have

$$E_{\rm sur} = \sum_k \left(6 - \frac{k}{2}\right) n_k, \tag{12.14}$$

where k is the number of nearest neighbors and n_k is the number of atoms with this number of nearest neighbors. As follows from this formula, internal atoms does not give a contribution to the cluster surface energy. Because the number of surface atoms is proportional to $n^{2/3}$, at large n we have

$$E_{\rm sur} = An^{2/3}. \tag{12.15}$$

Here the parameter A is the specific surface energy which, according to formulas (12.13) and (12.15), can be determined on the basis of the following relation if the total binding energy of atoms is known

$$A = 6n^{1/3} - E/n^{2/3}. \tag{12.16}$$

These relations can be a basis for determination of the energetic parameters of clusters of a different structure with a short-range interaction of atoms. We analyze this problem below.

12.4 Clusters with a Short-Range Interaction of Atoms

Let us construct clusters with a short-range interaction of atoms at zero temperature and determine the binding energy of atoms. We start from clusters of the face-centered cubic (fcc) structure, and these clusters can be by cutting out from the fcc lattice. The fcc structure has a high symmetry, and clusters of fcc structures can have surface atoms located on plane faces. The positions of atoms for all these planes, which include planes {100}, {110}, and {111} in crystallographic notations, are given in Fig. 12.3. We use the coordinate frame such that planes xy, xz, yz have the symmetry {100}. Then a high symmetry of the fcc structure is expressed by conservation of the system as a result of the transformations

$$x \longleftrightarrow -x, \quad y \longleftrightarrow -y, \quad z \longleftrightarrow -z, \quad x \longleftrightarrow y \longleftrightarrow z. \tag{12.17}$$

We now construct clusters of the fcc structure by the addition of atoms to a cluster or by cutting off the cluster from a crystal of the fcc structure. Take as the

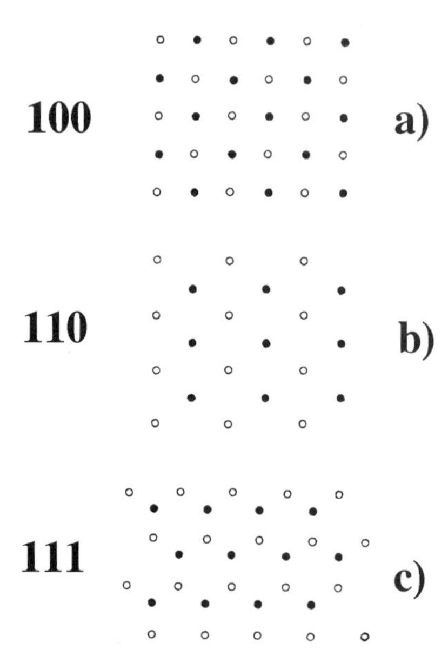

Figure 12.3. Positions of the centers of atoms on planes of the lattice of the fcc-structure. Black circles mark the centers of the atoms of the upper layer, and open circles indicate the positions of projections of the centers of atoms of the previous layer onto the surface. (a) Plane {100}, where each surface atom has 8 nearest neighbors (4 nearest neighbors of this layer and 4 nearest neighbors of the previous layer). (b) Plane {110}, where each surface atom has 7 nearest neighbors (2 nearest neighbors of this layer, 4 nearest neighbors of the previous layer, and 1 nearest neighbor from the second layer from the surface). (c) Plane {111}, where each surface atom has 9 nearest neighbors (6 nearest neighbors of this layer and 3 nearest neighbors of the previous layer).

origin of the coordinate frame any atom of this lattice or a center of a cell of the lattice so that the symmetry (12.17) takes place for cluster atoms. Then one can extract groups of atoms of the cluster such that the atoms of one group change their positions as a result of transformation (12.17). Such a group is called "the cluster shell," i.e., the fcc cluster has a shell structure. The maximum number of atoms of one shell corresponds to the total number of transformations (12.17) and is equal to 48. In clusters with filled shells the properties of each atom of a certain shell, in particular, a number of nearest neighbors, are the same for any atom of this shell. Hence, for a cluster description, it is enough to analyze the parameters of one atom of each shell.

Let us find the configuration of cluster atoms which correspond to the maximum binding energy of cluster atoms E (or a minimum surface energy E_{sur}) for a given number of atoms n at zero temperature. We assume the cluster shells to be filled or free and use the property of the fcc clusters that surface atoms can form plane faces when they are located in planes {100}, {110}, and {111} (see Fig. 12.3). Since there are six different planes of type {100}, 12 different planes of {110}, and eight different planes of type {111}, the maximum number of plane faces of fcc clusters is equal to 26. A cluster with such or other completed structure has the so-called magic number of atoms. In this case removal or addition of one atom to the cluster leads to an increase in its specific surface energy (12.16). Thus the magic numbers of clusters correspond to the profitable structures of clusters with respect to their specific energy. The appearance of the magic numbers of cluster atoms is the property of a system consisting of a finite number of atoms and results from the shell structure of clusters with a pair interaction of atoms.

We now construct the cluster geometrical figures with plane faces which correspond to the magic numbers of clusters of the fcc structure and evaluate the energy of such clusters with a short-range interaction of atoms. In order to calculate the surface energy of these clusters it is enough, in accordance with formula (12.14), to determine the number of bonds between nearest neighbors in this cluster. According to Fig. 12.3, the surfaces of the direction {111} are more profitable because a surface atom has nine nearest neighbors, while a surface atom of a plane of the direction {100} has eight nearest neighbors, and a surface atom of a plane of the direction {100} has seven nearest neighbors. From this it follows that the basis of the optimal structure of fcc clusters with a short-range interaction of atoms is an octahedron—a geometrical figure whose surface consists of eight equilateral triangles. Below we calculate the number of atoms and the surface energy of clusters with a short-range interaction of atoms for the family of octahedrons (see Fig. 12.4(a)). The number of an octahedron in the family is m so that each of 12 edges of this octahedron contains $m+1$ atoms. Then this octahedral cluster has six vertex atoms, $12(m-1)$ nonvertex edge atoms and $4(m-1)(m-2)$ surface atoms which are located inside eight surface triangles. Thus the total number of surface atoms of the mth octahedron is equal to $4m^2+2$, and we have the relation

$$n_m = n_{m-1} + 4m^2 + 2,$$

where n_m is the total number of atoms of the mth octahedral cluster, and this relation gives (for simplicity, below we omit the subscript)

$$n = \frac{2m^3}{3} + 2m^2 + \frac{7m}{3} + 1. \tag{12.18}$$

In order to calculate the surface energy, we note that each vertex atom has four nearest neighbors, each interior edge atom has seven nearest neighbors, and each interior atom of surface triangles has nine nearest neighbors. Then, from formula (12.14), we obtain, for the surface energy of this cluster,

$$E_{\text{sur}} = 6(m+1)^2. \tag{12.19}$$

Let us now consider a truncated octahedron (Fig. 12.5) which can be obtained from the mth octahedral cluster by cutting off six regular pyramids whose vertices are the octahedron vertices, and each edge of these pyramids contains k atoms. Each pyramid contains $k(k+1)(2k+1)/6$ atoms. The number of atoms of the

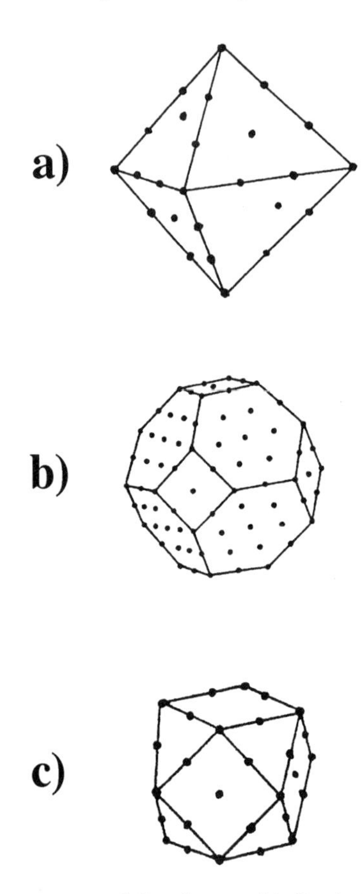

Figure 12.4. Symmetric structures of fcc-clusters: (a) Octahedron; (b) regular truncated octahedron—tetradecahedron; (c) cuboctahedron.

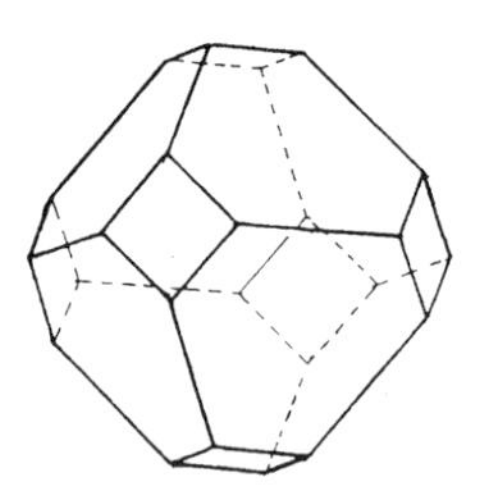

Figure 12.5. The icosahedron figure. Its 12 vertices are located on the surface of a sphere of radius R and the distances between neighboring vertices on the sphere are identical and equal to R_o, so that joining of the neighboring vertices gives 20 equilateral triangles.

truncated octahedron and its surface energy are equal to

$$n = \frac{2m^3}{3} + 2m^2 + \frac{7m}{3} + 1 - k(k+1)(2k+1), \qquad E_{\text{sur}} = 6(m+1)^2 - 6k(k+1). \tag{12.20}$$

The surface of a truncated octahedron includes six squares and eight hexagons. The regular truncated octahedron—tetradecahedron (Fig. 12.4(b)) has regular hexagons on its surface (see Fig. 12.4(b)). Let us consider a family of such figures which are characterized by a number p so that its edges contain $3p + 1$ atoms. Then from relations $m = 3p$ and $k = p$ we have, from formula (12.20),

$$n = 16p^3 + 15p^2 + 6p + 1; \qquad E_{\rm sur} = 48p^2 + 30p + 6. \tag{12.21}$$

One more symmetrical figure of truncated octahedra is the cuboctahedron (Fig. 12.4(c)). In this case three edges of the hexagon are equal to zero, and the surface of the cuboctahedron consists of six squares and eight equilateral triangles. Denote the number of the octahedron family by p so that each of its edges contains $2p+1$ atoms. Then $2(2p + 1) = m + 1$, so that from formula (12.20) we obtain, for the parameters of the cuboctahedral cluster,

$$n = \frac{10p^3}{3} + 5p^2 + \frac{11p}{3} + 1; \qquad E_{\rm sur} = 6(2p + 1)^2. \tag{12.22}$$

From the above expressions one can find the specific surface energy A of the clusters under consideration in accordance with formula (12.14). In particular, for large clusters with $m \gg 1$, $k \gg 1$, this value is equal to

$$A = 6 \cdot \frac{m^2 - k^2}{(2m^3/3 - 2k^3)^{2/3}}. \tag{12.23}$$

The specific surface energy A is equal to 7.86 for the octahedral cluster structure, 8.06 for the cuboctahedral cluster, 7.55 for the regular truncated octahedron—tetradecahedron and 7.60 ± 0.05 for fcc clusters with a maximum binding energy of atoms at a given number of atoms on average.

In order to determine the optimal cluster structures at low temperatures, it is enough to compare the values of the specific surface energies A for these structures at the same number of cluster atoms. This comparison shows that for large clusters $n \gg 100$ the fcc structure of clusters is more preferable than the hexagonal one. The icosahedral structure can be more favorable for these and larger sizes. Hence below we consider the icosahedral cluster structure. Note that the existence of the cluster icosahedral structure confirms a general peculiarity of cluster structures whose variety is wider than that of crystal structures. Indeed, crystals with a pair interaction of atoms can have two structures of close packing, fcc and hexagonal structures. Clusters consisting of these atoms along with these structures can have the icosahedral structure.

The icosahedron as a geometric figure has a surface consisting of 20 regular triangles (Fig. 12.6). Hence the distances between neighboring vertices of the icosahedron are identical as well as the distances from the icosahedral center to its vertices. This means that all the icosahedral vertices are located on a sphere which center is the icosahedral center. But the radius of this sphere (we denote it by R) differs from the distance between neighboring surface atoms R_o. Figure 12.6 gives projections of the icosahedron onto some planes and its developing view. We express below the different distances in the icosahedron and find the relation

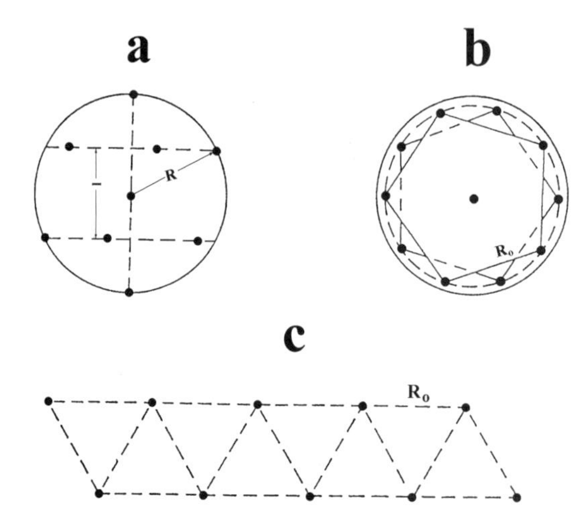

Figure 12.6. The icosahedral structure. (a) The projections of the icosahedron vertices onto the plane passing through its two axes. (b) The projections of the icosahedron vertices onto the plane which is perpendicular to one of its axes. (c) The developed view of the cylinder whose axis is one axis of the icosahedron, so that 10 icosahedron vertices are located on the surface of this cylinder. The projection of this cylinder onto the plane which is perpendicular to the cylinder axis is shown by a dashed line in the top view of the icosahedron (Fig. 12.6(b).

between the parameters R and R_o. The side length of the pentagon of Fig. 12.6(b) is equal to

$$R_o = 2r \sin\frac{\pi}{5}. \tag{12.24a}$$

The distance between nearest neighbors that are the vertices of different pentagons is

$$R_o = \sqrt{l^2 + \left(2r \sin\frac{\pi}{10}\right)^2}, \tag{12.24b}$$

where l is the distance between pentagons. The distance between a pole and an atom of a nearest pentagon equals

$$R_o^2 = r^2 + \left(R - \frac{l}{2}\right)^2. \tag{12.24c}$$

In addition, the following relation takes place between the sphere radius and the ring radius in which pentagons are inscribed

$$R^2 = r^2 + (l/2)^2. \tag{12.24d}$$

One can see that the first three equations give the relation between the icosahedral parameters, and the forth equation allows us to check the validity of the icosahedron definition. The first and second equations give

$$r = l = \left(\frac{1}{2} + \frac{1}{2\sqrt{5}}\right)^{1/2} R_o = 0.851 R_o. \tag{12.25a}$$

From the third equation we have

$$R = r\frac{\sqrt{5}}{2} = 0.951R_o. \tag{12.25b}$$

The last equation corroborates these relations. Thus, the distance from the center to the icosahedral vertices is approximately 5% less than between the nearest vertices of the icosahedron.

Let us construct the icosahedral cluster with m filled layers in the following way. Let us draw a sphere of radius mR and locate on it 12 vertices such that 20 regular triangles are formed as a result of joining the nearest vertices. Then we divide each edge of the icosahedron into m parts and draw lines through them which are parallel to the triangle sides. Atoms are located in both the vertices and crosses of these lines. Now let us divide each radius-vector joining the center with the vertices into m parts and draw $m - 1$ new spheres through the obtained points. Repeating the above operation with the triangles of each layer, we obtain the cluster of the icosahedral structure which contains m filled layers. Let us calculate the number of atoms of such a cluster. The surface layer of this cluster contains 12 vertex atoms, $m - 1$ nonvertex atoms in 30 of each of its edges, and $(m - 1)(m - 2)/2$ atoms inside each of the 20 surface triangles. This leads to the relation

$$n_m = n_{m-1} + 12 + 30(m - 1) + 10(m - 1)(m - 2),$$

where n_m is the number of atoms for the icosahedral, the cluster containing m filled layers. From this we have

$$n_m = \frac{10}{3}m^3 + 5m^2 + \frac{11}{3}m + 1. \tag{12.26}$$

In order to determine the binding energy of cluster atoms within the framework of a short-range interaction of atoms, it is necessary to calculate the number of bonds of different lengths between the atoms. We account for two lengths of R and R_o in the cluster and bonds of length R are realized between atoms of the neighboring layers, while the bonds of length R_o correspond to the nearest atoms of one layer. Let us present the total binding energy of cluster atoms in the form

$$E = -aU(R) - bU(R_o), \tag{12.27}$$

where a is the number bonds in the cluster of length R and b is the number bonds of length R_o. Now we calculate these numbers of bonds taking into account that each vertex atom has one bond of length R and five bonds of length R_o, each nonvertex edge atom has two bonds of length R and six bonds of length R_o, and each internal atom of a surface triangle has three bonds of length R and six bonds of length R_o. This leads to the following relations for the numbers of corresponding bonds for the icosahedral cluster with completed layers

$$a_m = a_{m-1} + 30m^2 - 30m + 12;$$

$$b_m = b_{m-1} + 30m^2.$$

This gives, for the number of corresponding bonds,

$$a_m = 10m^3 + 2m; \qquad b_m = 10m^3 + 15m^2 + 5m. \tag{12.28}$$

Taking into account that the lengths of bonds R and R_o are close to the equilibrium distance R_e in the diatomic molecule we obtain, from formula (12.28),

$$E = (a_m + b_m)D - \frac{1}{2}U''(R_e) \cdot [a_m(R - R_e)^2 + b_m(R_o - R_e)^2].$$

Taking the maximum of this value as a function of R, we obtain as a result of optimization of this expression,

$$E = (a_m + b_m)D - \frac{0.0012 a_m b_m}{0.904 a_m + b_m} \cdot R_e^2 U''(R_e). \tag{12.29}$$

The second term of this expression is small compared to the first term. For example, in the case of the truncated Lennard–Jones interaction potential, when $R_e^2 U''(R_e) = 72D$, the second term is 2.3% of the first term. This means that the binding energy is mainly determined by the number of bonds. But because the number of the nearest neighbors of internal atoms is the same for the icosahedral and close-packing structures, the second term of formula (12.29) is of importance for the choice of the optimal structure at a given number of cluster atoms. In particular, the asymptotic expression of the total binding energy of atoms E at large numbers of atoms n has the following form for the icosahedral cluster with filled layers

$$E = 5.864n - 6.56n^{2/3}. \tag{12.30}$$

Let us compare this expression with that of the fcc cluster where, according to formula (12.15), the asymptotic form of the total binding energy of atoms is $E = 6n - An^{2/3}$. One can see that the difference in the first term of expansion is determined by the second term in the right-hand side of formula (12.29), and formula (12.30) corresponds to the truncated Lennard–Jones interaction potential between the atoms. In this case the coincidence of these asymptotic formulas for the icosahedral and fcc structures takes place in the range $n = 400 \div 500$. This means competition of these structures in a wider region than this one, so that at some number of cluster atoms in this region the icosahedral structure is preferable, other regions clusters with the fcc structure are characterized by a higher binding energy of atoms. Of course, the position of the region of competition of the structures depends on the form of the pair interaction potential of atoms.

12.5 The Jelium Model of Metallic Clusters

Above we consider a system of many bound the atoms where atoms conserve their individuality. Now we present another case of atom interaction which relates to the formation of metallic particles from atoms. The related jelium model of clusters is suitable for clusters consisting of alkali metal atoms and assumes the charge of positive ions to be distributed uniformly over a cluster space. As a matter of fact, this model resembles plasma models with uniform distribution of the positive charge over a space. These models allow us to understand the principal plasma properties. In particular, electrons of a dense plasma can form the Wigner crystal at zero temperature which has an analogy with the jelium cluster model. Let us use the concept of the Wigner crystal for the jelium cluster model. Electrons in a field of a distributing positive charge have a specific potential energy whose depth is of the order of $e^2N^{-1/3}$, where N is the number density of electrons and ions. On the other hand, electrons of this degenerated electron gas have a typical kinetic energy of the order of $p_o^2/m \sim N^{2/3}$, where p_o is the Fermi momentum for electrons.

From this follows the form for the binding energy per electron

$$\varepsilon = aN^{2/3} + bN^{-1/3}, \tag{12.31}$$

where the parameters a, b have the order of a typical atomic value. Optimization of this formula gives the optimal value of the number density of electrons which is of the order of a typical atomic value. Thus the size of a metallic cluster within the framework of the jelium cluster model is established on the basis of competition of the electrostatic interaction between electrons and ions [the second term of formula (12.31)] and the exchange interaction between electrons due to the Pauli exclusion principle [the first term of formula (12.31)]. The result of this competition gives an optimal cluster size.

Along with the general properties of a dense plasma at zero temperature, the related cluster has specific properties which are determined by its finite size. Indeed, the form of the well which is created by the positive charge of the cluster influences the positions of electrons and, as a result of their interaction, a self-consistent field occurs which determines the quantum numbers of the electrons of the cluster. Let us consider this problem in a general form. The self-consistent field of the cluster has a spherical symmetry as follows from the problem symmetry (in reality it is valid strictly for clusters with filled electron shells).

Then the quantum numbers of electrons are the same as for atomic electrons. They are $nlm\sigma$, and in a general case we have $|m| \leq l$, $\sigma = \pm 1/2$. As for the condition $l + 1 \leq n$, which takes place for electrons located in the Coulomb field, this condition is now absent. This means that an electron with a certain n can have, in principle, any orbital momentum l. In particular, Fig. 12.7 yields the sequence of the filling of electron shells for clusters of alkali metals, as follows from the experiment. At the start, this sequence is the following $1s^2 1p^6 1d^{10} 2s^2 1f^{14} 2p^6 1g^{18} 2d^{10} 1h^{12} 3s^2$, and so on. The example of the jelium cluster shows the connection between physical concepts which were understood for physical objects of one type and can be suitable in a modified form for other physical objects. This relates to the connection between the jelium model of clusters of a dense plasma with a low temperature, with the electron behavior in a self-consistent field of ions and electrons, and with the physics of metals.

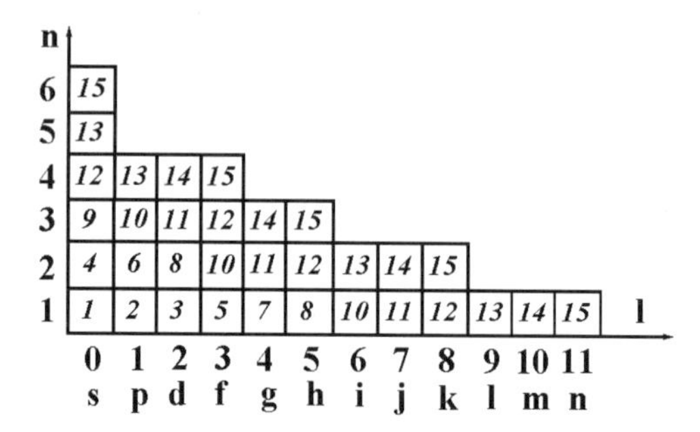

Figure 12.7. The sequence of shell filling for alkali metal clusters.

Problems

Problem 12.1. *Analyze the interaction potential of three atoms with one valent s-electron if they are arranged in one line.*

We use formulas (12.3) and (12.4) for the energy of the lowest state of the system under consideration. Let us take atom b as the middle atom and neglect the interaction between atoms a and c. Then the interaction potential depends on two distances, $x = R_{ab}$ and $y = R_{bc}$. According to formula (12.3), the interaction potential has the following form:

$$U(x, y) = V(x) + V(y) - \frac{1}{2\sqrt{2}}\sqrt{[\Delta(x) - \Delta(y)]^2 + \Delta^2(x) + \Delta^2(y)}, \quad (12.32)$$

where $V(x)$ is the long-range interaction potential between two atoms and $\Delta(x)$ is the exchange interaction potential. Note that $U(x, \infty) = V(x) - \frac{1}{2}\Delta(x)$ is the interaction potential for the diatomic molecule which is found in the even state. It is convenient to use this potential in a form which is similar to the Morse formula, so that $V(x) = A\exp[-\alpha(x - R_e)]$, $\Delta(x) = B\exp[-\beta(x - R_e)]$, where R_e is the equilibrium distance for the diatomic molecule. Introducing the dissociation energy D of the diatomic molecule we have, for the parameters of the long-range and exchange interaction potentials of two atoms,

$$A = \frac{\beta}{\alpha - \beta}D, \qquad B = \frac{2\alpha}{\alpha - \beta}D, \qquad \alpha > \beta.$$

These relations lead to the minimum of the pair interaction potential at distance R_e between the atoms with depth D of the potential well.

These relations allow us to present a general form of the electron energy which is called the potential energy surface for the system ABC of atoms. In this case the potential energy is presented as equipotential curves in the xy plane. In the frame of the xy-axes these curves are symmetric with respect to the line $x = y$. At large x or y they have the form of valleys which are separated by a potential barrier. Let us find the value of the barrier on the basis of the above formulas and approximations. Because of the symmetry, the barrier lies on the line $x = y$. The interaction potential of atoms on this line is equal to

$$U(x=y) - U(x, y=\infty) = V(x) - \frac{\Delta(x)}{2}$$
$$= 2A\exp[-\alpha(x-R_e)] - \frac{1}{2}B\exp[-\beta(x-R_e)].$$

Using the connection between the parameters of long-range and exchange interaction potentials, we find the position of the minimum $x_{\min}$ of this function and the barrier value $U_{\text{bar}} = D - U(x_{\min}, x_{\min})$:

$$x_{\min} = R_e + \frac{\ln 2}{\alpha - \beta}, \qquad U_{\text{bar}} = D - \frac{D}{2^{\beta/(\alpha-\beta)}}. \tag{12.33}$$

In particular, in the case of the Morse interaction potential $\alpha = 2\beta$ we have, for the barrier height,

$$U_{\text{bar}} = D/2. \tag{12.34}$$

Thus the potential energy surface of the system under consideration in the xy plane consists of two ravines separated by a barrier. These ravines are approximately located parallel to the x- and y-axes, so that the middle of each ravine is found at the distance R_e from the corresponding axis. In reality, transition from one ravine to the other corresponds to transition from the bound state of atoms AB to the bound state of atoms BC. Thus, the transition between these ravines corresponds to the chemical process $A + BC \to AB + C$. Therefore the ravine which is parallel to the x-axis is called the reagent valley, and the other ravine is called the product value. The line of an optimal transition from the reagent valley to the product valley is called the reaction way and the potential energy along this way is called the profile of the reaction way. In the case under consideration, the reaction way is a symmetric curve with respect to the line $x = y$. This is characterized by the maximum of the profile which is equal to U_{bar}.

Problem 12.2. *Evaluate the specific sublimation energy (the binding energy per atom) for a crystal of the face-centered cubic (fcc) structure with a pair interaction between atoms which has the form of the Lennard–Jones or Morse potential.*

It is convenient to introduce the unit of length $R_e/\sqrt{2}$, so that R_e is the equilibrium distance between atoms in the diatomic molecule, and we use the reduced energy units by expressing them in D units—the dissociation energy of the diatomic molecule or the energy per bond. Taking a test atom as the origin of the coordinate frame whose axes have the direction $\{100\}$ of the lattice, we account for the symmetry of the fcc structure (12.17),

$$x \longleftrightarrow -x, \quad y \longleftrightarrow -y, \quad z \longleftrightarrow -z, \quad x \longleftrightarrow y \longleftrightarrow z.$$

This allows us to place crystal atoms in shells so that all the atoms of one shell are located at identical distances from the central atom, and the atoms of one shell exchange their positions as a result of the above transformations. Therefore it is

Table 12.5. Parameters of atomic shells.

Shell	r_k^2/R_e^2	n_k		Shell	r_k^2/R_e^2	n_k
011	1	12		044	16	12
002	2	6		334	17	24
112	3	24		035	17	24
022	4	12		006	18	6
013	5	24		244	18	24
222	6	8		116	19	24
123	7	48		235	19	48
004	8	6		026	20	24
114	9	24		145	21	48
033	9	12		226	22	24
024	10	24		136	23	48
233	11	24		444	24	8
224	12	24		055	25	12
015	13	24		017	25	24
134	13	48		345	25	48
125	15	48		046	26	24

enough to use the parameters of one atom of each shell for the analysis of the contribution of this shell to the total binding energy. Below we characterize a shell by the coordinates of a test atom of a given shell xyz for which $0 \leq x \leq y \leq z$. Table 12.5 gives the number of atoms n_k for filled shells located at a given distance from the central atom. Note that for each atom of the crystal lattice, the value $x + y + z$ expressed in reduced units, is a whole even number.

On the basis of the data of Table 12.5 determine the binding energy of the crystal per atom which is equal to

$$\varepsilon = \frac{1}{2} \sum_k n_k U(r_k), \tag{12.35}$$

where $U(r_k)$ is the pair interaction potential of the central atom with an atom of a given shell, the factor 1/2 accounts for each bond relating to two atoms.

The general scheme of determination of the sum is the following. We present each term of the interaction potential in the form

$$\varepsilon = \frac{1}{2} \sum_{k=1}^{26} n_k U(r_k) + \int_{\sqrt{27}}^{\infty} 2\pi r_k^2 \, dr_k \sqrt{2} U(r_k),$$

Table 12.6. The total binding energy of the fcc crystal per one atom for a given pair interaction potential.

$U(r)$	r^{-6}	r^{-8}	r^{-12}	$\exp[2(r-R_e)]$	$\exp[4(r-R_e)]$	$\exp[8(r-R_e)]$
2ε	14.454	12.802	12.132	28.60	14.937	12.292
δ	0.002	$7\cdot 10^{-5}$	$1\cdot 10^{-7}$	0.01	$2\cdot 10^{-6}$	$2\cdot 10^{-13}$

and the error of this operation is estimated as $\delta = \pi\sqrt{26}U(\sqrt{26}) = 16U(\sqrt{26})$. Table 12.6 gives values of this sum for various forms of terms of interaction potentials and the error of the used scheme of calculation.

Let us formulate a general method for the calculation of the specific binding energy of crystal atoms on the basis of the data of Table 12.6 or data of such a type. We consider two methods for the presentation of the pair interaction potential so that in the first case it is the sum of two terms with an inverse power dependence on the distance between atoms and, in the second case it has a form of the Morse potential. In the first case the general form of the interaction potential is the following:

$$U(R) = D\left[\frac{l}{n-l}\left(\frac{R_e}{R}\right)^n - \frac{n}{n-l}\left(\frac{R_e}{R}\right)^l\right], \qquad l < n,$$

so that this dependence has a minimum $U(R_e) = -D$ at $R = R_e$. Denoting the sum $C_n = \sum_k (R_e/r_k)^n$, where r_k is the distance for atoms of the kth shell we have, for the specific binding energy of the crystal,

$$\frac{2\varepsilon}{D} = C_n\frac{l}{n-l}\left(\frac{R_e}{R}\right)^n - C_l\frac{n}{n-l}\left(\frac{R_e}{R}\right)^l.$$

This function has a minimum

$$\frac{2\varepsilon}{D} = -\frac{C_n^{n/(n-l)}}{C_l^{l/(n-l)}} \quad \text{at} \quad R_o = R_e\left(\frac{C_n}{C_l}\right)^{1/(n-l)}.$$

In particular, for the Lennard–Jones potential $l = 6$, $n = 12$, this formula yields, for the maximum binding energy of crystal atoms,

$$\frac{2\varepsilon}{D} = -C_6^2/C_{12} \quad \text{at} \quad R_o = R_e\left(\frac{C_{12}}{C_6}\right)^{1/6}$$

which coincides with formula (12.9).

In the case of the Morse pair potential

$$U(R) = D\left[e^{2\alpha(R-R_e)} - 2e^{\alpha(R-R_e)}\right]$$

we obtain, for the specific binding energy of the crystal per atom,

$$\frac{2\varepsilon}{D} = C_\alpha e^{2\alpha(R-R_e)} - 2C_{2\alpha}e^{\alpha(R-R_e)},$$

where $C_\alpha = \sum_k e^{\alpha(r_k - R_e)}$. This value has a minimum

$$\frac{2\varepsilon}{D} = C_\alpha^2 / C_{2\alpha} \quad \text{at} \quad R_o = R_e - \frac{1}{\alpha} \ln \frac{C_\alpha}{C_{2\alpha}}.$$

In particular, in the case $\alpha = 2$, this formula and the data of Table 12.6 give $R_o = 0.675 R_e$, $\varepsilon_{\min} = -27.4 D$. In the case $\alpha = 4$ we have $R_o = 0.951 R_e$, $\varepsilon_{\min} = -9.084 D$. Note that in both cases of this type of interaction potential the optimal distance between the nearest neighbors is less than the equilibrium distance between atoms of the diatomic molecule, and the specific binding energy exceeds the value $6D$ which corresponds to a short-range interaction potential.

CHAPTER 13

Elastic Collisions of Atomic Particles

13.1 Elastic Scattering of Classical Atomic Particles

The interaction of atomic particles determines the character of the collisional processes involving these particles. Hence the parameters of the scattering of these particles are expressed through the parameters of their interaction potentials or electron terms of the quasi-molecule consisting of colliding particles. We start from the elastic collisions of classical atomic particles, when their motion is governed by classical laws. Let $U(\mathbf{R})$ be the interaction potential of particles which depend on a distance $\mathbf{R}$ between them. We consider the elastic collisions of atomic particles when the internal states of the colliding particles do not vary. Then the positions of the colliding particles satisfy the following Newton equations:

$$m_1 \frac{d^2\mathbf{R}_1}{dt^2} = -\frac{\partial U}{\partial \mathbf{R}_1}, \qquad m_2 \frac{d^2\mathbf{R}_2}{dt^2} = -\frac{\partial U}{\partial \mathbf{R}_2} \tag{13.1}$$

where $\mathbf{R}_1$, $\mathbf{R}_2$ are the coordinates of the corresponding particles, m_1, m_2 are their masses and, because the interaction potential U for related particles depends only on the relative distance between particles, $\mathbf{R} = \mathbf{R}_1 - \mathbf{R}_2$, the forces which act on one particle from one another, are connected by the relation $\partial U/\partial \mathbf{R}_1 = -\partial U/\partial \mathbf{R}_2$. This form of interaction between particles allows us to divide the motion of two particles into the motion of their center of mass and their relative motion. Indeed, introducing the vector of the center of mass of particles $\mathbf{R}_c = (m_1\mathbf{R}_1 + m_2\mathbf{R}_2)/(m_1 + m_2)$, one can present the Newton equations (13.1) in the following form:

$$(m_1 + m_2)\frac{d^2\mathbf{R}_c}{dt^2} = 0, \qquad \mu \frac{d^2\mathbf{R}}{dt^2} = -\frac{\partial U}{\partial \mathbf{R}}, \tag{13.2}$$

where $\mu = m_1 m_2/(m_1 + m_2)$ is the reduced mass of the particles. Thus, the center of mass moves with a constant velocity, and the interaction of particles influences only their relative motion.

Therefore, we study below the motion of one particle with the reduced mass μ in the field of center $U(\mathbf{R})$, because the problem of the scattering of two particles is reduced to the problem of the scattering of one particle with mass μ in the field $U(\mathbf{R})$ of a scattering center. Note that the same situation takes place for the quantum character of the interaction of particles. Then the Schrödinger equation is separated as a result of using the same variables, and the scattering problem is described by the Schrödinger equation for the relative positions of particles.

We now introduce the parameters which characterize the scattering of particles. Under given initial conditions, the relative motion of particles is described by the certain trajectory of a scattering particle. Assuming the interaction potential to be spherical, $U = U(|\mathbf{R}_1 - \mathbf{R}_2|)$, one can characterize the trajectory of colliding particles by the following parameters (see Fig. 13.1): ρ is the impact parameter of collision, r_o is the distance of closest approach, and ϑ is the scattering angle which is the angle between the initial and final directions of motion of the particle. The parameter of the elementary act of the collision of particles is the cross section of this process, which is the ratio of the number of scattering acts per unit time to the flux of incident particles. Let us assume the dependence $\rho(\vartheta)$ to be monotonic in some range of ϑ and then find the differential cross section of collision as the ratio of the number of scattering acts per unit time and unit of solid angle to the flux of incident particles. In the case of a central force field the elementary solid angle is equal to $d\Omega = 2\pi\, d\cos\vartheta$, and particles are scattered in this angle element when the impact parameters range from ρ up to $\rho + d\rho$. Since the flux of particles is equal to Nv, where N is the number density of incident particles and $v = |\mathbf{v}_1 - \mathbf{v}_2|$ is the relative velocity of the colliding particles, the number of particles scattered per unit time into a given solid angle is equal to $2\pi\rho\, d\rho N v$, so that the differential

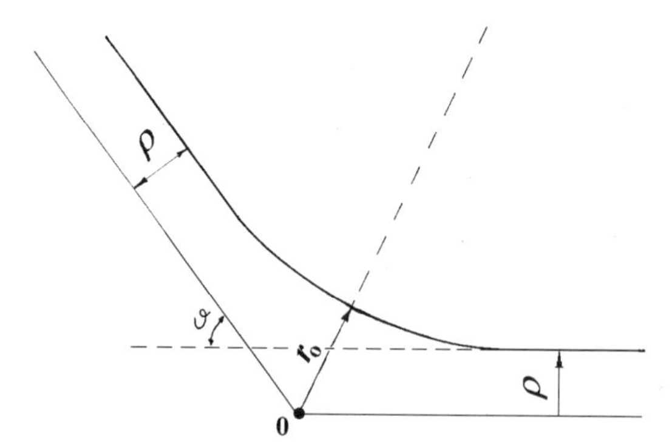

Figure 13.1. The parameters of the elastic scattering of particles. O is the position of the scattered force center, the solid curve is the trajectory of collision in the center-of-mass coordinate system, ρ is the impact parameter of the collision, r_o is the distance of closest approach, and ϑ is the scattering angle.

cross section is

$$d\sigma = 2\pi\rho\, d\rho. \tag{13.3}$$

The spherical symmetry of the field of the scattering center $U = U(|\mathbf{R}_1 - \mathbf{R}_2|)$ leads to a simple connection between the impact parameter of collision ρ and the distance of closest approach r_o. Use the conservation of the momentum of motion which is equal to $\mu\rho v$ at large distances between particles and $\mu v_\tau r_o$ at the distance of closest approach, where v_τ is the tangential component of the velocity at the distance of closest approach at which the normal component of the velocity is zero, so that the law of energy conservation gives $\mu v_\tau^2/2 = \mu v^2/2 - U(r_o)$. This leads to the following relationship:

$$1 - \frac{\rho^2}{r_o^2} = \frac{U(r_o)}{\varepsilon}, \tag{13.4}$$

where $\varepsilon = \mu v^2/2$ is the kinetic energy of particles in the center-of-mass coordinate system. These relations allow us to determine the angle of scattering ϑ. Indeed, the rotation momentum of particles $L = \mu v\rho = \mu v_\tau R$ is conserved in the course of collision. Here v is the relative velocity of particles at large distances between colliding particles when the interaction between them is weak. The tangential velocity is $v_\tau = v\rho/R = R\, d\vartheta/dt$, and this is the equation for the scattering angle. Next, the normal velocity of particles $v_R = dR/dt$ can be determined from the equation $\mu v_R^2/2 + \mu v_\tau^2/2 = \varepsilon = \mu v^2/2 - U(R)$. This yields

$$\frac{dR}{dt} = \pm v\sqrt{1 - \frac{\rho^2}{R^2} - \frac{U(R)}{\varepsilon}},$$

and the sign depends on the direction of the relative motion of particles. From this equation, on the basis of the symmetry of this problem [$R(t) = R(-t)$, t is time, and $t = 0$ at $R = r_o$], we obtain the following expression for the scattering angle in the classical case:

$$\vartheta = \pi - 2\int_{r_o}^{\infty} \frac{\rho\, dR}{R^2\sqrt{1 - \frac{\rho^2}{R^2} - \frac{U(R)}{\varepsilon}}}, \tag{13.5}$$

and according to formula (13.4) the root is zero if $R = r_o$. Of course, this formula gives $\vartheta = 0$ at $U(R) = 0$.

The scattering of particles at large angles is of importance for the various parameters of gases and plasmas. Parameters, such as transport coefficients and rates of relaxation for various degrees of freedom, are determined by the scattering of particles on large angles. In particular, the greatest averaging of the scattering cross section over the scattering angles has the form

$$\sigma^* = \int (1 - \cos\vartheta)\, d\sigma, \tag{13.6}$$

and this cross section is called the diffusion, or transport, cross section. An estimation for the scattering cross section at large angles $\vartheta \sim 1$ follows from formula

(13.4) and has the form

$$\sigma = \pi\rho_o^2, \quad \text{where} \quad U(\rho_o) \sim \varepsilon. \tag{13.7}$$

This estimate also corresponds to the diffusion cross section (13.6).

Thus, the problem of the elastic scattering of two atomic particles is reduced to the problem of the scattering of one particle in the central field. As follows from the above analysis, the parameters of scattering including the cross section of scattering, are determined by the interaction potential $U(R)$ between particles.

13.2 Scattering of Atoms in a Sharply Varied Interaction Potential

The interaction of atomic particles at small and moderate separations usually corresponds to repulsion (see Table 10.7) with a sharply varied interaction potential, because of the exchange interaction between the atomic particles and since the Coulomb interaction of atomic cores is partially shielded by atomic electrons at these separations. One can use a simple model for elastic collisions and an interaction potential which is called the hard sphere model. Then atoms are modeled by hard balls and their scattering is similar to the elastic scattering of billiard balls. The character of the scattering in this case is shown in Fig. 13.2, where R_o is the sum of the radii of these balls. According to Fig. 13.2, the following relation takes place between the collision and scattering parameters $\vartheta = \pi - 2\alpha$, $\sin\alpha = \rho/R_o$, which leads to $\rho = R_o\cos(\vartheta/2)$. Then on the basis of formula (13.2) we have the following expression for the differential cross section of scattering:

$$d\sigma = 2\pi\rho\, d\rho = (\pi R_o^2/2)\, d\cos\vartheta. \tag{13.8}$$

In particular, for the diffusion cross section of the collision of particles this gives

$$\sigma^* = \pi R_o^2, \tag{13.9}$$

which corresponds to formula (13.7).

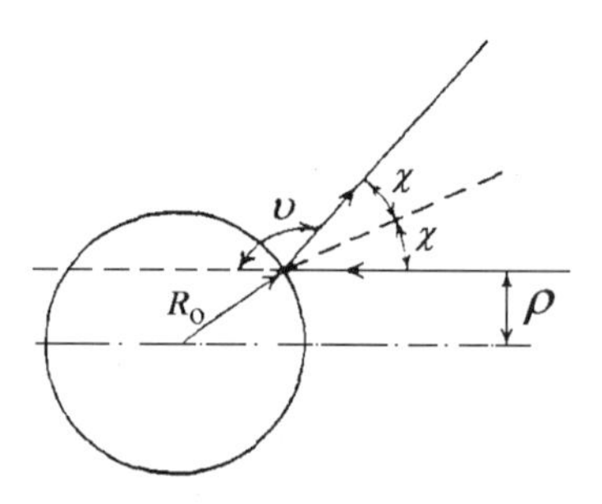

Figure 13.2. The trajectory of collision (solid lines with arrows) for the hard sphere model. R_o is the sphere radius, ρ is the collision impact parameter, and ϑ is the scattering angle.

As a matter of fact, this model relates to scattering of particles whose interaction is described by the potential of a hard wall:

$$U(R) = 0, \quad R > R_o; \qquad U(R) = \infty, \quad R \le R_o. \tag{13.10}$$

If we model a real interaction potential on this one, the question occurs as to what is the parameter R_o if the interaction potential is known. From general considerations it follows that this parameter can be found from the relation

$$U(R_o) \sim \varepsilon, \tag{13.11}$$

where ε is the energy of the particles in the center-of-mass coordinate system. We find below the numerical coefficient in this relation for a sharply varied interaction potential of particles as a result of the expansion over a small parameter $1/n$, where n is the logarithmic derivative of the interaction potential

$$n = -\frac{d \ln U(R)}{dR} \mid R = R_o. \tag{13.12}$$

Then the collision parameter may be constructed in the form of a series over a small parameter $\sim 1/n$. Formula (13.10) gives the first term of this expansion. The use of two expansion terms gives the possibility of determining the parameter R_o in the form

$$U(R_o) = a\varepsilon, \tag{13.13}$$

where the numerical constant a depends on the form of the collision cross section. The procedure of the determination of this parameter is fulfilled in Problems 13.1 and 13.2 for the scattering angle and the diffusion cross section. Now we determine the parameter a in formula (13.13) for the diffusion cross section $\sigma^* = \int(1 - \cos\vartheta)2\pi\rho\, d\rho$ and one more cross section for the scattering on large angles which is included in the expressions for the thermal conductivity and viscosity coefficients used in the kinetic theory of gases $\sigma^{(2)} = \int(1 - \cos^2\vartheta)2\pi\rho\, d\rho$. We use numerical calculations for the interaction potentials $U(R) = CR^{-n}$. Note that in this case we obtain, for a cross section on the basis of formula (13.9),

$$\sigma = \pi R_o^2 = \pi \left(\frac{C}{a\varepsilon}\right)^{2/n}. \tag{13.14}$$

Comparing the calculated cross section with this formula, one can find the value of the parameter a. This means that we can give the cross section for any n. In this way we find, for the related cross sections,

$$\sigma^* = \pi R_1^2, \quad \text{where } U(R_1) = 0.74\varepsilon;$$
$$\text{and} \quad \sigma^{(2)} = \frac{2}{3}\pi R_2^2, \quad \text{where } U(R_2) = 0.27\varepsilon. \tag{13.15}$$

Table 13.1 gives the comparison of the cross sections calculated on the basis of formulas (13.15) and the data of accurate calculations. The difference of these data characterizes the accuracy of formulas (13.15) for the real sharply varied interaction potentials.

Table 13.1. Comparison of formula (13.15) and the results of numerical calculations for the $\sigma^{(1)}$ and $\sigma^{(2)}$ cross sections of scattering in the interaction potential $U(R) = CR^{-n}$.

n	4	6	8	10	12	14
$\pi R_1^2/\sigma^*$	0.974	0.995	0.998	1.005	1.002	1.000
$2\pi R_2^2/(3\sigma^{(2)})$	1.040	1.001	0.985	0.984	0.983	0.985

13.3 Capture of Particles in an Attractive Interaction Potential

Another character of the motion of particles takes place for an attractive interaction potential of the colliding particles. Assume that the interaction potential of two atomic particles has a typical form of the potential well, and consider the limiting case of collision when the energy of particles ε in the center-of-mass system of coordinates is small compared to the potential well depth D. On the basis of the above expression we present the energy conservation law in the form $\varepsilon = \mu v_R^2/2 + \mu v_\tau^2/2 = \mu v^2/2 - U(R)$ and, since $v_\tau = v\rho/R$, this relation has the form

$$\frac{\mu v_R^2}{2} = \frac{\mu v^2}{2} - U(R) - \frac{\mu v^2 \rho^2}{2R^2} = \frac{\mu v^2}{2} - U_{\mathrm{ef}}(R), \tag{13.16}$$

where $U_{\mathrm{ef}}(R) = U(R) - \mu v^2\rho^2/(2R^2)$ is the effective interaction potential corresponding to the radial relative motion of particles. Since the effective interaction potential has a maximum, the trajectories of the particles can be divided into two groups. The boundary of these groups corresponds to the maximum of the effective interaction potential as a function of the distance of closest approach $r_o[U'_{\mathrm{ef}}(r_{\min}) = 0]$. Introducing the impact parameter ρ_c, which relates to this distance $r_{\min}$ of closest approach, we find that in the range $\rho > \rho_c$, i.e., $r_o > r_{\min}$, there takes place a monotonic dependence $r_o(\rho)$. In the range $\rho \le \rho_c$ the distance of closest approach transfers by a jump to zero. Hence, the capture of the colliding particles takes place at these impact parameters $\rho \le \rho_c$ and, in reality, particles approach distances of a strong repulsion between them.

Let us determine the capture cross section $\sigma_c = \pi\rho_c^2$ for the interaction potential of particles $U(R) = -C/R^n$. Then the impact parameter of capture ρ_c is determined as a minimum of the dependence $\rho(r_o)$ and, on the basis of formula (13.4) we find for the cross section of capture,

$$\sigma_c = \pi\rho_c^2 = \frac{\pi n}{n-2}\left[\frac{C(n-2)}{2\varepsilon}\right]^{2/n}. \tag{13.17}$$

As is seen, the dependence of the cross section on the parameters is similar to formula (13.7). In particular, in the case of the polarization interaction between the ion and atom $U(R) = -\alpha e^2/(2R^4)$ (α is the atom polarizability), the polarization cross section of capture resulting from ion–atom collisions is equal to, in the usual units,

$$\sigma_c = 2\pi\sqrt{\frac{\alpha e^2}{\mu v^2}}. \qquad (13.18)$$

As follows from this, the capture cross section of particles is mainly determined by the long-range part of the interaction potential. Indeed, according to formula (13.17), we have $-U(r_{\min})/\varepsilon = 2/(n-2)$ and, because of $\varepsilon \ll D$, this gives $|U(r_c)| \ll D$, i.e., the capture is determined by the long-range part of the attractive interaction. Next, since in reality $n \gg 1$, in the range $r_o > r_{\min}$ the interaction potential is small compared to the kinetic energy of particles. Hence, the capture cross section gives the main contribution to the cross section of scattering at large scattering angles.

For a realistic interaction potential of atomic particles, which is characterized by repulsion of the particles at small distances between them, the capture of particles leads to their strong approach. Indeed,

$$r_{\min} \sim R_e(D/\varepsilon)^{1/n} \gg R_e, \qquad (13.19)$$

where R_e is the equilibrium distance between particles.

As a result of capture, particles reach such distances between them where the attractive interaction potential remarkably exceeds their kinetic energy. A strong interaction of the particles in this region causes a strong scattering of particles. Therefore, one can assume that the capture of particles leads to their almost isotropic scattering and, hence, the diffusion cross section of scattering (13.6) is close to the cross section of capture (13.18). (In particular, in the case of the polarization interaction of particles, the diffusion cross section of the scattering of particles exceeds their cross section of capture by 10%). Thus, under the condition $\varepsilon \ll D$, the capture cross section can be used as the diffusion cross section of the scattering of particles.

13.4 The Scattering of Quantum Particles

Let us consider the elastic collisions of two particles when their motion is governed by the quantum laws. Then the Schrödinger equation for colliding particles, in the usual units, has the form

$$\left[-\frac{\hbar^2}{2m_1}\Delta_1 - \frac{\hbar^2}{2m_2}\Delta_2 + U(\mathbf{R}_1 - \mathbf{R}_2)\right]\Psi(\mathbf{R}_1, \mathbf{R}_2) = E\Psi(\mathbf{R}_1, \mathbf{R}_2), \qquad (13.20)$$

where we use the same notations as in the classical case described by the Newton equations (13.1) and where E is the total energy of particles. As above, we introduce new variables—the coordinate of the center of mass $\mathbf{R}_c = (m_1\mathbf{R}_1 + m_2\mathbf{R}_2)/(m_1 + m_2)$ and the relative distance between particles $\mathbf{R} = \mathbf{R}_1 - \mathbf{R}_2$. Using the relation for the sum of the Laplacians

$$\frac{\hbar^2}{2m_1}\Delta_1 + \frac{\hbar^2}{2m_2}\Delta_2 = \frac{\hbar^2}{2(m_1+m_2)}\Delta_{\mathbf{R}_c} + \frac{\hbar^2}{2\mu}\Delta_{\mathbf{R}}, \tag{13.21}$$

we separate the variables in the Schrödinger equation. Similar to the classical case, we obtain that the center of mass of the colliding particles moves with constant velocity, and information about the scattering of particles is contained in the Schrödinger equation for the wave function which describes the relative motion of particles. This Schrödinger equation has the form

$$-\frac{\hbar^2}{2\mu}\Delta_{\mathbf{R}}\psi(\mathbf{R}) + U(\mathbf{R})\psi(\mathbf{R}) = \varepsilon\psi(\mathbf{R}), \tag{13.22}$$

where $\varepsilon = \hbar^2 q^2/2\mu$ is the energy of particles in the center-of-mass coordinate system, so that q is the wave vector of the relative motion of particles. It is convenient to rewrite this equation in the form

$$(\Delta_{\mathbf{R}} + q^2)\psi(\mathbf{R}) = \frac{2\mu}{\hbar^2}U(\mathbf{R})\psi(\mathbf{R}). \tag{13.23}$$

The Grin function of this equation is

$$G(\mathbf{R}, \mathbf{R}') = -\frac{\exp(iq|\mathbf{R}-\mathbf{R}'|)}{4\pi|\mathbf{R}-\mathbf{R}'|},$$

so that the solution of this Schrödinger equation is

$$\psi(\mathbf{R}) = C\left[e^{i\mathbf{qR}} - \frac{\mu}{2\pi\hbar^2}\int \frac{\exp(iq|\mathbf{R}-\mathbf{R}'|)}{|\mathbf{R}-\mathbf{R}'|}U(\mathbf{R}')\psi(\mathbf{R}')\,d\mathbf{R}'\right]. \tag{13.24}$$

Formula (13.24) is an integral form of the Schrödinger equation. This form is convenient for obtaining the asymptotic expression for the wave function of the relative motion of particles which includes information about the scattering of particles. When R tends to infinity, so that $|\mathbf{R}-\mathbf{R}'| = R - \mathbf{R}'\mathbf{n}$, where $\mathbf{n}$ is the unit vector directed along $\mathbf{R}$, this equation gives

$$\psi(\mathbf{R}) = C\left[e^{i\mathbf{qR}} + f(\vartheta)\frac{e^{iqR}}{R}\right], \tag{13.25}$$

where the scattering amplitude is

$$f(\vartheta) = -\frac{\mu}{2\pi\hbar^2}\int \exp(-iq\mathbf{n}\mathbf{R}')U(\mathbf{R}')\psi(\mathbf{R}')\,d\mathbf{R}'. \tag{13.26}$$

Here ϑ is the angle between vectors $\mathbf{R}$ and $\mathbf{q}$, and the wave function is normalized such that it tends to $e^{i\mathbf{qR}}$ when $R \to \infty$. This formula does not allow us to determine the scattering amplitude in a general form because its expression includes the wave function of particles, but it is convenient for an approximate solution of the problem. In particular, in the Born approximation we neglect the interaction of the scattered particle with the force center, which gives $\psi(\mathbf{R}) = e^{i\mathbf{qR}}$.

Substituting this in formula (13.26) we obtain, for the scattering amplitude in the Born approximation,

$$f(\vartheta) = -\frac{\mu}{2\pi\hbar^2}\int e^{-i\mathbf{KR}}U(\mathbf{R}')\,d\mathbf{R}', \tag{13.27}$$

where $\mathbf{K} = q\mathbf{n} - \mathbf{q}$ is the variation of the wave vector for the relative motion of particles as a result of collision. This value is connected with the scattering angle ϑ by the relation $K = 2q\sin\vartheta/2$. The Born approximation is valid if the interaction potential is small compared to the kinetic energy at distances between particles $R \sim 1/q$ which are responsible for the scattering, i.e.,

$$U\left(\frac{1}{q}\right) \ll \varepsilon. \tag{13.28}$$

13.5 Phase Theory of Elastic Scattering

If the interaction potential does not depend on the angles $U = U(R)$, the scattering amplitude can be expressed through parameters of the Schrödinger equation for the radial wave function of the colliding particles. Indeed, let us expand the wave function over the spherical angular functions

$$\psi(\mathbf{R}) = \frac{1}{R}\sum_{l=0}^{\infty} A_l\varphi_l(R)P_l(\cos\vartheta),$$

$$e^{i\mathbf{qR}} = \sqrt{\frac{\pi}{2qR}}\sum_{l=0}^{\infty} i^l(2l+1)J_{l+1=2}(qR)P_l(\cos\vartheta),$$

$$f(\vartheta) = \sum_{l=0}^{\infty} f_l P_l(\cos\vartheta), \tag{13.29}$$

where $P_l(\cos\vartheta)$ is the Legendre polynomial and $J_{l+1/2}(qR)$ is the Bessel function which has the following form at large values of the argument:

$$J_{l+1=2}(x) = \sqrt{\frac{2}{\pi x}}\sin(x - \pi l/2).$$

The radial wave functions satisfy to the equation

$$\frac{d^2\varphi_l}{dR^2} + \left[q^2 - \frac{2\mu U(R)}{\hbar^2} - \frac{l(l+1)}{R^2}\right]\varphi_l = 0, \tag{13.30}$$

and have the following asymptotic form:

$$\varphi_l(R) = \frac{1}{q}\sin\left(qR - \frac{\pi l}{2} + \delta_l\right), \tag{13.31}$$

where δ_l are the scattering phases. These values include information about the scattering of particles. Indeed, substituting the expressions (13.29) into relation

(13.23) and equalizing the coefficients for the terms e^{iqR} and e^{-iqR} at large R, we obtain the following relations:

$$A_l = e^{i\delta_l}(2l+1)i^l, \qquad f_l = \frac{1}{2iq}(2l+1)(e^{2i\delta_l}-1). \tag{13.32}$$

Thus the scattering amplitude is expressed through the scattering phases. From this it follows that, for the total σ_t and diffusion σ^* cross sections of the scattering of atomic particles,

$$\sigma_t = \int_0^1 |f(\vartheta)|^2\, 2\pi\, d\cos\vartheta = \frac{4\pi}{q^2}\sum_{l=0}^{\infty}(2l+1)\sin^2\delta_l,$$

$$\sigma^* = \int_0^1 |f(\vartheta)|^2\,(1-\cos\vartheta)2\pi\, d\cos\vartheta = \frac{4\pi}{q^2}\sum_{l=0}^{\infty}(2l+1)\sin^2(\delta_l-\delta_{l+1}). \tag{13.33}$$

Relation (13.26) gives the following equation for the scattering phase:

$$\sin\delta_l = -\frac{\mu}{\hbar^2}\sqrt{2\pi q}\int_0^{\infty}\sqrt{R}\,J_{l+1/2}(qR)\varphi_l(R)U(R)\,dR. \tag{13.34}$$

In the Born approximation, when the radial wave functions

$$\varphi_l(R) = \sqrt{\frac{\pi R}{2q}}\,J_{l+1/2}(qR)$$

describe a free motion of particles, we obtain the following relation from equation (13.34):

$$\delta_l = -\frac{\pi\mu}{\hbar^2}\int_0^{\infty}U(R)\left[J_{l+1/2}(qR)\right]^2 R\,dR. \tag{13.35}$$

In particular, in the case of the electron–atom scattering and the polarization interaction potential between them, when $U(R) = -\alpha e^2/2R^4$, this formula gives

$$\delta_l = \frac{\pi\alpha q^2}{(2l-1)(2l+1)(2l+3)a_o}, \tag{13.36}$$

where $a_o = \hbar^2/(me^2)$ is the Bohr radius. This formula is valid if the main contribution to the integral gives large distances between an electron and atom $R \sim 1/q \gg a_o$, where the polarization interaction takes place. One can see that this is not valid for the zeroth scattering phase, because the integral (13.36) converges in this case.

Formulas (13.34) and (13.35) allow us to analyze the behavior of the scattering phases at small energies of the colliding particles. Since $J_{l+1/2}(x) \sim x^{l+1/2}$ at small x we obtain, from formulas (13.34) and (13.35), for a short-range interaction of particles $\delta_l \sim q^{2l+1}$ at small q. If the interaction potential has the dependence $U(R) \sim R^{-n}$ at large R we obtain, for $l \geq (n-3)/2$, the dependence of the scattering phases on the collision wave vector in the form $\delta_l \sim q^{n-2}$. In particular, for the polarization potential of the interaction $U(R) \sim R^{-4}$ this formula gives $\delta_l \sim q^2$ for $l \geq 1$ in accordance with formula (13.36).

Let us consider the behavior of the zeroth scattering phase at small energies of collision. Formula (13.34) gives $\delta_0 \sim q$ at small q. We present this relation in the form $\delta_0 = -Lq$, where the value L is called the scattering length. At small energies of collision the scattering amplitude is $f = -L$ so that, for the total σ_t and the diffusion σ^* cross sections of the collisions of atomic particles at small energies, we have

$$\sigma_t = \sigma^* = 4\pi L^2. \tag{13.37}$$

Let us obtain the quasi-classical limit for the collision of particles within the framework of the quantum formalism. Our goal is to establish a correspondence between the quasi-classical limit of the phase theory of the scattering and the classical scattering of particles. In the quasi-classical limit, large momenta of collision l give the main contribution to the cross section. Then we use the quasi-classical solution of the Schrödinger equation (13.30) which has the form

$$\varphi_l(R) = C \sin\left[q \int_{r_o}^{R} dR' \sqrt{1 - \frac{U(R')}{\varepsilon} - \frac{(l+1/2)^2}{q^2 R^2}} + \frac{\pi}{4}\right],$$

where r_o is the classical distance of closest approach which satisfies relation (13.4) and, in the above expression, we replace $l(l+1)$ by $(l+1/2)^2$. From this we obtain, for the scattering phase,

$$\delta_l = \lim_{R\to\infty}\left[q \int_{r_o}^{R} dR' \sqrt{1 - \frac{U(R')}{\varepsilon} - \frac{(l+1/2)^2}{q^2 (R^0)^2}} + \frac{\pi}{2}(l + \frac{1}{2}) - qR\right]. \tag{13.38}$$

At first we prove the relation

$$\sum_{l=0}^{\infty}(2l+1)P_l(\cos\vartheta) = \delta(1 - \cos\vartheta).$$

Indeed, from the definition of the product function for the Legendre polynomials

$$\frac{1}{\sqrt{1 - 2tx + t^2}} = \sum_{l=0}^{\infty} t^l P_l(x),$$

the relation follows

$$\sum_{l=0}^{\infty}(2l+1)t^l P_l(x) = \frac{1 - t^2}{(1 - 2tx + t^2)^{3/2}}.$$

This function is zero in the limit $t \to 1$ for any $x \neq 1$, and it equals ∞ at $x = 1$, $t = 1$. Next, the integral from this function over dx in the interval from $x = 0$ up to $x = 1$ is equal to unity in the limit $t \to 1$. This allows us to replace the above sum in the limit $t \to 1$ by the delta-function, so that the scattering amplitude (13.29) and (13.32) can be presented in the form

$$f(\vartheta) = \frac{1}{2iq}\sum_{l=0}^{\infty}(2l+1)e^{2i\delta_l}P_l(\cos\vartheta) - \frac{1}{2iq}\delta(1 - \cos\vartheta). \tag{13.39}$$

Considering the scattering of particles in a nonzero angle, we neglect the second term in formula (13.39). Then the differential cross section of the scattering per unit solid angle $d\Omega = 2\pi\, d\cos\vartheta$ has the form:

$$d\sigma = 2\pi\, d\cos\vartheta\, |f(\vartheta)|^2$$
$$= \frac{\pi d\cos\vartheta}{2q^2}\sum_{l=0}^{\infty}\sum_{n=0}^{\infty}(2l+1)(2n+1)P_l(\cos\vartheta)P_n(\cos\vartheta)\exp(2i\delta_l - 2i\delta_n).$$

Let us use the asymptotic expression for the Legendre polynomials at large values of the subscript (see Problem 6.2):

$$P_l(\cos\vartheta) = \frac{2\sin[(l+1/2)\vartheta + \pi/4]}{\sqrt{\pi(2l+1)\sin\vartheta}}, \qquad l\vartheta \gg 1.$$

This gives, for the differential cross section,

$$d\sigma = \frac{d\vartheta}{q^2}\sum_{l=0}^{\infty}\sum_{n=0}^{\infty}\sqrt{(2l+1)(2n+1)}$$
$$\times\left\{\cos[(l-n)\vartheta] - \cos\left[\frac{l+n+1}{2}\vartheta\right]\right\}\exp(2i\delta_l - 2i\delta_n).$$

Because, in the classical limit, the main contribution to these sums gives large collision momenta, we replace the sums by integrals. Due to oscillations of the cosines, the second term does not give a contribution to the result, while the integral for the first term converges near $l = n$. Thus, after integration over dn, we have

$$d\sigma = \frac{\pi\, d\vartheta}{q^2}\int_0^{\infty}(2l+1)\,dl\left[\delta\left(2\frac{d\delta_l}{dl} - \vartheta\right) + \delta\left(2\frac{d\delta_l}{dl} + \vartheta\right)\right]. \qquad (13.40)$$

Let us introduce the classical scattering angle on the basis of the relation

$$\vartheta_{cl} = \pm 2\frac{d\delta_l}{dl}, \qquad (13.41)$$

so that the $+$ corresponds to a repulsion interaction potential and the $-$ relates to an attractive one. Introducing the impact parameter of collision as $\rho = (l+1/2)/q$, rewrite the expression for the differential cross section of scattering

$$d\sigma = 2\pi\rho\, d\rho,$$

where the parameters ρ and $\vartheta_{\rm cl}$ are connected by a relation $\vartheta = \vartheta_{\rm cl}(\rho)$. As is seen, the obtained relation for the differential cross section coincides with the classical formula (13.3). In addition, from formula (13.41) and the quasi-classical expression of the scattering phase (13.38) follows the expression for the classical scattering angle

$$\frac{\pi \pm \vartheta_{\rm cl}}{2} = \int_{r_o}^{\infty}\frac{\rho\, dr}{\sqrt{1 - U(R)/\varepsilon - \rho^2/R^2}}. \qquad (13.42)$$

This result corresponds to the classical formula (13.5). Thus, the quantum formalism gives an accurate limiting transition to the classical scattering problem.

13.6 Total Cross Section of Scattering

The total cross section of the elastic scattering of particles determines the broadening of the spectral lines and is connected with a phase shift as a result of collision. This is given by formula 13.35):

$$\sigma_t = \int_0^1 |f(\vartheta)|^2 \, 2\pi \, d\cos\vartheta = \frac{4\pi}{q^2} \sum_{l=0}^{\infty} (2l+1) \sin^2 \delta_l. \tag{13.43}$$

In the classical limit the total cross section $\sigma_t = \int d\sigma$ is infinite because the classical particles interact and are scattered at any distances between them. Therefore, the classical total cross section must tend to infinity as the Planck constant $\hbar \to 0$.

Let us evaluate the total cross section of the collision of particles within the framework of the classical scattering theory. Because the scattering on small angles gives the main contribution to this cross section, we obtain below the dependence $\vartheta(\rho)$ at small scattering angles. In the zeroth approximation we assume the particles to move along straight classical trajectories. From formula (13.43) it follows that

$$\sigma_t = \int_0^{\infty} 8\pi\rho \, d\rho \sin^2 \delta(\rho), \tag{13.43a}$$

where $\rho = l/q$ is the impact parameter of collision, so that the total cross section can be evaluated according to the formula

$$\sigma_t = 2\pi\rho_t^2, \quad \text{where} \quad \delta(\rho_t) \sim 1. \tag{13.44}$$

The value ρ_t is called the Weiskopf radius. Take into account that the total cross section is determined by small scattering angles and a weak interaction during collision. Then, neglecting the interaction, we use formula (13.35) for the scattering phase

$$\delta_l = -\frac{\pi\mu}{\hbar^2} \int_{r_o}^{\infty} U(R) \left[J_{l+1/2}(qR) \right]^2 R \, dR = -\frac{1}{\hbar^2} \int_{r_o}^{\infty} \frac{U(R) \, dR}{\sqrt{q^2 - (l+1/2)^2/R^2}},$$

where the distance of closest approach is equal to $r_o = \rho = (l+1/2)/q$. Introducing the classical time from the relation $R^2 = \rho^2 + v^2 t^2$, we rewrite the above relation in the form

$$\delta_l = -\frac{1}{2\hbar} \int_{-\infty}^{\infty} U(R) \, dt. \tag{13.45}$$

Correspondingly, formula (13.44) can be rewritten in the form

$$\sigma_t = 2\pi\rho_t^2, \quad \text{where} \quad \frac{\rho_t U(\rho_t)}{\hbar v} \sim 1. \tag{13.46}$$

In particular, for the interaction potential $U(R) = CR^{-n}$, formula (13.45) gives

$$\delta = -\frac{C}{2\hbar} \int_{-\infty}^{\infty} \frac{dt}{(\rho^2 + v^2 t^2)^n} = -\frac{C\sqrt{\pi}}{2\hbar v \rho^{n-1}} \frac{\Gamma\left(\frac{n+1}{2}\right)}{\Gamma\left(\frac{n}{2}\right)}, \tag{13.47}$$

and we obtain the following expression for the total cross section of scattering

$$\sigma_t = \int_{-\infty}^{\infty} 8\pi\rho\, d\rho \sin^2\delta(\rho) = 2\pi\left(\frac{C}{\hbar v}\right)^{2/(n-1)}\left[\frac{\sqrt{\pi}\,\Gamma\left(\frac{n-1}{2}\right)}{\Gamma\left(\frac{n}{2}\right)}\right]^{2/(n-1)}\Gamma\left(\frac{n-3}{n-1}\right). \tag{13.48}$$

Since the total cross section is determined by the quantum effects, it tends to infinity in the classical limit. Indeed, from formulas (13.46) and (13.48) it follows that the total cross section tends to infinity in the limit $\hbar \to 0$.

Let us compare the diffusion and total cross sections of scattering in the classical limit. Then the kinetic energy of particles $\varepsilon = \mu v^2/2$ satisfies the relation

$$\varepsilon \gg \hbar/\tau, \tag{13.49}$$

where $\tau \sim \rho/v$ is a typical collision time. Formula (13.49) gives the criterion for the classical character of the particle motion $l = \mu\rho v/\hbar \gg 1$, where l is the collision momentum of particles. If this criterion is fulfilled, one can consider the motion of particles along classical trajectories. Next, in this case, according to formula (13.46) the total cross section of scattering σ_t remarkably exceeds a cross section σ of scattering on large angles which is estimated by formula (13.7). From formulas (13.7) and (13.46) it follows that, in the classical case, $U(\rho_o)/U(\rho_t) \sim \mu\rho_t v/\hbar \gg 1$, so that for a monotonic interaction potential $U(R)$ this gives $\rho_t \gg \rho_o$, i.e.,

$$\sigma_t \gg \sigma. \tag{13.50}$$

In particular, in the case of the polarization interaction potential $U(R) = -\alpha e^2/(2R^4)$, this ratio is equal to

$$\frac{\sigma_t}{\sigma_c} = \left(\frac{\pi}{4}\right)^{2/3}\Gamma\left(\frac{1}{3}\right)\left(\frac{\alpha e^2\mu^2 v^2}{\hbar^4}\right)^{1/6} = 2.3 l_c^{2/3},$$

where formula (13.18) is used for the capture cross section and the total cross section σ_t is given by formula (13.48). Here $l_c = \mu\rho_c v/\hbar$ is the collisional momentum of capture, so that the impact parameter ρ_c corresponds to the capture of particles. In the case of the classical character of capture we have $l_c \gg 1$, and the above ratio is large.

13.7 Oscillations in Quasi-Classical Cross Sections

Above we obtain that the quantum formalism gives classical formulas for the differential cross sections in the classical limit. But there is a principal difference between the classical and quantum descriptions of particle scattering in the classical limit. Indeed, according to formula (13.39), the quasi-classical expression for the scattering phase has the following structure:

$$f(\vartheta) = f_{\text{cl}}(\vartheta)e^{i\eta_l}, \tag{13.51}$$

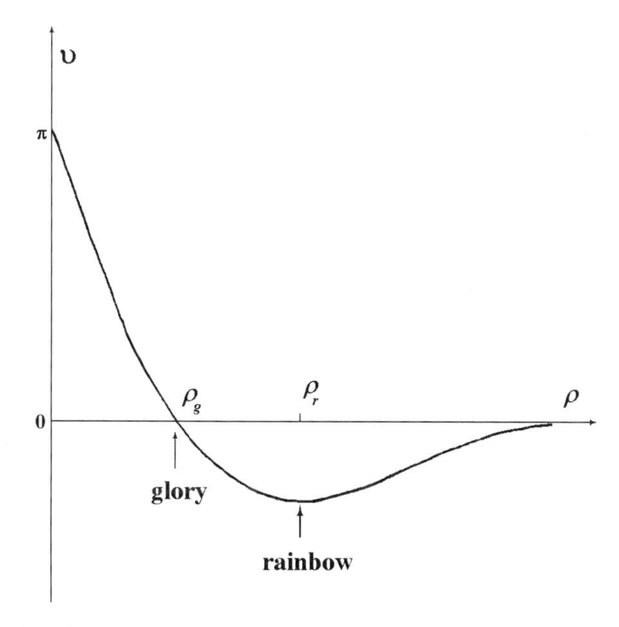

Figure 13.3. The dependence of the classical scattering angle on the collision impact parameter for a typical interaction potential of two atoms when an attraction at large distances between them is changed by repulsion at small distances. ρ_g refers to the glory point and ρ_r is the impact parameter of the rainbow.

where $f_{\text{cl}}(\vartheta)$ is the classical scattering amplitude and the square of its module is the classical differential cross section of scattering. The principal difference between the quasi-classical and classical scattering amplitudes consists of the quasi-classical phase $\eta_l = 2\delta_l$ which is a distinctive feature of the quantum values. This leads to specific peculiarities of the scattering of atomic particles.

Let us analyze the scattering of atomic particles for a realistic character of their interaction potential which corresponds to the attraction of particles at large distances between them and corresponds to their repulsion at small distances. Then the scattering angle, as a function of the collision impact parameter, has the form as given in Fig. 13.3. Indeed, frontal collisions correspond to the scattering angle $\vartheta = \pi$. At large impact parameters the scattering angle tends to zero, and for an attractive interaction potential it is negative. Thus the scattering angle, as a function of the impact parameter of collision, has a negative minimum. This is called the rainbow point. Near this point identical scattering angles correspond to two different impact parameters of collision in the region of negative angles of scattering. The scattering amplitude for negative scattering angles has the following structure, according to formula (13.51):

$$f(\vartheta) = f_1(\vartheta)e^{i\eta_1} + f_2(\vartheta)e^{i\eta_2},$$

where f_1, f_2 are the classical scattering amplitudes that are positive values. As is seen, the total scattering amplitude depends on the relative value of the corresponding phases. Indeed, the differential cross section is, in this case,

$$\frac{d\sigma}{d\Omega} = f_1^2 + f_2^2 + 2 f_1 f_2 \cos(\eta_1 - \eta_2),$$

and the phases η_1, η_2 are monotonic functions of both the impact parameter of collision and the collision velocity. Therefore, the differential cross section of the scattering has an oscillating structure in the range of negative scattering angles as a function of both the scattering angle and collision velocity.

The other effect corresponds to the zero scattering angle. Collisions with such impact parameters proceed without scattering because scattering in an attractive region of interaction is compensated for by scattering in a repulsive region. This point is called the glory point and, in accordance with formula (13.41), it corresponds to a maximum of the scattering phase as a function of the impact parameter. But such a character of scattering gives an oscillation structure of the differential and total cross section of scattering. Indeed, from formula (13.43a) it follows that if function $\delta_l(\rho)$ has a maximum, then the integral depends on this maximum value. Because the maximum phase varies with the collision energy, the cross section, as a function of the collision energy has an oscillation structure in spite of the classical character of scattering.

In order to take this effect into account, let us divide the scattering phase into two parts

$$\delta_l = \delta_l^{(\mathrm{reg})} + \alpha(l - l_g)^2,$$

where the first term corresponds to a regular part of the scattering phase and the second term accounts for its stationarity near the glory point. Here l_g is the collision momentum at the glory point, and $\alpha = \frac{1}{2}(d^2\delta_l/dl^2)$ at this point. The first term of this expression determines the total cross section of scattering (13.44), and the second term acts only near the glory point. From this we have, for the total cross section (13.43)

$$\sigma_t = \sigma_{\mathrm{LLS}} + \Delta\sigma_{\mathrm{NU}}, \tag{13.52}$$

where the first term is the so-called Landau–Lifshiz–Schiff cross section which accounts for the regular part of the scattering and, for the inverse power interaction potential, $U(R) = CR^{-n}$ is given by formula (13.48). The value $\Delta\sigma_{\mathrm{NU}}$, the Nikitin–Umanskij correction, includes oscillations of the cross section as a function of the collision velocity. For the above character of extraction of the phase stationarity, this correction to the total cross section is given by

$$\Delta\sigma_{\mathrm{NU}} = -\frac{8\rho_g}{\pi^{3/2} q^{1/2} \sqrt{\left|\frac{d^2\delta}{d\rho^2}\right|}} \cos\left(2\delta_g - \frac{\pi}{4}\right), \tag{13.53}$$

where the derivative is taken at the glory point. Since the scattering phase monotonically depends on the collision velocity, this part of the cross section oscillates. Because $\rho_g < \rho_t$ and $l_g = q\rho_g \gg 1$, the second term of the total cross section (13.52) is small compared to the first one. Oscillations in the scattering cross sec-

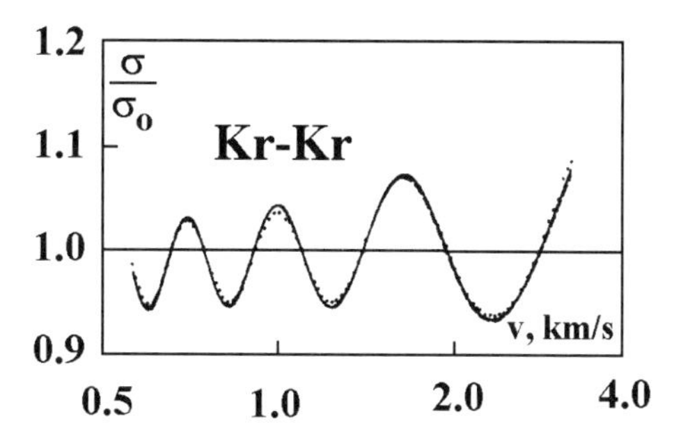

Figure 13.4. The relative values of the total cross section of collision of two thermal krypton atoms, as a function of the reduced collision velocity.

tion give information about the interaction potential of the colliding particles. As a demonstration of this phenomenon, Fig. 13.4 gives the relative values of the total cross section of the scattering of two thermal krypton atoms as a function of the collision velocity.

13.8 Elastic Collisions Involving Slow Electrons

Let us consider the collision of a slow electron with an atom. Then the character of the collision depends on the total spin of the electron–atom system, so that any cross section of the electron–atom collision is

$$\sigma = \frac{S}{2S+1}\sigma_- + \frac{S+1}{2S+1}\sigma_+, \tag{13.54}$$

where S is the atom spin, the cross sections σ_-, σ_+ correspond to the total spin $S - 1/2$ and $S + 1/2$ for the electron–atom system, and we assume that the electron–atom scattering proceeds in each channel independently. In particular, at small electron energies, this formula together with (13.37) gives, for the total or diffusion cross section of electron scattering on an atom with spin 1/2,

$$\sigma = \pi L_-^2 + 3\pi L_+^2, \tag{13.55}$$

where L_-, L_+ are the electron–atom scattering length for the corresponding total spin of the system. Because the exchange interaction of the electron and atom can strongly depend on their total spin, these parameters are different. For example, for the electron–hydrogen atom scattering we have $L_- = 5.8$ and $L_+ = 1.8$. Note that formulas (13.54) and (13.55) use the LS-scheme of the coupling of the electron and atom momenta into the total moment. Below we study the scattering for one channel and use atomic units.

Let us consider the scattering of a slow electron for a certain total spin. In the limit of its zero energy the diffusion and total cross section of the electron–atom scattering in this channel is given by formula (13.37) $\sigma = 4\pi L^2$, and the scattering length takes into account the exchange and other types of interaction when the electron is located near the atom. Below we also take into account the polarization interaction of the electron and atom which determines the dependence of the cross section on the electron energy at small collision energies. Within the framework of the phase theory of the scattering and the perturbation theory, we determine above (see formula (13.36)) the electron–atom phases of scattering due to the polarization interaction potential for all the scattering phases except δ_0. As for δ_0, in this case, the perturbation theory does not work and the integral (13.35) of the perturbation theory diverges. Indeed, the radial wave function of the s-electron at small electron energies is proportional to $r - L$, where r is the distance from the atom and L is the electron–atom scattering length, so that the polarization potential leads to divergence at small r. The physical reason of this divergence consists of an influence of the polarization interaction potential on the electron behavior in the internal region. But, if we assume that the polarization interaction potential acts only at large distances from the atom and does not influence the electron behavior in the internal region, one can separate the polarization potential from other interaction potentials. Then this potential acts on the electron only at large distances from the atom and we obtain formally that the corresponding integrals, which take into account the polarization potential, are determined by large electron–atom distances. Separating the action of the polarization potential from the other ones, we take $L = 0$ in expression (13.35) for the zero-scattering phase and further add to it the term with the scattering length. Then we obtain, for the zero scattering phase on the basis of formula (13.35), in atomic units

$$\delta_0 = -Lq - \frac{\pi\alpha q^2}{3}. \tag{13.56}$$

Joining this with expressions (13.36) for the other scattering phases, we find the scattering amplitude on the basis of formulas (13.29) and (13.32).

Using the above considerations, one can solve this task in a simpler way if we present the scattering amplitude in the following form by accounting for formula (13.27):

$$f(\vartheta) = -L - \frac{1}{2\pi}\int (1 - e^{-i\mathbf{K}\mathbf{r}})U(r)\,d\mathbf{r}, \tag{13.57}$$

Here we extract the scattering amplitude on the zero angle similar to formula (13.39). This formula gives, for the polarization interaction potential,

$$f(\vartheta) = -L + \frac{\pi\alpha}{4}K = -L - \frac{\pi\alpha q}{2}\sin\frac{\vartheta}{2}. \tag{13.58}$$

Thus the differential cross section of the electron–atom scattering is equal to, in this case,

$$\frac{d\sigma}{d\cos\vartheta} = 2\pi|f(\vartheta)|^2 = 2\pi L^2 + 2\pi^2\alpha q L\sin\frac{\vartheta}{2} - \frac{\pi^3\alpha^2 q^2}{4}(1 - \cos\vartheta).$$

From this we obtain, for the total and diffusion cross section of the electron–atom scattering at low energies,

$$\sigma_t = 4\pi \left(L^2 + \frac{2}{3}\pi\alpha q L + \frac{\pi^2}{8}\alpha^2 q^2 \right),$$
$$\sigma^* = 4\pi \left(L^2 + \frac{4}{5}\pi\alpha q L + \frac{\pi^2}{6}\alpha^2 q^2 \right). \qquad (13.59)$$

The important conclusion resulting from these formulas consists of a sharp minimum for the cross sections of the electron–atom scattering at small collision energies if L is negative. This is called the Ramsauer effect and is observed in the elastic scattering of electrons on argon, krypton and xenon atoms. One can expect this effect if the scattering length is negative. Then, according to formula (13.56), δ_0 vanishes at small energies when the contribution of the other phases to the cross section is relatively small. As follows from formula (13.59), the total cross section has the minimum $(4\pi/9)L^2$ at the electron wave number $q = -8L/3\pi\alpha$. The minimum of the diffusion cross section is equal to $(4\pi/25)L^2$ and corresponds to the electron wave vector $q = -12L/5\pi\alpha$. Thus, the scattering cross section drops by order of magnitude at small electron energies that is of importance for the processes in gases or plasmas involving electrons.

Problems

Problem 13.1. *Determine the scattering angle for the collision of particles with a sharply varied repulsive interaction potential.*

We are working on formula (13.5) for the scattering angle

$$\vartheta = \pi - 2\int_{r_o}^{\infty} \frac{\rho\, dR}{R^2\sqrt{1 - \frac{\rho^2}{R^2} - \frac{U(R)}{\varepsilon}}},$$

where ρ is the impact parameter of collision, and r_o is the distance of closest approach which are connected by relation (13.4), $1 - \rho^2/r_o^2 = U(r_o)/\varepsilon$. Below we expand the expression for the scattering angle over a small parameter $1/n$, where $n = -d\ln U(R)/dR$. The zeroth approximation corresponds to the use of the hard sphere model for an interaction potential. The hard sphere model corresponds to the interaction potential: $U(R) = 0$ for $R > r_o$, $U(R) = \infty$ for $R \le r_o$, where r_o is taken for a given impact parameter, i.e., this value depends on the impact parameter in accordance with formula (13.4). Thus we have, for the scattering angle in the zeroth approximation,

$$\vartheta = \pi - 2\arcsin\frac{\rho}{r_o}.$$

Then in the first approximation we have

$$\vartheta = \pi - 2\arcsin\frac{\rho}{r_o} + 2\Delta\vartheta,$$

where

$$\Delta\vartheta = \int_{r_o}^{\infty} \frac{\rho\, dR}{R^2} \left(\frac{1}{\sqrt{1 - \frac{\rho^2}{R^2}}} - \frac{1}{\sqrt{1 - \frac{\rho^2}{R^2} - \frac{U(R)}{\varepsilon}}} \right).$$

The above expression for the scattering angle is exact and $\Delta\vartheta \sim 1/n$, i.e., it is proportional to a small parameter of the theory. Our goal is to determine the first expansion term of $\Delta\vartheta$ over $1/n$. In order to avoid the divergence in the integral for $\Delta\vartheta$, we use the relation

$$-\frac{d}{d\rho}\int_{r_o}^{\infty} dR \left(\sqrt{1 - \frac{\rho^2}{R^2} - \frac{U(R)}{\varepsilon}} - \sqrt{1 - \frac{\rho^2}{R^2}} \right)$$
$$= -\frac{dr_o}{d\rho}\sqrt{1 - \frac{\rho^2}{R^2}} - \int_{r_o}^{\infty} \frac{\rho\, dR}{R^2} \left(\frac{1}{\sqrt{1 - \frac{\rho^2}{R^2}}} - \frac{1}{\sqrt{1 - \frac{\rho^2}{R^2} - \frac{U(R)}{\varepsilon}}} \right).$$

From this it follows that

$$\Delta\vartheta = -\frac{dr_o}{d\rho}\sqrt{1 - \frac{\rho^2}{R^2}} - \frac{d}{d\rho}\int_{r_o}^{\infty} dR \left(\sqrt{1 - \frac{\rho^2}{R^2} - \frac{U(R)}{\varepsilon}} - \sqrt{1 - \frac{\rho^2}{R^2}} \right).$$

Taking into account that the above integral converges near $R = r_o$ $(R - r_o \sim 1/n)$, one can calculate this integral with an accuracy of the order of $1/n$:

$$-\int_{r_o}^{\infty} dR \left(\sqrt{1 - \frac{\rho^2}{R^2} - \frac{U(R)}{\varepsilon}} - \sqrt{1 - \frac{\rho^2}{R^2}} \right)$$
$$\approx \sqrt{1 - \frac{\rho^2}{r_o^2}} \int_{r_o}^{\infty} dR \left[1 - \sqrt{1 - \frac{r_o^n}{R^n}} \right]$$
$$= \sqrt{r_o^2 - \rho^2} \int_0^1 \frac{1 - \sqrt{1 - x}}{n x^{1+1/n}} dx = \frac{2}{n}(1 - \ln 2)\sqrt{r_o^2 - \rho^2},$$

where

$$x = \frac{U(R)}{\varepsilon} \frac{1}{\sqrt{1 - \frac{\rho^2}{R^2}}} = \frac{r_o^n}{R^n}.$$

From this we have

$$\Delta\vartheta = \frac{dr_o}{d\rho}\sqrt{1 - \frac{\rho^2}{R^2}} + \frac{2}{n}(1 - \ln 2)\frac{d}{d\rho}\sqrt{r_o^2 - \rho^2}.$$

Let us introduce the value $u = U(r_o)/\varepsilon = 1 - \rho^2/r_o^2$. Since $n \gg 1$ and in the scattering region the value u is close to unity we obtain, from the above formulas,

$$\vartheta = 2 \arcsin \sqrt{u} + 2 \left[\frac{2 - (n - 2)\ln 2}{n} \right] \frac{\sqrt{u(1 - u)}}{1 + (n - 2)u/2}$$

$$\approx 2\arcsin\sqrt{u} - 2\ln 2\frac{\sqrt{u(1-u)}}{1+nu/2},$$

where $n = -d\ln u/dr_o$. We conserve 1 in the denominator of this formula in order to have the possibility for an expansion of this formula at small scattering angles.

Problem 13.2. *Determine the diffusion cross section of the elastic scattering of particles as a result of expansion over a small parameter for the scattering of particles with a sharply varied interaction potential.*

We are working on the result of the previous problem for the scattering angle. Taking into account that collisions with $u \sim 1$ give the main contribution to the diffusion cross section we have, for the scattering angle,

$$\vartheta = 2\arcsin\sqrt{u} - \frac{4\ln 2}{n}\sqrt{\frac{1-u}{u}}.$$

From this we find the diffusion cross section

$$\sigma^* = \int_0^\infty (1-\cos\vartheta)\pi\, d\rho^2$$
$$= 2\pi\int_0^1 u[(1-u)\,dr_o^2 - r_o^2\,du] + \pi\int_0^1 \frac{4\ln 2}{n}\sqrt{\frac{1-u}{u}}2\sqrt{u(1-u)}r_o^2\,du.$$

Here, since $u = 1-\rho^2/r_o^2$, we use the relation $d\rho^2 = (1-u)\,dr_o^2 - r_o^2\,du$. Taking into account that the first term is of the order of $1/n$ from the second term, we take the second integral on parts. We have

$$-2\pi\int_0^1 ur_o^2\,du = \pi\int_0^1 r_o^2\,du^2 = \pi r_o^2|_0^1 - \pi\int_0^1 u^2\,dr_o^2$$
$$= \pi R_o^2 + \frac{2\pi}{n}\int_0^1 r_o^2 u\,du = \pi R_o^2\left(1+\frac{1}{n}\right),$$

where $u(R_o) = 1$. We use that $u \sim r_o^{-n}$, so that $dr_o/r_o = -du/(nu)$. Note that $\rho = 0$ corresponds to $u = 1$, and that $\rho = \infty$ corresponds to $u = 0$. Repeating these operations for the other integral and keeping only terms of the order of $1/n$ we find, finally,

$$\sigma^* = \pi R_o^2\cdot\left(1+\frac{3-4\ln 2}{n}\right).$$

Let us represent the diffusion cross section in the form $\sigma^* = \pi R_1^2$, so that $R_1 = R_o + (3-4\ln 2)/2n$. We have

$$u(R_1) = (R_o/R_1)^n = \exp(-3/2 + 2\ln 2) = 4\exp(-3/2) = 0.89.$$

Thus the diffusion cross section has the form

$$\sigma^* = \pi R_1^2, \quad \text{where} \quad \frac{U(R_1)}{\varepsilon} = 0.89.$$

In a similar way one can determine the cross section

$$\sigma^{(2)} = \int_0^\infty (1 - \cos\vartheta) 2\pi\rho\, d\rho.$$

Repeating the above operations we find, finally,

$$\sigma^{(2)} = \int_0^\infty (1 - \cos\vartheta) 2\pi\rho\, d\rho = \frac{2}{3}\pi R_2^2, \quad \text{where} \quad \frac{U(R_2)}{\varepsilon} = 0.23.$$

Note that the values of the resultant constants differ from those obtained from a comparison with the numerical calculations for the interaction potentials $U = CR^{-n}$ which are given by formulas (13.15).

Problem 13.3. *Determine the diffusion cross section of the collision for particles with the polarization potential of interaction.*

For the polarization interaction potential $U(R) = -\alpha e^2/2R^4$, according to formula (13.5), the scattering angle is

$$\vartheta = \pi - 2\int_{r_o}^\infty \frac{\rho\, dR}{R^2\sqrt{1 - \frac{\rho^2}{R^2} - \frac{\alpha e^2}{2R^4\varepsilon}}}.$$

Here the distance of closest approach r_o is zero, if the impact parameter of the collision $\rho \le \rho_c$, and it is given by formula (13.4) for $\rho > \rho_c$, where the value ρ_c is determined by formula (13.17). So, we divide the impact parameter range into two parts, $\rho \le \rho_c$ and $\rho > \rho_c$. Let us introduce the reduced variables $x = \left(\alpha e^2/2\varepsilon R^4\right)^{1/4}$ and $y = \rho/\rho_c$. Then we have the following expression for the diffusion cross section

$$\sigma^* = \sigma_c(1 + J_1 + J_2),$$

where $\sigma_c = \pi\rho_c^2 = 2\pi\sqrt{\alpha e^2/\varepsilon}$ is the cross section of the capture of a particle in the polarization interaction potential, the integrals in the expression for the diffusion cross section are

$$J_1 = \int_0^1 2y\, dy \cos\int_0^\infty \frac{2y\sqrt{2}\, dx}{\sqrt{1 + x^4 - 2x^2y^2}},$$

$$J_2 = \int_1^\infty 2y\, dy \left(1 + \cos\int_0^{x_o} \frac{2y\sqrt{2}\, dx}{\sqrt{1 + x^4 - 2x^2y^2}}\right),$$

and the distance of closest approach is $x_o = y^2 - \sqrt{y^4 - 1}$. These integrals are calculated by numerical methods and are equal to $J_1 = -0.101$ and $J_2 = 0.207$. Thus the diffusion cross section of the scattering in the polarization interaction potential is

$$\sigma^* = 1.10\sigma_c,$$

i.e., the contribution to the diffusion cross section from collisions with impact parameters larger than ρ_c is approximately 10%.

Problem 13.4. *Present formula* (13.46) *for the total cross section of the scattering for a sharply varied interaction potential.*

Let us take the interaction potential of particles in the form $U(R) = CR^{-n}$ assuming $n \gg 1$. For the total cross section of the scattering we have formula (13.46) which we presented in the form

$$\sigma_t = 2\pi\rho_t^2, \quad \text{where} \quad \frac{\rho_t U(\rho_t)}{\hbar v} = g.$$

Our goal is to determine the value of the parameter g in this formula. For this we use expression (13.48) for the total cross section of scattering with this interaction potential and transfer to the limit $n \to \infty$. Then we have, for the parameter g,

$$g = \frac{\Gamma\left(\frac{n}{2}\right)}{\sqrt{\pi}\,\Gamma\left(\frac{n-1}{2}\right)}\left[\Gamma\left(\frac{n-3}{n-1}\right)\right]^{(n-1)/2}.$$

We have, for $n \gg 1$,

$$\left[\Gamma\left(\frac{n-3}{n-1}\right)\right]^{(n-1)/2} = \left[\Gamma(1) - \frac{2}{n}\Gamma'(1)\right]^{n/2} = e^{\psi(1)}\Gamma(1),$$

where $\psi(1) = \Gamma'(1)/\Gamma(1) = -C + 1 = 0.423$. Next, the Stirling formula gives

$$\frac{\Gamma\left(\frac{n}{2}\right)}{\Gamma\left(\frac{n-1}{2}\right)} = \frac{\left(\frac{n}{2}\right)^{n/2}}{\sqrt{e}\left(\frac{n-1}{2}\right)^{(n-1)/2}} = \sqrt{\frac{n}{2e}}\left(1 - \frac{2}{n}\right)^{n/2} = \sqrt{\frac{ne}{2}}.$$

Thus, finally, we have $g = \sqrt{ne/2\pi}\,e^{\psi(1)} \approx \sqrt{n}$, so that the formula for the total cross section of scattering has the form

$$\sigma_t = 2\pi\rho_t^2, \quad \text{where} \quad \frac{\rho_t U(\rho_t)}{\hbar v} = \sqrt{n}.$$

Problem 13.5. *Compare values of the diffusion cross section of scattering for similar attractive and repulsive interaction potentials of particles* $U(R) = \pm CR^{-n}$.

Take, as the diffusion cross section of the scattering of particles, the capture cross section σ_c which is given by formula (13.17). The difference of these cross sections is approximately 10% for $n = 4$ and these cross sections are coincident for $n \to \infty$. Use formula (13.14) for the diffusion cross section σ_{rep} of the scattering of particles with a repulsive interaction potential. Table 13.2 contains the ratio of these cross sections $\eta = \sigma_c/\sigma_{\text{rep}}$ for some n. As is seen, the scattering cross section in an attractive interaction potential exceeds that for a repulsive interaction potential, because the interaction of particles during collisions is stronger for the attractive interaction potential.

Problem 13.6. *Determine the contribution to the diffusion cross section from large impact parameters for the scattering of particles with an attractive potential* $U(R) = CR^{-n}$ *if* $n \gg 1$.

Table 13.2. The ratio of the cross sections of the scattering of particles $\eta = \sigma_c/\sigma_{\rm rep}$ for similar attractive and repulsive interaction potentials.

n	4	6	8	10	12	14
η	1.76	1.72	1.63	1.54	1.49	1.44

We assume that scattering on small angles takes place for collisions with impact parameters $\rho > \rho_c$, where ρ_c is the capture impact parameter. Then the scattering angle $\vartheta \sim \rho^{-n}$ for the related interaction potential, and that part of the diffusion cross section $\Delta\sigma$, which is determined by the impact parameters of collision $\rho > \rho_c$, is equal to

$$\Delta\sigma = \int_{\rho_c}^{\infty} 2\pi\rho d\rho(1-\cos\vartheta) = \int_{\rho_c}^{\infty} \pi\rho d\rho\vartheta^2 = \pi\rho_c^2\vartheta_c^2/(n-1) = \sigma_c\frac{\vartheta_c^2}{(n-1)},$$

where $\sigma_c = \pi\rho_c^2$ is the capture cross section and ϑ_c is the scattering angle at the impact parameter of collision ρ_c.

Let us calculate the scattering angle if it is small. Then we consider the motion of a scattering particle along straight trajectories and calculate the momentum $\Delta p_\perp$, which a scattering particle obtains from the force center in the direction perpendicular to its trajectory. The scattering angle is $\Delta p_\perp/p$, where p is the initial momentum of the scattering particle, and we have

$$\vartheta = \frac{\Delta p_\perp}{p} = \frac{1}{p}\int_{-\infty}^{\infty}\left|\frac{\partial U}{\partial R}\right|\frac{\rho}{R}dt = \frac{\rho}{\varepsilon}\int_{\rho}^{\infty}\left|\frac{\partial U}{\partial R}\right|\frac{dR}{\sqrt{R^2-\rho^2}} = \frac{U(\rho)}{\varepsilon}\frac{\sqrt{\pi}\,\Gamma\left(\frac{n}{2}\right)}{\Gamma\left(\frac{n-1}{2}\right)}.$$

Above we use the relation $R^2 = \rho^2 + v^2t^2$ for the free motion of the scattering particle and the dependence $U(R) \sim R^{-n}$ for the interaction potential. In this case we have

$$\frac{U(R_o)}{\varepsilon} = \frac{2}{n-2} \quad \text{and} \quad \frac{\rho_c^2}{R_o^2} = \frac{n}{n-2},$$

where R_o is the distance of closest approach to the impact parameter ρ_c. From this we obtain

$$\vartheta_c = \left(1-\frac{2}{n}\right)^{\frac{n}{2}}\frac{\sqrt{\pi}\,\Gamma\left(\frac{n}{2}\right)}{\Gamma\left(\frac{n-1}{2}\right)}$$

and

$$\frac{\Delta\sigma}{\sigma_c} = \frac{\vartheta_c^2}{(n-1)} = \frac{\pi}{n-1}\left(1-\frac{2}{n}\right)^{n}\frac{4}{(n-2)^2}\frac{\Gamma^2\left(\frac{n}{2}\right)}{\Gamma^2\left(\frac{n-1}{2}\right)}.$$

At large n we present this ratio in the form

$$\frac{\Delta\sigma}{\sigma_c} = \frac{2\pi}{(n-2)^2e^2},$$

so that this gives the correct result $\Delta\sigma/\sigma_c = 0.21$ for the polarization interaction potential $U(R) \sim R^{-4}$, in accordance with the result of Problem 13.3, and it has the correct asymptotic form in the limit of large n. As is seen, the contribution of the large impact parameters to the diffusion cross section drops sharply with an increase in n.

CHAPTER 14

Resonant Processes in Slow Atomic Collisions

14.1 Specifics of Slow Inelastic Atomic Collisions

Let us consider the slow collisions of atomic particles when the relative velocity of the nuclei v and a typical atomic velocity v_e for this process satisfy the relation

$$v \ll v_e. \tag{14.1}$$

This criterion corresponds to large energies of atomic particles. For example, in the case of H^+–H collisions this corresponds to collision energies $\varepsilon \ll 10\,\text{keV}$. In this chapter we focus on the processes which proceed at large and moderate distances between the colliding particles. Such processes are characterized by large cross sections compared to the typical atomic cross sections.

In the course of slow atomic collisions, when the criterion (14.1) is valid, the electron subsystem follows for the variations of atomic fields due to motion of the nuclei, and the equilibrium of the electron subsystem is established at each distance between the nuclei, because the time of the establishment of this equilibrium is small compared to a typical collision time. Hence, the electron state of this nonstationary electron system is close to that of motionless nuclei. This gives a method for the description of slow atomic collisions. Such a description is based on the parameters of the quasi-molecule which is the related system of colliding particles with motionless nuclei.

Let us express this statement in mathematical form. We have the Schrödinger equation for colliding atomic particles

$$i\hbar\frac{\partial\Psi(\mathbf{r},\mathbf{R},t)}{\partial t} = \hat{H}\Psi(\mathbf{r},\mathbf{R},t), \tag{14.2}$$

where $\mathbf{r}$ is the sum of electron coordinates, $\mathbf{R}$ describes the relative position of the atomic particles, and t is time. There is a system of eigenwave functions $\{\psi_i\}$ for each distance between the nuclei. These wave functions satisfy the Schrödinger equations

$$\hat{H}\psi_i(\mathbf{r}, R) = \varepsilon_i(R)\psi_i(\mathbf{r}, R), \tag{14.3}$$

where i denotes the quasi-molecule state. Thus, the electron state of the system of colliding atomic particles is close to a state of this system under motionless nuclei. The electron terms $\varepsilon_i(R)$ are called the adiabatic terms.

Let us use the adiabatic wave functions as a basis for the description of the evolution of the electron system during collisions. Assuming the motion of the nuclei to be governed by classical laws, we present the wave function of the system in the form

$$\Psi(\mathbf{r}, \mathbf{R}, t) = \sum_i c_i(t)\psi_i(\mathbf{r}, \mathbf{R}) \exp\left(-i\int_0^t \varepsilon_i(t')\,dt'\right). \tag{14.4}$$

Substituting this expression into the Schrödinger equation (14.2) and using the orthogonality of the wave functions $\psi_i(\mathbf{r}, R)$, we obtain the following set of equations as a result of multiplying (14.2) by $\psi_i(\mathbf{r}, R)$ and integrating over the electron coordinates

$$\dot{c}_i(t) = \sum_k c_k(t)\left(\frac{\partial}{\partial t}\right)_{ik} \exp\left(-i\int^t \omega_{ik}(t')dt'\right), \tag{14.5}$$

where $\hbar\omega_{ik}(t) = \varepsilon_i\left[\mathbf{R}(t)\right] - \varepsilon_k\left[\mathbf{R}(t)\right]$, and the matrix element is

$$\left(\frac{\partial}{\partial t}\right)_{ik} = \int \psi_i \frac{\partial \psi_k}{\partial t}\,d\mathbf{r}.$$

The set of equations (14.5) is called the adiabatic set of equations. As a matter of fact, this is a new form of the Schrödinger equation (14.2) which is equivalent to equation (14.2). But this form of the Schrödinger equation is more convenient for the analysis of transitions in slow atomic collisions. Let us construct the perturbation theory on the basis of this set of equations assuming, that at the beginning, the wave function of the system is $\psi_o(\mathbf{r}, \infty)$ and in the course of its evolution it is close to $\psi_o(\mathbf{r}, R)$ whose phase corresponds to the stationary state. Then the set of equations (14.5) can be rewritten within the framework of the perturbation theory by a change of amplitudes c_k, in the right-hand side of this equation and by δ_{ko} this transforms the set of equations (14.5) to the form

$$\dot{c}_i(t) = \left(\frac{\partial}{\partial t}\right)_{io} \exp(-i\int^t \omega_{io}(t')\,dt').$$

The probability of transition to state i after collision is $P_i = |\dot{c}_i(\infty)|^2$, where

$$c_i(t) = \int_{-\infty}^{\infty}\left(\frac{\partial}{\partial t}\right)_{io} \exp(-i\int_{-\infty}^t \omega_{io}(t')\,dt'). \tag{14.6}$$

One can see that the integral (14.6) is of the order of $\exp(-\Delta\varepsilon a/\hbar v)$, where a is a typical size which represents a shift in the distance between the nuclei corresponding to a remarkable change of the matrix element $(\partial/\partial t)_{io}$, $\Delta\varepsilon$ is a typical distance between levels o and i and v is the collision velocity. Thus the following parameter is of importance to understand the adiabatic character of slow atomic collisions

$$\xi = \frac{\Delta\varepsilon a}{\hbar v}. \tag{14.7}$$

This is called the Massey parameter. If this parameter is large, the probability of the transition between the corresponding levels is small in an exponential manner. Hence, analyzing the transitions between the electron states, one can be restricted by adiabatic states for which only the Massey parameter is not large. Other states may be excluded from this consideration.

14.2 Resonant Charge Exchange Processes

According to the above analysis, the transitions between the states of close energy are effective in slow collisions. The processes with the energy variation $\Delta\varepsilon = 0$ are called resonant processes and for quasi-resonant processes $\Delta\varepsilon$ is relatively small. We start with the analysis of resonant processes from the resonant charge exchange process which proceeds according to the scheme

$$A^+ + A \rightarrow A + A^+, \tag{14.8}$$

and as a result of this process a valent electron transfers from the field of one ion to the other. For simplicity, we consider the case of one state of an atom and ion, as this takes place in the case when an atom A has a valent s-electron and the ion electron shell is completed. The eigenstate of the quasi-molecule, consisting of the colliding ion and atom, can be even or odd depending on the property of the corresponding wave functions to conserve or change their sign as a result of electron reflection with respect to the symmetry plane. This plane is perpendicular to the axis joining the nuclei and passes through its middle. These electron eigenwave functions are determined by formula (10.17):

$$\psi_g = \frac{1}{\sqrt{2}}(\psi_1 + \psi_2), \qquad \psi_u = \frac{1}{\sqrt{2}}(\psi_1 - \psi_2), \tag{14.9}$$

where the wave functions ψ_1, ψ_2 are the molecular wave functions which correspond to the electron location in the field of the corresponding ion. The wave functions (14.9) are the eigenfunctions of the electron Hamiltonian, so that according to (10.18) we have

$$\hat{H}\psi_g = \varepsilon_g \psi_g, \qquad \hat{H}\psi_u = \varepsilon_u \psi_u. \tag{14.10}$$

For simplicity, we will be guided by the case when an ion A^+ has a filled electron shell, and its atom A has one valent s-electron above the filled shell, so that the process (14.8) corresponds to a transition of the valent electron from the field of one ion to the field of the other ion. Let us study the electron behavior in the absence of nonadiabatic transitions. At the beginning let the electron be located in the field of the first ion, i.e., the electron wave function is $\Psi(\mathbf{r}, \mathbf{R}, -\infty) = \psi_1(\mathbf{r})$. This means that at that time $c_1 = c_2 = 1/\sqrt{2}$. Then in the absence of nonadiabatic transitions we have, for the electron wave function according to formula (14.4), in atomic units

$$\Psi(\mathbf{r}, \mathbf{R}, t) = \frac{1}{\sqrt{2}}\psi_g(\mathbf{r}, \mathbf{R}) \exp\left[-i\int_{-\infty}^{t} \varepsilon_g(t')\,dt'\right] + \frac{1}{\sqrt{2}}\psi_u(\mathbf{r}, \mathbf{R}) \exp\left[-i\int_{-\infty}^{t} \varepsilon_u(t')\,dt'\right]. \tag{14.11}$$

Let us introduce the S-matrix of transition in the usual way such that if the initial state of a system is ψ, its state at time t is $\hat{S}\psi$. Then substituting (14.9) into (14.11), we find the element of the S-matrix, at time t,

$$S_{12}(t) = \exp\left[-i\int_{-\infty}^{t} \frac{(\varepsilon_g + \varepsilon_u)}{2}\,dt'\right] \cdot i \sin \int_{-\infty}^{t} \frac{(\varepsilon_g - \varepsilon_u)}{2}\,dt'. \tag{14.12}$$

This formula shows that in the absence of inelastic transitions between the states the electron transfers from the field of one ion to the field of the other ion due to the interference of the states. This is the nature of both the resonant charge exchange process and other resonant processes. Formula (14.2) gives, for the probability of the charge exchange process as a result of an ion–atom collision,

$$P_{12} = |S_{12}(\infty)|^2 = \sin^2 \int_{-\infty}^{\infty} \frac{(\varepsilon_g - \varepsilon_u)}{2}\,dt = \sin^2 \int_{-\infty}^{\infty} \frac{\Delta(R)}{2}\,dt, \tag{14.13}$$

and the exchange interaction potential between the ion and atom is introduced on the basis of formula (10.19), $\Delta(R) = \varepsilon_g(R) - \varepsilon_u(R)$.

The cross section of the charge exchange process, for this case of transition between two states, is equal to

$$\sigma_{\mathrm{res}} = \int_0^{\infty} 2\pi\rho\,d\rho \sin^2 \zeta(\rho), \qquad \zeta(\rho) = \int_{-\infty}^{\infty} \frac{\Delta(R)}{2}\,dt, \tag{14.14}$$

where $\zeta(\rho)$ is the charge exchange phase. We now calculate this integral by taking into account a strong dependence $\Delta(R)$. For this goal we take the charge exchange phase $\zeta(\rho)$ in the form $\zeta(\rho) = A\rho^{-n}$ and consider the result in the limit $n \to \infty$. We obtain

$$\sigma_{\mathrm{res}} = \int_0^{\infty} 2\pi\rho\,d\rho \sin^2 \zeta(\rho) = \frac{\pi}{2}(2A)^{2/n}\Gamma\left(1 - \frac{2}{n}\right)\cos\frac{\pi}{n}.$$

Let us present the result in the form

$$\sigma_{\mathrm{res}} = \frac{\pi R_o^2}{2} f_n,$$

where R_o is determined by the relation $\zeta(R_o) = a$, so that the function f_n is equal to

$$f_n = (2a)^{2/n}\Gamma\left(1 - \frac{2}{n}\right)\cos\frac{\pi}{n}.$$

We take the parameter a such that the second term of expansion of the function f_n over a small parameter $1/n$ would be zero. This gives

$$a = \frac{e^{-C}}{2} = 0.28,$$

where $C = 0.557$ is the Euler constant. Then the function f_n is equal to

$$f_n = \exp\left(-\frac{2C}{n}\right)\Gamma\left(1-\frac{2}{n}\right)\cos\frac{\pi}{n},$$

and its expansion at large n has the form

$$f_n = 1 - \frac{\pi^2}{6n^2}.$$

Table14.1 contains values of this function and its asymptotic expansion for some n. These data confirm the validity of the used expansion over a small parameter $1/n$.

Because of an exponential dependence $\zeta(\rho)$ at large $\rho[\zeta(\rho) \sim e^{-\gamma\rho}]$, i.e., $n = \gamma\rho \gg 1$, the expression for the cross section of the resonant charge exchange process can be represented in the form

$$\sigma_{\text{res}} = \frac{\pi R_o^2}{2} - \frac{\pi^3}{12\gamma^2}, \quad \text{where} \quad \zeta(R_o) = \frac{e^{-C}}{2} = 0.28. \tag{14.15}$$

In the deduction of this formula we suppose the parameter γR_o to be large. This is the basis of the asymptotic theory of resonant charge exchange which uses a small parameter $1/(\gamma R_o)$. Table14.2 contains the cross sections of the resonant charge exchange process for some ion–atom pairs with a transferring s-electron, and for the ground ion and atom states and the quantity of a large parameter γR_o in these cases. The data of Table 14.2 confirm the used assumption of a large value of the parameter γR_o in reality. Note that according to Fig. 10.3, within the framework of the asymptotic theory under consideration, the exchange interaction potential $\Delta(R)$ is determined with an accuracy up to $\sim 1/(\gamma R_o)^2$, because we do not take into account the contribution of the atomic regions of Fig. 10.3 into the exchange interaction potential. Hence, it is correct to account for two expansion terms over a small parameter $1/(\gamma R_o)$ in formula (14.15).

Let us determine the dependence of the cross sections of the resonant charge exchange process on the collision velocity. On the basis of formulas (10.20) and (10.23) we take the dependence of the exchange interaction potential Δ on the ion–atom distance R in the form $\Delta \sim e^{-\gamma R}$. This gives $\zeta(R_o) \sim e^{-\gamma R_o}/v$. From

Table 14.1. Values of the function f_n and its asymptotic expression.

n	2	4	6	8	∞
f_n	0.88	0.94	0.97	0.98	1.00
$1-\pi^2/(6n^2)$	0.59	0.90	0.95	0.97	1.00

Table 14.2. Parameters of the resonant charge exchange process at the ion collision energy of 1 eV in the laboratory coordinate system (an atom is at rest) for the ground atom and ion states.

Element	H	He	Li	Be	Na	Mg	K	Ca	Cu
$R_o\gamma$	10.5	10.5	13.6	12.7	14.9	13.8	15.7	14.7	14.3
σ_{res}, 10^{-15} cm^2	4.8	2.7	22	10	26	15	34	21	16
$-d\ln\sigma_{res}/d\ln v$, 10^{-2}	4.8	4.8	3.7	3.9	3.4	3.6	3.2	3.4	3.5
Element	Zn	Rb	Sr	Ag	Cd	Cs	Ba	Au	Hg
$R_o\gamma$	14.0	16.3	15.4	14.5	14.4	16.8	16.0	14.5	14.5
σ_{res}, 10^{-15} cm^2	12	38	25	17	14	44	29	14	12
$-d\ln\sigma_{res}/d\ln v$, 10^{-2}	3.6	3.1	3.2	3.4	3.5	3.0	3.1	3.4	3.4

this it follows that

$$R_o = \frac{1}{\gamma}\ln\frac{v_o}{v},$$

where the parameter $v_o \gg 1$. Thus we obtain the following dependence of the charge exchange cross section on the collision velocity:

$$\sigma_{res} = \frac{\pi R_o^2}{2} = \frac{\pi}{2\gamma^2}\ln^2\frac{v_o}{v}. \qquad (14.16a)$$

This dependence is weak enough, if the parameter γR_o is large. From this we have

$$\frac{d\ln\sigma_{res}}{d\ln v} = -\frac{2}{\gamma R_o},$$

which allows us to represent the velocity dependence for the cross section in the form

$$\frac{\sigma_{res}(v)}{\sigma_{res}(v_1)} = \left(\frac{v_1}{v}\right)^{\alpha}, \qquad \alpha = \frac{2}{\gamma R_o}. \qquad (14.16b)$$

The data of Table14.3, where the ratio $\sigma_{res}(0.3v)/\sigma_{res}(v)$ is given, show that the velocity dependence for the cross section is weaker, the larger the parameter γR_o is.

Let us express the cross section (14.15) through the exchange ion–atom interaction potential $\Delta(R)$. Because of the exponential dependence of the exchange interaction potential $\Delta \sim e^{-\gamma R}$ on a distance R between particles, the integral for the charge exchange phase $\zeta(\rho)$ converges near the distance of closest approach of the particles which coincides with the impact parameter of collision ρ. Then, taking $R = \sqrt{\rho^2 + z^2} = \rho + z^2/(2\rho)$ where $z = vt$, v is the velocity of collision, t is time, and accounting for $\Delta \sim e^{-\gamma R}$, we obtain

$$\zeta(\rho) = \int_{-\infty}^{\infty}\frac{\Delta(R)}{2}\,dt = \frac{\Delta(\rho)}{2v}\int_{-\infty}^{\infty}\exp\left(-\frac{z^2\gamma}{2\rho}\right)dz = \frac{\Delta(\rho)}{v}\sqrt{\frac{\pi\rho}{2\gamma}}. \qquad (14.17)$$

Table 14.3. The reduced cross sections of a resonant charge exchange; σ_o is the cross section of a resonant charge exchange for a transferring s-electron; σ_{10}, σ_{11} are the cross sections for a p-transferring electron for zero and one momentum projections on the impact parameter of collision; $\sigma_{\text{res}} = \sigma_{10}/3+2\sigma_{11}/3$; $\sigma_3, \sigma_4, \sigma_5$ are the average cross sections for the transition of a p-electron for elements of Groups 3, 8, Groups 4, 7 and Groups 5, 6 of the periodic table, respectively; and $\sigma_{1/2}, \sigma_{3/2}$ are the average cross sections for the total momentum 1/2 and 3/2 respectively, for a transferring p-electron between structureless cores. The asymptotic parameters of the transferring s- and p-electrons under consideration are identical.

$R_o\gamma$	6	8	10	12	14	16
$\sigma_o(v)/\sigma_o(0.3v)$	0.669	0.740	0.786	0.818	0.842	0.860
σ_{10}/σ_o	1.40	1.29	1.23	1.19	1.16	1.14
σ_{11}/σ_o	1.08	0.98	0.94	0.92	0.91	0.91
$\sigma_{\text{res}}/\sigma_o$	1.19	1.08	1.04	1.01	0.99	0.95
σ_3/σ_o	1.17	1.09	1.05	1.03	1.02	1.01
σ_4/σ_o	1.50	1.32	1.23	1.18	1.14	1.12
σ_5/σ_o	1.44	1.29	1.22	1.17	1.14	1.11
$\sigma_{1/2}/\sigma_o$	1.18	1.10	1.07	1.05	1.04	1.03
$\sigma_{3/2}/\sigma_o$	1.18	1.10	1.06	1.04	1.03	1.02

On the basis of this formula, it is convenient to present expression (14.15) for the cross section of the charge exchange process in the form

$$\sigma_{\text{res}} = \frac{\pi R_o^2}{2}, \quad \text{where} \quad \sqrt{R_o}\Delta(R_o) = 0.22v\sqrt{\gamma}, \tag{14.18}$$

Here the exchange potential of the ion–atom interaction is given by formulas (10.20) and (10.23).

In order to ascertain the accuracy of the asymptotic theory, we consider the charge exchange of the proton on the hydrogen atom at a collision energy of 1 eV in the laboratory coordinate system and analyze various versions of the asymptotic theory. In this case formula (14.15) has the form

$$\sigma_{\text{res}} = \frac{\pi R_o^2}{2}, \quad \text{where} \quad \zeta(R_o) = \frac{1}{v}\cdot\frac{4}{e}\sqrt{\frac{\pi}{2}}R_o^{3/2}\exp(-R_o) = 0.28. \tag{14.19a}$$

One can account for the next term of the expansion of the phase $\zeta(R_o)$ over the small parameter $1/R_o$. Then formula (14.15) has the form

$$\sigma_{\text{res}} = \frac{\pi R_o^2}{2}, \quad \text{where} \quad \zeta(R_o) = \frac{1}{v}\cdot\frac{4}{e}\sqrt{\frac{\pi}{2}}R_o^{3/2}\left(1+\frac{7}{8R_o}\right)\exp(-R_o) = 0.28. \tag{14.19b}$$

One can evaluate the exchange phase $\zeta(\rho)$ on the basis of the exchange interaction potential $\Delta(R)$ directly by the use of formula (14.15). This gives, for the charge exchange cross section,

$$\sigma_{\rm res} = \frac{\pi R_o^2}{2}, \quad \text{where} \quad \zeta(R_o) = \frac{4R_o^2}{v\exp(1)}\left[K_o(R_o) + \frac{1}{R_o}K_1(R_o)\right] = 0.28. \tag{14.19c}$$

Finally, one can find the charge exchange cross section directly

$$\sigma_{\rm res} = \int_0^\infty 2\pi\rho\, d\rho \sin^2\zeta(\rho), \tag{14.19d}$$

where the charge exchange phase is given by formulas (14.19a, b, and c). Calculation of the cross section in the hydrogen case at an energy of 1 eV in the laboratory coordinate system gives on the basis of the above formulas for the values of the charge exchange cross sections 172, 175, 175, respectively, in atomic units, if we use formulas (14.19a, b, and c), and also 170, 173, 174, if we use formula (14.19d) with the above expressions for the phase of charge exchange. The statistical treatment of these data gives 173 ± 2 for the average cross section, i.e., the error in this case, which can be considered as the best accuracy of the asymptotic theory, is approximately 1%.

The accuracy of the asymptotic theory is determined by the small parameter $1/(R_o\gamma)$ and the above accuracy is of the order of $1/(R_o\gamma)^2$. According to the data of Table 14.2, the best accuracy of the asymptotic theory is of the order of 1 under the parameters of the resonant charge exchange process with a transferring s-electron. In reality, the accuracy of the evaluated cross sections within the framework of the asymptotic theory is determined by the accuracy of the asymptotic coefficient A for the wave function (4.11) of a transferring electron far from the core. From formula (14.15) follows the relative accuracy of the cross section $\Delta\sigma$,

$$\frac{\Delta\sigma_{\rm res}}{\sigma_{\rm res}} = \frac{4}{R_o\gamma}\cdot\frac{\Delta A}{A}. \tag{14.20}$$

One can see that this error arises in the first approximation of the expansion of the charge exchange cross section over a small parameter. In particular, if the error in the asymptotic coefficient is $\Delta A/A = 10\%$, the error in the cross section is 3–4% for the cases of Table 14.2, as follows from formula (14.20). In particular, the asymptotic coefficient for the helium atom in the ground state is 2.8 ± 0.3 (see Chapter 4), if we are based on the hydrogenlike electron wave functions. According to formula (14.20) this corresponds to an accuracy of 4% at the collision energy of 1 eV. Thus, the accuracy of the asymptotic coefficients is of importance for the accuracy of the asymptotic theory for the cross section of a resonant charge exchange, and the accuracy of the cross sections of a resonant charge exchange with the transferring s-electron lies in reality between 1% and 5% at small collision energies. As a demonstration of the asymptotic theory for the transition of an s-electron, Figs. 14.1 and 14.2 give the cross sections of a resonant charge exchange for collisions of rubidium and cesium atoms with their ions depending on the collision energy.

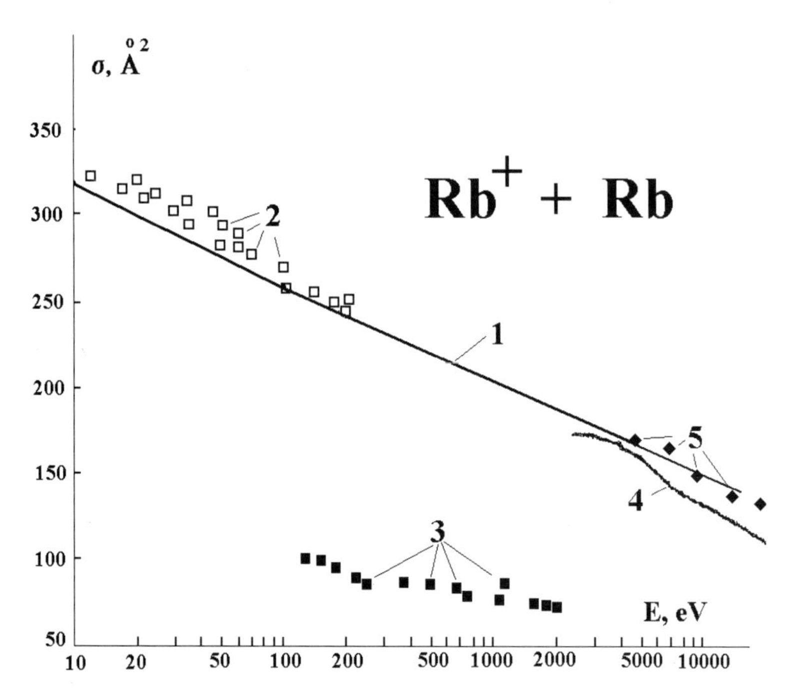

Figure 14.1. The cross sections of a resonant charge exchange for rubidium. Curve 1 corresponds to formula (5.8), signs relate to experimental data.

14.3 Resonant Charge Exchange with Transition of the p-Electron

The resonant charge exchange process with the transition of an electron of a nonzero momentum is entangled with the process of rotation of the electron momentum. Partially, these processes can be separated because the exchange interaction potential of the colliding particles strongly depends on a change of distance between the interacting particles. This means that a strong interaction takes place in a narrow range of distances between the colliding particles where the molecular axis turns on a small angle of the order of $1/\sqrt{R_o\gamma}$. Indeed, the range of the distances between the nuclei ΔR, where the phase of charge exchange ζ varies remarkably, is $\Delta R \sim 1/\gamma$, which corresponds to the angle of rotation $\vartheta \sim vt/R \sim 1/\sqrt{R\gamma} \ll 1$. Nevertheless, it is necessary to account for the rotation of the molecular axis in the course of electron transition.

Figure 14.3 shows the geometry of a collision in a center-of-mass coordinate system, when the configuration of colliding particles is close to that at the distance of closest approach. We have the following relation, which connects a current angle θ between the molecular and quantization axes and the angle ϑ between these axes at the distance of closest approach,

$$\cos\theta = \cos\vartheta\cos\alpha + \sin\vartheta\sin\alpha\cos\varphi, \tag{14.21}$$

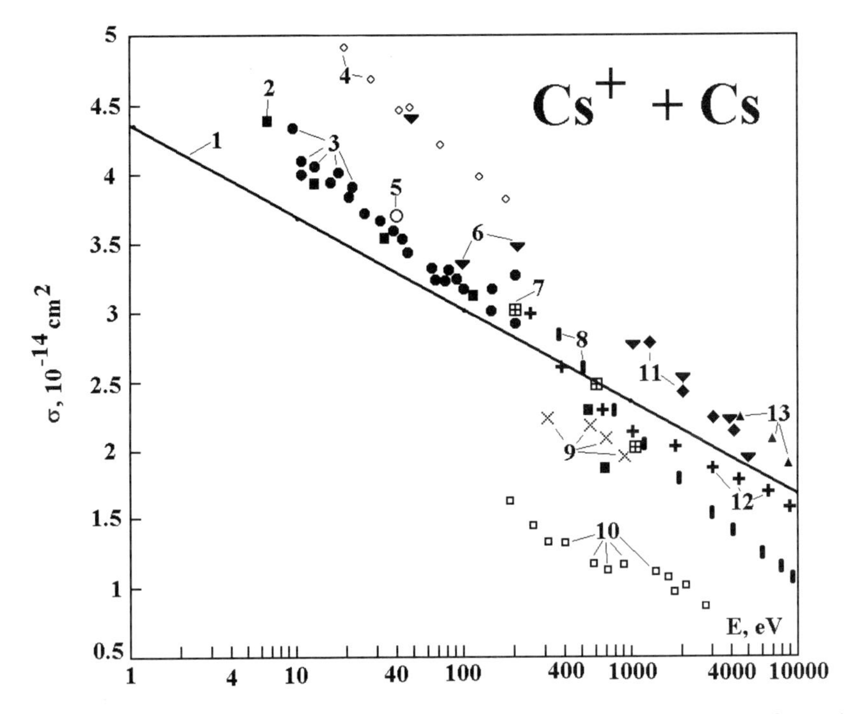

Figure 14.2. The cross sections of a resonant charge exchange for cesium. Curve 1 corresponds to formula (5.8), signs relate to experimental data.

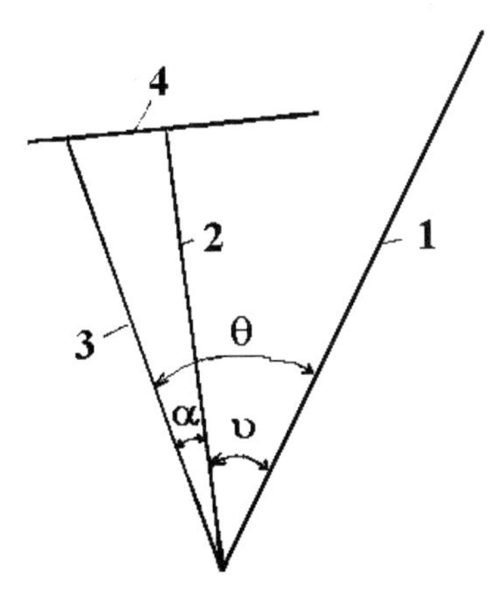

Figure 14.3. The geometry of the nuclear trajectory in the centre-of-mass coordinate system. 1, Quantization axis; 2, molecular axis at the distance of closest approach; 3, current molecular axis, 4, trajectory of nuclear motion. θ, ϑ are the angles between the quantization and molecular axes, current and at the distance of closest approach, α, φ are polar angles of a current molecular axis with respect to the distance of closest approach.

where α, φ are the polar angles of the molecular axis, so that $\sin\alpha = vt/R$, v is the collision velocity, t is time, and R is the distance between the colliding particles.

A small parameter of the theory $1/\rho\gamma$ simplifies determination of the phase of charge exchange and the cross section of this process. On the basis of formulas (10.58) the exchange interaction potentials of atoms and their ions with filling p-shells in neglecting the spin-orbital interaction one can find the charge exchange phase (14.14),

$$\zeta(\rho, \vartheta) = \int_{-\infty}^{\infty} \frac{\Delta(R, \theta)}{2} dt.$$

On the basis of this formula we have, for the charge exchange phase as a result of an expansion over a small parameter $1/\rho\gamma$ in the case of atoms of Groups 3, 5, 6, and 8 of the periodic table of elements,

$$\zeta(\rho, \vartheta, \varphi) = \zeta(\rho, 0)\left[\cos^2\vartheta - \frac{1}{\gamma\rho}\cos^2\vartheta + \frac{1}{\gamma\rho}\sin^2\vartheta(2+\cos^2\varphi)\right]. \quad (14.22)$$

This expression relates to the large impact parameters of collision, and $\zeta(\rho, 0)$ is the phase of the charge exchange process when a quantization axis has the same direction as the molecular axis at the distance of closest approach. The value $\zeta(\rho, 0)$ can be expressed through the charge exchange phase ζ_o which is given by formula (14.17) and relates to transition of an s-electron with the same asymptotic parameters γ, A. This connection for the resonant charge exchange process involving atoms of Groups 3 and 8 of the periodic table of elements has the form

$$\zeta(\rho, 0) = 3\zeta_o(\rho), \quad (14.23a)$$

and for atoms of Groups 5 and 6 this connection is

$$\zeta(\rho, 0) = 7\zeta_o(\rho), \quad (14.23b)$$

Note that our analysis relates to the ground state of the colliding atom and ion.

In the case of atoms of Groups 4 and 7 of the periodic table of elements the expression for the charge exchange phase at large impact parameters of collision has the form

$$\begin{aligned}\zeta(\rho, \vartheta, \varphi) &= 5\zeta_o(\rho)\{\sin^2\vartheta_1\sin^2\vartheta_2 + \frac{1}{\gamma\rho}[2\cos^2\vartheta_1 + 2\cos^2\vartheta_2 + \sin^2\vartheta_1\cos^2\vartheta_2 \\ &\quad + \cos^2\vartheta_1\sin^2\vartheta_2 - \sin^2\vartheta_1\sin^2\vartheta_2(\cos^2\varphi_1 + \cos^2\varphi_2) \\ &\quad + \sin 2\vartheta_1 \sin 2\vartheta_2 \cos\varphi_1\cos\varphi_2]\}, \quad (14.24)\end{aligned}$$

where ϑ_1, φ_1 and ϑ_2, φ_2 are the polar angles of the quantization axes of the atom and ion, respectively, with respect to the molecular axis at the distance of closest approach.

Thus, separating the depolarization of a colliding atom and ion from the charge exchange process, we average the cross section over directions of the molecular axis with respect to the quantization axis. Below we find the average cross section of the resonant charge exchange process $\overline{\sigma}$ according to the formula

$$\overline{\sigma}_{\mathrm{res}} = \frac{1}{2}\int_{-1}^{1} \sigma(\vartheta)\, d\cos\vartheta, \qquad (14.25)$$

where $\sigma(\vartheta)$ is the cross section of charge exchange at an angle ϑ between the impact parameter of collision and the quantization axis.

According to the exponential dependence of the charge exchange phase on the impact parameter of collision $\zeta(\rho, \vartheta) \sim \exp(-\gamma\rho)$ we have, in the case of atoms of Groups 3, 5, 6, 8 of the periodic table of elements, comparing the cross section with that at identical directions of the impact parameter of collision and quantization axis

$$R_o(\vartheta, \varphi) = R_o(0) + \frac{1}{\gamma}\ln\frac{\zeta(\rho, \vartheta, \varphi)}{\zeta(\rho, 0)}. \qquad (14.26)$$

This gives, for the average cross section of resonant charge exchange,

$$\overline{\sigma}_{\mathrm{res}} = \frac{1}{4}\int_0^1\int_0^{2\pi}\left[R_o(0) + \frac{1}{\gamma}\ln\frac{\zeta(R_o, \vartheta, \varphi)}{\zeta(R_o, 0)}\right]^2 d\cos\vartheta\, d\varphi. \qquad (14.27)$$

As a matter of fact, formula (14.27) means that the dependence of the exchange phase ζ on the impact parameter ρ of collision has the form $\zeta \sim \exp(-\gamma\rho)$. This formula is the basis for the determination of the average cross section of the resonant charge exchange process when this process results from the transition of a p-electron. This formula is valid for elements of Groups 3, 5, 6, 8 of the periodic table when atoms and ions are in the ground states, and one of these states is an S-state, so that the phase of charge exchange depends on two angles ϑ, φ. In the same manner, one can find the charge exchange phase for elements of Groups 4 and 7 which depend on four angles $\vartheta_1, \varphi_1, \vartheta_2, \varphi_2$. The data for the reduced cross sections of charge exchange for these cases are given in the Table 14.3.

Let us compare the cross sections of resonant charge exchange for transition of s and p-valent electrons if these electrons are characterized by the same asymptotic parameters γ and A. Supposing the dependence of the charge exchange phase $\zeta(\rho, \vartheta, \varphi)$ on the impact parameter ρ of collision to be exponential $\zeta(\rho, \vartheta, \varphi) \sim \exp(-\gamma\rho)$, and neglecting the momentum rotation during the electron transition we obtain, for the average cross section $\overline{\sigma}_{\mathrm{res}}$ of the resonant charge exchange process,

$$\overline{\sigma}_{\mathrm{res}} = \sigma_o \int_0^1 d\cos\vartheta \int_0^{2\pi}\frac{d\varphi}{2\pi}\left(1 + \frac{1}{\gamma R_o}\ln\frac{\zeta(\rho, \vartheta, \varphi)}{\zeta_o(\rho)}\right)^2, \qquad (14.28)$$

where σ_o is the cross section for a resonant charge exchange for a transferring s-electron with the same asymptotic parameters and $\zeta_o(\rho)$ is the charge exchange phase for an s-electron which is given by formula (14.15). Table 14.3 contains the ratios of the cross sections of resonant charge exchange for elements of Groups 3–8 of the periodic table to the cross sections with a transferring s-electron with the same asymptotic parameters as a function of a small parameter of the asymptotic theory. The use of additional assumptions for the evaluation of the charge exchange cross sections with a transferring p-electron decreases the accuracy of the asymptotic theory in this case. In addition to this, Figs. 14.3 and 14.3 give the cross section of the resonant charge exchange process for most elements of the periodic table. These cross sections are evaluated on the basis of the above formula. Figures 14.6–14.8 contain the cross sections as a function of the collision energy for neon, krypton, and xenon. The experimental data presented in these figures and Figs. 14.1 and 14.2 show that the accuracy of the contemporary experiment is lower than that of the asymptotic theory.

14.4 Coupling of Electron Momenta in the Resonant Charge Exchange Process

Analyzing the charge exchange process, we neglect above the relativistic effects in comparison with the electrostatic interactions. Returning to a general case of the coupling of electron moments during collisions, we consider six cases of the Hund coupling which are given in Table 11.1. The consideration is simplified for the resonant charge exchange process because the electron transition proceeds in a narrow range of distances between the colliding particles. Indeed, according to formulas (14.15) and (14.17), for typical impact parameters ρ of the charge exchange process

$$\Delta(\rho) \sim v\sqrt{\frac{\gamma}{\rho}}.$$

Because a typical rotational energy of colliding particles is $V_r \sim v/\rho$, we have

$$\Delta(\rho) \sim V_r\sqrt{\gamma\rho} \gg V_r.$$

Since the exchange interaction potential Δ is a part of the electrostatic energy V_e, from this it follows that only three cases of the Hund coupling, cases "a", "b", "c", are of interest for resonant charge exchange. Moreover, because the rotational axis rotates on a small angle during the electron transition, the difference of the cross sections for cases "a" and "b" of the Hund coupling is small and, practically, for the analysis of real charge exchange processes, one can restrict only by cases "a" and "c" of the Hund coupling for the analysis of the resonant charge exchange process. Above we consider this process for case "a" of the Hund coupling, and below we focus on case "c".

Note that because of a low atom symmetry in the case of the jj-coupling scheme, charge exchange in this case can be realized by the transition of only one electron, as follows from (11.8) and (11.9). Therefore, we consider below the transition of an electron with a total momentum 1/2 or 3/2 between two identical cores. We use formula (11.16) for the exchange interaction potentials

$$\Delta_{1/2} = \frac{1}{3}\Delta_{10} + \frac{2}{3}\Delta_{11},$$

PERIODICAL SYSTEM OF ELEMENTS
Cross sections of resonant charge exchange

Period \ Group	I	II	III	IV	V	VI	VII	VIII		
1	1.008 $1s\ ^2S_{1/2}$ $_1$H Hydrogen 1.000 2.00 6.12 4.82 3.65	$1s^2\ ^1S_0$ 4.003 $_2$He Helium 1.344 2.87 3.4 2.7 2.1								
2	6.491 $2s\ ^2S_{1/2}$ Li Lithium 0.630 0.82 26 22 18	9.012 $2s^2\ ^1S_0$ $_4$Be Berillium 0.828 1.6 13 10 8.2	$2p\ ^2P_{1/2}$ 10.81 $_5$B Boron 0.781 0.88 13 11 8.5	$2p^2\ ^3P_0$ 12.011 $_6$C Carbon 0.910 1.3 10 8.6 6.9	$2p^3\ ^4S_{3/2}$ 14.007 $_7$N Nitrogen 1.034 1.5 7.6 6.2 5.0	$2p^4\ ^3P_2$ 15.999 $_8$O Oxygen 1.000 1.3 8.1 6.6 5.3	$2p^5\ ^2P_{3/2}$ 18.998 $_9$F Fluorine 1.132 1.6 6.0 4.9 4.0	$2p^6\ ^1S_0$ 20.179 $_{10}$Ne Neon 1.259 1.8 4.1 3.3 2.6		
3	22.990 $3s\ ^2S_{1/2}$ $_{11}$Na Sodium 0.615 0.74 31 26 21	24.305 $3s^2\ ^1S_0$ $_{12}$Mg Magnesium 0.750 1.3 18 15 12	$3p\ ^2P_{1/2}$ 26.982 $_{13}$Al Aluminium 0.663 0.61 22 19 15	$3p^2\ ^3P_0$ 28.086 $_{14}$Si Silicon 0.774 1.1 18 15 12	$3p^3\ ^4S_{3/2}$ 30.974 $_{15}$P Phosphorus 0.878 1.6 13 11 9.2	$3p^4\ ^3P_2$ 32.06 $_{16}$S Sulfur 0.873 1.1 12 10 84	$3p^5\ ^2P_{3/2}$ 35.453 $_{17}$Cl Chlorine 0.976 1.8 10 8.4 6.9	$2p^6\ ^1S_0$ 39.948 $_{18}$Ar Argon 1.076 2.0 7.0 5.8 4.7		
4	39.098 $4s\ ^2S_{1/2}$ $_{19}$K Potassium 0.565 0.52 40 34 28	40.08 $4s^2\ ^1S_0$ $_{20}$Ca Calsium 0.670 0.95 25 21 17	44.956 $3d4s^2\ ^2D_{3/2}$ $_{21}$Sc Scandium 0.693 1.1 24 20 16	47.88 $3d^24s^2\ ^3F_2$ $_{22}$Ti Titanium 0.708 1.2 22 19 15	50.942 $3d^34s^2\ ^4F_{3/2}$ $_{23}$V Vanadium 0.704 1.2 23 19 16	51.996 $3d^54s\ ^7S_3$ $_{24}$Cr Chromium 0.705 1.1 22 19 15	54.938 $3d^54s^2\ ^6S_{5/2}$ $_{25}$Mn Manganese 0.739 1.3 20 17 14	55.847 $3d^64s^2\ ^5D_4$ $_{26}$Fe Iron 0.762 1.4 18 15 13	58.933 $3d^74s^2\ ^4F_{9/2}$ $_{27}$Co Cobalt 0.760 1.4 19 16 13	58.69 $3d^84s^2\ ^3F_4$ $_{28}$Ni Nickel 0.749 1.4 20 17 14
4	$3d^{10}4s\ ^2S_{1/2}$ 63.546 $_{29}$Cu Copper 0754 1.3 19 16 13	$4s^2\ ^1S_0$ 65.38 $_{30}$Zn Zinc 0.831 1.7 15 12 10	$4p\ ^2P_{1/2}$ 69.72 $_{31}$Ga Gallium 0.664 0.60 24 20 16	$4p^2\ ^3P_0$ 72.59 $_{32}$Ge Germanium 0.762 1.3 21 18 15	$4p^3\ ^4S_{3/2}$ 74.922 $_{33}$As Arsenic 0.850 1.6 16 13 11	$4p^4\ ^3P_2$ 78.96 $_{34}$Se Selenium 0.847 1.5 16 13 11	$4p^5\ ^2P_{3/2}$ 79.904 $_{35}$Br Bromine 0.932 1.8 12 10 8.6	$4p^6\ ^1S_0$ 83.80 $_{36}$Kr Krypton 1.014 2.1 9.0 7.5 6.2		
5 20 17 14	85.468 $5s\ ^2S_{1/2}$ $_{37}$Rb Rubidium 0.554 0.48 45 38 32	87.62 $5s^2\ ^1S_0$ $_{38}$Sr Strontium 0.647 0.86 29 25 20	88.906 $4d5s^2\ ^2D_{3/2}$ $_{39}$Y Yttrium 0.682 1.0 25 21 18	91.22 $4d^25s^2\ ^3F_2$ $_{40}$Zr Zirconium 0.709 1.2 23 20 16	92.906 $4d^45s\ ^6D_{1/2}$ $_{41}$Nb Niobium 0.711 1.2 23 20 16	95.94 $4d^55s\ ^7S_3$ $_{42}$Mo Molibdenum 0.722 1.2 22 19 15	[98] $4d^55s^2\ ^6S_{5/2}$ $_{43}$Tc Technetium 0.731 1.3 22 18 15	101.07 $4d^75s\ ^5F_5$ $_{44}$Ru Ruthenium 0.736 1.2 21 17 14	102.91 $4d^85s\ ^4F_{9/2}$ $_{45}$Rh Rhodium 0.741 1.2 20 17 14	106.42 $4d^{10}\ ^1S_0$ $_{46}$Pd Palladium 0.783 2.1 18 16 13
5	$4d^{10}5s\ ^2S_{1/2}$ 107.87 $_{47}$Ag Silver 0.:46 1.2 20 17 14	$5s^2\ ^1S_0$ 112.41 $_{48}$Cd Cadmium 0.813 1.6 16 14 11	$5p\ ^2P_{1/2}$ 114.82 $_{49}$In Indium 0.652 0.58 26 22 18	$5p^2\ ^3P_0$ 118.69 $_{50}$Sn Tin 0.735 1.0 23 19 16	$5p^3\ ^4S_{3/2}$ 121.75 $_{51}$Sb Antimony 0.797 1.7 20 17 14	$5p^4\ ^3P_2$ 127.60 $_{52}$Te Tellurium 0.814 1.6 19 16 13	$5p^5\ ^2P_{3/2}$ 126.90 $_{53}$I Iodine 0.876 1.9 16 13 11	$5p^6\ ^1S_0$ 131.29 $_{54}$Xe Xenon 0.944 2.2 12 10 8.2		
6	132.9? $6s\ ^2S_{1/2}$ $_{55}$Cs Cesium 0.535 0.41 51 44 36	137.33 $6s^2\ ^1S_0$ $_{56}$Ba Barium 0.619 0.78 35 29 25	138.90 $5d6s^2\ ^2D_{3/2}$ $_{57}$La Lanthanum 0.640 0.90 32 27 23	178.49 $5d^26s^2\ ^3F_2$ $_{72}$Hf Hafnium 0.740 1.3 22 18 15	180.95 $5d^36s^2\ ^4F_{3/2}$ $_{73}$Ta Tantalum 0.762 1.4 20 17 14	183.85 $5d^46s^2\ ^5D_0$ $_{74}$W Tungsten 0.766 1.4 20 17 14	186.21 $5d^56s^2\ ^6S_{5/2}$ $_{75}$Re Rhenium 0.761 1.4 20 17 14	190.2 $5d^66s^2\ ^5D_4$ $_{76}$Os Osmium 0.801 1.7 18 15 13	192.22 $5d^76s^2\ ^4F_{9/2}$ $_{77}$Ir Iridium 0.816 1.7 17 14 12	195.08 $5d^96s\ ^3D_3$ $_{78}$Pt Platinum 0.812 1.5 17 14 12
6	$5d^{10}6s\ ^2S_{1/2}$ 196.97 $_{79}$Au Gold C.823 1.6 16 14 11	$5d^{10}6s^2\ ^1S_0$ 200.59 $_{80}$Hg Mercury 0.876 1.9 14 12 10	$6p\ ^2P_{1/2}$ 204.38 $_{81}$Tl Thallium 0.670 0.55 24 21 17	$6p^2\ ^3P_0$ 207.2 $_{82}$Pb Lead 0.738 1.1 24 20 17	$6p^3\ ^4S_{3/2}$ 208.98 $_{83}$Bi Bismuth 0.732 1.4 26 22 19	$6p^4\ ^3P_2$ [209] $_{84}$Po Polonium 0.788 1.5 21 18 15	$6p^5\ ^2P_{3/2}$ [210] $_{85}$At Astatine 0.813 1.9 20 17 15	$6p^6\ ^1S_0$ [222] $_{86}$Rn Radon 0.889 2.3 15 12 10		
7	[223] $7s\ ^2S_{1/2}$ $_{87}$Fr Francium 0.542 0.49 53 45 38	226.02 $7s^2\ ^1S_0$ $_{88}$Ra Radium 0.623 0.78 35 30 25	227.03 $6d7s^2\ ^2D_{3/2}$ $_{89}$Ac Actinium 0.636 0.70 32 27 22							

Actinides

232.04 $6d^27s^2\ ^3F_2$ $_{90}$Th Thorium 0.675 0.94 28 24 20	231.04 $5f^26d7s^2\ ^4K_{11/2}$ $_{91}$Pa Protactinium 0.647 0.70 30 25 21	238.03 $5f^36d7s^2\ ^5L_6$ $_{92}$U Uranium 0.670 0.89 28 24 20	237.05 $5f^46d7s^2\ ^6L_{11/2}$ $_{93}$Np Neptunium 0.580 0.70 46 39 33	[244] $5f^67s^2\ ^7F_0$ $_{94}$Pu Plutonium 0.612 0.73 38 32 27	[243] $5f^77s^2\ ^8S_{7/2}$ $_{95}$Am Americium 0.569 0.70 49 42 36

Key: Atomic weight (137.33); Symbol (Ba); Atomic number (56); Element (Barium); Shell of valent electrons ($6s^2$); Electron term ($^1S_0$); Asymptotic parameters (0.619, 0.78); Cross section of resonant charge exchange (in 10^{-15} cm^2 at 0.1, 1, 10 eV in laboratory system) (35, 30, 25).

Figure 14.4. The cross sections of resonant charge exchange for the basic elements of the periodical table.

Lantanides

Element	Atomic number	Symbol	Atomic weight	Shell of valent electrons	Electron term	Asymptotic parameters		Cross section (0.1, 1, 10 eV)		
Cerium	58	Ce	140.12	$4f5d6s^2$	$^1G_4$	0.638	0.88	32	27	23
Praseodymium	59	Pr	140.91	$4f^36s^2$	$^4I_{9/2}$	0.634	0.84	32	28	23
Neodymium	60	Nd	144.24	$4f^46s^2$	$^5I_4$	0.637	0.85	32	27	23
Promethium	61	Pm	[145]	$4f^56s^2$	$^6H_{5/2}$	0.640	0.86	32	27	22
Samarium	62	Sm	150.36	$4f^66s^2$	$^7F_0$	0.644	0.88	31	26	22
Europium	63	Eu	151.96	$4f^76s^2$	$^8S_{7/2}$	0.646	0.89	31	26	22
Gadolinium	64	Gd	157.25	$4f^75d6s^2$	$^9D_2$	0.672	1.0	28	24	20
Terbium	65	Tb	158.92	$4f^96s^2$	$^6H_{15/2}$	0.657	0.93	30	25	21
Dysprosium	66	Dy	162.50	$4f^{10}6s^2$	$^5I_8$	0.661	0.94	29	25	20
Holmium	67	Ho	164.93	$4f^{11}6s^2$	$^5I_{15/2}$	0.665	0.96	29	24	20
Erbium	68	Er	167.26	$4f^{12}6s^2$	$^3H_6$	0.670	0.98	28	24	20
Thulium	69	Tm	168.93	$4f^{13}6s^2$	$^2F_{7/2}$	0.674	0.99	28	23	19
Ytterbium	70	Yb	173.04	$4f^{14}6s^2$	$^1S_0$	0.678	1.0	27	23	19
Lutetium	71	Lu	174.97	$4f^{14}5d6s^2$	$^2D_{3/2}$	0.632	0.92	34	29	24

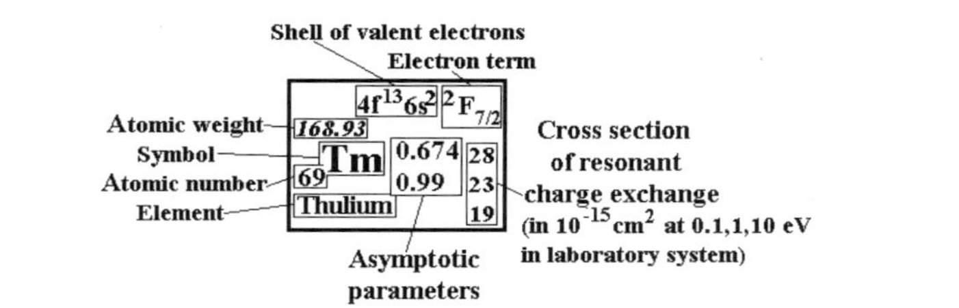

Figure 14.5. The cross sections of resonant charge exchange for lanthanides.

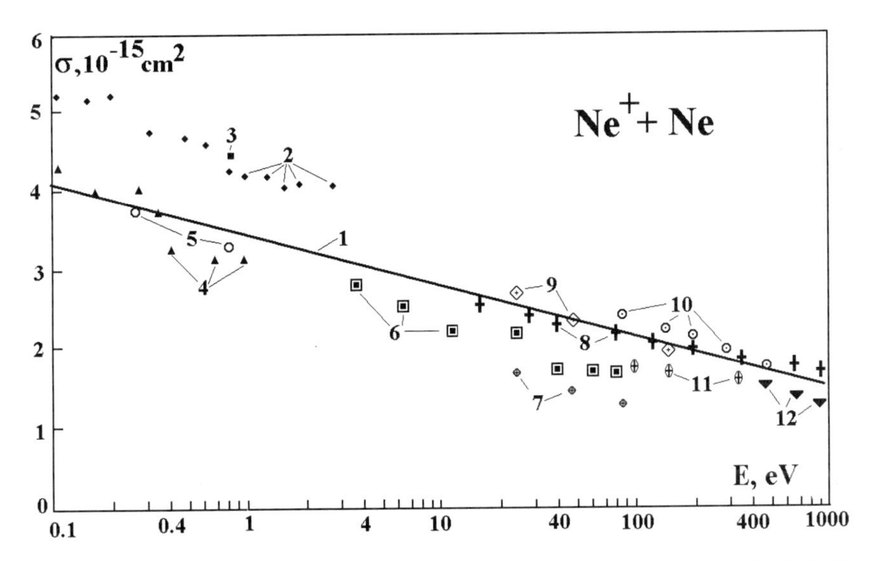

Figure 14.6. The cross sections of resonant charge exchange for neon. 1, Formulas (14.22), (14.27); 2–12, experiment.

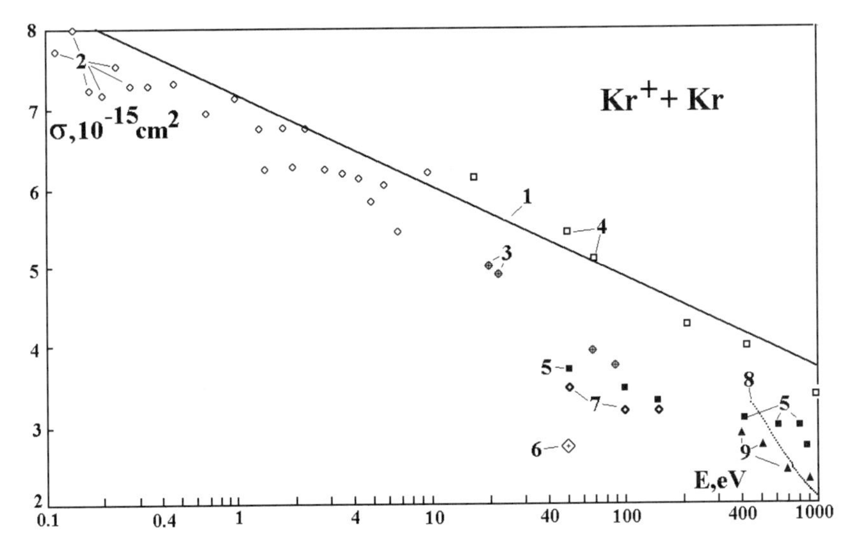

Figure 14.7. The cross sections of resonant charge exchange for krypton. 1, Formulas (14.22), (14.27); 2–9, experiment.

$$\Delta_{3/2}(\theta) = \left(\frac{1}{6} + \frac{1}{2}\cos^2\theta\right)\Delta_{10} + \left(\frac{1}{3} + \frac{1}{2}\sin^2\theta\right)\Delta_{11},$$

where θ is an angle between the molecular and quantization axes. Repeating the operation of the previous section we obtain from this, for the charge exchange

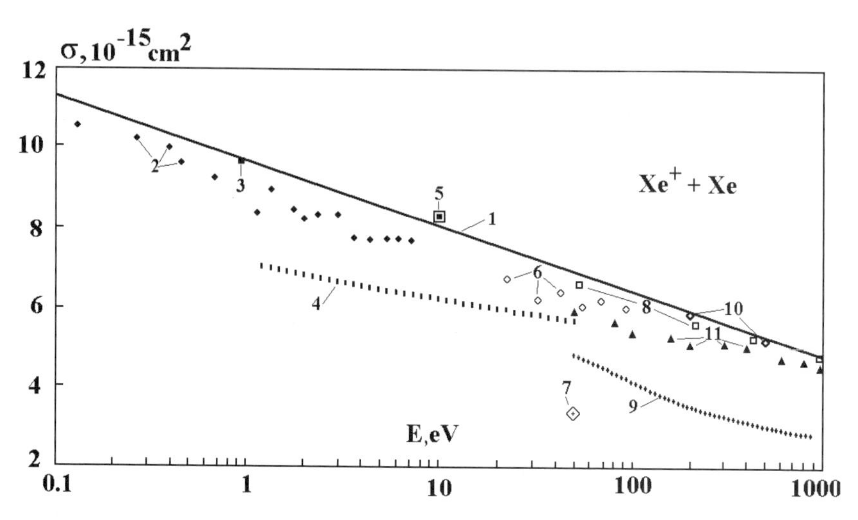

Figure 14.8. The cross sections of resonant charge exchange for xenon. 1, Formulas (14.22), (14.27), 2–11, experiment.

phase when the electron momentum projection is 1/2 and 3/2, respectively,

$$\frac{\zeta_{1/2}(\rho,\vartheta)}{\zeta_o(\rho)} = 1 + \frac{4}{\rho\gamma},$$

$$\frac{\zeta_{3/2}(\rho,\vartheta)}{\zeta_o(\rho)} = \frac{1}{2} + \frac{3}{2}\cos^2\vartheta + \frac{1}{\rho\gamma}\left(\frac{1}{2} + \frac{9}{2}\sin^2\vartheta + \frac{3}{2}\sin^2\vartheta\cos^2\varphi\right),$$

where $\zeta_o(\rho)$ is the charge exchange phase for a transferring s-electron with the same asymptotic parameters. On the basis of these phases of charge exchange, with the use of the method of the previous section, one can evaluate the cross sections for the resonant charge exchange process. The reduced quantities of the cross sections are given in Table 14.3 depending on the asymptotic parameter $R_o\gamma$. Table 14.4 represents the cross sections of this process for rare gases, i.e., $\overline{\sigma}$ is the average cross section for case "a" of the Hund coupling, $\sigma(\vartheta = 0)$ is the cross section of this process when the projection of the orbital momentum of the hole on the impact parameter direction is zero, and $\sigma_{1/2}$ and $\sigma_{3/2}$ are the charge exchange cross sections for the total ion momenta 1/2 and 3/2, respectively. We take the atom ionization potential with formation of the ion in different fine states to be identical, so that the difference in the cross sections under consideration is determined by the process dynamics. According to the data of Table 14.4, the difference of the average cross sections for both coupling schemes is small, and is lower than the accuracy of the cross sections evaluated which, in turn, is determined by the accuracy of the asymptotic coefficients of the wave function of an atomic valent electron. Hence, in spite of a significant dependence of the cross section of resonant charge exchange on the direction of the orbital momentum, the average cross section of this process is not sensitive to the scheme of coupling of electron momenta if the process is

Table 14.4. The resonant charge exchange cross sections for rare gases at the ion energy of 1 eV.

Element	Ne	Ar	Kr	Xe	Rn
γR_o	10.8	12.4	13.2	14.1	15.0
$\overline{\sigma}/\sigma(\vartheta = 0)$	0.85	0.867	0.87	0.87	0.88
$\sigma_{1/2}/\overline{\sigma}$	1.02	1.02	1.02	1.02	1.02
$\sigma_{1/2}/\sigma_{3/2}$	0.995	0.995	0.995	0.995	0.995
$\overline{\sigma}$, 10^{-15} cm^2	3.5	5.8	7.5	10	12

permitted in a one-electron approach. Hence, the cross sections averaged over the initial states weakly depend on the coupling scheme for atoms of Groups 3 and 8.

According to the data of Table 14.3, the difference between the cross sections of this process for different coupling schemes is significant for other groups of the periodic system of elements. This difference becomes dramatic for certain initial atomic and ionic states if the one-electron transition is forbidden according to the selection rules for one of these coupling schemes (see formula (10.60) and Tables 10.9, 11.8, and 11.9).

14.5 Spin Exchange

The spin exchange process is analogous to the resonant charge exchange process. We consider the case of the collision of two atoms with spin 1/2, as hydrogen or alkali metal atoms. If the spins of colliding atoms have different directions, the process corresponds to a change by valent electrons, i.e., the process proceeds according to the scheme

$$A\downarrow\, + B\uparrow\, \to A\uparrow\, + B\downarrow, \tag{14.29}$$

where the arrows indicate direction of spin. This process leads to transitions between the states of superfine structure. Indeed, let us consider a gas consisting of hydrogen or alkali metal atoms, and that the nuclear spin of the colliding atoms is equal to I. Then the total atom spin equals $F = I \pm 1/2$. Let the cross section of the spin exchange process be $\sigma_{\rm ex}$, and let the total momentum of both colliding atoms be equal to F. Then the cross section of the transition between the superfine states of atoms is

$$\sigma(F \to F') = 2 \cdot \frac{1}{2} \cdot \frac{2F'+1}{4I+2}\sigma_{\rm ex} = \frac{2F'+1}{4I+2}\sigma_{\rm ex}. \tag{14.30}$$

Here the factor 2 takes into account that two atoms with total spin F partake in the collisions, the next factor is the probability that the colliding atoms have opposite directions of electron spins, and the factor $(2F'+1)/(4I+2)$ is the probability of transition of each colliding atom in a new superfine state. Because this process leads to the decay of the initial superfine state of a colliding atom, it determines the number density of atoms in masers which operate on transitions between states of superfine structure (hydrogen and rubidium masers).

The physical nature of this process is similar to that of the resonant charge exchange process. Indeed, the wave function of the initial state in accordance with the notations of Chapter 10, has the form $\Psi_1 = \psi(1a)\psi(2b)$. This is the combination of the eigenwave functions of the quasi-molecule consisting of colliding atoms, i.e., it is a sum of the symmetrical and antisymmetric wave functions with respect to exchange of electrons. Therefore the interference of the corresponding eigenstates leads to exchange by electrons. By analogy with resonant charge exchange, the cross section of this process is given by formula (14.15),

$$\sigma_{\mathrm{ex}} = \frac{\pi R_o^2}{2}, \quad \text{where} \quad \zeta(R_o) = \int_{-\infty}^{\infty} \frac{\Delta(R) R\, dR}{2v\sqrt{R^2 - R_o^2}} dt = 0.28. \qquad (14.31)$$

Here the exchange interaction potential of atoms $\Delta(R)$ is determined by formula (10.28) as the difference between the energies of the symmetric and antisymmetric states of the systems of two atoms, and is given by formulas (10.31) and (10.32). Table 14.5 contains values of the spin exchange cross sections which correspond to the energy of collision of about 0.1 eV. Note that formula (14.31) for the cross section of the spin exchange process uses the small parameter $1/(2\gamma R_o) \ll 1$. Values of the parameter γR_o are given in Table 14.5 at a given collision energy.

14.6 Excitation Transfer

Excitation transfer from one excited atom to the other proceeds according to the scheme

$$A^* + B \to A + B^*. \qquad (14.32)$$

If an excited state is forbidden for the dipole radiation transition from the ground state, the cross section of this process is determined by the exchange interaction whose expression in the limiting case is given by formula (10.46). Below we consider in detail the following process:

$$A(P) + A(S) \to A(S) + A(P). \qquad (14.33)$$

Table 14.5. The cross sections of spin exchange σ_{ex} at the collision energy 0.1 eV in the laboratory coordinate system are expressed in 10^{-15} cm^2.

Colliding atoms	σ_{ex}	γR_o
$H - H$	2.0	6.8
$Na - Na$	11	10
$K - K$	15	10
$Rb - Rb$	17	11
$Cs - Cs$	19	11

This process is of interest for two reasons. First, the cross section of this process is high enough and, hence, it can influence the character of the interaction of resonant radiation with these atoms located in a gaseous system. Second, there is an interaction between this process and the rotation of the electron orbital momentum, so that in the example of this process one can ascertain peculiarities of this general problem.

The interaction operator in this case is determined by formula (10.7),

$$V = \frac{\mathbf{D}_1\mathbf{D}_2 - 3(\mathbf{D}_1\mathbf{n})(\mathbf{D}_2\mathbf{n})}{R^3},$$

where $\mathbf{D}_1$; $\mathbf{D}_2$ are the operators of the dipole moment for the first and second atoms, respectively, $\mathbf{n}$ is the unit vector directed along the molecular axis, and R is the distance between atoms. This transition proceeds without exchange by electrons and hence takes place at large distances between atoms. This character of the transition allows one to make an estimate of cross section. The probability of transition is determined by a shift of phases between the wave functions of different eigenstates similar to the case of the charge exchange process. Then the cross section of the transition is estimated as $\sigma_{\text{et}} \sim R_o^2$, where $\int \langle S|V|P\rangle dt \sim 1$. This gives

$$\sigma_{\text{et}} \sim \frac{d^2}{v}, \tag{14.34}$$

v is the collision velocity, $d = \langle S|D_z|P_z\rangle$, where P_z means the state with zero projection of the orbital electron momentum onto the z-axis. This cross section is high enough. In particular, for thermal collisions ($v \sim 10^{-4}, d \sim 1$) this cross section is estimated as $\sigma_{\text{et}} \sim 10^{-11}\,\text{cm}^2$.

For determination of the cross sections of partial transitions it is necessary to solve the Schrödinger equation for a suitable coordinate frame and basis wave functions. Let us use in this case the motionless frame of axes where the x-axis is directed along the velocity vector and the y-axis is directed along the impact parameter vector. It is convenient to take, as basis wave functions, the product of the wave functions of the noninteracting atoms ψ_k. This is similar to the representation of the Hamiltonian of the system of colliding atoms in the form $\hat{H} = \hat{H}_o + V$, where $\hat{H}_o$ describes the noninteracting atoms, so that $\hat{H}_o\psi_k = E_o\psi_k$ and E_o is the energy of the noninteracting atoms. Substituting the expansion of the wave function of the colliding atoms Ψ over the basis wave functions

$$\Psi = \sum_k c_k \psi_k e^{-iE_o t}$$

in the Schrödinger equation $i(\partial\Psi/\partial t) = \hat{H}\Psi$ and restricted by only the related states, we obtain the following set of equations for the amplitudes c_k:

$$i\dot{c}_k = \sum_m V_{mk} c_m. \tag{14.35}$$

In the related problem we have an additional symmetry with respect to the location of excitation. We separate the states of different symmetry by introducing symmetric basis functions by the formula

$$\Psi_{+} = \frac{1}{\sqrt{2}}\left[\varphi(1s)\psi(2p_i) + \psi(1p_i)\varphi(2s)\right],$$

where the wave functions φ, ψ correspond to the S- and P-atom states, respectively, the number in the wave function argument indicates to which atom this wave function corresponds, and the subscript i marks the projection of the P-atom momentum onto a given direction. In the same way we have, for the basis wave function of an antisymmetric state,

$$\Psi_{-} = \frac{1}{\sqrt{2}}\left[\varphi(1s)\psi(2p_i) - \psi(1p_i)\varphi(2s)\right].$$

The set of equations (14.35) is separated into two independent sets for c_i^+ and c_i^-; each of the obtained sets includes three equations. Let us consider the general properties of the S-matrix whose element S_{ik} is introduced as the probability amplitude $c_i(t=\infty)$ for a final state i under the initial condition $c_j(t=-\infty) = \delta_{jk}$. Since we have $V_{ik}^+ = -V_{ik}^-$, this gives $S_{ik}^- = \left(S_{ik}^+\right)^*$. The scattering matrix for excitation transfer is equal to

$$S_{ik}^{\text{et}} = \frac{1}{2}\left(S_{ik}^+ - S_{ik}^-\right) = i\,\text{Im}\,S_{ik}^+. \tag{14.36}$$

The scattering matrix for elastic scattering is

$$S_{ik}^{\text{el}} = \frac{1}{2}\left(S_{ik}^+ + S_{ik}^-\right) = \text{Re}\,S_{ik}^+.$$

Let us separate the state with zero momentum projection onto the direction which is perpendicular to both the velocity and impact parameter of collision. This state does not mix with other states, so that the equation for the probability amplitude of this state has the form

$$i\frac{dc_z^+}{dt} = \frac{d^2}{R^3}c_z^+,$$

where $d = \langle S|D_z|P_z\rangle$ and P_z means the state with zero projection of the orbital electron momentum onto the axis z. The solution of this equation gives

$$S_{zz}^+(\rho) = \exp\left(-i\int_{-\infty}^{\infty}\frac{d^2}{R^3}\,dt\right) = \exp\left(-i\frac{2d^2}{v\rho^2}\right).$$

This leads to the following expressions for the cross sections of the corresponding processes:

$$\sigma_{zz}^{\text{et}} = \int_0^\infty 2\pi\rho\,d\rho\left[\text{Im}\,S_{zz}^+\right]^2 = \frac{\pi^2 d^2}{v}, \qquad \sigma_{zz}^{\text{el}} = \int_0^\infty 2\pi\rho\,d\rho\left[\text{Re}\,S_{zz}^+\right]^2 = \frac{\pi^2 d^2}{v}. \tag{14.37}$$

In the case of the states with orbital momentum located in the motion plane, the processes of excitation transfer and depolarization are mixed. We give results from numerical solutions of the set of equations in this case. The cross sections of excitation transfer are equal to, in units $\sigma_o = \pi d^2/v$,

$$\sigma_{xx}^{\rm et} = 2.65\sigma_o, \qquad \sigma_{yy}^{\rm et} = 0.43\sigma_o, \qquad \sigma_{xy}^{\rm et} = \sigma_{yx}^{\rm et} = 0.56\sigma_o.$$

From this it follows that, for the average cross section of excitation transfer,

$$\sigma_{\rm et} = \frac{1}{3}\sum_{i,k}\sigma_{ik}^{\rm et} = 2.26\frac{\pi d^2}{v}, \tag{14.38a}$$

the average cross section of elastic scattering is equal to

$$\sigma_{\rm el} = 2.58\frac{\pi d^2}{v}. \tag{14.38b}$$

This yields, for the total cross section of the scattering of atoms, $\sigma_t = \sigma_{\rm et} + \sigma_{\rm el} = 4.8\pi d^2/v$.

The analysis of the above process allows us to formulate a general approach to resonance processes. The character of transitions is determined by the interference of some states, and the resonant processes are characterized by large cross sections. This means that the transition proceeds at large distances between the colliding particles. Then one can neglect the interaction with states which do not partake in the process. Thus, the problem is reduced to the analysis of interaction of several states at large distances between the colliding atomic particles where this interaction is weak. But the processes of transition between states of a related structure are mixed with the processes of rotation of momenta of colliding particles which is a general feature of the resonant processes of collisions of atomic particles with nonzero momenta.

14.7 The Matching Method

The above analysis shows that transitions between resonant states at large distances between colliding atomic particles are accompanied by the processes of rotation of momenta of colliding particles. Now we consider this problem in detail and give a commonly used method of an approximate solution to the problem. First, let us consider the above problem of excitation transfer from the P-atom to the parent S-atom. The set of equations (14.35), in the frame of axes which is connected to the molecular axis, has the form (in atomic units)

$$i\frac{dc_x^+}{dt} = -\frac{2d^2}{R^3}c_x^+ - i\frac{d\theta}{dt}c_y^+, \qquad i\frac{dc_y^+}{dt} = \frac{d^2}{R^3}c_y^+ + i\frac{d\theta}{dt}c_x^+,$$

where $d\theta/dt = -\rho v/R^2$ is the rate of rotation of the molecular axis. Introducing the new variables

$$a_{x,y}^+ = c_{x,y}^+ \exp\left(-i\int^t \frac{d^2}{2R^3}dt'\right),$$

we transform the set of equations to the symmetric form

$$i\frac{da_x^+}{dt} = -\frac{3d^2}{2R^3}a_x^+ + i\frac{\rho v}{R^2}a_y^+, \qquad i\frac{da_y^+}{dt} = \frac{3d^2}{2R^3}a_y^+ - i\frac{\rho v}{R^2}a_x^+. \tag{14.39}$$

Let us compare this set with the set of equations in the motionless frame of axes which includes vectors **v** and ρ (see Fig. 14.1) and can be obtained from this one by introducting the new probability amplitudes

$$b_x^+ = a_x^+ \cos\theta + a_y^+ \sin\theta, \qquad b_y^+ = -a_x^+ \sin\theta + a_y^+ \cos\theta.$$

The obtained set of equations has the form

$$i\frac{db_x^+}{dt} = -\frac{3d^2}{2R^3}\cos 2\theta \cdot b_x^+ - \frac{3d^2}{2R^3}\sin 2\theta \cdot b_y^+,$$
$$i\frac{db_y^+}{dt} = \frac{3d^2}{2R^3}\cos 2\theta \cdot b_y^+ - \frac{3d^2}{2R^3}\sin 2\theta \cdot b_x^+. \tag{14.40}$$

From the set of equations (14.39) it follows that at large distances between atoms $R \gg d^2/(\rho v)$ case "d " of the Hund coupling (see Table 11.1) is realized, so that the rotation energy of the molecular axis exceeds the Σ–Π splitting of levels. Hence, in this region one can neglect the Σ–Π splitting of levels, i.e., transitions between states are absent in the motionless frame of coordinates. On the contrary, in the region $R \ll d^2/(\rho v)$, where case "b" of the Hund coupling is valid, one can neglect the rotation energy compared to the Σ–Π splitting, so that in this region the probability amplitudes $b_{x,y}^+$ vary only their phases. If we suppose that the transition region between these limiting cases is narrow, one can join solutions for the probability amplitude in boundary points. As a result, the scattering matrix would be expressed through both Σ–Π splitting of levels and through the rotation angle of the molecular axis when case "b" of the Hund coupling is valid. This way of accounting for the rotation of molecular axis during the collisions of atomic particles is called the matching method. This approach is more suitable for a stronger dependence of the Σ–Π splitting of levels on distances between atoms than we have in the case under consideration. Below we consider the matching method in a general form for collision of the S- and P-atoms.

Let us start from the set of equations (14.35) for the probability amplitudes, and for symmetrization of this set of equations introduce the probability amplitudes by

$$a_{x,y} = c_{x,y} \exp\left(-i\int^t \frac{V_0 + V_1}{2}dt'\right),$$

where V_0, V_1 are the interaction potentials for the momentum projections 0 and 1 onto the molecular axis. Then we have the following set of equations by analogy with (14.39):

$$i\frac{da_x}{dt} = \Delta a_x - i\frac{d\theta}{dt}a_y, \qquad i\frac{da_y}{dt} = -\Delta a_y + i\frac{d\theta}{dt}a_x, \qquad i\frac{da_z}{dt} = -\Delta a_z, \tag{14.41}$$

where $\Delta = (V_0 - V_1)/2$. Here the x-axis is directed along the molecular axis and the y-axis is perpendicular to it and is found in the motion plane. It is convenient to

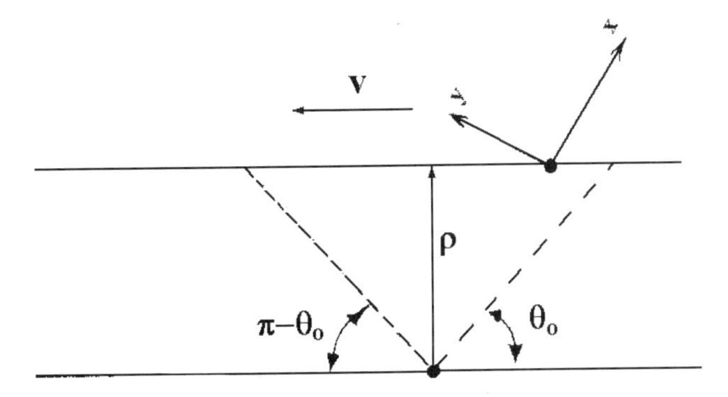

Figure 14.9. The geometry of collision of free moving atoms.

introduce the motionless frame of axes whose axes are directed along the collision velocity **v** and the impact parameter ρ of collision (see Fig. 14.9), so that in the related case of the P-atom the transition from one to the other frame of axes leads to the following transformations of the wave functions:

$$\psi_x = -\cos\theta\,\psi_v + \sin\theta\,\psi_\rho, \qquad \psi_y = \sin\theta\,\psi_v + \cos\theta\,\psi_\rho.$$

The inverse transformation has the form

$$\psi_v = -\cos\theta\,\psi_x + \sin\theta\,\psi_y, \qquad \psi_\rho = \sin\theta\,\psi_x + \cos\theta\,\psi_y.$$

This leads to the following relations between the probability amplitudes

$$a_x = -a_v\cos\theta + a_\rho\sin\theta, \qquad a_y = a_v\sin\theta + a_\rho\cos\theta, \tag{14.42a}$$

and

$$a_v = -a_x\cos\theta + a_y\sin\theta, \qquad a_\rho = a_x\sin\theta + a_y\cos\theta. \tag{14.42b}$$

From this we have the set of equations in the motionless frame of axes

$$i\frac{da_v}{dt} = \Delta\cos 2\theta a_v - \Delta\sin 2\theta a_\rho, \qquad i\frac{da_\rho}{dt} = -\Delta\sin 2\theta a_v - \Delta\sin 2\theta a_\rho. \tag{14.43}$$

Now let us solve the set of equations (14.41) within the framework of the matching method. Let us introduce the initial conditions $a_v(t = -\infty) = A$, $a_\rho(t = -\infty) = B$ which correspond to the probability amplitudes in the rotational frame of axes $a_x(t = -\infty) = -A$, $a_y(t = -\infty) = B$. Introduce the characteristic angle θ_o such that at this angle $\Delta = d\theta/dt$, i.e., the distance between atoms $R_m(\theta_o)$ at this angle satisfies the relation

$$\Delta(R_m) = \frac{\rho v}{R_m^2}. \tag{14.44a}$$

In the region $\theta \le \theta_o$ one can neglect the phases which are gained by the wave functions and probability amplitudes. Indeed, a typical phase shift resulting from this region according to equations (14.43) is estimated as $\eta \sim \int \Delta\, dt \ll \int (d\theta/dt)\, dt \sim \theta_o \sim 1$, i.e., $\eta \ll 1$. Thus we obtain, for the probability amplitudes in the motionless and rotational frames of axes on the boundary of this region $\theta = \theta_o$:

$$a_v = A, \quad a_\rho = B; \qquad a_x = -A\cos\theta_o + B\sin\theta_o, \qquad a_y = A\sin\theta_o + B\cos\theta_o.$$

In the next region, $\theta_o \le \theta \le \pi - \theta_o$, one can neglect the rotation so that the amplitudes a_x, a_y, in accordance with the set of equations (14.41), gain the phases $-\zeta_o$ and ζ_o, respectively, where $\zeta_o = \int \Delta\, dt$ and the integral is taken over this region. Below, for simplicity, we define the phase ζ_o as

$$\zeta_o = \int_{-\infty}^{\infty} \Delta\, dt, \tag{14.44b}$$

because times beyond this region do not contribute to this integral. Thus we have, on the basis of the second boundary at $\theta = \pi - \theta_o$,

$$a_x = -Ae^{-i\zeta_o}\cos\theta_o + Be^{-i\zeta_o}\sin\theta_o, a_y = Ae^{i\zeta_o}\sin\theta_o + Be^{i\zeta_o}\cos\theta_o.$$

This gives, for amplitudes in the motionless frame of axes at the second boundary,

$$a_v = A(i\sin\zeta_o - \cos\zeta_o\cos 2\theta_o) + B\sin 2\theta_o\cos\zeta_o,$$
$$a_\rho = -A\sin 2\theta_o\cos\zeta_o - B(i\sin\zeta_o + \cos\zeta_o\cos 2\theta_o).$$

The last region $\theta \ge \pi - \theta_o$ is symmetric with respect to the first region. In this region the amplitudes a_v, a_ρ do not vary.

Finally, we obtain, for the S-matrix by taking into account that the amplitudes c_i differ from amplitudes a_i by the factor $\exp\left[-i\int(V_o + V_1)\,dt/2\right]$,

$$S_{ik} = e^{-i\int^t \frac{V_o+V_1}{2}dt'}\begin{pmatrix} i\sin\zeta_o - \cos\zeta_o\cos 2\theta_o & -\sin 2\theta_o\cos\zeta_o \\ \sin 2\theta_o\cos\zeta_o & -i\sin\zeta_o - \cos\zeta_o\cos 2\theta_o \end{pmatrix}. \tag{14.45}$$

Now let us take into account the state with zero momentum projection onto the momentum of motion which is directed along the z-axis which is perpendicular to the motion plane. From the last equation of the set (14.41) it follows that this state develops independently of the other states and the corresponding probability amplitude is $a_z = a_z(-\infty)e^{i\zeta_o}$. Including this state in consideration we obtain, for the scattering matrix,

$$S_{ik} = e^{-i\int^t[(V_o+V_1)/2]\,dt'} \times \begin{pmatrix} i\sin\zeta_o - \cos\zeta_o\cos 2\theta_o & -\sin 2\theta_o\cos\zeta_o & 0 \\ \sin 2\theta_o\cos\zeta_o & -i\sin\zeta_o - \cos\zeta_o\cos 2\theta_o & 0 \\ 0 & 0 & e^{i\zeta_o} \end{pmatrix}. \tag{14.46}$$

One can see that, in the cases $\zeta_o = 0$ and $\theta_o = \pi/2$, the internal matrix contains only diagonal matrix elements which are equal to one, so that $S_{ik} = \delta_{ik}$. In this case transitions between the states are absent. In the case $\theta_o = \pi/2$ the internal matrix contains diagonal matrix elements $e^{-i\zeta_o}$ and $e^{i\zeta_o}$ which means that transitions between the states are absent but that the wave functions gain different phases depending on the energy of the corresponding state.

Above we use the peculiarity of the P-state whose basis can be composed from three states with zero momentum projection onto each axis of the frame of coordinates. Now let us formulate the above problem within the framework of the quantum numbers connected with one direction, which is the z-axis directed perpendicular to the motion plane. First we obtain the expression for the scattering matrix for the P-state on the basis of (14.46). Then we express the eigenwave function Φ_i corresponding to the z-axis through the above wave functions. Here i is the momentum projection onto the z-axis, and these wave functions are connected with the old functions by the relations

$$\Phi_0 = \psi_z, \qquad \Phi_1 = \frac{1}{\sqrt{2}}(-\psi_v + i\psi_\rho), \qquad \Phi_{-1} = \frac{1}{\sqrt{2}}(-\psi_v - i\psi_\rho).$$

From this we obtain the connection between elements of the S-matrix:

$$\begin{aligned} S_{00} &= S_{zz}, \\ S_{11} &= \frac{1}{2}(S_{vv} + S_{\rho\rho}) + \frac{i}{2}(S_{v\rho} - S_{\rho v}), \\ S_{-1,-1} &= \frac{1}{2}(S_{vv} + S_{\rho\rho}) - \frac{i}{2}(S_{v\rho} - S_{\rho v}), \\ S_{1,-1} &= \frac{1}{2}(S_{vv} - S_{\rho\rho}) + \frac{i}{2}(S_{v\rho} + S_{\rho v}), \\ S_{-1,1} &= \frac{1}{2}(S_{vv} - S_{\rho\rho}) - \frac{i}{2}(S_{v\rho} + S_{\rho v}), \end{aligned} \tag{14.47}$$

where the matrix element S_{ik} corresponds to the transition $i \to k$. Then the scattering matrix for the related quantum numbers are given in Table 14.6 where the initial states correspond to columns and the final states correspond to rows.

Let us find elements of the scattering matrix using the quantum numbers with respect to the z-axis. Let us analyze transformations of the wave function corresponding to one momentum projection on the z-axis resulting from the above operations. If the initial wave function is Φ_1, rotation of the frame of axes by angle θ_o transforms it, in a new frame of axes, into $\Psi = \Phi_1 e^{i\theta_o}$. Next, transfer to the molecular axis by using the relations

$$\Phi_1 = \frac{1}{\sqrt{2}}\left(\psi_x + i\psi_y\right), \qquad \Phi_{-1} = \frac{1}{\sqrt{2}}\left(\psi_x - i\psi_y\right),$$

Table 14.6. The elements of S-matrix for collisions of atoms in the S- and P-states within the framework of the matching method. The parameters θ_o and ζ_o are determined by formulas (14.44).

	-1	0	1
-1	$\cos\zeta_o \cdot e^{-2i\theta_o}$	0	$i\sin\zeta_o$
0	0	$e^{i\zeta_o}$	0
1	$i\sin\zeta_o$	0	$-\cos\zeta_o \cdot e^{2i\theta_o}$

where the wave functions ψ_x and ψ_y describe the states with zero momentum projection on the molecular x- and y-axes which are located in the motion plane and are perpendicular to the z-axis. The reversal transformation gives

$$\psi_x = \frac{1}{\sqrt{2}}(\Phi_1 + \Phi_{-1}), \qquad \psi_y = -i\frac{1}{\sqrt{2}}(\Phi_1 - \Phi_{-1}).$$

After rotation of the molecular axis from the angle θ_o up to $\pi - \theta_o$ the wave functions ψ_x, ψ_y gain phases $-\zeta_o$ and ζ_o, respectively, so that the wave function has the form

$$\Psi = \frac{1}{\sqrt{2}}\left(\psi_x e^{-i\zeta_o} + i\psi_y e^{i\zeta_o}\right)e^{i\theta_o} = \Phi_1 e^{i\theta_o}\cos\zeta_o - i\Phi_{-1}e^{i\theta_o}\sin\zeta_o.$$

The last operation which returns us to the initial frame of axes is the rotation around the z-axis by the angle $\pi - \theta_o$ in the opposite direction. Then, finally, we obtain, for the wave function,

$$\Psi = \Phi_1 e^{i(2\theta_o - \pi)}\cos\zeta_o + i\Phi_{-1}\sin\zeta_o.$$

This gives, for elements of the scattering matrix,

$$S_{1,1} = -e^{i2\theta_o}\cos\zeta_o, \qquad S_{1,-1} = i\sin\zeta_o, \tag{14.48}$$

in accordance with the data in Table 14.6.

Now let us formulate the matching method in the general form on the basis of the above transformation of wave functions. Let us introduce the rotation function $D^j_{mm'}(\alpha, \beta, \gamma)$ which describes the transformation of the wave function as a result of rotation on the Euler angles α, β, γ. Here j is the momentum and m, m' are the momentum projections onto the z-axis in a given frame of axes. The wave function ψ_{jm} in a new frame of axes obtains the form

$$\psi_{jm} = \sum_{m'} D^j_{mm'}(\alpha, \beta, \gamma)\psi_{jm'}.$$

The rotation function (or Wigner D-function) $D^j_{mm'}(\alpha, \beta, \gamma)$ describes the following sequence of rotations. The first operation is the rotation around the z-axis by angle α, the second operation is the rotation around the new y-axis by angle β, and the third transformation of the frame of axes is the rotation around the new z-axis by angle γ. Taking into account the change of phases for the wave functions during a strong coupling with the molecular axis we obtain, in this case for the scattering matrix,

$$S^j_{m\mu} = \sum_{\Lambda} D^j_{\mu\Lambda}(0, -\pi/2, -\pi + \theta_o)e^{i\zeta_\Lambda} D^j_{m\Lambda}(\theta_o, \pi/2, 0),$$

where ζ_Λ are the phases which are gained by the wave functions of states with the momentum projection Λ onto the molecular axis. The general properties of the rotation functions are

$$D^j_{mm'}(\alpha, \beta, \gamma) = \exp(-i\alpha m)d^j_{mm'}(\beta)\exp(-im'\gamma).$$

Using standard notations $d^j_{mm'}(\pi/2) = \Delta^j_{mm'}$ and taking into account the direction of rotations we obtain, for the scattering matrix in the case under consideration,

$$S^j_{m\mu} = e^{i(m+\mu)\theta_0 - i\mu\pi} \sum_{\Lambda} e^{i\zeta_\Lambda} \left(\Delta^j_{\Lambda\mu}\right)^* \Delta^j_{\Lambda m}. \tag{14.49}$$

14.8 Depolarization of Atoms in Collisions

The S-matrix of atomic transitions gives total information about the processes of collisions of the P-atom with the S-atom within the framework of the matching method. Let us use this S-matrix for determination of the cross section of the depolarization of the P-atom as a result of collision with the S-atom. We take the quantization axis onto which the momentum projection of this atom is zero, and this axis forms the polar angles α and φ with the motionless frame of axes related to the collision. Then the initial wave function of the P-atom is expressed in the following way, through the wave functions with zero projection of the atom momentum onto axes of the motionless frame of coordinates,

$$\Psi = \psi_z \cos\alpha - \psi_v \sin\alpha \cos\varphi + \psi_\rho \sin\alpha \sin\varphi.$$

This gives, for the probability amplitude of survival of the initial atom state,

$$\begin{aligned}\left\langle \Psi^* \hat{S} \Psi \right\rangle &= S_{zz} \cos^2\alpha + S_{xx} \sin^2\alpha \cos^2\varphi + S_{yy} \sin^2\alpha \sin^2\varphi \\ &\quad - (S_{xy} + S_{yx}) \sin^2\alpha \cos\varphi \sin\varphi.\end{aligned}$$

From this follows the expression for the probability averaging on the initial direction of the momentum with respect to collision (we take into account $S_{xy} = -S_{yx}$):

$$\begin{aligned}P = 1 - \left|\left\langle \Psi^* \hat{S} \Psi \right\rangle\right|^2 &= 1 - \frac{1}{5}\left(|S_{xx}|^2 + \left|S_{yy}\right|^2 + |S_{zz}|^2\right) - \frac{1}{15}|S_{yx} + S_{xy}|^2 \\ &- \frac{1}{15}\left[S^*_{xx}(S_{yy} + S_{zz}) + S^*_{yy}(S_{xx} + S_{zz}) + S^*_{zz}(S_{xx} + S_{yy})\right].\end{aligned} \tag{14.50}$$

This leads to a cumbersome expression for the probability of transition from a given state

$$P = \frac{8}{15} + \frac{4}{15}\cos^2\zeta_o + \frac{4}{15}\cos 2\theta_o \cos^2\zeta_o - \frac{8}{15}\cos^2 2\theta_o \cos^2\zeta_o. \tag{14.51}$$

As is seen, this expression includes the phases $\pm\zeta_o$ which are gained by the wave functions during a strong interaction of states with the molecular axis, and the rotation angle θ_o which characterizes a size of this region. Let us consider the case of a sharply varied interaction potential as a function of a distance between particles, because the matching method is valid only for this case. Then the phase ζ_o is high enough, so that averaging over this phase gives $\left\langle\cos^2\zeta_o\right\rangle = 1/2$, and the probability of transition from the initial state according to formula (14.51) is

$$P = \frac{2}{3} + \frac{2}{15}\cos 2\theta_o - \frac{4}{15}\cos^2 2\theta_o.$$

Taking into account a strong dependence $\Delta(R)$ in comparison with a power dependence and accounting for $\sin\theta_o = \rho/R_m$ we obtain, for the cross section of the depolarization process,

$$\sigma_{\text{dep}} = \int_0^\infty P \cdot 2\pi\rho\, d\rho = \frac{26}{45}\pi R_m^2, \tag{14.52a}$$

where the matching radius is given by the relation:

$$\left|\frac{V_o(R_m) - V_1(R_m)}{2}\right| = \frac{v}{R_m}. \tag{14.52b}$$

In particular, in the case when $V_o(R) - V_1(R) = C/R^6$ this formula gives

$$\sigma_{\text{dep}} = 0.76\pi\left(\frac{C}{v}\right)^{2/5}, \tag{14.53}$$

while the numerical evaluation gives the factor 0.78. Thus, along with the possibility of extracting the main features of the process, the matching method is capable of giving reliable numerical evaluations for the yield parameters of the processes.

14.9 Relaxation of Atomic Momentum in Isotropic Collisions

The above consideration includes only one aspect of the relaxation of an atom with nonzero momentum as a result of collisions with a gas of isotropically moving atoms. Let us consider this problem in a general form within the framework of the formalism of the density matrix which has the form, in this case,

$$\rho_{mm'}^j = \psi_{jm}^* \psi_{jm'},$$

where j is the atom moment, and m, m' are its projections onto a given direction. In particular, for the case $m = m'$, the density matrix $\rho_{mm}^j = |\psi_m|^2$ gives the population of a given state. The formalism of the density matrix allows one to analyze some coherent phenomena resulting from collisions involving the related atom.

Because of a random distribution of gaseous atoms on the directions of motion, the averaged density matrix is isotropic which simplifies the consideration of these processes. Then it is convenient to introduce the polarization atomic momenta

$$\rho_{\kappa q}^j = \sum_{m,m'} \begin{bmatrix} j & j & \kappa \\ m & -m' & q \end{bmatrix} (-1)^{j-m'} \rho_{mm'}^j,$$

so that the parameter κ varies from 0 to $2j$, and the parameter q varies from $-\kappa$ to κ. The value $\rho_{\kappa q}^j$ characterizes the multipole atom momentum of rank 2^κ, and q corresponds to a spherical projection of this momentum. For example, ρ_{00}^j determines the population of the related state, the values $\rho_{1,-1}^j$, $\rho_{1,0}^j$, and $\rho_{1,1}^j$ are components of a vector which characterizes the atom orientation, and five values of the second rank ρ^j_{2q} are responsible for the atom alignment and are proportional to components of the tensor of the atom quadrupole moment. Usage of this κq-representation of the density matrix is convenient for the isotropic character of collisions which leads to an independent evolution of the spherical components of tensors of different κ-ranks. This allows one to separate the kinetic equation for the density matrix in blocks each of them characterizing a certain rank of the spherical tensor. This means that the evolution for one rank of the tensor does not influence the tensor of another rank. In particular, we consider above the evolution of orientation by assuming the population of the considering state to be conserved. This leads to the relation between components of the scattering matrix $\sum_i |S_{ki}|^2 = 1$. But the obtained expressions for relaxation of orientation are valid for any evolution of the atom population. The other convenience of this description is such that evolution of the tensor of each rank is described by only one relaxation constant. In particular, in the case of the orientation relaxation, it is the depolarization cross section. Thus, the formalism of the spherical tensors of the density matrix is convenient for the understanding and analysis of relaxation processes resulting from isotropic collisions.

14.10 Transitions between States of Multiplet Structure

Rotation of the atomic angular momentum leads to transitions between the states of the atom's multiplet structure. Above we consider the process of spin exchange when the exchange by valent electrons leads to transition between the states of superfine structure. Below we consider the other example of such a type when the distance between the levels of multiplet structure is relatively small. The state of the multiplet structure results from summation of the angular and spin atomic momenta. Therefore a change of the angular momentum leads to transitions between the states of multiplet structure. As an example of such a process, we consider the transitions between states of the fine structure of the P-atom as a result of the rotation of the angular momentum which is described by the scattering matrices (14.36) and (14.39) and which is given in Table 14.4. Let us take the initial atom state to be related to the total momentum $j = 1/2$ and its projection $m = 1/2$ onto the z-axis which is perpendicular to the motion plane. Then the atom wave function before collision is

$$\Psi_{1/2,1/2} = -\sqrt{\frac{2}{3}}\psi_1\eta_- + \frac{1}{\sqrt{3}}\psi_0\eta_+ = \frac{1}{\sqrt{3}}\left(-\psi_x\eta_- - i\psi_y\eta_- + \psi_z\eta_+\right),$$

where η_+, η_- are the spin wave functions related to the atom spin projections $+\frac{1}{2}$ and $-\frac{1}{2}$ onto the z-axis, ψ_1, ψ_0 are the atom wave functions corresponding to the values 1 and 0 for the angular momentum projection onto the z-axis, and ψ_x, ψ_y, ψ_z are the wave functions with zero momentum projection onto an indicated axis. Note that the parity of states with respect to the motion plane is conserved during collision of the atoms. Hence transitions from the states 1/2, 1/2 are possible only in the states 3/2, 1/2 and 3/2, $-3/2$.

Let us introduce the scattering matrix $S(jm \to j'm')$ as the amplitude of transitions between the indicated states of colliding atoms. Because the transition in a new fine state is determined by a change in the space atom state, this matrix can be expressed through matrix (14.46). Indeed, the final atom state is described by the wave function

$$\Psi' = \frac{1}{\sqrt{3}}\left(-S_{xx}\psi_x\eta_- - S_{xy}\psi_y\eta_- - iS_{yy}\psi_y\eta_- - iS_{yx}\psi_x\eta_- + S_{zz}\psi_z\eta_+\right).$$

From this it follows that

$$S\left(\frac{1}{2},\frac{1}{2}\to\frac{1}{2},\frac{1}{2}\right) = \frac{1}{3}(S_{xx}+S_{yy}+S_{zz}) + \frac{i}{3}(S_{yx}-S_{xy}). \tag{14.54}$$

In the same manner we find the scattering matrix for the state $j = 1/2, m = -1/2$. We obtain the wave function of this state from the wave function of the state 1/2, 1/2 by changing the direction of the angular and spin momenta. Then, repeating the above operations we get, for the scattering matrix,

$$S(\frac{1}{2},-\frac{1}{2}\to\frac{1}{2},-\frac{1}{2}) = \frac{1}{3}(S_{xx}+S_{yy}+S_{zz}) - \frac{i}{3}(S_{yx}-S_{xy}).$$

From this we obtain, for the probability of transition from states with $j = 1/2$ to states with $j = 3/2$,

$$P = 1 - \frac{1}{2}\left|S\left(\frac{1}{2},\frac{1}{2}\to\frac{1}{2},\frac{1}{2}\right)\right|^2 - \frac{1}{2}\left|S\left(\frac{1}{2},-\frac{1}{2}\to\frac{1}{2},-\frac{1}{2}\right)\right|^2$$
$$= 1 - \frac{1}{9}\left|S_{xx}+S_{yy}+S_{zz}\right|^2 - \frac{1}{9}\left|S_{yx}-S_{xy}\right|^2.$$

Using expressions (14.46) for the scattering matrix, one can connect the probability of the transition from the state with $j = 1/2$ through the parameters of collision

$$P(1/2\to 3/2) = \frac{7}{9} - \frac{5}{9}\cos^2\zeta_o + \frac{4}{9}\cos 2\theta_o\cos^2\zeta_o. \tag{14.55}$$

In particular, for the case of a sharply varied interaction potential as a function of the distances between colliding particles, one can take an average over the phase ζ_o which is high enough. Then we have $\langle\cos^2\zeta_o\rangle = 1/2$, $\langle\cos\zeta_o\rangle = 0$, and the probability of transition between the states of fine structure is

$$P = \frac{2}{3} + \frac{2}{9}\cos 2\theta_o.$$

Taking into account $\sin\theta_o = \rho/R_m$ and neglecting the power dependence on R, in comparison with the exponential dependence of $\Delta(R)$ we find, for the cross section of the transition between the states of fine structure,

$$\sigma\left(\frac{1}{2}\to\frac{3}{2}\right) = \int_0^\infty P\cdot 2\pi\rho\,d\rho = \frac{2}{3}\pi R_m^2, \tag{14.56a}$$

where the matching radius R_m is determined by the relation

$$\left|\frac{V_o(R_m) - V_1(R_m)}{2}\right| = \frac{v}{R_m}. \tag{14.56b}$$

The cross section of the inverse transition, averaged over the momentum projections, is expressed through the above cross section on the basis of the principle of detailed balance

$$\sigma\left(\frac{3}{2} \to \frac{1}{2}\right) = \frac{1}{2}\sigma\left(\frac{1}{2} \to \frac{3}{2}\right) = \frac{1}{3}\pi R_m^2. \tag{14.57}$$

As one more example of the transitions between the states of fine structure, we consider depolarization of the state with $j = 1/2$. Introduce an angle α between the quantization axis and the z-axis which is perpendicular to the motion plane. If $\Phi_{1/2,1/2}$ is the initial wave function relating to the initial axis, and $\Psi_{1/2,1/2}$, $\Psi_{1/2,-1/2}$ are the wave functions quantized on the z-axis, they are connected by the relation

$$\Phi_{1/2,1/2} = \Psi_{1/2,1/2}\cos\frac{\alpha}{2} + \Psi_{1/2,-1/2}\sin\frac{\alpha}{2}.$$

The final atom state after collision is described by the wave function

$$\Phi' = S\left(\frac{1}{2},\frac{1}{2} \to \frac{1}{2},\frac{1}{2}\right)\Psi_{1/2,1/2}\cos\frac{\alpha}{2} + S\left(\frac{1}{2},-\frac{1}{2} \to \frac{1}{2},-\frac{1}{2}\right)\Psi_{1/2,-1/2}\sin\frac{\alpha}{2},$$

where the expression for the scattering matrix is given by formula (14.44). From this we find the probability of depolarization of the related fine state as a result of collision,

$$\begin{aligned} P &= \left|\langle\Phi' \mid \Phi_{1/2,-1/2}\rangle\right|^2 \\ &= \sin^2\frac{\alpha}{2}\cos^2\frac{\alpha}{2}\left|S\left(\frac{1}{2},\frac{1}{2} \to \frac{1}{2},\frac{1}{2}\right) - S\left(\frac{1}{2},-\frac{1}{2} \to \frac{1}{2},-\frac{1}{2}\right)\right|^2 \\ &= \frac{4}{9}\sin^2\frac{\alpha}{2}\cos^2\frac{\alpha}{2}\left|S_{xy} - S_{yx}\right|^2 = \frac{4}{9}\sin^2\alpha\,\sin^2 2\theta_o\cos^2\zeta_o. \end{aligned}$$

Averaging this expression over angle α and a large random phase ζ_o and integrating it over the impact parameter of collision similar to the above examples we obtain, for the depolarization cross section,

$$\sigma\left(\frac{1}{2},\frac{1}{2} \to \frac{1}{2},-\frac{1}{2}\right) = \frac{8}{81}\pi R_m^2. \tag{14.58}$$

The examples under consideration reflect the character of the transitions between the degenerated states of atoms resulting from collisions with other atoms. Then the rotation of the atom angular momentum as a result of the rotation of the molecular axis in the course of collision and a strong coupling of states with the molecular axis at certain distances between the colliding particles determine the variation of the initial wave function of the atom that causes transitions between the degenerated states. These factors are responsible for transitions in the above cases when particles are moving along straight trajectories according to the classical

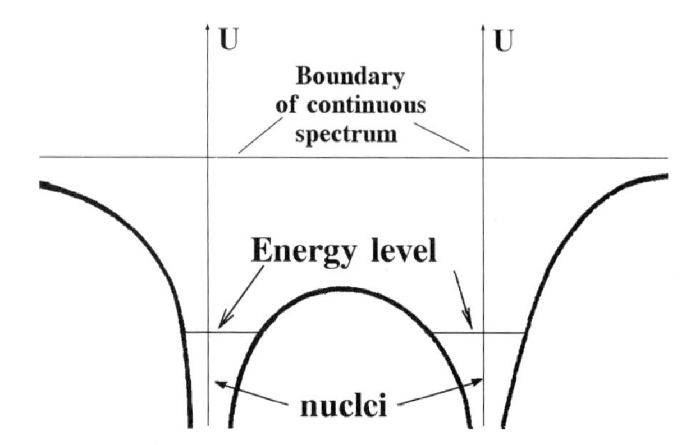

Figure 14.10. The character of a charge exchange process for the interaction of a highly excited atom and ion.

laws. Violation of such a character of the motion of particles can lead to additional possibilities of transitions between the states of multiplet structure.

Problems

Problem 14.1. *Determine the cross section of the charge exchange process for the collisions of slow highly excited atoms and ions. Assume the electron motion to be classical.*

Figure 14.10 gives the crosssection of the potential energy of an excited electron in the field of two Coulomb centers. The electron potential energy is equal to

$$U = -\frac{1}{r_1} - \frac{1}{r_2} + \frac{1}{R},$$

where r_1, r_2 are the electron distances from the corresponding nucleus and R is the distance between the nuclei. The electron wave function ψ satisfies the Schrödinger equation

$$-\frac{1}{2}\Delta\psi + U\psi = \varepsilon\psi,$$

so that at large ion–atom distances, when the fields of the individual ions are separated by a barrier, the electron energy is equal to $\varepsilon = -\gamma^2/2$, where $\gamma^2/2$ is the ionization potential for this atom state. Let us assume that the approach of the atoms and ions proceeds fast enough so that the electron energy does not vary as a result of their collision. Then the condition of disappearance of the barrier $U(r_1 = r_2 = R_o/2) = \varepsilon$ gives $R_o = 8/\gamma^2$. Assuming the transition of the electron in the field of the other atom to be a result of the disappearance of the barrier we obtain, for the cross section of the resonant charge exchange process in this case,

$$\sigma_{\text{res}} = \frac{\pi R_o^2}{2} = \frac{32\pi}{\gamma^4} = \frac{8\pi}{J^2},$$

where $J = \gamma^2/2$ is the atom ionization potential.

Note that this formula corresponds to the diabatic way of collisions so that the second ion does not change the electron energy in the field of the first ion in the course of collision. For the adiabatic way of approaching the atomic particles we obtain another cross section. Then as a result of a slow approach the ionization potential of the molecule is conserved in the course of collision, and the condition of disappearance of the barrier has the form $U(r_1 = r_2 = R_o/2) = 4/R_o = \varepsilon = -\gamma^2/2 + 1/R_o$. This gives, for the distance between the nuclei when the barrier vanishes, $R_o = 6/\gamma^2$. Considering the charge exchange process to be a result of the disappearance of the barrier we find, for the cross section of the resonant charge exchange process in this case,

$$\sigma_{\text{res}} = \frac{\pi R_o^2}{2} = \frac{18\pi}{\gamma^4} = \frac{9\pi}{2J^2}.$$

Problem 14.2. *Determine the contribution of the tunnel transitions to the cross section of the resonant charge process for a highly excited atom.*

Let us take the diabatic case of the approach of particles and calculate the part of the cross section due to tunnel transitions. The ion–atom exchange interaction potential has the following dependence on the distance between the nuclei in this case:

$$\Delta(R) \sim \psi^2\left(\frac{R}{2}\right) \sim \exp\left[-2\int_{z_o}^{R/2} dz\sqrt{2\left(U + \gamma^2/2\right)}\right],$$

where ε is the electron binding energy, $U(R)$ is the interaction potential of the electron with the nuclei, and $z_o = R_o/2$ corresponds to the disappearance of the barrier. Assuming $\Delta R = R - R_o \ll R_o$ and $R/2 - z \ll R/2$, we obtain

$$U = -\frac{1}{z} - \frac{1}{R - z} + \frac{1}{R} + \frac{\gamma^2}{2} = \frac{3\Delta R}{R_o^2} - \frac{16(R/2 - z)^2}{R_o^3}.$$

This leads to the following exponential dependence of the exchange interaction potential:

$$\Delta(R) \sim \exp\left[-\frac{3\pi\sqrt{2}}{8}\frac{R - R_o}{\sqrt{R_o}}\right] \sim \exp\left[-\frac{\pi\sqrt{3}(R - R_o)\gamma}{8}\right],$$

where we account for $R_o = 6/\gamma^2$.

From this we find the addition to the cross section of the resonant charge exchange process. This addition to the formula of the previous problem is determined by tunnel transitions, so that the cross section is given by formula (14.15): $\sigma_{\text{res}} = \pi\rho_o^2/2$, where ρ_o is the impact parameter of the collision for which $\int_{-\infty}^{\infty} \Delta(\rho_o)\,dt/2 = 0.28$. This gives the equation for the parameter ρ_o:

$$\frac{1}{v}\exp\left[-\frac{\pi\sqrt{3}(\rho_o - R_o)\gamma}{8}\right] = \text{const.}$$

From this it follows that

$$\rho_o - R_o = \frac{8}{\pi\sqrt{3}\gamma}\ln\frac{v_o}{v},$$

where the parameter v_o weakly depends on the collision velocity v. Representing the cross section in the form $\sigma_{\text{res}} = \sigma_o + \Delta\sigma$, where $\sigma_o = \pi R_o^2/2$ and $R_o = 6/\gamma^2$, we obtain

$$\Delta\sigma = \pi R_o(\rho_o - R_o) = \frac{8R_o}{\pi\sqrt{3}}\ln\frac{v_o}{v}.$$

From this we find that

$$\frac{\Delta\sigma}{\sigma} = \frac{16}{\pi\sqrt{3}}\frac{1}{\gamma R_o}\ln\frac{v_o}{v} = \frac{8\gamma}{3\pi\sqrt{3}}\ln\frac{v_o}{v} = 0.5\gamma\ln\frac{v_o}{v} \ll 1,$$

because for a highly excited atom $\gamma \ll 1$. Thus, tunnel transitions give a small contribution to the cross section of the charge exchange process in the slow collision of an ion and highly excited atom.

Problem 14.3. *Analyze a jump in the cross section of the resonant charge exchange for rare gases as a result of the transition between cases "a" and "c" of the Hund coupling.*

At low collision velocities when ions are found in the ground state ($j = 3/2$), the cross section is equal to $\sigma_{3/2}$ according to the used notations, and the transition into the ion state $j = 1/2$ is forbidden. At high collision velocities this channel is opened, and the resonant charge exchange process corresponds to case "a" of the Hund coupling. Let us assume that these coupling schemes lead to identical cross sections, if the ionization potential is identical to the ion formation in both fine states (see Table 14.4). Hence, a variation of the cross section in the course of the transition between cases "c" and "a" of the Hund coupling is due to different atom ionization potentials with the formation of different fine ion states. The jump in the cross sections due to this effect is

$$\Delta\overline{\sigma}_{\text{res}} = \frac{1}{3}\frac{\Delta I}{I}\overline{\sigma}_{\text{res}}, \tag{14.59}$$

where the first factor is the probability of the ion state $j = 1/2$ and the second factor accounts for the dependence (14.16a) of the cross section on the electron binding energy. According to formula (14.59), the relative variation of the cross section is approximately 0.4% for Ar, approximately 2% for Kr, and approximately 4% for Xe. A typical collision velocity v for this transition can be estimated from the expression for a typical time of the process

$$\tau \sim \frac{1}{v}\sqrt{\frac{R_o}{\gamma}} \sim \Delta_f,$$

as follows from formula (14.17), and Δ_f is the ion fine splitting of levels. A typical collisional energy for this transition is estimated as $\sim$ 10 eV for Ar, $\sim$ 100 eV for Kr, and $\sim$ 600 eV for Xe.

Problem 14.4. *Analyze the expression for the scattering matrix* (14.46) *in the case of the elastic scattering of colliding particles if the scattering angle of particles is equal to* ϑ.

Let us repeat the operations which we made to obtain expression (14.46) for the scattering matrix. A variation of the wave functions is absent until $R > R_m$, where this value is determined by the expression

$$\frac{d\theta}{dt}(R_m) = \Delta(R_m).$$

In this case the expression for the rotation angle during a weak coupling of states with the molecular axis is

$$\theta_o = -\int_{R_m}^{\infty} \frac{\rho \, dR}{R^2} \left(1 - \frac{\rho^2}{R^2} - \frac{U(R)}{E}\right)^{-1/2}$$

instead of $\sin\theta_o = \rho/R_m$ which takes place in the absence of scattering. Here $U(R)$ is the interaction potential which is assumed to be identical for all atom states, and E is the collision energy in the center of mass frame of the axes. As is seen, for $U(R) = 0$, both formulas give the same value of θ_o.

It is clear that a change of the motion trajectory leads to a change in the difference of the phases for the wave function of states related to different momentum projections because of the different times of a strong coupling of states with the molecular axis. But formally the expression for this value $\zeta_o = \int_{-\infty}^{\infty}(V_o - V_1)\,dt/2$ is conserved.

Next, the last operation in the determination of the S-matrix for returning to the initial frame of axes corresponded to rotation of the system by the angle $\pi - 2\theta_o$ around the z-axis. Now because of the scattering, this angle is equal to $\pi - 2\theta_o - \vartheta$. Finally, the expression for the scattering matrix has the following form instead of (14.46) (we must change $2\theta_o$ in formula (14.46) to $2\theta_o - \vartheta$):

$$S_{ik} = e^{-i\int^{t}[(V_o+V_1)/2]\,dt'}$$
$$\times \begin{pmatrix} i\sin\zeta_o - \cos\zeta_o\cos(2\theta_o - \vartheta) & -\sin(2\theta_o - \vartheta)\cos\zeta_o & 0 \\ \sin(2\theta_o - \vartheta)\cos\zeta_o & -i\sin\zeta_o - \cos\zeta_o\cos(2\theta_o - \vartheta) & 0 \\ 0 & 0 & e^{i\zeta_o} \end{pmatrix}.$$

CHAPTER 15

Inelastic Slow Atomic Collisions

15.1 Transitions in Two-Level Systems

According to the Massey criterion (14.7), the transitions between states with large distances between the electron terms are characterized by a small probability in slow collisions. This means that the transitions proceed between close by energy states only or they take place in a range of distances where the electron terms of these states are intersected or pseudointersected. The first case corresponds to quasiresonant processes which were considered in the previous chapter. Then the electron levels of transition are close or are coincident at infinite distances between the colliding atoms, so that the Massey parameter (14.7) is small in a wide range of distances. The other case corresponds to the approach of two electron terms and their intersection in some range of distances. Such distances between the colliding particles are responsible for the transition between states where the electron terms of these states are close. Just as in this region it is necessary to analyze the evolution of the system of colliding atoms, because other range of distances takes place the adiabatic development of this quasi-molecule without a change in the probability of its location in each of the related states. Hence, our task is to analyze the behavior of a quasi-molecule consisting of colliding atoms in the transition region where the electron terms of the states under consideration are close. Note that the system of colliding atoms goes through the transition region twice—during the approach and removal of atoms. The phases obtained by the wave function in an intermediate region can both lead to summation and subtraction of the amplitudes of the corresponding states. Thus, although we assume the colliding atoms to move along classical trajectories, an interference of states takes place and the transition has a quantum character. In order to analyze the character of the transition under related conditions, we are restricted by the case of two nondegenerated states which partake in the process under consideration.

We take the stationary wave functions as basis wave functions. This means that in contrast to the set of equations (14.5) for probability amplitudes, which is written in the adiabatic basis, we use a diabatic basis similar to the set of equations (14.35). So, let us choose a Hamiltonian $\hat{H}_o$ which is close to the Hamiltonian of the system of colliding atoms $\hat{H}$, and take as the basis the eigenfunctions ψ_1, ψ_2 of the Hamiltonian $\hat{H}_o$. The wave function of the colliding atoms we present in the form (in atomic units)

$$\Psi = (c_1\psi_1 + c_2\psi_2)\exp\left[-\frac{i}{2}\int^t (H_{11} + H_{22})\,dt'\right], \tag{15.1}$$

where $H_{11} = \langle\psi_1|\hat{H}|\psi_1\rangle$ and $H_{22} = \langle\psi_2|\hat{H}|\psi_2\rangle$. Substituting this expansion into the Schrödinger equation $i(\partial\Psi/\partial t) = \hat{H}\Psi$, multiplying it by ψ_1^* or ψ_2^*, and integrating over the electron coordinates, we obtain the following set of equations for the probability amplitudes

$$i\frac{dc_1}{dt} = \frac{\kappa}{2}c_1 + \frac{\Delta}{2}c_2, \qquad i\frac{dc_2}{dt} = \frac{\Delta}{2}c_1 - \frac{\kappa}{2}c_2, \tag{15.2}$$

where $\kappa(R) = H_{11} - H_{22}$ and $\Delta(R) = 2H_{12} - (H_{11} + H_{22})\langle\psi_2 \mid \psi_1\rangle$.

The solution of this set of equations is determined by the time dependence for the parameters κ and Δ. Below we consider the case when, in the transition region, $\kappa =$ const and $\Delta = 2\alpha e^{-t/\tau}$. This case corresponds to the resonant charge exchange if $\kappa = 0$. A general method of the solution of this set of equations is based on the asymptotic solutions. In region $\kappa \gg \Delta$ the solutions of the set of equations (15.2) have

$$c_1 = a\exp\left(\frac{i}{2}\int^t \kappa\,dt'\right), \qquad c_2 = b\exp\left(-\frac{i}{2}\int^t \kappa\,dt'\right), \tag{15.3a}$$

and the values $|c_1|$, $|c_2|$ are conserved in time. In the region $\kappa \ll \Delta$ the solutions of the set of equations (15.2) have the form

$$c_1 = \cos\left(\int^t \frac{\Delta}{2}\,dt'\right), \qquad c_2 = i\sin\left(\int^t \frac{\Delta}{2}\,dt'\right), \tag{15.3b}$$

and the values $|c_1 + c_2|$, $|c_1 - c_2|$ do not vary in time. Below we divide the time region into several parts and, solving the set of equations in each region, we join these solutions in common regions.

Let us take the initial conditions $c_1(-\infty) = 1, c_2(-\infty) = 0$, so that until $\kappa \gg \Delta$, we have $|c_1| = 1$, $c_2 = 0$. In the next time region the set of equations (15.2) has the form

$$i\frac{dc_1}{dt} = \frac{\kappa}{2}c_1 + \alpha e^{t/\tau}c_2, \qquad i\frac{dc_2}{dt} = \alpha e^{t/\tau}c_1 - \frac{\kappa}{2}c_2. \tag{15.4}$$

The solutions of this set of equations under the initial conditions $c_1(-\infty) = 1$, $c_2(-\infty) = 0$ are the following:

$$c_1 = \sqrt{\frac{\beta}{\mathrm{ch}\beta}} \cdot e^{t/(2\tau)} J_{-1/2-i\kappa\tau/2}(\alpha\tau e^{t/\tau}), \qquad c_2 = \sqrt{\frac{\beta}{\mathrm{ch}\beta}} \cdot e^{t/(2\tau)} J_{1/2-i\kappa\tau/2}(\alpha\tau e^{t/\tau}),$$

where $\beta = \pi\kappa\tau/2$. We assume the transition regions near the points $\Delta = \kappa$ to be divided by a large time interval. Hence, after passing the first transition region, the probability amplitudes reach the asymptotic expressions (15.3b) before the second transition region. In the region $\Delta \gg \kappa$, the above expressions have the following form:

$$c_1 = \sqrt{\frac{1}{ch\beta}} \cdot \cos\left(\alpha\tau e^{t/\tau} + i\frac{\beta}{2}\right), \qquad c_2 = \sqrt{\frac{1}{ch\beta}} \cdot \sin\left(\alpha\tau e^{t/\tau} + i\frac{\beta}{2}\right).$$

Comparing these solutions with (15.3b), rewrite them in the form

$$c_1 = \sqrt{\frac{1}{\mathrm{ch}\beta}} \cdot \cos\left(\int_{-\infty}^{t} \frac{\Delta}{2}\, dt' + i\frac{\beta}{2}\right), \quad c_2 = \sqrt{\frac{1}{\mathrm{ch}\beta}} \cdot \sin\left(\int_{-\infty}^{t} \frac{\Delta}{2}\, dt' + i\frac{\beta}{2}\right).$$

The set of equations for the probability amplitude in the region $\Delta \sim \kappa$, during the removal of atoms, can be obtained from the set of equations (15.4) by replacing the time sign in the expression of $\Delta(t)$, i.e.,

$$i\frac{dc_1}{dt} = \frac{\kappa}{2}c_1 + \alpha e^{-t'/\tau}c_2, \qquad i\frac{dc_2}{dt} = \alpha e^{-t'/\tau}c_1 - \frac{\kappa}{2}c_2,$$

and t' differs from t in (15.4) by an initial time. Joining the solutions of this equation at $t' = -\infty$ with the above asymptotic expressions we obtain, in the limit $t' \to \infty$ the following expression for the probability of the transition,

$$P = |c_2(\infty)|^2 = \sin^2\left(\int_{-\infty}^{\infty} \frac{\Delta}{2}\, dt\right) / ch^2\beta = \sin^2\left(\int_{-\infty}^{\infty} \frac{\Delta}{2}\, dt\right) / ch^2\frac{\pi\kappa\tau}{2}. \tag{15.5}$$

Averaging over the phase, we obtain

$$P = \frac{1}{2\mathrm{ch}^2\frac{\pi\kappa\tau}{2}}. \tag{15.6}$$

Formula (15.5) is called the Rosen–Zener formula, while the general formula (15.5) is the Demkov formula. Note that the Demkov formula (15.5) transfers to the expression for the probability of resonant charge exchange (14.13) in the limit $\kappa = 0$.

The above deduction shows the character of change of the probability amplitudes in the course of the evolution of the system. From this it follows that the scattering matrix, resulting from a single passage of the transition region, has the form

$$S = \begin{pmatrix} \sqrt{1-p}e^{-i\varsigma} & -pe^{-i\eta} \\ pe^{i\eta} & \sqrt{1-p}e^{i\varsigma} \end{pmatrix}. \tag{15.7}$$

Averaging over phases we have, for the transition probability resulting from the double passage of the transition region (during approach and removal),

$$P = 2p(1-p). \tag{15.8}$$

One can consider this value as the sum of two probabilities, so that the first one is the product of the probability of the transition during the first passage of the transition region by the probability of its absence during the second passage, i.e., it is equal to $p(1-p)$, and the second probability corresponds to the reversible character of the transition and is equal to $(1-p)p$.

The set of equations (15.2) does not have an analytic solution for any form of functions $\kappa(t)$ and $\Delta(t)$. We consider one more case when this set has an analytic solution. The case when

$$\kappa = F v_R t, \qquad \Delta = \text{const.} \tag{15.9}$$

in the transition region is called the Landau–Zener case. The Landau–Zener formula for a single passage of the transition region has the form

$$p = \exp\left(-\frac{\pi\Delta^2}{2Fv_R}\right). \tag{15.10}$$

There is a more general analytical solution for the probability of a single passage of the transition region when the parameters of the diabatic terms are given by the dependence

$$\begin{aligned} \kappa(t) &= H_{11} - H_{22} = \varepsilon(1 - \cos\vartheta e^{-t/\tau}), \\ \Delta(t) &= 2H_{12} - (H_{11} + H_{22})\langle\psi_2 \mid \psi_1\rangle = \varepsilon\sin\vartheta e^{-t/\tau}. \end{aligned} \tag{15.11}$$

Then the electron terms of the related states E_1 and E_2 are determined by the formulas

$$E_{1,2} = \frac{H_{11}+H_{22}}{2} \pm \Delta E, \qquad \Delta E = \sqrt{\kappa^2+\Delta^2} = \varepsilon\sqrt{1 - 2\cos\vartheta e^{-t/\tau} + e^{-2t/\tau}}. \tag{15.12}$$

The minimum of the difference of the level energies $\Delta E = \varepsilon \sin\vartheta$ corresponds to a time which follows from the relation $e^{-t/\tau} = \cos\vartheta$. The limiting case $\vartheta = \pi/2$ corresponds to the Rosen–Zener case, and the limit $\vartheta \to 0$ corresponds to the Landau–Zener case. The probability of transition in a general case is given by the Nikitin formula

$$p = \frac{\mathrm{sh}[\pi\varepsilon(1+\cos\vartheta)/(2\alpha]}{\mathrm{sh}(\pi\varepsilon/\alpha)} \exp\left[-\frac{\pi\varepsilon(1+\cos\vartheta)}{2\alpha}\right]. \tag{15.13}$$

The Nikitin formula includes two parameters, ε/α and ϑ. In the limiting case $\vartheta = \pi/2$ this gives

$$p = \frac{1}{\exp(\pi\varepsilon/\alpha) + 1},$$

which corresponds to the Rozen–Zener formula (15.6) accounting for relation (15.8) for the double passage of the transition region. In the other limiting case

$\vartheta \to 0$ formula (15.13) is transformed into the Landau–Zener formula (15.10),

$$p = \exp\left(-\frac{\pi \varepsilon \vartheta^2}{2\alpha}\right).$$

15.2 Cross Section of a Nonresonant Charge Exchange

Let us use the above results for the evaluation of the cross section for the nonresonant charge exchange process when two states of the system of colliding atoms participate in the process. This takes place if two levels are close at infinite distances between the atoms and the transition is determined by the exchange interaction potential which mixes these states. Then one can use the Demkov formula (15.5) for the probability of the transition by taking for these states $\kappa = \text{const}$ and accounting for a real dependence of the exchange interaction potential Δ on the distance between the colliding particles R. If we use this dependence in the form of formulas (10.20), (10.23), $\Delta(R) \sim e^{-\gamma R}$, the parameter α in the Demkov formula is equal to $\alpha = \gamma v_R$, where the radial velocity of the free moving particles is $v_R = v\sqrt{1 - \rho^2/R^2}$.

Let us introduce the characteristic distance between the atomic particles R_c such that

$$\Delta(R_c) = \kappa.$$

Using the new probability amplitudes $a_1 = c_1 e^{-i\kappa t}$, $a_2 = c_2 e^{i\kappa t}$, we obtain the set of equations (15.4) in the form

$$i\frac{da_1}{dt} = \frac{\Delta}{2} e^{i\kappa t} a_2, \qquad i\frac{da_2}{dt} = \frac{\Delta}{2} e^{-i\kappa t} a_1. \tag{15.14}$$

At the large impact parameters of collisions $\rho > R_c$, we solve this set of equations on the basis of the perturbation theory. This gives, for the transition probability,

$$P(\rho) = \left| \int_{-\infty}^{\infty} \frac{\Delta(R)}{2} e^{i\kappa t}\, dt \right|^2 = \zeta^2 \exp\left(-\frac{\rho\kappa^2}{\gamma v^2}\right), \qquad \rho \geq R_c, \tag{15.15}$$

where the exchange phase is

$$\zeta(\rho) = \int_{-\infty}^{\infty} \frac{\Delta}{2} dt = \sqrt{\frac{\pi\rho}{2\gamma}} \cdot \frac{\Delta(\rho)}{v}.$$

One can check the validity of the perturbation theory for the large impact parameters of collisions. The transition probability decreases with an increase in an impact parameter, so that $P(\rho) < P(R_c)$. Next, $P(R_c) = \zeta_c^2 \exp(-2\zeta_c^2/\pi)$, where $\zeta_c = \zeta(R_c)$. This value as a function of ζ_c has a maximum $P = \pi/(2e) < 1$ at $\zeta_c^2 = \pi/2$. Thus, one can consider this function to be small in a wide region of parameters ζ_c.

At impact parameters $\rho < R_c$ we use the Demkov formula (15.5) for the transition probability, which has the form

$$P(\rho) = \frac{\sin^2 \zeta}{\mathrm{ch}^2 \delta_R}, \quad \text{where} \quad \delta_R = \frac{\pi \kappa}{2\gamma v \sqrt{1 - \rho^2/R^2}}. \tag{15.16}$$

Let us evaluate the cross section of this process in the limiting cases. At large collision velocities one can take $\kappa = 0$, so that the cross section according to formula (15.16) is determined similarly to the cross section of the resonant charge exchange (14.15):

$$\sigma = \pi R_o^2/2, \quad \text{where} \quad \zeta(R_o) = \sqrt{\frac{\pi R_o}{2\gamma}} \cdot \frac{\Delta(R_o)}{v} = 0.28. \tag{15.17}$$

In the other limiting case of small collision velocities one can average over the phase in the Demkov formula (15.5), i.e., we transfer to the Rozen–Zener formula (15.6). Then the cross section equals

$$\sigma = \frac{\pi R_c^2}{2} F(\delta), \quad \text{where} \quad \delta = \frac{\pi \kappa}{2\gamma v} \quad \text{and} \quad F(\delta) = \int_0^1 \frac{2x dx}{\mathrm{ch}^2(\delta/x)}. \tag{15.18}$$

The values of the function $F(\delta)$ are given in Table 15.1. The asymptotic expressions of this function at large and small argument values are

$$F(0) = 1, \qquad F(\delta \to \infty) = \frac{4}{\delta} e^{-2\delta}.$$

Note that the large impact parameters give a small contribution to the cross section. This can be estimated on the basis of formula (15.15) for the transition probability at these impact parameters. We have

$$\Delta\sigma = \int_{R_c}^{\infty} 2\pi\rho \, d\rho \, P(\rho) = \frac{\pi^2 R_c^2}{2 + v^2\gamma^2/\kappa^2} \exp\left(-\frac{\kappa^2 R_c}{\gamma v^2}\right).$$

A maximum of this value $\Delta\sigma_{\max} = \pi^2 R_c/(e^2\gamma)$ corresponds to the collision velocity $v = \kappa\sqrt{R_c/\gamma}$. Since, usually, $\gamma R_c \gg 1$, this region gives a small contribution to the cross section of the process. Thus the cross section consists of two parts. At large collision velocities the resonant charge exchange cross section is given by formula (15.17) and at small collision velocities it is determined by the

Table 15.1. Values of the function $F(\delta)$ which is determined by formula (15.18).

δ	$F(\delta)$		δ	$F(\delta)$		δ	$F(\delta)$
0.2	0.859		1.2	0.131		2.2	0.0137
0.4	0.647		1.4	0.084		2.4	0.0087
0.6	0.456		1.6	0.0536		2.6	0.0055
0.8	0.308		1.8	0.0339		2.8	0.0035
1	0.203		2	0.0215		3	0.0023

Rozen–Zener cross section (15.8). The cross section has a flat maximum in a wide transition region where these limit cases are combined. According to the numerical calculations, the maximum cross section is approximately equal to $0.54\pi R_c^2$.

The conditions under consideration correspond to the nonresonant charge exchange process

$$A^+ + B \to A + B^+.$$

Then the parameter κ is equal to the difference of the ionization potentials of atoms A and B, and Δ is the exchange interaction potential corresponding to this electron transition. As an example, we consider the process of nonresonant charge exchange

$$\mathrm{O}^+ + \mathrm{H} \to \mathrm{O} + \mathrm{H}^+.$$

In this case the parameter κ is equal to 0.02 eV, and the exchange interaction potential is given by formula (10.38) for the zero momentum projection of the oxygen valence electron. The transitions proceed in a narrow range of distances between the colliding particles. At small collision velocities this takes place near separation R_c, at large collision velocities the main contribution to the exchange phase gives the range near the distance of closest approach. This allows one to account for the different projections of the electron momentum in a simple way. Figure 15.1 gives the experimental cross section of this process and is evaluated in a simple way by neglecting the electron momentum rotation during a collision.

15.3 Principle of Detailed Balance for the Excitation and Quenching Processes

The principle of detailed balance connects the cross sections of an inelastic collision process and an inverse process. The connection between the parameters of these processes follows from the time reversal operation. If the process proceeds

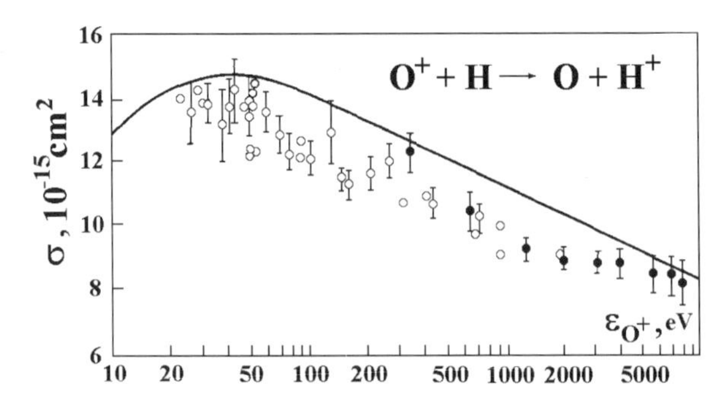

Figure 15.1. The cross section of charge exchange involving the oxygen ion and hydrogen atom. The curve is constructed on the basis of formulas (15.17), (15.18); signs refer to experiments.

according to the scheme

$$A + B_j \to A + B_f, \tag{15.19}$$

time reversal leads to the process

$$A + B_f \to A + B_j. \tag{15.20}$$

Denoting the cross section of the first process by σ_{jf} and the cross section of process (15.20) by σ_{fi}, we place one particle A and one particle B in a volume Ω. Particle B can be found only in states j or f and, due to equilibrium, the number of transitions $j \to f$ per unit time w_{jf} is equal to the number of transitions $f \to j$ per unit time w_{fj}. Next, introducing the interaction operator V which is responsible for these transitions we have, within the framework of the perturbation theory for the rates of these transitions,

$$w_{jf} = \frac{2\pi}{\hbar}\left|V_{jf}\right|^2 \frac{dg_f}{d\varepsilon}, \qquad w_{fj} = \frac{2\pi}{\hbar}\left|V_{fj}\right|^2 \frac{dg_j}{d\varepsilon}.$$

Here $dg_f/d\varepsilon$, $dg_i/d\varepsilon$ are the statistical weights per unit energy for states of the corresponding channel of processes and these formulas are written in the usual units. Using the definition of the cross sections of the processes

$$\sigma_{jf} = w_{jf}/(Nv_j) = \Omega w_{jf}/v_j, \qquad \sigma_{fj} = w_{fj}/(Nv_f) = \Omega w_{fj}/v_f.$$

where $N = 1/\Omega$ is the number density of particles and v_j, v_f are the relative velocities of particles for a given channel of the process. The time reversal leads to the connection between matrix elements of the interaction potential $V_{jf} = V_{fj}^*$. This gives the following relation between the cross sections of the direct and inverse processes

$$\sigma_{jf} = \sigma_{fj}\left(v_f \frac{dg_f}{d\varepsilon}\right)\left(v_j \frac{dg_j}{d\varepsilon}\right)^{-1}. \tag{15.21}$$

Let us use this relation for inelastic electron–atom collisions. Then v_j, v_f are the initial electron velocities for a given channel of the process, and the statistical weight of the corresponding channel is equal to $dg_f = \Omega[d\mathbf{p}_f/(2\pi\hbar)^3]g_f$, where g_f is the statistical weight of the atom state; formula (15.21) takes the form

$$\sigma_{jf} = \sigma_{fj}\frac{v_f^2 g_f}{v_j^2 g_j}. \tag{15.22}$$

In particular, near the threshold, the excitation cross section σ_{ex} has the form ($\sigma_{\text{ex}} = \sigma_{jf}$):

$$\sigma_{\text{ex}} = C\sqrt{E - \Delta\varepsilon},$$

where E is the energy of the incident electron and $\Delta\varepsilon$ is the excitation energy for this transition. Then, according to formula (15.22), the cross section of atom quenching $\sigma_q = \sigma_{fi}$, resulting from the collision with a slow electron of energy

$\varepsilon = E - \Delta\varepsilon \ll \Delta\varepsilon$, is

$$\sigma_q = C \frac{g_o \Delta\varepsilon}{g_f \sqrt{E - \Delta\varepsilon}}.$$

From this it follows that the rate constant of atom quenching by a slow electron

$$k_q = v_f \sigma_q = C \frac{g_o \Delta\varepsilon \sqrt{2}}{g_f \sqrt{m}} \tag{15.23}$$

does not depend on the electron energy (m is the electron mass). This means that in a plasma containing slow electrons the rate of quenching of the excited atoms does not depend on the distribution function of the electrons on energy.

15.4 Quenching of the Excited Atom States by Ion Impact

If an atom has two energetically nearby levels, and their states are mixed by an external electric field, the transitions between these states can result from the collision of the atom with an ion. Below we consider this problem starting from the transition between the $2s$- and $2p$-states of the hydrogen atom colliding with an ion. This process leads to quenching of the excited atom state and is of importance in astrophysics. The process is caused by the interaction potential between colliding particles (in atomic units)

$$V = \mathbf{rR}/R^3,$$

where $\mathbf{R}$ is the distance between nuclei and $\mathbf{r}$ is the electron coordinate; we use atomic units. Taking into account the free motion of particles $R^2 = \rho^2 + v^2t^2$ (ρ is the impact parameter of collision, v is the relative velocity of the colliding particles, and t is time), we present the wave function of the excited hydrogen atom colliding with an ion in the form,

$$\Psi = a_0 \psi_{p0} e^{-it\varepsilon_{p0}} + a_1 \psi_{p1} e^{-it\varepsilon_{p1}} + c\psi_s e^{-it\varepsilon_s},$$

where ε_s, ε_{p0}, ε_{p1} are the energies of the corresponding states and the subscripts $p0$, $p1$ indicate the momentum projection of the p-electron on the ρ-axis. We neglect the transitions to other states of the hydrogen atom because of a large transferring energy. The set of equations for the probability amplitudes takes the form

$$\begin{aligned} i\dot{c} &= \frac{d\rho}{R^3} a_{p0} + \frac{dvt}{R^3} a_{p1} - \frac{\omega}{2} c, \\ i\dot{a}_{p0} &= \frac{d\rho}{R^3} c + \frac{\omega}{2} a_{p0}, \\ i\dot{a}_{p1} &= \frac{dvt}{R^3} c + \frac{\omega}{2} a_{p1}, \end{aligned} \tag{15.24}$$

where ω is the difference between the energies of the p- and s-levels, and d is the matrix element from the electron coordinate r between the radial wave functions of the s- and p-states. We now neglect the fine splitting of p-levels. Next we account for the precise positions of these levels, so that the $2^2S_{1/2}$-level of the hydrogen atom lies lower than the $2^2P_{1/2}$-level by 0.035 cm^{-1} and is higher than the $2^2P_{3/2}$-level by 0.330 cm^{-1}. Let us first use the perturbation theory assuming in the zero-approximation that $c = 1$. Then we have

$$a_{p0} = \frac{2d\omega}{v^2} K_0\left(\frac{\omega\rho}{v}\right); \qquad a_{p1} = \frac{2d\omega}{v^2} K_1\left(\frac{\omega\rho}{v}\right),$$

where K_0, K_1, are the Macdonald functions and the perturbation theory is valid for $\rho \gg d/v$. Let us consider the case of large collision velocities

$$v^2 \gg d\omega,$$

when the cross section of the process is determined by the impact parameters

$$\frac{d}{v} \ll \rho \ll \frac{v}{\omega}.$$

Thus, within the framework of the perturbation theory we obtain, for quenching of the s-state,

$$P(\rho) = \frac{4d^2\omega^2}{v^4}\left[K_0^2\left(\frac{\omega\rho}{v}\right) + K_0^2\left(\frac{\omega\rho}{v}\right)\right], \qquad \rho \gg \frac{d}{v}.$$

Let us present the cross section of quenching of the s-state in the form

$$\begin{aligned}\sigma_{sp} &= \int_0^{\rho_o} P(\rho) 2\pi\rho\, d\rho + \int_{\rho_o}^{\infty} 2\pi\rho\, d\rho \cdot \frac{4d^2\omega^2}{v^4}\left[K_0^2\left(\frac{\omega\rho}{v}\right) + K_0^2\left(\frac{\omega\rho}{v}\right)\right] \\ &= \int_0^{\rho_o} P(\rho) 2\pi\rho\, d\rho + 8\pi\frac{d^2}{v^2}\ln\frac{2ve^{-C}}{\omega\rho_o},\end{aligned} \tag{15.25}$$

where the impact parameter ρ_o is taken such that $\omega\rho_o/v \gg 1$ and $P(\rho) \ll 1$; $C = 0.577$ is the Euler constant. The first term results from the solution of the set of equations (15.24). But one can estimate the parameter ρ_o in formula (15.25) from the relation $P(\rho_o) \sim 1$, which gives $\rho_o \sim v/\omega$, and

$$\sigma_{sp} = 8\pi\frac{d^2}{v^2}\ln\frac{av^2}{\omega d}.$$

The numerical coefficient a follows from the solution of the set of equations at small impact parameters of the collision where the perturbation theory is violated. If we find ρ_o from the relation $P(\rho_o) = 1$, we obtain $a = \exp(0.5 - C) = 0.9$. As is seen, under a given condition $v^2 \gg d\omega$, an error in the coefficient a weakly influences the result because of a large value under logarithm. Therefore, one can take $a = 1$ in the above formula, so that the quenching cross section is

$$\sigma_{sp} = 8\pi\frac{d^2}{v^2}\ln\frac{v^2}{\omega d}. \tag{15.26}$$

Let us use this formula for the $2s$-hydrogen atom, accounting for splitting of the levels of fine structure. Denote by $\omega_1 = 0.035\,\text{cm}^{-1}$ the distance between the $2S_{1/2}$- and $2P_{1/2}$-levels and by $\omega_1 = 0.330\,\text{cm}^{-1}$ the distance between the $2S_{1/2}$- and $2P_{3/2}$-levels. Then using the perturbation theory for each level separately, and accounting for the statistics of each level, we obtain formula (15.26) with the following value of the parameter $\omega = (\omega_1\omega_2^2)^{1/3} = 0.16\,\text{cm}^{-1}$. The cross section (15.26) is high enough. In particular, in the case of the collision of the proton at the collision energy in the center-of-mass coordinate frame $\varepsilon = 1\,\text{eV}$, formula (15.26) gives the cross section of quenching of the $2s$-hydrogen atom $\sigma_{sp} = 2.8\cdot 10^{-10}\,\text{cm}^2$.

We now consider quenching of a highly excited atom by ion impact on the basis of the above result. Then the transitions from a state with quantum numbers nl mostly proceed in states with quantum numbers n, $l+1$, where the distance between the levels is equal to $\omega = (\delta_l - \delta_{l+1})/n_*^3 = \Delta_l/n_*^3$. Here n_* is the effective principal quantum number and $n_* \gg l$, δ_l is the quantum defect for this level and $\Delta_l = \delta_l - \delta_{l+1}$. We have, for the quenching probability according to the above solution of the set of equations (15.24) within the framework of the perturbation theory,

$$P(\rho) = \frac{4\omega^2}{3v^4}\left|\mathbf{r}_{nl;n,l+1}\right|^2\left[K_0^2\left(\frac{\omega\rho}{v}\right) + K_1^2\left(\frac{\omega\rho}{v}\right)\right]. \tag{15.27}$$

Summing over the various momentum projections, we have

$$\left|\mathbf{r}_{nl;n,l+1}\right|^2 = \sum_{m'}\left|\langle nlm\,|\mathbf{r}|\,n,l+1,m'\rangle\right|^2 = \frac{9}{4}n_*^4\frac{l+1}{2l+1}[1-(l+1)^2/n_*^2].$$

Accounting for $n_* \gg l$ we obtain, for the quenching cross section,

$$\sigma_q = 6\pi\frac{n_*^4}{v^2}\frac{l+1}{2l+1}\ln\frac{av^2n_*}{\Delta_l},$$

where the numerical coefficient $a \sim 1$. If we obtain the parameter ρ_o from the relation $P(\rho_o) = 1/2$, we have $a = 0.5\sqrt{(2l+1)/(l+1)}$. If this parameter follows from the relation $P(\rho_o) = 1$, the above value of a is multiplied by $\sqrt{2}$. Thus, one can take $a = 1$, so that the quenching cross section is determined by the formula

$$\sigma_q = 6\pi\frac{n_*^4}{v^2}\frac{l+1}{2l+1}\ln\frac{v^2n_*}{\Delta_l}. \tag{15.28}$$

15.5 Quenching of Highly Excited Atoms in Atomic Collisions

Some examples of the effective processes with large cross sections corresponding to processes involving a highly excited atom because of a high density of its levels. Below we analyze the process of quenching of a highly excited state resulting from a collision with an atom in the ground state. This atom has a filled electron shell and hence the interaction between atoms has a short-range character. We take this interaction in the form (10.42), so that the interaction operator of the colliding atoms is given by the formula

$$V = 2\pi L\delta(\mathbf{r} - \mathbf{R}), \tag{15.29}$$

where L is the electron-incident atom scattering length, $\mathbf{r}$ is the electron coordinate, and $\mathbf{R}$ is the coordinate of the perturbed atom.

Let us determine the cross section of the process within the framework of the perturbation theory. We have, for the probability of the transition from the initial j-state to a final f-state,

$$P_{jf} = \left| \int_{-\infty}^{\infty} V_{jf}\, dt \right|^2 = 4\pi^2 L^2 \left| \int_{-\infty}^{\infty} \psi_j^*(\mathbf{R})\psi_f(\mathbf{R})\, dt \right|^2 ,$$

where ψ_j is the electron wave function of the corresponding state. This formula can be used to estimate the quenching cross section as a function of the principal quantum number. Since $|\psi(\mathbf{R})|^2 \sim a^{-3} \sim n^{-6}$, where $a \sim n^2$ is the size of a highly excited atom, and $\int dt \sim a/v_a$ we have, for the transition probability $P_{ik} \sim L^2/(v_a^2 n^8)$, where v_a is the collision velocity. This gives, for the transition cross section,

$$\sigma \sim a^2 P_{jf} \sim \frac{L^2}{v_a^2 n^4}, \quad \text{where} \quad P_{jf} \ll 1, \quad \text{i.e., } L/v_a \ll n^4. \tag{15.30}$$

At small n this process has an adiabatic character, and the quenching cross section sharply decreases with a decrease in n. Thus the cross section of this process as a function of n has a sharp maximum at $n_{\max} \sim (L/v_a)^{1/4}$, which corresponds to the cross section

$$\sigma_{\max} \sim \frac{L}{v_a} \sim n_{\max}^4. \tag{15.31}$$

15.6 Quenching of Metastable States in Collisions

Although this process is characterized by small cross sections, it can determine the lifetime of metastable atoms in gases. Interaction of the metastable atom and perturbed atom partially removes violation for the dipole radiation, so that the quasi-molecule consisting of the metastable and perturbed atoms emits dipole radiation. Denote the interaction operator for these atoms by V and assume that the main contribution to this process gives a resonantly excited state r which is nearest to the related metastable atom state m. Then the wave function of the quasi-molecule consisting of the metastable and perturbed atoms according to the first order of the perturbation theory can be presented in the form

$$\Psi = \psi_m + \frac{V_{mr}}{\varepsilon_m - \varepsilon_r}\psi_r,$$

where ψ_m, ψ_r are the wave functions for the metastable and nearest resonantly excited states, and ε_m, ε_r are the energies of these states. According to formula

(1.23) the probability per unit time for radiation of the quasi-molecule equals

$$w(R) = \frac{V_{mr}^2(R)}{(\varepsilon_m - \varepsilon_r)^2 \tau_r}, \tag{15.32}$$

where τ_r is the radiative lifetime of the resonantly excited state and, for simplicity, we neglect the difference of the photon energy and the excitation energy of the resonantly excited state.

Let us consider a general problem when a transition with a small probability takes place in the course of collision, and our goal is to evaluate the process cross section. Then the probability of this process, for the given parameters of collision, is equal to

$$P = \int_{-\infty}^{\infty} w(R)\, dt = \frac{2}{v} \int_{r_o}^{\infty} \frac{w(R) \rho \, dR}{R^2 \sqrt{1 - \rho^2/R^2 - U(R)/\varepsilon}}, \tag{15.33}$$

where r_o is the distance of closest approach, ε is the collision energy in the center-of-mass system, $U(R)$ is the interaction potential of the colliding atoms, and we use the connection between the parameters of collision on the basis of formulas (13.4) and (13.5). From this we have, for the cross section of this process,

$$\begin{aligned} \sigma &= \int_0^{\infty} 2\pi \rho \, d\rho \cdot \frac{2}{v} \int_{r_o}^{\infty} \frac{w(R) \rho \, dR}{R^2 \sqrt{1 - \rho^2/R^2 - U(R)/\varepsilon}} \\ &= \frac{4\pi}{v} \int_{R_o}^{\infty} R^2 \, dR \cdot w(R) \sqrt{1 - \frac{U(R)}{\varepsilon}}, \end{aligned} \tag{15.34}$$

where R_o is determined by the relation $U(R_o) = \varepsilon$. If this process proceeds in a gas, it is convenient to introduce the rate constant of the process,

$$k = \langle v\sigma \rangle = 4\pi \int_0^{\infty} \exp\left[-\frac{U(R)}{T}\right] \cdot w(R) R^2 \, dR, \tag{15.35}$$

where the averaging is made over the Maxwell distribution of atoms, and T is the gaseous temperature expressed in energetic units. Formulas (15.34) and (15.35) allow us to evaluate the cross section and rate constant for various processes of this type. In particular, in the related case the expression for the rate constant has the form

$$k = \frac{4\pi}{(\epsilon_m - \epsilon_r)^2 \tau_r} \int_0^{\infty} \exp\left[-\frac{U(R)}{T}\right] \cdot V_{mr}^2(R) R^2 \, dR. \tag{15.36}$$

If this process proceeds mostly in a repulsive region of the interaction of atoms, this gives a strong temperature dependence because of the sharp dependence $V_{mr}(R)$.

Problems

Problem 15.1. *Evaluate the position and value of the maximum cross section of the nonresonant charge exchange process for transition between two states.*

We work on the set of equations (15.14) for the probability amplitudes. According to the above analysis, the maximum of the cross section of this process is expected at small values of the parameter $\kappa/(\gamma v)$. Therefore, we will expand the transition probability over this small parameter. At the impact parameters of collision $\rho \leq R_c$ the transition probability is given by the Demkov formula which has the form, at $\kappa/(\gamma v) \ll 1$,

$$P(\rho) = \sin^2 \zeta \cdot \left[1 + \frac{1}{2}\left(\frac{\pi\kappa}{2\gamma v}\right)^2 \frac{1}{\sqrt{1-\rho^2/R_c^2}}\right]^{-1}, \qquad R_c - \rho \gg \frac{1}{\gamma}.$$

The transition probability for large impact parameters $\rho \geq R_c$ is determined by formula (15.15) of the perturbation theory which, at $\kappa/(\gamma v) \ll 1$, can be presented in the form

$$P(\rho) = \sin^2 \zeta \cdot \left(1 - \frac{\rho\kappa^2}{2\gamma v^2}\right), \qquad \rho \geq R_c.$$

From this it follows that for the transition cross section,

$$\sigma = \int_0^\infty 2\pi\rho\, d\rho\, P(\rho) = \sigma_{res} - \frac{\pi R_c^2}{2}\left(\frac{\pi\kappa}{2\gamma v}\right)^2 \ln(a\gamma R_c),$$

where $\sigma_{\text{res}} = \int_0^\infty \sin^2 \zeta(\rho) \cdot 2\pi\rho\, d\rho$ is the cross section of the resonant charge exchange, and a parameter $a \sim 1$, because the used Demkov formula for the probability transition is valid up to $R_c - \rho \sim 1/\gamma$. Because $\sigma_{\text{res}} \sim \pi R_c^2$, the second term in the expression for the transition cross section is small compared to the first one.

Let us analyze the obtained expression for the cross section. Its maximum is close to $\pi R_c^2/2$ because of a weak dependence of the cross section of resonant charge exchange on the collision velocity, and the velocity $v_{\max}$, which corresponds to the cross section maximum, is given by the relation

$$\frac{\kappa}{\gamma v_{\max}} = \frac{2}{\pi}\sqrt{\frac{2\sigma_{\text{res}}}{\pi R_c^2} \cdot \frac{1}{R_o\gamma} \cdot \frac{1}{\ln(aR_c\gamma)}},$$

where $R_o = \sqrt{2\sigma_{\text{res}}/\pi}$. Since $\sigma_{\text{res}} \approx \pi R_c^2/2$, we have

$$\frac{\kappa}{\gamma v_{\max}} = \frac{2}{\pi}\sqrt{\frac{1}{R_o\gamma} \cdot \frac{1}{\ln(aR_c\gamma)}} < 1/\sqrt{\gamma R_o} \ll 1,$$

which was used above.

Let us compare the velocities $v_{\max}$ and v_o which are given by the relation $\sigma_{\text{res}}(v_o) = \pi R_c^2/2$. According to the definition of the resonant charge exchange cross section (14.15), we have

$$\zeta(R_c) = \frac{1}{v_o}\sqrt{\frac{\pi R_c}{2\gamma}}\Delta(R_c) = 0.28,$$

and because of $\Delta(R_c) = \kappa$, we obtain

$$\frac{\kappa}{\gamma v_o} = \frac{0.22}{\sqrt{\gamma R_c}}.$$

This gives

$$\frac{v_{\max}}{v_o} = 0.35\sqrt{\ln(a\gamma R_c)}.$$

From this it follows that the values v_o and $v_{\max}$ have the same order of magnitude.

Problem 15.2. *Determine the cross section of the detachment of a negative ion with the external electron shell s^2 in collision with a fast ion or electron within the framework of the perturbation theory.*

Let us take into account one binding state of the negative ion and use formula (15.27) for the probability of the related transition which has the form

$$P(\rho) = \frac{4\omega^2}{3v^4} \sum_k |\mathbf{r}_{ok}|^2 \left[K_0^2\left(\frac{\omega\rho}{v}\right) + K_1^2\left(\frac{\omega\rho}{v}\right) \right],$$

where subscript o refers to the negative ion state and subscript k refers to the system of an atom and free electron. In the limit $\rho \ll v/\omega$ this formula gives

$$P(\rho) = \frac{8}{3v^2\rho^2} \sum_k |\mathbf{r}_{ok}|^2 = \frac{8}{3v^2\rho^2} \left(r^2\right)_{oo}, \qquad \rho \ll \frac{v}{\omega}.$$

Using the asymptotic wave function (7.8) for the s-electron of the negative ion

$$\psi_o = B\sqrt{\frac{\gamma}{2\pi}} e^{-\gamma r}/r,$$

and accounting for two valent electrons of the negative ion we obtain, for the matrix element by analogy with Problem 9.5,

$$\left(r^2\right)_{oo} = \frac{B^2}{\gamma^2},$$

where $\gamma^2/2$ is the electron binding energy in the negative ion and B is the asymptotic coefficient. From this it follows that, for the cross section of the related process,

$$\sigma_{\text{det}} = \int_0^\infty 2\pi\rho\, d\rho\, P(\rho) = \frac{16\pi}{3} \frac{B^2}{v^2\gamma^2} \ln \frac{\rho_{\max}}{\rho_{\min}}.$$

As the upper limit of integration we take $\rho_{\max} \sim v/\omega \sim v/\gamma^2$, because at the larger impact parameters of collision the used expansion of the Macdonald functions is not valid. As the lower limit of integration we take $\rho_{\min} \sim (v\gamma)^{-1}$ from the relation $P(\rho_{\min}) \sim 1$ where the used perturbation theory is violated. From this it follows that, for the cross section of detachment of the negative ion,

$$\sigma_{\text{det}} = \frac{16\pi}{3} \frac{B^2}{v^2\gamma^2} \ln \frac{av^2}{\gamma},$$

where the parameter $a \sim 1$. This formula corresponds to the detachment of the negative ion by both electron and ion impact. This is valid for the corresponding high-collision velocities

$$v \gg \sqrt{\gamma},$$

at which $\rho_{\max} \gg \rho_{\min}$.

CHAPTER 16

Collisions with Transitions in States of Continuous and Quasi-Continuous Spectra

16.1 Transitions in States of Continuous Spectra

The ionization of atomic particles in slow collisions results from the intersection of a corresponding electron term with the boundary of continuous spectra. Examples of such behavior of the electron terms were given in Chapter 10 for the interaction of a negative ion with an atom, and in Chapter 11 for the interaction of two helium atoms. Now we consider methods for the analysis of such transitions. If the related discrete level which is denoted by i is found above the boundary of continuous spectra, we have an autoionization or autodetachment state, and the corresponding discrete level is characterized by a width Γ, so that the energy of this state is equal to $E_j = \varepsilon(R) + i\Gamma(R)$, where

$$\Gamma = 2\pi \sum_f \left|H_{jf}\right|^2 \frac{dg_f}{d\varepsilon}, \tag{16.1}$$

where the subscript f corresponds to the states of continuous spectra, g_f is the statistical weight of this state, and this formula gives the probability per unit time for transitions from a given state which corresponds to the perturbation theory. Since the wave function of this state is $\Psi \sim \exp(-iEt)$, the probability of the survival of a state is equal to $P = |\Psi|^2 = \exp(-\Gamma t)$. But the description of the state on the basis of its width demands the fulfillment of two conditions. First, the energy above the boundary of the continuous spectra must remarkably exceed its width $\varepsilon \gg \Gamma$. Second, the rate of variation of a distance between atoms must be small enough, so that $d\varepsilon/dt \ll \varepsilon^2$. Below we analyze the behavior of an electron term near the continuous spectra boundary where these conditions can be violated.

If an electron term corresponding to two atoms intersects the boundary of continuous spectra, it intersects first the infinite number of terms of excited states. Let us evaluate the probability for the survival of this term when it goes into continuous spectra. Assume that all terms of the excited states are parallel each to other and that the interaction with each term is small, i.e., the regions of transition in each of the intersected states are separated. The probability of the survival of a related j-state, after the passage of the intersection region with a state f, is given by the Landau–Zener formula (15.10):

$$p_f = \exp\left(-\frac{2\pi}{\beta}\left|H_{jf}\right|^2\right), \quad \text{where} \quad \beta = Fv_R = d(E_j - E_f)/dt.$$

Hence the probability of survival of the related term, after the passing of regions of intersections with n parallel terms, is equal to

$$P_n = \prod_{f=1}^{n} p_f = \exp\left(-\frac{2\pi}{\beta}\sum_f \left|H_{jf}\right|^2\right).$$

As follows from the above formulas, the probability of survival of the related term can be presented in the form

$$P_n = \exp\left(-\int^t \Gamma dt'\right), \quad \text{where} \quad \Gamma = \frac{2\pi}{\beta dt}\sum_f \left|H_{jf}\right|^2 = 2\pi\frac{dn}{d\varepsilon}\sum_f \left|H_{jf}\right|^2, \tag{16.2}$$

and $dn/d\varepsilon$ is the number of levels per unit energy. As is seen, this consideration leads to the concepts of a level width in accordance with (16.1) although the related term is found under the boundary of continuous spectra.

Using the concepts of the width of an autoionization or an autodetachment level we obtain, for the cross section of the electron release resulting from the collision of two atoms or an atom and a negative ion,

$$\sigma = \int_0^{R_c} 2\pi\rho\, d\rho \left[1 - \exp\left(-\int^t \Gamma dt'\right)\right], \tag{16.3}$$

where R_c is the distance of intersection between an electron term of the initial electron state with the boundary of continuous spectra, and the integral over time is taken for a time when this term is above the boundary of continuous spectra.

Let us consider the process of detachment of a negative ion as a result of atomic collisions within the framework of the above concepts of an autodetachment state of the quasi-molecule consisting of a colliding negative ion and atom. The behavior of the corresponding electron terms for these atomic particles is given in Fig. 16.1, and the electron release takes place at distances between the colliding particles less than R_c. A simple result corresponds to the case of slow collisions, if decay of the negative ion takes place for collisions whose distance of closest approach is less than R_c. The cross section of this process is equal to $\pi\rho_c^2$, where ρ_c is the impact parameter of the collision with the closest approach distance R_c. Using relationship (13.4) between the impact parameter of collision and the distance of

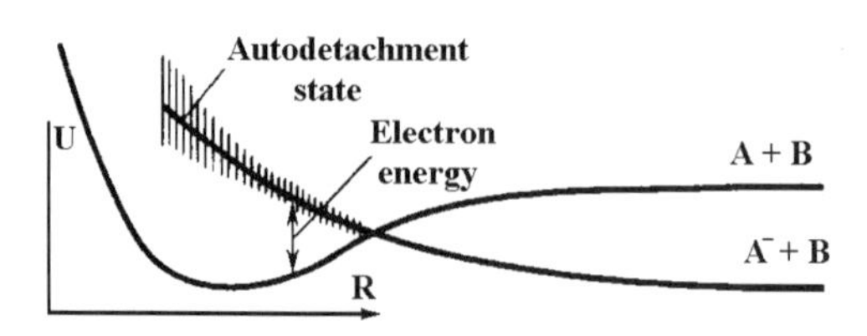

Figure 16.1. The character of detachment of a negative ion resulting from collision with an atom. The width of the autodetachment level is shown.

closest approach we obtain, for the detachment cross section,

$$\sigma_{\rm det} = \pi R_c^2 \left[1 - \frac{U(R_c)}{\varepsilon}\right], \tag{16.4}$$

where $U(R)$ is the interaction potential of colliding particles at a distance R between them, and ε is the collision energy in the center-of-mass frame of axes.

One can use formulas (15.33), (15.34), (15.35) for the analysis of this and identical processes, resulting in decay of the autodetachment or autoionization state of a system of interacting atomic particles. We consider the limit when the interaction between colliding particles is weak, so that the decay probability during one collision is small. Then taking into account that decay proceeds at distances less than R_c we obtain, by analogy with (15.33) for the decay probability

$$P = \int_{-\infty}^{\infty} \Gamma(R)\, dt = \frac{2}{v} \int_{r_o}^{R_c} \frac{\Gamma(R) \rho\, dR}{R^2 \sqrt{1 - \rho^2/R^2 - U(R)/\varepsilon}},$$

where r_o is the distance of closest approach, and ε is the collision energy in the center-of mass system. From this we have, for the cross section of this process,

$$\sigma = \int_0^{\infty} 2\pi \rho\, d\rho \cdot \frac{2}{v} \int_{r_o}^{R_c} \frac{\Gamma(R) \rho\, dR}{R^2 \sqrt{1 - \rho^2/R^2 - U(R)/\varepsilon}}$$

$$= \frac{4\pi}{v} \int_{R_o}^{R_c} R^2 \Gamma(R) \sqrt{1 - \frac{U(R)}{\varepsilon}}, \tag{16.5}$$

where R_o is determined by $U(R_o) = \varepsilon$. We use this relation below to evaluate some processes of this type.

16.2 Mutual Neutralization at Collisions of Positive and Negative Ions

This process proceeds according to the scheme

$$A^- + B^+ \to A + B^*. \tag{16.6}$$

Process (16.6) can be responsible for the recombination of charges in a weakly ionized gas and, by nature, it is analogous to the charge exchange process because of a tunnel transition of a valence electron in the field of the positive ion. But the

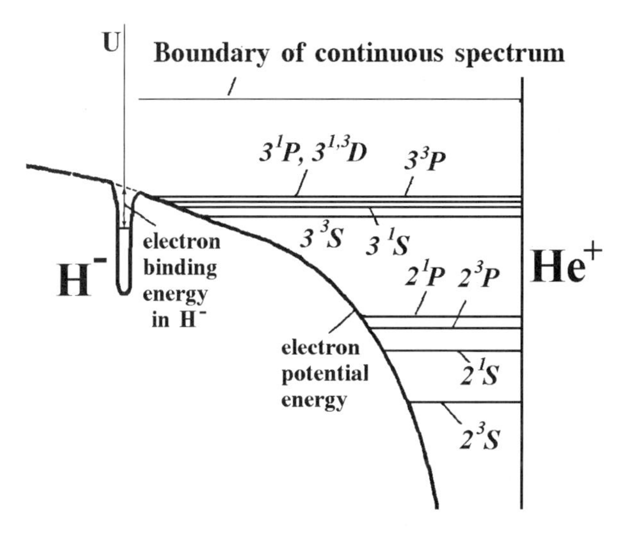

Figure 16.2. Positions of levels of the interacting hydrogen negative ion and helium atom at the distance between nuclei 2 nm (the cross section of the electron potential surface is shown). The process $H^- + He^+ \to H + He^*$ corresponds to formation of excited helium atoms with the main quantum number $n = 3$.

difference compared to a typical charge exchange corresponds to the possibility of the electron transition on many excited levels (see, e.g., Fig. 16.2). This allows us to use the quasi-continuous model for this process. Within the framework of this model we assume that the final electron state is similar to a state of continuous spectra, so that the transition character is the same as for the decay of an atomic particle in an electric field with the release of an electron. In contrast to the charge exchange process, a reverse electron transition in the atom field is absent under conditions of the quasi-continuous model because of a large number of occupied states in the ion field. Thus the negative ion term is characterized by a certain width, although its nature differs from that considered above. Now our goal is to evaluate this width.

We use the same method as in the case of atom decay in an electric field (see Chapters 2, 6, and 11), i.e., we evaluate the rate of the electron tunnel transition by accounting for the form of the barrier. Let the valent electron be in the s-state of the negative ion. Then the asymptotic expression of its wave function is given by formula (7.8):

$$\psi(r) = \frac{B}{r}\sqrt{\frac{\gamma}{2\pi}}e^{-\gamma r},$$

where r is the electron distance from the negative ion nucleus. In the transition region the electron wave satisfies the following Schrödinger equation:

$$\left(-\frac{1}{2}\Delta - \frac{1}{|\mathbf{R}-\mathbf{r}|}\right)\psi = \left(-\frac{\gamma^2}{2} - \frac{1}{R}\right)\psi,$$

where R is the distance between nuclei, and $\mathbf{r}$ is the electron distance from the positive ion. The equation can be separated into the ellipticity coordinates $\xi = (|\mathbf{R}-\mathbf{r}|+r)/R$, $\eta = l(|\mathbf{R}-\mathbf{r}|-r)/R$:

$$\psi(\mathbf{r}) = X(\xi)Y(\eta),$$
$$\frac{d}{d\xi}(\xi^2-1)\frac{dX}{d\xi} + \left(R\xi - \frac{R^2\gamma^2}{4}\xi^2 - \frac{R}{2}\xi^2 + C\right)X = 0;$$
$$\frac{d}{d\eta}(1-\eta^2)\frac{dY}{d\eta} + \left(-R\eta + \frac{R^2\gamma^2}{4}\eta^2 + \frac{R}{2}\eta^2 - C\right)Y = 0. \qquad (16.7)$$

Because of the large distance between nuclei (or a small rate of electron tunnel leakage) the electron tunneling takes place near the axis joining the nuclei. The electron wave function (7.8) has the following form near the axis ($\xi - 1 \ll 1$) and far from the center of the negative ion

$$\psi = \frac{2B\sqrt{2\gamma}}{R(1-\eta)\sqrt{\pi}}\exp\left[-\frac{R\gamma}{2}(1-\eta) - \frac{R\gamma}{2}(\xi-1)\right], \qquad \theta = 2\sqrt{\frac{\xi-1}{1-\eta}} \ll 1. \qquad (16.8)$$

Substituting this into equations (16.7), we find the separation constant

$$C = \frac{R^2\gamma^2}{4} + R\gamma - \frac{R}{2}.$$

From this one can obtain the quasi-classical electron wave function and the rate of tunnel transition. The wave function $Y(\eta)$ has the turning point $\eta_o = -(R\gamma^2/2-1)/(R\gamma^2/2+1)$, so that left from it the wave function has an exponential dependence on η and right from the turning point the wave function as a function of η oscillates. In the quasi-classical regions left and right of the turning point the electron wave function has the form

$$Y(\eta) = -\frac{iD}{\sqrt{|p|}\sqrt{1-\eta^2}}\exp\left(-\int_{\eta_o}^{\eta}|p|\,d\eta'\right), \qquad \eta < \eta_o;$$
$$Y(\eta) = \frac{D}{\sqrt{p}\sqrt{1-\eta^2}}\exp\left(i\int_{\eta_o}^{\eta}p\,d\eta' - i\frac{\pi}{4}\right), \qquad \eta > \eta_o,$$

where

$$p = \sqrt{\frac{R^2\gamma^2}{4} - \frac{R\gamma}{1-\eta^2} - \frac{R}{2}\frac{1-\eta}{1+\eta}}.$$

Joining this solution with expression (16.8) for the negative ion wave function, we find

$$D = -\frac{iB\gamma\sqrt{2}}{R\sqrt{\pi}}\exp\left[-R\gamma + \sqrt{2R}f\left(\frac{R\gamma^2}{2}\right)\right], \qquad (16.9)$$

$$\text{where} \quad f(x) = \frac{\ln(\sqrt{x}+\sqrt{1+x})}{\sqrt{1+x}}.$$

Now let us evaluate the rate of the electron tunneling. The probability of the tunnel transition per unit time is equal to $P = \int_S \mathbf{j}\, d\mathbf{S}$, where the electron current density is $\mathbf{j} = (i/2)(\psi \nabla \psi^* - \psi^* \nabla \psi)$, and the element of the surface near the axis is $dS = \rho\, d\rho\, d\varphi = (R/2)^2(1-\eta^2)\xi\, d\xi\, d\varphi$, so that the electron current density right of the turning point is equal to

$$j = \frac{2D^2 X^2(\xi)}{R\sqrt{(1-\eta^2)(\xi^2-\eta^2)}}.$$

From this we obtain, for the probability of the electron tunnel transition per unit time,

$$\Gamma(R) = \frac{\pi D^2}{\gamma} = \frac{B^2}{2R^2} \exp\left[-R\gamma + \sqrt{2R} f\left(\frac{R\gamma^2}{2}\right)\right]. \tag{16.10}$$

In the limit, $R\gamma^2 \ll 1$, this gives

$$\Gamma(R) = \frac{B^2}{2R^2} \exp\left(-\frac{2}{3} R^2 \gamma^3\right). \tag{16.11a}$$

In the other limit case, $R\gamma^2 \gg 1$, we get

$$\Gamma(R) = \frac{B^2}{2R^2} \exp(-2R\gamma). \tag{16.11b}$$

The cross section of the exchange process is determined by formula (16.3) which we use in the form

$$\sigma_{\text{ex}} = \int_0^\infty 2\pi\rho\, d\rho \left[1 - \exp(-F)\right], \quad \text{where} \quad F(\rho) = \int_{-\infty}^\infty \Gamma\, dt. \tag{16.12}$$

Let us assume the cross section of the elastic scattering of two ions to be relatively small, i.e., suppose the free motion of ions and account for a strong dependence $F(\rho)$. Then we divide the integral (16.12) into two parts

$$\sigma_{\text{ex}} = \int_0^{\rho_o} 2\pi\rho\, d\rho + \int_{\rho_o}^\infty 2\pi\rho\, d\rho \left[1 - \exp(-F)\right],$$

so that $F(\rho_o) \ll 1$ and the second integral converges in a narrow range of ρ. At these ρ we have $F(\rho) = F(\rho_o)\exp[-\omega(\rho - \rho_o)]$, where $\omega = d\ln F/d\rho$ and $\omega\rho_o \gg 1$. Then we obtain

$$\sigma_{\text{ex}} = \pi\rho_o^2 + \frac{2\pi\rho_o}{\omega} \int_0^{F(\rho_o)} \frac{dx}{x}(1 - e^{-x}) = \pi\rho_o^2 + \frac{2\pi\rho_o}{\omega}[C + \ln F(\rho_o)],$$

where $C = 0.577$ is the Euler constant. With accuracy of the order of $1/(\omega\rho_o)^2$ this can be represented in the form

$$\sigma_{\text{ex}} = \pi R_o^2, \quad \text{where} \quad F(R_o) = e^{-C} = 0.56. \tag{16.13}$$

In particular, in the case (16.11b), $R_o\gamma^2 \gg 1$ when the action of a field of a positive ion on the transferring electron is analogous to the action of a constant electric field, this formula gives

$$\sigma_{\text{ex}} = \frac{3\pi}{2\gamma^3} \ln \frac{v_o}{v}, \quad \text{where} \quad v_o = \frac{3.9A^2}{R_o^2 \gamma^3}. \tag{16.14}$$

One can account for the elastic scattering of nuclei when it is essential. This scattering is determined by the Coulomb interaction of nuclei. Assuming that the electron tunnel transition proceeds mainly near the distance of the closest approach of atoms, we use the relation between the impact parameter ρ and the distance of closest approach R_o:

$$1 - \frac{\rho^2}{R_o^2} = -\frac{1}{R_o \varepsilon},$$

where ε is the collision energy. This gives, for the cross section of the related process,

$$\sigma_{\text{ex}} = \pi R_o^2 + \pi R_o/\varepsilon, \quad \text{where} \quad F(R_o) = 0.56. \tag{16.15}$$

As is seen, the cross section of the mutual neutralization of ions slowly depends on the collision energy at large energies ($\varepsilon \gg 1/R_o$) and is inverse to the collision energy at small energies ($\varepsilon \ll 1/R_o$). As a demonstration of this method, Fig. 16.3 gives values of the cross sections of the mutual neutralization process N^+, $O^+ + O^- \rightarrow N, O + O$, which are evaluated on the basis of formulas (16.13), (16.14), (16.15), and also the experimental values of these cross sections.

Let us analyze the criterion of validity of the above quasi-continuous model for the considered charge exchange process. Since the electron transfers in many excited levels of the positive ion field, the perturbation theory is valid for the transition on each of these levels. Hence we consider the set of equations (15.14) within the framework of the perturbation theory which now has the form

$$i\frac{dc_n}{dt} = \frac{\Delta_n}{2} e^{i\omega_n t} a, \tag{16.16}$$

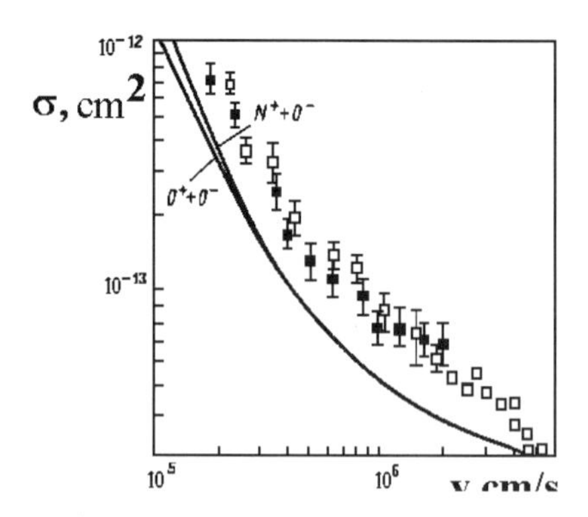

Figure 16.3. The cross section of charge exchange for processes $N^+ + O^- \rightarrow N + O$, $O^+ + O^- \rightarrow 2O$. Curves are calculated on the basis of formulas (16.13)–(16.15), signs refer to experiments.

where n is the number of a state for the electron in the field of the positive ion, so that c_n is the probability amplitude for this state, a is the probability amplitude for the negative ion state, and Δ_n, ω_n are the exchange interaction potential and difference diagonal matrix elements of the Hamiltonian for these states. Solving this equation on the basis of the perturbation theory ($a = 1$) we obtain, for the transition probability,

$$P_n(\rho) = \left| \int_{-\infty}^{\infty} \frac{\Delta_n}{2} e^{i\omega_n t}\, dt \right|^2 = \frac{1}{v^2} \cdot \frac{2\pi\rho}{\alpha} \Delta_n^2(\rho) \exp\left(-\frac{\omega_n^2 \rho}{\alpha v^2}\right),$$

where $\alpha = -d \ln \Delta_n / d\rho$. The total probability of the electron transfer is equal to

$$P(\rho) = \sum_n P_n(\rho) = \frac{1}{v^2} \cdot \frac{2\pi\rho}{\alpha} \sum_n \Delta_n^2(\rho) \exp\left(-\frac{\omega_n^2 \rho}{\alpha v^2}\right).$$

We use formula (10.44) for the exchange interaction between negative and positive ions through the state of an excited atom

$$\Delta_n = 2B\sqrt{2\pi\gamma}\,\psi_n(\mathbf{R}),$$

where B is the asymptotic coefficient for the electron wave function of the negative ion and $\psi_n(\mathbf{R})$ is the electron wave function for the corresponding excited atom state at the point of location of a perturbed atom. This gives

$$P(\rho) = \frac{1}{v^2} \cdot \frac{16\pi^2 B^2 \gamma \rho}{\alpha} \sum_n |\psi_n(\rho)|^2 \exp\left(-\frac{\omega_n^2 \rho}{\alpha v^2}\right).$$

Taking into account the high density of excited levels, we substitute the summation in this formula by integration. Then the difference of energies for the negative ion and excited atom is

$$\omega_n = \frac{\gamma^2}{2} - \frac{1}{2n^2} + \frac{1}{R},$$

where $\gamma^2/2$ is the electron binding energy in the negative ion and n is the principal quantum number of an excited atom state. The integration of the exponent gives

$$\int dn \exp\left(-\frac{\omega_n^2 \rho}{\alpha v^2}\right) = \sqrt{\frac{2\pi\alpha v^2}{\rho}} \cdot \left(\sqrt{\frac{d^2\omega_n^2}{dn^2}\Big|_{n_o}}\right)^{-1} = \frac{v\sqrt{\pi\alpha}}{\sqrt{\rho}} \left(\frac{d\omega_n}{dn}\Big|_{n_o}\right)^{-1}$$
$$= v n_o^3 \sqrt{\frac{\pi\alpha}{\rho}},$$

where n_o is defined by the relation $\omega_{n_o} = 0$. Thus we obtain

$$P(\rho) = \frac{16\pi^2 B^2 \gamma \rho}{v} \sqrt{\frac{\pi\rho}{\alpha}} n_o^3 \sum_{l,m} |\psi_{nlm}(\rho)|^2 . \qquad (16.17)$$

From this one can determine the criteria of validity of the quasi-continuous model. This requires that the exponent varies continuously with the change of the principal quantum number, so that

$$\frac{\rho}{\alpha v^2}\left|\omega_n^2-\omega_{n+1}^2\right| \ll 1.$$

Since $d\omega_n/dn = n^{-3}$ and the main contribution into the integral gives $\omega_n \sim v\sqrt{\alpha/\rho}$, this criterion has the form

$$v \gg \frac{1}{n^3}\sqrt{\frac{\rho}{\alpha}}. \tag{16.18a}$$

This condition provides nonadiabatic conditions for the electron transitions on neighboring levels of an excited atom. The other condition requires the process to be slow, so that a typical process time $\tau_{col} \sim (1/v)\sqrt{\rho/\alpha}$ must be large compared to a typical atomic time $\sim 1/\gamma^2$. This gives

$$v \ll \gamma^2\sqrt{\frac{\rho}{\alpha}}. \tag{16.18b}$$

A comparison of criteria (16.18) shows that the quasi-continuous model is valid for states with a small electron binding energy.

Let us assume that the electron state in the positive ion field is similar to a highly excited hydrogen atom. Then it is convenient to use parabolic quantum numbers for the electron excited states, so that summation in formula (16.17) has the form

$$\sum_{n_1,n_2,m}\left|\psi_{nn_1,n_2m}(\mathbf{R})\right|^2 = \pi n^4\sum_{n_1=0}^{n}\left|F\left(-n,1,\frac{2R}{n}\right)\right|^2.$$

If the electron is located in a classically prohibited region under the barrier, we obtain

$$\sum_{n_1,n_2,m}\left|\psi_{nn_1,n_2m}(\mathbf{R})\right|^2 = \frac{1}{32\pi n^3[p_n(R)]^2}\exp\left(-2\int_{r_n}^{R}p_n\,dr\right),$$

where $p_n(r)=\sqrt{1/n^2-2r}$ and the turning point r_n is determined by the relation $p_n(r_n)=0$. The above expression is valid in the quasi-classical region $\int_{r_n}^{R}p_n\,dr \gg 1$. From this we obtain

$$P(\rho)=\frac{1}{v}\sqrt{\frac{\pi\rho}{\alpha}}\frac{B^2}{2\gamma\rho^2}\exp\left(-2\int_{r_n}^{R}\sqrt{\gamma^2+\frac{2}{R}-\frac{2}{r}}dr\right).$$

Write this expression in the form

$$P(\rho)=\frac{1}{v}\sqrt{\frac{\pi\rho}{\alpha}}\Gamma(\rho),\quad \text{where}\quad \Gamma(\rho)=\frac{B^2}{2\gamma\rho^2}\exp\left(-2\int_{r_n}^{R}\sqrt{\gamma^2+\frac{2}{R}-\frac{2}{r}}dr\right). \tag{16.19}$$

Since

$$\alpha=-\frac{d\ln\Delta(\rho)}{d\rho}=-\frac{d\ln\Gamma}{2d\rho},$$

one can rewrite the above formula for the transition probability in the form $P(\rho) = \int_0^\infty \Gamma(\rho)\,dt$, where the time t is introduced from the relation $R^2 = \rho^2 + v^2t^2$. This shows the analogy of this consideration on the basis of the perturbation theory and the above consideration within the framework of the concept of an autodetachment state. Let us expose expression (16.19) for the level width $\Gamma(\rho)$ in the limiting cases. In the limit $R\gamma^2 \ll 1$ this gives

$$\Gamma(R) = \frac{B^2}{2\gamma R^2} \exp\left(-\frac{2}{3}R^2\gamma^2\right), \qquad R\gamma^2 \ll 1,$$

and

$$\Gamma(R) = \frac{B^2}{2\gamma R^2} \exp(-2R\gamma), \qquad R\gamma^2 \gg 1,$$

in accordance with (16.11). Thus, the coincidence of expressions for the process rates, which are obtained by different methods, demonstrates both the physical nature of this process and the conditions of validity of the result.

16.3 Collisional Transitions Involving Multicharged Ions

The character of the electron transition resulting from the collision of an atom and multicharged ion is similar to the case of the exchange process of the collision of negative and positive ions. In both cases the electron transfers to a group of levels of highly excited states of the electron in the field of a Coulomb center. Hence the above formulas can be used for charge exchange during the collision of atoms and multicharged ions in a certain range of collision parameters. Below we use the results of the previous section for the charge exchange process involving multicharged ions. In the case of the charge exchange process involving an atom and a multicharged ion of a charge Z, the criteria of validity of the quasi-continuous model (16.18) have the form

$$\gamma^2\sqrt{\frac{\rho}{\alpha}} \gg v \gg \frac{Z^2}{n^3}\sqrt{\frac{\rho}{\alpha}}, \tag{16.20}$$

and the energy difference for the transition states at a distance R between nuclei is equal to

$$\omega_n = \frac{\gamma^2}{2} - \frac{Z^2}{2n^2} + \frac{Z}{R}.$$

In the case $\rho\gamma^2 \gg Z$ we have $\alpha \approx \gamma$, $n \approx Z/\gamma$, so that the criteria (16.20) are the following, in this case,

$$\gamma\sqrt{\rho\gamma} \gg v \gg \gamma, \qquad \rho\gamma^2 \gg Z. \tag{16.21a}$$

In the other limit we obtain

$$\sqrt{Z\gamma} \gg v \gg \gamma, \qquad \rho\gamma^2 \ll Z. \tag{16.21b}$$

From this it follows that the quasi-continuous model is working for some range of parameters if $\gamma \sim 1$, $Z \gg 1$.

Let us demonstrate this model in the case of the electron transition from the ground state of the hydrogen atom in the highly excited states of a multicharged ion. This model is especially simple, if the main contribution to the cross section gives impact parameters ρ which satisfy the relation $\rho\gamma^2 \ll Z$. Then an action of the multicharged ion on the hydrogen atom is determined by its electric field whose strength is equal to $E = Z/R^2$, where R is the distance between nuclei. The rate of hydrogen atom decay in a constant electric field of strength E is (see Problem 6.3):

$$w = \frac{4}{E} \exp\left(-\frac{2}{3E}\right).$$

This decay corresponds to the electron tunnel transition in the field of the multicharged ion and is similar to the tunnel electron transition in a constant electric field. Hence this probability of decay per unit time is analogous to the level width $\Gamma(R)$ which is equal to, according to this formula ($E = Z/R^2$),

$$\Gamma(R) = \frac{4R^2}{Z} \exp\left(-\frac{2R^2}{3Z}\right).$$

Then, on the basis of formulas (16.12), (16.13) we obtain, for the charge exchange cross section,

$$\sigma_{\text{ex}} = \pi R_o^2, \quad \text{where} \quad \int_{-\infty}^{\infty} \Gamma(R)\, dt|_{R_o} = \frac{2}{v}\sqrt{\frac{6\pi}{Z}} \cdot R_o^2 \exp\left(-\frac{2R_o^2}{3Z}\right) = 0.56. \tag{16.22}$$

This cross section can be represented in the form

$$\sigma_{\text{ex}} = \frac{3}{2}\pi Z \Phi\left(\frac{v}{\sqrt{Z}}\right), \qquad \Phi \ll Z,$$

where the function $\Phi(x)$ satisfies the relation

$$\Phi \exp(-\Phi) = \frac{0.56x}{3\sqrt{6\pi}} = 0.043x.$$

As a matter of fact, the cross section (16.22) is the upper boundary for the charge exchange cross section, because it gives the cross section in the region $R_o^2 \gg Z$ and gives an excessive result at the collision parameters where this condition is violated. The lower boundary for the cross section one can determine as $\sigma_{\text{ex}} = \pi R_*^2$, where R_* is the distance between nuclei when the barrier disappears between fields of the multicharged ion and proton. Since $R_* = 2\sqrt{2Z}$, the lowest boundary for the cross section is

$$\sigma_{\min} = 8\pi Z. \tag{16.23}$$

Figure 16.4 contains cross sections (16.22) and (16.23) between which are found the real values of this process.

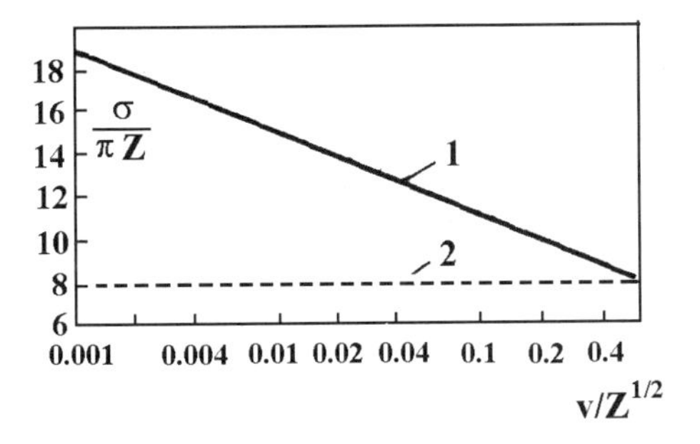

Figure 16.4. The cross section of charge exchange involving the hydrogen atom and a multicharged ion of a charge $Z \gg 1$. Curve 1 corresponds to formula (16.22), curve 2 refer to formula (16.23).

16.4 Associative Ionization and the Penning Process

Collisional ionization involving excited atoms can be determined by different mechanisms. Each of them corresponds to the decay of an autoionization state of the quasi-molecule consisting of colliding atoms. One such process, the associative ionization process, proceeds according the scheme

$$A^* + B \rightarrow AB^+ + e. \tag{16.24}$$

As a demonstration of the peculiarity of this process, Fig. 16.5 explains the character of the associative ionization process in the case of the collisions of an excited helium atom with a helium atom in the ground state. At thermal collisions, the cross section of this process is approximately $2 \cdot 10^{-15}\,\mathrm{cm}^2$ for the 3^1D-excited state and essentially less for other excited states. According to the data of Fig. 16.5, a profitable position of terms of this state is due to the interaction of the corresponding electron term with the repulsive term corresponding to $\mathrm{He}(2^1P) + \mathrm{He}$. In the case of repulsion at the distance R_c of the term intersection with the boundary of continuous spectra, $U(R_c) > 0$, this process has a threshold at the collision energy $\varepsilon_o = U(R_c)$. As follows from formula (16.7), the threshold form of the cross section of associative ionization is the following:

$$\sigma_{\mathrm{as}} = C(\varepsilon - \varepsilon_o)^{3/2}, \quad \text{where} \quad C = \frac{4\pi}{3}\sqrt{2\mu}R_c^2\Gamma(R_c)/\left|\frac{dU}{dR}\right|_{R=R_c},$$

where μ is the reduced mass of colliding particles and we use the threshold behavior ($\Gamma = \mathrm{const}$) for $\Gamma(R)$.

The other mechanism of ionization takes place in the case that the excitation energy of one colliding atom exceeds the ionization potential of its partner (see also Chapter 6). This process is called the Penning process if the excited state is a metastable state. The Penning process can be responsible for the ionization of atoms in a gas discharge if the gaseous atoms have a high ionization potential and if there is an admixture of atoms with a small ionization potential. Examples of

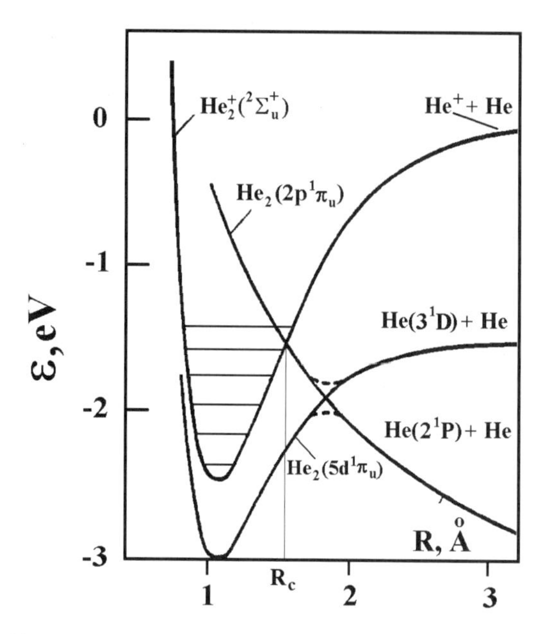

Figure 16.5. Positions of electron terms of the excited helium molecule and molecular ion which are responsible for the process $He(3^1D) + He \to He_2^+ + e$.

these processes are

$$He(2^3S) + Ar(Kr, Xe) \to He(1^1S) + Ar^+(Kr^+, Xe^+). \tag{16.25}$$

In this case the quasi-molecule consisting of colliding atoms is found in the autoionization state at any distance between atoms, but the width of the autoionization level decreases strongly with an increase in the distance between atoms. Indeed, let us represent the electron Hamiltonian in the form $\hat{H} = \hat{H}_o + V$, where the perturbation potential V is responsible for the transition of interacting electrons. According to formula (16.1), the level width is equal to

$$\Gamma = 2\pi \sum_f |V_{jf}|^2 \frac{dg_f}{d\varepsilon}, \tag{16.26}$$

where the subscript j corresponds to the initial state of the system and the subscript f correspond to a final state of the process which relates to the formation of a free electron. Note that this process corresponds to the transition of two electrons, because an excited atom transfers to the ground state, and a valence electron of the second atom releases.

Let us evaluate the dependence of the cross section of the Penning process on the collision energy (see Fig. 16.6). At small collision energies (the energy range 1 of Fig. 16.6) the cross section decreases with an increase in the collision energy because of a decrease in the time of location of colliding particles in the attractive zone. The cross section reaches its minimum at energies $\varepsilon \sim D$, where D is the depth of an attractive well for colliding particles. At higher collision energies

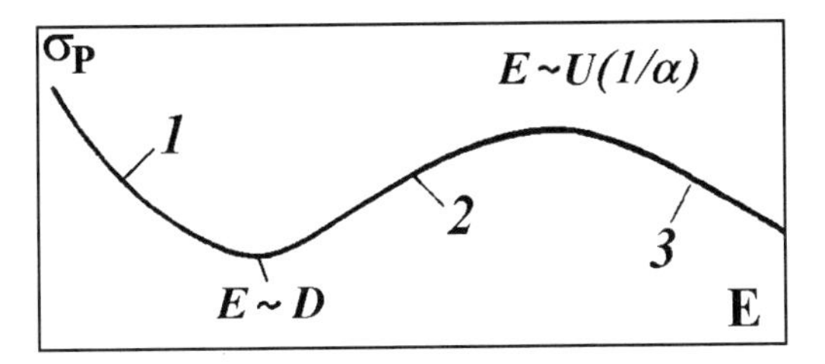

Figure 16.6. Schematic form of the dependence of the cross section for the Penning process on the collision energy.

(the energy range 2 of Fig. 16.6) the cross section increases with an increase in the collision energy because of a stronger interaction at small distances between atoms. Let us analyze the behavior of the cross section in this region, assuming $\Gamma(R)$ to be the strongest distance dependence in formula (16.5). Then, introducing $\alpha = -d\ln\Gamma(R_o)/dR$, where $U(R_o) = \varepsilon$ we obtain, from formula (16.5),

$$\sigma = \frac{4\pi}{v}\int_{R_o}^{\infty} R^2\Gamma(R)\sqrt{1-\frac{U(R)}{\varepsilon}}\,dR = \frac{2}{v}\cdot\left(\frac{\pi}{\alpha}\right)^{3/2}\cdot R_o^2\Gamma(R_o)\sqrt{\frac{|U'(R_o)|}{\varepsilon}}.$$

This formula is valid for a strong dependence $\Gamma(R)$, and the cross section reaches a maximum at $\alpha R_o \sim 1$. The next energy range 3 of Fig. 16.6 corresponds to a slow decrease in the cross section with an increase in the collision energy due to factors in expression (16.5) which depend on the collision velocity or energy. Figure 16.7 contains examples of the Penning process which confirm a general dependence of its cross section on the collision energy in accordance with the data of Fig. 16.6.

One more specific of this process refers to the spectrum of the released electron. The electron energy in each case is equal to the difference of the energies for the corresponding states of colliding atoms, so that the spectrum of released electrons gives information about both the distances between colliding atomic particles where the process takes place, and about the final states of formed particles.

16.5 Ionization in Collisions with Resonantly Excited Atoms

Although the Penning process with the partaking of a metastable atom is more spread, the ionization process involving a resonantly excited atom is more effective. Below we evaluate the cross section of this process. The interaction operator in this case is determined by formula (10.5),

$$V = \frac{1}{R^3}\left[\mathbf{D}_1\mathbf{D}_2 - 3(\mathbf{D}_1\mathbf{n})(\mathbf{D}_2\mathbf{n})\right],$$

where the subscripts 1, 2 refer to the corresponding atom and $\mathbf{n}$ is a unit vector directed along the molecular axis. We assume the cross section in this case to be large compared to a typical atomic value, so that the above expansion of the

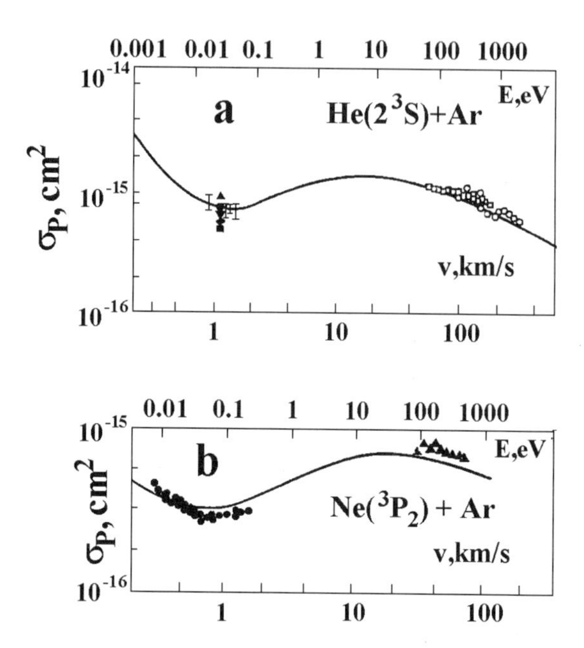

Figure 16.7. The dependence of the cross section of the Penning process on the collisional velocity for ionization of the argon atom in collisions with (a) He(2^3S) and (b) Ne($^3P_2$). The solid curve corresponds to calculations, signs refer to experiments.

interaction operator on a small parameter $1/R$ is valid. Let us take the motionless frame of axes, whose axes x, y, z are directed along the collision velocity $\mathbf{v}$, the impact parameter of collision ρ, and the motion moment respectively. Write the interaction operator in the form

$$V = \frac{1}{R^3}\left[D_{1x}D_{2x}(1 - 3\cos^2\theta) + D_{1y}D_{2y}(1 - 3\sin^2\theta) \right.$$
$$\left. + D_{1z}D_{2z} - 3(D_{1x}D_{2y} + D_{1y}D_{2x})\right]\sin\theta\cos\theta,$$

where θ is the angle between the vectors $\mathbf{v}$ and ρ. Thus the level width is expressed through the matrix elements of the dipole moment operators of atoms. The photoionization cross section is expressed through the same matrix elements. Hence, the value $\Gamma(R)$ can be expressed through the photoionization cross sections

$$\sigma_{\rm ph}^x = \frac{4\pi^2\omega}{c}\left|(D_{2x})_{jf}\right|^2 g_f,$$
$$\sigma_{\rm ph}^y = \frac{4\pi^2\omega}{c}\left|\left(D_{2y}\right)_{jf}\right|^2 g_f,$$
$$\sigma_{\rm ph}^z = \frac{4\pi^2\omega}{c}\left|(D_{2z})_{jf}\right|^2 g_f.$$

We use atomic units; the superscripts x, y, z denote the polarization of an incident photon whose frequency ω coincides with the atom excitation energy. Below, for simplicity, we assume that the photoionization cross sections do not depend on a photon polarization (the second atom has a filled electron shell). Then the level width or the rate of electron release resulting from decay of the quasi-molecule is equal to

$$\Gamma(R) = \frac{c\sigma_{\text{ph}}}{2\pi\omega R^6}\Big[\left|(D_{1x})_{jf}\right|^2(1+3\cos^2\theta) + \left|(D_{1y})_{jf}\right|^2(1+3\sin^2\theta) + \left|(D_{1z})_{jf}\right|\Big], \tag{16.27}$$

where the matrix element $(D_{1x})_{jf}$ is taken between the ground and excited states of atom A. The autonization term width $\Gamma(R)$ has an inverse power dependence on separations, while this dependence is an exponential one for the processes of type (16.24). Assuming the motion of colliding atoms to be free, from formula (16.27) we have

$$\int_{-\infty}^{\infty}\Gamma(R)dt = \frac{9c\sigma_{\text{ph}}}{32\omega\rho^5 v}\left[\left|(D_{1x})_{jf}\right|^2 + \frac{7}{3}\left|\left(D_{1y}\right)_{jf}\right|^2 + \frac{2}{3}\left|(D_{1z})_{jf}\right|\right].$$

This gives, for the ionization cross section according to formula (16.3),

$$\sigma_{\text{ion}} = \pi\Gamma\left(\frac{5}{3}\right)\left\{\frac{9c\sigma_{\text{ph}}}{32\omega v}\left[\left|(D_{1x})_{jf}\right|^2 + \frac{7}{3}\left|\left(D_{1y}\right)_{jf}\right|^2 + \frac{2}{3}\left|(D_{1z})_{jf}\right|\right]\right\}^{2/5}. \tag{16.28}$$

Let us average this cross section over the direction of the excited atom momentum. First we consider the case when this process corresponds to an $S - P$ transition for the first atom. Then we have, for the ionization cross section, if the P-state has a zero momentum projection on an axis with polar angles θ, φ:

$$\sigma_{\text{ion}} = \sigma_o\left(1 - \frac{1}{4}\cos^2\theta + \frac{3}{4}\sin^2\theta\cos^2\varphi - \frac{1}{2}\sin^2\theta\sin^2\varphi\right),$$

where

$$\sigma_o = \pi\Gamma\left(\frac{5}{3}\right)\left[\frac{9c\sigma_{\text{ph}}}{32\omega v}\left|(\mathbf{D}_1)_{ik}\right|^2\right]^{2/5} = 3.72\left(\frac{cf\sigma_{\text{ph}}}{\omega^2 v}\right)^{2/5}, \tag{16.29}$$

the oscillator strength for this transition is $f = \frac{2}{3}\omega\left|(\mathbf{D}_1)_{ik}\right|^2$, and the coordinate frame is such that x is the polar axis, $\varphi = 0$ corresponds to the plane xy. Then the average cross section for an $S - P$ transition is equal to $0.986\sigma_o$, and the partial cross sections for the corresponding directions of the atom momentum are equal to

$$\sigma_{\text{ion}}^x = 0.996\sigma_o, \qquad \sigma_{\text{ion}}^y = 0.969\sigma_o, \qquad \sigma_{\text{ion}}^z = 0.982\sigma_o.$$

As is seen, the difference of these partial cross sections is small, and the average cross section of ionization is

$$\sigma_{\text{ion}} = 0.986\sigma_o = 3.16\left[\frac{c\sigma_{\text{ph}}}{\omega v}\left|(\mathbf{D}_1)_{ik}\right|^2\right]^{2/5}. \tag{16.30}$$

This formula is valid if, in accordance with the concept of the autoionization level width, slow collisions give the main contribution to the cross section, so that the level width is small compared to the excess of the energy of an excited atom above the boundary of continuous spectra. This gives $\omega - J \gg 1/\tau \sim v/\rho$, where

J is the ionization potential of the second atom ($\omega - J$ is the energy of a released electron), τ is a typical collision time, v is the collision velocity, and $\rho \sim \sqrt{\sigma_{\text{ion}}}$ is a typical impact parameter. Thus the criterion of validity for formula (16.30) has the form

$$v^2 \ll (\omega - J)^2 \sigma_{\text{ion}}. \tag{16.31}$$

16.6 Collisional Ionization of Highly Excited Atoms

This process is determined by the interaction of a highly excited electron and an incident atom. Assuming the electron motion to be classical, one can consider this process as a result of the scattering of the excited electron on an incident atom. If the energy obtained by the electron exceeds its binding energy, the ionization process takes place. Within the framework of this model we have, for the ionization probability,

$$P(\rho) = \int dt\, |\psi(\mathbf{R})|^2\, |\mathbf{v} - \mathbf{v}_a| \int_{\Delta\varepsilon \geq J} d\sigma,$$

where $\mathbf{v}$ is the electron velocity, $\mathbf{v}_a$ is the relative velocity of nuclei, $\psi(\mathbf{R})$ is the electron wave function at the point of location of an incident atom, $d\sigma$ is the differential cross section of the electron–atom scattering, $\Delta\varepsilon$ is the energy change as a result of collision, and J is the ionization potential of the highly excited atom, i.e., the electron binding energy, so that $|\psi(\mathbf{R})|^2\, |\mathbf{v} - \mathbf{v}_a| \int_{\Delta\varepsilon \geq J} d\sigma$ is the ionization probability per unit time. Integration of this expression over the impact parameters of collision leads to the following expression for the ionization cross section:

$$\sigma_{\text{ion}} = \int P(\rho)\, d\rho = \int d\rho \frac{dz}{v_a}\, |\psi(\mathbf{R})|^2\, |\mathbf{v} - \mathbf{v}_a| \int_{\Delta\varepsilon \geq J} d\sigma = \left\langle \frac{|\mathbf{v} - \mathbf{v}_a|}{v_a} \int_{\Delta\varepsilon \geq J} d\sigma \right\rangle, \tag{16.32}$$

where an average is made over the electron distribution in a highly excited atom and we use that the collision trajectory is straight, i.e., $dt = dz/v_a$.

Formula (16.32) can be obtained in another way. Indeed, the probability per unit time for the ionization of a highly excited atom by incident atoms is equal to $N\langle |\mathbf{v} - \mathbf{v}_a| \int_{\Delta\varepsilon \geq J} d\sigma \rangle$, where N is the number density of incident atoms. Division of this value on the flux Nv_a of incident atoms leads to formula (16.32).

Let us show formula (16.32). We have, for a change of the electron energy resulting from a collision with an atom,

$$\Delta\varepsilon = \frac{P^2}{2\mu} - \frac{(\mathbf{P} - \Delta\mathbf{p})^2}{2\mu} = \mathbf{v}_a \Delta\mathbf{p} - \frac{\Delta\mathbf{p}^2}{2\mu},$$

where $\mathbf{P} = \mu\mathbf{v}_a$ is the nuclear momentum before collision, μ is the nuclear reduced mass, and $\Delta\mathbf{p}$ is the change of the electron momentum. The last value is equal to

$$\Delta p = 2\, |\mathbf{v} - \mathbf{v}_a| \sin\frac{\vartheta}{2},$$

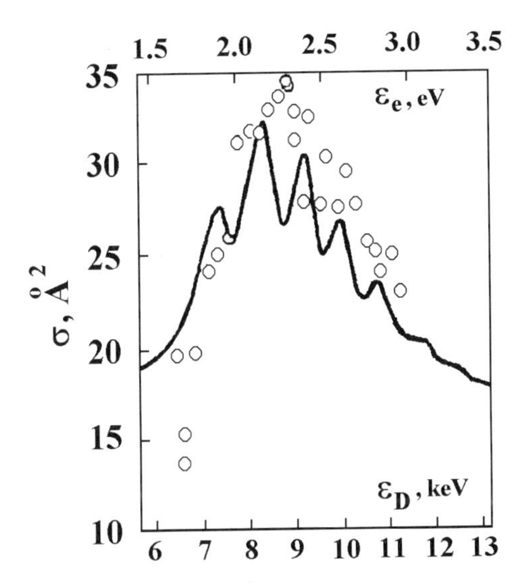

Figure 16.8. The cross section of the electron–nitrogen molecule elastic scattering (solid curve) and the ionization cross section of a highly excited deuterium atom with nitrogen molecule (open circles) at identical collisional velocities of electrons.

where ϑ is the scattering angle for electron–atom collision. Let us consider the case

$$v_a \gg v \sim 1/n,$$

where n is the principal quantum number of the highly excited electron [$J = 1/(2n^2)$]. In this case the energy change is equal to $\Delta\varepsilon = \mathbf{v}_a \Delta\mathbf{p} = v_a^2(1 - \cos\vartheta)$. Then the relation $\Delta\varepsilon \geq J$ is fulfilled for scattering angles which give the main contribution to the electron–atom elastic cross section, and the ionization cross section is equal to

$$\sigma_{\text{ion}} = \sigma_{\text{ea}}, \qquad v_a \gg 1/n, \tag{16.33}$$

where $\sigma_{\text{ea}} = \int d\sigma$ is the electron–atom elastic cross section. As a demonstration of this result, Fig. 16.8 compares the cross section of a highly excited deuterium atom resulting from collision with a nitrogen molecule and the cross section of the electron–nitrogen molecule elastic scattering. One can see the same character of the dependence on the collision energies for these cross sections.

16.7 Electron Attachment to Molecules

This process has a resonant character and results from electron capture on an autodetachment term of the negative ion. Hence, this process is effective at the favorable positions of electron terms of the molecule and its negative ion. As a demonstration of this, Table 16.1 contains the values of the rate constants for the attachment of thermal electrons to halomethane molecules. Although these

Table 16.1. The rate constants of electron attachment to halomethane molecules at room temperature.

Molecule	Rate constant, cm^3/s	Molecule	Rate constant, cm^3/s
CH_3Cl	$< 2 \cdot 10^{-15}$	CF_2Cl_2	$1.6 \cdot 10^{-9}$
CH_2Cl_2	$5 \cdot 10^{-12}$	$CFCl_3$	$1.9 \cdot 10^{-7}$
$CHCl_3$	$4 \cdot 10^{-9}$	CF_3Br	$1.4 \cdot 10^{-8}$
$CHFCl_2$	$4 \cdot 10^{-12}$	CF_2Br_2	$2.6 \cdot 10^{-7}$
CH_3Br	$7 \cdot 10^{-12}$	$CFBr_3$	$5 \cdot 10^{-9}$
CH_2Br_2	$6 \cdot 10^{-8}$	CF_3I	$1.9 \cdot 10^{-7}$
CH_3I	$1 \cdot 10^{-7}$	CCl_4	$3.5 \cdot 10^{-7}$
CF_4	$< 1 \cdot 10^{-16}$	CCl_3Br	$6 \cdot 10^{-8}$
CF_3Cl	$1 \cdot 10^{-13}$		

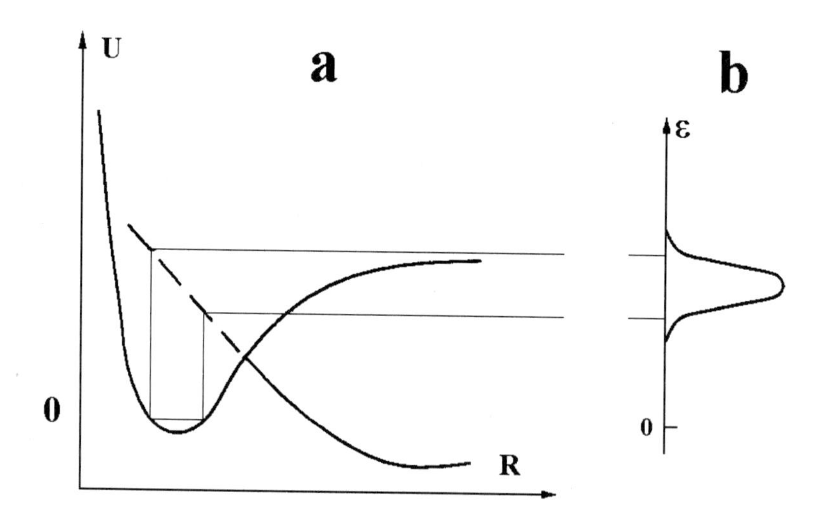

Figure 16.9. Positions of the molecule and negative ion electron terms during electron attachment to (a) a molecule, and (b) the spectrum of captured electrons.

molecules have a similar or close structure, the difference of these rate constants shows their dependence on the fine elements of interaction in these molecules and their negative ions that determine the positions of their electron terms.

Figure 16.9 represents the typical positions of the electron terms which partake in the electron attachment process. This process proceeds in three stages which include the electron capture in an autodetachment state, development of the forming negative ion, and the decay of the autodetachment state. Positions of the electron terms of Fig. 16.9 are typical for diatomic molecules and lead to the following expression for the electron attachment cross section

$$\sigma_{\rm at} = \frac{\pi\hbar^2}{2m\varepsilon}\int \frac{\Gamma^2(R)\,|\varphi_o(R)|^2\,dR}{[\varepsilon - E(R)]^2 + \Gamma^2(R)/4}\exp\left[-\int \Gamma(R)\,dt/\hbar\right], \qquad (16.34)$$

where R is the nuclear coordinate which is responsible for the electron attachment process, ε is the electron energy, m_e is the electron mass, $\Gamma(R)$ is the width of the autodetachment state, $E(R)$ is the excitation energy for the autodetachment state at a given nuclear coordinate, and $\varphi_o(R)$ is the nuclear wave function. The value $|\varphi_o(R)|^2\,dR$ is the probability of finding the nuclear coordinate in an interval from R to $R + dR$.

The basis of formula (16.34) is the Breight–Wigner formula for the resonant process, and the last factor $\exp\left[-\int \Gamma(R)\,dt/\hbar\right]$ is the probability of survival for the autodetachment state in the course of its evolution. In the case of diatomic molecules, when the rate of this process is small in thermal collisions, the survival probability is small. For example, for the process

$$e + H_2 \to H + H^-,$$

when the resonance in the electron attachment corresponds to the electron energy is at 4 eV, the survival probability at this electron energy is equal to $3 \cdot 10^{-6\pm0.2}$.

The small probability of survival of the autodetachment state in the course of its evolution leads to a strong dependence of the electron attachment cross section on the vibrational temperature of the molecule. In addition, the electron energy at which the cross section has a maximum, decreases with excitation of the molecule. For example, the process

$$e + O_2 \to O + O^-$$

is characterized by the resonance at the electron energy 6.7 eV at room temperature of the molecules, and the maximum cross section is $1.2 \cdot 10^{-18}\,\mathrm{cm}^2$. An increase in the molecular temperature up to 1930 K shifts the energy of the cross section maximum to about 5.7 eV, and the maximum cross section becomes about $4 \cdot 10^{-18}\,\mathrm{cm}^2$ at this molecular temperature. Note that only 21% of molecules are excited vibrationally at this molecular temperature. In the case when the oxygen molecule is found in the metastable electron state $^1\Delta_g$ whose excitation energy is 1 eV, the maximum cross section of electron attachment is approximately $6 \cdot 10^{-18}\,\mathrm{cm}^2$ and corresponds to the electron energy 5.7 eV. The resonance width in the electron attachment cross section increases with excitation of the molecule.

Let us consider the case when the electron attachment process is effective in thermal collisions. The molecules for which it takes place are used for the protection of power electric systems from electrical breakdown. Therefore, the electron attachment to such molecules is well studied. The effective electron attachment is possible if the molecular electron term and the electron term of the autodetachment negative ion are intersected near the bottom of the molecular potential well. This takes place for some complex molecules (see Table 16.1), in particular, the rate constant of electron attachment to the SF_6 molecule at room temperature is equal to $2.5 \cdot 10^{-7}\,\mathrm{cm}^3/s$. Because of the profitable positions of electron terms in these cases, the survival probability of the autodetachment term during its evolution to a stable state is close to unity, and the electron attachment cross section is equal to the electron capture on an autodetachment term, that is,

$$\sigma_{\text{at}} = \frac{\pi \hbar^2}{2m_e \varepsilon} \int \frac{\Gamma^2(R)\, |\varphi_o(R)|^2\, dR}{[\varepsilon - E(R)]^2 + \Gamma^2(R)/4}. \tag{16.35}$$

We now consider the limiting case when $\varepsilon \gg \Gamma(R)$ and the energy difference for the transition terms is linear with respect to the reaction coordinate R, i.e.,

$$\varepsilon - E(R) = E'_R(R - R_c),$$

where R_c is the point of intersection of terms. Then in the limit of small Γ we have for the capture cross section

$$\sigma_{\text{at}} = \frac{\pi^2 \hbar^2}{m_e \varepsilon} \frac{\Gamma(R_o)\, |\varphi_o(R_o)|^2}{E'_R}, \tag{16.36}$$

where the resonance point R_o is given by $\varepsilon = E(R_o) = E'_R(R_o - R_c)$. In the region of a weak dependence of the formula (16.36) parameters on the electron energy we have

$$\sigma_{\text{at}} \sim 1/\varepsilon.$$

At small electron energies according to the Wigner threshold law we have, for attachment of the s-electron, $\Gamma \sim \sqrt{\varepsilon}$, $\sigma_{\text{at}} \sim 1/\sqrt{\varepsilon}$, i.e., the rate constant of electron attachment $k_o = v\sigma_{\text{at}}$ (v is the electron velocity) does not depend on the electron energy in this limiting case. Combining these energy dependencies for the electron attachment cross section we have, for the rate constant of the electron attachment process,

$$k_{\text{at}} = \frac{k_o}{\sqrt{1 + \frac{\varepsilon}{\varepsilon_1}}}. \tag{16.37}$$

In particular, in the case of electron attachment to the SF_6-molecule the parameters of this formula are

$$k_o = (4.6 \pm 0.6) \cdot 10^{-7}\,\text{cm}^3/\text{s}, \qquad \varepsilon_1 = 6 \pm 2\,\text{meV}.$$

Analyzing the character of this process, we consider symmetric or almost symmetric molecules such as SF_6 or halomethane such molecules as CX_4, CX_kY_{4-k} where X, Y are halogen atoms. These molecules are characterized by a certain symmetry with respect to the transposition of halogen atoms. Evidently, the electron wave function of the ground molecular state is symmetric with respect to such transpositions, as well as the ground state of the negative ion of this molecule. This means that the capture of an s-electron leads to the formation of the ground electron state of the negative ion. Usually these molecules have stable negative ions. Because the electron capture results in the formation of an autodetachment state of the negative ion, one can conclude that compression of this negative ion leads to a decrease in the electron binding energy up to the intersection of its electron term and its molecular electron term.

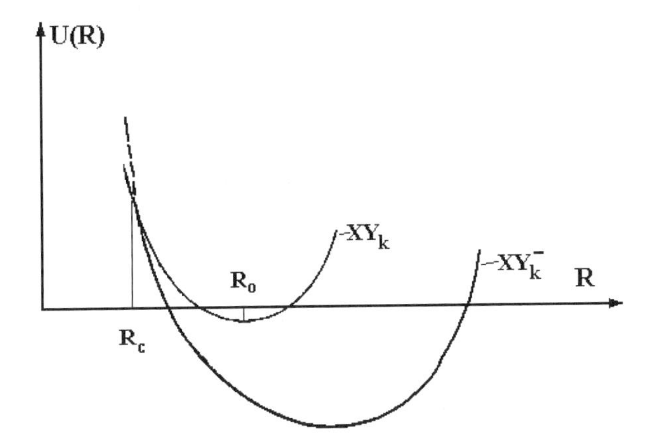

Figure 16.10. Positions of the electon terms for complex molecules which form a stable negative ion and are characterized by a high electron attachment rate in thermal collisions.

Possible positions of the electron terms in this case are given in Fig. 16.10. The evolution of a forming autodetachment state of the negative ion proceeds such that all the distances between the central atom and halogen atom remain identical, i.e., the reaction coordinate R is the distance from the central atom to a halogen atom in this case. Next, because the intersection of the electron terms of the molecule and negative ion ground states proceeds left from the bottom of the potential well, the efficiency of this process can increase with an increase in the molecular temperature. For example, the rate constant of attachment of an ultraslow electron to the SF_6 molecule is $k_o = (1.7 \pm 0.3) \cdot 10^{-7}\,\mathrm{cm^3/s}$ at low vibrational temperatures below 100 K, when the molecule is found in the ground vibrational state, and $k_o = (10 \pm 2) \cdot 10^{-7}\,\mathrm{cm^3/s}$ at vibrational temperatures above 400 K when the temperature dependence of the rate constant becomes weak.

Problems

Problem 16.1. *Determine the width of the autodetachment electron term near the intersection point of the boundary of continuous spectra for the electron term of a quasi-molecule consisting of an atom and a negative ion. Use the conditions of Problem 10.9 of a short-range interaction of the valent electron with each atom.*

Use the relation for the electron binding energy, which is obtained in Problem 10.9,

$$\alpha - \kappa - \frac{1}{R}\exp(-\alpha R) = 0,$$

where $\alpha^2/2$ is the electron binding energy and κ is the logarithmic derivation of the electron wave function near each atom. The intersection with the boundary of continuous spectra ($\alpha = 0$) takes place at a distance R_c between nuclei $R_c = 1/\kappa$. Let us expand the above relation over a small parameter αR near the intersection point. Expanding $e^{-\alpha R}$, and accounting for terms up to $(\alpha R)^3$, we obtain

$$-\kappa + \frac{1}{R} + \frac{\alpha^2 R}{2} - \frac{\alpha^3 R^2}{6} = 0.$$

Then, introducing the energy of the autodetachment state as $E_a = -\alpha^2/2$, we obtain in the first approximation $E_a = (R_c - R)/R_c^3$. In the next approximation, taking into account the term $\sim (\alpha R)^3$ in the above relation we have, for the energy of the autodetachment state,

$$E = E_a + i\Gamma, \quad \text{where} \quad \Gamma = \frac{\sqrt{2}}{3} R_c^2 E_a^{3/2}.$$

As is seen in this case, $\Gamma \ll E_a$.

Problem 16.2. *Determine the spectrum of released electrons as a result of the detachment of a negative ion in slow collisions with atoms.*

We use that the decay at a given distance between nuclei leads to the formation of a free electron whose energy is equal to the excitation energy of the autodetachment state E_a, i.e., it is the difference between the energy of a given electron term and the boundary of continuous spectra. Let us introduce the probability of decay of the autodetachment state up to time t:

$$P = 1 - \exp\left(-\int_0^t \Gamma \, dt'\right),$$

where $\Gamma(R)$ is the width of the autodetachment term. From this we have, for the energy distribution function of released electrons,

$$f(\varepsilon)\, d\varepsilon = \frac{dP}{dt} d\varepsilon \Big/ \frac{dE_a}{dt} = \frac{\Gamma \, d\varepsilon}{v_R \, |dE_a/dR|} \exp\left(-\int_0^\varepsilon \frac{\Gamma d\varepsilon'}{v_R \, |dE_a/dR|}\right),$$

where $v_R = dR/dt$. In particular, using the results of the previous problem ($\varepsilon = E_a, \Gamma = (\sqrt{2}/3) R_c^2 E_a^{3/2}$), we have $(dE_a/dR = -1/R_c^3)$,

$$f(\varepsilon)\, d\varepsilon = \frac{5}{2} C \varepsilon^{3/2} \, d\varepsilon \exp(-C\varepsilon^{5/2}), \quad \text{where} \quad C = \frac{2\sqrt{2}}{15} R_c^5 / v_R.$$

Problem 16.3. *Determine the cross section of ionization of a highly excited atom in collisions with a resonantly excited atom.*

Based on formula (16.30) for the ionization cross section, we use the Kramers formula (Problem 6.5) for the photoionization cross section

$$\sigma_{\text{ph}} = \frac{16\pi}{3\sqrt{3}} \cdot \frac{1}{n^5 \omega^3},$$

where n is the principal quantum number of a highly excited atom and ω is the photon frequency. From this we have, for the ionization cross section,

$$\sigma_{\text{ion}} = \frac{9}{\omega^2 n^2} \cdot \left(\frac{f}{v}\right)^{2/5},$$

where f is the oscillator strength for transition between the ground and resonantly excited states of the first (quenched) atom.

Problem 16.4. *Analyze the role of elastic scattering in the ionization of atomic particles resulting from collisions with an* He(2^1P)*-atom.*

At small collision energies the capture of particles leads to their approach. Because of a strong interaction at small distances between colliding particles, this can lead to an effective decay of an autoionization state. Let us evaluate this part of the process by presenting the decay cross section in the form

$$\sigma_{\text{ion}} = \xi \sigma_{\text{cap}},$$

where ξ is the probability of decay at a close approach and the capture cross section is given by formula (13.17). We take the interaction potential in the form $U(R) = CR^{-6}$, and the van der Waals constant is equal to, in this case according to formula (10.12), $C = \alpha \cdot \left(r^2\right)_{oo}$ where α is the polarizability of the ionizing atom, the value $\left(r^2\right)_{oo}$ relates to the excited electron. Thus we have, for the capture cross section,

$$\sigma_{\text{cap}} = \frac{3\pi}{2}\left[\frac{2\alpha \cdot \left(r^2\right)_{oo}}{\varepsilon}\right]^{1/3},$$

where ε is the collision energy. Averaging this cross section over the Maxwell distribution of atoms, we obtain

$$\sigma_c = \frac{\langle v\sigma_{cap}\rangle}{\langle v\rangle} = 5.36\left[\frac{2\alpha \cdot \left(r^2\right)_{oo}}{T}\right]^{1/3}.$$

Table 16.2 contains the comparison of this cross section and the ionization cross section (16.30) which is obtained by neglecting the elastic scattering of the colliding particles. This comparison is made at the gaseous temperature 1000 K. This demonstrates that both mechanisms of ionization must be taken into account under related conditions.

Table 16.2. The cross sections of some processes at the temperature 1000 K. Here σ_{ph} is the photoionization cross section which is included in formula (16.30), σ_{ion} is the ionization cross section of the process $\text{He}(2^1P) + A \to \text{He}(1^1S) + A^+ + e$ in accordance with formula (16.30) which is averaged over the Maxwell distribution of atoms and σ_c is the averaging cross sections of capture for these particles.

A	Ar	Kr	Xe	H_2	N_2	O_2
σ_{ph}, 10^{-17} cm^2	3.5	3.8	3.2	0.6	2	2
σ_{ion}, 10^{-15} cm^2	8.6	8.9	8.4	2.6	6.9	6.9
σ_c, 10^{-15} cm^2	10.8	12.4	14.5	8.5	11	10.6

References

[1] N.W. Ashcroft and N.D. Mermin, *Solid State Physics*, Holt, Rinehart, and Wilson, New York, 1976.

[2] S. Bashkin, J.D. Stoner. *Atomic energy levels and Grotrian diagrams*. vol.1,2. North Holland, Amsterdam, 1975-1978.

[3] P.F. Bernath, *Spectra of Atoms and Molecules*, Oxford University Press, New York, 1995.

[4] H.A. Bethe, *Intermediate Quantum Mechanics*, Benjamin, New York, 1964.

[5] P. Burke, *Theory of Electron-Atom Collisions*, Cambridge University Press, Cambridge, 1995.

[6] E.U. Condon and G.H. Shortly, *Theory of Atomic Spectra*, Cambridge University Press, Cambridge, 1953.

[7] N.B. Delone and V.P. Krainov, *Fundamentals of Nonlinear Optics of Atomic Gases*, Wiley, New York, 1988.

[8] Yu.N. Demkov and V.N. Ostrovskii, *Zero-range Potentials and their Applications in Atomic Physics*, Plenum, New York, 1988.

[9] W. Demtrader. *Laser Spectroscopy*. Springer, Berlin, 1980.

[10] G.F. Drukarev, *Collisions of Electrons with Atoms and Molecules*, Plenum, New York, 1987.

[11] T.F. Gallagher, *Rydberg Atoms*, Cambridge University Press, Cambridge, 1994.

[12] H. Haken, *The Physics of Atoms and Quanta: Introduction to Experiments and Theory*, Springer-Verlag, Berlin, 1994.

[13] F. Harald, *Theoretical Atomic Physics*, Springer-Verlag, Berlin, 1991.

[14] R.K. Janev, L.P. Presnyakov, and V.P. Schevelko, *Physics of Highly Charged Ions*, Springer-Verlag, Berlin, 1985.

[15] Ch. Kittel, *Introduction to Solid State Physics*, Wiley, New York, 1986.

[16] L.D. Landau and E.M. Lifshitz, *Quantum Mechanics*, Pergamon, Oxford, 1965.
[17] I. Lindgren, J. Morrison. *Atomic Many-Body Theory*. Springer, Berlin, 1982.
[18] V.S. Lisitsa, *Atoms in Plasmas*, Springer, Berlin, 1994.
[19] H.S.W. Massey, *Negative Ions*, Cambridge University Press, Cambridge, 1976.
[20] H.S.W. Massey, *Atomic and Molecular Collisions*, Taylor and Francis, New York, 1979.
[21] I.E. McCarthy and E. Weigold, *Electron–Atom Collisions*, Cambridge University Press, Cambridge, 1995.
[22] E.W. McDaniel, J.B.A. Mitchell, and M.E. Rudd. *Atomic Collisions*. Wiley, New York, 1993.
[23] C.E. Moore. *Atomic Energy Levels*, vol. 1,2,3. Nat. Bur. Std., Washington, 1949-1958.
[24] E.E. Nikitin and S.Ya. Umanskii, *Theory of Slow Atomic Collisions*, Springer-Verlag, Berlin, 1984.
[25] S.H. Patil, K.T. Tang. *Asymptotic Methods in Quantum Mechanics*. Springer, Berlin, 2000.
[26] B.M. Smirnov, *Negative Ions*, McGraw Hill, New York, 1982.
[27] I.I. Sobelman, *Atomic Spectra and Radiative Transitions*, Springer-Verlag, Berlin, 1979.

Index

Graduate Texts in Contemporary Physics

F.T. Vasko and A.V. Kuznetsov: **Electronic States and Optical Transitions in Semiconductor Heterostructures**

A.M. Zagoskin: **Quantum Theory of Many-Body Systems: Techniques and Applications**

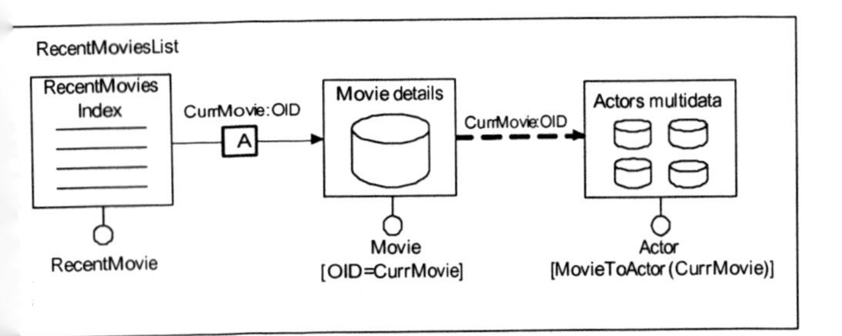

Figure 9.5. Example of automatic and transport links.

the example in Figure 9.5, extending the content of the page **ies List** shown in Figure 9.4. The link between the index and : has been defined as *automatic*: When the page is accessed, the e first movie appearing in the index will be shown to the user, need for her interaction. A multidata unit has been added to mes of the actors playing in the selected movie. A *transport* link ss the OID of the current movie to the multidata unit. This OID he multidata unit in a parametric selector associated with the **or** relationship defined between the entities **Movie** and **Actor** to y the actors associated with the current movie. Note that the nk admits the user's interaction for selecting a different movie endered as an anchor; conversely, the output link of the data unit ıble any selection and thus is defined as transport and is not n anchor.

parameters. In some cases, contextual information is not point to point during navigation but can be set as globally all the pages of the site view. This is possible through *global* which abstract the implementation-level notion of session-ta.

rs can be set through the *Set unit* and consumed within a page *t unit*. The visual representation of such two units is reported in n example of use of the get unit will be shown in the next

s. In addition to the specification of read-only Web sites, where on is limited to information browsing, WebML also supports tion of services and content management operations requiring over the information hosted in a site (e.g., the filling of a ley or an update of the users' personal information). WebML nal primitives for expressing built-in update operations, such as ting, or modifying an instance of an entity (represented through *elete,* and *modify* units, respectively) or adding or dropping a

Table 9.1. The Five Predefined Content Units in WebML

Data Unit	Multidata Unit	Index Unit	Scroller Unit	Entry Unit
Data unit Entity [conditions]	Multidata unit Entity [conditions]	Index unit Entity [conditions]	Scroller unit Entity [conditions]	Entry unit

Page composition. Pages are made of *content units*, which are the elementary pieces of information, possibly extracted from data sources, published within pages. Table 9.1 reports the five WebML predefined content units, representing the elementary information elements that may appear in the hypertext pages.

Units represent one or more instances of entities of the structural schema, typically selected by means of queries over the entity attributes or over relationships. In particular, *data units* represent some of the attributes of a given entity instance; *multidata units* represent some of the attributes of a set of entity instances; *index units* present a list of descriptive keys of a set of entity instances and enable the selection of one of them; *scroller units* enable the browsing of an ordered set of objects. Finally, *entry units* do not draw content from the elements of the data schema, but publish a form for collecting input values from the user.

Data, multidata, index, and scroller units include a *source* and a *selector*. The source is the name of the entity from which the unit's content is retrieved. The selector is a predicate, used for determining the actual objects of the source entity that contribute to the unit's content. The previous collection of units is sufficient to logically represent arbitrary content on a Web interface (Ceri et al., 2002). However, some extensions are also available, for example, the *multichoice* and the *hierarchical* indexes reported in Table 9.2. These are two variants of the index unit that allow one to choose multiple objects and organize a list of index entries defined over multiple entities hierarchically.

Link definition. Units and pages are interconnected by links, thus forming a hypertext. Links between units are called *contextual*, because they carry some information from the source unit to the destination unit. In contrast, links between pages are called *noncontextual*.

Table 9.2. Two Index Unit Variants

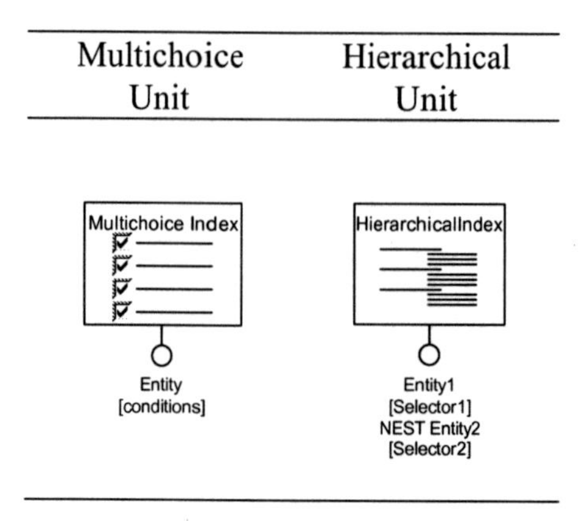

Multichoice Unit	Hierarchical Unit
Multichoice Index Entity [conditions]	HierarchicalIndex Entity1 [Selector1] NEST Entity2 [Selector2]

In contextual links, the binding between the source unit and the destination unit of the link is formally represented by link parameters, associated with the link, and by parametric selectors, defined in the destination unit. A *link parameter* is a value associated with a link between units, which is transported as an effect of the link navigation, from the source unit to the destination unit. A *parametric selector* is, instead, a unit selector whose condition contains one or more parameters.

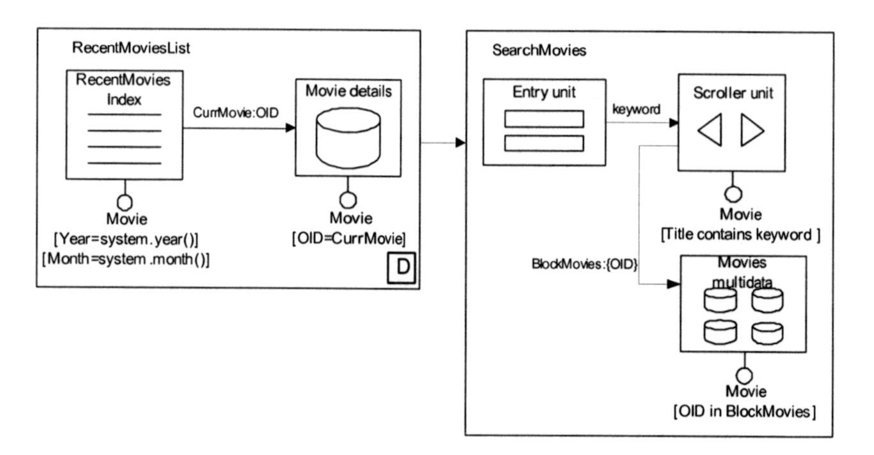

Figure 9.4. Example of contextual and noncontextual navigation.

As an example of page composition and unit linking, Figure 9.4 reports a simple hypertext, containing two pages of the **Movies** Area. The page **Recent Movies List** contains an index unit defined over the **Movie** entity, which shows the list of movies shown in the last month, and a data unit also

defined over the **Movie** entity, which d
selected from the index. Two sel
[Month=system.month()]) are defined t
movies of the current month and year. Th
contextual link, carrying the parameter
identifier (OID) of the selected item. Th
selector (**[OID=CurrMovie]**), which uses t
the data of the specific movie.

OIDs of the objects displayed or c
considered the default context associate
parameters over links and parametric sele
omitted and simply inferred from the diag

An example of a noncontextual link i
List page to the **Search Movies** pag
parameter, because the content of the de
the content of the source page.

The page **Search Movies** shows a
contains three units: an entry unit denotin
of the title to be searched, a scroller unit
having a selector for retrieving only the
their titles (**[Title contains keyword]**
scrollable block of search results. Throu
move to the first, previous, next, and last

Automatic and transport links. In son
to differentiate a specific link behavior, w
displayed as soon as the page is accessed
its incoming link. This effect can be ach
automatic link, graphically represented b
is "navigated" in the absence of a use
contains the source unit of the link is acc

Also, there are cases in which a link
information from one unit to another an
This type of link is called a *transport lin*
only parameter passing and not interact
represented as dashed arrows.

Conside
Recent Mo
the data un
details of t
without the
show the n
is used to p
is used by
MovieToAc
retrieve on
automatic l
and is thus
does not e
rendered as

Global
transferred
available to
parameters,
persistent d

Paramete
through a G
Table 9.3.
subsection.

Operatio
user interac
the specific
write acces
shopping tr
offers additi
creating, de
the *create,*

relationship between two instances (represented through the *connect* and *disconnect* unit, respectively). The visual representation of such units is reported in Table 9.4.

Table 9.3. The WebML Global Parameter Units

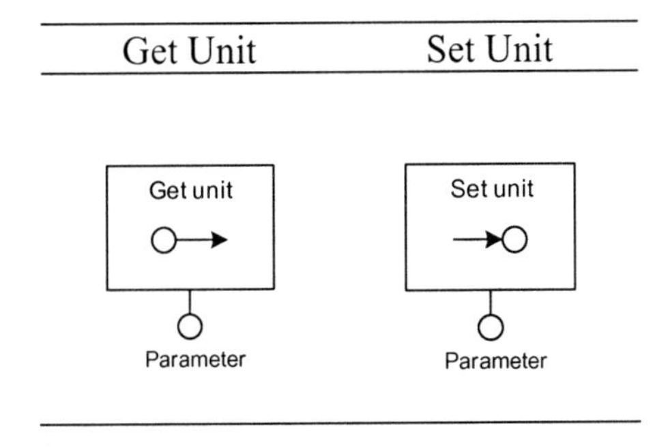

Table 9.4. The WebML Operation Units

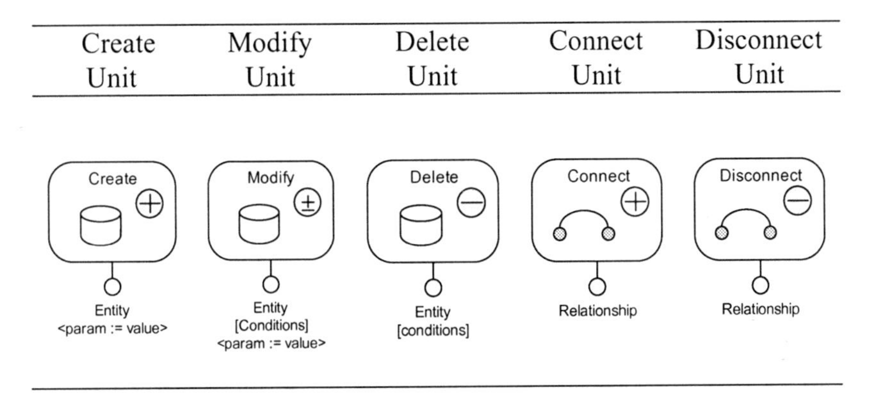

Other utility operations extend the previous set. For example, *login* and *logout* units (see Table 9.5) are respectively used (1) for managing access control and verifying the identity of a user accessing the application site views and (2) for closing the session of a logged user.

Operation units do not publish the content to be displayed to the user but execute some processing as a side effect of the navigation of a link. Like content units, operations may have a source object (either an entity or a relationship) and selectors, may receive parameters from their input links, and may provide values to be used as parameters of their output links. The result of executing an operation can be displayed in a page by using an appropriate content unit, for example, a data or multidata unit, defined over the objects updated by the operation.

Table 9.5. Login and Logout Operations, Supporting Site View Access Control

Login Unit	Logout Unit
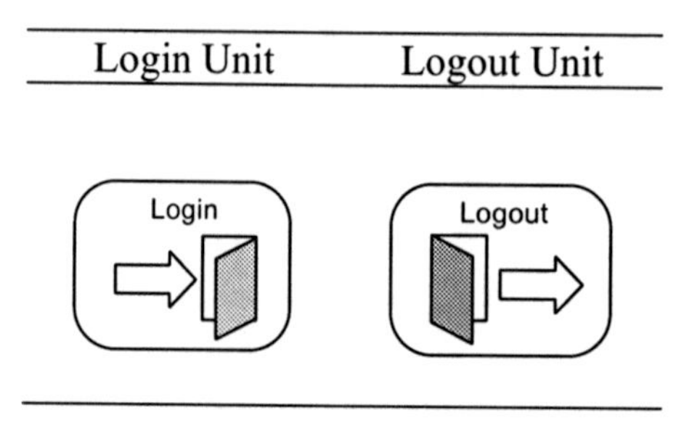	

Regardless of their type, WebML operations may have multiple incoming contextual links, which provide the parameters necessary for executing the operation. One of the incoming links is the activating link (the one followed by the user for triggering the operation), while the others just transport contextual information and parameters, for example, the identifiers of some objects involved in the operation.

Two or more operations can be linked to form a chain, which is activated by firing the first operation. Each operation can have two types of output links: one *OK link* and one *KO link*. The former is followed when the operation succeeds; the latter when the operation fails. The selection of the link to follow (OK or KO) is based on the outcome of the operation execution and is under the responsibility of the operation implementation.

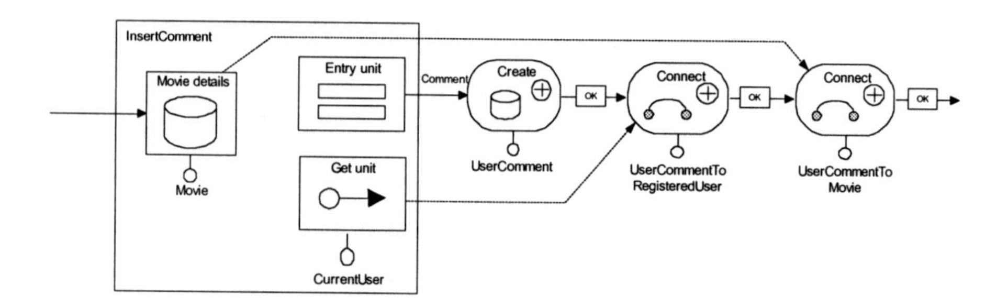

Figure 9.6. Example of content management.

The example in Figure 9.6 shows the content of the `Insert Comment` page in the `Movies` area. Through the entry unit the user can insert a comment for the movie currently displayed by the `Movie details` data unit. A get unit is defined to retrieve the data of the currently logged user, which have been stored in a global parameter after the login. When the user submits a comment, a chain of operations is triggered and executed: First, a new comment instance is created in the `UserComment` entity, containing the text inserted by the user; then, the new comment is associated to the current user (by creating a new instance of the relationship `UserCommentToRegisteredUser`) and to the current movie (relationship `UserCommentToMovie`). In the example, KO links are not explicitly drawn: By default, they lead the user to the page from which the operation chain has been triggered.

9.2.3 Other Development Phases

The phases following conceptual modeling consist of implementing the application, testing and evaluating it in order to improve its internal and external quality, deploying it on top of a selected architecture, and maintaining and possibly evolving the application once it has been deployed.

As described in more details in Section 9.3, the WebRatio development environment (WebModels, 2006) largely assists the implementation phase. First of all, it offers a visual environment for drawing the data and hypertext conceptual schemas. Such visual specifications are then stored as XML documents, which are the inputs for the WebML code generator, which then produces the data and hypertext implementation.

For space reasons, the remaining phases of the application life cycle are only hinted at in this chapter, but they are nonetheless well supported by WebML and WebRatio. In particular:

- The model-driven approach benefits the systematic testing of applications, thanks to the availability of the conceptual model and the model transformation approach to code generation (Baresi et al., 2005). With respect to the traditional testing of applications, the focus shifts from verifying individual Web applications to assessing the correctness of the code generator. The intuition is that if one could ensure that the code generator produces a correct implementation for all legal and meaningful conceptual schemas (i.e., combinations of modeling constructs), then testing Web applications would reduce to the more treatable problem of validating the conceptual schema. The research work conducted in this area has shown that it is possible to quantitatively evaluate the confidence in the correctness of a model-driven code generator, by formally measuring the coverage of a given test set (that is, of a set of sample conceptual schemas) with respect to the entire universe of syntactically admissible schemas. Different notions of coverage have been proposed, and heuristic rules have been derived for minimizing the number of test cases necessary to reach the desired coverage level of the testing process.
- Model-driven development also fosters innovative techniques for quality assessment. The research in this area has led to a framework for the model-driven and automatic evaluation of Web application quality (Fraternali et al., 2004; Lanzi et al., 2004; Meo and Matera, 2006). The

framework supports the *static* (i.e., compile-time) analysis of conceptual schemas and the *dynamic* (i.e., run-time) collection of Web usage data to be automatically analyzed and compared with the navigation dictated by the conceptual schema. The static analysis is based on the discovery in the conceptual schema of design patterns and on their automatic evaluation against quality attributes encoded as rules. Conversely, usage analysis consists of the automatic examination and mining of enriched Web logs, called *conceptual logs* (Fraternali et al., 2003), which correlate common HTTP logs with additional data about (1) the units and link paths accessed by the users, and (2) the database objects published within the viewed pages.

- In a model-driven process, maintenance and evolution also benefit from the existence of a conceptual model of the application. Requests for changes can in fact be turned into changes at the conceptual level, either to the data model or to the hypertext model. Then, changes at the conceptual level are propagated to the implementation. This approach smoothly incorporates change management into the mainstream production life cycle and greatly reduces the risk of breaking the software engineering process due to the application of changes solely at the implementation level.

9.3 IMPLEMENTATION

Application development with WebML is assisted by WebRatio (WebModels, 2006), a commercial tool for designing and implementing Web applications. The architecture of WebRatio (shown in Figure 9.7) consists of two layers: a *design layer*, providing functions for the visual editing of specifications, and a *run-time layer*, implementing the basic services for executing WebML units on top of a standard Web application framework.

The design layer includes a graphical user interface (shown in Figure 9.8) for data and hypertext design, which produces an internal representation in XML of the WebML models. A data mapping module, called Database Synchronizer, maps the entities and relationships of the conceptual data schema to one or more physical data sources, which can be either created by the tool or pre-existing. The Database Synchronizer can forward- and reverse-engineer the logical schema of an existing data source, propagate the changes from the conceptual data model to the physical data sources, and vice versa.

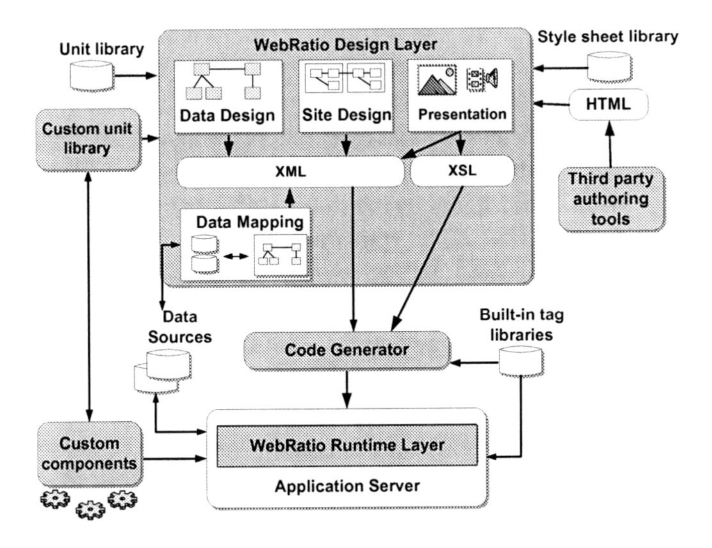

Figure 9.7. The WebRatio architecture.

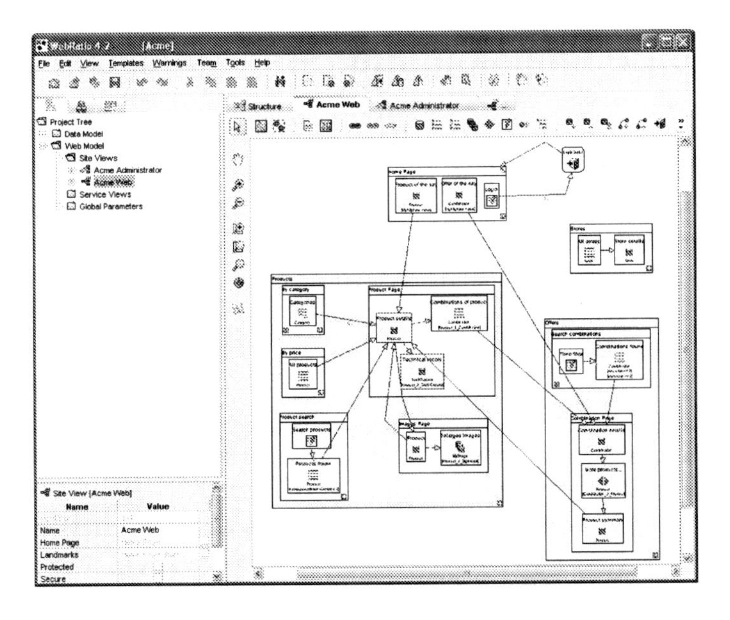

Figure 9.8. WebRatio's graphical user interface.

A third module (called *EasyStyler Presentation Designer*) offers functionality for defining the presentation style of the application, allowing the designer to create XSL stylesheets from XHTML mock-ups, associate XSL styles with WebML pages, and organize page layout, by arranging the relative position of content units in each page.

The design layer is connected to the run-time layer by the WebRatio code generator, which exploits XSL transformations to translate the XML specifications visually edited in the design layer into application code executable within the run-time layer, built on top of the Java2EE platform. In particular, a set of XSL translators produces a set of *dynamic page templates* and *unit descriptors*, which enable the execution of the application in the run-time layer. A dynamic page template (e.g., a JSP file) expresses the content and markup of a page in the markup language of choice (e.g., in HTML, WML, etc.). A unit descriptor is an XML file that expresses the dependencies of a WebML unit from the data layer (e.g., the name of the database and the code of the SQL query computing the population of an index unit).

The design layer, code generator, and run-time layer have a plug-in architecture: New software components can be wrapped with XML descriptors and made available to the design layer as custom WebML units, the code generator can be extended with additional XSL rules to produce the code needed for wrapping user-defined components, and the components themselves can be deployed in the run-time application framework. As described in the following section, such a plug-in architecture has been exploited to extend WebRatio to support new WebML constructs that have been recently defined for covering advanced modeling requirements.

9.4 ADVANCED FEATURES

The core concepts of WebML have been extended to enable the specification of complex applications, where Web services can be invoked, the navigation of the user is driven by process model specifications, and page content and navigation may be adapted (like in a multichannel, mobile environment). In the next subsections we briefly present the extensions that have been integrated in the WebML model for designing service-enabled, process-enabled, and context-aware Web applications.

9.4.1 Service-Enabled Web Applications

Web services have emerged as essential ingredients of modern Web applications: They are used in a variety of contexts, including Web portals for collecting information from geographically distributed providers or B2B applications for the integration of enterprise business processes.

To describe Web services interactions, WebML has been extended with Web service units (Manolescu et al., 2005), implementing the WSDL (W3C, 2002) classes of Web service operations.

We start by recalling some basic aspects of WSDL, providing the foundation of the proposed WebML extensions. A *WSDL operation* is the basic unit of interaction with a service and is performed by exchanging messages.

Two categories of operations are initiated by the client:

- *One-way* operations consist of a message sent by the client to the service.
- *Request-response* operations consist of one request message sent by the client and one response message built by the service and sent back to the client.

Two other operation categories are initiated by the service:

- *Notification operations* consist of messages sent to the service.
- *Solicit* and *response* operations are devised for receiving request messages sent to the service and providing messages as responses to the client.

WebML supports all four categories of operations. In particular, we interpret the operations initiated by the service as a means for *Web services publishing*. Therefore, we assume that these operations will not be used within the traditional hypertext schemas representing the Web site, but within appropriate *Service views*, which contain the definition of published services. The operations initiated by the client are instead integrated within the specification of the Web application. In the following subsections we will see how they can be specified in WebML and present some examples applied to the Movie database running case.

9.4.1.1 Modeling Web Applications Integrated with Web Services

The specification of Web service invocation from within a Web application exploits the request-response and one-way operations. Here we show an example of a request-response operation. Suppose we want to extend the Movie database Web application with the possibility of retrieving books related to a particular movie from a remote Web service (e.g., the Amazon

Web service). Assume that the request-response operation `SearchBooks` allows one to obtain a list of books meeting search criteria provided as input to the service (e.g., keywords contained in the title). The remote Web service responds with the list of books meeting the given search criteria.

The WSDL request-response operation is modeled through the request-response unit, whose graphical notation is shown in Figure 9.9. This operation involves two messages: the message sent to the service and the message received from the service. The corresponding unit is labeled with the Web service operation name and includes two arrows that represent the two messages. This operation is triggered when the user navigates one of its input links; from the parameters transferred by these links, a message is composed and then sent to a remote service as a request. The user waits until the arrival of the response message from the invoked service; then she can resume navigation from the page reached by the output link of the Web service operation unit.

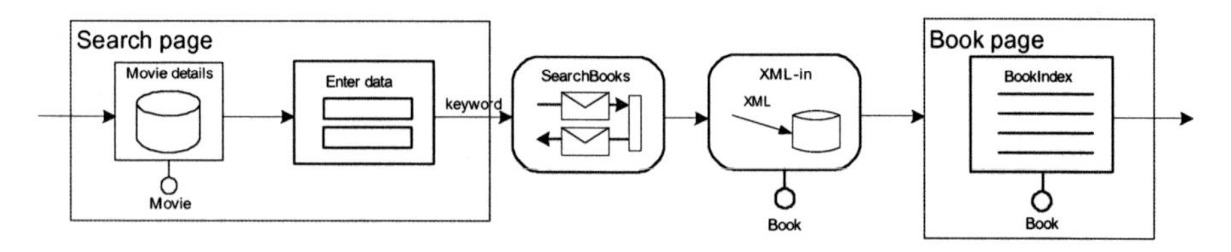

Figure 9.9. Example of usage of the request-response operation.

In the example in Figure 9.9, the user can browse to the `Search page`, where an entry unit permits the input of search criteria, preloaded from the currently selected movie. From this information, a request message is composed and sent to the `SearchBooks` operation of the Web service exposed by the service provider. The user then waits for the response message, containing a list of books satisfying the search criteria. From these options, a set of instances of the `Book` entity is created through the XML-in operation unit (which receives as input XML data and transforms them into relational data) and displayed to the user by means of the `Book Index` unit; the user may continue browsing, e.g., by choosing one of the displayed books. Further details about data transformations and about the storage of data retrieved from Web services can be found in recent publications (Manolescu et al., 2005).

One-way operations are modeled in a similar way: The main difference is that the service will not provide any response. Therefore, once the message is sent to the service, the user continues navigation without waiting for the response.

9.4.1.2 Modeling Web Services Publishing

WebML also supports the publication of Web services that can be invoked by third-party applications. From the application point of view, no user interaction is required in a published Web service. The actions to be performed when the notification or the solicit-response operations are triggered are not specified through pages, but as a chain of operations (e.g., for storing or retrieving data, or for executing generic operations such as sending emails). Therefore, the publishing of Web services can be specified separately from the site view of a Web application. We introduce the following concepts:

- *Service view*: a collection of ports that expose the functionality of a Web service through WSDL operations
- *Port*: the individual service, composed by a set of WSDL operations; each individual WSDL operation is modeled through a chain of WebML operations starting with a solicit-response and/or notification operation

Therefore, the business logic of a WSDL operation is described by a chain of WebML operations, specifying the actions to be performed as a consequence of the invocation of the service, and possibly building the response message to be sent back to the invoker. Each WSDL operation starts with a *solicit unit*, which triggers the service, and possibly ends with the *response unit*, which provides a message back to the service. Here we show an example of a solicit-response operation.

Suppose we want to extend the Movie database application with the publication of a service providing the list of movies satisfying search criteria. The WSDL operation is modeled through a chain of WebML operations starting with the solicit unit (`SearchSolicit`), shown in Figure 9.10. The solicit unit receives the SOAP message from the requester and decodes the search keywords, passing them as parameters to the next WebML operation in the sequence. This is a so-called XML-out (Manolescu et al., 2005) operation unit, which extracts from the database the list of movies that correspond to the specified conditions and formats it as an XML document. After the XML-out operation, the composition of the response message is performed through the *response unit* (`SearchResponse`).

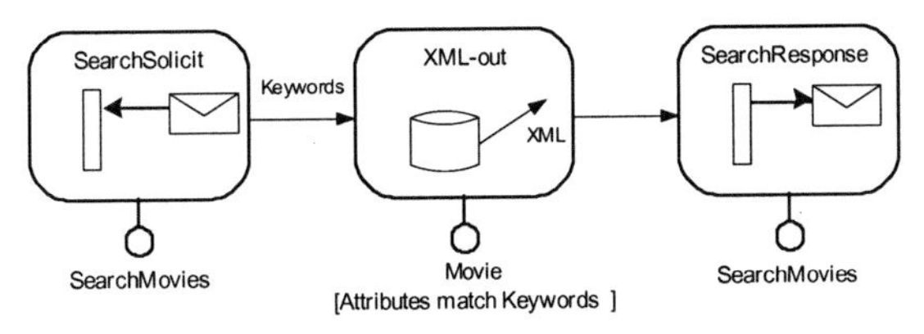

Figure 9.10. Example of usage of the solicit-response operation.

Notice that the schema of Figure 9.10 can be seen as the dual specification of the *SearchBooks* service invocation pattern, represented in Figure 9.9.

In addition to the above-mentioned examples, WebML also supports the exchange of asynchronous messages (Brambilla et al., 2004) and complex Web service conversations (Manolescu et al., 2005).

From the implementation standpoint, the deployment and publishing of Web services required the extension of the run-time WebRatio with a SOAP listener able to accept SOAP requests.

9.4.2 Process-Enabled Web Applications

Today the mission of Web applications is evolving from the support of online content browsing to the management of full-fledged collaborative workflow-based applications, spanning multiple individuals and organizations. WebML has been extended for supporting lightweight Web-enabled workflows (Brambilla, 2003; Brambilla et al., 2003, 2007), thus transferring the benefits of high-level conceptual modeling and automatic code generation also to this class of Web applications.

Integrating hypertexts with workflows means delivering Web interfaces that permit the execution of business activities and embodying constraints that drive the navigation of users. The required extensions to the WebML language are the following:

- *Business process model:* A new design dimension is introduced in the methodology. It consists of a workflow diagram representing the business process to be executed, in terms of its activities, the precedence constraints, and the actors/roles in charge of executing each activity.
- *Data model:* The data model representing the domain information is extended with a set of objects (namely, entities and relationships) describing the meta-data necessary for tracking the execution of the business process, both for logging and for constraints evaluation purposes.
- *Hypertext model:* The hypertext model is extended by specifying the business activity boundaries and the workflow-dependent navigation links.

Besides the main models, the proposed extension affects the following aspects of the WebML methodology:

- *Development process:* Some new phases are introduced in the development process, to allow the specification of business processes and their integration in the conceptual models (see Figure 9.11).

- *Design tools:* A new view shall be introduced for supporting the design of the workflow models within the WebML methodology.
- *Automatic generation tools:* A new transformer is needed for translating workflow diagrams into draft WebML specifications of the Web applications implementing the process specification.

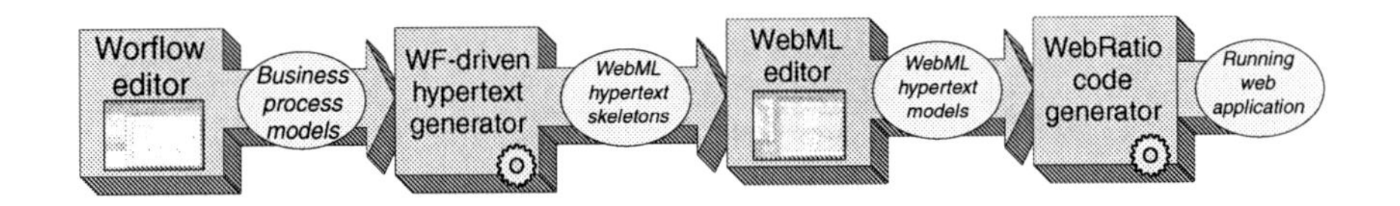

Figure 9.11. Steps of the proposed methodology: Square boxes represent the design steps and the involved tools; bubbles represent the expected results of each step.

The following sections present the details of the process-related extensions, by referring to a specific aspect of the Internet movie database case study, namely the subscription process. Details will be provided about the new features of the development process, the business process modeling, and the data and hypertext modeling.

9.4.2.1 Extensions to the Development Process

The development process is enriched by a set of new design tasks and automatic transformations that addresses the workflow aspects of the application. Figure 9.11 shows the expected steps of the development, the results of each steps, and the involved tools: Through a visual workflow editor, the analyst specifies the business process model to be implemented; the designed workflow model can be processed by an automatic transformation that generates a set of hypertext skeletons implementing the specified behavior; the produced skeletons can be modified by designers by means of CASE tools for conceptual Web application modeling; the resulting models can be processed by automatic code generators that produce the running Web application.

9.4.2.2 Workflow Model and Design Tool

Many standard notations have been proposed to express the structure of business processes. For our purposes, we adopt the Business Process Management Notation (BPMN), which covers the basic concepts required by WfMC (Workflow Management Coalition) and is compatible with Web service choreography languages (e.g., BPEL4WS) and standard business process specification languages (e.g., XPDL). A visual design tool for business processes has been implemented for covering this design phase. The tool is an Eclipse plug-in and allows one to specify BPMN diagrams.

Figure 9.12 shows a subscription process that could apply to the Movie database scenario (the case study has been extended to avoid a simplistic example): The user specifies whether he is a private customer or a company, then he alternatively submits the company or his own personal information, and finally a user manager accepts the subscription.

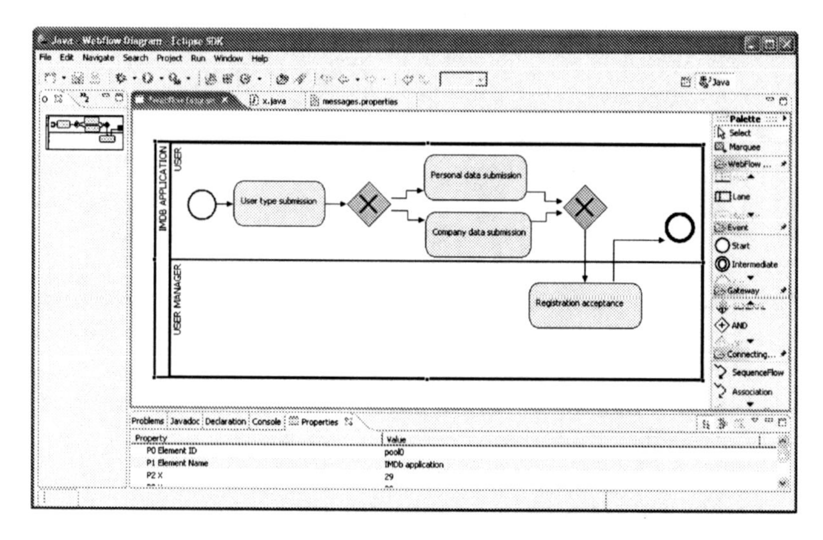

Figure 9.12. Subscription process represented in BPMN in the BP design tool.

9.4.2.3 Data Model Extensions: Workflow Meta-Data

The extensions to the data model include some standard entities for recording activities instances and process cases, thus allowing one to store the state of the business process execution and enacting it accordingly. The adopted meta-model is very simple (see Figure 9.13): The `Case` entity stores the information about each instantiation of the process, while the `Activity` entity stores the status of each activity instance executed in the system. Each activity belongs to a single case. Connections to user and application data can be added, for the purpose of associating domain information to the process execution. Typical requirements are the assignment of application objects to activity instances and the tracking of the relation between an activity and its executor (a user).

Notice that the proposed meta-model is just a guideline. The designer can adopt more sophisticated meta-data schemas or even integrate with underlying workflow engines through appropriate APIs (e.g., Web services) for tracking and advancing the process instance.

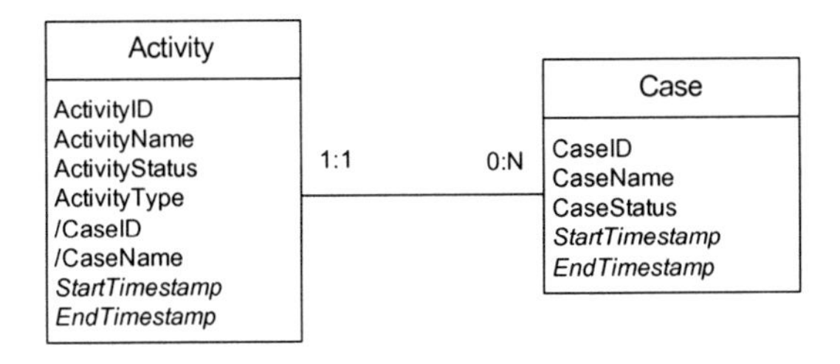

Derived attributes:
/CaseID {Self.Activity2Case.CaseID}
/CaseName {Self.Activity2Case.CaseName}

Figure 9.13. Workflow meta-data added to the data model.

9.4.2.4 Hypertext Model Extensions: Activities and Workflow Links

The hypertext model is extended with two new primitives:

- *Activity:* An activity is represented by an area tagged with a marker "A." The whole hypertext contained in the area is the implementation of the activity.
- *Workflow link:* Workflow links are links that traverse the boundary of any activity area. They are used for hypertext navigation, but their behavior includes workflow logic, which is not explicitly visible in the hypertext. Every link entering an activity represents the start of the execution of the activity; every outgoing link represents the end of the activity. The actual behavior of the workflow links is specified by a category associated with the link.

Incoming links can be classified as *Start link*, allowing an existing activity to start from scratch; *Start case link,* allowing one to create a new case and a new activity and to start them; *Create link,* allowing one to create a new activity and start it; *Resume link*, allowing one to resume the execution of an activity once it has been suspended.

Outgoing links can be classified as *Complete link*, which closes the activity and sets its status to completed; *Complete case link*, which closes the activity and the whole case, setting their status to completed; *Suspend link,* which suspends the execution of an activity (that can be resumed later through a resume link); *Terminate link,* which closes the activity and sets its status to terminated (e.g., for exception management).

Notice that *if* and *switch* units can be used to express navigation conditions. Moreover, a specific approach has been studied for managing exceptions within workflow-based Web applications (Brambilla et al., 2005; Brambilla and Tziviskou, 2005), but it is not discussed here for the sake of brevity. Moreover, by combining workflows and Web services extensions, the design of distributed processes can be obtained (Brambilla et al., 2006).

9.4.2.5 Mapping Workflow Schemas to Hypertext Models

Workflow activities are realized in the hypertext model by suitable configurations of pages and units, enclosed within an activity area. Workflow constraints must be turned into navigation constraints among the pages of the activities and into data queries on the workflow meta-data for checking the status of the process, thus ensuring that the data shown by the application and user navigation respect the constraints described by the process specification. The description of how the precedence and synchronization constraints between the activities can be expressed in the hypertext model is specified in Brambilla et al. (2003), which describes the mapping between each workflow pattern and the corresponding hypertext.

A flexible transformation, depending on several tuning and style parameters, has been included in the methodology for transforming workflow models into skeletons of WebML hypertext diagrams.

The produced WebML model consists of an application data model, workflow meta-data, and hypertext diagrams. The transformation supports all the main WfMC precedence constraints, which include sequences of activities, AND-, OR-, XOR- splits and joins, and basic loops.

Since no semantics is implied by the activity descriptions, the generated skeleton can only implement the empty structure of each activity and the hypertext and data queries that are needed for enforcing the workflow constraints. The designer remains in charge of implementing the interface and business logic of each activity. Additionally, it is possible to annotate the activities with a set of predefined labels (e.g., create, update, delete, browse), thus allowing the transformer tool to map the activity to a coarse hypertext that implements the specified behavior.

Once the transformation has been accomplished, the result can be edited with WebRatio (WebModels, 2006), thus allowing the designer to refine the generated hypertext and to implement the internal behaviour of each activity.

9.4.2.6 Workflow-Based Hypertext Example

Figure 9.14 shows the hypertext diagram for the `Personal Data Submission` activity, which is part of the example process depicted in Figure 9.12. Notice that the shown implementation is the final result of the two steps of automatic hypertext skeleton generation and of hypertext refinement by the designer. The link marked with the "..." label may come from any hypertext fragment in the site view.

Before starting the activity, a condition is checked for verifying that the **`Company data submission activity`** is not started yet, since it is defined as mutually exclusive with respect to the **`Personal Data Submission`** activity (a corresponding XOR-split decision gateway is shown in Figure 9.14). Hence, the condition to be checked before starting **`Personal Data Submission`** is that the instance of **`Company data submission`** activity within the current case has a status not yet *Active*. Notice that we assume an ordered set of possible values for the status (*Created* < *Inactive* < *Active* < *Suspended* < *Resumed* < *Completed*), and at most one instance of the activity **`Company data submission`** exists within a case, because of the construction rules of the instances of the workflow. Therefore, the condition extracts the activity of type **`Company data submission`** not yet started. If this instance exists, the *Start* link is followed and the **`Personal Data Submission`** activity is started (i.e., its status in the database is set to *Active*). The user submits his own information and the Modify unit updates the database, then the *Complete* link closes the activity and redirects the user to the home page.

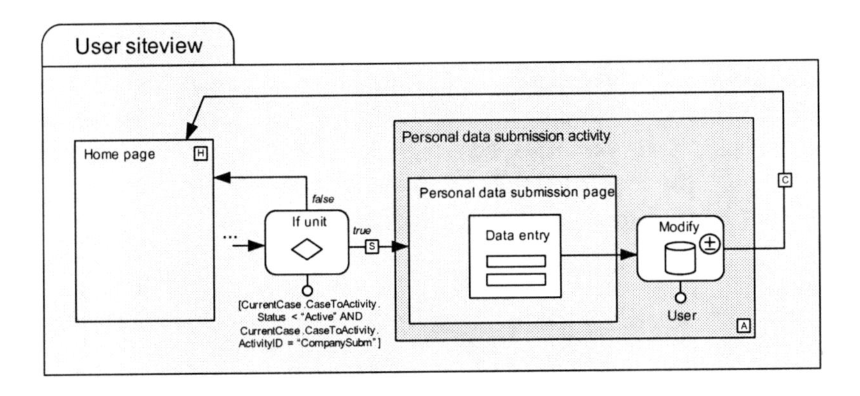

Figure 9.14. Example of hypertext representing the Personal data submission activity.

9.4.3 Context-Aware Web Applications

WebML has also been applied to the design of adaptive, context-aware Web applications (Ceri et al., 2003, 2006, 2007). The overall design process for context-aware applications follows the activity flow typically used for conventional Web applications. However, some new issues must be considered for modeling and exploiting the context at the data level and for modeling adaptive behaviors in the hypertext interface.

9.4.3.1 Modeling User and Context Data

During data design, the user and context requirements can be translated into three different subschemas complementing the application data (see Figure 9.15):

- The *User subschema*, which clusters data about users and their access rights to application data. In particular, the entity **User** provides a basic profile of the application's users, the entity **Group** allows access rights for a group of users to be managed, and the entity **SiteView** allows users (and user groups) to be associated with specific hypertexts. In the case of adaptive context-aware applications, users may require different interaction and navigation structures, according to the varying properties of the context.
- The *Personalization subschema*, which consists of entities from the application data associated with the **User** entity by means of relationships expressing user preferences for some entity instances, or the user's ownership of some entity instances. For example, the relationship between the entities **User** and **UserComment** in Figure 9.15 enables the selection and the presentation to the user of the comments she has posted. The relationship between the entities **User** and **Movie** represents the preferences of the user for specific movies. The role of this subschema is to support the customization of contents and services, which is one relevant facet of adaptive Web applications.

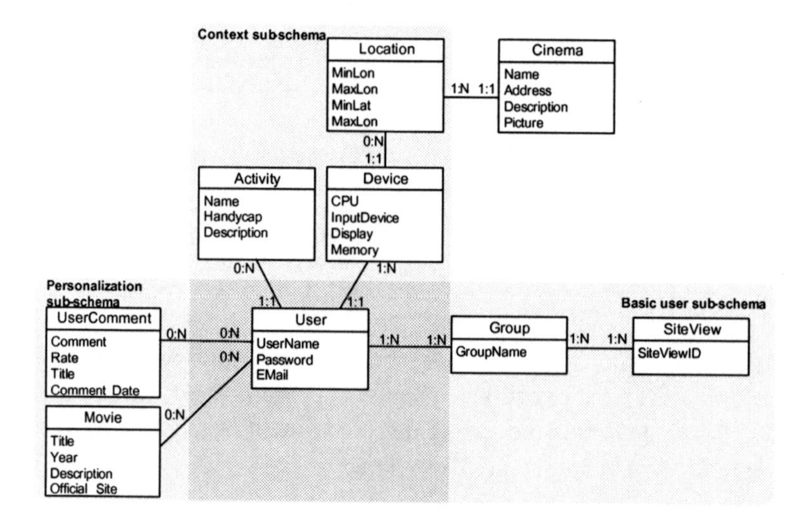

Figure 9.15. Three subschemas representing context data.

- The *Context subschema*, including entities such as **`Device`**, **`Location`**, and **`Activity`**, which describe context properties relevant for providing adaptivity. Context entities are connected to the entity **`User`** to associate each user with his (personal) context.

9.4.3.2 Identifying Context-Aware Pages

During hypertext design, adaptive requirements are considered to augment the application's front end with reactive capabilities. As illustrated in Figure 9.16, context-awareness in WebML can be associated with selected pages, and not necessarily with the whole application. Location-aware applications, for example, adapt "core" contents to the position of a user, but typical "access pages" (including links to the main application areas) might not be affected by the context of use.

We therefore tag adaptive pages with a *C* label (standing for "Context-aware") to distinguish them from conventional pages. This label indicates that some adaptivity actions must be associated with the page. During application execution, such actions must be evaluated prior to the computation of the page, since they can serve to customize the page content or to modify the navigation flow defined in the model.

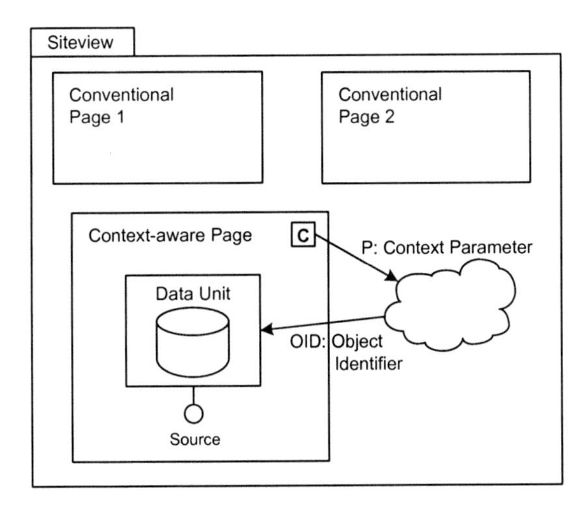

Figure 9.16. Hypertext schema highlighting context-aware pages. Context-aware pages are labeled with a "C" and are associated with a context cloud.

As shown in Figure 9.16, adaptivity actions are clustered within a *context cloud*. The cloud is external to the page, and the adaptivity actions that it clusters are kept separate from the page specification. Such a notation highlights the different roles played by pages and context clouds: The former act as providers of content and services, the latter act as modifiers of such content and services.

In order to monitor the state of the context and execute adaptivity actions, C-pages must be provided with autonomous intervention capabilities. The standard HTTP protocol underlying most of today's Web applications implements a strict pull paradigm. In the absence of a proper push mechanism, reactive capabilities can therefore be achieved by periodically refreshing the viewed page and by triggering the execution of adaptivity actions before the computation of the page content. This polling mechanism "simulates" the active behavior necessary for making pages sensitive to the context changes.

9.4.3.3 Specifying Adaptivity Actions in Context Clouds

Context clouds contain adaptivity actions expressed as sequences of WebML operations and are associated with a page by means of a directed arrow, i.e., a link, exiting the C label. This link ensures communication between the page logic and the cloud logic, since it can transport parameters derived from the content of the page, useful for computing the actions specified within the cloud. Vice versa, a link from the cloud to the page can transport parameters computed by the adaptivity actions, which might affect the page contents with respect to a new context.

The specification of adaptivity actions relies both on the use of the standard WebML primitives and on a few novel constructs, related to the acquisition and use of context data:

1. *Acquisition and management of context data.* This may consist of the retrieval of context data from the context model stored within the data source, or of the acquisition of fresh context data provided by device- or client-side-generated URL parameters, which are then stored in the application data source. These are the first actions executed every time a C-page is accessed, for gathering an updated picture of the current context.
2. *Condition evaluation.* The execution of some adaptivity actions may depend on some conditions, e.g., evaluating whether the context has changed and hence triggering some adaptivity actions.
3. *Page content adaptivity.* Parameters produced by context data acquisition actions and by condition evaluation can be used for page computation. They are sent back to the page by means of a link exiting the context cloud and going to the page. The result is the display of a page where the content is adapted to the current context.
4. *Navigation adaptivity.* The effect of executing the adaptivity actions within the context cloud can be the redirection to a different page. The

specification of context-triggered navigation just requires a link exiting the context cloud to be connected to pages other than the cloud's source page.

5. *Adaptivity of the hypertext structure.* To deal with coarse-grained adaptivity requirements, e.g., the change of device, role, or activity, the adaptivity actions may lead to the redirection toward a completely different site view.
6. *Adaptivity of presentation properties.* To support finer-grained adjustments of the interface, the adaptivity actions may induce the run-time modification of the presentation properties (look and feel, content position and visibility, and so on).

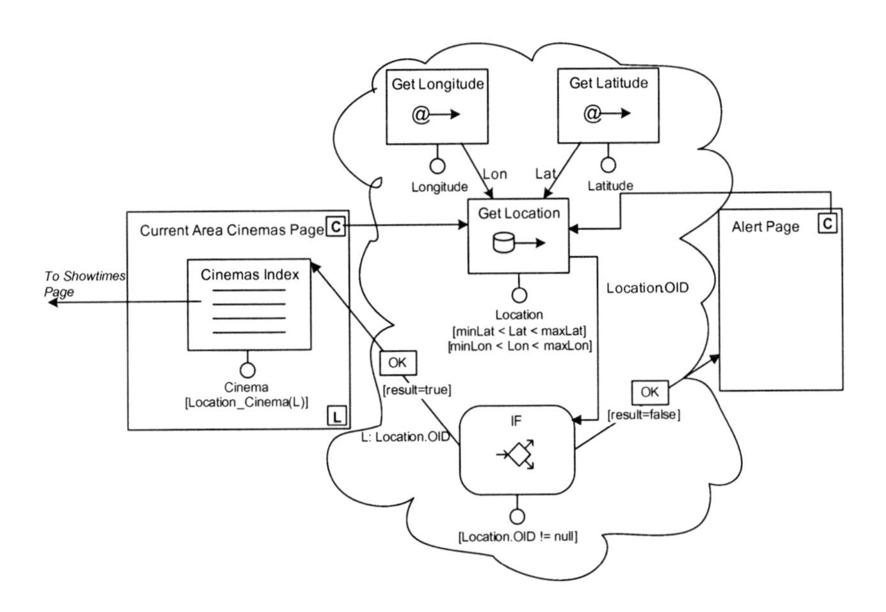

Figure 9.17. The WebML specification of adaptivity actions providing users with context-aware information about cinemas.

Figure 9.17 illustrates an example of adaptivity actions, applied to the **Current Area Cinemas** page. Upon page access, some adaptivity actions in the cloud are executed, which may change the content of the page based on the geographical position of the user. Specifically, the user's **Latitude** and **Longitude** are retrieved by the **Get Longitude** and **Get Latitude** units, which are examples of the *GetClientParameter* operation unit, introduced in WebML to access context data sensed at the client side. In the example, the two parameters **Longitude** and **Latitude** represent the position coordinates sensed through a user's device equipped with a GPS module. The retrieved position values are used by the **Get Location** unit to identify a (possible) location stored in the database for the current user's position. `Get Location` is a *Get Data* unit, a content unit for retrieving values (both scalars and sets) from an entity of the data model without displaying them on a page. The location OID is evaluated through an *If* unit: If it is not null (i.e., the sensed coordinates fall into a location stored in the application data source), the list of cinemas in that location is visualized in the `Current Area Cinemas` page; otherwise, the user is automatically redirected to the `Alert` page, where a message notifies of the absence of information about cinemas in the current area.

Figure 9.17 also models the `Alert` page as context-aware; in particular, this page shares its adaptivity actions with the `Current Area Cinemas` page. Therefore, as soon as an automatic refresh of the `Alert` page occurs, the shared actions are newly triggered and the application is adapted to the user's new position.

More details on the WebML extensions for adaptivity and context-awareness and on their implementation in WebRatio can be found in Ceri et al. (2003, 2006, 2007).

9.5 INDUSTRIAL EXPERIENCE

We conclude the illustration of WebML with an overview of the most significant aspects of transferring model-driven development to industrial users. The reported activities are based on WebML and WebRatio, but we deem that the achieved results demonstrate the effectiveness and economic sustainability of MDD in a more general sense. As a case study, we focus on the applications developed by Acer EMEA, the Europe, Middle East, and South Africa branch of Acer, for which five years of experience and data are available. In particular, we will review some of the realized projects, highlighting their functional and nonfunctional requirements, their dimensional parameters, and the key aspects of their development, deployment, evolution, and economic evaluation. The experience started with the first version of the Acer-Euro application (http://www.acer-euro.com), which aimed at establishing a software infrastructure for managing and Web-deploying the marketing and communication content of an initial group of 14 countries out of the 31 European Acer national subsidiaries. The content of Acer-Euro 1.0 included the following main areas: *About Acer*, *Products*, *News*, *Service & Support*, *Partner Area*, and *Where to buy*.

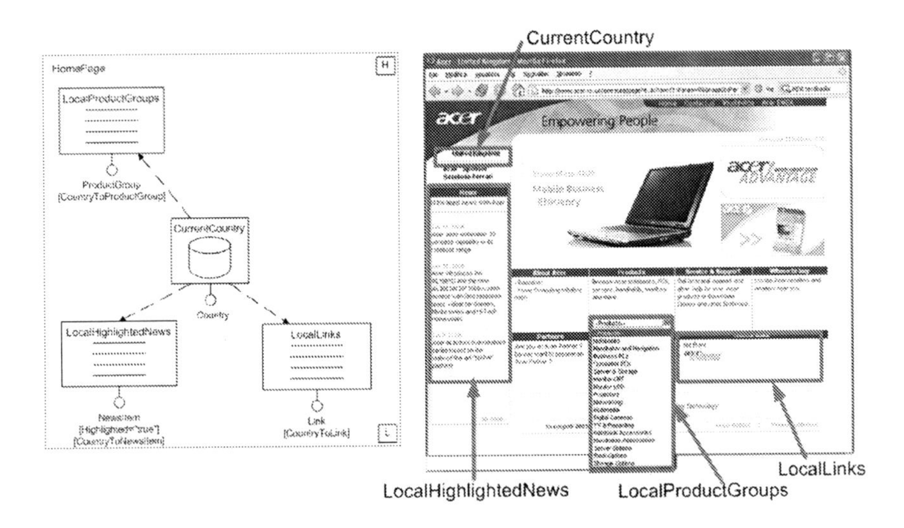

Figure 9.18. The WebML specification of the home page of a national site of Acer-Euro (left) and its rendition in HTML (right).

Figure 9.18 shows the home page of a national site of Acer-Euro (left) and its rendition in HTML generated by WebRatio. The Acer-Euro 1.0 system supported two main functions:

1. *Content publishing:* comprising the architecture, tools, and processes to make content about the Acer European Web sites available on the Web to the users of the target countries.
2. *Content management:* comprising the architecture, tools, and processes needed to gather, store, update, and distribute to the destination countries the content related to the Acer European Web sites.

Figure 9.19 shows the schedule and milestones of the Acer-Euro 1.0 project. Only 7 weeks elapsed from the approval of the new site map and visual identity to the publishing of the 14 national Web sites and to the delivery of the CMS to Acer employees. In this period, two distinct prototypes were formally approved by the management: Prototype 1, with 50% of functionality, was delivered at the end of week 2; prototype 2, with 90% of functionality, at week 5. Overall, nine prototypes were constructed in six weeks: two formal, seven for internal assessment.

The development team consisted of four persons: one business expert and one junior developer from Acer, and one analyst and one Java developer from Politecnico di Milano.

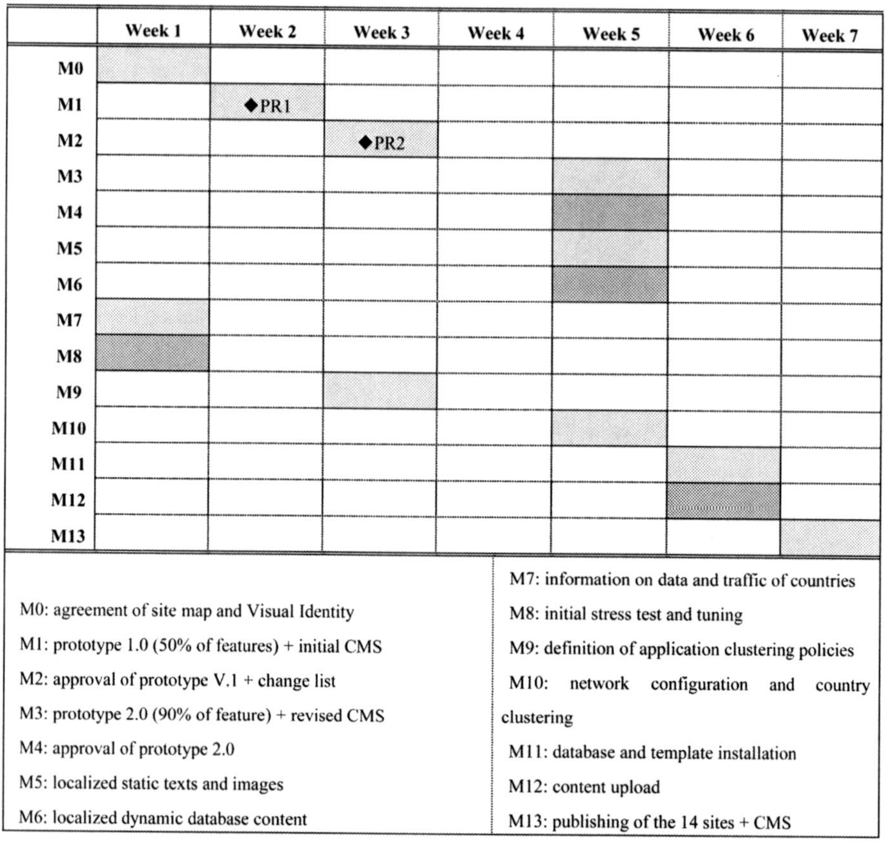

	Week 1	Week 2	Week 3	Week 4	Week 5	Week 6	Week 7
M0							
M1		◆PR1					
M2			◆PR2				
M3							
M4							
M5							
M6							
M7							
M8							
M9							
M10							
M11							
M12							
M13							

M0: agreement of site map and Visual Identity
M1: prototype 1.0 (50% of features) + initial CMS
M2: approval of prototype V.1 + change list
M3: prototype 2.0 (90% of feature) + revised CMS
M4: approval of prototype 2.0
M5: localized static texts and images
M6: localized dynamic database content
M7: information on data and traffic of countries
M8: initial stress test and tuning
M9: definition of application clustering policies
M10: network configuration and country clustering
M11: database and template installation
M12: content upload
M13: publishing of the 14 sites + CMS

Figure 9.19. The schedule and milestones of the Acer-Euro 1.0 project.

Figure 9.19 shows the most relevant figures of the project: only six weeks of development plus one week of testing were sufficient for analyzing, designing, implementing, verifying, documenting, and deploying a set of midsized, functionally complex, multilingual Web applications. As illustrated by the dimensional and economic parameters reported in Table 9.6, such result has to be ascribed to

1. The high degree of automation brought to the process by the use of the model-driven approach: More than 90% of the application and database code were synthesized automatically by the WebRatio development environment from the WebML models of the applications, without the need to manually intervene on the produced code.
2. The overall productivity of the development process: The productivity value is obtained by counting the number of function points (FPs) of the project and dividing this value by the number of staff-months

employed in the development. The result is an average productivity rate of 131.5 FP/staff month, which is 30% greater than the maximum value expected for traditional programming languages in the Software Productivity Research Tables (SPR, 2006). This latter result is a consequence of the former: High automation implies a substantial reduction of the manually written repetitive code and a high reuse of design patterns.

Table 9.6 Main Dimensional and Economic Parameters of the Acer-Euro Project

Class	Dimension	Value
Time & effort	Number of elapsed workdays	49
	Number of development staff-months (analysts and developers)	6 staff-months (6 weeks × 4 persons)
	Total number of prototypes	9
	Average elapsed man days between consecutive prototypes	5,4
	Average number of development man days per prototype	15,5
Size	Number of localized B2C Web sites	14
	Number of localized CMS applications	4 (Admin, News, Product, Other)
	Number of supported languages	12 for B2C Web sites, 5 for CMS
	Number of data entry masks	39
	Number of automatically generated database tables	46
	Number of automatically generated database views	82
	Number of automatically generated database queries	279 for extraction, 89 for update
	Number of automatically generated JSP page templates	48
	Number of automatically generated or reused Java classes	250
	Number of automatically generated Java lines of code	12,500 Noncommented lines of code
Degree of automation	Number of manually written SQL statements	17 (SQL constraints)
	Percentage of automatically generated SQL code	96%
	Number of manually written/adapted Java classes /JSP templates	10% JSP templates manually adapted
	Percentage of automatically generated Java and JSP code	90% JSP templates, 100% Java classes
Productivity	Number of function points	177 (B2C web site) + 612 (CMS) = 789
	Average number of FP delivered per staff-month	131.5

Another critical success factor has been the velocity in focusing the requirements, thanks to the rapid production of realistic prototypes. At the end of week 2, the top management could already evaluate an advanced prototype, which incorporated 50% of the requested functionality, and this initial round of requirement validation proved essential to the delivery of a compliant solution in such a limited time. With respect to traditional prototyping, which exploits a simplified architecture, WebRatio generates code directly for the actual delivery platform; in this way, stress test and architecture tuning could already start at week 1 on the very first prototype, greatly improving the parallelism of work and further reducing time to market.

The benefits of MDD were manifested not only in the development of the first version, but were even more sensible in the maintenance and evolution phase. Figure 9.20 shows the timeline of the additional releases and spin-off projects of Acer-Euro. Four major releases of Acer-Euro were delivered between 2001 and 2006, and the number of applications grew from the initial 5 to 13 intranet and Internet applications, serving more corporate roles and supporting more sophisticated workflow rules.

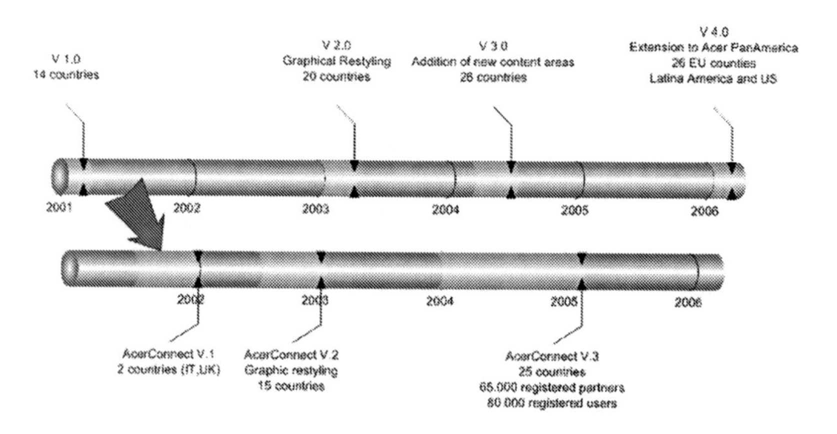

Figure 9.20. The evolution of the Acer-Euro project in five years.

At the end of 2005, Acer-Euro was rolled out in 26 European countries and extended also to the Acer Pan-American subsidiaries, including Latin America and the United States. As early as June 2001, an extension of the Acer-Euro platform was scheduled, to address the delivery and management of content for the channel operators (Acer partners). This spin-off project, called Acer Connect, is a multi-actor extranet application targeted to Acer partners, characterized by the following features:

1. the segmentation of the users accessing the site into a hierarchy of groups corresponding to both Acer's and partners' business functions
2. the definition of different access privileges and information visibility levels to groups

3. the provision of an Acer European administration role, able to dynamically perform via the Web all administrative and monitoring tasks
4. the provision of an arbitrary number of nation-based and partner-based administration roles, with responsibility for local content creation and publishing, and local user administration
5. a number of group-tailored Web applications (e.g., sales, marketing) targeting content to corporate-specific or partner-specific user communities
6. the management of administrative and business functions in multiple languages flexibly set by administrators and users
7. a security model storing group and individual access rights into a centrally managed database, to enforce global control over a largely distributed application
8. content personalization based on group-specific or user-specific characteristics, for ensuring one-to-one relationships with partners
9. advanced communication and monitoring functions for the effective tracking of partners' activity and of Acer's quality of services

The first version of Acer Connect was deployed in Italy and the UK in December 2001, after only seven months of development and with an effort of 24 staff-months. Today, Acer Connect is rolled out in 25 countries and hosts 65,000 registered partners, delivering content and services to a community of over 80,000 users. Acer Connect and Acer-Euro share part of the marketing and communication content, and therefore the former project was realized as an evolution of the latter; starting from the data model of Acer-Euro, the specific functions of Acer Connect were added, and new applications were modeled and automatically generated. The model-driven approach greatly reduced the complexity of integration, because the high-level models of the two systems were an effective tool for reasoning about the functionality to reuse and develop.

Besides Acer Connect, several other projects were spun off, to exploit the customer and partner communities gathered around these two portals. Figure 9.21 overviews the delivered B2C projects, which collectively total over 10,800,000 visits per month.

As a remark on the long-term sustainability of MDD, we note that, despite their complexity and multinational reach, both Acer-Euro and Acer Connect are maintained and evolved by one junior developer each, working on the project at part time. In total, only 5 junior developers, allocated to the projects at part time, maintain the 56 mission-critical Web applications implemented by Acer with WebML.

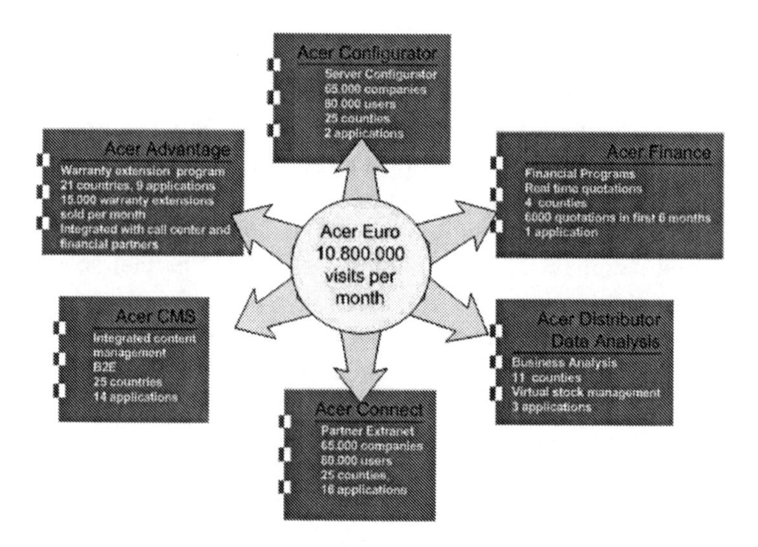

Figure 9.21. The main applications developed in Acer with WebML.

On the negative side of MDD, the initial training and switching costs have been reported as the most relevant barrier. MDD requires nontechnical knowledge on the modeling of software solutions, which must be acquired with a mix of conventional and on-the-job training. Furthermore, developers have their own previous consolidated skills and professional history, and switching to a completely new development paradigm is felt to be a potential risk. Acer estimates that it takes developers from 4 to 6 months to become fully acquainted and productive with MDD, WebML, and WebRatio. However, Acer's figures demonstrate that the initial investment in human capital required by MDD pays off in the mid-term. The number of applications developed and maintained per unit of development personnel increases with the developers' expertise and exceeds 10 fully operational, complex, and distributed Web applications per developer.

9.6 CONCLUDING REMARKS

In this chapter we have described the Web Modeling Language, a conceptual notation for specifying the design of complex, distributed, multi-actor, and adaptive applications deployed on the Web and on service-oriented architectures using Web services. WebML was born in academia but soon spun off to the industrial battlefield, where it faced the development of complex systems with requirements often exceeding the expressive power of the language. This fruitful interplay of academic design and industrial experience made the language evolve from a closed notation for data-centric

Web applications to an open and extensible framework for generalized component-based development. The core capability of WebML is expressing application interfaces as a network of collaborating components, which sit on top of the core business objects. WebML incorporates a number of built-in, off-the-shelf components for data-centric, process-centric, and Web service-centric applications and lets developers define their own components, by wrapping existing software artifacts and reverse-engineering them. In other words, the essence of WebML boils down to a standard way of describing components, their interconnection and passage of parameters, their exposition in a user interface, and the rules for generating code from their platform-independent model.

This flexibility allowed several extensions of the language, in the direction of covering both new application requirements and deployment architectures. The ongoing work is pursuing a number of complementary objectives:

1. Extending the model-driven approach to all the phases of the application life cycle: WebML is being used as a vehicle to investigate the impact of MDD on development activities like business requirement elicitation and reengineering, cost and effort estimation, testing, quality evaluation, and maintenance.
2. Extending the capability of the user interface beyond classical hypertexts: The expressive power of WebML is presently inadequate to express Rich Internet Applications and classical client-server applications; research is ongoing to identify the minimal set of concepts needed to capture the Web interfaces of the future.
3. Broadening the range of deployment platforms: WebML and WebRatio are being extended to target code generation for nonconventional infrastructures. A version of WebRatio for digital television has been already built, and experimentation is ongoing for deploying applications on top of embedded systems and mobile appliances for the DVB-H standard.

REFERENCES

Baresi, L., Fraternali, P., Tisi, M., and Morasca, S., 2005, Towards model-driven testing of a Web application generator. *Proceedings 5th International Conference on Web Engineering* (ICWE'05), Sydney, Australia, pp. 75–86.

Beck, K., 1999, Embracing change with extreme programming. *IEEE Computer*, **32**(10): 70–77.

Boehm, B., 1988, A spiral model of software development and enhancement. *IEEE Computer*, **21**(5): 61–72.

Booch, G., Rumbaugh, J., and Jacobson, I., 1999, *The Unified Modeling Language User Guide (Object Technology Series),* Addison-Wesley, Reading, MA.

Brambilla, M., 2003, Extending hypertext conceptual models with process-oriented primitives. *Proceedings Conceptual Modeling* (ER 2003), Chicago, IL, pp. 246–262.

Brambilla, M., Ceri, S., Comai, S., Fraternali, P., and Manolescu, I., 2003, Specification and design of workflow-driven hypertexts. *Journal of Web Engineering*, **1**(2): 163–182.

Brambilla, M., Ceri, S., Comai, S., and Tziviskou, C., 2005, Exception handling in workflow-driven Web applications. *Proceedings World Wide Web International Conference* (WWW'05), Chiba, Japan, May 10–13, pp. 170–179.

Brambilla, M., Ceri, S., Fraternali, P., and Manolescu, I., 2007, Process modeling in Web applications. *ACM Transactions on Software Engineering and Methodology*. In print.

Brambilla, M., Ceri, S., Passamani, M., and Riccio, A., 2004, Managing asynchronous Web services interactions. *Proceedings ICWS 2004*, pp. 80–87.

Brambilla, M., and Tziviskou, C., 2005, Fundamentals of exception handling within workflow-based Web applications. *Journal of Web Engineering*, **4**(1): 38–56.

Ceri, S., Daniel, F., Facca, F., Matera, M., and the MAIS Consortium, 2006, Front-end methods and tools for the development of adaptive applications. In *Mobile Information Systems. Infrastructure and Design for Flexibility and Adaptivity*, B. Pernici, ed., Springer-Verlag, pp. 209–246.

Ceri, S., Daniel, F., and Matera, M., 2003, Extending WebML for modelling multi-channel context-aware Web applications. *Proceedings WISE '03 Workshops*, IEEE Press, pp. 225–233.

Ceri, S., Daniel, F., Matera, M., and Facca, F., 2007, Model-driven development of context-aware Web applications. *ACM Transactions on Internet Technology*, **7**(1), Article No. 2.

Ceri, S., Fraternali, P., and Bongio, A., 2000, Web Modeling Language (WebML): A modeling language for designing Web sites. *Computer Networks*, **3**(1–6): 137–157.

Ceri, S., Fraternali, P., Bongio, A., Brambilla, M., Comai, S., and Matera, M., 2002, *Designing Data-Intensive Web Applications,* Morgan Kaufmann, San Francisco.

Conallen, J., 2000, *Building Web Applications with UML (Object Technology Series),* Addison-Wesley, Reading, MA.

Fraternali, P., Lanzi, P.L., Matera, M., and Maurino, A., 2004, Model-driven Web usage analysis for the evaluation of Web application quality. *Journal of Web Engineering*, **3**(2): 124–152.

Fraternali, P., Matera, M., and Maurino, A., 2003, Conceptual-level log analysis for the evaluation of Web application quality. *Proceedings LA-WEB 2003*, IEEE Press, pp. 46–57.

Garzotto, F., Paolini, P., and Schwabe, D., 1993, HDM—A model-based approach to hypertext application design. *ACM Transactions on Information Systems*, **11**(1): 1–26.

Kruchten, P., 1999, *The Rational Unified Process: An Introduction*, Addison-Wesley, Reading, MA.

Isakowitz, T., Sthor, E.A., and Balasubranian, P., 1995, RMM: A methodology for structured hypermedia design. *Communications of the ACM*, **38**(8): 34–44.

Lanzi, P.L., Matera, M., and Maurino, A., 2004, A framework for exploiting conceptual modeling in the evaluation of Web application quality. *Proceedings ICWE 2004*, Springer-Verlag, pp. 50–54.

Manolescu, I., Brambilla, M., Ceri, S., Comai, S., and Fraternali, P., 2005, Model-driven design and deployment of service-enabled Web applications. *ACM Transactions on Internet Technology,* **5**(3): 439–479.

Meo, R., and Matera, M., 2006, Designing and mining Web applications: A conceptual modeling approach. In *Web Data Management Practices: Emerging Techniques and Technologies*, A. Vakali and G. Pallis, eds., Idea Group Publishing, Hershey, PA.

SPR (Software Productivity Research), 2006, SPR Programming Language Table—Version PLT2005a. Retrieved February 2006 from http://www.spr.com.

WebModels, 2006. WebRatio Tool Suite. Retrieved October 2006 from http://www.webratio.com.

W3C, 2006, WSDL Web Service Description Language. Retrieved October 2006 from https://www.w3.org/2002/ws/desc.

Chapter 10

HERA

Geert-Jan Houben,[1,2] Kees van der Sluijs,[1] Peter Barna,[1] Jeen Broekstra,[1,3] Sven Casteleyn,[2] Zoltán Fiala,[4] Flavius Frasincar[5]

[1]*Technische Universiteit Eindhoven, PO Box 513, 5600 MB Eindhoven, The Netherlands, {g.j.houben, k.a.m.sluijs, p.barna, j.broekstra}@tue.nl*

[2]*Vrije Universiteit Brussel, Pleinlaan 2, 1050 Brussels, Belgium, {Geert-Jan.Houben, Sven.Casteleyn}@vub.ac.be*

[3]*Aduna, Prinses Julianaplein 14b, 3817 CS Amersfoort, The Netherlands, jeen@aduna.biz*

[4]*Technische Universität Dresden, Mommsenstr. 13, D-01062, Dresden, Germany, zoltan.fiala@inf.tu-dresden.de*

[5]*Erasmus Universiteit Rotterdam, PO Box 1738, 3000 DR Rotterdam, The Netherlands, frasincar@few.eur.nl*

10.1 INTRODUCTION

This chapter illustrates a method for Web information systems (WIS) design that found its origins in an approach for hypermedia presentation generation. It was also this focus on hypermedia presentation generation that gave the first engine complying with this method its name, HPG (Frasincar, 2005). The method distinguishes three main models that specify the generation of hypermedia presentations over available content data. With a model for the content, a model for the hypermedia navigation construction, and a model for the presentation construction, the method enables the creation of a hypermedia-based view over the content. Originally, in the first generation of the method and its toolset, the models specified a transformation from the content to the presentation. The engine that was compliant with this definition was based on XSLT and is therefore known as HPG-XSLT.

One of the characteristic aspects that HPG-XSLT supported was adaptation. As an illustrative example, we show in Figure 10.1 how the engine could produce different presentations from a single design in which the "translation" to formats such as HTML, SMIL, and WML was dealt with generically.

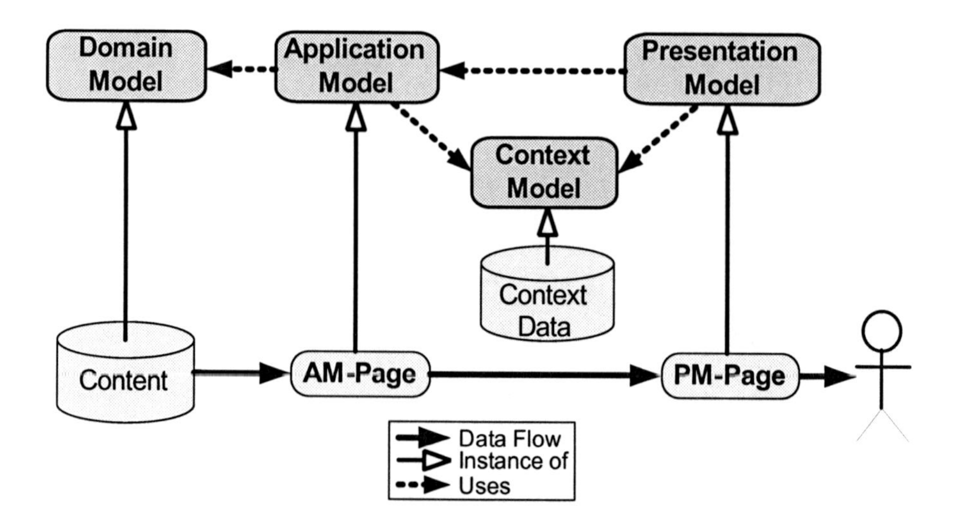

Figure 10.1. Hera models.

Characteristic for the Hera models was not only their focus on user- and context-adaptation support, but also the choice to base the models on the resource description framework (RDF) (Klyne and Carrol, 2004) and RDF schema (RDFS) (Brickley and Guha, 2004). The use of Web standards such as RDF and RDFS as a modeling paradigm facilitates easy deployment on very heterogeneous data sources: The only assumption made is that a semistructured description (in RDF) of the domain is available for processing. Not only is such a representation less costly to develop than any alternative, but it also enables the reuse of existing knowledge and flexible integration of several separate data sources into a single hypermedia presentation.

During further research into the development of the method, support was extended for more advanced dynamics. Whereas the first XSLT-based approach primarily transformed the original content data into a hypermedia document, with which the user could interact by following links with a Web browser, the subsequent engine version allowed the inclusion of form processing, which led to the support of other kinds of user interaction while retaining the hypermedia-based nature. Out of this effort, a Java-based version of the engine became available that used RDF queries to specify the data involved in the forms.

The experience from these HPG-based versions and the aim for further exploitation of the RDF-based nature of the models have led to a further refinement of the approach in what is now termed Hera-S. The Hera-S-compliant models do combine the original hypermedia-based spirit of the Hera models with more extensive use of RDF querying and storage. Realizing this, RDF data processing using the Sesame framework (Broekstra et al., 2002) and its query language SeRQL (Broekstra, 2005) caters for extra flexibility and interoperability.

In the current version of the Hera method that we present in this chapter, we aim to exemplify the characteristic elements included in the method. As we mentioned before, there is the RDF-based nature of the models. There is certainly also the focus on the support for adaptation in the different model elements. Adapting the data processing to the individual user and the context that the user is in (in terms of application, device, etc.) is a fundamental element in WIS design and one that deserves the right attention: Managing the different design aspects and thus controlling the complexity of the application design is crucial for an effective design and implementation.

In this chapter we first address the main characteristics of the method and then we explain the models, i.e., the main design artifacts, for the book's running example. We present the implementation of the hypermedia presentation generation process induced by the models. We also consider some extensions to the basic approach that can help the design process in certain scenarios.

10.2 METHOD

We discuss the Hera approach and illustrate it by means of examples from the Hera models for the running example (in this case we use Hera-S-compliant versions of those models). In this section we will capture the main elements of the key models used in the example before we go into details in the next section.

The purpose of Hera is to support the design of applications that provide navigation-based Web structures (hypermedia presentations) over semantically structured data in a personalized and adaptive way. The design approach centers on *models* that represent the core aspects of the application design. Figure 10.2 gives an overview of these models. With the aid of a tool for executing those models (e.g., HPG-XSLT or Hera-S), we can also generate the application, as depicted in this figure. Thus, the appropriate pipeline of models captures the entire application design, leaving room for the designer to change or extend the implementation where desired. In this section we give a short overview over the different models and associated modeling steps, while each of them is presented in more detail in subsequent sections.

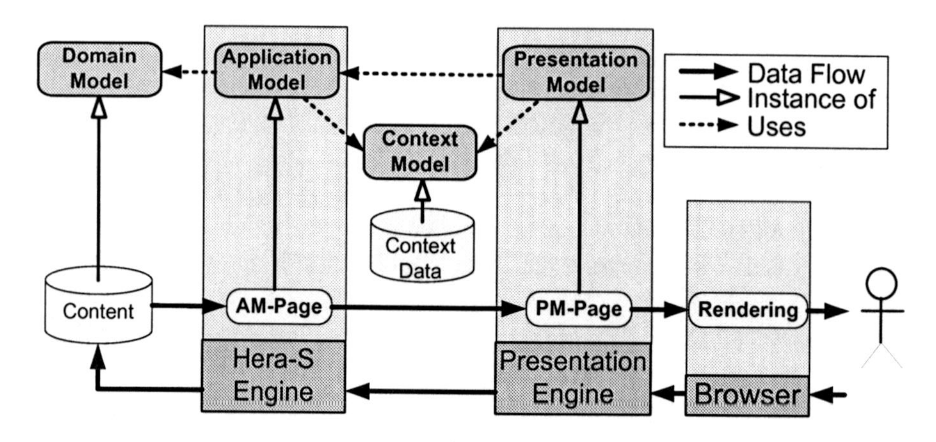

Figure 10.2. Hera-S models and tool pipeline.

Before we can create a model to specify the core design of the application, we need as a starting point in Hera a *domain model* (DM) that describes the structure of the content data. The sole purpose of the DM is to define how the designer perceives the semantical structure of the content data: It tells us what we need to know about the content over which we want the application to work. Based on this DM, the designer creates an *application model* (AM) that describes a hypermedia-based navigation structure over the content. This navigation structure is devised for the sake of delivering and presenting the content to the user in a way that allows for (semantically) effective access to the content.

In turn, this effective access can imply the personalization or adaptation that is deemed relevant. Hera allows dynamic personalization and adaptation of the content. For this purpose, context data are maintained (under control of the application) in a so-called context model (CM). These context data are typically updated based on the (inter)actions of the user as well as on external information.

Thus, on the basis of DM and CM, the AM serves as a recipe that prescribes how the content is transformed into a navigational structure. To be more precise, instantiating the AM with concrete content results in AM (instance) *pages* (AMP). These AMPs can be thought of as pages that contain content to be displayed and navigation primitives (based on underlying semantic relations from the DM) that the user can use to navigate to other AMPs and thus to semantically "move" to a different part of the content. An AMP itself is not yet directly suitable for a browser, but can be transformed into a suitable presentation by a presentation generator, i.e., an engine that executes a specification, for example, a presentation model (PM) of the concrete presentation design in terms of layout and other (browser-specific) presentation details. In Section 10.7 we demonstrate that both proprietary and external engines can be used for this task. For the Hera method, this presentation-generation phase itself is not specific and may be done in whatever way is preferred. So, the AM specifies the (more conceptual or semantical) construction of the navigational structure over the content, while the subsequent presentation phase, possibly specified by a PM, is responsible for the transformation of this structure into elements that fit the concrete browsing situation.

AMP creation is conceptually *pull-based*, meaning that a new AMP is constructed in the Hera pipeline only upon request (in contrast to constructing the whole instantiation of the AM at once, which was done, for example, in the implementation by the HPG-XSLT engine). Through navigation (link-following) and forms submission, the user triggers the feedback mechanism, which results in internally adapting (updating) the Web site navigation or context data and the creation of a new AMP.

As indicated in the introduction, Hera models use RDF(S) to represent the relevant data structures. In the next sections we will see this for the specification of the data in DM, CM, and AM. In the engines these RDF(S) descriptions are used to retrieve the appropriate content and generate the appropriate navigation structures over that content. In HPG-XSLT the actual retrieval was directly done by the engine itself, whereas in HPG-Java this was done with the aid of expressions that are based on SeRQL (Broekstra, 2005) queries. In Hera-S the actual implementation exploits the fact that we have chosen to use RDF(S) to represent the model data and allows us to use native *RDF querying* to access data, for example, the content (DM) and context (CM) data. For this Hera-S allows the application (AM) to connect to the content and context data through the Sesame RDF framework. This solution, combining Hera's navigation design and Sesame's data processing by associating SeRQL queries to all navigation elements, allows us to effectively apply existing Semantic Web technology and a range of its solutions that is becoming available. We can thus include background knowledge (e.g., ontologies, external data sources), and we can connect to third-party software (e.g., for business logic) and to services through the RDF-based specifications. We can also use the facilities in Sesame for specific adaptation to the data processing and to provide more extensive interaction processing (e.g., client-side, scripting). The dynamics and flexibility required in modern Web information systems can thus be met by accommodating the requirements that evolve from an increasing demand for personalization, feedback, and interaction mechanisms. We point out that

Hera-S models in principle are query- and repository-independent. We only require a certain type of functionality; if a repository fulfills these requirements, it can be used for implementation of Hera-S models.

10.3 DATA MODELING

Before the application design can consider the personalized navigation over the domain content, the relevant data need to be specified. As a necessary first step in the approach, the *data modeling* step leads to the construction of the data models for the domain content and the context of the user.

The modeling of the domain content uses RDF(S) and is primarily targeted toward capturing the semantical structure of the domain content. With the Hera-S engine we even allow the model to be an OWL (Dean and Schreiber, 2004) ontology (without restrictions). If we look at the UML representation for the IMDb example that is used throughout the book, this can be easily modeled as an RDFS or OWL definition. In this case this could be done by using a UML-to-OWL conversion process (several papers have been written on the relations between UML and OWL) (Hart et al., 2004). We could, however, also create this model ourselves, e.g., by using an ontology editor like Protégé.[1] Figure 10.3 contains a screenshot of the UML model translated into an RDFS hierarchy together with its properties and OWL restrictions in Protégé. We divided the UML model into four parts. One part, with the prefix imdb, contains the "core" of the movie domain, describing the movies and the persons involved with those movies. Another part, with the prefix cin, models the cinemas that show the movies that are modeled in the imdb part.

In this figure we also see prefixes starting with cm. They relate to the *context* modeling. The CM is modeled and implemented in a similar way as the DM. The main difference between the two is that the content data are meant to be presented to the user, while the context data are meant to support the context-dependent adaptation of the application. So, the content data typically contain the information that in the end is to be shown to the user, while the context data typically contain information used (internally) for personalization and adaptation of content delivery. This distinction might not always be strict, but as it is only a conceptual distinction in Hera-S, the designer may separate content and context in whatever way he desires. As a consequence, we assume that context data are under direct control of the engine, while the content often is not. In the IMDb example, the context model is modeled in the same way as the domain. We first maintain a model,

[1] http://protege.stanford.edu.

with the prefix cm1, that contains users and their comments on movies in the imdb part. The second part, with the prefix cm2, contains a description of tickets bought by the user for a particular movie showing in a particular cinema.

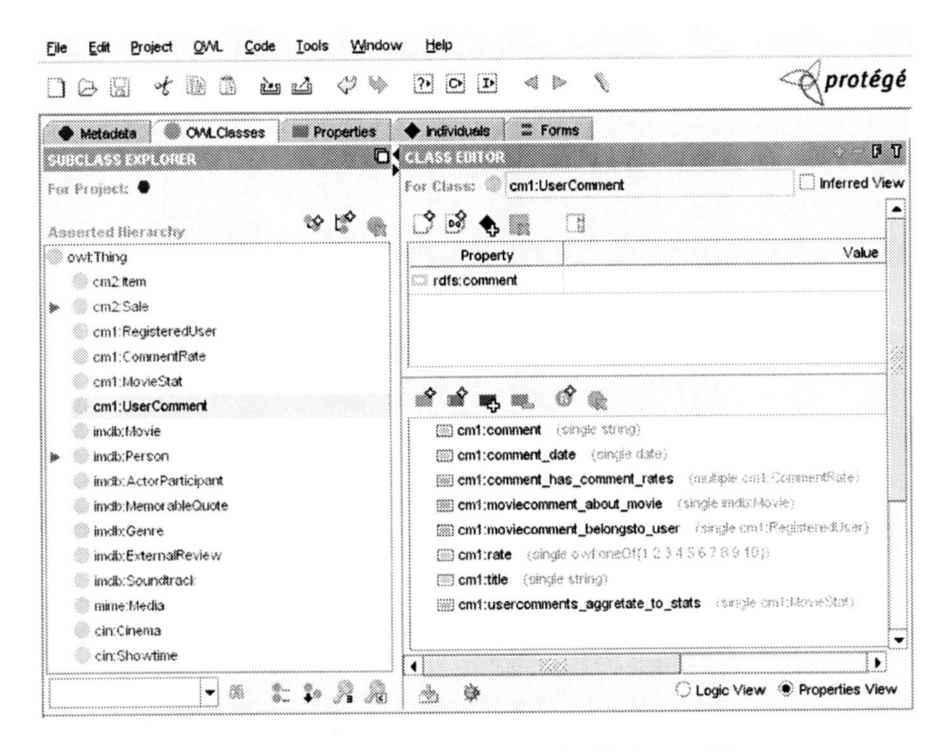

Figure 10.3. Protégé screenshot for the IMDb data modeling.

Considering the role and function of the context data, we can identify different aspects of context. We will come back to context data later when we discuss adaptation in the AM, but we now address the context data modeling. Even though the designer is free to choose any kind of context data, we discern three types in general: session data, user data, and global data.

- *Session data* are relevant to a certain session of a certain user. An example of such data is the current browsing context, such as the device that is used to access the Web application or the units browsed in the current session.
- *User data* are relevant to a certain user over multiple sessions (from initial user data to data collected over more than one session). User (profile) data can be used for personalization (even at the beginning of a

new session). Note that for maintaining these user data over time, the application needs some authentication mechanism.

- *Global data* are usage data relevant to all users over all sessions. Global data typically consist of aggregated information that gives information about groups of people. Examples include "most visited unit" or "people that liked item x, also browsed item y."

In Figure 10.4 we show part of an RDF-graph representation of the domain data model that we will use as a basis for the examples in this chapter. It shows the main classes and a selection of relationships between those classes, while omitting their data type properties.

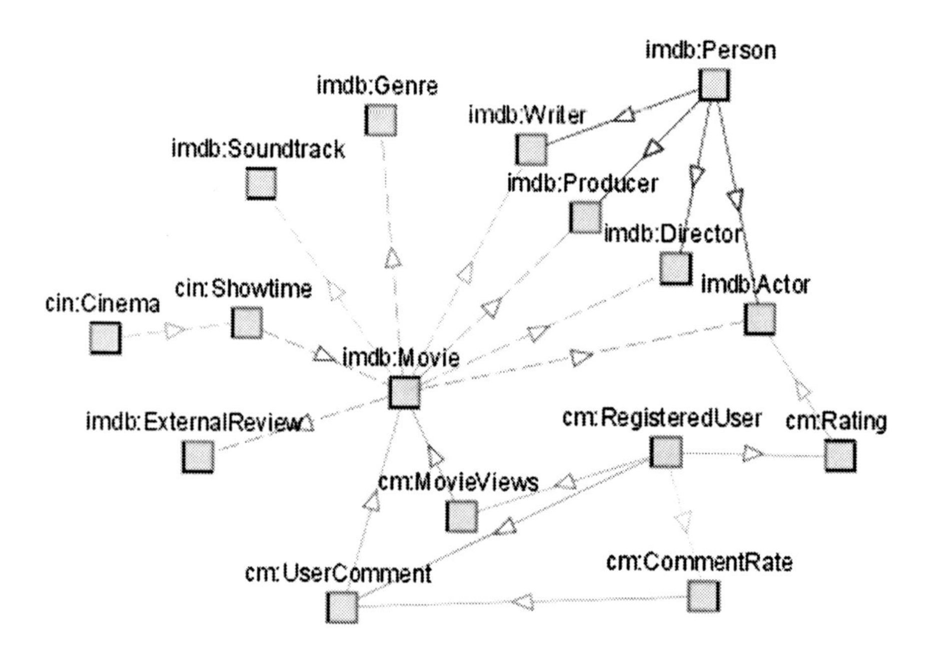

Figure 10.4. RDF-graph representation of IMDb domain and context data.

Both the DM and CM data are implemented in Hera-S using a Sesame repository. For the CM, which is under direct control of the application, this allows the application to manage and update the context as it perceives this context. Next to this, it also provides the means for other processes to use and update (parts of) this information. The context data could, for instance, be manipulated by business logic software for the sake of adaptation, or by external user-profiling software.

Another great advantage of using Sesame is the possibility to combine several data sources (both content and context data) at the same time. In this way, designers can couple additional data sources to the already existing ones and can thus easily extend the domain content. This also offers possibilities to exploit additional knowledge when performing a search. Currently, we are involved in the exploitation of the WordNet ontology[2] (Miller, 1995), time ontologies (Hobbs and Pan, 2004), and geographic ontologies[3] (Chipman et al., 2005). In this way, a keyword search can be extended with synonyms extracted from the WordNet ontology, or a search for a city can be extended with surrounding cities from a geographic ontology. By supporting unrestricted RDFS or OWL DMs, Hera-S is particularly suited to (re-)use existing domain ontologies. Moreover, many existing data sources that are not yet available in RDFS or OWL format can be used via Semantic Web wrapping techniques[4] (Thiran et al., 2005). In the latter case Sesame can be used as a mediator between such a data source and Hera-S.

As we will see in detail in the next section, the access to the data from the DM or CM is part of the application definition. It means that the access to the RDF data is part of the model. In principle, we assume that the concepts from the DM and CM are associated with concrete data elements in the data storage structure. As we use the Sesame RDF framework as our back-end repository, these data can be exploited and reasoned upon. Accessing the content data in Hera-S will be done via explicit SeRQL queries; by making them explicit in the models, we support customizable access via customizable SeRQL queries. Thus, the full potential of the SeRQL query language can later be used in the AMP creation. For the purpose of defining the content and context, we can abstract from the SeRQL queries, but for the support of different types of adaptation, we benefit from making this SeRQL access explicit.

10.4 APPLICATION MODELING

Based on the domain definition, application modeling results in the application model (AM) that specifies the navigational behavior of the Web application. The AM enables designers to specify how the (navigational) access to the data (dynamically retrieved from the domain) is structured by describing which data are shown to the user and what Web pages the user can navigate to. At the same time, the AM allows this specification to be dynamic, such that the navigational access to the data can be personalized to a user and adapted for a specified context.

---

[2] http://www.semanticweb.org/library/.

[3] http://reliant.teknowledge.com/DAML/Geography.owl.

[4] http://simile.mit.edu/RDFizers/.

Since in the AM we use Turtle and SeRQL syntax, we first highlight in Section 10.4.1 the most relevant elements from those languages. In Section 10.4.2 we present the basic AM constructs and exemplify them based on the IMDb example as discussed in the previous section. In Section 10.4.3 we give examples of adaptation expressed in the AM. Section 10.4.4 contains a number of more advanced modeling primitives. In Section 10.4.10 we illustrate a model builder that offers designers a visual tool to help create the domain models and the AM and produce the correct RDF serialization for those graphical representations.

10.4.1 Queries and Syntax

10.4.1.1 Turtle

Turtle (Terse RDF Triple Language) (Beckett, 2004) is an RDF syntax format designed to be compact, easy to use, and easy to understand by humans. Although not an official standard, it is widely accepted and implemented in RDF toolkits.

In Turtle, each part of the triple is written down as a full URL or as a qualified name (using namespace prefixes). In our examples, we will mostly use the latter form, for brevity. For example,

```
my:car rdf:type cars:Volvo
```

denotes the RDF statement "my car is of type Volvo." "my:" and "cars:" in this example are namespace prefixes that denote the vocabulary/ontology from which the term originates. Turtle also introduces the predicate "a" as a shortcut for the "rdf:type" relation.

```
my:car a cars:Volvo
```

denotes the same RDF statement as the first example.

In order to list several properties of one particular subject, we can use the semicolon to denote branching:

```
my:car a cars:Volvo ;
        my:color "Red"
```

denotes two statements about the car: that it is of type "Volvo" and that its color is red (denoted by a string literal value in this example).

When the value of a property is itself an object with several properties, we can denote this by using square brackets ([and]). In RDF terms, such brackets denote a blank node (which can be thought of as an existential qualifier). For example:

```
my:car a cars:Volvo ;
    my:hasSeat [
                    a my:ComfortableSeat ;
        my:material "Leather"
    ]
```

denotes that the car has a seat that is “something” (a blank node) that has as its type “my:ComfortableSeat” and that has leather as the material.

In the next sections, we will regularly use Turtle syntax forms in various examples.

10.4.1.2 SeRQL

SeRQL (Broekstra et al., 2002; Broekstra, 2005) (Sesame RDF Query Language) is an RDF query and transformation language that uses graph templates (in the form of path expressions) to bind variables to values occurring in the queried RDF graph. It is an expressive language with many features and useful constructs.

An SeRQL query consists of a set of clauses (SELECT, FROM, and WHERE). As in SQL, the SELECT clause describes the projection, i.e., the ordered set of bound values that is to be returned as a query result. The FROM clause describes a graph template that is to be matched against the target graph, and the WHERE clause specifies additional Boolean constraints on matching values. A simple example query that selects all instances of the class “Volvo” is

```
SELECT aCar
FROM {aCar} rdf:type {cars:Volvo}
```

As one can see, the path expression syntax bears a strong resemblance to Turtle syntax, except that in SeRQL, each node in the graph is surrounded by braces ({ and }). In the above query, “aCar” is a variable that is to be matched in the target graph against all statements that conform to the pattern, i.e., those that have “rdf:type” as their predicate and “cars:Volvo” as their object.

As in Turtle, SeRQL paths can be branched (using a semicolon) as well as chained. For example, in the following query we use both chaining and branching to select a car, its color, its owner, and the address of that owner:

```
SELECT aCar, color, owner, address
FROM {aCar} rdf:type {cars:Volvo} ;
                my:color {color} ;
                my:owner {owner} my:address {address}
```

Additionally, we can use the WHERE clause to specify additional constraints on the results. To adapt the above query to return results only for those cars whose color is red, we could add a WHERE clause:

```
WHERE color = "Red"
```

10.4.2 Units, Attributes, and Relationships

Now we will discuss the constructs that we provide in our AM. We will start in this section with the basic constructs that are sufficient to build basic Web applications and then move on to more complex constructs for realizing richer behavior.

The AM is specified by means of *navigational units* (shorthand: units) and *relationships* between those units. The instantiation of units and relationships is defined by (query) expressions that refer to the (content and context) data, as explained in Section 10.3.

The unit can be used to represent a "page." It is a primitive that (hierarchically) groups elements that will be shown to the user together. Those elements shown to the user are called "attributes," and so units build hierarchical structures of attributes.

An attribute is a single piece of information that is shown to the user. This information may be constant (i.e., predefined and not changing), but usually it is based on information inside the domain data. If we have a unit for a concept *c*, then typically an attribute contained in this unit is based on a literal value that is directly associated with *c* (for example, as a data type property). Note that literals may denote not only a string type, but also other media by referring to a URL. Furthermore, we offer a built-in media class (denoted as hera:Mime) that can be used to specify an URL and the MIME type of the object that can be found at the URL. This can be used if the media type is important during later processing.

Below we give an example of the definition of a simple unit, called "MovieUnit," to display information about a movie. We mention two elements in this definition:

- From the second until the seventh line, we define which data instantiate this unit. These data are available as input (am:hasInput) from the environment of this unit, e.g., passed on as a link parameter or available as a global value. In this case we have one variable, "M": The fourth line specifies the (literal) name of the variable, while the fifth line indicates the type of the variable. In this case, a value from the imdb:Movie class concept from the domain will instantiate this unit.

- In the 8th until the 14th line, starting with am:hasAttribute, we decide to display an attribute of this movie, namely a title. We label this attribute "Title" (so that later we can refer to it), and we indicate (with "am:hasQuery") how to get its value from the data model. This query uses the imdb:movieTitle (datatype) property applied to the value of M. Note that in the query "$M" indicates that M is a Hera variable, i.e., outside the scope of the SeRQL query itself. The output of the SeRQL query result is bound to the Hera variable T (implicitly derived from the SELECT list).

In our RDF/Turtle syntax the definition looks like the following:

```
:MovieUnit a am:NavigationUnit ;
  am:hasInput [
    am:variable [
      am:varName "M" ;
      am:varType imdb:Movie
    ]
  ] ;
  am:hasAttribute [


    rdfs:label "Title" ;
    am:hasQuery
      "SELECT T
        FROM {$M} rdf:type {imdb:Movie};
                  imdb:movieTitle {T}"
  ]
```

For the attribute value instead of the simple expression (for which we can even introduce a shorthand abbreviation), we can use a more complicated query expression, as long as the query provided in "am:hasQuery" returns a data type property value.

Relationships can be used to link units to each other. We can use relationships to contain units within a unit, thus hierarchically building up the "page" (we call these *aggregation relationships*), but we can also exploit these relationships for navigation to other units (we call these *navigation relationships*).

As a basic example, we include in the unit for the movie not only its title but also a (sub)unit with the name and photo of the lead actor and a navigational relationship that allows us to navigate from the lead-actor information to the full bio-page (unit) for that actor. Note that from now on we omit namespaces in our text when they appear to be obvious.

- We have separated here the definitions of "MovieUnit" and "ActorUnit" (which allows later reuse of the ActorUnit), but we can also define subunits inside the unit that contains them.
- In the definition of the MovieUnit, one can notice, compared to the previous example, that we have an additional subunit with its label LeadActor, with its type ActorUnit, and with the query that gives the value with which we can instantiate the subunit.
- In the definition of the ActorUnit. one can notice its input variable, two attributes, and a navigation relationship. This navigation relationship has a label "Actor-Bio," targets a "BioUnit," and, with the query based on the "imdb:actorBio" property, determines to which concrete BioUnit this ActorUnit offers a navigation relationship. Note that in this case the variable $B is passed on with the navigational relationship (it is also possible to specify additional output variables that are passed on with the relationship).

```
:MovieUnit a am:NavigationUnit ;
  am:hasInput [ am:variable [ am:varName "M";
                              am:varType imdb:Movie]] ;
  am:hasAttribute [ rdfs:label "Title" ; ... ] ;
  am:hasUnit [
    rdfs:label "LeadActor" ;
    am:refersTo :ActorUnit ;
    am:hasQuery
      "SELECT L
       FROM {$M} rdf:type {imdb:Movie};
                 imdb:movieLeadActor {L} rdf:type {imdb:Actor}"
]

:ActorUnit a am:NavigationUnit ;
  am:hasInput [ am:variable [ am:varName "A" ;
                              am:varType imdb:Actor]] ;
  am:hasAttribute [
    rdfs:label "Name";
    am:hasQuery
      "SELECT N
       FROM {$A} rdf:type {imdb:Actor};
                 imdb:actor_name {N}" ] ;
```

```
am:hasAttribute [
  rdfs:label "Photo" ;
  am:hasQuery
    "SELECT P
     FROM {$A} rdf:type {imdb:Actor};
                imdb:actorPhoto {P}" ] ;
am:hasNavigationRelationship [
  rdfs:label "Actor-Bio" ;
  am:refersTo :BioUnit ;
  am:hasQuery
    "SELECT B
     FROM {$A} rdf:type {imdb:Actor};
                imdb:actorBio {B}"
]
```

Thus, in these examples we see that each element contained in a unit, whether it is an attribute, a subunit, or a navigational relationship, has a query expression (hasQuery) that determines the value used for retrieving (instantiating) the element.

Sometimes we know that in a unit we want to contain subunits for each of the elements of a set. For example, in the MovieUnit we might want to provide information for all actors from the movie (and not just the lead actor). Below we show a different definition for MovieUnit that includes a *set*-valued subunit element (am:hasSetUnit). In its definition, one can notice

- the label "Cast" for the set unit
- the indication that the elements of the set unit are each an "ActorUnit"
- the query that determines the set of concrete actors for this movie to instantiate this set unit, using the imdb:movie_actor object property

```
:MovieUnit a am:NavigationUnit ;
  am:hasInput [am:variable [am:varName "M" ;
                            am:varitype imdb:Movie]] ;
  am:hasAttribute [rdfs:label "Title" ; … ] ;
  am:hasSetUnit [
    rdfs:label "Cast" ;
    am:refersTo ActorUnit ;
    am:hasQuery
      "SELECT A
       FROM {$M} rdf:type {imdb:Movie} ;
                  imdb:movieActor {A}"
  ]
```

So, we see that a set unit is just like a regular unit, except that its query expression will produce a set of results, which will cause the application to arrange for displaying a set of (in this example) ActorUnits.

Likewise, we can have set-valued query expressions in navigational relationships, and with am:tour and am:index we can construct guided tours and indexes, respectively. With these the order of the query determines the order in the set, and with the index an additional query is used to obtain anchors for the index list.

10.4.3 Adaptation Examples

Adaptation and personalization are important aspects within the Hera methodology. For this purpose, the query expressions can be used to include conditions that provide control over the instantiation of the unit. Typically, these conditions use data from the context model (CM) and thus depend on the current user situation. For example, we can use U as a (global) variable that denotes the current (active) user for this browsing session (typically, this gets instantiated at the start of the session). Let us assume that in the CM for each Actor there is a cm:actorRating property that denotes U's rating of the actor (from 1 to 5 stars) and that the user has indicated to be interested only in actors with more than 3 stars. We could then use this rating in adapting the Cast definition in the last example:

```
am:hasSetUnit [
  rdfs:label "Cast";
  am:refersTo ActorUnit ;
  am:hasQuery
    "SELECT A
     FROM {$U} cm:actorRating {} cm:stars {V} ;
                cm:ratingOnActor {A} imdb:playsIn {$M}
    WHERE V > 3"
]
```

Here we see how we can influence (personalize) the input to an element (in this case a set) by considering the user context in the query that determines with which values the element is constructed. To be precise, in this example we state inside the movie unit what actors of the cast this user will be provided with, i.e., which values we "pass on."

Another user adaptation example would be that the user has indicated not to be interested in photos from actors. We could then change the query for the photo attribute accordingly:

```
:ActorUnit a am:NavigationUnit ;
  am:hasInput [ am:variable [ am:varName "A" ;
                              am:varType imdb:Actor]] ;
  ...
  am:hasAttribute [
    rdfs:label "Photo" ;
    am:hasConditionalQuery [
      am:if "SELECT *
             FROM {$U} cm:showElement {}
                       cm:showAbout {imdb:actorPhoto}"

      am:then "SELECT P
               FROM {$A} imdb:actorPhoto {P}"
    ]
  ];
  ....
```

Here we see that with "am:hasConditionalQuery" the attribute becomes "conditional," i.e., the photo attribute is only shown when the condition (am:if) query produces a non-empty result. We can also add an "am:else" part here and display, for example, the string "no photo displayed." We point out that this query can be written in one single (nested) SeRQL query, but for the clarity of adaptation specification we use this syntax sugaring.

Finally, we present a more complex example of adaptation. Consider again the ActorUnit from the previous section, which showed an actor's name, his picture, and a link to his bio. Now imagine we would like to add the list of movies in which the actor played. However, because some movies are age-restricted, we would like to restrict this list so that adult-rated movies are only shown to registered users that are 18 or older. As in the previous adaptation examples, this adaptation can be achieved by tweaking the SeRQL query that computes the list of movies:

```
:ActorUnit a am:NavigationUnit ;
  am:hasInput [ am:variable [ am:varName "A" ;
                    am:varType imdb:Actor]] ;
  ...
    am:hasSetUnit [
    rdfs:label "Movies Played In";
    am:refersTo MovieUnit ;
    am:hasConditionalQuery [
    am:if "SELECT *
        FROM        {$U} cm:age {G}
        WHERE G > 17"
```

```
am:then "SELECT M
         FROM {$A} imdb:actorMovie {M},
                {M} rdf:type {imdb:Movie}"
am:else "SELECT M
    FROM {$A} imdb:actorMovie {M},
           {M} rdf:type {imdb:Movie}; imdb:mpaaRating {R}
                WHERE R != "NC-17""
]]
```

First, notice in the code excerpt the am:hasSetUnit, which represents the list of movies for the active actor (A). This list is defined by a conditional query, in which it is verified whether the active user (U) is registered and whether his age is over 17 (am-if). If this condition holds (am:then), all movies of the particular actor are computed. If the condition does not hold (am:else), the computed movie list is restricted to movies that are not MPAA (Motion Picture Association of America)-rated as "NC-17" (No Children Under 17 Admitted).

10.4.4 Other Constructs

Earlier we explained the basic constructs. In this section we will look at some additional features of Hera-S that also allow designers to use some more advanced primitives in order to construct richer applications.

10.4.5 Update Queries

For the sake of adaptation, we need to maintain an up-to-date context model. In order to do so, we need to perform updates to this data. For this, we have the functionality to specify an am:onLoad update query and an am:onExit update query within every unit; these are executed on loading (navigating to) and exiting (navigating from) the unit. Furthermore, we allow an update query to be attached to a navigation relationship so that the update query is executed when a link is followed. In all cases, the designer may also specify more than one update query.

In our example we could, for instance, maintain the number of page views (visits) of a certain movieUnit and update this information if the movieUnit is loaded using the "onLoad" query:

```
:MovieUnit a am:NavigationUnit ;
  …
  am:onLoad [
    am:updateQuery
```

```
        "UPDATE {V} cm:amount {views+1}
         FROM {$U} rdf:type {cm:RegisteredUser};
                   cm:userMovieViews {V} cm:amount {views};
                   cm:viewsOfMovie {$M}"
    ];
    ....
```

10.4.6 Frame-Based Navigation

We explained earlier that units can contain other units. The root of such an aggregation hierarchy is called a *top-level unit*. The default semantics of a navigational relationship (that is defined somewhere inside the hierarchy of a top-level unit) is that the user navigates from the top-level unit to the top-level unit that is the target of the relationship. In practice, this often means that in the browser the top-level unit is replaced by the target unit. However, we also allow specify that the navigation should only consider the (lower-level) unit in which the relationship is explicitly specified, so that only that unit is replaced while the rest of the top-level unit remains unchanged.

This behavior is similar to the frame construct from HTML. We specify this behavior by explicitly indicating the source unit for the relationship. Inspired by the HTML frame construct, we allow the special source indications "_self" (the unit that contains the relation), "_parent" (the unit that contains the unit with the relation), and "_top" (the top-level unit—the default behavior). Alternatively, relations may also indicate another containing unit by referring to the label of the contained unit. An example of a navigational relationship with a source indication looks like

```
am:hasNavigationRelationship [
  ...
  am:source am:_self ;
  ... ]
```

10.4.7 Forms

Besides using relationships (links) for navigation, we also support applications that let the user provide more specific feedback and interact. For this we provide the form unit. A form unit extends a normal unit with a collection of input elements (that allow the user to input data into the form) and an action that is executed when the form is submitted. In a form a navigational relationship typically has a button that activates the submission.

Below we give an example of a form that displays the text "Search Movie:" (line 3) with one text input field (lines 5 to 11) to let the user enter the movie she wants to browse to. If the user enters a value in this field, it is bound to the variable movieName (line 9). After submitting the form via a button with the text "Go" (lines 14 and 15), the user navigates to the MovieUnit (line 14) that will display the movie for which the name was entered in the input field, which is specified in the query (starting in line 20) using the variable movieName.

```
:MovieSearchForm a am:FormUnit ;
  am:hasAttribute [


    am:hasValue "Search Movie: "
  ];
  am:formElement [
    rdfs:label "Search Input";
    am:formType am:textInput;
    am:binding[
          am:variable [am:varName "movieName" ;
                       am:varType xsd:String ]]
  ];
  am:formElement [
    rdfs:label "Submit Button";
    am:formType am:button;
    am:buttonText "Go";
    am:hasNavigationRelationship [
      rdfs:label "Search Form-Movie" ;
      am:refersTo :MovieUnit ;
      am:hasQuery
        "SELECT M
         FROM {M} rdf:type {imdb:Movie};
                  imdb:movieTitle {X}
         WHERE X = $movieName"
    ]
  ]
```

10.4.8 Scripting Objects

Current Web applications offer users a wider range of client-side functionality by different kinds of scripting objects, like Javascript and VBscript, stylesheets, HTML+TIME timing objects, etc. Even though WIS methods like Hera concentrate more on the creation of a platform-independent hypermedia presentation than a data domain, and these scripts are often (but not always) browser-/platform-specific, we still provide the designer a hook to insert these kinds of scripting objects.

The designer can specify within a scripting object whatever code she wants, as this will be left untouched in generating the AMPs out of the AM. Furthermore, the designer can add an am:hasTargetFormat property to specify one or more target formats for format-specific code, e.g., HTML or SMIL. This allows us later in the process to filter out certain format-specific elements if these are not wanted for the current presentation. The scripting objects can use the variables that are defined within the scope of the units. Scripting objects can be defined as an element within any other element (i.e., units and attributes). Furthermore, it can be specified whether or not the script should be an attribute of its superelement (e.g., similar to elements in HTML that have attributes and a body). The need to place some specific script on some specific place is, of course, decided by the designer.

10.4.9 Service Objects

An application designer might want to use additional functionality that cannot be realized by a client-side object but that involves the invocation of external server-side functionality. Therefore, we provide so-called service objects (am:serviceObject) to support Web services in the AM. The use of a service object and the reason to provide support for it are similar to that of scripting objects. The designer is responsible for correctness and usefulness of the service object.

Think of utilizing a Web service from an online store selling DVDs in order to be able to show on a movie page an advertisement for buying the movie's DVD. A service object needs three pieces of information:

- a URL of the Web service one wants to use
- a SOAP message that contains the request to the Web service
- a definition of the result elements

A service object declaration can be embedded as a part of every other element. If a unit is navigated to ("created"), first the service objects will be executed. The results of the service object either will be directly integrated into the AM and treated as such, or the result can be bound to variables. Service objects can use unit variables in their calls.

10.4.10 Model Builders

In most RDF serializations it can become difficult to see which structures belong together and what the general structure of the document is, especially as the documents get larger. This also applies to the Hera models and has the consequence that manually creating them can become error-prone. It is therefore beneficial to offer the designer tool support for creating those models graphically. Based on a given HPG version, Hera Studio (Figure 10.5) contains a domain, context, and application model editor in which the designer can specify the models in a graphical way. All these models can subsequently be exported to an RDF serialization that can be used by Hera.

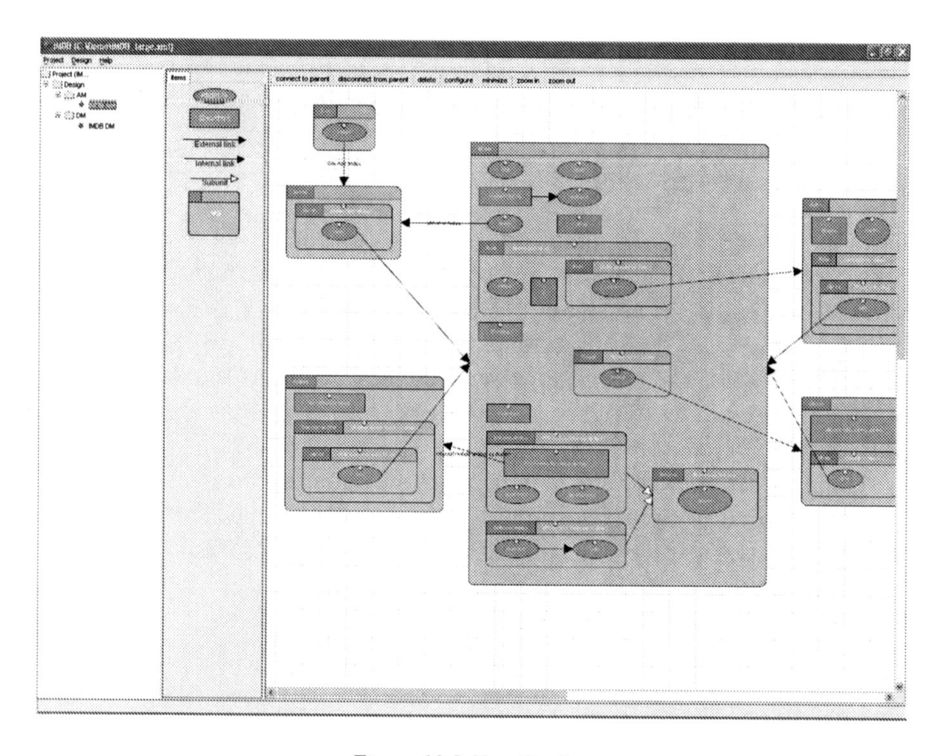

Figure 10.5. Hera Studio.

Note that Hera Studio is not a general-purpose OWL or RDF(S) editor such as Protégé, for instance; rather, it is a custom-made version specialized for Web applications designed through Hera models.

In the DM editor (Figure 10.6), designers can define classes and object properties between those classes. For every class, a number of data type properties can be given that have a specified media type (e.g., String, Image, etc.). Furthermore, inheritance relations for classes and properties can be denoted. In addition, instances of the classes and properties can be specified. Note that if more complex constructs are needed, the designer could also use a general-purpose OWL/RDF editor like Protégé.

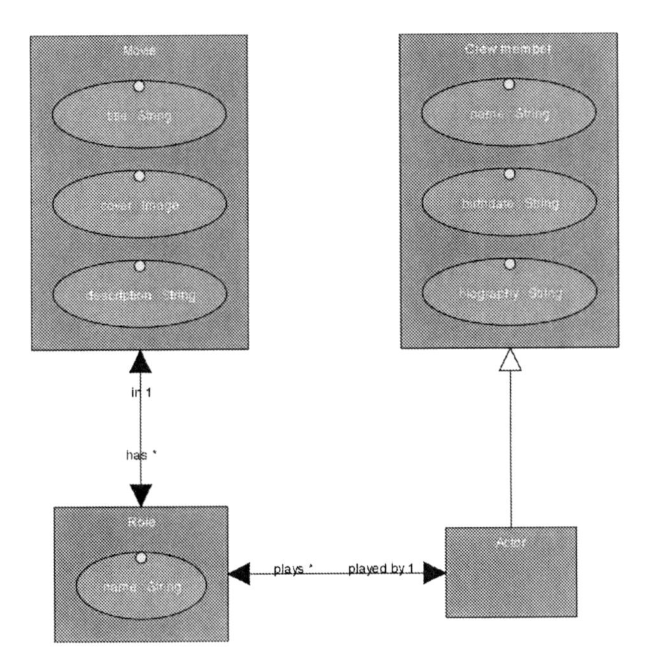

Figure 10.6. DM example.

The AM editor provides a graphical way for specifying an AM (Figure 10.7 gives an example). It specifically allows organizing the units and the relationships between them. Per unit, elements can be defined and displayed. Detailed information like queries is hidden in the graphical view and can be configured by double-clicking the elements. For the simpler constructs, the editor provides direct help: For example, when defining a data type property, the editor gives the designer a straightforward selection choice from the (inherited) data type properties of the underlying context and domain models. However, for the more complex constructs, the designer has the freedom to express his own queries and element properties. In addition, the designer can control the model's level of detail to get a better overview of the complete model.

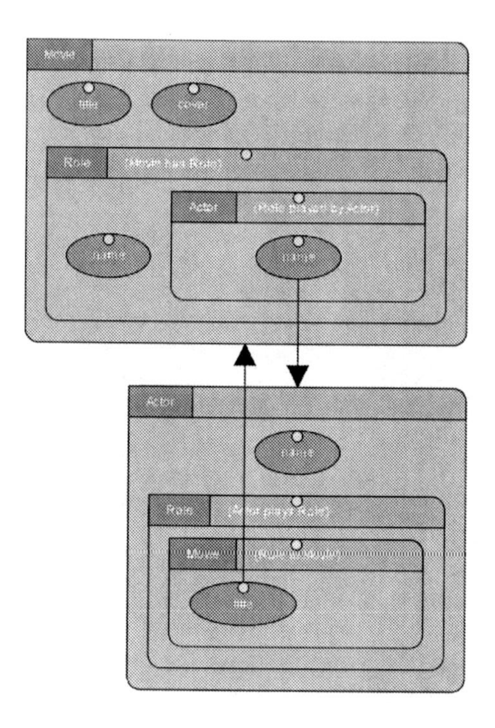

Figure 10.7. AM example.

Currently, the AM editor is being extended to support the more specific Hera-S constructs mentioned in this section. In this process, we will also extend the model-checking functionality that allows the designer to check if the Web application fulfills certain requirements. Furthermore, we plan to extend the builders with an optional lightweight presentation component.

10.5 ASPECT ORIENTATION IN MODEL DESIGN

Before we continue with presentation design and implementation, we make a side step and turn to an element of design support that we are currently working on and that uses principles from aspect orientation.

As described in the previous sections, Hera-S provides conceptual WIS design support on the basis of data contained in an RDF repository like Sesame and that are accessed and manipulated through an RDF query language like SeRQL. In this setting, adaptation is specified by SeRQL queries that, based on (DM and) CM data, conditionally instantiate navigational units in the AM. Examples of such an adaptation can be found in Section 10.4.3. In most cases, the desired adaptation is expressed by expressions that are embedded in the SeRQL query and that have the explicit purpose of restricting the set of instances; we can call these expressions *adaptation conditions*. We observe that often these adaptation conditions can conceptually be detached from the rest of the SeRQL queries and explicitly specified at (AM) model level. In this way, with each AM modeling element (i.e., units, relationships) we can nicely associate its adaptation conditions that explicitly denote the restriction of this element according to the user model's attributes/values.

Typically, in a Web application several adaptation issues need to be taken into account in parallel (e.g., age-group restriction, accessibility, device dependence). Adaptation engineering thus constitutes a significant effort in specifying the application's functionality. Moreover, although the adaptation conditions for an adaptation issue can occur at one specific place in the design (e.g., to restrict adult-rated material on a certain page), it is (more) often the case that they cannot be pinpointed to one particular element (e.g., when one does not want anything on the site to show adult-rated material to minors) and need to be applied at different places in the design (models). Concretely, consider the last example from Section 10.4.3, which restricts the list of movies starring a particular actor. Obviously, the designer may be required to specify other lists of movies (e.g., in the am:CinemaUnit, to denote the movies played in a particular cinema) also at other places in the design. To enforce the age-group restriction policy throughout the application or Web site, the designer thus needs to incorporate the necessary conditions in all SeRQL queries involving the selection of movies (or any other content that may be age-restricted, e.g., adult actors).

A similar observation was made in (regular) software development, when considering different design concerns of an application: Some concerns cannot be localized to a particular class or module; instead they are inherently distributed over the whole application. Such a concern is called a *cross-cutting* concern. To cleanly separate the programming code addressing this concern from the regular application code, aspect-oriented programming (Kiczales et al., 1997) was introduced. Inspired by the principles of aspect orientation, Hera-S provides (adaptation) design support to specify, in an aspect-oriented way, the different cross-cutting adaptation concerns by means of an aspect-oriented adaptation specification.

Applying aspect orientation to extend an AM with different additional adaptation concerns is thus done by modeling each concern as an *aspect*. Each aspect is composed of a number of advice-pointcut pairs. In this setting, the notions of advice and pointcut are as follows:

- *Advice*: Advice specifies a particular transformation in terms of modifications to the different (navigational) elements of the AM. In most cases, a single modification will add a single adaptation condition to certain navigational units or relationships in the form of an SeRQL query.

- *Pointcut*: A pointcut defines a query on the set of navigational units and relationships of an application model, which specifies exactly the elements to which certain advice should be applied.

These advice-pointcut pairs can thus be used to inject adaptation conditions to (certain elements of) the AM. It is a current research topic to investigate the limitations of this process of transforming these adaptation conditions in the corresponding SeRQL queries, a restricted form of what is called *weaving* in aspect terminology.

To exemplify this approach, we illustrate how two additional adaptation concerns, age-group (restriction) and device dependence, can be specified in an aspect-oriented way over an (existing, in this case non-adaptive) AM. For the first adaptation aspect, namely age-group (restriction), let us express the motivating example mentioned earlier in this section: Restrict visibility of all adult-rated, i.e., NC-17-rated, content throughout the application, and only show it when the user's age has been confirmed to be above 17. In an aspect-oriented way, this adaptation strategy is specified as follows:

```
POINTCUT SET WITH PARENT cm:movie
ADVICE
  SELECT M
  FROM {M} am:MPAA-rating {R}; rdf:type {imdb:Movie}
  WHERE R != 'NC-17'
  OR EXISTS
    (SELECT * FROM  {$U} cm:age {G}
     WHERE G > 17)
```

This pointcut-advice pair specifies first the pointcut: wherever in the AM a navigational unit is used that represents a set of movie elements (i.e., the pointcut part). In all these places (in the advice) a condition is added in the form of an SeRQL expression, which denotes that the age (an attribute from the CM) should be over 17 to view NC-17-rated material. Similar pointcut-advice pairs can be specified to restrict visibility of items with other MPAA ratings. Note that any movie set, wherever it appears in the AM, is restricted: The adaptation is not localized to one particular navigational unit and is thus truly cross-cutting.

The semantics of this condition addition is that this query expression is performed after the one that was defined originally for this element. Concretely, interpretation and execution of the above aspect result in modification of the SeRQL queries instantiating (a set of) movies. Note that the last example of Section 10.4.3 is one particular occurrence of such a set of movies, for which the adaptation was manually specified by the designated SeRQL query. However, to achieve automatic *weaving* of aspects, a more generic *pipeline approach* is best suited. In this approach, the original (non-adaptive) query expression Q producing the set of instances (i.e., movies in this case) is taken as a starting point. Subsequently, each adaptation condition C_i specified for this set gives rise to an SeRQL query Q_i that takes as input the result of the previous query $Q_{(i-1)}$ and filters from this result the element according to the adaptation condition C_i. For adaptation conditions $C_1 \ldots C_n$ specified for a set, and possibly originating from different adaptation issues, the resulting query will thus be of the form $Q_n \circ Q_{(n-1)} \circ \ldots \circ Q_1 \circ Q$. Evidently, other approaches, such as query rewriting or query merging, are possible.

A second example concerns device dependence: In order not to overload small-screen users (e.g., PDA users), we decided not to show pictures. Therefore, we can specify the following pointcut-advice pair:

```
POINTCUT ATTRIBUTE
ADVICE
  SELECT P
  FROM {P} hera:Mime {} hera:mimeType {T}
  WHERE T != 'image\*'
  OR EXISTS
    (SELECT * FROM {$U} cm:device {D}
                        WHERE D != 'pda')
```

In the pointcut, all attributes from the AM are selected. For these attributes, in the advice part, the picture attributes (denoted by the mime-type specification as mentioned in Section 10.4.2) are filtered out for PDA users, restricting their visibility for these PDA users. Note that, once again, our example addresses a truly cross-cutting concern: Anywhere in the AM where a picture attribute is used, it will be filtered out for PDA users.

To conclude this section on aspect-oriented adaptation support, we would like to point out that the primary way of defining adaptation, namely to manually specify it in the AM by means of SeRQL queries, as was illustrated in Section 10.4, is still available to the designer. The aspect-oriented support presented here merely offers the designer an additional and alternative means of specifying, in a straightforward and distributed way, the adaptation conditions for (sets of) AM elements.

10.6 IMPLEMENTATION

After illustrating the AM design, we now turn to the engine implementing the model, i.e., generating the hypermedia views over the content according to what is specified in the AM. As explained in Section 10.1, we distinguish three implementations for the Hera models. The Hera-S implementation is based on our previous experiences with the Hera Presentation Generator (HPG) (Frasincar, 2005), an environment that supports the construction of WIS using the Hera methodology. Considering the technologies used to implement Hera's data transformations, we distinguish two variants of the HPG: HPG-XSLT, which implements the data transformations using XSLT stylesheets, and HPG-Java, which implements the data transformations using Java. In HPG-XSLT we employed as an XSLT processor Saxon, one of the most up-to-date XSLT processors implementing XSLT 2.0. In HPG-Java we used two Java libraries, the Sesame library for querying the Hera models and the Jena library for building the new models. Hera-S resembles in many ways HPG-Java, but it is based on the revision of the Hera models presented in this paper.

10.6.1 HPG-XSLT and HPG-Java

First we take a look at the tools for HPG-XSLT and HPG-Java that build the foundation for Hera-S. HPG-XSLT has an intuitive designer's interface, visualizing the Hera pipeline for the development of a Web application (see Figure 10.8). The user is guided in a sequence of steps to create the complete Web application, which, generated with XSLT stylesheets, results in a concrete, but static, Web site for a given platform. In the interface we see that each step in this advanced HPG view is associated with a rectangle labeled with the step's name (e.g., Conceptual Model, Unfolding AM, Application Adaptation, etc.). In each step a number of buttons are connected with within-step arrows and between-step arrows that express the data flow. Such a button represents a transformation or input/output data depending on the associated label (e.g., Unfold AM is a transformation, Unfolding sheet AM is an input, and Unfolded AM is an output). The arrows that enter into a transformation (left, right, or top) represent the input, and the ones that exit from a transformation (bottom) represent the output. The last step is the generation of the presentation in the end format (e.g., HTML).

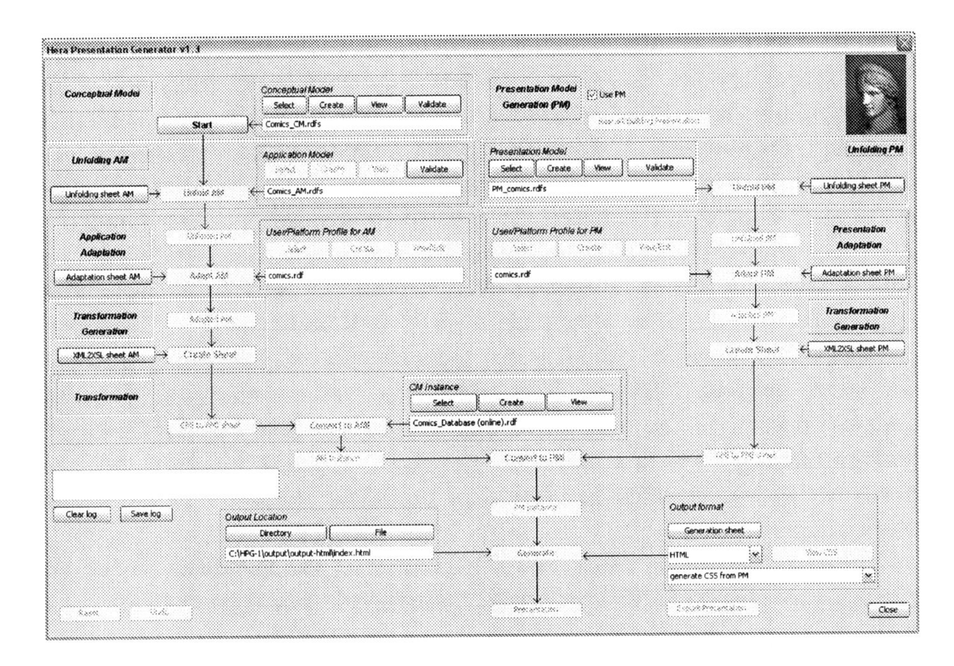

Figure 10.8. Screenshot from HPG-XSLT.

The models for HPG-XSLT can be created by hand but also with the help of Microsoft Visio templates, which provide a graphical environment to aid the correct construction of the different models (see Figure 10.9 for an example of the corresponding AM builder).

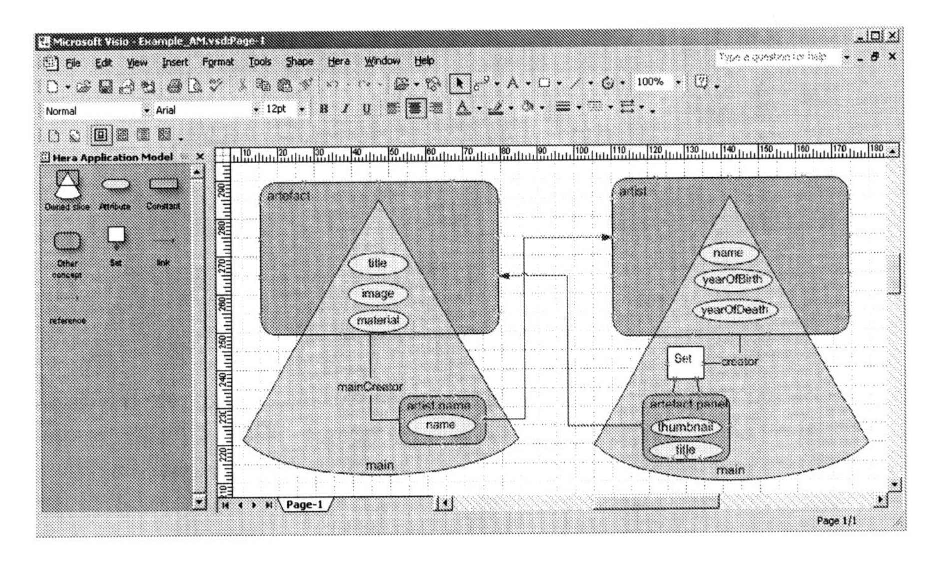

Figure 10.9. Visio templates HPG-XSLT.

HPG-XSLT is an effective demonstration tool and sufficient for simple Web site applications. However, WIS may require more flexibility and more dynamics, for the application to be able to dynamically change in an ever-changing environment. A standalone client creating static Web pages was not enough for this purpose, so we created a server-side engine that evaluates every page request and dynamically creates an adapted (e.g., personalized) page; this engine was called HPG-Java. HPG-Java is a Java Servlet that can be run within a Servlet container Web server like Apache Tomcat. The application can be configured by the Hera models and a basic configuration file that indicates where on the server these models are provided (and some additional database settings). The server-side version allows data to be updated based on the user behavior, providing data for the sake of personalization. HPG-Java does not provide a designer platform, but the designer can use an adapted version of the Visio templates to create the models.

In order to give the reader some indication of the dynamics (based on queries) provided by HPG-Java, we use the example from Figure 10.10.

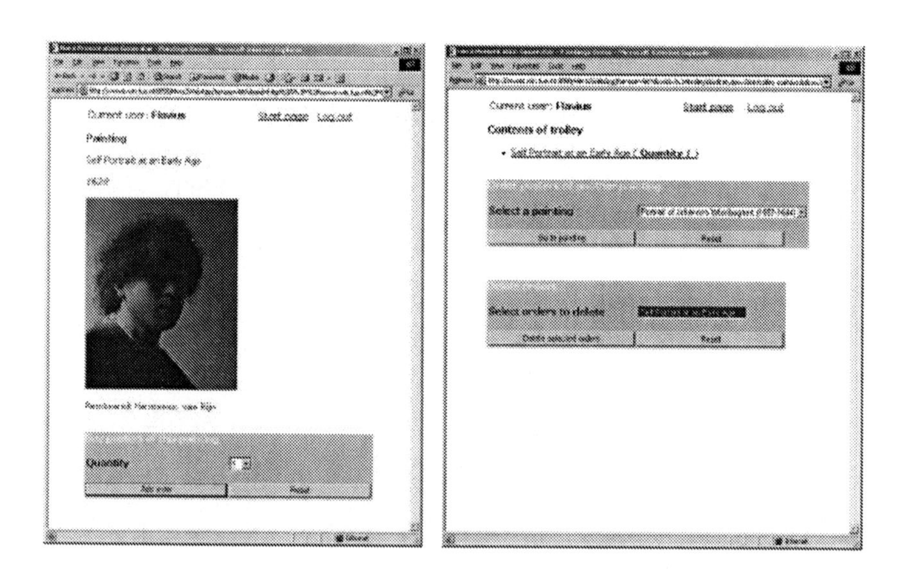

Figure 10.10. Web application served by HPG-Java.

On the left-hand side of the figure is the current page, while on the right-hand side is the next page that needs to be computed. When the user presses the "Add order" button, two queries are executed. The first query builds a new order, and the second query adds the currently created order to the trolley. Based on this newly computed data, the next page is generated. This new page displays the list of ordered paintings (based on the orders in the trolley) and also provides two forms. The first form allows the user to select the next painting, and the second form enables the user to delete an order previously added to the trolley.

10.6.2 Hera-S

As we have explained earlier, the architecture of the Hera-S version is similar to that from the earlier HPG-Java version, but obviously it accommodates the newer Hera-S AMs with their SeRQL queries. A major difference is that the AM is less tightly coupled to the domain, in the sense that the designer has more freedom to select elements and concepts in the domain by the use of the SeRQL queries. Furthermore, we allow the domain to be any repository of RDF data (i.e., no proprietary data model). Note that this does not only apply to the domain, but also to the context. The implementation is again a Java servlet; however, storage and manipulation of all the meta-models are now handled by different Sesame repositories.

Furthermore, the Hera-S implementation concentrates on the application-model level only. Several presentation modules exist (Section 10.7 will go into more details) that can configure the presentation of AM data, each with outstanding features that might be more appropiate in different situations. Having separate implementations allows a presentation module to be plugged into the pipeline that fits the situation, thus also making Hera-S more platform-independent.

In Figure 10.11 we see the main components that make up the Hera-S implementation architecture. The domain model (DM), application model (AM), and all context data are realized as Sesame repositories, exploiting Sesame's capability that enables storage, reasoning, and querying of RDF and OWL data.

The content is interfaced to the rest of the system through the DM repository. A major advantage of this approach is that integrated querying of both schema and instance data becomes possible. To enable this, the content has to be represented as RDF statements. For non-RDF content repositories, this can be achieved in various ways. The simplest and most straightforward case is an offline translation of the data to RDF and simply storing that RDF in a Sesame repository. However, this approach has a drawback for certain cases by duplicating data, which means that updates to the data need to happen in two places. An alternative way of realizing the link between DM and data is by creating a wrapper component around the actual data source that does online back-and-forth translation. The Sesame architecture caters for this scenario by having a storage abstraction layer called the SAIL API (Broekstra et al., 2002). A simple wrapper component around virtually any data source can be realized as a SAIL implementation and then be integrated effortlessly into the rest of the Sesame framework and thus into our Hera-S environment.

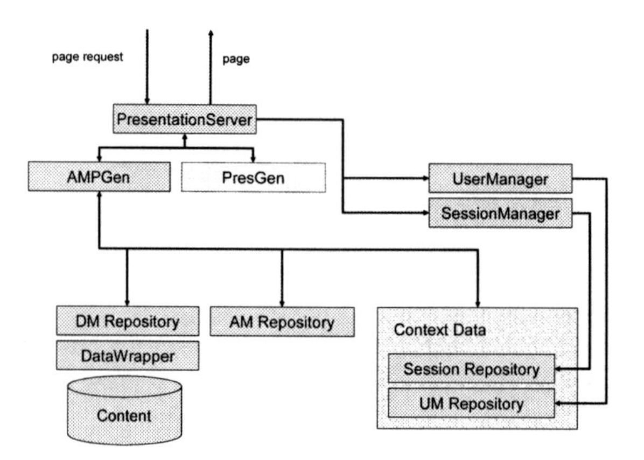

Figure 10.11. Hera-S architecture.

The entire system has an event-driven architecture. When a request for a certain page comes in at the PresentationServer component, the request is translated to a request for an AMP. At the same time, the UserManager and SessionManager components are informed of the request. These two manager components can then take appropriate actions in updating the context data repositories [specifically, the UM (user model) Repository, and the Session Repository].

Independently from this, the so-called AMPGen component retrieves the requested part of the AM that contains the conceptual specification that is the basis for the next AMP. It then starts the AMP creation process by following that specification.

The actual AMP is internally implemented as a volatile (in-memory) Sesame repository, which means that all transformation operations on it can simply be carried out as RDF queries and graph manipulations using the SeRQL query language. When the AMP has been fully constructed, it is sent back to the presentation-generation component. This presentation-generation component can then transform the AMP into an actual page (in terms that a thin client such as a Web browser can understand, e.g., XHTML). The result is then finally sent back to the client (as the response).

In the Hera-S system, SeRQL query expressions are extensively used to define mappings and filters between the different data sources and the eventual AMP. Since all these data are expressed as RDF graphs, an RDF query/transformation language is a natural choice as a mapping tool.

In Section 10.7 we will describe a presentation-generation process that uses the AMP as input to generate a presentation for different user platforms. It will also show a screenshot from an application that can be generated with the Hera-S engine in combination with that presentation-generation solution. Via the feedback mechanism, built-in as parameters in the links, user actions will trigger subsequent actions in the Hera-S engine (i.e., presentation generation typically does not interfere with this process).

10.7 PRESENTATION DESIGN

In this section we address one particular approach to presentation design. The presentation design step of Hera bridges the gap between the logical level from the AM and the actual implementation. If needed, the presentation model (PM) can specify the details of this transformation. Complementary to the AM, where the designer is concerned with the structure of the information and functionality as it needs to be presented to the user (by identifying navigational units and relationships), the PM specifies how the content of those navigation units is displayed. According to these specifications, AMPs can be transformed to a corresponding Web presentation in a given output format, e.g., XHTML, cHTML, WML, etc. We do stress that we foresee multiple alternative ways to render AMPs in specific output formats, and the designer is free to choose a way to configure this transformation of AMPs into output. In this section we illustrate one particular way, which uses a PM to detail the presentation design and which is implemented using a particular document format for adaptive Web presentations.

10.7.1 Presentation Model Specification

The PM is defined by means of so-called regions and relationships between regions. Regions are abstractions for rectangular parts of the user display and thus satisfy browsing platform constraints. They group navigational units from the AM; like navigation units, regions can be defined recursively. They are further specified by a layout manager, a style, and references to the navigational units that they aggregate. We note that the usage of layout managers was inspired by the AMACONT project's component-based document format (Fiala et al., 2003), adopting its abstract layout manager concept in the Hera PM. As will be explained in Section 10.7.2, this enables us to use AMACONT's flexible presentation capabilities for the generation of a Web presentation.

Figure 10.12 shows an excerpt of the PM for the running example, i.e., the regions associated to the MovieUnit navigational unit and its subregions.

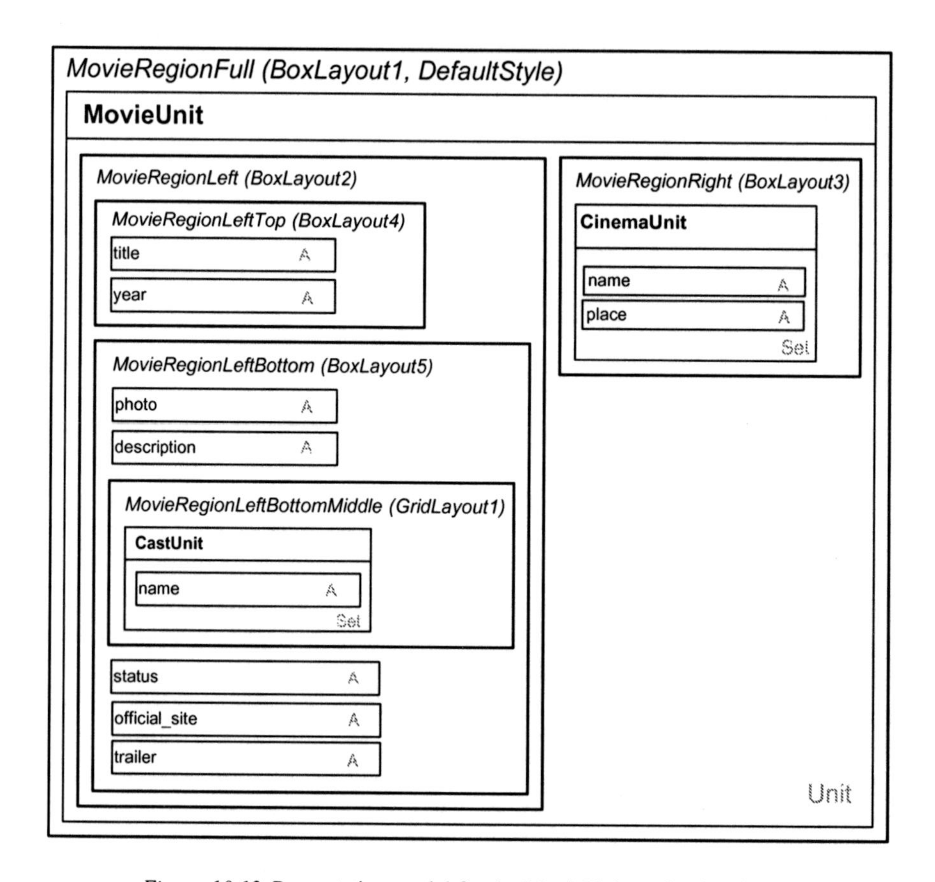

Figure 10.12. Presentation model for the MovieUnit navigational unit.

The MovieUnit navigational unit is associated with the region called MovieRegionFull. It uses the layout manager BoxLayout1 for the arrangements of its subregions (MovieRegionLeft and MovieRegionRight) and the style given by DefaultStyle. BoxLayout1 is an instance of the layout manager class BoxLayout that allows us to lay out the subregions of a region either vertically or (as in this case) horizontally. The style describes the font characteristics (size, color), background (color), hyperlink colors, etc. to be used in a region. The definition of styles was inspired by Cascading Style Sheets (CSS) (Bos et al., 2005). We chose to abstract the CSS formatting attributes because (1) not every browser supports CSS at the current moment and (2) we would like to have a representation of the style that can be customized based on user preferences.

Both MovieRegionLeft and MovieRegionRight use BoxLayout's with a vertical organization of their inner regions. For the title and year attributes, BoxLayout4 is used, which specifies a horizontal arrangement. For photo, description (the region containing the names in the cast), status, official_site, and trailer, BoxLayout5 is used, which indicates a vertical arrangement. The names in the cast are organized using GridLayout1 (an instance of the layout manager class GridLayout), a grid with four columns and an unspecified number of rows. The number of rows was purposely left unspecified, as one does not know a priori (i.e., before the presentation is instantiated and generated) how many names the cast of a movie will have. The regions that do not have a particular style associated with them inherit the style of their container region. Note that in Figure 10.12 we have omitted constant units [e.g., (,), Cast, etc.] in order to simplify the explanation.

Besides the layout manager classes exemplified in Figure 10.12, the definition of PM supports additional ones. BorderLayout arranges subregions to fit in five directions: north, south, east, west, and center. OverlayLayout allows us to present regions on top of each other. FlowLayout places the inner regions in the same way as words are placed on a page: The first line is filled from left to right, and the same is done for the second line, etc. TimeLayout presents the contained regions as a slide show and can be used only on browsers that support time sequences of items, e.g., HTML+TIME (Schmitz et al., 1998) or SMIL (Bulterman et al., 2005). Due to the flexibility of the approach, this list can be extended with other layout managers that future applications might need.

The specification of regions allows us to define the application's presentation in an implementation-independent way. However, to cope with users' different layout preferences and client devices, Hera-S also supports different kinds of adaptation in presentation design. As an example, based on the capabilities of the user's client device (screen size, supported document formats, etc.), the spatial arrangement of regions can be adapted. Another adaptation target is the corporate design (the "look-and-feel") of the resulting Web pages. According to the preferences and/or visual impairments of users, style elements like background colors, fonts (size, color, type), or buttons can be varied. For a thorough elaboration of presentation-layer adaptation, the reader is referred to Fiala et al. (2004).

Turning back to our running example, we now consider how to adapt the PM for the MovieUnit navigational unit to the typical small display size and horizontal resolution of a handheld device. In this respect, we aim at replacing the layout managers BoxLayout1 and GridLayout1 with BoxLayout's specifying vertical arrangements for their containment elements. Note that the PM facilitates specifying such adaptations by the assignment of multiple layout or style alternatives (variants) as simple conditions attached to "region-layout manager assignments," in correspondence with the adaptation conditions of the AMACONT document format. These conditions are simple Boolean expressions consisting of constants, arithmetic and logical operators, as well as references to context model parameters.

10.7.2 Presentation Generation Implementation

After the specification of the PM, we now turn to its implementation. As mentioned above, we illustrate it by using the AMACONT project's component-based document model, which is perfectly suited for this task as this PM was based on AMACONT presentation principles in the first place. This approach aims at implementing personalized ubiquitous Web applications by aggregating and linking configurable document components. These are instances of an XML grammar representing adaptable content on different abstraction levels. Media components encapsulate concrete media assets (text, structured text, images, videos, HTML fragments, CSS) by describing them with technical meta-data. Content units group media components by declaring their layout in a device-independent way. Document components define a hierarchy from content units to fulfill a semantic role. Finally, the hyperlink view defines links that are spanned over components. For more details on the AMACONT document model, the reader is referred to Fiala et al. (2003).

Whereas the AMACONT document model provides different adaptation mechanisms, in this chapter we focus on its presentation support. For this purpose it allows us to attach XML-based abstract layout descriptions (layout managers) to components. Document components with such abstract layout descriptions can be automatically transformed to a Web presentation in a given Web output format. As mentioned above, the PM was specified by adopting AMACONT's layout manager concept to the model level. This enables the automatic translation of AMPs to a component-based Web presentation based on a PM specification. The corresponding presentation generation pipeline is illustrated in Figure 10.13.

In a first transformation step (AMP to component) the AMPs are translated to hierarchical AMACONT document component structures. In doing so, both the aggregation hierarchy and the layout attributes of the created AMACONT components are configured according to the PM configuration. Beginning at top-level document components and visiting their subcomponents recursively, the appropriate AMACONT layout descriptors (with adaptation variants) are added to each document component. This transformation can be performed in a straightforward way and was already described in detail by Fiala et al. (2004). The automatically created AMACONT documents are then processed by AMACONT's document generation pipeline. In a first step, all adaptation variants are resolved according to the current state of the context model. Second, a Web presentation in a given Web output format (e.g., XHTML, XHTML Basic, cHTML, or WML) is created and delivered to the client.

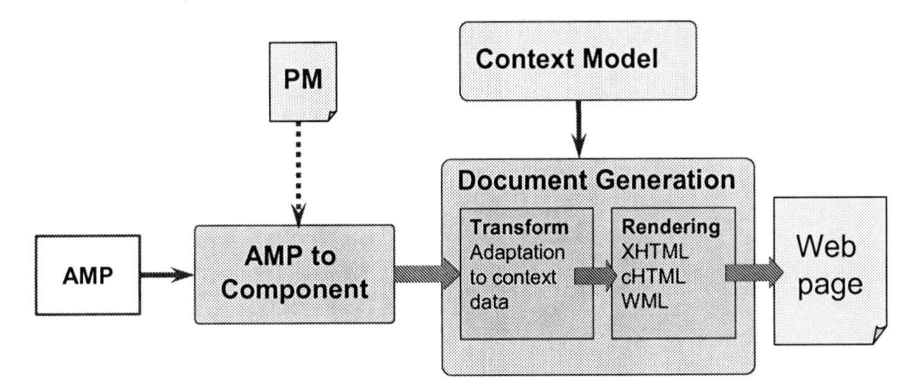

Figure 10.13. Presentation generation with AMACONT.

Figure 10.14 illustrates the XHTML page generated for the PC version of our running example. It represents an instantiation of the Movie navigational unit with data used for "The Matrix" movie. It also shows the cinemas that are currently playing this movie.

Note that the presentation of content elements corresponds to the PM specification that was illustrated in Figure 10.12.

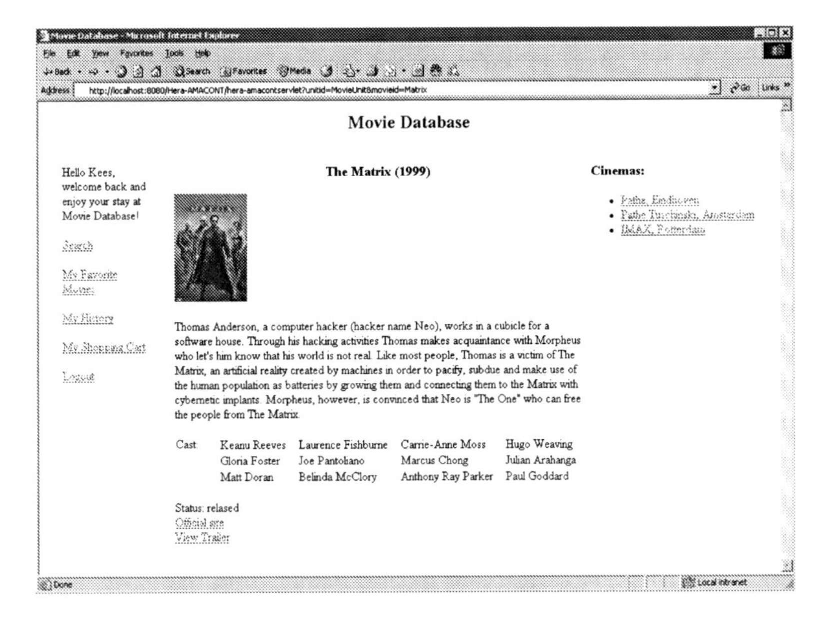

Figure 10.14. Presentation on a PC.

Figure 10.15. Presentation on a PDA.

As an alternative, Figure 10.15 shows the same page as presented on a PDA, exemplifying the layout adaptation. As can be seen, the resulting presentation is in correspondence with the PM adaptation specified above, i.e., all content elements are displayed below each other.

10.8 SUMMARY

In this chapter we have discussed on the basis of the running example the models and tools that make up the Hera approach to WIS design. This approach is characterized by a focus on adaptation in the navigation design, and a number of the facilities are motivated by the goals of this adaptation support. The most characteristic element of the approach is the choice to use RDF as the main language for expressing the domain and context data and the application model (AM) that defines the context-based navigation over and interaction with the content. Since the storage and retrieval of the RDF data involve the manipulation of RDF data, we have chosen to use a Sesame-based approach, i.e., making the different RDF data models available as Sesame repositories. Consequently, we use SeRQL query expressions in the definition of the AM. With the RDF and SeRQL expressions, we have models that allow a more fine-grained specification of adaptation and context dependency. Also, we can more extensively exploit the interoperability of RDF data, for example, when integrating data sources (e.g., for background knowledge) and interfering with the data processing independently from the navigation. This enables a clean separation of concerns that helps in personalization and adaptation and in the inclusion of external data sources.

REFERENCES

Beckett, D., 2004, *Turtle: Terse RDF Triple Language.* Technical report. http://www.dajobe.org/2004/01/turtle/.

Bos, B., Çelik, T., Hickson, I., and Lie, H.W., 2005, *Cascading Style Sheets, level 2 revision 1*, CSS 2.1 specification, W3C Working Draft, June 13.

Brickley, D., and Guha, R.V., 2004, *RDF Vocabulary Description Language 1.0: RDF Schema.* W3C Recommendation, Feb. 10. http://www.w3.org/TR/rdf-schema/.

Broekstra, J., Kampman, A.. and van Harmelen, F., 2002, Sesame: An architecture for storing and querying RDF and RDF schema. *Proceedings First International Semantic Web Conference, ISWC2002*, Sardinia, Italy, June 9-12, Springer-Verlag Lecture Notes in Computer Science (LNCS) no. 2342, pp. 54–68. http://www.openrdf.org/.

Broekstra, J., 2005, SeRQL: A second-generation RDF query language. Chapter 4 in Storage, Querying and Inferencing for Semantic Web Languages, PhD thesis, Vrije Universiteit Amsterdam, ISBN 90-9019-236-0. http://www.openrdf.org/doc/SeRQLmanual.html.

Bulterman, D., Grassel, G., Jansen, J., Koivisto, A., Layaïda, N., Michel, T., Mullender, S., and Zucker, D., 2005, *Synchronized Multimedia Integration Language (SMIL 2.1)*, W3C Recommendation Dec. 13.

Chipman, A., Goodell, J., Harpring, P., Beecroft, A., Johnson, R., and Ward, J., 2005, *Getty Thesaurus of Geographic Names: Editorial Guidelines*. Available at http://www.getty.edu/research/conducting_research/vocabularies/guidelines/tgn_1_contents_intro.pdf.

Dean, M., and Schreiber, G., 2004, *The OWL Web Ontology Language Reference.* W3C Recommendation, Feb. 10. http://www.w3.org/TR/owl-ref/.

Fiala, Z., Frasincar, F., Hinz, M., Houben, G.J., Barna, P., and Meissner, K., 2004, Engineering the presentation layer of adaptable Web information systems. *International Conference on Web Engineering, ICWE 2004*, July 28-30, Munich, Germany, Springer-Verlag Berlin Heidelberg, LNCS 3140, pp. 459–472.

Fiala, Z., Hinz, M., Meißner, K.. and Wehner, F., 2003, A component-based approach for adaptive, dynamic Web documents. *Journal of Web Engineering,* **2**(1&2): 58–73.

Frasincar, F., 2005, Hypermedia Presentation Generation for Semantic Web Information Systems, PhD thesis, Chapter 4 (Hera Presentation Generator), Eindhoven University of Technology Press Facilities, ISBN 90-386-0594-3, pp. 67–87.

Hart, L., Emery, P., Colomb, B., Raymond, K., Taraporewalla, S., Chang, D., Ye, Y., Kendall, E., and Dutra, M., 2004, OWL full and UML 2.0 compared. http://www.itee.uq.edu.au/~colomb/Papers/UML-OWLont04.03.01.pdf.

Hobbs, J.R., and Pan, F., 2004, An ontology of time for the Semantic Web. *ACM Transactions on Asian Language Information Processing, TALIP,* **3**(1): 66–85.

Kiczales, G., Lamping, J., Mendhekar, A., Maeda, C., Lopes, C., Loingtier, J.M., and Irwin, J., 1997, Aspect-oriented programming. *Proceedings 11th European Conference on Object-Oriented Programming, ECOOP'97*, Jyväskylä, Finland, pp. 220–242.

Klyne, G.. and Carrol, J., 2004, *Resource Description Framework (RDF): Concepts and Abstract Syntax.* W3C Recommendation, Feb. 10. http://www.w3.org/TR/rdf-concepts/.

Miller, G.A., 1995, Wordnet: A lexical database for English. *Communications of the ACM,* **38**(11): 39–41.

Schmitz, P., Yu, J., and Santangeli, P., 1998, *Timed Interactive Multimedia Extensions for HTML (HTML+TIME)*, W3C Note Sept. 18.

Thiran, P., Hainaut, J.L., and Houben, G.J., 2005, Database wrappers development: Towards automatic generation. *Ninth European Conference on Software Maintenance and Reengineering, CSMR'05*, Manchester, UK, IEEE CS Press, pp. 207–216.

Chapter 11

WSDM: WEB SEMANTICS DESIGN METHOD

Olga De Troyer, Sven Casteleyn, Peter Plessers
Research Group WISE, Department of Computer Science, Vrije Universiteit Brussel, Pleinlaan 2, 1050 Brussels, Belgium

11.1 WSDM OVERVIEW

WSDM was introduced by De Troyer and Leune in 1998 (De Troyer and Leune, 1998). At that time the acronym stood for **W**eb **S**ite **D**esign **M**ethod and only targeted information-providing Web sites. In the meantime, the method has evolved a great deal and now also allows traditional Web applications as well as Semantic Web applications to be designed, hence the renaming of the method into **W**eb **S**emantics **D**esign **M**ethod.

More than other Web design methods, WSDM is a methodology, i.e., it not only provides modeling primitives that allow a Web developer to construct models that describe the Web site/application from different perspectives and at different levels of abstraction, but it also provides a systematic way to develop the Web application. Developing a Web site/application with WSDM starts with the formulation of the so-called mission statement and follows a well-defined design philosophy that offers the designer the necessarily support to structure the Web site. The method consists of a sequence of phases. Each phase has a well-defined output. For each phase, a (sub)method describing how to derive the output from its input is provided. The output of one phase is the input of a following phase. As already indicated, currently the method allows the development of Web sites as well as Web applications. For the sake of simplicity, we will use the term "Web systems" to indicate both Web sites and Web applications. It is also important to notice that WSDM allows us to develop Web systems that are semantically annotated, in this way effectively enabling the Semantic Web. Content-related (semantic) annotations as well as structural annotations can be generated. Content-related annotations make the semantics of the content explicit. Structural annotations are annotations that explicitly describe the semantics of the different structural elements of the Web systems. Structural annotations can be exploited by third parties to transcode the Web system to a different format, for example to formats appropriate for a screen reader, or they can be exploited by search engines for their page segmentation [see, e.g., Deng et al. (2004) for an overview of page segmentation by search engines].

WSDM follows an *audience-driven* design approach. An audience-driven design philosophy means that the different target audiences (visitors) and their requirements are taken as starting point for the design and that the main structure of the Web site is derived from this. Concretely, this results in different navigation paths (called audience tracks) offered from the home page, one for each different kind of visitor.

Figure 11.1 shows an overview of the different phases of WSDM. The different phases are shown sequentially. However, in practice, the design process is rather iterative. In this section a brief description of the different phases is given. In the following sections the different phases are explained into more detail and illustrated with examples from this book's common example, the Internet Movies Database Web site,[1] or IMDb for short. Note that it is not realistic and also not the purpose to redesign the entire system here. Simplifications made were necessarily to reduce the size of the drawings and the examples.

In the first phase of the method, the mission statement specification, the mission statement for the Web system is formulated. The goal of this phase is to identify the purpose of the Web system, as well as the subject and the target users. Without giving due consideration to the purpose, there is no proper basis for making design decisions or for evaluating the effectiveness of the Web system. The target users are the users whom we want to address or who will be interested in the Web system. The subject of the Web system is, of course, related to the purpose and the target users of the Web system. The subject must allow fulfilling the purpose of the Web system, and it must be adapted for the target users. The output of this phase is the *mission statement*. It is formulated in natural language and must describe the purpose, subject, and target users of the Web systems. In fact, the mission statement establishes the borders of the design process. It allows (in the following phases) deciding which information or functionality to include or exclude, how to structure it, and how to present it.

---

[1] http://www.imdb.com.

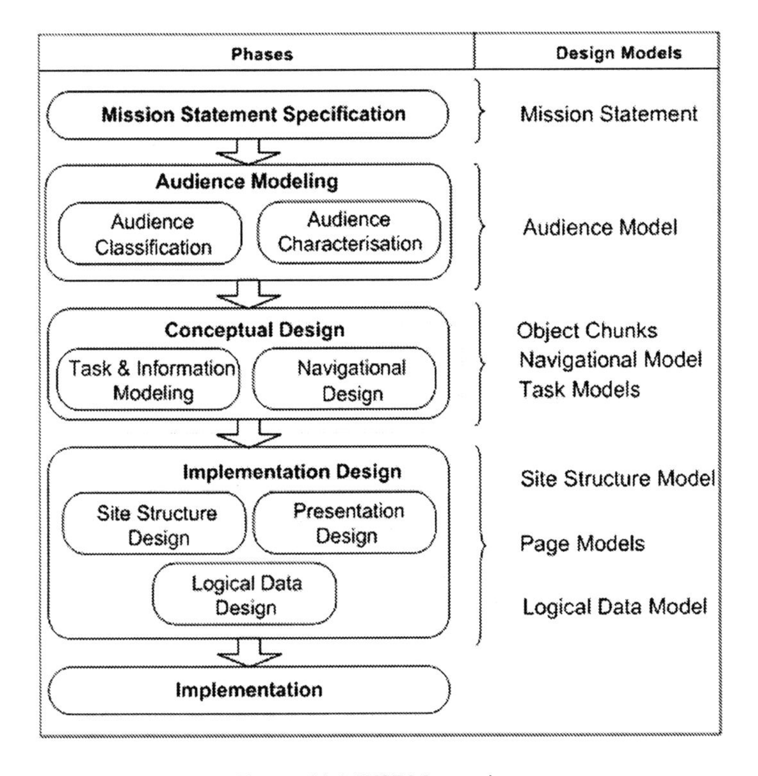

Figure 11.1. WSDM overview.

The next phase is the audience modeling phase. The target users identified in the mission statement are refined into *audience classes*. This means that the different types of users are further identified and classified into audience classes. The classification is based on the requirements of the different users. Users with the same information and functional requirements become members of the same audience class. Users with additional requirements form audience subclasses. In this way a hierarchy of audience classes is constructed. For each audience class, relevant characteristics (e.g., age, experience level) are given.

The next phase, the conceptual design, is used to specify the information, functionality, and structure of the Web system at a conceptual level. A conceptual design makes an abstraction from any implementation technology or target platform. The information and functionality are specified during the task & information modeling subphase. The overall conceptual structure including the navigational possibilities for each audience class is specified during the navigational design subphase.

During the implementation design phase, the conceptual design models are complemented with information required for an actual implementation. It consists of three subphases: site structure design, presentation design, and logical data design. During site structure design, the conceptual structure of the Web system is mapped onto pages, i.e., is the designer decides which components (representing information and functionality) and links will be grouped onto Web pages. For the conceptual design, different site structures can be defined, targeting different devices, contexts, or platforms. The presentation design is used to define the look and feel of the Web system as well as the layout of the pages. The logical data design is only needed for data-intensive Web systems. In case the data will be maintained in a database, a database schema is constructed (or an existing one can be used), and the mapping between the conceptual data model and the actual data source is created. Evidently, other types of data sources are possible (XML, RDF, OWL, etc.).

The last phase is the implementation. The actual implementation can be generated automatically from the information collected during the previous phases.

11.2 MISSION STATEMENT SPECIFICATION

In the first phase of the method, the mission statement for the Web system should be formulated. To develop a successful Web system, it is necessary to first reflect on the purpose of the Web system; otherwise, there will be no proper basis for making design decisions or for evaluating the effectiveness of the Web system. For example, for a company, the purpose may range from simply "having an identity on the Web," to "advertising some of its products," to "provide a full-fledged e-shop"; for public and local authorities, it may range from providing general information to a full-fledged e-government system that allows users to arrange official matters (e.g., apply for official documents) using the Web. The purpose should be established in consultation with the different stakeholders.

The different stakeholders should also agree on the topics that should be covered by the Web system. Even if the purpose is clear, it may be necessary to explicitly name the topics the system will deal with. For example, a company may decide to offer online information about products but only for their products in a higher price range. Another example is a high school that decides only to offer information about its educational system and courses, but not about its research activities.

Furthermore, the target users should also be identified. In principle, everybody can visit a public Web site, including people for whom the Web site is not relevant. However, it is impossible to satisfy the expectations of each possible visitor. It is better to focus on the users needed to make the Web system successful, called the *target users*. For example, consider a company that only sells very specialized technical items. In this case, the

Web site should focus on the people who need these products. These people are likely not the general public but instead probably very technical and specialized people. When building a Web system for a university, an important group of users you probably want to address are potential students.

It is clear that the purpose, subject (topics), and the target users are highly related. The subject and the target users of the Web system must allow the purpose to be fulfilled, and the subject must be suitable for the target users. So, one may argue that if the purpose is stated, the topics and target users are implicitly stated as well. However, to avoid misunderstandings, it is better to explicitly identify all three aspects of the mission statement. The mission statement is formulated in natural language.

Later on, in the following phases, the mission statement serves as the basis to decide what information or functionality to include (or not), how to structure it, and how to present it. Information or functionality that is not needed for the purpose or is not covered by the subject should not be considered. How to present and structure information and functionality is highly dependent on the target users, e.g., information should be presented and organized differently to professionals than to a common public. In addition, the mission statement can be used during validation to check if the Web system has indeed achieved the formulated purpose.

To illustrate this phase, we have formulated a mission statement for the example Web system. Currently, imdb.com mainly focuses on movies, but there is also a part about games. Therefore, the mission statement is formulated as follows:

> To be the biggest and best movie and game site on earth. For movies, this will be achieved by providing as much information as possible on movies, including their actors, directors, and producers, as well as to provide news, allow exploring show times, buy tickets in selected cinemas, and to share personal opinions about movies. For games, information about games is offered as well as news, and game lovers should be able to exchange information.

This mission statement defines the purpose, subject, and target users as follows:

- **Purpose**: to be the biggest and best movie and game site by (1) providing information and news on movies and games, (2) allowing users to explore show times of movies and to buy tickets, and (3) allowing movie lovers and game lovers to share personal opinions and exchange information.
- **Subject**: movies and related information such as actors, directors, and producers; selected cinemas; games.
- **Target users**: movie lovers and game lovers.

As you may have noticed, the purpose may involve multiple goals. Here, the stakeholders want to realize their long-term purpose (to be the biggest and best movie and game site) by means of three (related) goals. The subject may also deal with different topics. Here, movies as well as cinemas and games are considered. Also, the target users may be composed of different groups. Here, two different groups of users are involved: movie lovers and game lovers.

11.3 AUDIENCE MODELING

The mission statement formulated in the first phase is only a first and very incomplete description of the system that should be developed. Because WSDM is an audience-driven design method, the first concern that is elaborated is the set of target users. The target users identified in the mission statement are refined into *audience classes*. This is done by means of two subphases: the audience classification and the audience characterization. During audience classification, the different types of users are identified in more detail and classified into so-called audience classes. During audience characterization, relevant characteristics are specified for each audience class. We describe each subphase in more detail in the following subsection, but first we discuss why it is important to distinguish between different types of users.

In general, a Web system has different types of visitors who may have different needs. Consider as an example a university's Web site. Typical users of such a Web site are potential students, enrolled students, and researchers. Potential students are looking for general information about the university and the content of the different programs of study. The enrolled students need detailed information about the different courses, timetables, and contact information of the lecturers (telephone extension, room number, and contact hours). Researchers look for information on research projects and publications and general information on the researchers (full address, research interests, and research activities). This illustrates that different types of users (WSDM uses the term *audience class*) may have different information and/or functional requirements. To ensure good usability, this should be reflected in the Web system. For example, a student should be able to follow a navigation path that leads him to the information he is interested in without having to travel through pages of other (for him) nonrelevant information. If this is not the case and all information is provided to all users, a user has to scan the page(s) in order to find the links, the pages, and the pieces of information or functionality that are relevant for him. Providing too much nonrelevant information enhances the lost-in-hyperspace syndrome.

Next to the fact that different types of users may have different informational and functional requirements, it may be necessary to represent the (same) information or functionality in different ways to different kinds of users. This depends on the characteristics of the users. As an example, we again consider the university example. Potential students, especially secondary school students, are not familiar with the university jargon and should be addressed in a young and dynamic way. Also, by preference, the information should be offered in the native language. The enrolled students are familiar with the university jargon. They also prefer to have the information in the native language; however, for foreign students (e.g., who follow exchange programs) English should be used as the communication language. For researchers, it may be sufficient to use English. When the information and functionality are grouped based on the requirements of the different types of users, it is also possible to adapt the presentation to the characteristics of the different type of users without the need to rely on adaptive Web systems or personalization (Brusilovsky, 1996). Although for some situations, personalization may be undoubtedly the best solution, in other situations it may be less appropriate.

11.3.1 Audience Classification

The target users informally identified in the mission statement are the input for the audience classification. These target users are refined and classified into audience classes based on differences in their informational and functional requirements. All members of an audience class must have the same set of informational and functional requirements.

Sometimes, the set of informational and functional requirements of one audience class is a subset of the set of requirements of another audience class. To accommodate such situations, the concept of *audience subclass* is used. Figure 11.2 gives the graphical representation of an audience class and an audience subclass. Audience class B is an audience subclass of audience class A, which means that audience class B has all the same informational and functional requirements as audience class A but also some extra requirements. In other words, the set of requirements of audience class A is a subset of the set of requirement of audience (sub)class B. From the point of view of their populations, the population of the audience subclass is a subset of the population of the audience superclass. Indeed, the members of the audience subclass have more requirements than the members of the audience superclass, so these can only be fewer people.

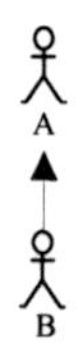

Figure 11.2. Graphical representation of audience class and subclass.

WSDM prescribes two alternative methods for discovering the audience classes. The first method uses the activities of the organization for which the Web system needs to be developed and the role people play in these activities. Only activities that are related to the subject of the Web system are considered. Each activity involves people, who may be potential users for the Web system. These people should be identified. If they belong to the target users of the Web system, their requirements (informational as well as functional, usability, and navigational requirements) are formulated (in an informal way and at a high level). Users with the same informational and functional requirements become members of the same audience class. Whenever the informational and functional requirements of a set of users differ, a new audience class is defined or, if possible, an audience subclass is introduced. If possible, the activities are decomposed in order to refine the audience classes. This may introduce audience subclasses. The decomposition stops if no new subclasses emerge or if decomposition is not possible anymore. In summary, the method is as follows:

Step 1: Consider the activities of the organization related to the purpose of the Web system.
Step 2: For each activity,

1. Identify the people involved.
2. Restrict them to the target users.
3. Identify their requirements.
4. Divide them into audience classes based on different informational or functional requirements.
5. Decompose the activity if possible, and repeat Step 2.

In this way a hierarchy of audience classes is constructed. At the top of this *audience class hierarchy* we always place the audience class *Visitor*. The Visitor audience class represents all target users. The requirements associated with the Visitor class are the requirements that are common to all users.

The second method starts with identifying all possible informational and functional requirements of the target users, without wondering how to classify them into audience classes. The different audience classes and the subclass relations between them are derived from a matrix that is constructed using the different requirements as dimensions of the matrix. The cells in the matrix answer the question, "Does every user who has the requirement of this row also (always) have the requirement of this column?" Informally, every row *i* of such a matrix characterizes the users having the requirement associated with this row in terms of the other requirements they have. From this requirements matrix, the audience class hierarchy can be (automatically) derived as follows.

If two or more rows are exactly the same, this means that the users represented by these rows are the same in terms of possible requirements, and thus the users having these requirements should belong to the same audience class. If the set of "Y" entries of a row *k* is a subset of the "Y" entries of another row *l*, this means that the users represented by row *l* have the same requirements as the users represented by row *k* and some extra requirements. So, the audience class represented by row *l* is an audience subclass of the audience class represented by the row *k*.

This algorithm is formally specified in Casteleyn and De Troyer (2001). In summary, it looks as follows:

Step 1. Construct the audience class matrix based on all the requirements formulated for the Web system to be built.
Step 2. Determine the equivalence rows. Each set of equivalence row represents an audience class. The user requirements for this audience class are the requirements associated with the rows. Meaningful names should be given to the audience classes.
Step 3. Identify subset relations between the rows of the different equivalence classes. These subset relations result in audience subclass relations.
Step 4. Construct the integrated audience class hierarchy (using the audience classes and the audience subclass relations).

We now illustrate the audience classification phase for the IMDb example. We first illustrate the activity-based method and then the method based on the requirements matrix.

We suppose that IMDb is run by a separate organization. In this case, the activities of this organization could be

- Provide information about movies and games.
- Sell cinema tickets.
- Maintain information about movies and games.

The people (and other organizations) involved in these activities are movie lovers, game lovers, cinemas, and the organization's database administrators. Figure 11.3 illustrates this.

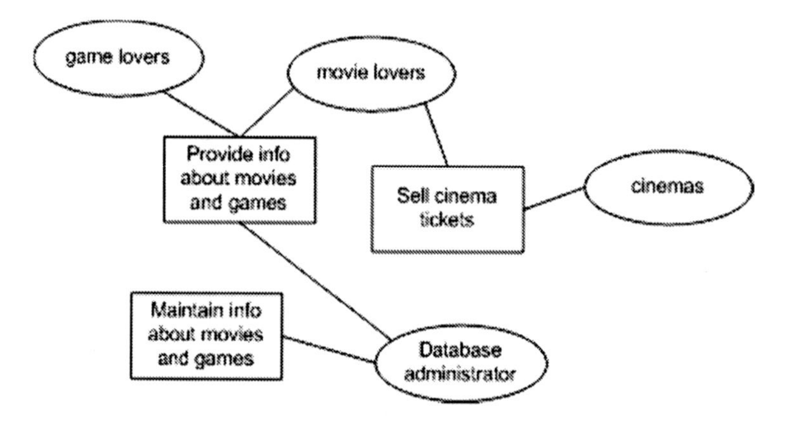

Figure 11.3. Activity diagram for the IMDb example.

The organization's database administrators are not part of the target users, and maintaining the information about movies and games is also not part of the purpose of the Web system. Therefore, the database administrators are not considered in the audience classification. The cinemas are involved in the activity "Sell cinema tickets" because the information about the movies, show times, and available seats must be synchronized with the ticket information of the cinemas. However, we suppose that this synchronization is done by means of interactions with the cinemas' information systems, for which no manual intervention is needed. Therefore, cinemas can also be discarded. This leaves us with the movie lovers and the game lovers. The requirements for these users can be specified as follows:

Movie lovers

1. To get an overview of the movies currently playing
2. To get an overview of the movies coming soon
3. To find a movie by means of a search function
4. To browse the movie database
5. To get an overview of new DVDs
6. To get an overview of DVDs coming soon
7. To obtain information about a movie
8. To obtain information about a person involved in a movie
9. To read news about movies
10. To have links to other interesting sites in relation to movies

11. To explore show times at different cinemas of movies currently playing
12. To buy tickets for a show time in a selected cinema
13. To manage a personal movie list
14. To post messages on the movie message boards
15. To enter a comment about a movie

Game lovers

1. To get an overview of games
2. To get information about a game
3. To read news about games
4. To have links to other interesting sites in relation to games
5. To post messages on the game message boards

The requirements for these two groups are sufficiently different to put them in separate audience classes. This results in two different audience classes: Movie Lover and Game Lover. Note that these classes don't need to be disjoint: A person may be a movie lover as well as a game lover. However, such a person will in general not want to look for movie information and game information at the same time. Also, in general, a movie lover is not necessarily a game lover, and vice versa.

There is no useful further decomposition for these activities. The resulting audience class hierarchy is shown in Figure 11.4. Note that Visitor is the top of the hierarchy.

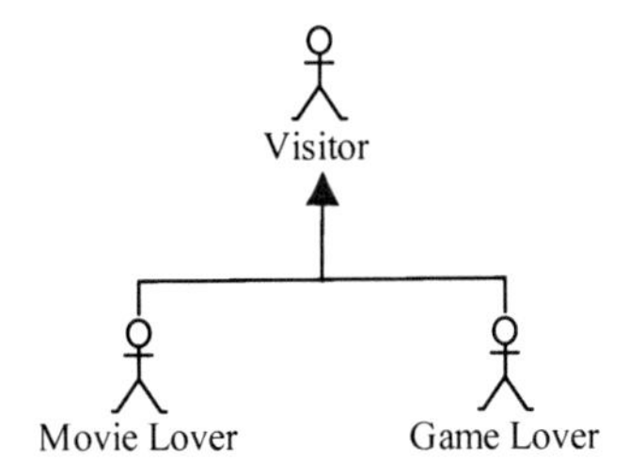

Figure 11.4. Audience class hierarchy for the IMDb example.

Some parts in the IMDb system require authorization to access and are protected by a login. Therefore, additional (authorization) requirements are needed. A user must be able to register; once he is registered, he is able to log in. Also logout must be possible. These are requirements common to the movie lovers and the game lovers and are therefore assigned to the Visitor class.

Visitor

1. To register
2. To log in
3. To log out

To use the matrix method to derive the (final) audience class hierarchy, we first have to list all informational and functional requirements of the target users, here being movie lovers and game lovers. These are defined as follows:

1. To get an overview of the movies currently playing
2. To get an overview of the movies coming soon
3. To find a movie by means of a search function
4. To browse the movie database
5. To get an overview of new DVDs
6. To get an overview of DVDs coming soon
7. To obtain information about a movie
8. To obtain information about an actor
9. To obtain information about a director
10. To read news about movies
11. To obtain links to other interesting sites in relation to movies
12. To explore show times at different cinemas of movies currently playing
13. To buy tickets for a show time in a selected cinema
14. To manage a personal movie list
15. To post messages on the movie message boards
16. To enter a comment about a film
17. To get an overview of games
18. To get information about a game
19. To read news about games
20. To obtain links to other interesting sites in relation to games
21. To post messages on the game message boards
22. To register
23. To log in
24. To log out

Then the requirements matrix is constructed. For each requirement, a row and a column are created. Then the cells are filled by answering the question, "Does every user who has the requirement of this row also (always) have the requirement of this column?" For example, for row 1 (requirement 1), we obtain the following answers:

- For columns 1 to 16: "Y"
- For columns 17 to 21: "N"
- For columns 22 to 24: "Y"

The other cells are filled in a similar way. Table 11.1 shows a reduced version of the matrix where equal rows and columns are displayed as a single row or column.

Table 11.1. Reduced Requirements Matrix

	1–16	17–21	22–24
1–16	Y	N	Y
17–21	N	Y	Y
22–24	N	N	Y

The following audience classes can be derived from this matrix (equal rows):

- Rows 1 to 16: Movie Lover
- Rows 17 to 21: Game Lover
- Rows 22 to 24: Visitor (this audience class only has the requirements that are common to all users)

The following subclass relations can be derived (subset between rows):

- Movie Lover (rows 1–16) is an audience subclass of Visitor (rows 22–24).
- Game Lover (rows 17–21) is an audience subclass of Visitor (rows 22–24).

This result is the same audience hierarchy as the one given in Figure 11.4.

Once the audience classes are identified, it should be investigated if the members of those audience classes have special usability or navigational requirements. Different examples of navigational requirements can be found in the IMDb example. For example, for the Movie Lover audience class, we can formulate the following navigational requirements:

- The user should be able to navigate directly from the information of a movie to the show times and the ordering of tickets when the movie is currently played.
- If more information about some item shown is available, then the user should always be able to directly navigate to this information; e.g., from the movie information to the information about its directors, its actors, its genre, etc.

11.3.2 Audience Characterization

In the second subphase of the audience modeling, the audience characterization, relevant characteristics should be specified for each audience class. Examples of characteristics are level of experience with Web sites in general, frequency of use, language issues, education/intellectual abilities, age, income, lifestyle, etc. Some of the characteristics may be translated into usability requirements, while others may be used later on in the implementation design phase to guide the design of the "look and feel" of the navigation tracks of the different audience classes, e.g., choice of colors, fonts, graphics, etc.

The target users of the IMDb example are a very broad audience; they don't have very specific characteristics. There are also no differences worth mentioning between the characteristics of the audience class Game Lover and the audience class Movie Lover. Therefore, the characteristics for all audience classes are specified as follows:

Characteristics for all audience classes in the IMDb example :
- Able to communicate in English
- Have reasonable experience with the Web
- Are young people or adults

11.4 CONCEPTUAL DESIGN

So far in the method, the informational, functional, usability, and navigational requirements as well as the characteristics of the potential visitors have been identified and different audience classes have been defined. The goal of the conceptual design is to turn these informal requirements into high-level, formal descriptions that can be used later on to generate (automatically or semiautomatically) the Web system.

During conceptual design, we concentrate on the **conceptual** "what and how" rather than on the visual "what and how." The conceptual "what" is covered by the task & information modeling subphase and deals with the modeling of the content and functionality of the Web system; the conceptual "how" is covered by the navigational design subphase and specifies the conceptual structure of the Web system and the navigation. We describe these two subphases in more detail in the next subsections.

11.4.1 Task and Information Modeling

Instead of starting with an overall conceptual data model, like most Web design methods do, WSDM starts by analyzing the requirements of the different audience classes. This will result in a number of tiny conceptual descriptions, called *object chunks*, which model the information and functionality needed to satisfy these requirements. These conceptual descriptions are integrated into an overall conceptual model. This approach is used because WSDM follows the audience-driven design philosophy. It has the following advantages:

- The developer is forced to concentrate on the actual needs of the users rather than on the information (and functionality) already available in the organization. In this way, the chance that information is missing in the actual system will be less than in a data-driven approach where the available data are taken as the starting point. In addition, the information and functionality already available in the organization are not necessarily what the users need. Also, the way the information is organized and structured in the organization is not necessarily the way external users need it.
- It gives consideration to the fact that different types of users may have different requirements and that it may be necessary to use different structures or terminology for different types of users. By modeling the requirements for each audience class separately, we can give due consideration to this.

The output of the task & information modeling subphase is a collection of *task models* and associated *object chunks*. We first explain the task modeling and afterwards the information modeling.

11.4.1.1 Task Modeling

The purpose of task modeling is to model in detail the different tasks the members of each audience class need to be able to perform and to formally describe the data and functionality that are needed for those tasks. The tasks that a member of an audience class needs to be able to perform are based on the requirements formulated for the audience class during audience classification, i.e., for each informational and functional requirement formulated for an audience class, a task is defined that should allow one to satisfy this requirement. Each task is modeled into more details using an adapted version of the task-modeling technique CTT (Paterno et al., 1997; Paterno, 2000). Essentially, in CTT, tasks are decomposed into subtasks until elementary tasks are obtained. In addition, temporal relationships between subtasks are specified to indicate the order in which the subtasks need to be performed. The result is a *task model*.

CTT was developed in the context of human–computer interaction to describe user activities. CTT looks like hierarchical task decomposition, but it distinguishes four different categories of tasks (user tasks, application tasks, interaction tasks, and abstract tasks). CTT also has an easy-to-grasp graphical notation. However, we do not completely follow the original specifications of CTT, but have adopted them slightly to better satisfy the particularities of the Web and Web design:

1. WSDM does not consider user tasks. User tasks are tasks performed by the user without using the application (such as thinking on or deciding

about a strategy). They are not useful to consider at this stage of the design. This means that we only use the following categories of tasks (see Figure 11.5 for the graphical notation):

- *Application tasks*: tasks executed by the application. Application tasks can supply information to the user or perform some calculations or updates, e.g., checking username and password is typically an application task.
- *Interaction tasks*: tasks performed by the user by interaction with the system, e.g., entering information using a form.
- *Abstract tasks*: tasks that consist of complex activities and thus require decomposition into subtasks, e.g., ordering tickets for a movie.

2. A (complex) task is decomposed into (sub)tasks. The same task can be used in different subtrees. Tasks are identified by means of their name. CTT prescribes that if the children of a task are of different categories, then the parent task must be an abstract task. WSDM does not follow this rule. We use the category of the task to explicitly indicate who will be in charge of performing the task. For an interaction task, the user will be in charge; for an application task, the application will be in charge. In this way, we can indicate at a conceptual level who will initiate a subtask, or who will make a choice between possible subtasks.
3. CTT has a number of operators to express temporal relations among tasks. For some of the operations, we have changed the meaning slightly and an extra operator for transactions has been added:

- *Order-independent* (T1 |=| T2): The tasks can be performed in any order.
- *Choice* (T1 [] T2): One of the tasks can be chosen and only the chosen task can be performed.
- *Concurrent* (T1 ||| T2): The tasks can be executed concurrently.
- *Concurrent with information exchange* (T1 |[]| T2): The tasks can be executed concurrently, but they have to synchronize in order to exchange information.
- *Deactivation* (T1 [> T2): The first task is deactivated once the second task is started.
- *Enabling* (T1 >> T2): The second task is enabled when the first one terminates.
- *Enabling with information exchange* (T1 []>> T2): The second task is enabled when the first one terminates, but, in addition, some information is provided by the first task to the second task.
- *Suspend-resume* (T1 |> T2): This indicates that T1 can be interrupted to perform T2; when T2 is terminated, T1 can be reactivated from the state reached before the interruption.

- *Iteration* (T*): The task can be performed repetitively. In CTT the meaning is that the action is performed repetitively: When the action terminates, it is restarted automatically until the task is deactivated by another task. The interpretation in WSDM is that the task can be repeated several times and ends when the one in charge decides not to repeat the task (e.g., the user who decides not to redo the task but to continue with the next task).
- *Finite iteration* (T(n)): indicates if the task has to be repeated a fixed number of times (number known in advance).
- *Optional* ([T]): indicates that the performance of the task is optional.
- *Recursion*: occurs when the subtree that models a task contains the task itself. This means that performing the task can be a recursive activity.
- *Transaction* (-> T <-): The task must be executed as a transaction. This means that if the task, or in case of a complex task one of the tasks in the task's subtree, is not completed successfully, the whole task will not be successful and all activities should be rolled back (i.e., "all or nothing").

4. In WSDM, the level of detail provided in the task model is less than in the original CTT method. The reason for this is the use of the object chunks. As we will explain, with each elementary task, an object chunk can be associated that further describes the task in terms of informational and functional needs.

Figure 11.5. Graphical notation for the different types of tasks.

We illustrate the task modeling for some of the requirements formulated for the IMDb example. Figure 11.6 shows the task model for the tasks defined for the requirement "To find a movie by means of a search function" of the audience class Movie Lover. For this requirement, the task "Search IMDB" is defined. This abstract task is decomposed into three sequential tasks: "Specify Query," "View Results," and "Show Movie." The task "Show Movie" is optional and is further decomposed into "Show Movie Info" (to show information like title, director, etc.) and "Provide Extras." This task allows the user to choose among "Show Photos" (to display photos of the movie), "Add to My List" (to add the movie to the user's personal movie list), and "Post Message" (to post messages on the message boards associated with the movie). The task "Add to My List" is composed of an application task "Update My List" that will add the movie to the user's personal movie list and to the task "Manage My List." In order not to overload the figure, the abstract tasks "Manage My List" and "Post Message" are not further elaborated in this CTT.

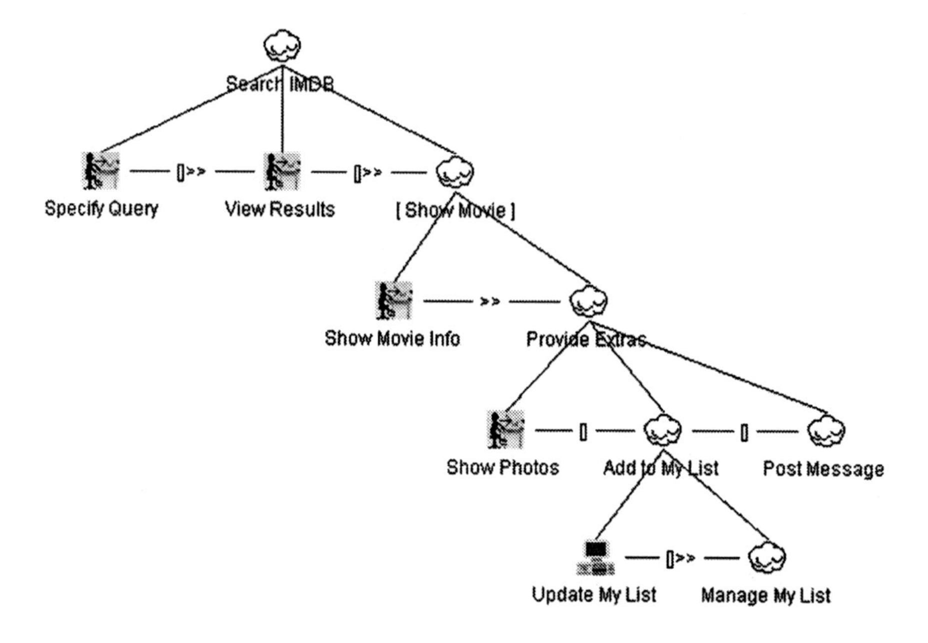

Figure 11.6. Task model for the task "Search IMDB."

Figure 11.7 shows the task model for the task defined for the requirements "To obtain show times of movies currently playing" and "To buy tickets for a show time in a selected cinema" of the audience class Movie Lover. We decided to support both requirements by a single task, as buying tickets requires knowing the show time, and offering the possibility to buy tickets while exploring show times may stimulate the sale of tickets. The task "Showtimes & Buy Tickets" is decomposed into two sequential tasks. First, the location must be specified using the "Specify Location" task. This is done by means of an interaction task to enter the location and the movie(s) (task "Enter Location Movie"), optionally followed by an interaction task to choose the location if more than one location exists for the information entered (task "Choose Location"). Next, the user can explore the show times and optionally buy tickets by means of the task "Select Showtime & Buy Tickets." This task is decomposed into the task "Explore Show Times," which can be repeated, followed by the optional task "Buy Tickets." For the task "Explore Show Times," first the show times associated with the requested location, movie(s), and date are showed ("View

Showtimes"), and then the user may change these parameters ("Change Parameters") and obtain the show times again.

The task models created in this way allow a first-level description of the functionality to be provided by the Web system (i.e., they describe a kind of workflow). More details are given by means of the object chunks. The object chunks are described in the next section.

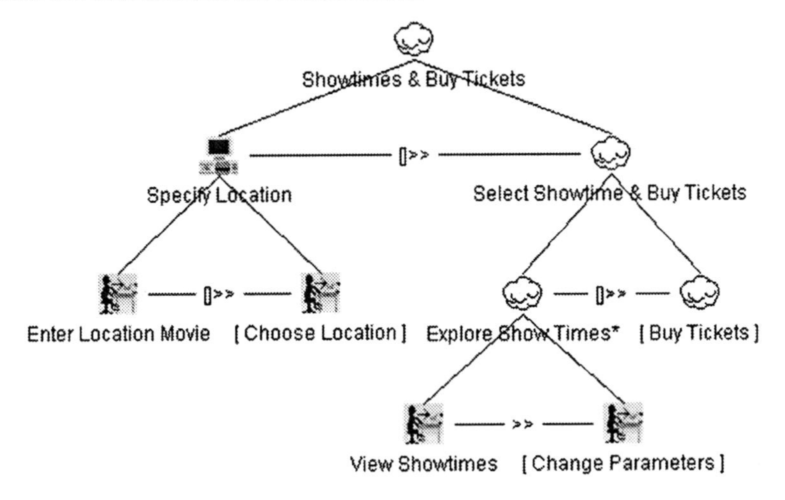

Figure 11.7. Task model for the task "Show Times & Buy Tickets."

11.4.1.2 Information Modeling

When a task model is completed, an object chunk is created for each elementary task in this model. The main purpose of an object chunk is to formally describe the information and functionality needed by the user when he has to perform the associated task. If the requirement associated with the task is a pure informational requirement (i.e., the user is only looking for information; she doesn't need to perform actions), then the object chunk can be considered as the conceptual description of the information that will be displayed on (a part of) the screen. For this purpose, a standard conceptual modeling language is sufficient. However, to be able to deal with functionality (e.g., filling in a form and processing it), a data manipulation language is also needed. We first discuss the modeling of informational requirements and then discuss what is needed to allow modeling functionality.

WSDM uses OWL[2] to model the information needs. OWL is becoming the standard for the Semantic Web. Its use as a specification language for the object chunks allows an easy integration with and use of existing domain ontologies and allows making the object chunks available as local application

[2] www.w3.org/TR/owl-features.

ontology in case no relevant domain ontology already exists. It also provides the basis for the generation of semantic annotations (see Section 11.6). Note that, as there isn't yet a generally used and compact graphical notation for OWL, we use the ORM graphical notation (Halpin, 2001) to give a graphical representation of OWL. We have opted for the ORM notation because ORM is very close to OWL, and therefore the mapping from ORM to OWL is straightforward. In addition, because of the purpose of the object chunks, there is no need to specify any of the advanced types of restrictions supported by OWL. An ORM data type is graphically represented as a dotted circle, an ORM object type is represented as a solid circle, an ORM subtype is connected to its supertype object type by means of an arrow, roles are represented as boxes connected to their respective data type or object type, a mandatory constraint on a role is represented as a black dot on its connection, and a uniqueness constraint is represented as an arrow over one or two role boxes. See Figure 11.8 for some examples. The mapping from ORM to OWL is sketched in Table 11.2. Suppose L is an ORM data type; N, N', N1, and N2 are ORM object types; and r and r' are roles. Informally, we can state that an ORM object type is mapped onto an OWL class; an ORM role connected to a data type is mapped onto an OWL data type property; and an ORM role connected to an object type onto an OWL object property.

As already indicated, the use of OWL allows an easy way of coupling the concepts used in the object chunks to concepts in existing (external) ontologies. This coupling is later on (in the implementation phase—see Section 11.6) used to generate semantic annotations. The namespace mechanism of OWL is used to refer in an object chunk to concepts defined in ontologies. To refer to a concept in an ontology, the identifying prefix of the ontology is used to qualify the names of the concept. For example, "FOAF:Person" refers to the class Person defined in the ontology identified by the prefix FOAF.

Figure 11.8 shows an example object chunk "ShowMovie." This object chunk is associated with the elementary task "Show Movie Info" of the task model "Search IMDB" given in Figure 11.6. In this object chunk the use of two external ontologies is illustrated: a basic IMDB ontology[3] (prefixed with "IMDB") and the well-known FOAF ontology[4] (prefixed with "FOAF" and used to describe persons). For example, the classes "IMDB:Movie," "FOAF:Image," "IMDB:Genre," and "FOAF:Person" refer to classes from these ontologies. Also, properties can refer to properties defined in ontologies; e.g., "IMDB:genres" refers to such a property.

[3] http://www.csd.abdn.ac.uk/~ggrimnes/dev/imdb/IMDB.rdfs.
[4] http://www.foaf-project.org/.

Table 11.2. Mapping Between ORM and OWL

ORM	OWL
N	<Class rdf:ID=“N”/>
N’ subtype of N	<Class rdf:ID=“N’”> <subClassOf rdf:resource=“#N”/> </Class>
(N1, r, L)	<DatatypeProperty rdf:ID=“r”/> <Class rdf:about=“#N1”> <subClassOf> <Restriction> <onProperty rdf:resource=“#r”/> <allValuesFrom rdf:resource=“L”/> </Restriction> </subClassOf> </Class>
(N1, r’, N2)	<ObjectProperty rdf:ID=“r”/> <Class rdf:about=“#N1”> <subClassOf> <Restriction> <onProperty rdf:resource=“#r’”/> <allValuesFrom rdf:resource=“N2”/> </Restriction> </subClassOf> </Class>
(r,r’)	<Property rdf:about=“#r”> <inverseOf rdf:resource=“#r’/> </Property> (for object properties only)
Mandatory role r	… <Restriction> <onProperty rdf:resource=“#r”/> <someValuesFrom rdf:resource=“…”/> </Restriction> …
Uniqueness constraint of role r	… <Restriction> <onProperty rdf:resource=“#r”/> <maxCardinality> 1 </maxCardinality> </Restriction> …

To allow communication between tasks, parameters (input as well as output parameters) can be specified for object chunks. For example, the object chunk “ShowMovie” (Figure 11.8) has an instance of type “IMDB:Movie” as input parameter (represented by *m). Input parameters are used in general to restrict the information that should be presented to the user. For instance, the input parameter *m of type IMDB:Movie is used to express that only the information related to this particular movie should be shown. This is graphically represented by putting this parameter in the corresponding class. In fact, placing the parameter *m in the class "IMDB:Movie" restricts the complete schema to a view.

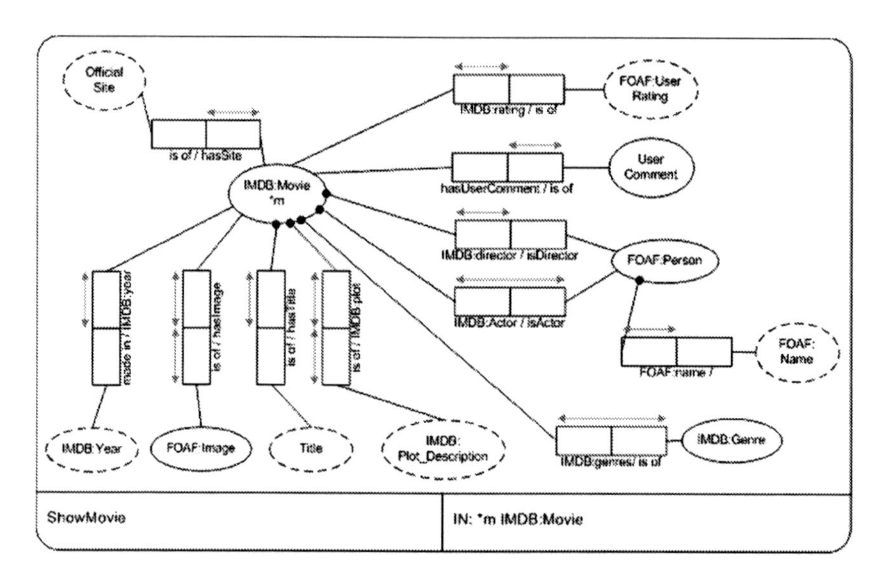

Figure 11.8. Object chunk "ShowMovie."

To be able to deal with functionality (e.g., fill in a form or update, add, or delete information), a (conceptual) data manipulation language is needed. WSDM provides a graphical conceptual data manipulation language. However, notice that this language has limited expressive power and does not intend to allow complex functionality to be specified. We don't believe that a graphical language is appropriate for this. However, using the primitives provided by the language, most commonly used functionalities for Web systems, such as adding, deleting, and changing information, can be specified. For more complex functionality, WSDM supports the use of (external) Web services. More in particular, the conceptual data manipulation language of WSDM provides support for

- specifying that the user can select one or several instances from a class or a property (e.g., to allow the user to select an actor from the list of available actors in the IMDb example). For this, the symbols "!" (single selection) and "!!" (multiple selection) are used.
- expressing interactive input (e.g., needed to fill in a form). The symbol "?" is used for the input of a single value, while "??" is used for

expressing interactive input of more than one value. Note that these symbols can only be applied to value types (data type properties). Class instances cannot be entered directly. They should be created using the "NEW" operator (see next bullet).

- To manipulate the data itself, a number of primitive operators are available. "NEW" indicates the creation of a new class instance; "REMOVE" is used to indicate the removal of one or more class instances; the symbol "+" above a relation indicates the addition of a property (and its inverse), and the symbol "-" indicates the removal of properties.
- Furthermore, an assignment operator ("=") is available to assign values to variables (also called *referents*), as well as several built-in functions and default referents. For example, *USER is available to refer to the current user of a session.

A more detailed description of this graphical language can be found in De Troyer and Casteleyn (2001) and De Troyer et al. (2005).

Figure 11.9 shows an object chunk involving functionality: adding a movie to the personal movie list of the user. This object chunk is created for the elementary (application) task "Update My List" in the task model "Search IMDB" given in Figure 11.6. The movie instance that needs to be added to the personal list is given by means of the input parameter *m (and passed to this task after the user has opted for the task "Add to My List" in the task "Show Movie" where the user was viewing a certain movie). The list to which the movie needs to be added is denoted by the referent *l and refers to the "My List" instance that "belongs to" the "User" instance *USER (which is the predefined referent used to refer to the current user). The "+" below the object properties "is in" and "has" specified the addition of the relationship (i.e., instantiation of both object properties for *l and *m). Note that *l is returned as output parameter because in the task "Add to My List" this information need to be passed to the task "Manage My List."

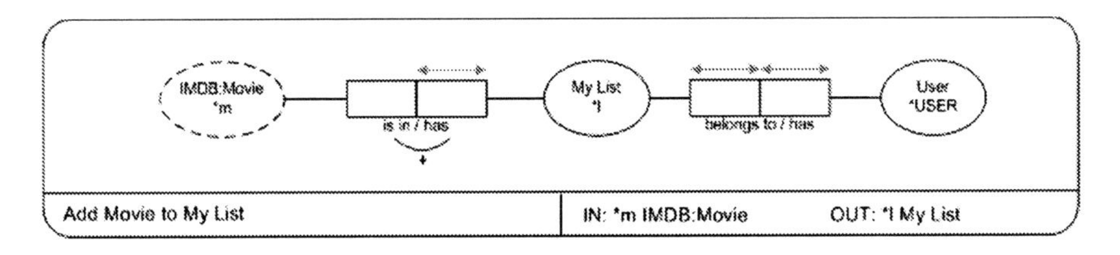

Figure 11.9. Object chunk "Add Movie to My List."

11.4.2 Navigational Design

The goals of the navigational design are to define the conceptual structure of the Web system and to model how the members of the different audience classes can navigate through the Web system and perform their tasks. Because of the audience-driven approach of WSDM, a *navigation track* is created for each audience class. Such an audience track can be considered as a subsystem containing all and only the information and functionality needed by the members of the associated audience class. The internal structure of an audience track is derived from the task models made for this audience class. In addition, navigational requirements formulated during audience modeling are also taken into account. In the next subsection we explain in detail how a navigational track is created. Next, all audience tracks are combined into a basic conceptual navigation structure by means of *structural links* (see Section 11.4.2.2). In WSDM, the structure defined between the audience tracks corresponds to the hierarchical structure defined between the audience classes in the audience class hierarchy.

Once the main conceptual navigation structure has been derived, *semantic* and *navigational aid links* are added (see also Section 11.4.2.2). Semantic links are navigational links based on semantic relationships that exist between objects in the domain and that are modeled in the object chunks by means of object properties. Semantic links express task-independent navigation (which may have been expressed in the form of navigational requirements during audience modeling). Navigational aid links are links that enhance navigation, such as home links, landmarks, quick links, etc. In contrast to structural links and semantic links, navigational aid links are strictly speaking not necessary but are added to enhance the usability of the Web system. A more detailed discussion on the different types of links used in WSDM can be found in De Troyer and Casteleyn (2003a, 2003b).

The output of the navigational design phase is the *navigational model*. The navigational model is expressed in term of *components* and *links* between components. Components can be considered as (conceptual) navigation units that group the information/functionality conveyed in one or more object chunks. As indicated, WSDM distinguishes between structural links, process logic links, semantic links, and navigational aid links. The process logic links express part of a workflow or the invocation of (external) functionality (e.g., a Web service). In general, a link may be defined from one component to one (other) component (one-to-one link), but also from one component to a set of components (one-to-many link), or from a set of components to one single component (many-to-one link), or from a set of components to a set of components (many-to-many link). As typical implementation formats, such as HTML, do not support one-to-many, many-to-one, or many-to-many links, these kinds of links need to be implemented as a collection of one-to-one hyperlinks when generating (HTML) output. However, these kinds of links are useful to consider during conceptual design because they allow abstracting from the current implementation limitations and provide more semantics. For example, in the presentation design (see Section 11.5.2) a one-to-many link can be represented as a single menu, and structural annotations can be generated to indicate the semantics of the links (see Section 11.6).

Note that the links specified in the navigational model are actually link types. That means that even a one-to-one link may result in different hyperlinks in the actual Web system. For example, if we specify in the navigational model that the user can navigate from a "Movie" component to an "Actor" component, this is modeled by means of a one-to-one link between these two components. However, a movie may involve several actors; therefore, for each individual movie, the one-to-one link may give rise to several hyperlinks, one to each of its actors.

Links can have parameters to indicate that relevant information should be passed from the source component to the target component when a user follows the link. A parameter is usually an output parameter from an object chunk connected to the source component.

Next to parameters, conditions can also be specified for links. A condition allows restricting the availability of the link to different users, devices, or timeframes. For example, a link may become unavailable after a certain date, or only users who are logged in are presented certain links. To some extent, conditional links allow for adaptation of the Web system.

The graphical representation of components and the different kinds of links are given in Figure 11.10. An external component refers to an external system or a Web service.

Note that the navigational model only provides the *conceptual* structure (including navigation) of the Web system. The mapping of this conceptual structure onto (Web) pages and hyperlinks is specified during site structure design, which is part of the implementation design phase (see Section 11.5.1).

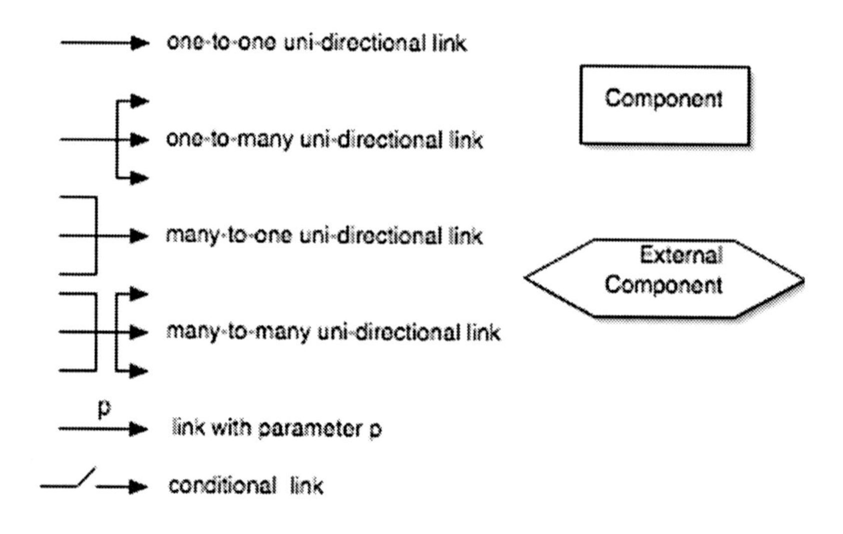

Figure 11.10. Graphical representation of components and links.

11.4.2.1 Creating the Navigational Tracks

To create a navigational track for an audience class, for each task a *task navigational model* is created. A task navigational model is the translation of the task model (represented by means of a CTT) into a navigational structure.

A task navigational model is created by defining a component for each elementary interaction task defined in the hierarchical decomposition of the task. The object chunk that was created for the elementary task is connected to this component. See Figure 11.11 for the graphical representation. Also, object chunks created for application tasks can be attached to components. In fact, components are a kind of placeholder for the object chunks (which represents the actual information and/or functionality). By linking components instead of object chunks, it is possible to use the same object chunk in different task navigational models and even in different navigational tracks without losing the modeling context of the different links.

Next, process logic links between components are used to express the workflow or process logic, which is expressed in the task model by means of the temporal CTT relations. In fact, the temporal CTT relations are translated into links (one-to-one, one-to-many, many-to-one, or many-to-many links; conditional or nonconditional links). Components and process logic links can be grouped into a *transaction* to indicate that they constitute a conceptual unit (for which the all-or-nothing property holds). To avoid complex diagrams, complex subtasks may be modeled separately, in *subtask navigational models*.

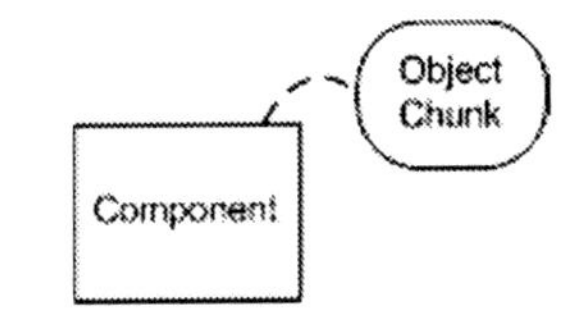

Figure 11.11. An object chunk connected to a component.

Until now, we permitted the designer to neglect the fact that not all information and functionality should be freely available. In many Web systems, parts of the information and functionality need to be protected in some way. This can be modeled during the navigational design by means of a *protection area*. Graphically, this is represented by including the component(s) (together with their associated object chunks) that need to be protected into a named box labeled with a key symbol (an example can be found in Figure 11.12). In a similar way, it is possible to indicate that some information transfer needs to be secure. This protection area concept for expressing security and validation allows abstracting (in the different navigational models) from how to achieve the validation or security. If relevant, how this must be achieved can be specified by means of a separated navigational model. An example is given in Figure 11.14 and is explained later on.

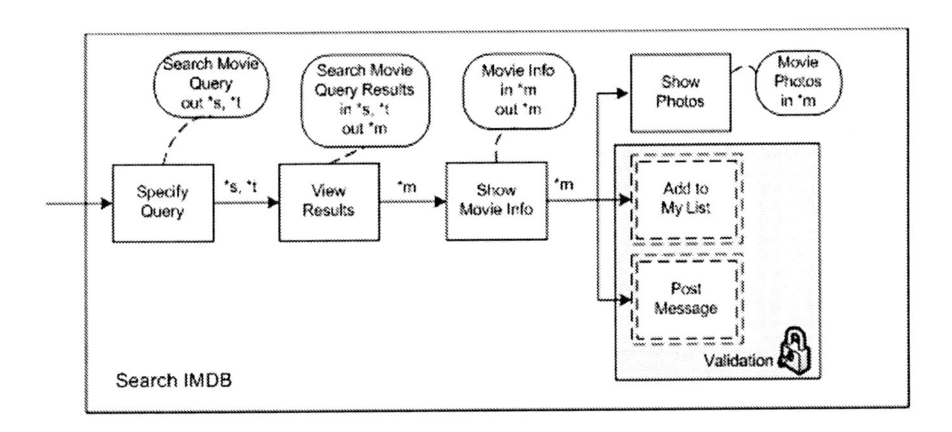

Figure 11.12. Task navigational model for the task "Search IMDB."

We illustrate the task navigational models by means of some examples. Figure 11.12 shows the task navigational model for the task "Search IMDB." Its task model was given in Figure 11.6. The navigational subtask models "Add to My List" and "Post Message" are protected by the "Validation" protection area. These parts of the Web system can only be accessed after the user has been authorized. How this must be done is specified in the task navigation model given in Figure 11.14, which has been derived from the task model given in Figure 11.13.

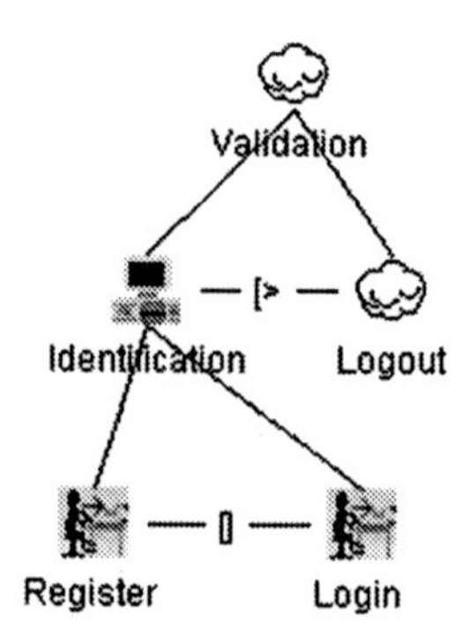

Figure 11.13. Task model "Validation."

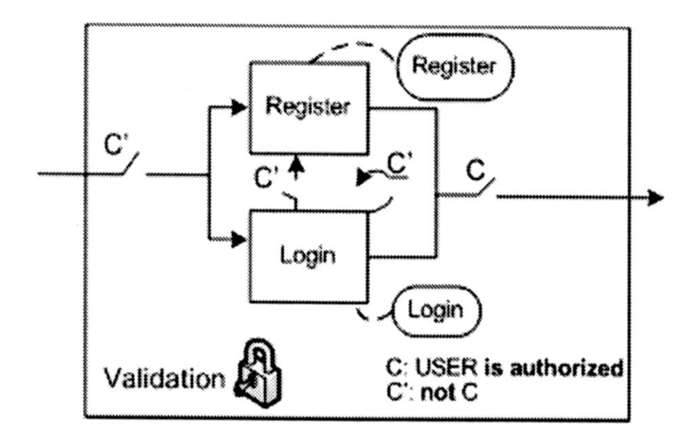

Figure 11.14. Task navigational model for "Validation."

Figure 11.15 gives the task navigational model for the task "Show Time & Buy Tickets" for which the task model was given in Figure 11.7. Note the use of an external component "Buy Tickets" to indicate that this is handled by an external service. Link parameters as well as input and output parameters are omitted in this diagram.

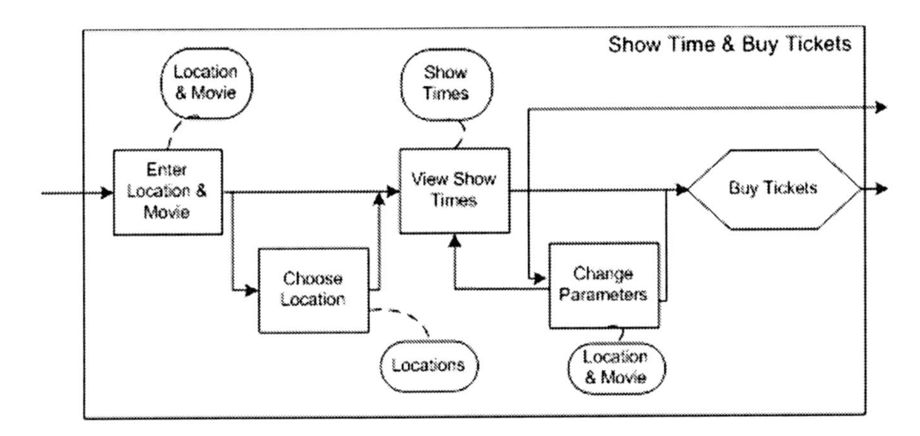

Figure 11.15. Task navigational model for the task "Show Time & Buy Tickets."

Once the task navigational models are constructed for the different tasks of an audience class, they should be composed into a navigational track using structural links. This can be done by first making a task model of how the members of the audience class are allowed to select the different tasks and then translating this task into a task navigational model. In the simple case that, at any moment in time, a member of the audience class can freely select between the different tasks, this modeling process can be reduced to the introduction of a new component that is linked to the different task navigational models by means of a one-to-many link. This principle is illustrated in Figure 11.16, which shows the navigational track for the Game Lover audience class of the IMDb example. For the sake of simplicity, the navigational models are represented by means of their shorthand notation (dotted double-lined rectangles). The newly introduced component is not connected to an object chunk, because it does not provide any information or functionality itself. Instead, its sole role is to allow navigation to the different tasks. In the actual Web system, this may result in a menu that provides links to the different tasks. However, if there are a lot of tasks for an audience track, it may be better (from a usability point of view) to structure the tasks in some way and to provide groups of tasks (which may result in groups of menus).

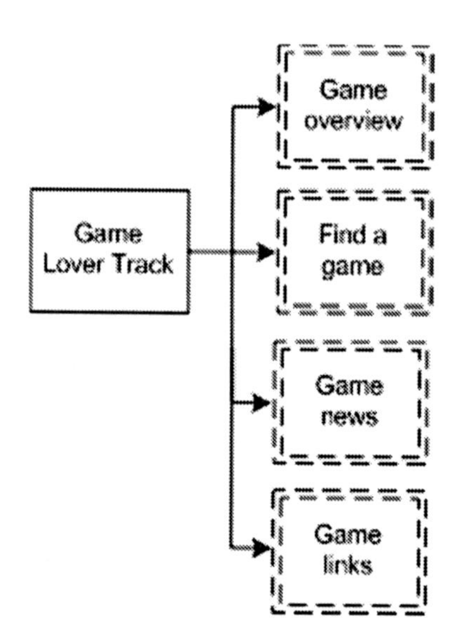

Figure 11.16. Navigational track for the Game Lover audience class.

11.4.2.2 Creating the Navigational Model

Once the navigational tracks for the different audience classes are constructed, they need to be composed into a single structure. This will be the main conceptual structure of the Web system. Because WSDM is audience-driven, the structure between the audience tracks must correspond to the hierarchical structure defined between the audience classes in the audience class hierarchy. We illustrate this with the IMDb example. The navigational model for the IMDb example is given in Figure 11.17. Note that for space limitations the different task navigational models are given by means of their shorthand notation and that the navigational track for the Movie Lover is also given by means of its shorthand notation (a double-lined rectangle). Note the correspondence with the audience class hierarchy given in Figure 11.4.

During navigational design, we also define semantic links that will enhance the navigation. Semantic links are based on semantic relationships that exist between objects in the application domain and that are modeled in the object chunks by means of object properties. For example, in the IMDb example, there are semantic relationships between movie and actor, between movie and director, and between movie and cinema. Also, in the navigational requirements for the Movie Lover audience class, it was stated

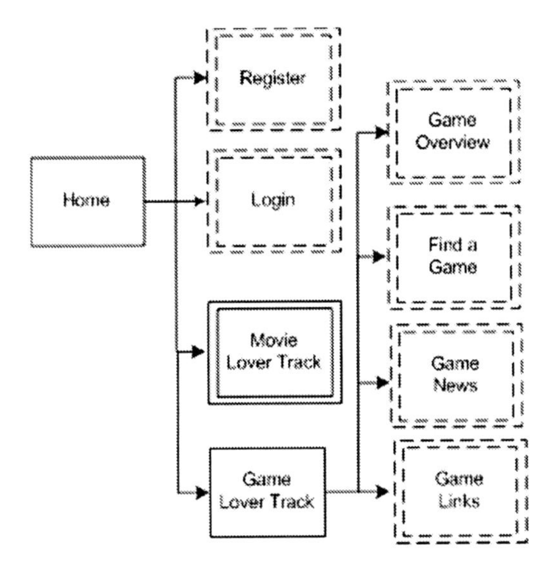

Figure 11.17. Main conceptual navigation structure for the IMDb example.

that the user should be able to navigate directly from the movie to the information of each of these related items. This facility can be modeled by means of semantic links. A semantic link is a link between two object chunks and must be based on the existence of a semantic relationship between two classes (e.g., movie having director). The source object chunk should contain the object property that expresses this semantic relationship between the two classes (movie having director). The target object chunk should provide the request information (information about director). We illustrate this with an example. Consider the object chunk "ShowMovie" as shown in Figure 11.8. For a movie we decided to provide the name of the director (modeled in "ShowMovie"). Suppose that the object chunk "Director Info" models the information provided for a director. Then, based on the object property "IMDB:director," we can define a semantic link from the object chunk "ShowMovie" to the object chunk "Director Info." See Figure 11.18 for a graphical representation. If needed, the link can be labeled with the name of the property used. Semantic links are independent of a particular task. This means that in each task where the chunk "ShowMovie" is used, the link will be available. Therefore, they don't need to be represented in the different task navigation models, and they will not overload these diagrams.

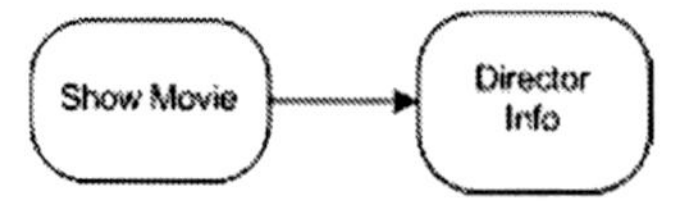

Figure 11.18. Graphical representation of a semantic link.

In the IMDb example many semantic links are possible. In Figure 11.19, a selection of possible semantic links is given. Note that in this example all links are bidirectional because the user must always be able to navigate to more information if this is available. For example, a link from the movie page to the director page must be available, and vice versa.

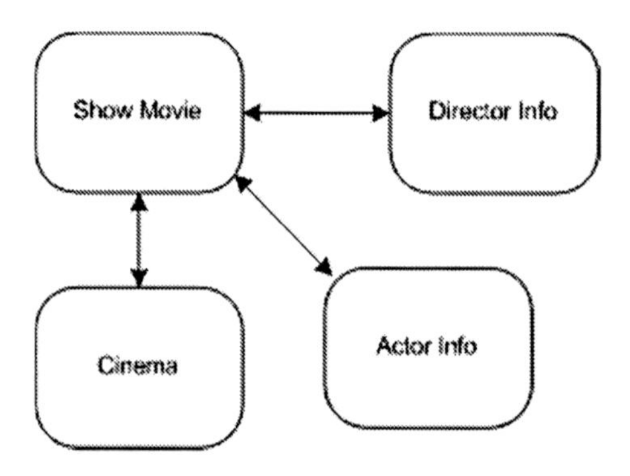

Figure 11.19. Some example semantic links for the IMDb example.

On top of the conceptual structure defined by means of the structural links, and the navigation possibilities defined by the process logic links and the semantic links, navigational aid links can be added to ease the navigation even more and to enhance the usability. From the viewpoint of being able to reach information and functionality, they are strictly speaking not needed; all information and functionality should also be reachable by means of the navigational tracks. Navigational aid links can be compared to adding an index and post-it pointers to chapters in a book: The information in and the structure of the book stay the same, but the user is provided with shortcuts to access the information more easily. A typical example of a navigational aid link is the home link, which can often be found on each page of a Web site. Also, landmarks are examples of navigational aid links. Not to overload the diagrams, the home component and the landmark components are represented by means of a symbol. Later on, during the implementation phase, home links and landmark links can be generated. In Figure 11.20 the conceptual structure of the IMDb example is enhanced with navigational aid links (home and landmarks and a link from the "Login" navigational task model to the "Register" navigational task model).

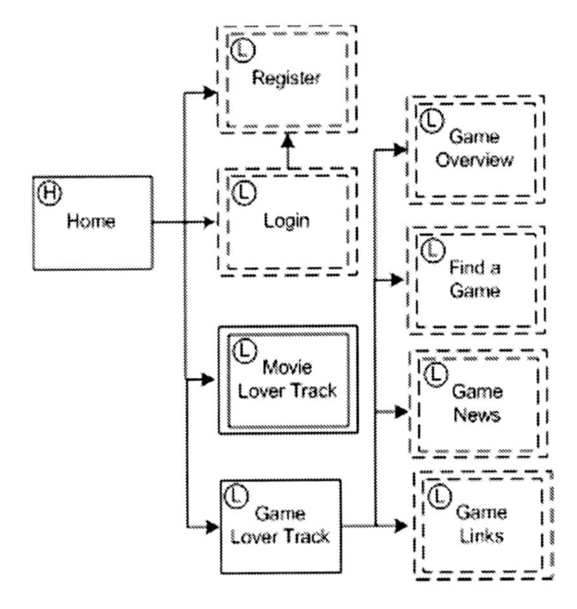

Figure 11.20. Navigational aid links.

11.5 IMPLEMENTATION DESIGN

The goal of the implementation design is to complement the conceptual design with the necessary details for the implementation. In principle, it would be possible to generate an implementation from the conceptual design, but this is not realistic for several reasons. First, the Web is very visually oriented, and the standards for presentation have become very high in recent years. Professional Web systems need to have a professional look and feel, and graphical designers are usually involved to achieve this. If the Web system is generated directly from the conceptual design, only standard and rather simplistic presentations can be obtained. Therefore, a presentation design is needed. Second, the information provided on the Web system may already be available from some data source (e.g., a relational database). In this case, no new data source needs to be created, but a mapping should be defined from the conceptual description of the information (the object chunks) to the actual data source. Third, Web users don't like to make unnecessary clicks (clicks that don't lead to new information), but, on the other hand, too much information on a single page will overload the page and also decrease the usability. Therefore, information and functionality should be grouped onto pages in such a way that a good balance is reached between the amount of information on a page and the number of clicks needed to reach information.

For these reasons, WSDM has an implementation design phase consisting of three subphases: the site structure design, the presentation design, and the logical data design. The following subsections describe these subphases into more detail.

11.5.1 Site Structure Design

During site structure design, the designer decides how the components from the navigational model will be grouped into pages. The characteristics of the different audience classes may be taken into account when deciding which information to group on a page. For example, for an audience class with the characteristic that the average age is over 50, the designer might want to limit the amount of information on a single page. It is also possible to define different site structures for the same design, each supporting a different device. For a device with a small screen (e.g., a PDA), it may even be necessary to distribute the information related to a single component onto different pages.

By default, each component (with its associated object chunks and links) is placed on a single page. However, the designer can decide to group different components on a single page or to use different pages for a single component. When components are grouped on a page, the designer should respect the conceptual structure expressed by means of the different links; e.g., if a component cannot be reached by a link from another component, these two components cannot be placed on the same page.

The output of this phase is the *site structure model.* The site structure is graphically presented by drawing pages over the components that should be grouped. Figure 11.21 illustrates a part of the site structure design for the IMDb example. The Home component and the links from this component are placed on a single page; Register and Login are each on a different page; and the components "Find a Game" and "Game Links" and all links coming from the "Game Lover Track" component are also put together on a single page. The rest has been left unspecified in this figure.

Note that the pages defined during the site structure design are abstract pages. Each abstract page will give rise to one or a set of concrete pages when the actual implementation is generated. For example, the page containing the "ShowMovie" object chunk will result (in the case of a static Web site) in many different concrete pages, one for each movie.

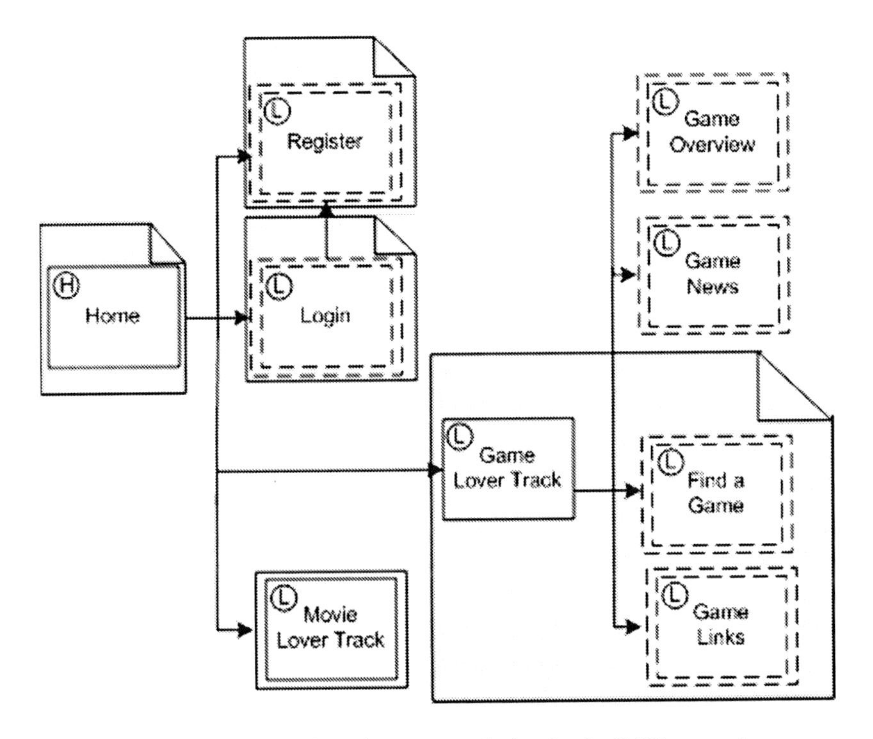

Figure 11.21. Part of the site structure design for the IMDb example.

11.5.2 Presentation Design

During presentation design, the look and feel of the Web system, as well as the layout of the pages (i.e., positioning of page elements), is defined. To enhance a consistent look and feel, templates are used. Therefore, page templates are defined. Typically, a Web system may require different kinds of pages, e.g., a home page, a title page, leaf page. For each of these page types, a template can be created. These templates are subsequently used in the page design, where the layout is defined for each of the pages defined in the site structure model. The layout describes how the information and functionality (modeled by means of the object chunks) and assigned to a page (by means of the components) should be laid out on the page.

For both the template and page design, a number of presentation-modeling concepts are available. To position information, the concept of a *grid* is used. A grid contains *rows* with *cells*. A cell contains a *multimedia value*, another grid (nesting of grids), or a *referent* from an object chunk assigned to the page [remember that a referent refers to an instance/value (or a set of instances/values) from a class or a property]. Absolute and relative height and width can be specified for grids, rows, and cells. By nesting grids, specifying the width and height of the different grids, rows, and cells, information can be positioned on the page.

The value of a cell can be associated with a *hyperlink* (which must be based on a link contained in the page and defined during navigational modeling). Furthermore, when a grid or cell is associated with a referent that represents a set (of object instances or values), then, in the actual implementation, the grid will be repeated for each instance of this set.

To display multimedia values correctly, additional properties may be required. For instance, an image and a video require a height and width property.

Furthermore, a number of high-level presentation-modeling concepts are also provided. The high-level presentation-modeling concepts are more powerful and more intuitive for the designer. They also are useful to capture the semantics of a presentation element. Most of the concepts have a well-known meaning: *OrderedBulletList*, *BulletList*, *Table*, *Menu*, *TableOfContent*, *BreadcrumbTrail*, *Section*, *Banner*, *Copyright*, *Advertisement*, *Figure*, *Icon*, *Logo*, *Marquee* (a string or an image that scrolls horizontally across the screen).

Forms are widely used in Web systems. To support them the following control concepts are available: a select control to model that a selection can be made out of multiple options, an input control to model that a value can be entered, and an action control to specify that an action should be performed. These controls can be associated with a presentation concept. Types of select controls are a *RadioButton*, a *CheckBox*, a *ListBox,* and a *DropDown* box. An input control is either a *TextBox* or a *SecretTextBox* (typically used for entering passwords). A typical action control is a *PushButton*. The behavior associated with an action control is defined by associating an event and an action to the control. It expresses the fact that when the specified event occurs for the associated presentation concept, the specified action will be performed. Possible events are *OnClick*, *OnLoad*, and *onHoover*; possible actions are *PopUp*, *Show*, *Scroll*, *Reset*, *Submit,* and *Cancel*. A popup menu, for instance, can be defined using the Menu presentation concept with associated *OnClick* event and associated *PopUp* action; an expandable menu can be defined using the Menu presentation concept, where the elements of the menu are associated with the event *OnClick* and the action *Show*.

Templates are specified using the presentation-modeling concepts mentioned. A template can also be composed out of a *Header*, *Footer*, *SideBar*, and/or *ContentPane*. Each template furthermore contains at least one *editable region*. An editable region denotes an area that one needs to specify further when the template is used for a page design. An editable region can be placed anywhere in a grid.

To specify style, WSDM currently relies on Cascading Style Sheets (CSS).[5] This allows style specification for any particular element and has enough expressive power to describe most styles commonly found in Web systems.

For each page, the designer chooses a template and then specifies how the links and the information (specified by means of the object chunks) will be positioned in the editable regions of the template. This is done using the presentation-modeling concepts mentioned. For each object chunk connected to a component included in the page, a grid is constructed. Each data type property of an object chunk is placed in a cell of the grid. For functionality, control concepts are used. If needed, multimedia values can be added to enhance the presentation (e.g., titles, labels, graphics, etc.).

For each link contained in the page, the designer needs to specify the anchor. This is done by adding the link to the relevant cell of the grid. Note how this linking mechanism does not differentiate between the type of anchors (e.g., a text element, an image, a table): A link is uniformly specified on a cell of a grid, no matter what its content is.

The characteristics and usability requirements of the audience classes should be taken into account when designing the different templates and pages.

The output of this phase is the *presentation model* consisting of a set of *templates*, a set of *styles,* and, for each page defined in the site structure model, a *page model*.

Figure 11.22 shows a simple example template and Figure 11.23 a simple example page model.

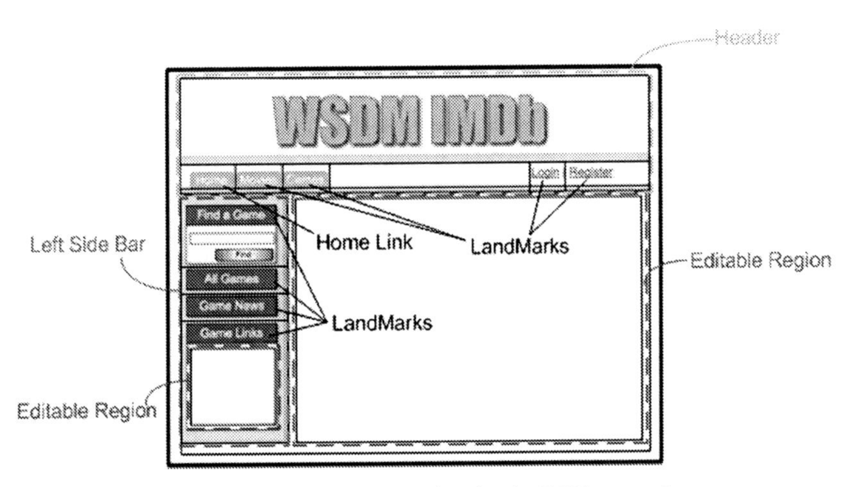

Figure 11.22. Example template for the IMDb example.

[5] www.w3.org/Style/CSS/.

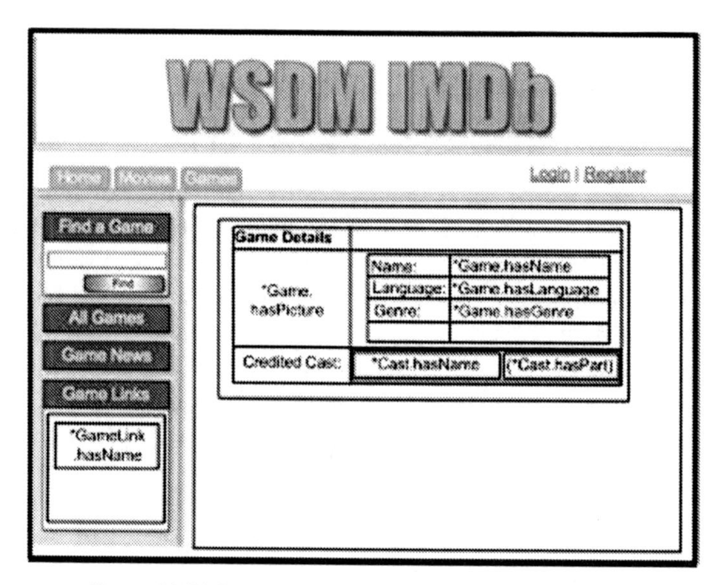

Figure 11.23. Template page model for the IMDb example.

11.5.3 Logical Data Design

The information provided by the Web system is described by means of the different object chunks made during task & information modeling. The different object chunks are related by means of the *reference ontology* that contains the different concepts used in the different object chunks. The object chunks are views on the reference ontology that is incrementally constructed during information modeling. The reference ontology may be based on one or more external ontologies or created from scratch. This reference ontology can be considered as the conceptual schema for the data to be provided by the Web system. In case no data storage is already available, a logical data schema needs to be created from this conceptual schema. This is comparable to the creation of a relational database schema from a conceptual ER schema or UML schema. The logical data schema can be a relational database schema, an XML schema, an RDF schema, or even the OWL schema of the reference ontology itself. While generating the logical data schema, it is important to keep track of the mapping between the reference ontology and the logical data schema, because later on (in the implementation phase) the conceptual queries and updates expressed in the object chunks need to be translated into queries and updates onto the logical database schema. Because of space limitations, it is not possible to describe in this chapter how a logical data schema can be generated from a reference ontology and how the mappings can be expressed. Normally, this process should be supported by a CASE tool, in which case the designer is not burdened with the creation of the logical data schema and the mappings.

In a second scenario, an existing data store is available. In this case it is only needed to define the mapping between the reference ontology and this data store.

The output of this phase is a *logical data schema* and a *data source mapping*.

11.6 IMPLEMENTATION

The actual implementation can be generated automatically from the information collected during the different design phases by means of the different design models. As proof-of-concept, a transformation pipeline (using XSLT) has been defined, which takes as input the object chunks (with corresponding data source mapping), navigational, site structure, style & template, and page models and outputs the actual implementation for the chosen platform and implementation language. This transformation is performed fully automatically. A description of this transformation pipeline is out of the scope of this chapter. An overview can be found in Plessers et al. (2005b). An example of a page (showing game details) is given in Figure 11.24.

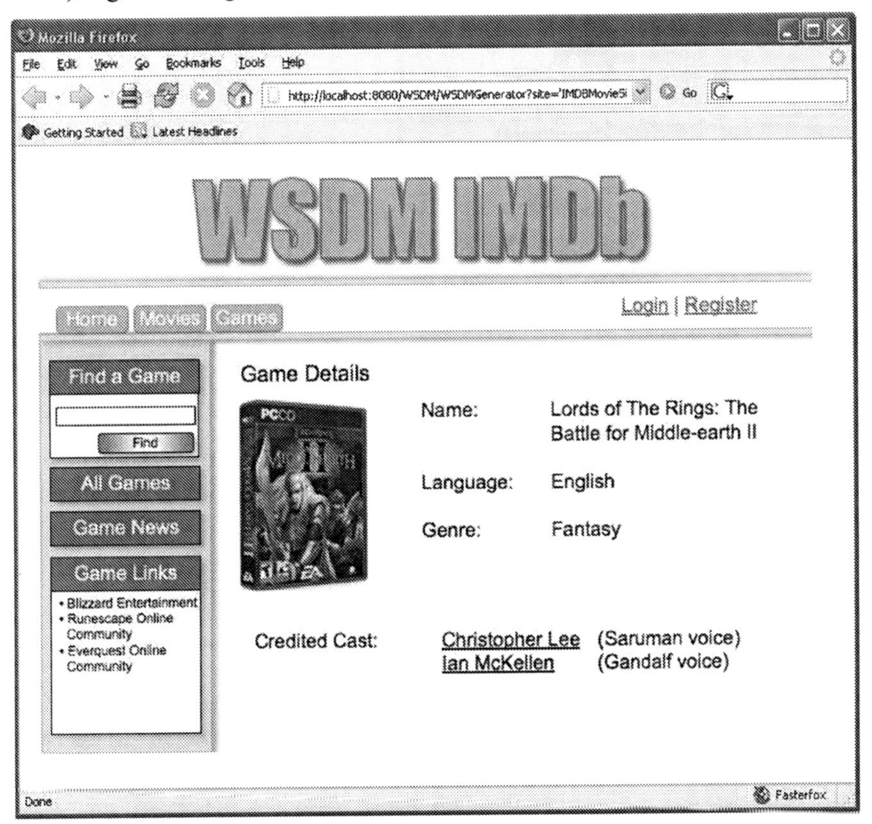

Figure 11.24. Example page from the IMDb example.

Important to notice is that, based on the design information collected, semantic annotations can be generated. More in particular, the use of one or more external ontologies during the conceptual design allows expressing explicitly the semantics of the presented data by means of content-related semantic annotations using these ontologies. However, it is also possible to annotate the Web system such that the semantics of its structure are also made explicit. Dedicated ontologies [e.g., the WAfA ontology (Yesilada et al., 2004) developed in the context of accessibility for visually impaired user] can be used to make the semantics of the different structural elements (e.g., a navigation menu, a logo, an advertising banner) explicit. These so-called structural annotations can be exploited by third parties that require specific knowledge about the Web system's structure: page transcoders to transcode a Web page in, for example, a format more appropriate for screen readers or search engine indexers. Structural annotations can be generated without any additional effort from the designer by exploiting the design information captured by means of the design models. How the content-related and structural annotations are generated is outside the context of this chapter. A description of this can be found in Plessers et al. (2005a, b) and Plessers and De Troyer (2004a, b).

As an example, consider (a part of) the generated content-related semantic annotations for a Web page showing the movie details for "The Terminator" movie, based on the object chunk "ShowMovie" (recall Figure 11.8):

```
<IMDB:Movie rdf:ID="23">
<hasTitle>The Terminator</hasTitle>
<IMDB:year>1984</IMDB:year>
<FOAF:plot>A human-looking cyborg from the future …<FOAF:plot>
…
</IMDB:Movie>
```

When generating the actual data on a Web page, span tags enclose the data originating from the data source. For the example above, the generated movie title and year are as follows:

```
<span id="1">The Terminator</span>
<span id="2">1984</span>
```

Finally, exploiting the mapping between the reference ontology and the actual data source (defined in the logical data design subphase), the (HTML) code and the generated annotations are linked together:

page.html#xpointer(id("1"))<=>page.owl#xpointer(id("23")/hasTitle)
page.html#xpointer(id("2"))<=>page.owl#xpointer(id("23")/IMDB:year)

11.7 FURTHER ISSUES

WSDM has been extended in different directions. The most important ones are the extensions to support localization (De Troyer and Casteleyn, 2004) and the extensions for adaptation (Casteleyn, 2005). We will briefly describe the principles used for these extensions.

11.7.1 Localization of Web Systems

Public Web systems are accessible from all over the world. This offers opportunities for companies and organizations to attract visitors from across international borders and to do business with them. Two different approaches are possible to address this issue: Develop one single Web system to serve everyone, or develop "localized" Web systems for particular localities. The "one-size-fits-all" approach may be appropriate for particular communities (e.g., researchers), but in general it will be less successful. In general, it is recommended to localize a global Web system, i.e., to create different versions and adapt those versions to the local communities they target. Members of a community share not only a common language, but also common cultural conventions. Since measurement units, keyboard configurations, default paper sizes, character sets, and notational standards for writing time, dates, addresses, numbers, currency, etc. differ from one culture to another, it is self-evident that local Web systems should address these issues. Some jokes, symbols, icons, graphics, or even colors may be completely acceptable in one country but trigger negative reactions in another country. Sometimes even the style or tone of the site's text might be considered offensive by a particular cultural entity, resulting in the text's needing to be rewritten rather than merely translated. Next to cultural differences, it may also be necessary to adapt the content to regional differences, such as differences in the services and products offered, differences in price, and differences in regulations.

As for classical software, Web system localization is often done once the Web system is completely developed and available for a particular community. We believe that the globalization process[6] could benefit from

[6] According to LISA (Localization Industry Standards Association; http://www.lisa.org), localization of a thing is adapting it to the needs of a given locality. Globalization is about spreading a thing to several different countries, and making it applicable and usable in those countries.

taking localization requirements into consideration while designing the Web system. Then it may be easier to actually realize globalization because the internationalization activities[7] may already be considered and prepared for during the design process. For this reason, WSDM was extended to support Web localization. We shortly explain how the different (sub)phases have been adapted.

11.7.1.1 The Mission Statement Specification

To be able to take localization into account during the design process, the mission statement should also mention the different localities for which the Web system needs to be developed. A locality describes a particular place, situation, or location. Localities are identified by means of a name and a label. Examples of localities are the Unied States, Japan, and the Flemish community in Belgium.

As an example, suppose that next to the English version (which is targeted to Americans), we also want localized versions of the IMDb Web system for France and Germany. Then, the mission statement can be reformulated as follows:

> To be the biggest and best movie and game site on earth. For movies, this will be achieved by providing as much information as possible on movies including their actors, directors, and producers, as well as to provide news, allow exploring show times, buy tickets in selected cinemas in the United States, and to share personal opinions about movies. For games, information about games is offered as well as news, and game lovers should be able to exchange information. Next to the U.S. version, localized versions for France and Germany should be offered; information dependent on the country, such as the movies currently playing, should be adapted for each version. Exploring show times and buying tickets are only available for the United States.

Here, the localities that are targeted are the United States, France, and Germany.

11.7.1.2 Audience Modeling

To support localization, a distinction is made between requirements and characteristics typical for an audience class and those typical for a locality.

The requirements and characteristics that are typical for a locality are related to the language, culture, habits, or regulations of the locality. Some examples of locality requirements are that an address should always include the state; for each price it should be indicated if tax is included and, if not,

[7] Internationalization consists of all preparatory tasks that will facilitate subsequent localization.

the percentage of tax that needs to be added should be mentioned. Locality characteristics will typically deal with issues such as language use, reading order, use of color, and use of symbols.

Then the localities are linked to the different audience classes. An audience class may span different localities. For example, in the IMDb example all classes identified so far are applicable for all localities, but in fact only people within the United States should be able to explore show times and buy tickets. Therefore, we can refine the audience class hierarchy and introduce a new subclass for the people who can explore show times and buy tickets. Then this class only needs to be supported in the U.S. locality.

11.7.1.3 Conceptual Modeling

During task modeling, a task model is defined for each requirement. Now, we also have requirements formulated for the different localities. These requirements also need to be considered during task modeling. When constructing the task models, we need to inspect the locality requirements to check if additional or different steps are needed when decomposing a task. If a task must be completely different for a specific locality (which is rarely the case), a different CTT must be created and labeled with this locality. If only some additional steps are needed, then these steps are labeled with the localities for which they are needed.

Also, when constructing the object chunks, we need to inspect the locality requirements to check if additional information is needed. If this is the case, this information is added to the object chunk and labeled with the locality for which it is needed. In the object chunks, we should also indicate which information is dependent on the locality. For example, in the IMDb example, the description of a movie needs to be given in the language of the locality, and the movies currently played and the movies coming soon will be different for each locality. Labeling the classes and properties that are locality-dependent indicates this. In the navigational design, the audience tracks are labeled with all the localities for which they are applicable.

11.7.1.4 Implementation Design

Usually, the site structure design will be independent of the locality, i.e., for each locality the site structure will be the same. However, if some task models are very different for different localities, a different site structure may be needed.

During the presentation design, the localization characteristics formulated during audience modeling need to be taken into consideration. Different templates should be created for different localities if this is needed (e.g., different colors, different labels).

When creating a logical data schema, we need to take into account that the information may be different for different localities, as indicated by the labeling of and within the object chunks. Depending on the situation, different data sources for each locality may be needed or only different fields for some properties. Many different solutions are possible; we will not go into details here. More information can be found in De Troyer and Casteleyn (2004).

11.7.2 Adaptation

WSDM provides flexible design support for the specification of (automatic) reorganization of structure and content of the Web system (at run time), based on the way users access and use the Web system. Note that this type of adaptation, also called *optimization* (Perkowitz and Etzioni, 1997), differs from personalization (which is what is usually intended when the term "adaptation" is used): Optimization improves the Web system as a whole (for all users), whereas personalization adapts the Web system for a single user (i.e., the current user). The possibility to take into account and anticipate during design the actual use of the Web system at run time offers the following advantages:

1. **Anticipate and react on run-time browsing behavior**: For example, make popular pages more directly available (add navigational aids links).
2. **Evaluate and use design alternatives automatically**: For example, merge audience tracks if their separation seems to be less useful.
3. **Detect and correct design flaws**: For example, detect and correct misplaced information.
4. **Better tailor the Web system to satisfy business goals:** For example, add or replace strategic business information in such a way that it appears on the most popular pages.

To specify this type of adaptive behavior, a dedicated language, called the Adaptation Specification Language (ASL), was introduced. ASL is a high-level, rule-based adaptation specification language that allows the designer to specify adaptation strategies (i.e., *which* adaptation needs to be done) and adaptation policies (i.e., *when* adaptation needs to be done). ASL is event-based: User-generated *events* (e.g., clicking a link, visiting a page, starting a session) will trigger the adaptation strategies. The strategies themselves are specified using *rules* (e.g., iterations, conditional execution of an action, predefined transformations on the relevant design models). By allowing the designer to specify which event(s) need(s) to be tracked, and when and how adaptation should be performed based upon these events, the designer has a powerful mechanism to specify how the organization and structure of the Web system should be improved (at run time) based on the actual use of the Web system. The remainder of this section explains an example of a useful adaptation strategy (and policy) and highlights some interesting features of ASL. For an in-depth discussion of ASL (including formal specification, example strategies, experimentation results), we refer to Casteleyn (2005).

Consider as an example adaptation strategy the *promotion* strategy, discussed in the context of WSDM in Casteleyn et al. (2003). Promotion of a component makes the component easier to find by moving it closer to the root (e.g., the home page) of the Web system. Here, the promotion is based on the popularity of the components: The most popular component(s) [the component(s) with the highest number of accesses] are promoted. ASL allows specifying for which components the number of accesses needs to be tracked and how this should be done (i.e., per session, per load, per click). For the IMDb example, a useful adaptation strategy might be to promote the movie and game visited most often (overall) during the past month. Therefore, the accesses to each individual movie and game need to be counted. A general script for counting the access to the elements of some set (of design elements) is used for this. This script can be used in different adaptation strategies:

```
script trackAmountOfAccesses(Set) :
    forEach element in Set
    begin
        addTrackingVariable element.amountOfAccesses ;
        monitor load on element do element.amountOfAccesses :=
                                    element.amountOfAccesses + 1
    end
```

Intuitively, the for-each rule in this script states that a tracking variable *amountOfAccesses* is declared (i.e., addTrackingVariable) and attached to each element of the given set. Furthermore, load events on the elements (i.e., for all users and for all sessions) will give rise to the increment of the *amountOfAccesses* tracking variable of that particular element.

The actual promotion in this case consists of linking the most popular movie and game to the root of the Web system. First, a script implementing the general principle of promotion is given. Here, the original links to the promoted component are kept (so only a navigational aids link is added). Alternative promotion strategies can be defined. The promotion script is specified as follows in ASL (note that in ASL the shorter term "node" is used instead of "component"):

```
script promoteNode(Set, promoteTo) :
    begin
    let promoteNodeMaxAccesses be
        max(Set [MAP on element: element.amountOfAccesses]);
    forEach node in Set :
        if node.amountOfAccesses = promoteNodeMaxAccesses
        then addLink (navigationAid, promoteTo, node)
    end
```

Having defined the adaptation strategy, we are able to specify the adaptation policy, i.e., *when* the adaptation should be performed. In this example, we collect the accesses to movies/games for one month and perform a promotion once every month:

```
when initialization do
    begin
            call trackAmountOfAccesses(ALL MovieDetailNode);
            call trackAmountOfAccesses(ALL GameDetailNode)
    end

when 1 month from now do
    begin
            call promoteNode(ALL MovieDetailNode, root);
            call promoteNode(ALL GameDetailNode, root)
            reset(ALL MovieDetailNode
                    [MAP on element: element.amountOfAccesses]);
            reset(ALL GameDetailNode
                    [MAP on element: element.amountOfAccesses]);
    end
```

Note that the ALL keyword is used to obtain a set of all instances of a particular (conceptual) component. In this case, all concrete Movie and Game Detail nodes (i.e., each individual movie or game detail page) are the subjects of the adaptation strategy. The first part of the adaptation policy specifies that when the Web system is initialized, the number of accesses to Movie and Game Detail nodes is initialized. The second policy specifies that after one month (note that this will be repeated each month), the promotion strategy is applied, and the components containing the most popular movie and game are promoted to the root. Finally, all tracking variables are reset for the next month of tracking.

11.8 SUMMARY

WSDM is a Semantic Web design method based on an audience-driven design philosophy. This means that the requirements of the target audience, rather than the data available in the organization or its internal organization, are the starting point of the modeling process. The different audience classes and their different requirements are also reflected in the actual structure of the Web system. This approach is used to offer the designer a well-defined method to identify the information and functionality needed for a Web system and to structure it in an appropriate way. This must prevent the developers from only providing information that happened to be available and structuring it in a way that is obvious only for them. The method is based on the principle that a Web system should be designed for and adapted to its target audiences.

The method also makes a clear distinction between the conceptual design and the implementation design. Issues like grouping of information and functionality in pages and graphical presentation and layout are not considered to be conceptual issues but implementation design issues, because more than one grouping into pages or more than one presentation design is possible for the same conceptual design.

Last but not least, WSDM allows developing Web systems that are semantically annotated by means of one or more ontologies. Next to the regular content-related semantic annotations, structural annotations can also be generated. These are annotations that describe the semantics of the different structural elements of the Web system and can be exploited by other applications to transcode the Web system to formats more suitable for purposes other than human reading.

Furthermore, the clear separation of design concern by means of different design concepts and models as well as a clear separation between conceptual issues and implementation issues have shown to pay off: The modeling of a new design concern can easily been added. This has been demonstrated for adding localization and adaptation.

REFERENCES

Brusilovsky, P., 1996, Methods and techniques of adaptive hypermedia. *User Modeling and User-Adapted Interaction*, **6**(2–3), Springer Science+Business Media B.V., New York, ISSN 0924-1868, pp. 87–129.

Casteleyn, S., 2005, Designer specified self re-organizing Web sites, PhD thesis, Vrije Universiteit, Brussels.

Casteleyn, S., and De Troyer, O., 2001, Structuring Web sites using audience class hierarchies. *Conceptual Modeling for New Information Systems Technologies, ER 2001*

Workshops, HUMACS, DASWIS, ECOMO, and DAMA, Lecture Notes in Computer Science, **2465**, Springer-Verlag, ISBN 3-540-44-122-0, pp. 198–211.

Casteleyn, S., De Troyer, O., and Brockmans, S., 2003, Design time support for adaptive behavior in Web sites. *Proceedings 18th ACM Symposium on Applied Computing*, ACM, ISBN 1-58113-624-2, pp. 1222–1228.

Deng, C., Yu, S., Wen, J.-R., and Ma, W.-Y., 2004, Block-based Web search. *SIGIR 2004: Proceedings 27th Annual International ACM SIGIR Conference on Research and Development in Information Retrieval*, Sheffield, UK, July 25-29, M. Sanderson, K. Järvelin, J. Allan, and P. Bruza, eds., ACM, ISBN 1-58113-881-4, pp. 456–463.

De Troyer, O., and Casteleyn, S., 2001, The conference review system with WSDM. *First International Workshop on Web-Oriented Software Technology, IWWOST'01* (also http://www.dsic.upv.es/~west2001/iwwost01/), O. Pastor, ed., Valencia University of Technology, Spain.

De Troyer, O., and Casteleyn, S., 2003a, Modeling complex processes for Web applications using WSDM. *Proceedings Third International Workshop on Web-Oriented Software Technologies (held in conjunction with ICWE2003), IWWOST2003* (also http://www.dsic.upv.es/~west/iwwost03/articles.htm), D. Schwabe, O. Pastor, G. Rossi, and L. Olsina, eds., Oviedo, Spain.

De Troyer, O., and Casteleyn, S., 2003b, Exploiting link types during the conceptual design of Web sites. *International Journal of Web Engineering Technology (IJWT),* **1**(1): 17–40.

De Troyer, O., and Casteleyn, S., 2004, Designing localized Web sites. *Proceedings 5th International Conference on Web Information Systems Engineering (WISE2004)*, X. Zhou, S. Su, M.P. Papazoglou, M.E. Orlowska, and K.G. Jeffery, eds., Springer-Verlag, Brisbane, Australia, ISBN 3-540-23894-8, pp. 547–558.

De Troyer, O., Casteleyn, S., and Plessers, P., 2005, Using ORM to model Web systems. *On the Move to Meaningful Internet Systems 2005: OTM 2005 Workshops, International Workshop on Object-Role Modeling (ORM'05), Lecture Notes in Computer Science,* **3762**, Springer, ISBN 3-540-29739-1, pp. 700–709.

De Troyer, O., and Leune, C., 1998, WSDM: A user-centered design method for Web sites. *Computer Networks and ISDN Systems, Proceedings 7th International World Wide Web Conference*, Elsevier, Brisbane, Australia, pp. 85–94.

Halpin, T., 2001, *Conceptual Schema and Relational Database Design: From Conceptual Analysis to Logical Design*, Morgan Kaufmann, San Francisco.

Paterno F., 2000, *Model-Based Design and Evaluation of Interactive Applications*, Springer-Verlag, New York.

Paterno, F., Mancini, C., and Meniconi, S., 1997, ConcurTaskTrees: A diagrammatic notation for specifying task models. *Proceedings INTERACT 97*, Chapman & Hall, pp. 362–366.

Perkowitz, M., and Etzioni, O., 1997, Adaptive Web sites: An AI challenge. *Proceedings 15th International Joint Conference on Artificial Intelligence*, Morgan Kaufmann, pp. 16–23.

Plessers, P., Casteleyn, S., and De Troyer, O., 2005a, Semantic Web development with WSDM. *Proceedings 5th International Workshop on Knowledge Markup and Semantic Annotation (SemAnnot 2005)*, Galway, Ireland.

Plessers, P., Casteleyn, S., Yesilada, Y., De Troyer, O., Stevens, R., Harper, S., and Goble, C., 2005b, Accessibility: A Web engineering approach. *Proceedings 14th International World Wide Web Conference (WWW2005)*, A. Ellis and T. Hagino, eds., ACM, Chiba, Japan, ISBN 1-59593-046-9, pp. 353–362.

Plessers, P., and De Troyer, O., 2004a, Web design for the Semantic Web. *Proceedings WWW2004 Workshop on Application Design, Development and Implementation Issues in the Semantic Web, CEUR Workshop Proceedings, Vol. 105 Web, WWW2004 Workshop*, C. Bussler, S. Decker, D. Schwabe, and O. Pastor, eds., ISBN 1613-0073, New York.

Plessers, P., and De Troyer, O., 2004b, Annotation for the Semantic Web during Web site development. *Proceedings ICWE 2004 Conference, Lecture Notes in Computer Science,* **3140**, N. Koch, P. Fraternali, and M. Wirsing, eds., Springer, Munich, Germany, ISBN 3-540-22511-0, pp. 349–353.

Yesilada, Y., Harper, S., Goble, G., and Stevens, R., 2004, Screen readers cannot see (ontology-based semantic annotation for visually impaired Web travelers). *Web Engineering, 5th International Conference, ICWE 2005, Sydney, Australia, July 27-29, 2005, Proceedings*, *Lecture Notes in Computer Science*, **3579**, Springer, ISBN 3-540-27996-2, pp. 445–458.

Chapter 12

AN OVERVIEW OF MODEL-DRIVEN WEB ENGINEERING AND THE MDA

Nathalie Moreno,[1] José Raúl Romero,[2] Antonio Vallecillo[1]
[1]*Dept. Lenguajes y Ciencias de la Computación, University of Málaga, Spain*
[2]*Dept. Informática y Análisis Numérico, University of Córdoba, Spain*

12.1 INTRODUCTION

Model-Driven Software Development (MDSD) is becoming a widely accepted approach for developing complex distributed applications. MDSD advocates the use of models as the key artifacts in all phases of development, from system specification and analysis to design and implementation. Each model usually addresses one concern, independently from the rest of the issues involved in the construction of the system. Thus, the basic functionality of the system can be separated from its final implementation; the business logic can be separated from the underlying platform technology, etc. The transformations between models enable the automated implementation of a system from the different models defined for it.

Web Engineering is a specific domain in which MDSD can be successfully applied. Most of the technology is here to implement systems that exploit the Web paradigm, but the effective design of Web applications is still a concern: The complexity and requirements of Web applications are constantly growing, while the supporting technologies and platforms rapidly evolve.

Existing model-driven Web Engineering (MDWE) approaches already provide excellent methodologies and tools for the design and development of most kinds of Web applications. They address different concerns using separate models (navigation, presentation, data, etc.) and are supported by model compilers that produce most of the application's Web pages and logic based on the models. However, these proposals also present some limitations, especially when it comes to modeling further concerns, such as architectural styles or distribution. Furthermore, current Web systems need to interoperate with other external applications, something that requires their integration with third-party Web services, portals, and also with legacy systems. Finally, many of these Web Engineering proposals do not fully exploit all the potential benefits of MDSD, such as complete platform independence, model transformation and merging, or meta-modeling. Miller and Mukerji (2003) from the Object Management Group (OMG™) have introduced a new approach for organizing the design of an application into (yet another set of) separate models so that portability, interoperability, and reusability can be obtained through architectural separation of concerns. MDA covers a wide spectrum of topics and issues (MOF-based meta-models, UML profiles, model transformations, modeling languages and tools, etc.) and also promises the interoperability required between models and tools from separate vendors. On the other camp, Software Factories (Greenfield and Short, 2004) provide effective concepts and resources for the model-based design and development of complex applications, and it is our belief that they can be successfully used for Web Engineering, too.

In this chapter we will introduce the main concepts involved in MDWE and discuss its current strengths, weaknesses, and major challenges, especially in the context of the MDA initiative.

12.2 DOMAIN-SPECIFIC MODELING

Domain-specific modeling (DSM) is a way of designing and developing systems that involves the systematic use of domain-specific languages (DSLs) to represent the various facets of a system. Such languages tend to support higher-level abstractions than general-purpose modeling languages and are closer to the problem domain than to the implementation domain. Thus, a DSL follows the domain abstractions and semantics, allowing modelers to perceive themselves as working directly with domain concepts. Furthermore, the rules of the domain can be included in the language as constraints, thereby disallowing the specification of illegal or incorrect models.

DSLs play a cornerstone role in DSM. In general, defining a modeling language involves at least two aspects: the domain concepts and rules (abstract syntax), and the notation used to represent these concepts (concrete syntax—let it be textual or graphical). Each model is written in the language of its meta-model. Thus, a meta-model will describe the concepts of the language, the relationships between them, and the structuring rules that constrain the model elements and combinations in order to respect the domain rules. We normally say that a model conforms to its meta-model (Bézivin, 2005).

Meta-models are also models, and therefore they need to be written in another language, which is described by its meta-meta-model. This recursive definition normally ends at that level, since meta-meta-models conform to themselves.

A typical example of a meta-model-defined DSL is ATL (Jouault and Kurtev, 2006b), which is a transformation language. A large library of ATL transformations is available from the Eclipse meta-model open source library. The interested reader can consult the work by Bézivin (2005) for a more complete and detailed introduction to these topics.

DSM often also includes the idea of code generation: automating the creation of executable source code directly from the DSM models. Being free from the manual creation and maintenance of source code implies significant improvements in developer productivity, reduction of both defects and errors in programs, and a better resulting quality. Moreover, working with models of the problem domain instead of models of the code raises the level of abstraction, hiding unnecessary complexity and implementation-specific details, while putting the emphasis on already familiar terminology.

A DSM environment may be thought of as a meta-modeling tool, i.e., a modeling tool used to define a modeling tool or CASE tool. The domain expert only needs to specify the domain-specific constructs and rules, and the DSM environment provides a modeling tool tailored for the target domain. The resulting tool may either work within the DSM environment or, less commonly, may be produced as a separate standalone program. Using a DSM environment can significantly lower the cost of obtaining tool support for a DSM language, since a well-designed DSM environment will automate the creation of program parts that are costly to build from scratch, such as domain-specific editors, browsers, and components.

Examples of DSM environments include commercial ones such as MetaEdit+; open source environments, such as the Generic Eclipse Modeling System; or academic ones such as the Generic Modeling Environment (GME; http://www.isis.vanderbilt.edu/projects/gme/). The increasing popularity of DSM has led to DSM frameworks being added to existing integrated development environments, such as the Eclipse Modeling Project (EMP) and Microsoft's DSL Tools for Software Factories.

12.3 MDA

One of the best known MDSD initiatives is called Model-Driven Architecture (MDA®), which is an approach to software development produced and maintained by the OMG, a consortium that produces and maintains computer industry specifications for interoperable enterprise applications. MDA is a registered trademark of the OMG, together with its related acronym, model-driven development (MDD), another OMG trademark.

The goal of MDA is one that is often sought: to separate business and application logic from its underlying execution platform technology so that (1) changes in the underlying platform do not affect existing applications and (2) business logic can evolve independently from the underlying technology. A tool that implements the MDA concepts will allow developers to produce models of the application and business logic and also generate code for a target platform by means of transformations.

The major benefit of this approach is that it raises the level of abstraction in software development. Instead of writing platform-specific code in some high-level language, software developers focus on developing models that are specific to the application domain but independent of the platform. In a nutshell, MDA is a broad conceptual framework that describes an overall approach to software development.

MDA is not to be confused with MDSD. MDA is the OMG implementation of MDSD, using the set of tools and standards defined by OMG. These OMG standards include UML® (Unified Modeling Language), MOF (Meta-Object Facility), XMI (XML Metadata Interchange), and MOF/QVT (Query/View/Transformations), among others. All these standards can be obtained from the OMG Web site (www.omg.org).

12.3.1 The MDA Framework

The MDA framework is basically organized around the so-called platform-independent models (PIMs) and platform-specific models (PSMs) and on the model transformations between them. The PIM is a specification of a system in terms of domain concepts. These domain concepts exhibit a specified degree of independence of different platforms (e.g., CORBA, .NET, and J2EE). The system can then be compiled using any of those platforms as a target by transforming the PIM to a platform-specific model (PSM). Thus, the PSM specifies how the system uses a particular type of platform. Finally, the application's code is considered a form of PSM (at the lowest level).

In MDA, a platform is a set of subsystems and technologies that provides a set of functionality through interfaces and specified usage patterns, which any application supported by that platform can use without concern for the details of how the functionality provided by the platform is implemented (Miller and Mukerji, 2003). As in MDSD, each model in MDA conforms to a meta-model, which in MDA can be defined using MOF.

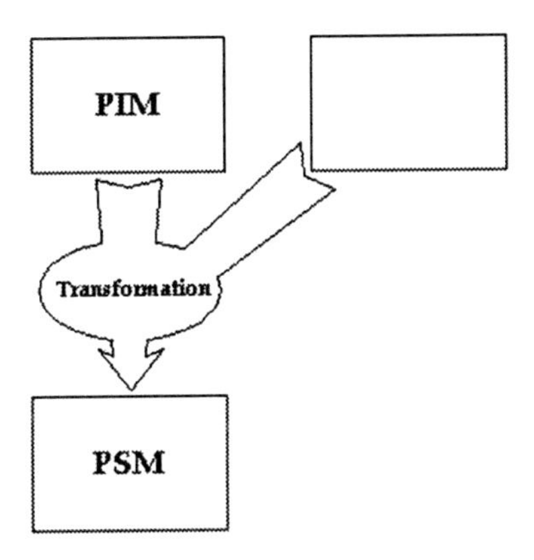

Figure 12.1. The MDA pattern.

In addition to models, transformations are also at the heart of MDA. Model transformation is the process of converting one model to another model of the same system (see Figure 12.1). Such transformations can be done following many ways: using types, marks, templates, etc. In MDA, software development becomes an iterative model transformation process: Each step transforms one PIM of the system at one level into one PSM at the next level, until a final system implementation is reached, with the particularity that each PSM of a transformation can become the PIM of the next transformation (within another level of abstraction). In this context, the implementation is just another model, which provides all the information necessary to construct the system and put it into operation.

12.3.2 OMG Approaches for Defining DSLs

Both PIMs and PSMs are models and are therefore defined using modeling languages. Although in theory MDA's models can be defined using any modeling language, OMG strongly suggests that models are specified using UML or any other MOF-compliant language (i.e., whose meta-meta-model is MOF). This interest for being MOF- and UML-compliant arises from the increasing need to be able to interoperate with other notations and tools, and to exchange data and models, thus facilitating and improving reuse.

OMG defines three main possible approaches for defining domain-specific languages. The first solution is to develop a meta-model that is able to represent the relevant domain concepts. This means creating a new domain language, an alternative to UML, using the MOF meta-modeling facilities provided by OMG for defining object-based visual languages (i.e., the same mechanisms that have been used for defining UML and its meta-model). In this way, the syntax and semantics of the elements of the new language are defined to faithfully match the domain's specific characteristics. The problem is that standard UML tools will not be able to deal with such a new language (to edit models that conform to the meta-model, compile them, etc.). This approach is the one followed by languages such as the CWM (Common Warehouse Metamodel) or the W2000 (Baresi et al., 2006b) notations, since the semantics of some of these languages' constructs do not match the semantics of the corresponding UML model elements.

The second and third solutions are based on extending UML. Extensions of the UML can be either heavyweight or lightweight. The difference between lightweight and heavyweight extensions comes from the way in which they extend the UML meta-model. Heavyweight extensions are based on a modified UML meta-model with the implication that the original semantics of modeling elements is changed, and therefore the extension might no longer be compatible with UML tools.

Lightweight extensions are called UML *profiles* and are based on the extension mechanisms provided by UML (OMG, 2005b; Fuentes and Vallecillo, 2004) (stereotypes, tag definitions, and constraints) for specializing its meta-classes, but without breaking their original semantics. UML profiles may impose new restrictions on the extended meta-classes, but they should respect the UML meta-model without modifying the original semantics of the UML elements (i.e., the basic features of UML classes, associations, properties, etc. will remain the same, only new constraints can be added to the original elements). Syntactic sugar can also be defined in a profile, in terms of icons and symbols for the newly defined elements. One of the major benefits of profiles is that they can be handled in a natural way by UML tools.

In UML profiles, stereotypes define particularizations of given UML elements, adding some semantics to them. For instance, we can define the stereotype <<persistent>> that extends UML classes to represent persistent elements in a particular domain. Tag definitions specify the possible attributes of stereotypes (e.g., the name of the table where the persistent element should be stored). Finally, constraints define the domain rules that the stereotyped UML elements should obey in order to make up correct models (e.g., suppose that we do not want abstract classes to be stereotyped as persistent). Figure 12.2 graphically shows the UML specification of this example stereotype.

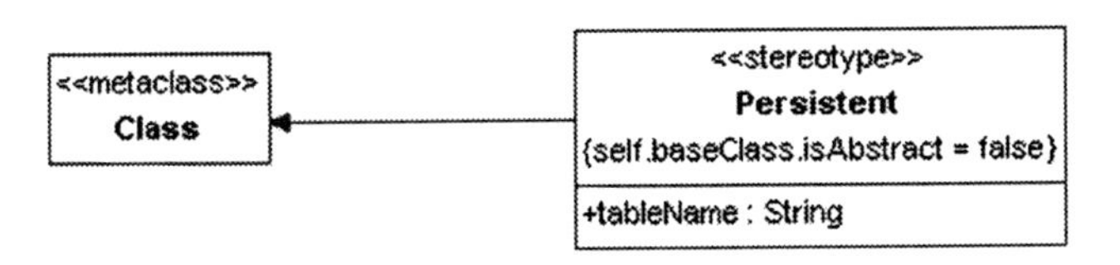

Figure 12.2. An example of a UML 2.0 stereotype specification.

Constraints on stereotypes are normally specified using OCL (Object Constraint Language) (OMG, 2006), whose current version (2.0) is fully aligned with UML. Constraints can be either directly attached to the modeling elements (as shown in the figure) or separately specified and then related to the element to which they apply by identifying their context:

```
context Persistent inv:
   self.baseClass.isAbstract = false
```

Perhaps the best-known example of customizing UML for a specific domain is SysML, a DSL for systems engineering (www.sysml.org). In addition, there is a whole set of UML profiles that customize UML to deal with the specific concepts required in several relevant application domains (e.g., real-time, business process modeling, finance, etc.) or implementation technologies (such as .NET, J2EE, or CORBA).

The main advantage of UML profiles is probably not the extension of the UML meta-model (which is already too large and complex to be used in full) but that they allow “restricting” the set of UML elements that need to be used in a given domain, particularizing the semantics of those elements in order to capture the semantics and structuring rules of the domain-specific elements they represent. It is important to repeat that such a particularization can only be done by refinement, and without changing the original semantics of UML elements.

Finally, meta-transformations that transform back and forth from the profile definition to the meta-model definition can also be specified, as shown in Figure 12.3.

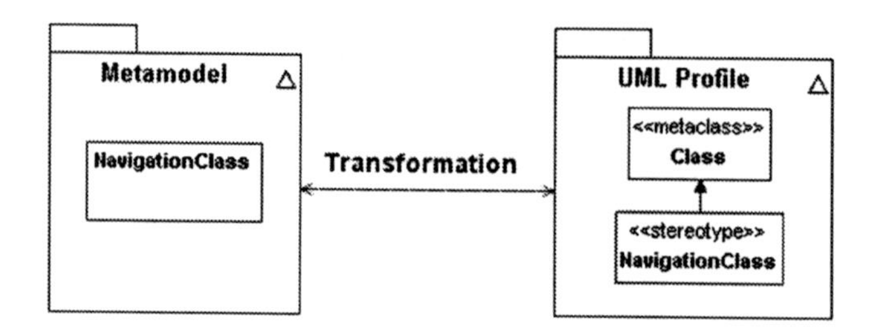

Figure 12.3. Example of transformation between a "profilable" meta-model and a profile.

12.3.3 Model Transformations

A model transformation can be viewed as a transformation between two models that describes how elements in the source model are converted into elements in the target model. This is done by relating the appropriate meta-model elements in the source and target meta-models and defining constraints and guards on such relations (e.g., the preconditions on the transformation to take place). It is important to notice that model transformations are also models, and therefore they conform to a meta-model that describes the language in which they are expressed.

MDA describes a wide variety of models and transformations between models. While there are many kinds of transformations, they can fit broadly into two main categories:

- Vertical mappings (or *refinements*), which relate system models at different levels of abstraction—such as PIM to PSM mappings, or reverse-engineering mappings. Until now, vertical transformations have in most cases been developed within modeling tools using Web tool-specific proprietary languages. For the same reason that domain know-how should not be tied to a particular platform, it is thus critical that model transformations are not dependent on a given CASE tool.
- Horizontal mappings, which relate or integrate models covering different aspects or domains within a system, but at the same level of abstraction. Horizontal mappings maintain the consistency between levels, guaranteeing that an entity needs to be consistent with what is said about the same entity in any other specification at the same level of abstraction. This includes the consistency of that entity's properties, structure, and behavior.

In MDA, OMG proposes MOF-QVT (Query/View/Transformation) (OMG, 2005a) as the standard language for specifying model transformations. Many other model transformation languages, like VIATRA by the University of Budapest, ATL by INRIA, RubyTL (Sánchez and García-Molina, 2006) by the University of Murcia, etc., are also available, with different levels of compliance to the QVT standard (Jouault and Kurtev, 2006a). The interested reader can visit the "Model Transformation" Web site (www.model-transformation.org) for a complete listing of model transformation languages and tools.

12.4 MODEL-DRIVEN WEB ENGINEERING PROPOSALS

As mentioned in the introduction, Web Engineering is a specific domain in which MDSD can be successfully applied, due to its particular characteristics: There is a precise set of concerns that need to be addressed (navigation, presentation, business processes, etc.); the basic kinds of Web applications is well known (Kappel et al., 2006) (document-centric, transactional, workflow-based, collaborative, etc.); and the set of architectural patterns and structural features used in Web systems is reduced and precisely defined. In fact, existing model-based Web Engineering approaches—most of which have been described in this book—already provide excellent methodologies and tools for the design and development of most kinds of Web applications.

These approaches come basically from two main areas. First, a few proposals are based on hypermedia design methods, introducing the required expressiveness and mechanisms to capture relevant Web-specific elements, such as navigation. Prominent examples of these initiatives are HDM (Garzotto et al., 1993), RMM (Frasincar, 2001), WebML (Ceri et al., 2002), W2000 (Baresi et al., 2006b), WSDM (De Troyer and Leune, 1998), Hera (Vdovjak et al., 2003), and Webile (Di Ruscio, 2004), the majority of which are based on the classic E/R model or on extensions of it. Another group of more recent approaches emerged as extensions of conventional object-oriented development techniques, adapting them to cope with the particular characteristics of Web systems. In this group we can find methods such as EORM (Lange, 1994), OOHDM (Schwabe et al., 1999), UWE (Koch, 2001), OOWS (Pastor et al., 2006), OO-Method (Pastor et al., 2001), OO-H (Gómez and Cachero, 2003), and MIDAS (De Castro et al., 2006).

These proposals are model-driven because they address the different concerns involved in the design and development of a Web application using separate models (such as content, navigation, and presentation) and then are supported by model compilers that produce most of the application's Web pages and logic from the original models. Furthermore, most of them count with development processes that support their notations and tools and have been successfully used in commercial environments for building many different kinds of Web systems. And although all methodologies adopt different notations and propose their own constructs, they all share a common ground of concepts—and thus they might be considered as somehow based on a common meta-model, as suggested by Koch and Kraus (2003).

However, as the complexity of Web applications grows (to be able to deliver, e.g., large e-commerce, e-learning, or e-government applications), and new requirements are imposed on Web systems, most of these proposals show some limitations:

- They are usually tied to particular architectural styles and technologies, i.e., they do not allow the parameterizable construction of Web applications using different platform technologies and architectural styles—they typically build client-server applications only, and based on very specific platform technologies (PHP, ASP, EJB, or JSP). The problem is that these architectural styles and target technologies are no longer relevant when, for example, mobility and nomadic features are required for some types of Web applications.
- Most of these proposals were originally conceived to deal with particular kinds of Web applications (such as Web information systems, hypermedia applications, or adaptive Web applications), so they deal with a fixed set of common concerns (navigation, presentation, etc.). Therefore, they are very good at modeling certain aspects, but very weak at modeling others. In addition, they are difficult to extend to model further aspects (such as internal processes, distribution, and some other extra-functional concerns) in a natural, modular, and independent way.

Finally, Web applications currently need to interoperate with other external systems. This requires their integration with third-party Web services, portals, and legacy systems—meaning, among other things, that their processes, choreography, and part of their business logic must be explicitly available for integration with these external systems (Moreno and Vallecillo, 2005a). Not all MDWE proposals address this issue at the model level; the integration is mostly achieved at the implementation level.

Solving all these limitations is not a trivial task. We are currently observing how some Web Engineering proposals are evolving to cope with some of these issues. For instance, some of them are developing extensions to address more and more aspects. Examples include UWE and OO-H, which have incorporated a process model into their original approaches (Koch et al., 2004) and are working to deal with the architectural style of the final application, too (Cáceres et al., 2006). WebML has also evolved to be able to deal with legacy systems and for context-awareness (Ceri et al.,

2007). The problem with these incremental extensions is that, unless their efforts to include new concerns are made in a very well-organized and interoperable manner, we may end up with proposals that have grown by adding too many new features in an unnatural and artificial way, and therefore may become too complex and brittle.

Another problem that some of these proposals are also facing is their use of proprietary notations and tools. This forces customers and developers to buy and use "yet-another" modeling tool (with the learning costs and efforts involved in the process) if they want to take advantage of them. Even worse, these proprietary tools do not interoperate with the rest of the tools being used by the customer, which forces him to work with a whole set of isolated development environments, each one different (and incompatible) from the rest—something that the customer is not going to tolerate.

Thus, we are witnessing how the Web Engineering community considers the use of standard UML notation, techniques, and supporting tools for modeling Web systems, including the adaptation of their own modeling languages, representation diagrams, and development processes to UML. There is a need to be able to be compatible and to interoperate with other notations and tools, and to seamlessly exchange data and models with them. This is the case for WebML, which is defining UML-based representations of its modeling language so that the WebML notation and its development process can be smoothly integrated into standard UML development environments (Moreno et al., 2006; Schauerhuber et al., 2006).

The advent of the model-driven architecture (MDA) initiative may also bring significant benefits here and may help to address most of the limitations cited above in a natural way. As mentioned in the preceding section, MDA provides an approach for organizing the design of an application into separate models so that portability, interoperability, and reusability can be achieved through architectural separation of concerns. In addition, the new modeling notation UML 2.0 incorporates a whole new set of diagrams and concepts that are more appropriate for modeling the specific structure and behavior of software systems, in particular of Web applications (e.g., the new structuring mechanisms, or the improved specification and semantics of state machines and activities).

Of course, the use of UML and MDA for model-driven Web Engineering is not free from problems. As any other initiative, it brings along both benefits and drawbacks and also has both supporters and detractors. The next two sections are dedicated to explaining these ideas in detail.

12.5 MDA-BASED WEB ENGINEERING

MDA provides several interesting opportunities to improve current Web Engineering approaches, helping them to overcome some of the limitations cited above.

12.5.1 Becoming UML- and MOF-Compliant

As previously mentioned, there is an increasing need to be able to interoperate and be compatible with other notations and tools, and to integrate with already existing modeling environments—in particular, with the UML tools that today are commonplace in many customer settings. Of course, other DSM environments are already being developed—some of them probably much better than those supporting the UML notation—but the problem is that they have not reached the level of acceptance and are not as widespread as UML modeling tools are today. And we are faced with the need to be able to offer a solution to our customers today.

In this sense, a very promising approach is the definition of UML profiles for representing proprietary Web Engineering modeling languages. This is the case with WebML, which has recently defined a meta-model and a UML profile (Moreno et al., 2006; Schauerhuber et al., 2006) for its notation. This allows the WebML language and its development process (supported by the WebRatio tool) to be smoothly integrated with standard UML development environments.

In addition, counting on a meta-model for WebML will allow its integration with other MDA tools as soon as they are available (editors, validators, metric evaluators, etc.) and also with other MDSD approaches and tools (using model transformations that allow the conversion of MOF-meta-models to other meta-modeling approaches, such as KM3 or Ecore).

12.5.2 Organizing Models According to the MDA Principles

We are also witnessing how other approaches that were originally UML-based are making use of the new MDA principles to reorganize their models in a modular manner, in such a way that each model focuses on one specific concern and then formulates its development processes in terms of model transformations and model merges.

Probably the most representative example is UWE, which has successfully restructured its original set of models (which represented the different concerns involved in the design and development of a Web application) in terms of meta-models, and the UWE development process in terms of transformations between them (Koch, 2006; Kraus, 2007). This has significantly enhanced the original proposal with better modularity, expressiveness, and reuse. Furthermore, the use of specification techniques for the transformations will allow UWE to redefine and improve many of the aspects of its development process, especially those that were originally hard-coded in the UWE supporting CASE tool, in order to benefit from model transformation rules defined at a higher abstraction level, e.g., using graph transformations or transformation languages.

Another interesting outcome of the work done by the UWE group when adopting the MDA principles into their proposal is the analysis of the models (and model transformations) that comprise the MDSD process for Web applications, focusing on the classification of the model transformations in terms of type, complexity, number of source models, involvement of marking models, implementation techniques, and execution type (Koch, 2006). This analysis could be very useful to other model-based Web Engineering methods if they decide to reformulate their proposals in terms of independent models and transformations between them. Other proposals, such as MIDAS, have also started to adopt such an approach by specifying the development process of Web information systems in terms of (meta)models and transformations between them (Cáceres et al., 2006).

12.5.3 Adding New Concerns

This reformulation of model-based Web Engineering proposals is also proving other benefits, such as the modular addition of further aspects into their designs. Most of these concerns were not contemplated originally, and integrating them was difficult because of the (usually ad hoc) internal structure of their supporting processes and tools.

One representative example is OO-H, whose authors realized that they had to be able to deliver Web applications with different software architectures and to different platforms, depending on the customers' specific requirements—in this case the customers were the ones demanding such features. The OO-H team managed to successfully reformulate part of their internal structure and methods, making the representation of the software architecture of the system a separate concern that could be captured as a separate model, and then merged (using QVT transformations) with the rest of the models of the system (such as navigation, presentation, etc.) (Meliá and Gómez, 2006).

UWE and OO-H have also investigated the explicit representation of the business processes of a Web application, as separate models (Koch et al., 2004). Their joint findings are very encouraging, because they managed to define a common way for modeling them for both proposals. This shows that reuse of meta-models across Web Engineering proposals is feasible.

Finally, UWE has also shown recently how other concerns, such as the user requirements (Koch et al., 2006), can be expressed as UML models and connected to the approach. This is one of the benefits they obtained once they fully reorganized their proposal as a set of separate models, related through model transformations (Kraus, 2007).

All these findings support the thesis that a common meta-model is possible for Web Engineering, as originally proposed by Koch and Kraus (2003). Furthermore, in the next section we will see how the existence of a common meta-model could allow the definition of a framework for building Web applications, which in the context of the MDA would also enable the exchange of models and tools between MDWE proposals.

12.6 WEI: A MODEL-BASED FRAMEWORK FOR BUILDING WEB APPLICATIONS

In this section we shall identify a general set of common concerns involved in the development of Web applications and present a model-driven Web architectural framework (WEI) for organizing and relating the different models that represent these concerns. Each WEI model focuses on one particular concern (navigation, presentation, architectural style, distribution) and at different levels of abstraction (platform-independent, platform-specific). The set of meta-models that define such models can be considered as a common meta-model for WEI.

MDWAF is also supported by a development methodology for building Web applications, which conforms to the MDA principles—in the sense that it is defined in terms of models and the relationships between them, so transformations can be easily formalized among the models until the final implementation is reached.

12.6.1 Identifying Reference Models for Web Applications

In general, the kinds of concerns involved in the development of a Web application will directly depend on the type of Web application being designed and also on the project requirements. Web applications have already been classified by complexity and development history (Kappel et al., 2006):

1. *Document-centric Web sites*, which are hierarchical collections of static HTML documents (basically, plain text and images) that offer read-only information based on a set of structured content, navigation patterns, and presentation characteristics designed and stored a priori.

The simplicity and stability of these systems limit the scope of Web modeling to three models: a **user interface structure** model that deals with the content of the information delivered to the client; a **navigation** model that points out the network of paths within the Web application; and a **presentation** model that refers to the visual elements that comprise the Web pages.

2. *Transactional Web applications*, which incorporate support for persistent data store, information location, concurrency control, failure, and configuration management. In addition to the navigation aspects of any hypermedia application, development of transactional Web applications implies the need for an effective **information structure** model, which is capable of capturing the processes of inserting, updating, and deleting data, and also a **distribution** model, which enables the establishment of alternatives for carrying out transactions. A clearer separation between data design, among behavioral aspects of the application, and from the user interface concerns is required.
3. *Interactive Web applications*, which are browser-based applications that allow dynamic content of Web pages, hence providing users with personalized information. This feature requires a **process** model that describes how business classes manage the information stored (i.e., the elements of the information structure model) and also requires that the navigation and presentation models are parameterizable to provide tailor-made information to individual users according to their preferences, goals, and knowledge. Furthermore, this type of application emphasizes modeling not only the information structure itself and its future consumers (i.e., the **users** model), but also the relationships or bridges between the information structure model, navigation model, and **business** model.
4. *Workflow-based Web applications*, which provide support for modeling structured business processes, activity flows, business rules, interactions among actors, roles, and a high-performance infrastructure for data storage (content management). Information is needed not only for the system actors but also for its processes. For this kind of Web applications, at a minimum the following models are required: a user interface structure model, a navigation model, a presentation model, an information structure model, a business model (i.e., the description of how functionality is encapsulated into business components and services), a process model (with a description of the behavior of the internal processes), and a **software architecture** model identifying the subsystems, components, and connectors (software and hardware) the application should have.
5. *Collaborative Web applications*, which are those executed by different groups of users that access Web resources to accomplish a specific task. They entail a modeling decomposition of the Web application

design into **views** or workspaces based on different **user roles**. For each group of users, the functional requirements, task, and activities to be performed must be specified. These issues involve **modularity** and **distribution** requirements on the process model. Finally, the information assets to be manipulated by views must also be modeled.

6. *Portal-oriented Web applications*, which integrate resources (data, applications, and services) from different sources in a single point. From an end-user perspective, a portal is a Web site with pages that are organized by some form of navigation. Pages can display either static HTML content or complex Web services. Personalization, behavior tracking of users, as well as message flows in Web service collaborations are extremely relevant in portal-oriented Web applications. Therefore, a **choreography** model needs to express the expected behavior of both the system processes and the **external services** in order to check their compatibility and interoperability to compose them to build the portal aggregated.
7. *Ubiquitous Web applications*, which need to be accessible at any time, from anywhere, and in any media, i.e., they must run on a variety of platforms, including mobile phones, personal digital assistants (PDAs), desktop computers, etc. This implies that their **presentation** and **navigational** models should be **adaptable** not only to different kinds of users, but also to different kinds of platforms and contexts. Consequently, this kind of application requires modeling the separation between **platform-independent** and **platform-specific** concerns.

Based on the set of concerns identified above, each one represented by one model, we have built an architectural framework for model-driven Web application development (WEI). Its basic structure is depicted in Figure 12.4. It is organized in three main layers (User Interface, Business Logic, and Data), each one corresponding to a viewpoint. In turn, each layer is composed of a set of models, which specify the entities relevant to each concern.

Far from being "yet another Web methodology," the aims of WEI can be summarized as follows:

1. to be able to represent, in terms of models and relationships between them, the concerns required for designing and developing Web applications—following an architectural separation of concerns as prescribed by MDA
2. to integrate and harmonize the models and practices proposed by existing approaches, addressing their concerns
3. to be extensible so that new concerns could be easily added

4. to provide as a common framework (and meta-model) in which current proposals could be integrated and formulated in terms of the MDA principles, hence allowing them to smoothly interoperate (by defining, e.g., interoperability bridges between compatible models coming from different proposals, whenever this is possible) and complement each other, share tools, etc.

At a high architectural design level, the whole WEI concept space is captured by 13 meta-models, organized in 3 main packages as shown in Figure 12.4. It is important to note that the models that comprise the framework have not been arbitrarily chosen but are based on the concerns covered by existing Web Engineering proposals (see also Table 12.1 later on) and our previous experience with the development of large distributed applications.

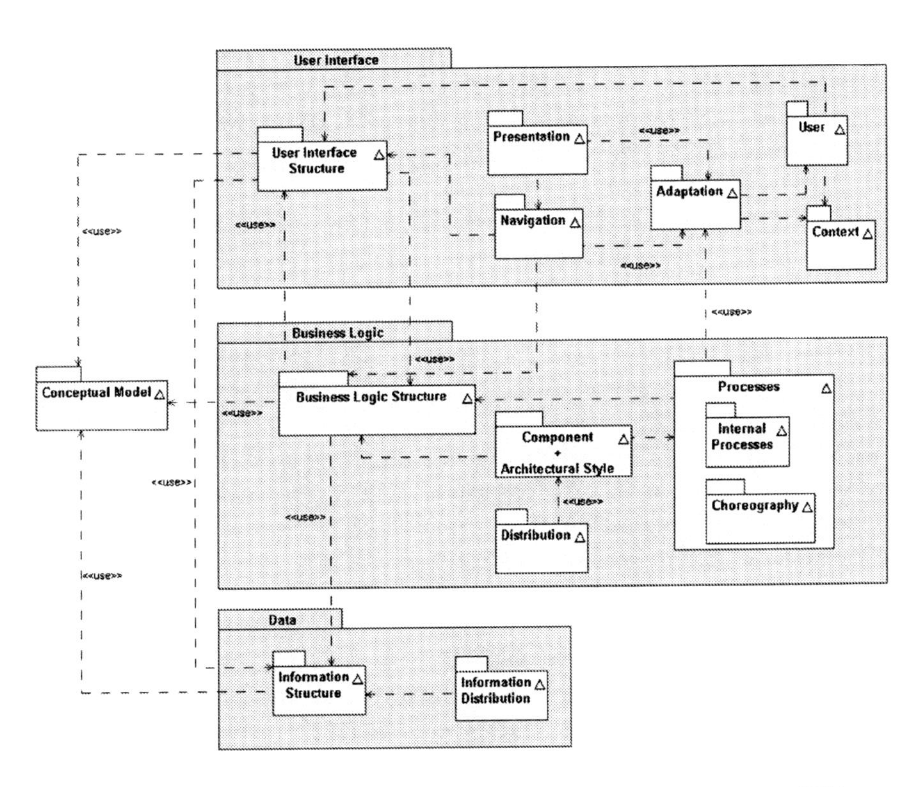

Figure 12.4. Models representing the concerns involved in the development of Web applications.

At the bottom level, the **Data Structure** package describes the organization of the information managed by the application (by means of, e.g., a database system) and provides a mechanism for storing it persistently. Information is depicted in terms of the data elements that constitute its information base and the semantic relationships between these elements. This level is organized in two models:

(i) The **Information Structure** model deals with the information that has to be made persistent, i.e., stored in a database.
(ii) The **Information Distribution** model describes the distribution and replication of the data being modeled, since information can be fragmented in *nodes* or replicated in different *locations*.

Then, the **User Interface** focuses on the facilities provided to the end user for accessing and navigating through the information managed by the application and describes how this information is presented depending on the context and the user profile. The User Interface level is responsible for accepting persistent, processed, or structured data from the Process and Data viewpoints, in order to interact with the end user and deliver the application contents in a suitable format. Originally, Web applications were specifically conceived to deal mainly with navigation and presentation concerns, but currently they also need to address other relevant issues:

(i) The **User Interface Structure** model encapsulates the information that the rest of the models at this level have about the information handled by the system (i.e., it is the *view* of such information from this viewpoint).
(ii) The **Navigation** model represents the application navigational requirements in terms of access structures that can be accessed via navigational links.
(iii) Navigational objects are not directly perceived by the user; rather, they are accessed via the **Presentation** model. This model captures the presentational requirements in terms of a set of *PresentationUnits*.
(iv) The **User** model describes and manages the user characteristics with the purpose of adapting the content and the presentation to the users' needs and preferences.
(v) The **Context** model deals with *Device*, *Network*, *Location,* and *Time* aspects and describes the environment of the application. These are needed to determine how to achieve the required customization.
(vi) The **Adaptation** model captures context features and user preferences to obtain the appropriate Web content characteristics (e.g., the number of embedded objects in a Web page, the dimension of the base-Web page without components, or the total dimension of the embedded components). Adaptation policies are usually specified in terms of ECA rules.

Finally, the **Business Logic** package encapsulates the business logic of the application, i.e., how the information is processed and how the application interacts with other computerized systems.

(i) The **Business Logic Structure** model describes the major classes or component types representing services in the system (*BusinessProcessInformation*), their attributes (*Attributes*), the signature of their operations (*Signature*), and the relationships between them (*Association*). The design of the Structure model is driven by the needs of the processes that implement the business logic of the system, taking into account the tasks that users can perform.

(ii) The **Internal Processes** model specifies the precise behavior of every *BusinessProcessInformation* or component as well as the set of activities that are executed in order to achieve a business objective. For a complete description of a business process, apart from the Structure model, we need information related to the *Activities* carried out by the *BusinessProcessInformation*, expressing their behavior and the *Flows* that pass around objects or data.

(iii) The **Choreography** model defines the valid sequences of messages and interactions that the different objects of the system may exchange. The choreography may be individually oriented, specifying the contract a component exhibits to other components (*PartialChoreography*), or it may be globally oriented, specifying the flow of messages within a global composition (*GlobalChoreography*).

(iv) The **Distribution** model describes how its basic entities, the *nodes*, are connected by means of point to point connections or *links*. While the Information Distribution model of the Data layer specifies the distribution of the data, this model describes the distribution of the processes that achieve the business logic of the system.

(v) The **Component+Architectural Style** model defines the fundamental organization of a system in terms of its components, their relationships, and the principles guiding its design and evolution, i.e., how functionality is encapsulated into business components and services.

The emphasis in each of these levels will depend on the kind of Web application being modeled (data-intensive, user interface-oriented, etc.).

A central model of the WEI framework is the **Conceptual Model**, which can be used for specifying the basic structure and contents on the Web application (so the rest of the "views" can relate to the elements of that model) and also for maintaining the consistency of the model specifications, establishing how the different viewpoints merge and complement each other.

Note that, in addition to the models, the framework predefines some dependencies between the models which determine those cases in which the definition of a model requires the previous specification of some other models. At a different level, the dependencies may also imply how the framework instantiation process should be carried out. Furthermore, these dependencies also specify correspondences between the elements from different models of the framework, especially when they may have been independently developed by different parties, or when they represent the system from different viewpoints, and therefore the same element is specified in different ways in different models (each one offering a partial view of the whole). In these cases, correspondences between model elements may also be subject to certain consistency rules, which check that the views do not impose contradictory requirements on the elements they share.

12.6.2 Modeling These Concerns

In order to formally define the framework, we have built an MOF meta-model for each model, which describes its entities and their relationships (http://www.lcc.uma.es/~nathalie/WEI/). MOF was selected as a meta-modeling language because of our interest in being MDA-compliant. Other alternatives were, of course, possible (using, e.g., KM3 or Ecore), but it was important for us to try to use OMG's notations and tools, to exercise the MDA approach. MagicDraw was selected as a modeling tool. The selection of a UML tool is really important, because they do not interoperate well, and therefore the tool you use may greatly condition your project.

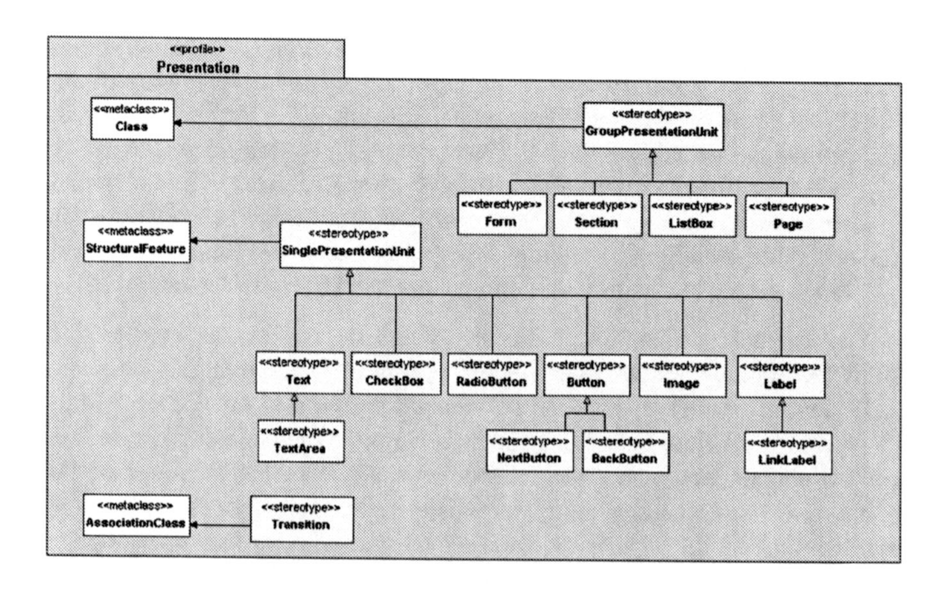

Figure 12.5. The WEI Presentation profile.

But the meta-models are just one part of the puzzle. Unlike other approaches, OMG does not provide a solution for directly building correct models from meta-models. Instead, you have to define your own DSL associated to these meta-models. In our case we defined lightweight extensions of UML, i.e., UML profiles, for representing these models (Moreno et al., 2005). As an example of it, Figure 12.5 shows the profile for the WEI presentation meta-model.

12.6.3 How the Framework is Used

WEI can be instantiated to build Web applications both from scratch and based on existing models (including those defined using other methodologies, e.g., UWE, WebML, or OO-H).

12.6.3.1 Building Applications from Scratch with WEI

The straightforward application of the framework in the context of MDA to develop a Web system from scratch has already been documented in detail (Moreno et al., 2005a, b; Moreno and Vallecillo, 2005c) and successfully applied to define and implement several kinds of Web applications such as the *Conference Review System* or the *Travel Agency Application*.

As a brief summary, the WEI methodology process involves the definition of at least three PIMs, each one corresponding to a viewpoint, as illustrated in Figure 12.6(b). Each PIM is composed of the set of models described in the previous section and is developed following the process depicted in Figure 12.6(a).

Once the three top-level PIMs have been appropriately defined, we need to mark them using the appropriate profile(s) for the target platform(s) and technologies. Once marked, we need to follow the MDA transformation process from PIMs to PSMs, applying a set of mapping rules (one for each mark and for each marked element). The result of the application of such mapping rules is a set of UML models of the application according to the target technologies (e.g., Java, JSP, Oracle, etc.). Finally, the PSMs are translated to code, applying a transformation process again.

It is important to note that *bridges* should be specified among the three PIMs and among their corresponding PSMs, and for which transformations are also required. Bridges are the key elements to maintain consistency between the different models at the same level of abstraction and to be able to provide links between them. A very interesting work by the group of Alfonso Pierantonio at the University of L'Aquila (Chicchetti et al., 2006) shows how model weaving can be effectively used to specify and implement such bridges, being able to connect the different artifacts and models produced during the development of Web applications—in particular, the models describing the data, navigation, and presentation aspects, whose connections are usually defined in an ad hoc manner, and their consistency manually maintained. Although their work is carried out using non-OMG notations and standards, it can be easily ported to the MDA context, using MOF meta-models and QVT transformations for establishing correspondences between elements from different views.

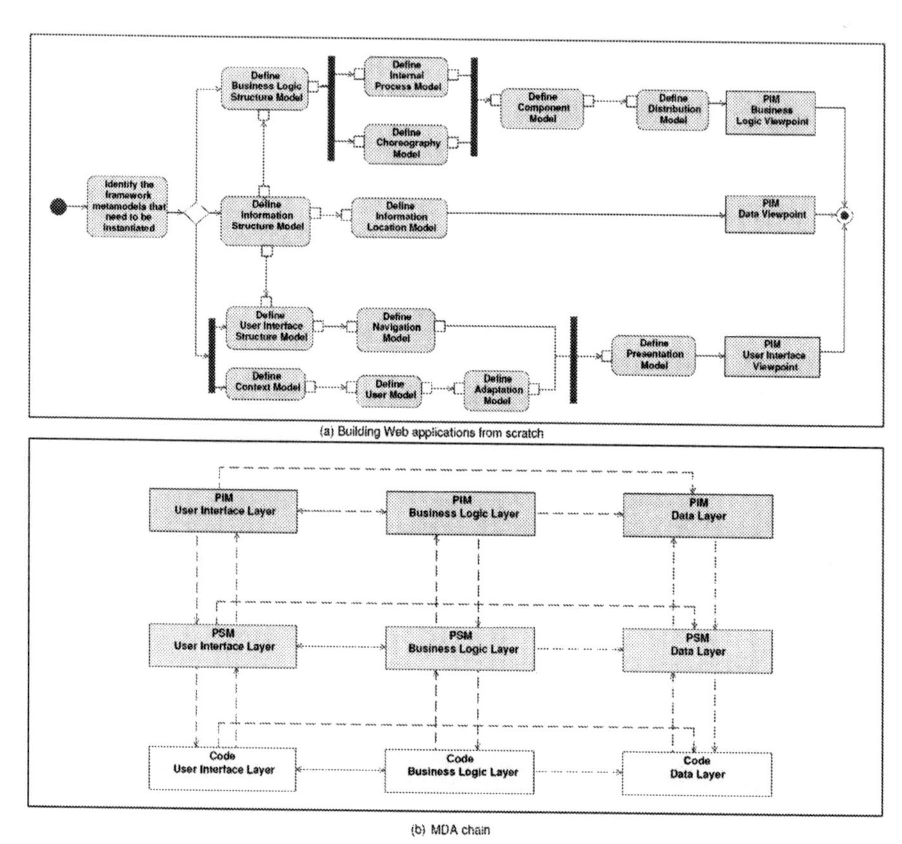

Figure 12.6. The WEI process.

12.6.3.2 Designing Web Applications by Reusing Models from Other Methodologies

One of the major advantages of our proposal is its ability to design and implement applications reusing both models and tools (e.g., model compilers) defined by other Web methodologies. Thus, a Web application developer could use, for instance, UWE or OO-H for designing the models of the User Interface layer, and WebML for designing the Data layer, or vice versa. Furthermore, models could be already defined for other applications and reused here for building fast prototypes.

Reusing models conforming to other Web methodologies requires the definition of interoperability bridges between "compatible" models coming from different methodologies and the appropriate models of our framework. Usually, the source and target entities defined in different approaches do not differ much. In addition, neither the models nor the entities described in our framework were arbitrarily chosen: Instead, they try to generalize the entities and models defined by most Web Engineering proposals (see Table 12.1). Thus, the interoperability bridges between models from different proposals are a priori feasible and even quite straightforward using WEI as a reference framework.

Table 12.1 Concerns and Models Covered by Current Web Engineering Proposals

Layer	Model	OOHDM	W2000	UWE	WebML	WSDSM	OOWS	OOH
User Interface	Structure	√	~	√	~	~	√	√
	User	~		√	√	√	√	
	Context	~		√	√			
	Adaptation			√	√		√	
	Navigation	√	√	√	√	√	√	√
	Present.	√	√	√	√	√	√	√
Business Logic	Structure		~	~		√	√	~
	Processes		√	√			√	√
	Choreogr.						√	
	Architect.			√	√			√
	Distribution							
Data	Inf. Struct.	√	√	√	√	√		√
	Inf. Distrib.							

There are, however, some issues that need to be addressed, which are similar to the traditional problems that appear when integrating models that represent different views of the same system. First, we may find models using different names to refer to the same elements. Second, we may find that one model may assume the existence of other models that either provide some services (e.g., the precise behavior that needs to be executed when a navigation link is traversed) or represent external systems or legacy applications that our Web system should be able to work with (by, e.g., exchanging data or invoking services). Third, the majority of Web Engineering proposals apply (almost the same) separation of concerns, but the problem is that their levels of abstraction and granularity do not always coincide. Fourth, some of the models that we want to reuse may deal with more than one of our framework concerns. And finally, we may find some aspects and concerns that have not been modeled, because they are implicitly assumed in the proposals' models (the most typical example is behavior).

The way in which we address the first four issues is by specifying bridges (either correspondences or transformations) between the elements living in different models. Such bridges have been defined using QVT relations. The last issue, i.e., the lack of models for representing some concerns, needs to be addressed by the explicit specification of such elements, in order to supply the "missing" information. This case currently happens when models to be reused come from methodologies that do not have all their information explicitly modeled, but that is hardwired into their supporting CASE tools. Thus, the models to be reused assume some information and semantics that are not available if we try to use them in a different environment. This problem is alleviated by the explicit representation of all concerns in the WEI framework, because all the information has to be supplied there.

12.7 ISSUES AND CHALLENGES FOR MDWE AND MDA

So far we have discussed how MDA and its related concepts and mechanisms can help in the effective design and development of Web applications. This section describes the major challenges faced by the introduction of MDA in the Web Engineering domain.

12.7.1 Maturity of MDA Standards and Tools

One of the major problems that any person approaching MDA discovers is the lack of maturity of the current standards and tools. For example, some standards considered key to MDA are not currently supported by tools, and some others have not even been finalized. Probably the most representative example is QVT, for which there is not a complete implementation available as of today. This is really frustrating and needs to be urgently addressed in order to avoid the dissatisfaction it produces to potential users.

12.7.2 Lack of Interoperability Between UML Tools

Despite the interoperability goals of the OMG, current UML modeling tools cannot properly interoperate, and exchanging models and diagrams between them is almost impossible. XMI is supposed to provide the solution to this problem, but most UML tool vendors fail to generate fully XMI-compliant specifications of the models they produce. What we currently see is that most vendors add proprietary extensions to the XMI tags, which cannot be understood by other tools. This is another sign of the current immature status of the MDA initiative, which we expect can be resolved soon (otherwise the vendors may kill this opportunity with their incompatibilities).

12.7.3 Need to Improve the Support for DSLs

As mentioned above, UML profiles are a very interesting option to define DSLs, not only because they are relatively simple to define, but also because, once defined, they can be (in theory) used by any UML tool to produce models that conform to that profile.

The current situation is not so bright, however. Actually, most UML tools provide support for defining UML profiles (in terms of their corresponding stereotypes, tag definitions, and constraints) but fail to be able to guarantee the constraints on the models because they do not support OCL checks. Therefore, you can specify a UML profile that represents a given application domain (that is, a DSL for that domain), but then there is no way of checking that the models users produce respect the structuring rules of that DSL, i.e., users can easily create wrong models. It is similar to defining a language but without providing a compiler that could check the grammar of the programs produced.

Another improvement that is also required is a better support for relating MOF meta-models with profiles, i.e., to map the meta-model of a DSL to its corresponding profile, as suggested at the end of Section 12.3.2. This would allow meta-models to be imported from other sources as well as our being able to use standard UML tools to easily draw models that conform to them. There are some academic proposals in this respect (Abouzahra et al., 2005), although this kind of mechanism should be implemented in most UML tools as part of their profiling facilities.

12.7.4 The Complexity of UML

The size and technical complexity of UML have been held responsible for hampering its wide adoption in many industrial environments. UML is a general-purpose modeling language for software-intensive systems that is designed to support many kinds of applications. Consequently, in contrast to specific DSM languages, UML is used for a wide variety of purposes across a broad range of domains. Thus, it counts with many modeling elements and diagrams, and even provides support to cope with different semantic variants, through the *semantic variation points* defined for some of its elements. This mechanism increases the potential adoption of UML in many different kinds of environments, but at the high cost of increasing its complexity and introducing a lack of focus and precision ("maximizing reuse minimizes use"). This kind of mechanism also has a strong impact on the learning curve of UML, and on the efforts required by system modelers to master and effectively use the UML notation.

12.7.5 The Ways in Which Modelers Work

Many of today's modelers are still casual in their approach; MDSD (and in particular MDA) requires increased rigor to produce models that are amenable to automatic generation of code. This means that users need to be very precise when designing their models—which in MDA implies plenty of training in UML modeling.

Note that this issue and the previous one could be greatly alleviated by the use of UML profiles that restricted the set of UML elements that can be used to model a domain-specific application and only allowed users to draw correct models with regard to the DSL meta-model (i.e., the profile). This is why very compact, precise, and specific UML-based DSLs, with a reduced number of elements and strong structuring rules, are being perceived as a key factor to the success of MDSD (Bézivin et al., 2005). However, current UML tools do not provide complete support for UML profiles (including the validation of their OCL constraints) as mentioned above. In addition, the use that average modelers make of UML stereotypes and profiles is not always correct, especially because this extension mechanism is not as simple as it might seem at first sight. Different studies have tried to analyze the way in which stereotypes are currently used, and the most common mistakes made by modelers when defining and using them (Atkinson et al., 2003; Henderson-Sellers and González-Pérez, 2006).

Another tendency that we also perceive in normal modelers is the use of DSLs that support agile methodologies and rapid prototyping for designing and developing Web applications. For instance, the use of Ruby is gaining acceptance in many areas (Schwabe, 2006), and experience shows that the increase in development performance and reduction in costs might be worth its use, especially when combined with frameworks such as Rails (Thomas et al., 2006).

12.7.6 MDA is not Just About Modeling

It is unrealistic to expect 100% code generation for every computing problem, and no vendor today can realistically offer a complete MDA solution. Thus, if you expect too much of MDA, it will fail. What MDA offers is just a way of approaching the design and development of systems, using a set of standard notations and tools to achieve interoperability and reuse across vendors and platform independence. But to realize the full benefits of MDA, organizations should not just introduce some modeling practices in their development processes; they must support the full software lifecycle development process, from analysis and requirements management through design, development, implementation, deployment, and maintenance. Otherwise, the full advantages of MDA will be lost.

12.7.7 Modeling Further Concerns

Finally, and especially in the case of more data-intensive Web applications (usually called Web-based information systems), we see a trend toward the incorporation of emerging initiatives like the Semantic Web, with supporting technologies such as (Semantic) Web services, and (Semantic) Web rule languages, which aim at fostering application interoperability. Semantic Web languages [e.g., RDF(S) or OWL] facilitate the description of models for such domains. However, the integration of all these models with the rest of the model-based Web Engineering approaches is still unresolved. This is not only a problem for MDA, but for any MDSD approach.

Further concerns, such as user requirements, as well as the role that the computation-independent model (CIM) defined by MDA plays in MDWE, need to be investigated, too.

12.8 CONCLUSIONS

In this chapter we have presented an overview of the current state of model-driven software development, and of model-driven Web Engineering in particular, especially in the context of MDA. We have analyzed the key concepts and mechanisms that these approaches provide and how the development of Web systems can benefit from them. Apart from introducing the advantages and opportunities that MDA can bring to MDWE, we have also discussed the current problems and threats that MDA faces for its successful adoption in industrial settings. Addressing and resolving them properly is possibly the major challenge for MDA today.

In summary, we have seen that there is a real need to integrate with UML environments, which are the ones currently demanded in many customer settings, and that MDA can help reformulate and reorganize current Web Engineering proposals in terms of models and transformations between them. MDWE can significantly benefit from the facts that each model can address a concern, that these concerns can be explicitly represented, and that they can be specified in a platform-independent manner—hence achieving the modularity, portability, reusability, and interoperability required for any competitive Web Engineering proposal. MDWE solutions cannot survive isolated any longer; they need to interoperate among themselves and be integrated into the customers' development environments. And these are precisely the issues that MDA can help them address in a very successful way.

ACKNOWLEDGMENTS

We would like to acknowledge the work of many MDSD, MDA, and MDWE experts who have been involved in investigating and addressing the problems of model-Web Engineering. Although the views in this chapter are the authors' solely responsibility, they could not have been formulated without the many long and clarifying discussions with these experts. In particular, we would like to thank Nora Koch, Jaime Gómez, Vicente Pelechano, Piero Fraternali, Oscar Pastor, Daniel Schwabe, Gustavo Rossi, Geert-Jan Houben, Joaquin Miller, Jean Bézivin, Alfonso Pierantonio, Bryan Wood, and many others too numerous to be named here. We would also like to thank both the organizers and the participants of the past editions of the Model-Driven Web Engineering (MDWE) workshop at the last ICWE conferences, where some of the issues presented here were originally raised and discussed.

This work has been supported by Spanish Projects TIN2005-25886-E and TIN2005-09405-C02-01.

REFERENCES

Abouzahra, A., Bézivin, J., Del Fabro, M.D., and Jouault, F., 2005, A practical approach to bridging domain specific languages with UML profiles. *Proceedings Best Practices for Model Driven Software Development* (OOPSLA'05), San Diego, CA.

Atkinson, C., Kühne, T., and Henderson-Sellers, B., 2003, Systematic stereotype usage. *Software and Systems Modelling*, **2**(3): 153–163.

Baresi, L., Colazzo, S., Mainetti, L., and Morasca, S., 2006a, Model-based Web application development. In *Web Engineering*, Springer, Heidelberg, pp. 303–334.

Baresi, L., Colazzo, S., Mainetti, L., and Morasca, S., 2006b, W2000: A modelling notation for complex Web applications. In *Web Engineering: Theory and Practice of Metrics and Measurement for Web Development*, Springer, New York, pp. 335–408.

Bézivin, J., Jouault, F., Rosenthal, P., and Valduriez, P., 2005, Modelling in the large and modelling in the small. *Proceedings European MDA Workshops: Foundations and Applications* (MDAFA 2003 and MDAFA 2004), Springer, LNCS 3599, pp. 33–46.

Bézivin, J., 2005, On the unification power of models. *Software and Systems Modelling* (SoSym), **4**(2): 171–188.

Cáceres, P., De Castro, V., Vara, J.M., and Marcos, E., 2006, Model transformations for hypertext modelling on Web information systems. *Proceedings ACM/SAC 2006 Track on Model Transformations* (MT2006), Dijon, France, pp. 1256–1261.

Ceri, S., Fraternali, P., Bongio, A., Brambilla, M., Comai, S., and Matera, M., 2002, *Designing Data-Intensive Web Applications*, Morgan Kaufmann, San Francisco.

Ceri, S., Daniel, F., Matera, M., and Facca, F., 2007, Model-driven development of context-aware Web applications. *ACM Transactions on Internet Technology* (TOIT), **7**(1).

Chicchetti, A., Di Ruscio, D., and Pierantonio, A., 2006, Weaving concerns in model-based development of data-intensive Web applications. *Proceedings ACM/SAC 2006 Track on Model Transformations* (MT2006), Dijon, France, pp. 1256–1261.

De Castro, V., Marcos, E., and López Sanz, M., 2006, A model-driven method for service composition modelling: A case study. *International Journal of Web Engineering and Technology*, **2**(4): 335–353.

De Troyer, O., and Leune, C.J., 1998, WSDM: A user centered design method for Web sites. *Proceedings 7th International Conference on World Wide Web*, Amsterdam, Elsevier Science Publishers B.V., pp. 85–94.

Di Ruscio, D., Muccini, H., and Pierantonio, A., 2004, A data modelling approach to Web application synthesis. *International Journal of Web Engineering and Technology*, 1(3): 320–337.

Frasincar, F., Houben, G., and Vdovjak, R., 2001, An RMM-based methodology for hypermedia presentation design. *Proceedings 5th East European Conference on Advances in Databases and Information Systems* (ADBIS '01), London, Springer-Verlag, pp. 323–337.

Fuentes, L., and Vallecillo, A., 2004, An introduction to UML profiles. *UPGRADE, The European Journal for the Informatics Professional*, **5**(2): 5–13.

Garzotto, F., Paolini, P., and Schwabe, D., 1993, HDM—A model-based approach to hypertext application design. *ACM Transactions on Information Systems*, **11**(1): 1–26.

Gómez, J., and Cachero, C., 2003, *OO-H Method: Extending UML to Model Web Interfaces*, Idea Group Publishing, Hershey, PA, pp. 144–173.

Greenfield, J., and Short, K., 2004, *Software Factories: Assembling Applications with Patterns, Frameworks, Models & Tools*, Wiley, New York.

Henderson-Sellers, B., and González-Pérez, C., 2006, Uses and abuses of the stereotype mechanism in UML 1.x and 2.0. *Proceedings MODELS 2006*, Italy.

Jouault, F., and Kurtev, I., 2006a, On the architectural alignment of ATL and QVT. *Proceedings ACM Symposium on Applied Computing*, Dijon, France, ACM Press.

Jouault, F., and Kurtev, I., 2006b, Transforming models with ATL. *Proceedings Model Transformations in Practice Workshop at MoDELS 2005*, Montego Bay, Jamaica. Springer, LNCS 3844, pp. 128–138.

Kappel, G., Pröll, B., Reich, S., and Retschitzegger, W., 2006, *Web Engineering—The Discipline of Systematic Development of Web Applications*, Wiley, New York.

Koch, N., 2001, Software engineering for adaptive hypermedia systems: Reference model, modelling techniques and development process. *Softwaretechnik—Trends*, **21**(1).

Koch, N., and Kraus, A., 2003, Towards a common metamodel for the development of Web applications. *Proceedings 3rd International Conference on Web Engineering* (ICWE 2003). Springer, LNCS 2722, pp. 497–506.

Koch, N., Kraus, A., Cachero, C., and Meliá, S., 2004, Integration of business processes in Web applications models. *Journal of Web Engineering* (JWE), **3**(1): 22–49.

Koch, N., Zhang, G., and Escalona, M.J., 2006, Model transformations from requirements to Web system design. *Proceedings 6th International Conference on Web Engineering* (ICWE 2006), Palo Alto, CA, ACM Press, pp. 281–288.

Koch, N., 2006, Transformation techniques in the model-driven development process of UWE. *Proceedings 2nd Model-Driven Web Engineering Workshop* (MDWE 2006), Palo Alto, CA.

Kraus, A., 2007, Model-driven software engineering for Web applications, PhD Thesis. Institut für Informatik, Ludwig-Maximilians-Universität, Munich.

Lange, D.B., 1994, An object-oriented design method for hypermedia information systems. *Proceedings 27th Annual Hawaii International Conference on System Sciences* (HICSS-27), Maui, IEEE Computer Society, pp. 366–375.

Meliá, S., and Gómez, J., 2006, The WebSA approach: Applying model driven engineering to Web applications. *Journal of Web Engineering* (JWE), **5**(2): 121–149.

Miller, J., and Mukerji, J., 2003, The MDA Guide. Draft v. 2.0, OMG doc. ab/2003-01-03.

Moreno, N., and Vallecillo, A., 2005a, A model-based approach for integrating third-party systems with Web applications. *Proceedings 5th International Conference on Web Engineering* (ICWE 2005), Springer, LNCS 3579, pp. 441–452.

Moreno, N., Romero, J.R., and Vallecillo, A., 2005b, Incorporating cooperative portlets in Web application development. *Proceedings 1st Model-Driven Web Engineering Workshop* (MDWE 2005), Sydney, Australia, pp. 70–79.

Moreno, N., and Vallecillo, A., 2005c, Modelling interactions between Web applications and third-party systems. *Proceedings IWWOST 2005,* Porto, Portugal, pp. 441–452.

Moreno, N., Fraternalli, P., and Vallecillo, A., 2006, A UML 2.0 profile for WebML modelling. *Proceedings 2nd Model-Driven Web Engineering Workshop* (MDWE 2006), Palo Alto, CA.

OMG, 2005a, MOF QVT Final Adopted Specification, OMG doc. ptc/05-11-01.

OMG, 2005b, UML 2.0 Superstructure Specification v. 2.0, OMG doc. formal/05-07-04.

OMG, 2006, OCL 2.0, OMG doc. ptc/06-05-01.

Pastor, O., Gómez, J., Insfran, E., and Pelechano, V., 2001, The OO-Method approach for information systems modelling: From object-oriented conceptual modelling to automated programming. *Information Systems*, **26**(7): 507–534.

Pastor, O., Fons, J., Abrahao, S., and Pelechado, V., 2006, Conceptual modelling of Web applications: The OOWS approach. In *Web Engineering*, E. Mendes and N. Mosley, eds., Springer, New York, pp. 277–302.

Sánchez, J., and García-Molina, J., 2006, A plugin-based language to experiment with model transformation. *Proceedings 9th International Conference MoDELS 2006*, Genova, Italy, Springer, LNCS 4199, pp. 336–350.

Schauerhuber, A., Wimmer, M., and Kapsammer, E., 2006, Bridging existing Web modelling languages to model-driven engineering: A metamodel for WebML. *Proceedings 2nd Model-Driven Web Engineering Workshop* (MDWE 2006), Palo Alto, CA.

Schwabe, D., 2006, Rapid prototyping of Web applications combining domain specific languages and model driven design. *Proceedings 6th International Conference on Web Engineering* (ICWE 2006), Palo Alto, CA, ACM Press.

Schwabe, D., Pontes, R.A., and Moura, I., 1999, OOHDMWeb: An environment for implementation of hypermedia applications in the WWW. *SigWEB Newsletter*, **8**(2).

Thomas, D., and Heinemeier, D., 2006, *Agile Web Development with Rails: A Pragmatic Guide*, 2nd ed., Pragmatic Bookshelf, Raleigh, NC.

Vdovjak, R., Frasincar, F., Houben, G., and Barna, P., 2003, Engineering Semantic Web information systems in Hera. *Journal of Web Engineering* (JWE), **2**(1–2): 3–26.

PART III

QUALITY EVALUATION AND EXPERIMENTAL WEB ENGINEERING

Chapter 13

HOW TO MEASURE AND EVALUATE WEB APPLICATIONS IN A CONSISTENT WAY

Luis Olsina, Fernanda Papa, Hernán Molina
GIDIS_Web, Engineering School, Universidad Nacional de La Pampa, Calle 9 y 110, (6360) General Pico, LP, Argentina, {olsinal,pmfer,hmolina}@ing.unlpam.edu.ar

13.1 INTRODUCTION

A recurrent challenge many software organizations face is to have a clear establishment of a measurement and evaluation of a conceptual framework useful for quality assurance processes and programs. While many useful approaches for and successful practical examples of software measurement programs exist, the inability to clearly and consistently specify measurement and evaluation concepts (i.e., the meta-data) could unfortunately hamper the progress of the software, and Web Engineering as a whole, and could hinder their widespread adoption.

Software and Web organizations introducing a measurement and evaluation program—maybe as part of a measurement and analyses process area and quality assurance strategy (CMMI, 2002)—need to establish a set of activities and procedures to specify, collect, store, and use trustworthy measurement and indicator data sets and meta-data. Moreover, to ensure, for analysis purposes, that measurement and indicator data sets are repeatable and comparable among different measurement and evaluation projects, appropriate meta-data of metrics and indicators should be adapted and recorded.

Therefore, in the present chapter we argue that at least three pillars are necessary to build, i.e., to design and to implement, a robust and sound measurement and evaluation program:

1. a process for measurement and evaluation, i.e., the main managerial and technical activities that might be planned and performed
2. a measurement and evaluation framework that must rely on a sound conceptual (ontological) base
3. specific model-based methods and techniques in order to carry out the specific project's activities

A measurement or evaluation process prescribes or informs a set of main phases, activities, and their input and output that might be considered. Usually, it says what to do but not how to do it; that is, it says nothing about the particular methods and tools in order to perform the specific activities' descriptions. Regarding measurement and evaluation processes for software, the *International Standard Organization* (ISO) published two standards: the ISO 15939 document issued in 2002 (ISO, 2002), which deals with the software measurement process, and the ISO 14598-5 issued in 1998 (ISO, 1998), which deals with the process for evaluators in its part 5. On the other hand, the CMMI (*Capability Maturity Model Integration*) initiative is also worthy of mention as another source of knowledge, in which specific support process areas such as measurement and analyses, decision analyses and resolution, among others, are specified. The primary aim of these documents was to reach a consensus about the issued models, processes, and practices. However, in Olsina and Martin (2004) we observe that very often a lack of consensus exists about the used terminology among the ISO standards.

Considering our second statement, we argue that in order to design and implement a robust measurement and evaluation program, a sound measurement and evaluation conceptual framework is necessary. Very often organizations start measurement programs from scratch more than once because they did not pay too much attention to the way metrics and indicators should be designed, recorded, and analyzed.

A well-established framework has to be built on a sound conceptual base, that is, on an ontological base. In fact, an ontology explicitly and formally specifies the main concepts, properties, relationships, and axioms for a given domain. In this direction, we have built an explicit specification of measurement and indicator meta-data, i.e., an ontology for this domain (Olsina and Martin, 2004). The sources of knowledge for this ontology stemmed from different software-related ISO standards (ISO, 1999, 2001, 2002) and recognized research articles and books (Briand et al., 2002; Kitchenham et al., 2001; Zuse, 1998), in addition to our own experience backed up by previous works on metrics and evaluation processes and methods (Olsina et al., 1999; Olsina and Rossi, 2002).

However, the metrics and indicators ontology itself is not sufficient to model a full-fledged measurement and evaluation framework but rather is the ground and rationale to building it. In Olsina et al. (2006b), the INCAMI framework (Olsina et al., 2005) is thoroughly analyzed in the light of its ontological roots. INCAMI is an organizational purpose-oriented measurement and evaluation framework that enables consistently saving not only meta-data of metrics and indicators but also values (data sets) for concrete real-world measurement and evaluation projects. It is made up of five main conceptual components, namely: the requirement, measurement, and evaluation of projects definition; the nonfunctional requirements definition and specification; the measurement design and execution; the evaluation design and execution; and the conclusion and recommendation components. We argue that this framework can be useful for different qualitative and quantitative evaluation methods and techniques with regard to the requirements, measurement, and evaluation concepts and definitions (Olsina et al., 2008).

On the other hand, the growing importance the Web currently plays in such diverse application domains as business, education, government, industry, and entertainment have heightened concerns about the quality and quality of delivered Web applications. It is necessary to have not only robust development methods to improve the building process (one of the main aims of this book) but also consistent ways to measure and evaluate intermediate and final products as well. In this sense measurement and evaluation methods and tools that are grounded on the quoted conceptual framework are the third pillar of our proposal.

There are different categories of methods (e.g., inspection, testing, inquiry, simulation, etc.) and specific types of evaluation methods and techniques such as the heuristic evaluation technique (Nielsen et al., 2001), the Web Quality Evaluation Method (WebQEM) (Olsina and Rossi, 2002) as a concept model-centered evaluation methodology for the inspection category, to name just a few. We argue that a method or technique is usually not enough to assess different information needs for diverse evaluation purposes. In other words, it is true that one size does not fit all needs and preferences, but an organization might at least adopt a method or technique in order to know the state of its quality and quality in use for understanding and improving purposes.

In order to illustrate the above three main points, this chapter is organized as follows. In Section 13.2 we present an abridged overview of the state-of-the-art of measurement and evaluation processes as well as a basic process that is akin to our framework. In Section 13.3 we analyze the main components of the INCAMI framework regarding the metrics and indicators ontological base; at the same time, as proof of these concepts, an external quality model to measure and evaluate the shopping cart component of a typical e-commerce site is employed. In Section 13.4, using the specific models, procedures, and processes, the WebQEM inspection methodology is illustrated with regard to the previous case study. Finally, additional discussions about the flexibility of the framework as well as concluding remarks are drawn in Section 13.5.

13.2 OVERVIEW OF MEASUREMENT AND EVALUATION PROCESSES

As previously mentioned, a measurement or evaluation process specifies a set of main phases, activities, their input and output, and sometimes control points that might be considered. Usually, a process says what to do but not how to do it.

For instance, the ISO 14598-5 standard prescribes an evaluation process to assess software quality which is a generic abstract process customizable for different evaluation needs; however, it does not prescribe or inform about specific evaluation methods and tools in order to perform the activities' descriptions.

On the other hand, it is important to remark that no unique ISO standard that integrates in one document the measurement and evaluation process as a whole exists. Instead, there are two separate standards: one for the evaluation process, issued in 1998 (ISO, 1998), and another for the measurement process, issued in 2002 (ISO, 2002). Regarding the former, in an introductory paragraph it says, "The primary purpose of software product evaluation is to provide quantitative results concerning software product quality that are comprehensible, acceptable to and can be dependable on by any interested party"; it continues, "This evaluation process is a generic abstract process that follows the model defined in ISO/IEC 9126."

In the ISO 14598-5 standard, the evaluation process comprises the five activities listed in Figure 13.1 (see ISO, 1998, for a detailed description):

1. *establishment of evaluation requirements*
2. *specification of the evaluation* based on the evaluation requirements and on the product provided by the requester
3. *design of the evaluation,* which produces an evaluation plan on the basis of the evaluation specification
4. *execution of the evaluation plan,* which consists of inspecting, modeling, measuring, and testing the products and/or its components according to the evaluation plan
5. *conclusion of the evaluation,* which consists of the delivery of the evaluation report

Figure 13.1. The main activities specified in the ISO 14598-5 evaluation process standard.

The ISO 15939 standard that deals with the measurement process says, "Software measurement is also a key discipline in evaluating the quality of software products and the capability of organizational software processes"; in addition,

> Continual improvement requires change within the organization. Evaluation of change requires measurement. Measurement itself does not initiate change. Measurement should lead to action, and not be employed purely to accumulate data. Measurement should have a clearly defined purpose. . . , This standard defines the activities and tasks necessary to implement a software measurement process ... each activity is comprised of one or more tasks. This International Standard does not specify the details of how to perform the tasks included in the activities.

In this standard two activities (out of four) are considered to be the core measurement process, namely: *plan the measurement process*, and *perform the measurement process*. These two activities are comprised of the following tasks (see Figure 13.2 and also ISO, 2002, for a detailed description):

1. Plan the Measurement Process:
   - *1.1 Characterize organizational unit*
   - *1.2 Identify information needs*
   - *1.3 Select measures*
   - *1.4 Define data collection, analysis, and reporting procedures*
   - *1.5 Define criteria for evaluating the information products and the measurement process*
   - *1.6 Review, approve, and provide resources for measurement tasks*
   - *1.7 Acquire and deploy supporting technologies*
2. Perform the Measurement Process:
   - *2.1 Integrate procedures*
   - *2.2 Collect data*
   - *2.3 Analyze data and develop information products*
   - *2.4 Communicate results*

Figure 13.2. The two core measurement processes specified in the ISO 15939 measurement process standard.

Lastly, the CMMI (CMMI, 2002) initiative[1] is also worthy of mention. This initiative specifies support process areas such as *measurement and analyses*, among others. It says, "The purpose of measurement and analysis is to develop and sustain a measurement capability that is used to support management information needs", Figure 13.3 shows the two specific goals

[1] There is a related ISO 15504 initiative named SPICE (*Software Process Improvement and Capability dEtermination*).

for this process area and its specific practices (which can be considered as activities or specific actions).

1 Align Measurement and Analysis Activities
   1.1 Establish measurement objectives
   1.2 Specify measures
   1.3 Specify data collection and storage procedures
   1.4 Specify analysis procedures
2 Provide Measurement Results
   2.1 Collect measurement data
   2.2 Analyze measurement data
   2.3 Store data and results
   2.4 Communicate results

Figure 13.3. The two specific goals and related practices for the CMMI Measurement and Analyses process area.

As the reader could observe in the previous figures, there is in principle no clear integrated proposal about measurement and evaluation activities even though both are closely intertwined, as we discuss in our framework later on. However, a common denominator between activities and tasks outlined in the previous figures can be observed. For instance, there are the definition and specification of requirements, e.g., activities 1 and 2 in Figure 13.1 deal with the establishment and specification of evaluation requirements; tasks 1.1 and 1.2 in Figure 13.2 are about measurement requirements, as is practice 1.1 in Figure 13.3. There are also design activities, i.e., defining, specifying, or ultimately planning activities; then, execution or implementation activities of the designed evaluation or measurement; and lastly, activities about the conclusion and communication of results.

On the other hand, we have been developing the WebQEM methodology since the late 1990s (Olsina et al., 1999; Olsina and Rossi, 2002). The underlying WebQEM process integrates activities for requirements, measurement, evaluation, and recommendations. Figure 13.5 shows the evaluation process, including the phases, main activities, input, and output. This model followed to some extent the ISO's process model for evaluators (ISO, 1998). The main activities are grouped into the following four major technical phases (see Figure 13.4):

1. *Nonfunctional Requirements Definition and Specification*
2. *Measurement and Elementary Evaluation* (both Design and Implementation stages)
3. *Global Evaluation* (both Design and Implementation stages)
4. *Conclusion and Recommendations*

Figure 13.4. The four phases underlying the WebQEM methodology and the INCAMI framework. Note that the specific activities are not listed in the figure.

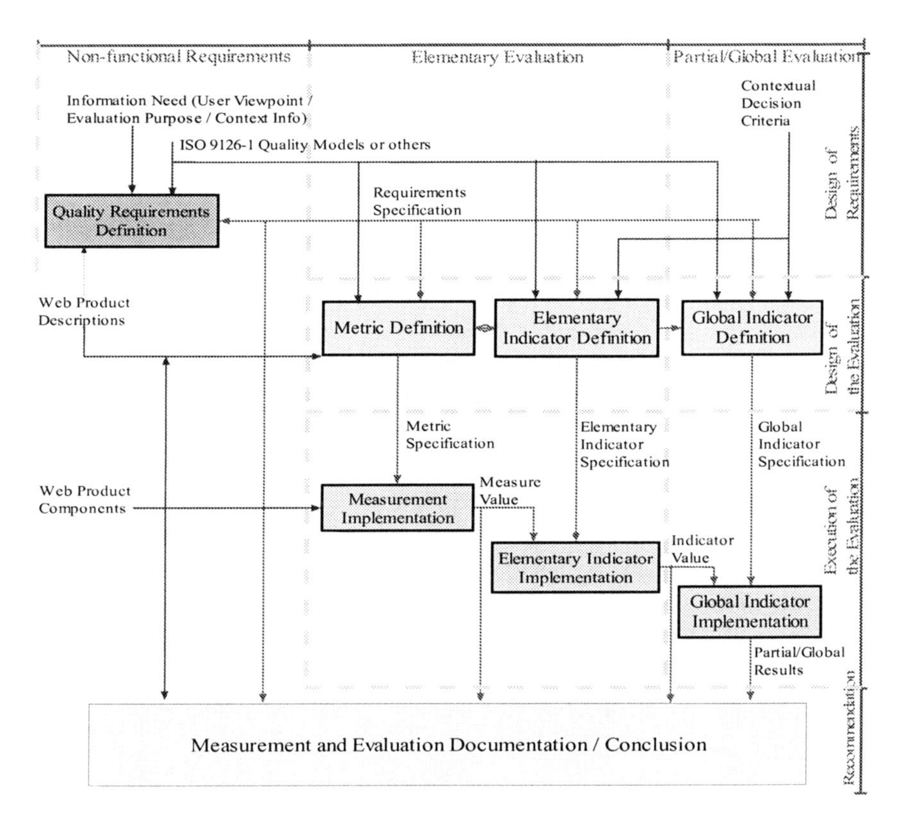

Figure 13.5. The basic measurement and evaluation process underlying the WebQEM methodology. The technical phases, main activities, and their input and output are represented (it might be assumed that some activities are iterative).

In the next section we thoroughly discuss the measurement and evaluation framework (the second pillar proposed in Section 13.1) in the light of its conceptual root and the above measurement and evaluation process. As an additional remark, in Olsina and Martin (2004) we observed that very often there is a lack of consensus about the used terminology among the quoted ISO standards, and some terms used mainly for the evaluation domain are missing.

13.3 FRAMEWORK FOR MEASURING AND EVALUATING NONFUNCTIONAL REQUIREMENTS

The proposed INCAMI (*Information Need*, *Concept model*, *Attribute*, *Metric,* and *Indicator*) framework (Molina et al., 2004; Olsina et al., 2005) is based upon the assumption that for an organization to measure and evaluate in a purpose-oriented way it must first specify nonfunctional requirements starting from information needs, then it must design and select the specific set of useful metrics for measurement purpose, and lastly it must interpret the metrics values by means of contextual indicators with the aim of evaluating or estimating the degree to which the stated requirements have been met and, ultimately, to draw conclusions and give recommendations.

As aforementioned, the conceptual framework is made up of five main components: the nonfunctional requirements definition and specification; the measurement design and execution; the evaluation design and execution; the conclusion and recommendation component; and the project definition itself. Currently, most of the components are supported by many of the ontological concepts, properties, and relationships defined in previous works (Olsina and Martin, 2004). For instance, to the nonfunctional requirements definition component, concepts such as *Information Need*, *Calculable Concept*, *Concept Model*, *Entity*, *Entity Category,* and *Attribute* intervene (all these terms are defined and illustrated in Section 13.3.4.1). Some other concepts were added to the framework in order to design and implement it as a Web application (the INCAMI_Tool).

In Sections 13.3.1 to 13.3.3 we give an abridged description of the first three components listed above. In Section 13.3.4 we thoroughly discuss the main terms for these components; in addition, each term is illustrated using as an example the external quality model to assess the shopping cart feature of the www.amazon.com site.

13.3.1 Information Need, Concept Model, and Attribute

First, for the nonfunctional requirements definition and specification component, the *Information Need* to a measurement and evaluation *Project* must be agreed upon. Information need is defined as the insight necessary to manage objectives, goals, risks, and problems. Usually, information needs come from two organizational project-level sources: goals that decision makers seek to achieve, or obstacles that hinder reaching the goals; e.g., obstacles involve basically risks and problems. The *InformationNeed* class (see Figure 13.6) has three properties: the *purpose*, the user *viewpoint*, and the *contextDescription*. (Note that from the process standpoint, outlined in the previous section, and particularly for the *Nonfunctional Requirements Definition and Specification* phase, we can represent an activity named *Identify Information Needs* and in turn tasks such as *Establish measurement/evaluation purpose*; *Establish the user viewpoint*; and *Specify the context of the measurement/evaluation.*)

Additionally, the *InformationNeed* class has two main relationships with the *CalculableConcept* and the *EntityCategory* classes, respectively. A calculable concept can be defined as an abstract relationship between attributes of entities' categories and information needs; in fact, internal quality, external quality, cost, etc. are instances of a calculable concept. In turn, a calculable concept can be represented by a *ConceptModel*; for example, ISO 9126-1 specifies quality models for the internal quality, external quality, and quality in use, respectively.

On the other hand, a common practice is to assess quality by means of the quantification of lower abstraction concepts such as *Attributes* of entities' categories. The attribute term can be defined in brief as a measurable property of an *EntityCategory* (e.g., categories of entities of interest to software and Web Engineering are resource, process, product, service, and project as a whole). An entity category may have many attributes, though only some of them may be useful just for a given measurement and evaluation project's information needs.

In summary, this component allows the definition and specification of nonfunctional requirements in a sound and well-established way. It has an underlying organizational strategy that is purpose-oriented by information needs and is concept model-centered and evaluator-driven by domain experts and users.

13.3.2 Metrics and Measurement

Regarding the measurement component, purposeful metrics should be selected in the process. In general, each attribute can be quantified by one or more metrics, but in practice just one metric should be selected for each attribute of the requirements tree, given a specific measurement project.

The *Metric* concept contains the definition of the selected *Measurement* or *Calculation Method* and the *Scale* (see Figure 13.8). For instance, the measurement method is defined as the particular logical sequence of operations and possible heuristics specified for allowing the realization of a metric description by a measurement; while the scale is defined as a set of values with defined properties. Thus, the metric m represents a mapping $m: A \rightarrow X$, where A is an empirical attribute of an entity category (the empirical world), X is the variable to which categorical or numerical values can be assigned (the formal world), and the arrow denotes a mapping. In order to perform this mapping, a sound and precise measurement activity definition is needed by explicitly specifying the metric's method and scale. We can apply an *objective* or *subjective* measurement method for *Direct Metrics*; conversely, we can perform a calculation method for *Indirect Metrics*, that is, when a *Formula* intervenes.

Once the metric has been selected, we can perform (execute or implement) the measurement process, i.e., the activity that uses a metric definition in order to produce a measure's value (see Figure 13.5). The *Measurement* class allows the date/time stamp, the information of the owner in charge of the measurement activity, and the actual or estimated yielded value to be recorded.

However, since the value of a particular metric will not represent the elementary requirement's satisfaction level, we need to define a new mapping that will produce an elementary indicator value. One fact worthy of mention is that the selected metrics are useful for a measurement process as long as the selected indicators are useful for an evaluation process in order to interpret the stated information need.

13.3.3 Indicators and Evaluation

For the evaluation component, contextual indicators should be selected. Indicators are ultimately the foundation for the interpretation of information needs and decision making. There are two types of indicators: *elementary* and *global indicators* (see Figure 13.9).

In Olsina and Martin (2004) the indicator is described as "the defined calculation method and scale in addition to the model and decision criteria in order to provide an estimate or evaluation of a calculable concept with respect to defined information needs." In particular, we define an elementary indicator as one that does not depend upon other indicators to evaluate or estimate a concept at a lower level of abstraction (i.e., for associated attributes to a concept model). On the other hand, we define a partial or global indicator as one that is derived from other indicators to evaluate or estimate a concept at a higher level of abstraction (i.e., for subconcepts and concepts). Therefore, the elementary indicator represents a new mapping coming from the interpretation of the metric's measured value of an attribute (the formal world) into the new variable to which categorical or numerical values can be assigned (the new formal world). In order to perform this mapping, elementary and global model and decision criteria for a specific user information need should be designed.

Therefore, once we have selected a scoring model, the aggregation process follows the hierarchical structure defined in the concept model, from bottom to top. Applying a stepwise aggregation mechanism, we obtain a global schema; this model lets us compute partial and global indicators in the execution stage. The global indicator's value ultimately represents the global degree of satisfaction in meeting the stated requirements (information need) for a given purpose and user viewpoint.

13.3.4 Definition and Exemplification of the INCAMI Terms

In this section (from Sections 13.3.4.1 to 13.3.4.3) we define the main terms that intervene in the above INCAMI framework's components, i.e., the requirement, measurement, and evaluation components. Each one is modeled by a class diagram (Figures 13.6, 13.8, and 13.9), where many (but not all) terms in the diagrams come from the metrics and indicators ontology. Note that for space reasons, we do not define each class attribute and relationships among classes, as is done in Olsina and Martin (2004).

In addition, for illustration purposes, we use an external quality model with associated attributes specified to the shopping cart of Web sites. We have conducted a case study in order to assess the shopping cart feature of the www.amazon.com site (details of this study will be given in Section 13.4).

13.3.4.1 Requirements Definition and Specification Model

As shown in Figure 13.6, this model includes all the necessary concepts for the definition and specification of requirements for measurement and evaluation projects. Nonfunctional requirements are the starting point of the measurement and evaluation process, so that a *requirement project* should be defined.

Definition 13.1. *RequirementProject* is a project that allows us to specify nonfunctional requirements for measurement and evaluation activities.

In our example the project *name* is "ExternalQuality_Amazon_05"; the *description* is "requirements for evaluating the external quality for the shopping cart feature of the www.amazon.com site"; with a starting date "2005/12/19" and an ending date "2005/12/30" and in charge of "Fernanda Papa" with the "pmfer@ing.unlpam.edu.ar" *contact* email.

Next, the *information need* should be specified. For this study, a basic information need may be "understand the external quality of the shopping cart component of a typical e-store, for a general visitor viewpoint, in order to incorporate the best features in a new e-bookstore development project."

Definition 13.2. *InformationNeed* is the insight necessary to manage objectives, goals, risks, and problems.

In our example the *information need* is stated by the *purpose* (i.e., to understand), the *user viewpoint* (i.e., a general visitor), in a given *context* of use (e.g., bandwidth constraints, among other contextual descriptions). In addition, an *entity category*, which is the *object* under analysis, and the *calculable concept*, which is the *focus* of *the information need,* must be defined.

Definition 13.3. *Entity Category* is the object category that is to be characterized by measuring its attributes.

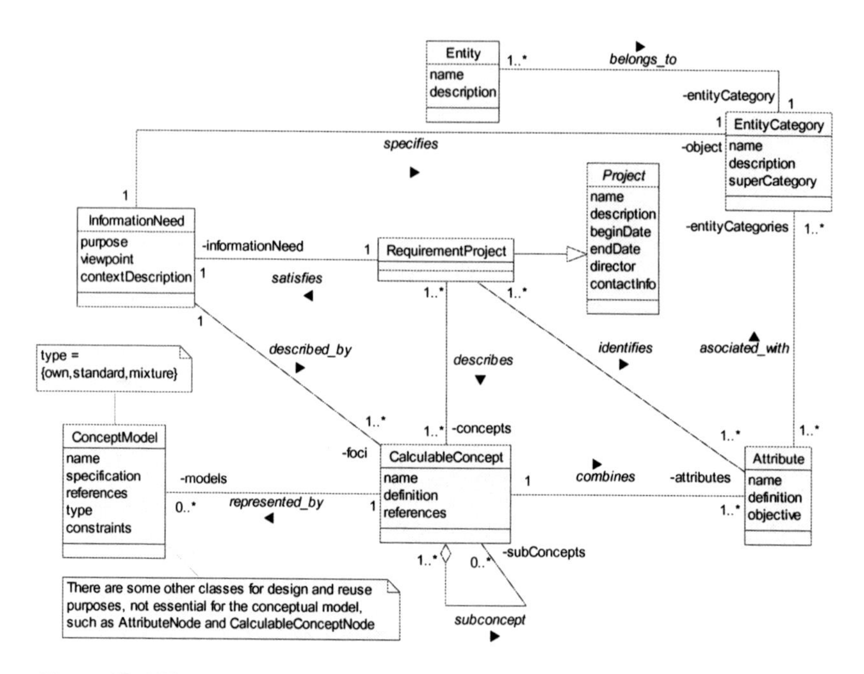

Figure 13.6. Key terms and relationships that intervene in the INCAMI requirements component for the definition and specification of nonfunctional requirements.

Definition 13.4. *Entity,* synonym Object, is a concrete object that belongs to an entity category.

Therefore, given the *entity category* (i.e., an e-commerce application, of which *superCategory* is a product), a concrete object *name* that belongs to this category is the "Amazon's shopping cart" Web component.

Definition 13.5. *CalculableConcept,* synonym Measurable Concept in ISO (2002), defines the abstract relationship between attributes of entity categories and information needs.

In the example the *calculable concept name* is "external quality" and its *definition* is "the extent to which a product satisfies stated and implied needs when used under specified conditions" (ISO, 1999). The external quality concept has *subconcepts* such as "usability", "functionality", "reliability", "efficiency", "portability", and "maintainability".

For instance, the "functionality" subconcept is defined in ISO (2001) as "the capability of the software product to provide functions which meet stated and implied needs when the software is used under specified conditions". In turn, the calculable concept (characteristic) "functionality" is split into five subconcepts (subcharacteristics): "suitability", "accuracy", "interoperability", "security", and "functionality compliance." Suitability is defined as "the capability of the software product to provide an appropriate set of functions for specified tasks and user objectives"; and accuracy as "the capability of the software product to provide the right or agreed results or effects with the needed degree of precision." See Figure 13.7, where these two subconcepts in the requirements tree are included as "Function Suitability" and "Function Accuracy", respectively (we used the name "function suitability" instead of "suitability" alone, in order to distinguish it from the name "information suitability", which is a subconcept of the Content characteristic).

On the other hand, the calculable concept can be *represented by* a *concept model.*

Definition 13.6. *ConceptModel,* synonym Factor or Feature Model, is the set of subconcepts and the relationships between them, which provide the basis for specifying the concept requirement and its further evaluation or estimation.

As mentioned earlier, INCAMI is a concept model-centered approach; the concept model *type* can be either a standard-based model or an organization own-defined model, or a mixture of both. The concept model used in the example is of the "mixture" *type* that is based mainly on the ISO external quality model (*reference* "(ISO, 1999)"), and the *specification* is shown in Figure 13.11 (note that the model also shows *attributes* combined to the *subconcepts*).

Definition 13.7. *Attribute,* synonym Property, Feature, is a measurable physical or abstract property of an entity category.

Note that the selected attributes are those properties relevant to the agreed-upon information need. The abridged representation in Figure 13.7 shows attribute *names* such as "Capability to delete items" (2.1.2) and "Precision to recalculate after deleting items" (2.2.2), among others.

2. Functionality
- 2.1. Function Suitability
  - *2.1.1. Capability to add items from anywhere*
  - *2.1.2. Capability to delete items*
  - *2.1.3. Capability to modify an item quantity*
  - *2.1.4. Capability to show totals by performed changes*
  - *2.1.5. Capability to save items for later/move to cart*
- 2.2. Function Accuracy
  - *2.2.1. Precision to recalculate after adding an item*
  - *2.2.2. Precision to recalculate after deleting items*
  - *2.2.3. Precision to recalculate after modifying an item quantity*

Figure 13.7. An excerpt (taken from Figure 13.11) of an instance of the external quality model with associated attributes specified for measurement and evaluation of the shopping cart component; for instance, the 2.1 and 2.2 codes represent specific calculable concepts and subconcepts; and the rest (in italic) are associated attributes to the above subconcepts. The model as a whole is depicted as a requirements tree.

For instance, the "Capability to delete items" attribute is defined (see the field *definition* in the Attribute class in Figure 13.6) as "the capability of the shopping cart to provide functions in order to delete appropriately items one by one or to the selected group at once."

The INCAMI_Tool, which is a prototype tool that supports this framework, currently implements concept models in the form of requirements trees. It also allows partially or totally previously edited requirements trees to be imported for a new project.

13.3.4.2 Measurement Design and Execution Model

The measurement model (see Figure 13.8) includes all the necessary concepts for the design and implementation of the measurement as a part of the *Measurement and Elementary Evaluation* phase shown in Figure 13.4. First, a *measurement project* should be defined.

Definition 13.8. *MeasurementProject* is a project that allows us, starting from a requirement project, to select the metrics and record the values in a measurement process.

Once the measurement project has been created, with similar information as that of a requirement project, the attributes in the requirements tree can be quantified by *direct* or *indirect metrics*.

Consider that for a specific measurement project just one metric should be selected for each attribute of the concept model. In the INCAMI_Tool, each metric is selected from a catalogue (Molina et al., 2004).

On the other hand, note that many measurement projects can rely on the same requirements, for instance, in a longitudinal analysis. In this case the starting and ending dates should change for each project.

Definition 13.9. *Metric*[2] is the defined measurement or calculation method and the measurement scale.

Definition 13.10. *DirectMetric*, synonym Single, Base Metric, is a metric of an attribute that does not depend on a metric of any other attribute.

---

[2] The "metric" term is used in ISO (1999, 2001) but not in ISO (2002). Furthermore, ISO (1999, 2001) uses the terms "direct measure" and "indirect measure" (instead of "direct" or "indirect metric"), while ISO (2002) uses "base measure" and "derived measure." In some cases we could state that they are synonymous terms, but in others such as "metric", which is defined in ISO (1999) as "the defined measurement method and the measurement scale", there is no term with exact matching meaning in ISO (2002). Furthermore, we argue that the measure term is not synonymous with the metric term. The measure term is defined in ISO (1999) (the meaning we adopted) as "the number or category assigned to an attribute of an entity by making a measurement" or in ISO (2002) as the "variable to which a value is assigned as the result of measurement" reflects the fact of the measure as the resulting value or output for the measurement activity (or process). Thus, we argue that the metric concept represents the specific and explicit definition of the measurement activity.

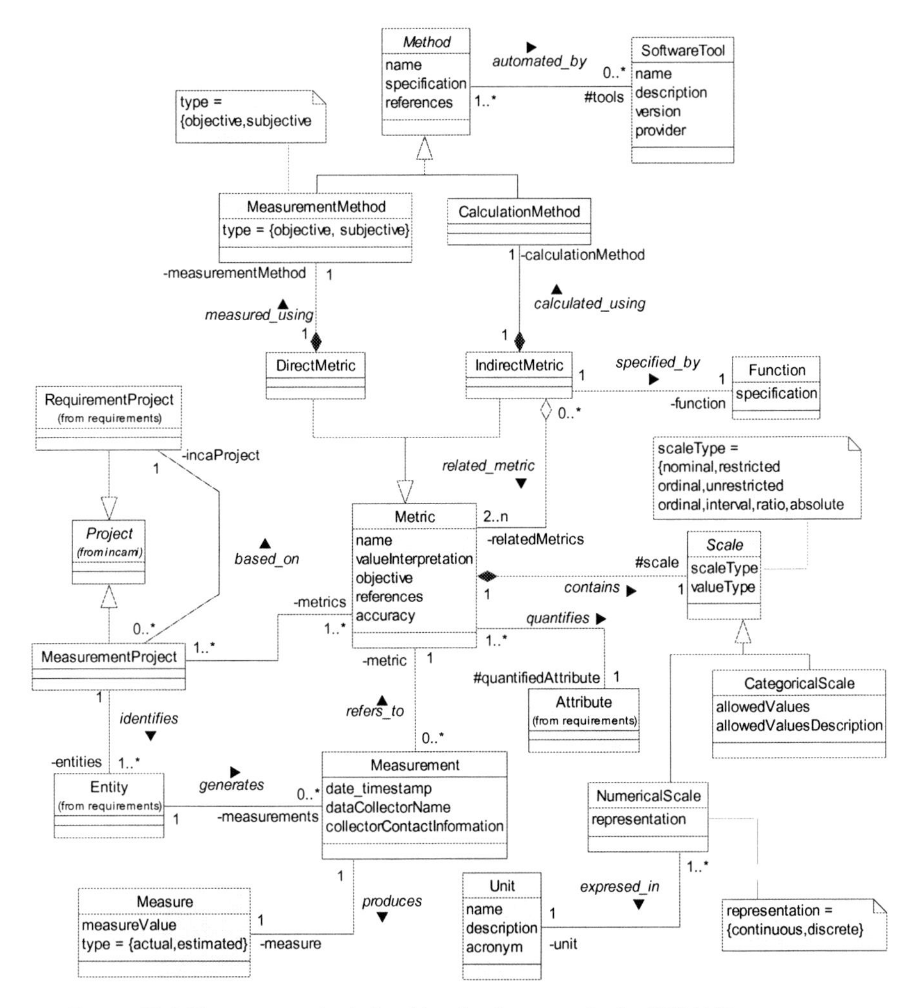

Figure 13.8. Key terms and relationships that intervene in the INCAMI measurement component for the definition of metric and measurement concepts.

For example, to the "Capability to delete items" attribute (coded 2.1.2 in Figure 13.7) we designed a direct metric named "Degree of the capability to delete items" that specifies four categories, namely:

0. Does not delete items at all
1. Delete just all at once
2. Delete one by one
3. Delete one by one or delete the selected group at once

Definition 13.11. *IndirectMetric,* synonym Hybrid, Derived Metric, is a metric of an attribute that is derived from the metrics of one or more other attributes.

Definition 13.12. *Function,* synonym Formula, Algorithm, Equation, is an algorithm or formula performed to combine two or more metrics.

There are two key terms in Definition 13.9: *Method* and *Scale*. For the latter, two types of scales have been identified: *Categorical* and *Numerical Scales:*

Definition 13.13. *Scale* is a set of values with defined properties.

The type of scales (*scaletype* attribute in the Scale class in Figure 13.8) depends on the nature of the relationship between values of the scale. The types of scales commonly used in software and Web Engineering are classified into nominal, ordinal (both restricted and unrestricted), interval (and quasi-interval), ratio, and absolute. The scale type[3] of measured and calculated values affects the sort of arithmetical and statistical operations that can be applied to values, as well as the admissible transformations among metrics.

Definition 13.14. *CategoricalScale* is a scale where the measured or calculated values are categories and cannot be expressed in units, in a strict sense.

Definition 13.15. *NumericalScale* is a scale where the measured or calculated values are numbers that can be expressed in units, in a strict sense.

Definition 13.16. *Unit* is a particular quantity defined and adopted by convention, with which other quantities of the same kind are compared in order to express their magnitude relative to that quantity.

The *scale type* of the above direct metric (see the example in Definition 13.10) is "ordinal" represented by a *categorical scale* with a "symbol" *value type.* The *allowedValues* for the ordinal categories are from 0 to 3, and the *allowedValuesDescription* are the names of the categories such as "Delete just all at once." Note that because the type of the scale is ordinal, a mapping of categories to numbers can be made, whereas the order is preserved.

As stated earlier, two key terms appear in the metric definition: *method* and *scale*. In the sequel, the method-related terms are defined.

Definition 13.17. *Method,* synonym Procedure, is a logical sequence of operations and possible heuristics, specified generically, for allowing the realization of an activity description.

Definition 13.18. *SoftwareTool,* synonym Software Instrument, is a tool that partially or totally automates a measurement or calculation method.

For example, the INCAMI_Tool, the current prototype tool that supports the WebQEM methodology, allows us to calculate indirect metrics (from direct metrics and parameters) in addition to calculating elementary and global indicators from elementary and global models. A previous tool for WebQEM was the WebQEM_Tool (Olsina et al., 2001). Different

[3] See a deeper discussion about type of scales in Chapter 14, Section 14.2.

commercial tools for data collection of direct metrics are widely well known and available for download.

Definition 13.19. *MeasurementMethod,* synonym Counting Rule, Protocol, is the particular logical sequence of operations and possible heuristics specified for allowing the realization of a metric description by a measurement.

To the exemplified direct metric (see the example in Definition 13.10), the counting rule was clearly specified as well as the measurement method *type*. The type of method can be either "subjective" i.e., where the quantification involves human judgment, or "objective" i.e., where the quantification is based on numerical rules. Generally, an objective measurement method type can be automated or semiautomated by a software tool. Nevertheless, for our example of a direct metric, even though the type is objective, no tool can automate the collection of data, and so a human must perform the task.

Definition 13.20. *CalculationMethod* is the particular logical sequences of operations specified for allowing the realization of a formula or indicator description by a calculation.

Definition 13.21. *Measurement* is an activity that uses a metric definition in order to produce a measure's value.

Definition 13.22. *Measure* is the number or category assigned to an attribute of an entity by making a measurement.

A *measurement* activity must be performed for each metric that intervenes in the project. It allows the *date/time stamp*, the *collector information* in charge of the measurement activity, and the *measure,* the "actual" or "estimated" value *type,* and the yielded *value* itself to be recorded.

Ultimately, for a specific measurement project, at least all the above concepts and definitions of the measurement model are necessary in order to specify, collect, store, and use trustworthy metrics' values and meta-data.

13.3.4.3 Evaluation Design and Execution Model

As introduced in Section 13.3.2, the value of a particular metric will not represent the elementary requirement's satisfaction level. Thus, we need to define a new mapping that will produce an elementary indicator value.

As aforementioned, the selected metrics are useful for designing and performing the measurement process as long as the selected indicators are useful for designing and executing the evaluation process for the stated information need, which is represented specifically in the concept model. The main concepts involved in the elementary and global evaluation are depicted in the model in Figure 13.9.

Definition 13.23. *EvaluationProject* is a project that allows us, starting from a measurement project and a concept model of a requirement project, to select the indicators and perform the calculations in an evaluation process.

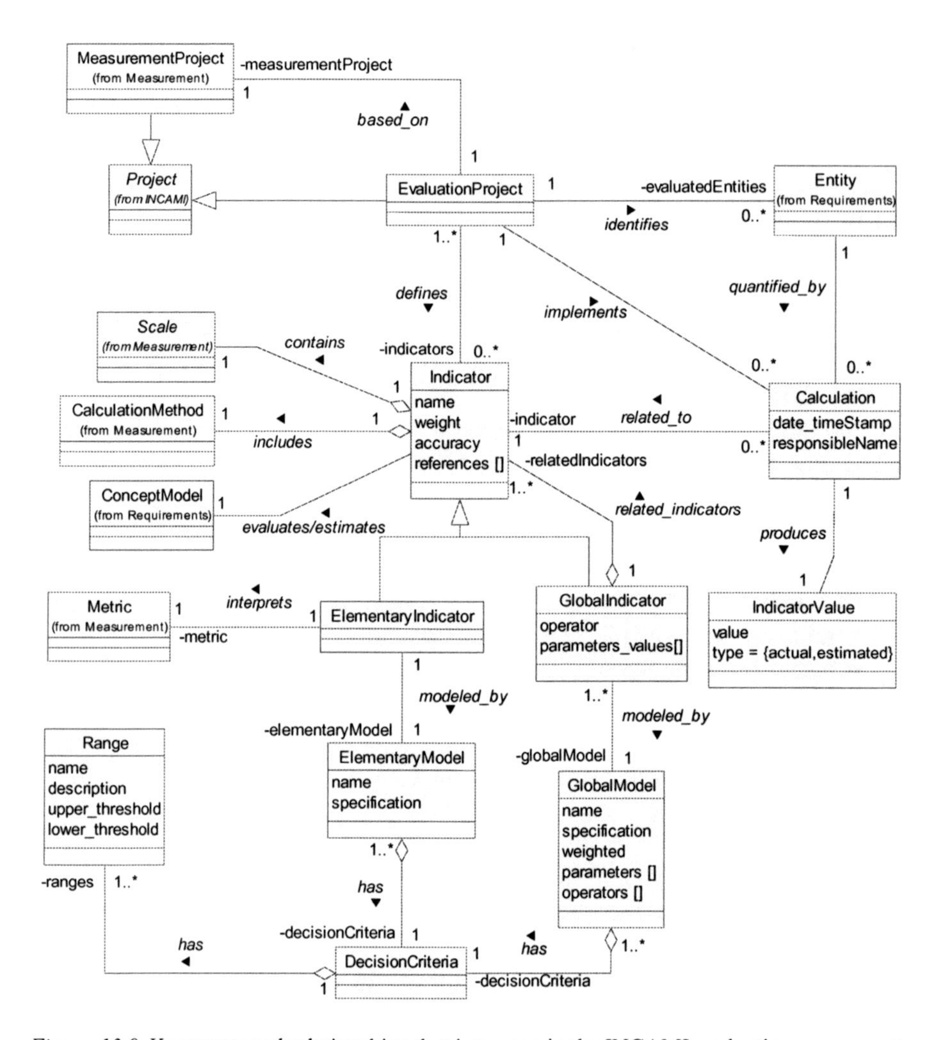

Figure 13.9. Key terms and relationships that intervene in the INCAMI.evaluation component for the definition of indicators and related concepts.

Once a measurement project has been created, one or more evaluation projects can in turn be created, relying on the recorded measurement values and meta-data, by adding related information with *indicators*.

Definition 13.24. *Indicator,* synonym Criterion, is the defined calculation method and scale in addition to the model and decision criteria in order to provide an estimate or evaluation of a calculable concept with respect to defined information needs.

Definition 13.25. *ElementaryIndicator,* synonym Elementary Preference, Elementary Criterion, is an indicator that does not depend upon other indicators to evaluate or estimate a calculable concept.

Therefore, an elementary indicator for each attribute of the concept model, i.e., for each leaf of the requirements tree, can be defined. For instance, to the 2.1.2 attribute of Figure 13.7, the *name* of the elementary indicator is "Performance Level of the Capability to Delete Items" (CDI_PL).

The elementary indicator interprets the metric's value of the attribute. To this end, an *elementary model* is needed.

Definition 13.26. *ElementaryModel,* synonym Elementary Criterion Function, is an algorithm or function with associated decision criteria that model an elementary indicator.

The *specification* of the elementary model can look like this: CDI _PL = (0.33 * CDI) * 100; where CDI is the direct metric for the *Capability to Delete Items* attribute (see Definition 13.10).

Note that, like a metric, an indicator has a *Scale* (see Definition 13.13). To the above example we considered a *numerical scale* where the *Unit* (see Definition 13.16) can be a normalized "percentage" scale. As mentioned, the elementary indicator interprets the metric's value of an attribute (an attribute as an elementary requirement). Then, the above elementary model interprets the percentage of the satisfied elementary requirement.

Definition 13.27. *DecisionCriteria,* synonym Acceptability Levels, are the thresholds, targets, or patterns used to determine the need for action or further investigation, or to describe the level of confidence in a given result.

Definition 13.28. *Range* is the threshold or limit values that determine the acceptability levels.

The decision criteria that a model of an indicator may have are the agreed-upon acceptability levels in given ranges of the scale; for instance, it is "unsatisfactory" if the *range* (regarding *lower_threshold* and *upper_threshold*) is "0 to 45", respectively; "marginal" if it is "greater than 45 and less or equal than 70"; otherwise, "satisfactory." A *description* or interpretation for "marginal" is that a score within this range indicates a need for improvement. An "unsatisfactory" rating means change actions must take high priority.

Definition 13.29. *GlobalIndicator,* synonym Global Preference, Global Criterion, is an indicator derived from other indicators to evaluate or estimate a calculable concept.

Definition 13.30. *GlobalModel,* synonym Aggregation Model, Scoring Model, or Function, is an algorithm or function with associated decision criteria that model a global indicator.

In order to enact the concept model (see Definition 13.6) for elementary, partial, and global indicators, an *aggregation model* and *decision criteria* must be selected. The quantitative aggregation and scoring models aim at making the evaluation process well structured, objective, and comprehensible to evaluators. For example, if our procedure is based on a "linear additive scoring model," the aggregation and computing of partial/global indicators (P/GI), considering relatives *weights* (W), is based on the following *specification*:

$$\text{P/GI} = (W_1\,\text{EI}_1 + W_2\,\text{EI}_2 + \ldots + W_m\,\text{EI}_m); \tag{13.1}$$

such that if the elementary indicator (EI) is in the percentage scale and unit, the following holds:

$$0 \leq \text{EI}_i \leq 100;$$

and the sum of weights for an aggregation block must fulfill

$$(W_1 + W_2 + \ldots + W_m) = 1$$

if $W_i > 0$; for i = 1, . . ., m, where m is the number of subconcepts at the same level in the tree's aggregation block (see Figure 13.11).

The basic arithmetic aggregation *operator* for input in Eq. (13.1) is the plus (+) connector. Besides, this model lets us compute partial and global indicators in the execution stage. Other nonlinear aggregation models or functions can be used such as logic scoring of preference (Dujmovic, 1996), fuzzy model, and neural models, among others.

Definition 13.31. *Calculation,* synonym Computation, is an activity that uses an indicator definition in order to produce an indicator's value.

Definition 13.32. *Indicator Value,* synonym Preference Value, is the number or category assigned to a calculable concept by making a calculation.

As a final remark, for a specific evaluation project, all the above concepts and definitions of the evaluation model are necessary in order to specify, calculate, store, and use trustworthy indicator values and meta-data. When the execution of the measurement and evaluation activities for a given project has been performed, decision makers can analyze the results and draw conclusions and recommendations with regard to the established information need. Ultimately, we argue that this framework can be useful for different qualitative and quantitative evaluation methods and techniques with regard to the requirements, measurement, and evaluation concepts and definitions discussed previously.

13.4 ASSESSING WEB QUALITY USING WEBQEM: A CASE STUDY

In Section 13.1, we stated that in order to build a robust and clear measurement and evaluation program, at least three pillars are necessary, namely (1) a process for measurement and evaluation, which is outlined in Section 13.2, (2) a measurement and evaluation framework based on an ontological base, which is analyzed in Section 13.3, and (3) specific model-based methods and techniques in order to perform the specific program or project's activities, which are the aim of this section.

While a measurement or evaluation process specifies what to do (i.e., a clear specification of activities' descriptions, input and output, etc.), a method specifies how to do and perform such activities' descriptions relying on specific models and criteria.

As mentioned, there are different categories of methods (for example, categories for inspection, testing, inquiry, simulation, etc.) and specific types of evaluation methods and techniques such as the heuristic evaluation technique, analyses of log files, or concept model-centered evaluation methods, among many others.

In this section we present the Web Quality Evaluation Methodology (WebQEM) (Olsina and Rossi, 2002) as a model-centered evaluation methodology for the inspection category, that is, inspection of concepts, subconcepts, and attributes stemming from a quality or quality-in-use requirement model, among others. In addition, WebQEM relies on the metric and indicator concepts for measurement and evaluation in order to draw conclusions and give recommendations. We have been developing the WebQEM methodology since the late 1990s. It has been used to evaluate Web sites in several domains, as documented elsewhere (Olsina et al., 1999, 2000, 2006a), in addition to evaluating some industrial Web sites.

In order to illustrate WebQEM and its applicability, we conducted an e-business case study by evaluating the external quality of the shopping cart feature of the Amazon Web site, taking into account a general visitor standpoint. Note that users are redirected to the Amazon Web site (www.amazon.com) from the IMDb, the Internet Movie Database Web site (www.imdb.com), when trying to buy a DVD.

13.4.1 External Quality Requirements Specification

Many potential attributes, both general and domain-specific, can contribute to the Web's external quality. However, as mentioned earlier, evaluation must be organizational, purpose-oriented for an identified information need. Let us establish that the purpose in the present study is to understand the external quality of the shopping cart component of a typical e-store, for a general visitor viewpoint, in order to incorporate the best features in a new e-bookstore development project. For this aim, a successful international site such as Amazon was chosen. On the other hand, recall that the ISO 9126-1 standard models the software quality from three related approaches, which can be summarized as follows:

- *Internal quality*, which is specified by a quality model (ISO, 2001; prescribing a set of six characteristics and a set of subcharacteristics for each characteristic) and can be measured and evaluated by static attributes of documents such as specification of requirements, architecture, or design; pieces of source code, and so forth. In the early phases of a software or Web life cycle, we can evaluate and control the internal quality of these early products, but assuring internal quality is not usually sufficient to assure external quality.
- *External quality*, which is specified by a quality model (likewise as in the previous cited model) and can be measured and evaluated by dynamic properties of the running code in a computer system, i.e., when the module or full application is executed in a computer or network simulating the actual environment as closely as possible. In the late phases of a software life cycle (mainly in different kinds of testing, or even in the acceptance testing, or furthermore in the operational state of a software or Web application), we can measure, evaluate, and control the external quality of these late products, but assuring external quality is usually not sufficient to assure quality in use.
- *Quality in use*, which is specified by a quality model (ISO, 2001; prescribing a set of four characteristics) and can be measured and evaluated by the extent to which the software or Web application meets a specific user's needs in an actual, specific context of use.

A point worthy of mention is the important difference between measuring and evaluating external quality and quality in use; see Olsina et al. (2006a) for an in-depth discussion on Web quality and these ISO models. The former generally involves no real users but rather experts, as long as the latter always involves real end users. The advantage of using expert evaluation without extensive user involvement is minimizing costs and time, among other features. Deciding whether or not to involve end users should be carefully planned and justified. On the other hand, without the end user's participation, it is unthinkable to conduct a task testing in a real context of use for quality-in-use evaluation. Nielsen et al. (2001) indicate that it is common for three to five subjects in the testing process for a given audience to produce meaningful results that minimize costs; however, they recommend running as many small tests as possible.

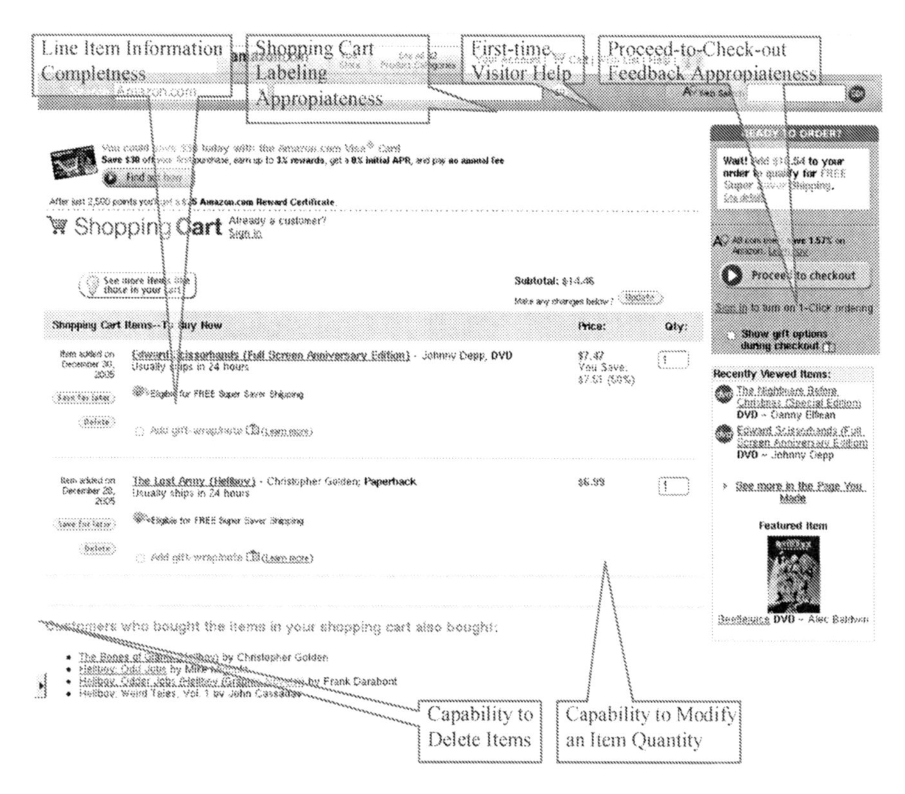

Figure 13.10. A screenshot of Amazon's shopping cart page with several attributes highlighted.

Considering the present study, Figure 13.10 shows a screenshot of Amazon's shopping cart page with several highlighted attributes, which intervene in the quality requirements tree of Figure 13.11.

To the external quality requirements definition, we considered 4 main characteristics: *Usability* (1), *Functionality* (2), *Content* (3), and *Reliability* (4), and 32 attributes related to them (see Figure 13.11). For instance, the Usability characteristic splits into subcharacteristics such as *Understandability* (1.1), *Learnability* (1.2), *Operability* (1.3), and *Attractiveness* (1.4).

Instead of previous quoted case studies, we now consider two separate characteristics: *Functionality* and *Content*. The *Functionality* characteristic splits into *Function Suitability* (2.1) and *Accuracy* (2.2), while the *Content* characteristic splits into *Information Suitability* (3.1) and *Content Accessibility* (3.2). As the reader can observe in Figure 13.11, we relate five measurable attributes to the *Function Suitability* subcharacteristic and three to *Function Accuracy*. In the latter subcharacteristic, we mainly consider precision attributes to recalculate values after making supported operations. On the other hand, in Olsina et al. (2006a) we also justified the inclusion of the *Content* characteristic for assessing the Web.

The following categories can help to evaluate information quality requirements of Web sites and applications (see also Lee et al., 2002):

1. **Usability**
    1.1. Understandability
        1.1.1. Shopping cart icon/label ease to be recognized
        1.1.2. Shopping cart labeling appropriateness
    1.2. Learnability
        1.2.1. Shopping cart help (for first-time visitor)
    1.3. Operability
        1.3.1. Shopping cart control permanence
        1.3.2. Shopping cart control stability
        1.3.3. Steady behavior of the shopping cart control
        1.3.4. Steady behavior of other related controls
    1.4. Attractiveness
        1.4.1. Color style uniformity (links, text, etc.)
        1.4.2. Aesthetic perception
2. **Functionality**
    2.1. Function Suitability
        2.1.1. Capability to add items from anywhere
        2.1.2. Capability to delete items
        2.1.3. Capability to modify an item quantity
        2.1.4. Capability to show totals by performed changes
        2.1.5. Capability to save items for later/move to cart
    2.2. Function Accuracy
        2.2.1. Precision to recalculate after adding an item
        2.2.2. Precision to recalculate after deleting items
        2.2.3. Precision to recalculate after modifying an item quantity
3. **Content**
    3.1. Information Suitability
        3.1.1. Shopping Cart Basic Information
            3.1.1.1. Line item information completeness
            3.1.1.2. Product description appropriateness
        3.1.2. Shopping Cart Contextual Information
            3.1.2.1. Purchase Policies Related Information
                3.1.2.1.1. Shipping and handling costs information completeness
                3.1.2.1.2. Applicable taxes information completeness
                3.1.2.1.3. Return policy information completeness
            3.1.2.2. Continue-buying feedback appropriateness
            3.1.2.3. Proceed-to-check-out feedback appropriateness
    3.2. Content Accessibility
        3.2.1. Readability by Deactivating the Browser Image Feature
            3.2.1.1. Image title availability
            3.2.1.2. Image title readability
        3.2.2. Support for text-only version
4. **Reliability**
    4.1. Nondeficiency (Maturity)
        4.1.1. Link Errors or Drawbacks
            4.1.1.1. Broken links
            4.1.1.2. Invalid links
            4.1.1.3. Reflective links
        4.1.2. Miscellaneous Deficiencies
            4.1.2.1. Deficiencies or unexpected results dependent on browsers
            4.1.2.2. Deficiencies or unexpected results independent of browsers

Figure 13.11. Specifying the external quality requirements tree to the shopping cart component from a general visitor standpoint.

- *Information accuracy.* This subcharacteristic addresses the very intrinsic nature of the information's quality. It assumes that information has its own quality per se. Accuracy is the extent to which information is correct, unambiguous, authoritative (reputable), objective, and verifiable. If a Web site becomes famous for inaccurate information, the Web site will likely be perceived as having little added value and will result in reduced visits.
- *Information suitability.* This subcharacteristic addresses the contextual nature of the information quality. It emphasizes the importance of conveying the appropriate information for user-oriented goals and tasks. In other words, it highlights the quality requirement that contents must be considered within the context of use and the intended audience. Therefore, suitability is the extent to which information is appropriate (appropriate coverage for the target audience), complete (relevant amount), concise (shorter is better), and current (see the specified attributes in Figure 13.11).
- *Accessibility.* It emphasizes the importance of technical aspects of Web sites and applications in order to make Web contents more accessible for users with various disabilities (see the specified attributes in Figure 13.11).
- *Legal compliance.* The capability of the information product to adhere to standards, conventions, and legal norms related to contents and intellectual property rights.

The INCAMI_Tool records all the information for a measurement and evaluation project. Besides the data in the project itself, it also saves to the *InformationNeed* class (see Figure 13.6) the purpose, user viewpoint, and context description meta-data; for the *CalculabeConcept* and *Attribute* classes, it saves all the names and definitions, respectively.

The *ConceptModel* class permits us to instantiate a specific model, that is, the external quality model in our case, allowing evaluators to edit and relate specific concepts, subconcepts, and attributes (the whole instantiated model looks like that in Figure 13.11, and an INCAMI_Tool screenshot of it appears in Figure 13.12).

13.4.2 Designing and Executing the Measurement and Elementary Evaluation

As mentioned in Section 13.2, the evaluators should design, for each measurable attribute of the instantiated external quality model, the basis for the measurement and elementary evaluation process by defining each specific metric and elementary indicator for the information needed accordingly.

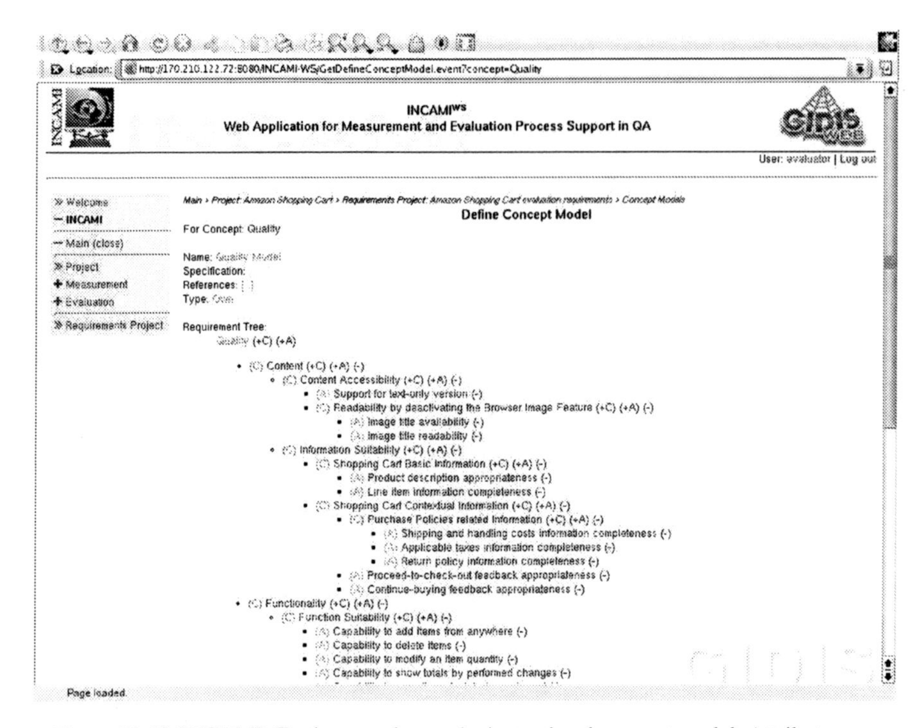

Figure 13.12. INCAMI_Tool screenshot to the instantiated concept model. Attributes are labeled with "A" on the left side of the tree; concepts and subconcepts with "C." In addition, "+C" and "+A" mean adding concepts or attributes, respectively, and "-" removing them.

In the design phase we record all the information for the selected metrics and elementary indicators regarding the conceptual schema of the *Metric* and *Elementary Indicator* classes shown in Figures 13.8 and 13.9, respectively. In addition, in Sections 13.4.2 and 13.4.3 the metric and indicator meta-data for the "Capability to delete items" attribute were illustrated. Finally, Figure 13.13 shows the name of the attributes and the name of each metric that quantifies them. Note that we can assign a metric for a given attribute by selecting it from the semantic catalogue (Molina et al., 2004); see the "Assign Metric" link in the figure.

Lastly, in the execution phase, we record for the *Measurement* and *Calculation* classes' instances the yielded final values for each metric and indicator. The data collection for the measurement activity was performed from December 19 to 30, 2005. From the metrics' values, the elementary indicators' values were calculated according to the respective elementary models.

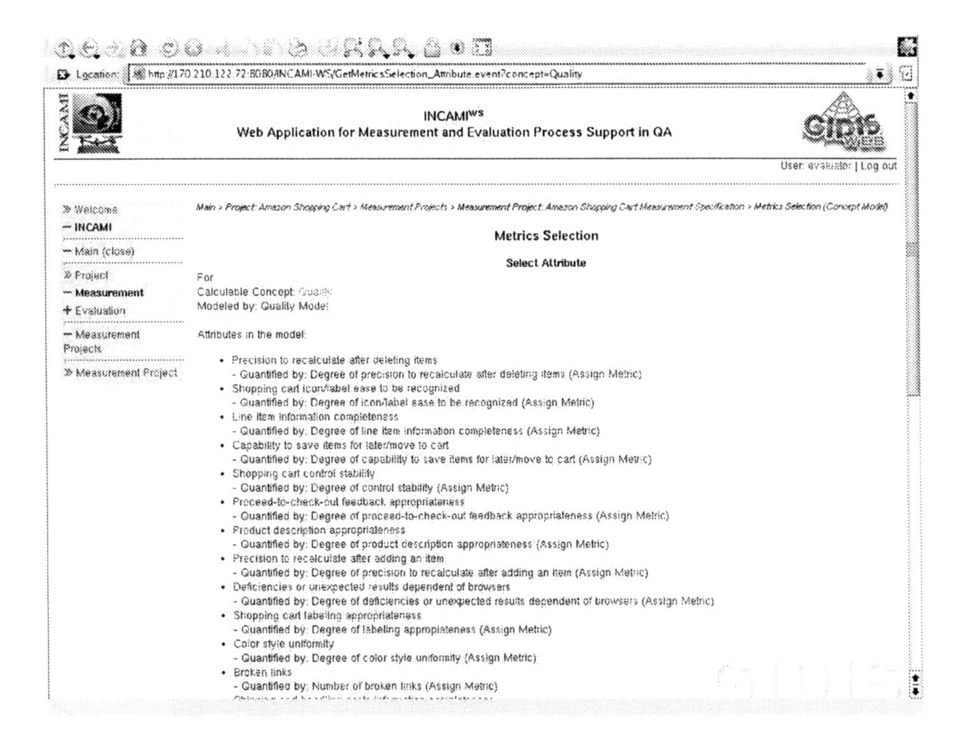

Figure 13.13. INCAMI_Tool screenshot of the metric selection process.

Figure 13.14 shows the selection process of a measurement value from a specific measurement project, which will be the input to the respective elementary indicator function in order to produce the indicator value (recall that for the same measurement project we can record measurement values at different times).

Once evaluators have designed and implemented the elementary evaluation, they should consider not only each attribute's relative importance but also whether the attribute (or subcharacteristic) is mandatory, alternative, or neutral. For this global evaluation task, we need a robust aggregation and scoring model, described next.

13.4.3 Designing and Executing the Partial/Global Evaluation

In the design of the global evaluation phase we select and apply an aggregation and scoring model (see *GlobalModel* class in Figure 13.9). Arithmetic or logic operators will then relate the hierarchically grouped attributes, subconcepts, and concepts accordingly.

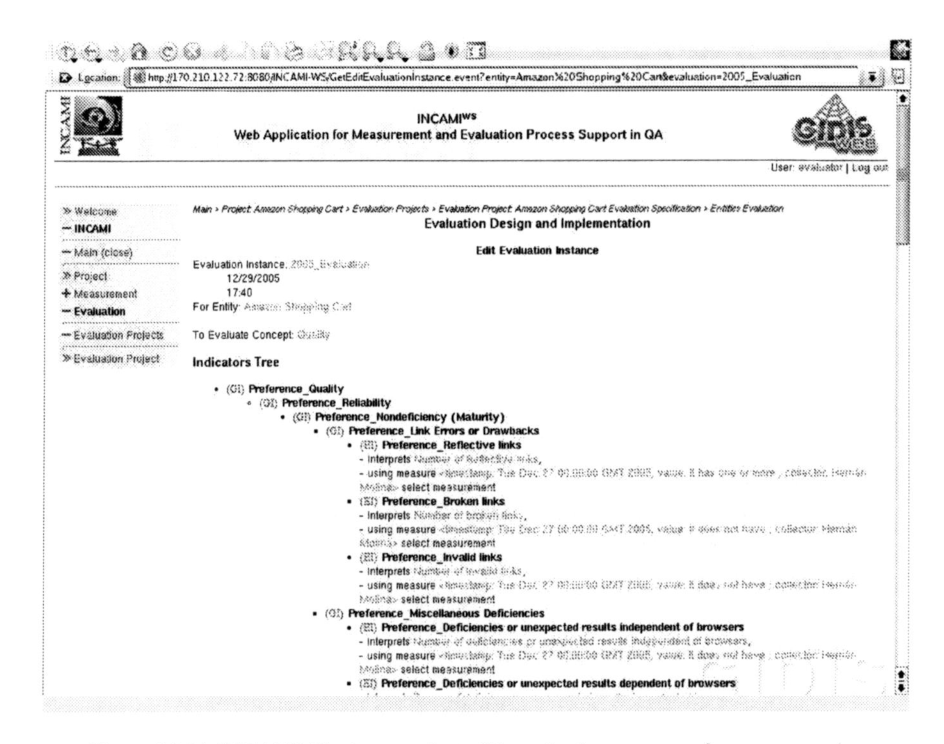

Figure 13.14. INCAMI_Tool screenshot of the selection process of a measure value for a given elementary indicator.

As mentioned earlier (see Definition 13.30), the INCAMI_Tool supports a linear additive or a nonlinear multicriteria scoring model (even other models can be used for designing the global evaluation such as fuzzy logic or neural networks not supported currently by the tool). We cannot use the additive scoring model to model input simultaneity (an *and* relationship among inputs) or replaceability (an *or* relationship), however, because it cannot express, for example, simultaneous satisfaction of several requirements as input. Additivity assumes that the insufficient presence of a specific attribute (in an input) can always be compensated for by the sufficient presence of any other attribute. Furthermore, additive models cannot model mandatory requirements; that is, a necessary attribute's or subcharacteristic's total absence cannot be compensated for by another's presence.

A nonlinear multicriteria scoring model lets us deal with simultaneity, neutrality, replaceability, and other input relationships using aggregation operators based on the weighted-power-means mathematical model. This model, called Logic Scoring of Preference (LSP) (Dujmovic, 1996), is a generalization of the additive scoring model and can be expressed as follows:

$$P/GI(r) = (W_1\, EI^r{}_1 + W_2\, EI^r{}_2 + \ldots + W_m\, EI^r{}_m)^{1/r}, \qquad (13.2)$$

where

$$-\infty \le r \le +\infty \; ; \; P/GI(-\infty) = \min(EI_1, EI_2, \ldots, EI_m),$$

$$P/GI(+\infty) = \max(EI_1, EI_2, \ldots, EI_m).$$

The power r is a parameter selected to achieve the desired logical relationship and polarization intensity of the aggregation function. If $P/GI(r)$ is closer to the minimum, such a criterion specifies the requirement for input simultaneity. If it is closer to the maximum, it specifies the requirement for input replaceability. Equation (13.2) is additive when $r = 1$, which models the neutrality relationship; that is, the formula remains the same as in the first additive model. Equation (13.2) is supra-additive for $r > 1$, which models input disjunction or replaceability, and it's sub-additive for $r < 1$ (with $r! = 0$), which models input conjunction or simultaneity.

For our case study (as in previous ones), we selected this last model and used a 17-level approach of conjunction–disjunction operators, as defined by Dujmovic. Each operator in the model corresponds to a particular value of the r parameter. When $r = 1$, the operator is tagged with A (or the + sign). The C conjunctive operators range from weak (C–) to strong (C+) quasi-conjunction functions, i.e., from decreasing r values, starting from $r < 1$.

In general, the conjunctive operators imply that low-quality input indicators can never be well compensated for by a high quality of some other input to output a high-quality indicator (in other words, a chain is as strong as its weakest link). Conversely, disjunctive operators (D operators) imply that low-quality input indicators can always be compensated for by the high quality of some other input.

Designing the LSP aggregation schema requires answering the following key basic questions (which are part of the *Global Indicator Definition* task in Figure 13.5):

- What is the relationship among this group of related attributes and subconcepts: conjunctive, disjunctive, or neutral [for instance, when modeling the attributes' relationship for the *Function Suitability* (2.1) subcharacteristic, we can agree they are neutral or independent of each other]?
- What is the level of intensity of the logic operator, from a weak to strong conjunctive or disjunctive polarization?
- What is the relative importance or weight of each element in the aggregation block or group?

Figure 13.15 shows some details of the enacted requirements tree for amazon.com as generated by our tool. Particularly, in the top part of Figure 13.15 we can see LSP operators, weights, and final values for elementary, partial, and global indicators; the bottom part shows only the indicator values and the respective colored bars in a percentage scale.

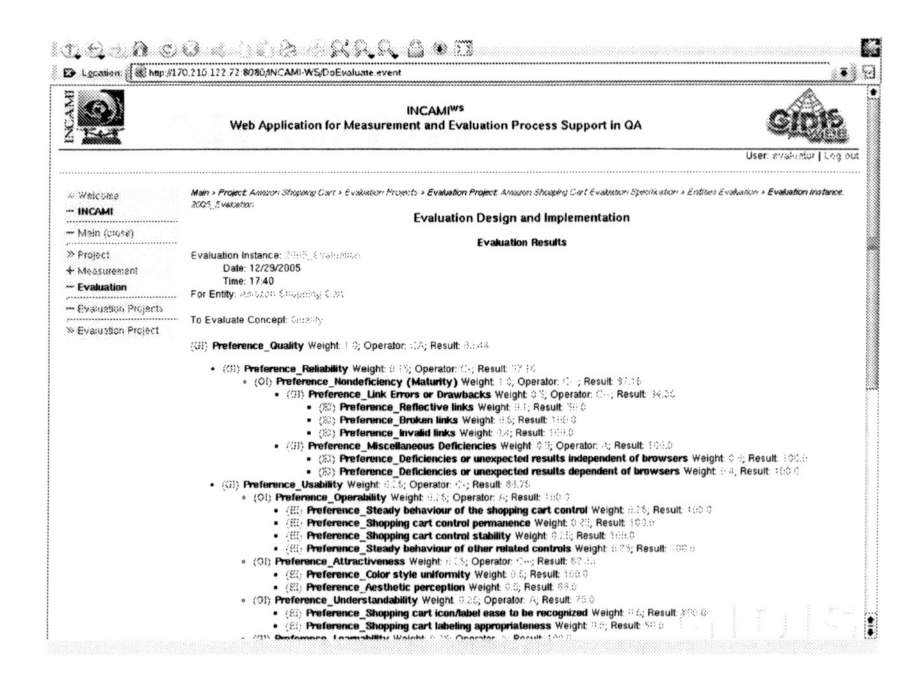

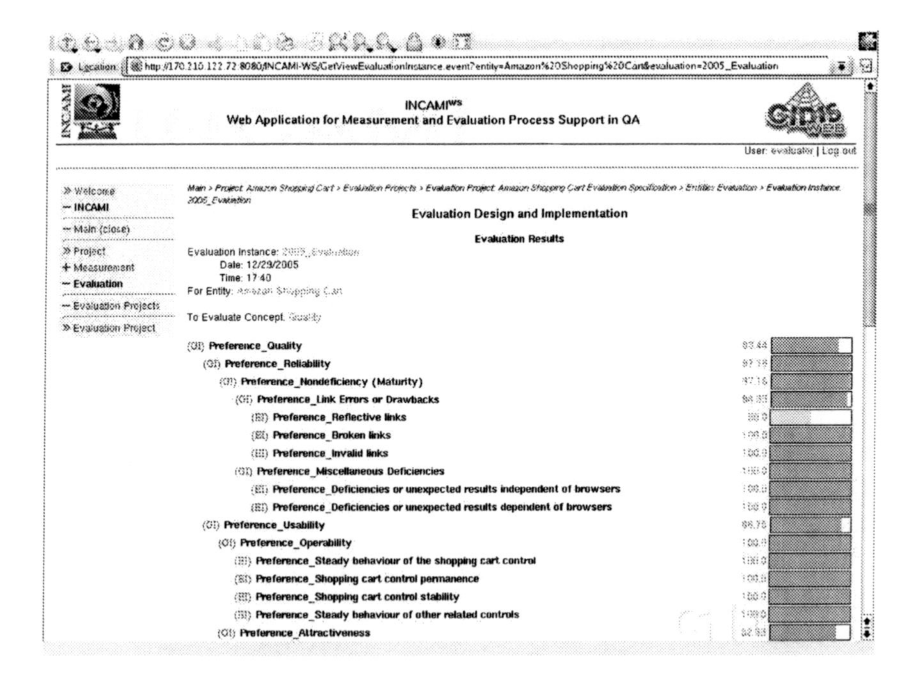

Figure 13.15. Once the weights and operators (in this case for the LSP aggregation model) were agreed on, the INCAMI_Tool yields elementary partial and global indicators in the execution phase, as highlighted in the figures. The top figure shows details of weights and operators, while the bottom figure shows just indicator values and the respective colored bars in the percentage scale.

13.4.4 Analyzing and Recommending

Once we have performed the final execution of the evaluation, decision makers can analyze the results and draw conclusions and recommendations. As stated in Section 13.4.1, we established (for illustration reasons) that the purpose in this study is to understand the external quality of the shopping cart component of a typical e-store, for a general visitor viewpoint in order to incorporate the best features in a new e-bookstore development project. The underlying hypothesis is that at the level of calculable concepts (characteristics in the ISO 9126 vocabulary) they accomplish at least the satisfactory acceptability range.

Table 13.1 shows the final results for the *Usability*, *Functionality*, *Content*, and *Reliability* characteristics and subcharacteristics, as well as partial and global indicator values for the amazon.com shopping cart.

Table 13.1. Summary of Partial and Global Indicators' Values for the Amazon.com Shopping Cart

Code	**Concept/Subconcept Name**	**Indicator Value**
	External Quality	*83.44*
1	Usability	88.75
1.1	Understandability	75.00
1.2	Learnability	100.00
1.3	Operability	100.00
1.4	Attractiveness	82.33
2	Functionality	87.61
2.1	Function Suitability	76.40
2.2	Function Accuracy	100.00
3	Content	71.40
3.1	Information Suitability	81.21
3.1.1	Shopping Cart Basic Information	81.70
3.1.2	Shopping Cart Contextual Information	80.47
3.1.2.1	Purchase Policies related Information	77.89
3.2	Content Accessibility	56.79
3.2.1	Readability by Deactivating the Browser Image Feature	67.75
4	Reliability	97.16
4.1	Nondeficiency (Maturity)	97.16
4.1.1	Link Errors or Drawbacks	94.35
4.1.2	Miscellaneous Deficiencies	100

The colored quality bars in the bottom part of Figure 13.15 indicate the acceptability ranges and clearly show the quality level each shopping cart feature has reached. In fact, the final indicator value to the external quality of the Amazon shopping cart was satisfactory getting a rank of 83.44 [that is a similar global indicator value for the study made in late 2004 (Olsina et al., 2006), using the same requirements and criteria, which ranked 84.32%]. Notice that a score within a yellow bar (marginal) indicates a need for improvement actions. An unsatisfactory rating (red bar) means change actions must take high priority. A score within a green bar indicates satisfactory quality of the analyzed feature.

Looking at the *Usability, Functionality, Content,* and *Reliability* characteristics, we can see that the scores fall in the satisfactory level, so that we can emulate these features in a new development project. However, none of them is 100%. For instance, if we look at the *Functionality* characteristic and particularly at the *Function Suitability* subconcept, which ranked 76.40, we can observe that the reason for this score is in part due to the *Capability to Delete Items* (2.1.2) attribute, which is not totally suitable (the indicator value was 66%).

In order to make a thorough causal analysis, we must look at the elementary indicator and metric specification. Regarding the INCAMI_Tool, the following elementary indicator model specification (see Definition 13.26) was edited: CDI _PL = (0.33 * CDI) * 100, where CDI is the direct metric for the *Capability to Delete Items* attribute.

In the example of Definition 13.10, the scale of the direct metric was specified in this way:

1. Does not delete items at all.
2. Delete just all at once.
3. Delete one by one.
4. Delete one by one or delete the selected group at once.

Thus, the resulting indicator value in the execution phase was 66% because the Amazon shopping cart allows users to delete only one item at once, but does not allow the selected group to be deleted at once.

Ultimately, we observe that the state-of-the-art of the shopping cart quality of this typical site is rather high, but the wish list is not empty, because of some weak-designed attributes. Notice that elementary, partial, and global indicators reflect results of these specific requirements for this specific audience and should not be regarded as generalized rankings. Moreover, results themselves from a case study are seldom intended to be interpreted as generalizations (in the sense of external validity).

13.5 DISCUSSION AND FINAL REMARKS

Our experience suggests that it is necessary to select metrics for purpose-oriented attributes as well as to identify contextual indicators in order to start and guide a successful measurement and evaluation program. In fact, organizations must have sound specifications of metric and indicator meta-data associated consistently to data sets, as well as a clear establishment of frameworks and programs in order to make measurement and analyses and quality assurance useful support processes to software and Web development and maintenance projects. Ultimately, the underlying hypothesis is that without appropriate recorded meta-data of information needs, attributes, metrics, and indicators, it is difficult to ensure that measure and indicator values are repeatable and comparable among an organization's projects; consequently, analyses and comparisons can be carried out in an inconsistent way as well.

Throughout this chapter we have stated that in order to build a robust and flexible measurement and evaluation program, at least three pillars are necessary: (1) a process for measurement and evaluation (outlined in Section 13.2); (2) a measurement and evaluation framework based on an ontological base (analyzed in Section 13.3); and (3) specific model-based methods and techniques for the realization of measurement and evaluation activities (a particular inspection method was illustrated in Section 13.4).

As a matter of fact, in the present chapter we have emphasized the importance of counting with a measurement and evaluation conceptual framework. The discussed INCAMI framework is based on the assumption that for an organization to measure and evaluate in a purpose-oriented way, it must first specify nonfunctional requirements starting from information needs, then it must design and select the specific set of metrics for measurement purposes, and last it must interpret the metric values by means of contextual indicators with the aim of evaluating or estimating the degree to which the stated information need has been met. Thus, consistent and traceable analyses, conclusions, and recommendations can be drawn.

Regarding other initiatives, the GQM (*Goal-Question-Metrics*) paradigm (Basili and Rombach, 1989) is a useful, simple, purpose-oriented measurement approach that has been used in different measurement projects and organizations. However, as Kitchenham et al. pointed out (2001), GQM is not intended to define metrics at a level of detail suitable to ensure that they are trustworthy, in particular, whether or not they are repeatable. Contrary to our approach, which is based on an ontological conceptualization of metrics and indicators, GQM lacks this conceptual base, and so it cannot assure that measurement values (and the associated meta-data like scale, unit, measurement method, and so forth) are trustworthy and consistent for ulterior analysis among projects.

On the other hand, GQM is a weak framework for evaluation purposes, i.e. GQM lacks specific concepts for evaluation in order to interpret attributes' measures. For instance, elementary and global indicators and related terms are essential for evaluation as shown in the previous sections. Conversely, GQM is more flexible than INCAMI in the sense that it is not always necessary to have a concept model specification in order to perform a measurement project.

In our humble opinion, an interesting improvement to the GQM approach that considers indicators has recently been issued as a technical note (Goethert and Fisher, 2003). This approach uses both the *Balance Scorecard* technique (Kaplan and Norton, 2001) and the *Goal-Question-Indicator-Measurement* method in order to purposely derive the required enterprise goal-oriented indicators and metrics. It is a robust framework for specifying enterprise-wide information needs and deriving goals and subgoals and then operationalizing questions with associated indicators and metrics. It says, "The questions provide concrete examples that can lead to statements that identify the type of information needed. From these questions, displays or indicators are postulated that provide answers and help link the measurement data that will be collected to the measurement goals" (Goethert and Fisher, 2003). However, this approach is not based on a sound ontological conceptualization of metrics and indicators as ours; furthermore, the terms "measure" and "indicator" are sometimes used ambiguously, which can result in data sets and meta-data being recorded inconsistently.

On the other hand, there exist other close initiatives to our research, such as the Kitchenham et al. (2001) conceptual framework as well as the cited ISO standards related to software measurement and evaluation processes. In summary, we tried to strengthen these contributions not only from the conceptual modeling point of view, but also from the ontological point of view, including a broader set of concepts.

Lastly, we argue that the INCAMI framework can be a useful conceptual base and approach for different qualitative and quantitative evaluation methods and techniques with regard to the requirement, measurement, and evaluation concepts and definitions analyzed in Section 13.3. Apart from inspection or *feature analyses* methods (like WebQEM), this framework can be employed for some other methods, such as neural networks and fuzzy logic, when they are intended to measure and evaluate quality, quality in use, and cost, among other calculable concepts.

Finally, due to the importance of managing the acquired enterprise-wide contextual knowledge during measurement and evaluation and during quality assurance projects, a semantic infrastructure that embraces contextual information and organizational memory management is currently being considered in the INCAMI framework. This will be integrated to the

INCAMI_Tool and framework, also making sure that ontologies and the Semantic Web are enabling technologies for our previous (Molina et al., 2004) and current research aims as well.

ACKNOWLEDGEMENTS

This research is supported by Argentina's UNLPam-09/F037 project, as well as the PICTO 11-30300 and PAV 127-5 research projects.

REFERENCES

Basili, V., and Rombach, H.D., 1989, The TAME project: Towards improvement-oriented software environments. *IEEE Transactions on Software Engineering*, **14**(6): 758–773.

Briand, L., Morasca, S., and Basili, V., 2002, An operational process for goal-driven definition of measures. *IEEE Transactions on Software Engineering*, **28**(12): 1106–1125.

CMMI, 2002, Capability Maturity Model Integration, Version 1.1, CMMI^{SM} for Software Engineering (CMMI-SW, V. 1.1) Staged Representation CMU/SEI-2002-TR-029, CMMI Product Team, SEI Carnegie Mellon University (available online).

Dujmovic, J., 1996, A method for evaluation and selection of complex hardware and software systems. *Proceedings 22nd International Conference for the Resource Management and Performance Evaluation of Enterprise CS* (CMG 96), Vol. 1, pp. 368–378.

Goethert, W., and Fisher, M., 2003, Deriving enterprise-based measures using the balanced scorecard and goal-driven measurement techniques. Software Engineering Measurement and Analysis Initiative, CMU/SEI-2003-TN-024 (available online).

ISO/IEC 14598-5, 1998, Information technology—Software product evaluation—Part 5: Process for evaluators.

ISO/IEC 14598-1, 1999, International standard, information technology—Software product evaluation—Part 1: General overview.

ISO/IEC 9126-1, 2001, International standard, software engineering—Product quality—Part 1: Quality model.

ISO/IEC 15939, 2002, Software engineering—Software measurement process.

Kaplan, R., and Norton, D., 2001, *The Strategy-Focused Organization, How Balanced Scorecard Companies Thrive in the New Business Environment*. Harvard Business School Press, Boston.

Kitchenham, B.A., Hughes, R.T., and Linkman, S.G., 2001. Modeling software measurement data. *IEEE Transactions on Software Engineering*, **27**(9): 788–804.

Lee, Y.W., Strong, D.M., Kahn, B.K., and Wang, R.Y., 2002, AIMQ: A methodology for information quality assessment. *Information & Management*, **40**(2): 133–146.

Molina, H., Papa, F., Martín, M., and Olsina, L., 2004, Semantic capabilities for the metrics and indicators cataloging Web system. In *Engineering Advanced Web Applications*, M. Matera and S. Comai, eds., Rinton Press Inc., Princeton, NJ, pp. 97–109, ISBN 1-58949-046-0.

Nielsen, J., Molich, R., Snyder, C., and Farrell, S., 2001, E-Commerce User Experience, NN Group.

Olsina, L., Godoy, D., Lafuente, G., and Rossi, G., 1999, Assessing the quality of academic Websites: A case study. *New Review of Hypermedia and Multimedia (NRHM) Journal*, **5**: 81–103.

Olsina, L., Lafuente, G., and Rossi, G., 2000, E-commerce site evaluation: A case study. *Proceedings 1st International Conference on Electronic Commerce and Web Technologies* (EC-Web 2000), London, Springer LNCS 1875, pp. 239–252.

Olsina, L., Papa, M.F., Souto, M.E., and Rossi, G., 2001, Providing automated support for the Web quality evaluation methodology. *Proceedings 4th Workshop on Web Engineering, at the 10th International WWW Conference*, Hong Kong, pp. 1–11.

Olsina, L., and Rossi, G., 2002, Measuring Web application quality with WebQEM. *IEEE Multimedia*, **9**(4): 20–29.

Olsina, L., and Martin, M., 2004, Ontology for software metrics and indicators. *Journal of Web Engineering,* **2**(4): 262–281, ISSN 1540-9589.

Olsina, L., Papa, F., and Molina, H., 2005, Organization-oriented measurement and evaluation framework for software and Web Engineering projects. *Proceedings International Congress on Web Engineering* (ICWE05), Sydney, Australia, Springer, LNCS 3579, pp. 42–52.

Olsina, L., Covella, G., and Rossi, G., 2006, Web quality. Chapter 4 in *Web Engineering,* E. Mendes and N. Mosley, eds., Springer, New York, ISBN 3-540-28196-7.

Olsina, L., Papa, F., and Molina, H., 2008, Ontological support for a measurement and evaluation framework. To appear in the *Journal of Intelligent Systems.*

Zuse, H., 1998, *A Framework of Software Measurement*, Walter de Gruyter, Berlín.

Chapter 14

THE NEED FOR EMPIRICAL WEB ENGINEERING: AN INTRODUCTION

Emilia Mendes
WETA Research Group, Computer Science Department, The University of Auckland, Private Bag 92019, Auckland, New Zealand

14.1 INTRODUCTION

The World Wide Web (Web) was originally conceived in 1989 as an environment to allow for the sharing of information (e.g., research reports, databases, user manuals) among geographically dispersed individuals. The information itself was stored on different servers and was retrieved by means of a single user interface (a Web browser). The information consisted primarily of text documents interlinked using a hypertext metaphor[1] (Offutt, 2002).

Since its original inception, the Web has changed into an environment employed for the delivery of many different types of applications. Such applications range from small-scale information-dissemination-like applications, typically developed by writers and artists, to large-scale commercial,[2] enterprise-planning and scheduling, collaborative-work applications. The latter are developed by multidisciplinary teams of people with diverse skills and backgrounds using cutting-edge, diverse technologies (Gellersen and Gaedke, 1997; Ginige and Murugesan, 2001; Offutt, 2002).

---

[1] http://www.zeltser.com/web-history/.

[2] The increase in the use of the Web to provide commercial applications has been motivated by several factors, such as the possible increase of an organization's competitive position, and the opportunity for small organizations to project their corporate presence in the same way as that of larger organizations.

Numerous current Web applications are fully functional systems that provide business-to-customer and business-to-business e-commerce, and numerous services to numerous users (Offutt, 2002).

Industries such as travel and hospitality, manufacturing, banking, education, and government have utilized Web-based applications to improve and increase their operations (Ginige and Murugesan, 2001). In addition, the Web allows for the development of corporate intranet Web applications, for use within the boundaries of their organizations (Houghton, 2000). The remarkable spread of Web applications into areas of communication and commerce makes it one of the leading and most important branches of the software industry (Offutt, 2002).

To date the development of industrial Web applications has been in general ad hoc, resulting in poor-quality applications that are difficult to maintain (Murugesan and Deshpande, 2001). The main reasons for such problems are unawareness of suitable design and development processes, and poor project management practices (Ginige, 2002). A survey on Web-based projects, published by the Cutter Consortium in 2000, revealed a number of problems with outsourced, large Web-based projects (Ginige, 2002):

- Eighty-four percent of surveyed delivered projects did not meet business needs.
- Fifty-three percent of surveyed delivered projects did not provide the required functionality.
- Seventy-nine percent of surveyed projects presented schedule delays.
- Sixty-three percent of surveyed projects exceeded their budget.

As the reliance on larger and more complex Web applications increases, so does the need for using methodologies/standards/best practice guidelines to develop applications that are delivered on time and within budget, have a high level of quality, and are easy to maintain (Lee and Shirani, 2004; Ricca and Tonella, 2001; Taylor et al., 2002). To develop such applications, Web development teams need to use sound methodologies, systematic techniques, quality assurance, rigorous, disciplined, and repeatable processes, better tools, and baselines. Web Engineering[3] aims to meet such needs (Ginige and Murugesan, 2001).

Web Engineering is described as (Murugesan and Deshpande, 2001) "the use of scientific, **engineering**, and management principles and systematic

[3] The term "Web Engineering" was first published in 1996 in a conference paper by Gellersen et al. (1997). Since then this term has been cited in numerous publications, and numerous activities devoted to discussing Web Engineering have taken place (e.g., workshops, conference tracks, entire conferences).

approaches with the aim of successfully developing, deploying and maintaining high quality Web-based systems and applications."

Engineering is widely taken as a disciplined application of scientific knowledge for the solution of practical problems. A few definitions taken from dictionaries support that:

> Engineering is the application of science to the needs of humanity. This is accomplished through knowledge, mathematics, and practical experience applied to the design of useful objects or processes. (Wikipedia, 2004)
>
> Engineering is the application of scientific principles to practical ends, as the design, manufacture, and operation of structures and machines. (Houghton, 1994)
>
> The profession of applying scientific principles to the design, construction, and maintenance of engines, cars, machines, etc. (mechanical engineering), buildings, bridges, roads, etc. (civil engineering), electrical machines and communication systems (electrical engineering), chemical plant and machinery (chemical engineering), or aircraft (aeronautical engineering). (Harper, 2000)

In all of the above definitions, the need for "the application of scientific principles" has been stressed, where scientific principles are the result of applying a scientific process (Goldstein and Goldstein, 1978). A process in this context means that our current understanding, i.e., our theory (hypothesis) of how best to develop, deploy, and maintain high-quality Web-based systems and applications, may be modified or replaced as new evidence is found through the accumulation of data and knowledge. This process is illustrated in Figure 14.1 and described below (Goldstein and Goldstein, 1978):

- *Observation*: To observe or read about a phenomenon or set of facts. In most cases the motivation for such observation is to identify cause-and-effect relationships between observed items, since these entail predictable results. For example, we can observe that an increase in the development of new Web pages seems also to increase the corresponding development effort.
- *Hypothesis*: To formulate a hypothesis represents an attempt to explain an *observation*. It is a tentative theory or assumption that is believed to explain the behavior under investigation (Fenton and Pfleeger, 1997). The items that participate in the *observation* are represented by variables (e.g., number of new Web pages, development effort), and the hypothesis indicates what is expected to happen to these variables (e.g., there is a linear relationship between the number of Web pages and the

development effort, showing that as the number of new Web pages increases, so does the effort to develop these pages). These variables first need to be measured; to do so, we need an underlying measurement theory.

- *Prediction*: To predict means to predict results that should be found if the rationale used in the hypothesis formulation is correct (e.g., Web applications with a larger number of new Web pages will use a larger development effort).
- *Validation*: To validate requires experimentation to provide evidence to either support or refute the initial hypothesis. If the evidence refutes the hypothesis, then the hypothesis should be revised or replaced. If the evidence is in support of the hypothesis, then many more replications of the experiment need to be carried out in order to build a better understanding of how variables relate to each other and their cause-and-effect relationships.

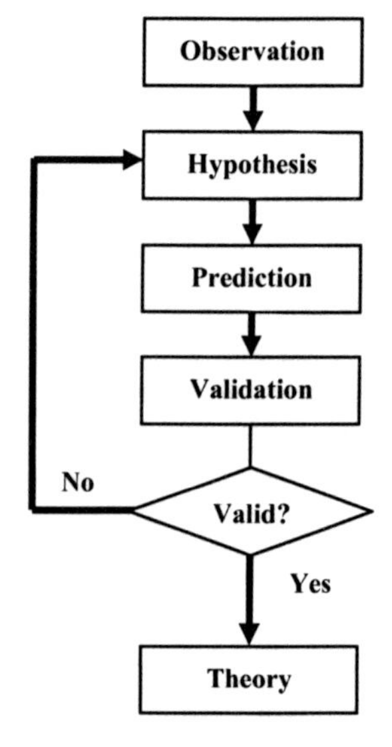

Figure 14.1. WSDM overview.

The scientific process supports knowledge building, which in turn involves the use of empirical studies to test hypotheses previously proposed and to assess if current understanding of the discipline is correct. Experimentation in Web Engineering is therefore essential (Basili, 1996; Basili et al., 1999).

The extent to which scientific principles are applied to developing and maintaining Web applications varies among organizations. More mature organizations generally apply these principles to a larger extent than less mature organizations, where maturity reflects an organization's use of sound development processes and practices (Fenton and Pfleeger, 1997). Some organizations have clearly defined processes that remain unchanged regardless of the people who work on the projects. For such organizations, success is dictated by following a well-defined process, where feedback is constantly obtained using product, process, and resource measures. Other organizations have processes that are not so clearly defined (ad hoc), and therefore the success of a project is often determined by the expertise of the development team. In such a scenario, product, process, and resource measures are rarely used, and each project represents a potential risk that may lead an organization, if it gets it wrong, to bankruptcy (Pressman, 1998).

The variables used in the formulation of hypotheses represent the attributes of real-world entities that we observe. An entity represents a process, product, or resource. A process is defined as a software-related activity. Examples of processes are Web development, Web maintenance, Web design, Web testing, and Web project. A product is defined as an artifact, deliverable, or document that results from a process activity. Examples of products are Web application, design document, testing scripts, and fault reports. Finally, a resource represents an entity required by a process activity. Examples of resources are Web developers, development tools, and programming languages (Fenton and Pfleeger, 1997).

In addition, for each entity's attribute that is to be measured, it is also useful to identify if the attribute is *internal* or *external*. Internal attributes can be measured by examining the product, process, or resource on its own, separate from its behavior. External attributes can only be measured with respect to how the product, process, or resource relates to its environment (Fenton and Pfleeger, 1997). For example, usability is in general an external attribute since its measurement often depends upon the interaction between user and application. An example of classification of entities is presented in Table 14.1.

The measurement of an entity's attributes generates quantitative descriptions of key processes, products, and resources, enabling us to understand behavior and results. This understanding lets us select better techniques and tools to control and improve our processes, products, and resources (Pfleeger et al., 1997).

The measurement theory that has been adopted in this chapter is the representational theory of measurement (Fenton and Pfleeger, 1997). It drives the definition of measurement scales, presented in Section 14.2, and the measures presented in Chapter 13.

Table 14.1. Classification of Process, Product, and Resources for Tukutuku[4] Data Set

Entity	Attribute	Description
PROCESS ENTITIES		
PROJECT		
	TYPEPROJ	Type of project (new or enhancement)
	LANGS	Implementation languages used
	DOCPROC	If project followed defined and documented process
	PROIMPR	If project team is involved in a process improvement program
	METRICS	If project team is part of a software metrics program
	DEVTEAM	Size of project's development team
WEB DEVELOPMENT		
	TOTEFF	Actual total effort used to develop the Web application
	ESTEFF	Estimated total effort necessary to develop the Web application
	ACCURACY	Procedure used to record effort data
PRODUCT ENTITY		
WEB APPLICATION		
	TYPEAPP	Type of Web application developed
	TOTWP	Total number of Web pages (new and reused)
	NEWWP	Total number of new Web pages
	TOTIMG	Total number of images (new and reused)
	NEWIMG	Total number of new images your company created
	HEFFDEV	Minimum number of hours to develop a single function/feature by one experienced developer that is considered high (above average)
	HEFFADPT	Minimum number of hours to adapt a single function/feature by one experienced developer that is considered high (above average)
	HFOTS	Number of reused high-effort features/functions without adaptation
	HFOTSA	Number of adapted high-effort features/functions
	HNEW	Number of new high-effort features/functions
	FOTS	Number of low-effort features off the shelf
	FOTSA	Number of low-effort features off the shelf adapted
	NEW	Number of new low-effort features/functions
RESOURCE ENTITY		
DEVELOPMENT TEAM		
	TEAMEXP	Average team experience with the development language(s) employed

14.2 MEASUREMENT SCALES

When we gather data associated with the attributes of entities we wish to measure, they can be collected using different scales of measurement. The characteristics of each scale type determine the choice of methods and

[4] The Tukutuku project collects data on industrial Web projects, for the development of effort estimation models and to benchmark productivity across and within Web companies. (See http://www.cs.auckland.ac.nz/tukutuku.)

statistics that can be used to analyze the data and how to interpret their corresponding measures. In this section we describe the five main scale types (Fenton and Pfleeger, 1997):

- Nominal
- Ordinal
- Interval
- Ratio
- Absolute

14.2.1 The Nominal Scale Type

The Nominal scale type represents the most primitive form of measurement. It identifies classes or categories where each category groups a set of entities based on their attribute's value. Here entities can only be organized into classes or categories, and there is no notion of ranking between classes. Classes can be represented as symbols or numbers; however, if we use numbers, they do not have any numerical meaning. Examples using a Nominal scale are given in Table 14.2.

Table 14.2. Examples of Nominal Scale Measures

Entity	Attribute	Categories
Web application	Type	e-Commerce, academic, corporate, entertainment
Programming language	Type	ASP (VBScript, .Net), Coldfusion, J2EE (JSP, Servlet, EJB), PHP
Web project	Type	New, enhancement, redevelopment
Web company	Type of service	1, 4, 5, 7, 9, 34, 502, 8

14.2.2 The Ordinal Scale Type

The Ordinal scale supplements the Nominal scale with information about the ranking of classes or categories. As with the Nominal scale, it also identifies classes or categories, where each category groups a set of entities based on their attribute's value. The difference between an Ordinal scale and a Nominal scale is that here there is the notion of ranking between classes. Classes can be represented as symbols or numbers; however, if we use numbers, they do not have any numerical meaning and represent ranking only. Therefore addition, subtraction, and other arithmetic operations cannot be applied to classes. Examples using an Ordinal scale are given in Table 14.3.

Table 14.3. Examples of Ordinal Scale Measures

Entity	Attribute	Categories
Web application	Complexity	Very low, low, average, high, very high
Web page	Design quality	Very poor, poor, average, good, very good
Web project	Priority	1,2,3,4,5,6,7

14.2.3 The Interval Scale Type

The Interval scale supplements the Ordinal scale with information about the size of the intervals that separate the classes or categories. As with the Nominal and Ordinal scales, it also identifies classes or categories, where each category groups a set of entities based on their attribute's value. As with the Ordinal scale, there are ranks between classes or categories. The difference between an Interval scale and an Ordinal scale is that here there is the notion that the size of intervals between classes or categories remains constant. Although the Interval scale is a numerical scale and numbers have a numerical meaning, the class zero does not mean the complete absence of the attribute we measured. To illustrate that, let's look at temperatures measured using the Celsius scale. The difference between 1 °C and 2 °C is the same as the difference between 6 °C and 7 °C: exactly 1°. There is a ranking between two classes; thus, 1 °C has a lower rank than 2 °C, and so on. Finally, the temperature 0 °C does not represent the complete absence of temperature, where molecular motion stops. In this example, 0 °C was arbitrarily chosen to represent the freezing point of water. This means that operations such as addition and subtraction between two categories are permitted (e.g., 50 °C − 20 °C = 70 °C − 40 °C; 5 °C + 25 °C = 20 °C + 10 °C); however, calculating the ratio of two categories (e.g., 40 °C/20 °C) is not meaningful (40 °C is not twice as hot as 20 °C), so multiplication and division cannot be calculated directly from categories. If ratios are to be calculated, they need to be based on the differences between categories. Examples using an Interval scale are given in Table 14.4.

Table 14.4. Examples of Interval Scale Measures

Entity	Attribute	Categories
Web project	Number of days relative to start of project	0,1,2,3,4,5,...
Human body	Temperature (Celsius or Fahrenheit)	Decimal numbers

14.2.4 The Ratio Scale Type

The Ratio scale supplements the Interval scale with the existence of a zero element, representing the total absence of the attribute measured. As with the Interval scale, it also provides information about the size of the intervals that separate the classes or categories. As with the Interval and Ordinal scales, there are ranks between classes or categories. As with the Interval, Ordinal, and Nominal scales, it also identifies classes or categories, where each category groups a set of entities based on their attribute's value. The difference between a Ratio scale and an Interval scale is the existence of an absolute zero. The Ratio scale is also a numerical scale, and numbers have a numerical meaning. This means that any arithmetic operations between two categories are permitted. Examples using a Ratio scale are given in Table 14.5.

Table 14.5. Examples of Ratio Scale Measures

Entity	Attribute	Categories
Web project	Effort	Decimal numbers
Web application	Size	Integer numbers
Human body	Temperature in Kelvin	Decimal numbers

14.2.5 The Absolute Scale Type

The Absolute scale supplements the Ratio scale with restricting the classes or categories to a specific unit of measurement. As with the Ratio scale, it also has a zero element, representing the total absence of the attribute measured. As with the Ratio and Interval scales, it also provides information about the size of the intervals that separate the classes or categories. As with the Interval and Ordinal scales, there are ranks between classes or categories. As with the Ratio, Interval, Ordinal, and Nominal scales, it also identifies classes or categories, where each category groups a set of entities based on their attribute's value.

The difference between an Absolute scale and the Ratio scale is the existence of a *fixed unit of measurement* associated with the attribute being measured. For example, using a Ratio scale, if we were to measure the attribute *effort* of a *Web project,* we could obtain an effort value that could represent effort in number of hours, or effort in number of days, and so on. In case we want all effort measures to be kept using number of hours, we can convert effort in number of days to effort in number of hours, or effort in number of weeks to effort in number of hours. Thus, an attribute measured using a given unit of measurement (e.g., number of weeks) can have its class converted into another using a different unit of measurement, but keeping the meaning of the obtained data unchanged. Therefore, assuming a single developer, a Web project's effort of 40 hours is equivalent to a Web project effort's of a week. Thus, the unit of measurement changes, but the data that have been gathered remain unaffected. If we were to measure the attribute *effort* of a *Web project* using an *Absolute scale,* we would need to determine in advance the unit of measurement to be used. Therefore, once the unit of measurement is determined, it is the one used when effort data are being gathered. Using our example on *Web project's effort,* if the unit of measurement associated with the *attribute effort* had been *number of hours,* then all the effort data gathered would have represented effort in number of hours only. Finally, as with the Ratio scale, operations between two categories, such as addition, subtraction, multiplication, and division, are also permitted. Examples using an Absolute scale are given in Table 14.6.

Table 14.6. Examples of Absolute Scale Measures

Entity	Attribute	Categories
Web project	Effort, in number of hours	Decimal numbers
Web application	Size, in number of HTML files	Integer numbers
Web developer	Experience developing Web applications, in number of years	Integer numbers

14.2.6 Summary of Scale Types

Table 14.7 presents one of the summaries we are providing regarding Scale types. It has been adapted from Maxwell (2005). It is also important to note that the Nominal and Ordinal scales do not provide classes or categories that have numerical meaning, and for this reason their attributes are called Categorical or Qualitative. Conversely, given that the Interval, Ratio, and Absolute scales provide classes or categories that have numerical meaning, their attributes are called Numerical or Quantitative (Maxwell, 2005).

Table 14.7. Summary of Scale Type Definitions

Scale Type	Is Ranking Meaningful?	Are Distances Between Classes the Same?	Does the Class Include an Absolute Zero?
Nominal	No	No	No
Ordinal	Yes	No	No
Interval	Yes	Yes	No
Ratio	Yes	Yes	Yes
Absolute	Yes	Yes	Yes

In relation to the statistics relevant to each measurement scale type, Table 14.8 presents a summary adapted from Fenton and Pfleeger (1997).

Table 14.8. Summary of Scale Type Definitions

Scale Type	Examples of Suitable Statistics	Suitable Statistical Tests
Nominal	Mode, frequency	Nonparametric
Ordinal	Median, percentile	Nonparametric
Interval	Mean, standard deviation	Nonparametric and parametric
Ratio	Mean, geometric mean, standard deviation	Nonparametric and parametric
Absolute	Mean, geometric mean, standard deviation	Nonparametric and parametric

14.3 OVERVIEW OF EMPIRICAL ASSESSMENTS

Validating a hypothesis or research question encompasses experimentation, which is carried out using an empirical investigation. This section details the three different types of empirical investigation that can be carried out, which are survey, case study, or formal experiment (Fenton and Pfleeger, 1997):

- *Survey*: a retrospective investigation of an activity in order to confirm relationships and outcomes (Fenton and Pfleeger, 1997). It is also known as "research-in-the-large", as it often samples over large groups of projects. A survey should always be carried out after the activity under focus has occurred (Kitchenham et al., 1995). When performing a survey, a researcher has no control over the situation at hand, i.e., the situation can be documented, compared to other similar situations, but none of the variables being investigated can be manipulated (Fenton and Pfleeger, 1997). Within the scope of software and Web Engineering, surveys are often used to validate the response of organizations and developers to a new development method, tool, or technique, or to reveal trends or relationships between relevant variables (Fenton and Pfleeger, 1997). For example, a survey can be used to measure the success of changing from Sun's J2EE to Microsoft's ASP.NET throughout an organization, because it can gather data from numerous projects. The downside of surveys is time. Gathering data can take many months or even years, and the outcome may only be available after several projects have been completed (Kitchenham et al., 1995).
- *Case study*: an investigation that examines the trends and relationships using as its basis a typical project within an organization. It is also known as "research-in-the-typical" (Kitchenham et al., 1995). A case study can investigate a retrospective event, but this is not the usual trend. A case study is the type of investigation of choice when wishing to examine an event that has not yet occurred and for which there is little or no control over the variables. For example, if an organization wants to investigate the effect of a programming framework on the quality of the resulting Web application but cannot develop the same project using numerous frameworks simultaneously, then the investigative choice is to use a case study. If the quality of the resulting Web application is higher than the organization's quality baseline, it may be due to many different reasons (e.g., chance, or perhaps bias from enthusiastic developers). Even if the programming framework had a legitimate effect on quality, no conclusions outside the boundaries of the case study can be drawn, i.e., the results of a case study cannot be generalized to every possible situation. Had the same application been developed several times, each

time using a different programming framework[5] (as in a formal experiment), then it would have been possible to have had a better understanding of the relationship between framework and quality, given that these variables were controlled. A case study samples *from the variables*, rather than over them. This means that, in relation to the variable programming framework, a value that represents the framework usually used on most projects will be the one chosen (e.g., J2EE). A case study is easier to plan than a formal experiment, but its results are harder to explain and, as previously mentioned, cannot be generalized outside the scope of the study (Kitchenham et al., 1995).

- *Formal experiment*: rigorous and controlled investigation of an event where important variables are identified and manipulated such that their effect on the outcome can be validated (Fenton and Pfleeger, 1997). It is also known as "research-in-the-small" since it is very difficult to carry out formal experiments in software and Web Engineering using numerous projects and resources. A formal experiment samples *over the variable that is being manipulated*, such that all possible variable values are validated, i.e., there is a single case representing each possible situation. If we apply the same example used when explaining case studies above, this means that several projects would be developed, each using a different object-oriented programming language. If one aims to obtain results that are largely applicable across various types of projects and processes, then the choice of investigation is a formal experiment. This type of investigation is most suited to the Web Engineering research community. However, despite the control that needs to be exerted when planning and running a formal experiment, its results cannot be generalized outside the experimental conditions. For example, if an experiment demonstrates that J2EE improves the quality of e-commerce Web applications, one cannot guarantee that J2EE will also improve the quality of educational Web applications (Kitchenham et al., 1995).

Other concrete issues related to using a formal experiment or a case study may impact the choice of study. It may be feasible to control the variables, but at the expense of a very high cost or a high degree of risk. If replication *is* possible, but at a prohibitive cost, then a case study should be used (Fenton and Pfleeger, 1997). A summary of the characteristics of each type of empirical investigation is given in Table 14.9.

[5] The values for all other attributes should remain the same (e.g., developers, programming experience, development tools, computing power, and type of application).

Table 14.9. Summary Characteristics of the Three Types of Empirical Investigations

Characteristic	Survey	Case Study	Formal Experiment
Scale	Research-in-the-large	Research-in-the-typical	Research-in-the-small
Control	No control	Low level of control	High level of control
Replication	No	Low	High
Generalization	Results representative of sampled population	Only applicable to other projects of similar type and size	Can be generalized within the experimental conditions

A set of steps broadly common to all three types of investigations is described below.

Define the goal(s) of your investigation and its context. Goals are crucial for the success of all activities in an investigation. Thus, it is important to allow enough time to fully understand and set the goals so that each is clear and measurable. Goals represent the research questions, which may also be presented by a number of hypotheses. By setting the research questions or hypotheses, it becomes easier to identify the dependent and independent variables for the investigation (Fenton and Pfleeger, 1997). A dependent variable is a variable whose behavior we want to predict or explain. An independent variable is believed to have a causal relationship with, or have influence upon, the dependent variable (Wild and Seber, 2000). Goals also help determine what the investigation will do, and what data are to be collected. Finally, by understanding the goals we can also confirm if the type of investigation chosen is the most suitable type to use (Fenton and Pfleeger, 1997).

Each hypothesis of an investigation will later be either supported or rejected. An example of hypotheses is given below (Wild and Seber, 2000):

H_0 *Using J2EE produces the same quality of Web applications as using ASP.NET.*

H_1 *Using J2EE produces a different quality of Web applications than using ASP.NET.*

H_0 is called the null hypothesis and assumes the quality of Web applications developed using J2EE is similar to that of Web applications developed using ASP.NET. In other words, it assumes that data samples for both groups of applications come from the same population. In this instance, we have two samples, one representing quality values for Web applications developed using J2EE, and the other, quality values for Web applications developed using ASP.NET. Here, quality is our dependent variable, and the choice of programming framework (e.g., J2EE or ASP.NET) is the independent variable.

H_1 is called the alternative or research hypothesis and represents what is believed to be true if the null hypothesis is false. The alternative hypothesis assumes that samples do not come from the same sample population. Sometimes the direction of the relationship between dependent and independent variables is also presented as part of an alternative hypothesis. If H_1 also suggested a direction for the relationship, it could be described as

H_1 Using J2EE produces a better quality of Web applications than using ASP.NET.

To confirm H_1 it is first necessary to reject the null hypothesis and, second, to show that quality values for Web applications developed using J2EE are significantly higher than quality values for Web applications developed using ASP.NET.

We have presented both null and alternative hypotheses since they are both equally important when presenting the results of an investigation, and, as such, both should be documented.

To see if the data justify rejecting H_0 we need to perform a statistical analysis. Before carrying out a statistical analysis it is important to decide the level of confidence we have that the data sample we gathered truly represents our population of interest. If we have 95% confidence that the data sample we are using truly represents the general population, there still remains a 5% chance that H_0 will be rejected when, in fact, it truly represents the current situation. Rejecting H_0 incorrectly is called the *Type I error*, and the probability of this occurring is called the *Significance level* (α). Every statistical analysis test uses α when testing whether or not H_0 should be rejected.

14.4 ISSUES TO CONSIDER WITH EMPIRICAL ASSESSMENTS

In addition to defining the goals of an investigation, it is also important to document the context of the investigation (Kitchenham et al., 2002). One suggested way to achieve this is to provide a table, similar to Table 14.1, describing the entities, attributes, and measures that are the focus of the investigation.

14.4.1 Prepare the Investigation

It is important to prepare an investigation carefully to obtain results from which one can draw valid conclusions, even if these conclusions cannot be scaled up. For case studies and formal experiments, it is important to define the variables that can influence the results and, once these are defined, decide how much control one can have over them (Fenton and Pfleeger, 1997).

Consider the following case study, which would represent a *poorly prepared investigation*.

The case study aims to investigate, within a given organization, the effect of using the programming framework J2EE on the quality of the resulting Web application. Most Web projects in this organization are developed using ASP.NET, and consequently all members of the development team have experience with this language. The type of application representative of the majority of applications this organization undertakes is in electronic commerce (e-commerce), and a typical development team has two developers. Therefore, as part of the case study, an e-commerce application is to be developed by two developers using J2EE. Because we have stated that this is a poorly executed case study, we will assume that no other variables have been considered or measured (e.g., developers' experience, development environment).

The e-commerce application is developed, and the results of the case study show that the quality of the delivered application, measured as the number of faults per Web page, is worse than that for the other similar Web applications developed using ASP.NET. When questioned as to why these were the results obtained, the investigator seemed puzzled, and without a clear explanation.

What is missing?

The investigator should have anticipated that other variables can also affect the results of an investigation and should therefore also be taken into account. One such variable is the developers' programming experience. Without measuring experience prior to the case study, it is impossible to discern if the lower quality is due to J2EE or to the effects of learning J2EE as the investigation proceeds. It is possible that one or both developers did not have experience with J2EE and that lack of experience interfered with the benefits of its use.

Variables such as developers' experience should have been anticipated and, if possible, controlled, or risk obtaining results that will be incorrect.

To control a variable is to determine a subset of values for use within the context of the investigation from the complete set of possible values for that variable. For example, using the same case study presented above, if the investigator had measured developers' experience with J2EE (e.g., low, medium, high) and was able to control this variable, then he could have determined that two developers experienced with J2EE should have participated in the case study. If there were no developers with experience in J2EE, two would have been selected and trained.

When conducting a case study, if it is not possible to control certain variables, they should still be measured, and the results documented. If, however, all variables are controllable, then the type of investigation to use is a formal experiment.

Another important issue is to identify the population being studied and the sampling technique used. For example, if a survey was designed to investigate the extent to which project managers use automatic project management tools, then a data sample of software programmers is not going to be representative of the population that has been initially specified.

With formal experiments, it is important to describe the process by which experimental subjects and objects are selected and assigned to treatments (Kitchenham et al., 2002), where a treatment represents the new tool, programming language, or methodology you want to evaluate. The experimental object, also known as experimental unit, represents the object to which the treatment is to be applied (e.g., development project, Web application, code). The control object does not use or is not affected by the treatment (Fenton and Pfleeger, 1997). In software and Web Engineering it is difficult to have a control in the same way as in, say, formal medical experiments. For example, if you are investigating the effect of a programming framework on quality, and your treatment is J2EE, you cannot have a control that is "no programming framework" (Kitchenham et al., 2002). Therefore, many formal experiments use as their control a baseline representing what is typical in an organization. Using the example given previously, our control would be ASP.NET, since it represents the typical programming framework used in the organization. The experimental subject is the "who" applying the treatment (Fenton and Pfleeger, 1997).

As part of the preparation of an investigation we also include the preparation and validation of data collection instruments. Examples are questionnaires, automatic measurement tools, timing sheets, etc. Each has to be prepared carefully such that it clearly and unambiguously identifies what is to be measured. For each variable it is also important to identify its measurement scale and measurement unit. So, if you are measuring effort, then you should also document its measurement unit (e.g., person hours, person months) or else obtain incorrect and conflicting data. It is also important to document at which stage during the investigation the data collection takes place. If an investigation gathers data on developers' programming experience (before they develop a Web application), size and effort used to design the application, and size and effort used to implement the application, then a diagram, such as the one in Figure 14.2, may be provided to all participants to help clarify what instruments to use and when to use them.

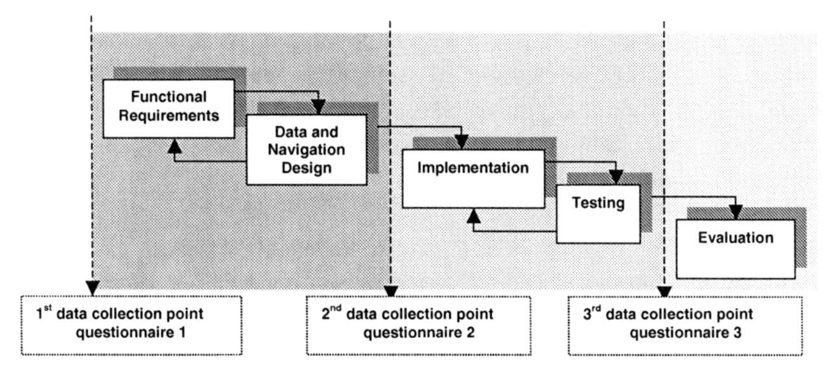

Figure 14.2. Plan detailing when to apply each project.

It is usual for instruments to be validated using pilot studies. A pilot study uses similar conditions to those planned for the real investigation, such that any possible problems can be anticipated. It is highly recommended that those conducting any empirical investigations use pilot studies, as they can provide very useful feedback and reduce or remove any problems not previously anticipated.

Finally, it is also important to document the methods used to reduce any bias.

14.4.2 Analyzing the Data and Reporting the Results

The main aspect of this final step is to understand the data collected and to apply statistical techniques that are suitable for the research questions or hypotheses of the investigation. For example, if the data were measured using a nominal or ordinal scale, then statistical techniques that use the mean cannot be applied, as this would violate the principles of the representational theory of measurement. If the data are not normally distributed, then it is possible to use nonparametric or robust techniques, or transform the data to conform to the normal distribution (Fenton and Pfleeger, 1997). Further details on data analysis are presented later in this chapter.

When interpreting and reporting the results of an empirical investigation, it is also important to consider and discuss the validity of the results obtained. There are three types of threats to the validity of empirical investigations (Kitchenham et al., 1995; Porter et al., 1997): *construct validity*, *internal validity,* and *external validity*. Each is described below.

Construct validity represents the extent to which the measures you are using in your investigation really measure the attributes of entities being investigated. For example, if you are measuring the size of a Web application using IFPUG function points, can you say that the use of IFPUG function points is really measuring the size of a Web application? How valid will the results of your investigation be if you use IFPUG function points to measure a Web application's size? As another example, if you want to measure the experience of Web developers developing Web applications and you use as a measure the number of years they worked for their current employer, it is unlikely that you are using an appropriate measure since your measure does not also take into account their previous experience developing Web applications.

Internal validity represents the extent to which external factors not controlled by the researcher can affect the dependent variable. Suppose that, as part of an investigation, we observe that larger Web applications are related to more productive teams, compared to smaller Web applications. We must make sure that team productivity is not being affected by using, for example, highly experienced developers to develop larger applications and less experienced developers to develop smaller applications. If the researcher is unaware of developers' experience, it is impossible to discern whether the results are due to developers' experience or due to legitimate economies of scale. Typical factors that can affect the internal validity of investigations are variations in human performance, learning effects where participants' skills improve as the investigation progresses, and differences in treatments, data collection forms used, or other experimental materials.

External validity represents the extent to which we can generalize the results of our investigation to our population of interest. In most empirical investigations in Web Engineering the population of interest often represents industrial practice. Suppose you carried out a formal experiment with postgraduate students to compare J2EE to ASP.NET, using as experimental object a small Web application. If this application is not representative of industrial practice, you cannot generalize the results of your investigation beyond the context in which it took place. Another possible problem with this investigation might be the use of students as subject population. If you have not used Web development professionals, it will also be difficult to generalize the results to industrial practice. Within the context of this example, even if you had used Web development professionals in your investigation, if they did not represent a random sample of your population of interest, you would also be unable to generalize the results to your entire population of interest.

14.5 DETAILING FORMAL EXPERIMENTS

A formal experiment is considered the most difficult type of investigation to carry out since it has to be planned very carefully such that all the important factors are controlled and documented, enabling its further replication. Due to the amount of control that formal experiments use, they can be further replicated and, when replicated under identical conditions, if results are repeatable, they provide a better basis for building theories that explain our current understanding of a phenomenon of interest. Another important point related to formal experiments is that the effects of uncontrolled variables upon the results must be minimized. The way to minimize such effects is to use randomization. Randomization represents the random assignment of treatments and experimental objects to experimental subjects.

In this section we are going to discuss the typical experimental designs used with formal experiments (Wohlin et al., 2005); for each typical design, we will discuss the types of statistical analysis tests that can be used to examine the data gathered from such experiments.

14.5.1 Typical Design 1

There is one independent variable (factor) with two values and one dependent variable. Suppose you are comparing the productivity between Web applications developed using J2EE (treatment) and Web applications developed using ASP.NET (control). Fifty subjects are participating in the experiment, and the experimental object is the same for both groups. Assuming other variables are constant, subjects are randomly assigned to J2EE or ASP.NET (see Figure 14.3).

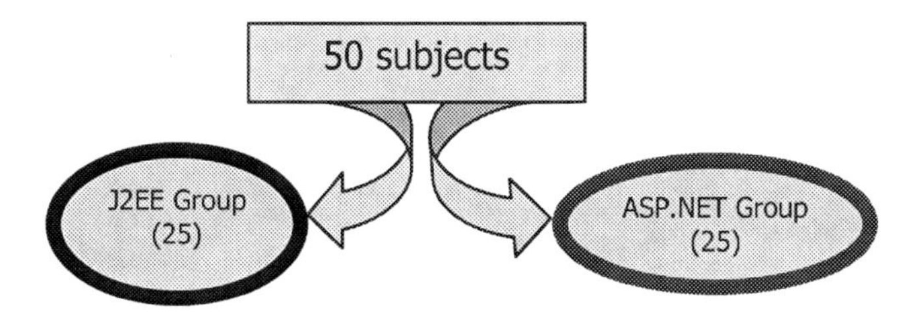

Figure 14.3. Example of one-factor design.

Once productivity data are gathered for both groups the next step is to compare the productivity data to check if productivity values for both development frameworks come from the same population (H_0) or from different populations (H_1). If the subjects in this experiment represent a large random sample or the productivity data for each group are normally distributed, you can use the independent samples t-test statistical technique to compare the productivity between both groups. This is a parametric test; as such, it assumed that the data are normally distributed or that the sample is large and random. Otherwise, the statistical technique to use would be the independent samples Mann–Whitney test, a nonparametric equivalent to the t test. Nonparametric tests make no assumptions related to the distribution of the data, and that is why they are used if you cannot guarantee that your data are normally distributed or represent a large random sample.

14.5.2 Typical Design 1: One Factor and One Confounding Factor

There is one independent variable (factor) with two values and one dependent variable. Suppose you are comparing the productivity between Web applications developed using J2EE (treatment) and Web applications developed using ASP.NET (control). Fifty subjects are participating in the experiment, and the experimental object is the same for both groups. A second factor (confounding factor)—gender—is believed to have an effect on productivity; however, you are only interested in comparing different development frameworks and their effect on productivity, not the interaction between gender and framework type on productivity. The solution is to create two blocks (see Figure 14.4), one with all the female subjects, and another with all the male subjects, and then, within each block, randomly assign a similar number of subjects to J2EE or ASP.NET (balancing).

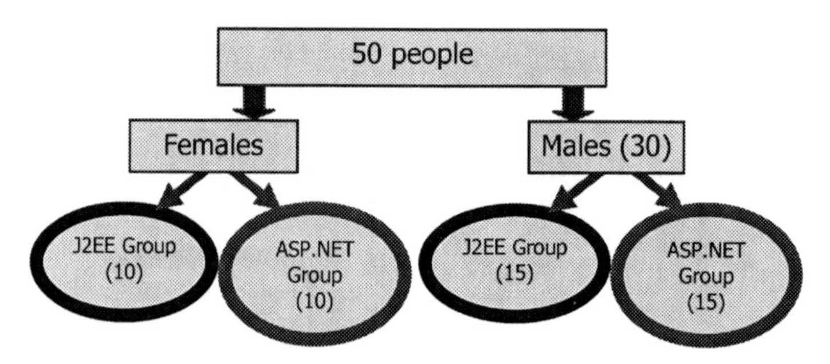

Figure 14.4. Example of blocking and balancing with one-factor design.

Once productivity data have been gathered for both groups, the next step is to compare the productivity data to check if productivity values for both groups come from the same population (H_0) or from different populations (H_1). The mechanism used to analyze the data would be the same one presented previously. Two sets of productivity values are compared, one containing productivity values for the 10 females and the 15 males who used J2EE, and the other containing productivity values for the 10 females and the 15 males who used ASP.NET. If the subjects in this experiment represent a large random sample or the productivity data for each group are normally distributed, you can use the independent samples t-test statistical technique to compare the productivity between both groups. Otherwise, the statistical technique to use would be the independent samples Mann–Whitney test, a nonparametric equivalent to the t test.

14.5.3 Typical Design 2

There is one independent variable (factor) with two values and one dependent variable. Suppose you are comparing the productivity between Web applications developed using J2EE (treatment) and Web applications developed using ASP.NET (control). Fifty subjects are participating in the experiment using the experimental object. You also want every subject to be assigned to both the control and the treatment. Assuming other variables are constant, subjects are randomly assigned to the control or the treatment and then swapped around (see Figure 14.5).

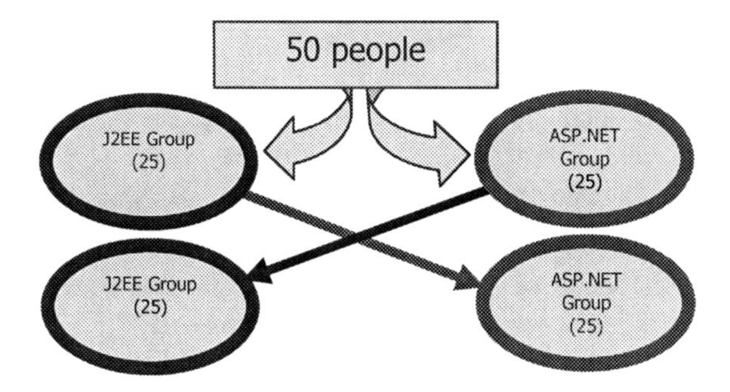

Figure 14.5. Example of Typical Design 2.

Once productivity data have been gathered for both groups, the next step is to compare the productivity data to check if productivity values for both groups come from the same population (H_0) or from different populations (H_1). Two sets of productivity values are compared: The first contains productivity values for 50 subjects when using J2EE; the second contains productivity values for the same 50 subjects when using ASP.NET. Given that each subject was exposed to both control and treatment, you need to use a paired test. If the subjects in this experiment represent a large random sample or the productivity data for each group are normally distributed, you can use the paired samples t-test statistical technique to compare the productivity between both groups. Otherwise, the statistical technique to use would be the two related samples Wilcoxon test, a nonparametric equivalent to the paired samples t test.

14.5.4 Typical Design 3

There is one independent variable (factor) with more than two values and one dependent variable. Suppose you are comparing the productivity among Web applications designed using Methods A, B, and C. Sixty subjects are participating in the experiment, and the experimental object is the same for all groups. Assuming other variables are constant, subjects are randomly assigned to one of the three groups (see Figure 14.6).

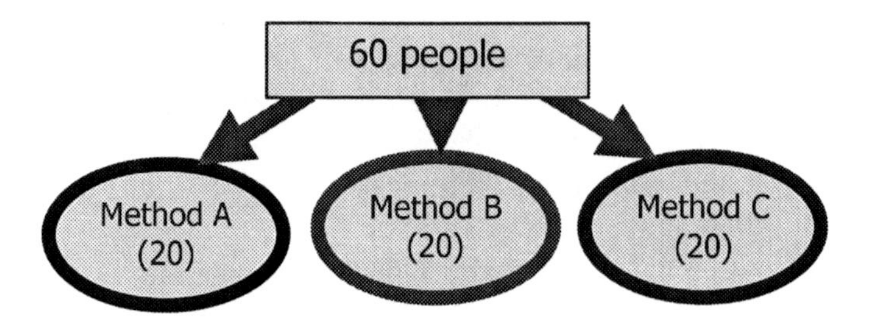

Figure 14.6. Example of Typical Design 3.

Once productivity data have been gathered for all three groups, the next step is to compare the productivity data to check if productivity values for all groups come from the same population (H_0) or from different populations (H_1). Three sets of productivity values are compared: The first contains productivity values for 20 subjects when using Method A; the second contains productivity values for another 20 subjects when using Method B; the third contains productivity values for another 20 subjects when using Method C. Given that each subject was exposed to only a single method, you need to use an independent samples test. If the subjects in this experiment represent a large random sample or the productivity data for each group are normally distributed, you can use the one-way ANOVA statistical technique to compare the productivity among groups. Otherwise, the statistical technique to use would be the Kruskal–Wallis H test, a nonparametric equivalent to the one-way ANOVA technique.

14.5.5 Typical Design 4

There are at least two independent variables (factors) and one dependent variable. Suppose you are comparing the productivity between Web applications developed using J2EE (treatment) and Web applications developed using ASP.NET (control). Sixty subjects are participating in the experiment, and the experimental object is the same for both groups. A second factor—gender—is believed to have an effect on productivity, and you are interested in assessing the interaction between gender and framework type on productivity. The solution is to create four blocks (see Table 14.10) representing the total number of possible combinations. In this example each factor has two values; therefore, the total number of combinations would be given by multiplying the number of values in the first factor by the number of values in the second factor (2 multiplied by 2, which is equal to 4). Then, assuming that all subjects have similar experience using both frameworks, within each gender block, subjects are randomly assigned to J2EE or ASP.NET (balancing). In this scenario each block will provide 15 productivity values.

Table 14.10. Example of Typical Design 4

		Gender	
		Female	Male
Framework	J2EE	Female, J2EE (15) *Block 1*	Male, J2EE (15) *Block 2*
	ASP.NET	Female, ASP.NET (15) *Block 3*	Male, ASP.NET (15) *Block 4*

Once productivity data have been gathered for all four blocks, the next step is to compare the productivity data to check if productivity values for males come from the same population (H_0) or from different populations (H_1), and the same has to be done for females. Here productivity values for blocks 2 and 4 are compared; and productivity values for blocks 1 and 3 are compared. If the subjects in this experiment represent a large random sample or the productivity data for each group are normally distributed, you can use the independent samples t-test statistical technique to compare the productivity between groups. Otherwise, the statistical technique to use would be the Mann–Whitney test, a nonparametric equivalent to the independent samples t test.

14.5.6 Summary of Typical Designs

Table 14.11 summarizes the statistical tests to be used with each of the typical designs previously introduced. Each of these tests is explained in detail in statistical books, such as Wild and Seber (2000).

Table 14.11. Examples of Statistical Tests for Typical Designs

Typical Design	Parametric Test	Nonparametric Test
Design 1: no explicit confounding factor	Independent samples t test	Independent samples Mann–Whitney test
Design 1: explicit confounding factor	Independent samples t test	Independent samples Mann–Whitney test
Design 2	Paired samples t test	Two-related samples Wilcoxon test
Design 3	One-way ANOVA	Kruskal–Wallis H test
Design 4	Independent samples t test	Mann–Whitney test

14.6 DETAILING CASE STUDIES

It is often the case that case studies are used in industrial settings to compare two different technologies, tools, or development methodologies. One of the technologies, tools, or development methodologies represents what is currently used by the company, and the other represents what is being compared to the company's current situation. Three mechanisms are suggested to organize such comparisons to reduce bias and enforce internal validity (Wohlin et al., 2005):

- To compare the results of using the new technology, tool, or development methodology to a company's baseline. A baseline generally represents an average over a set of finished projects. For example, a company may have established a productivity baseline against which to compare projects. This means that productivity data have been gathered from past finished projects and used to obtain an average productivity (productivity baseline). If this is the case, then the productivity related to the project that used the new technology, tool, or development methodology is compared against the existing productivity baseline, to assess if there was productivity improvement or decline. In addition to productivity, other baselines may also be used by a company, e.g., usability baseline or defect rate baseline.
- To compare the results of using the new technology, tool, or development methodology to a company's sister project, which is used as a baseline. This means that two similar and comparable projects will be carried out, one using the company's current technology, tool, or development methodology, and the other using the new technology, tool, or development methodology. Once both projects are finished, measures such as productivity, usability, and actual effort can be used to compare the results.

- Whenever the technology, tool, or development methodology applies to individual application components, it is possible to apply at random the new technology, tool, or development methodology to some components and not to others. Later measures such as productivity and actual effort can be used to compare the results.

14.7 DETAILING SURVEYS

There are three important points to stress here. The first is that, similarly to formal experiments and case studies, it is very important to define beforehand what is it that we wish to investigate (hypotheses) and what is the population of interest. For example, if you plan to conduct a survey to understand how Web applications are currently developed, the best population to use would be the one of Web project managers, as they have the complete understanding of the development process used. Interviewing Web developers may lead to misleading results, as it is often the case that they do not see the forest for the trees.

The second point is related to piloting the survey. It is important to ask different users, preferably representative of the population of interest, to read the instrument(s) to be used for data collection to make sure questions are clear and no important questions are missing. It is also important to ask these users to actually answer the questions in order to have a feel for how long it will take them to provide the data being asked for. This should be a similar procedure if you are using interviews.

Finally, the third point relates to the preparation of survey instruments. It is generally the case that instruments will be either questionnaires or interviews. In both cases instruments should be prepared with care and avoid misleading questions that can bias the results. If you use ordinary mail to post questionnaires to users, make sure you also include a self-addressed prepaid envelope to be used to return the questionnaire. You can also alternatively have the same questionnaire available on the Web. Unfortunately, the use of email as a means to broadcast a request to participate in a survey has been impaired by the advent of spam emails. Many of us today use filters to stop the receipt of unsolicited junk emails; therefore, many survey invitation requests may end up being filtered and deleted.

14.8 CONCLUSIONS

This chapter discussed the need for empirical investigations in Web Engineering and introduced the three main types of empirical investigation—surveys, case studies, and formal experiments. Each type of investigation was described, although greater detail was given to formal experiments as they are the most difficult type of investigation to conduct.

REFERENCES

Basili, V.R., 1996, The role of experimentation in software engineering: Past, current, and future. *Proceedings 18th International Conference on Software Engineering*, March 25–30, pp. 442–449.

Basili, V.R., Shull, F., and Lanubile, F., 1999, Building knowledge through families of experiments. *IEEE Transactions on Software Engineering*, July–Aug., **25**(4): 456–473.

Fenton, N.E., and Pfleeger, S.L., 1997, *Software metrics: A rigorous and practical approach*, 2nd ed., PWS Publishing Company, Boston.

Gellersen, H., Wicke, R., and Gaedke, M., 1997, WebComposition: An object-oriented support system for the Web engineering lifecycle. *Journal of Computer Networks and ISDN Systems*, September, *29*(8–13): 865–1553. Also (1996) in *Proceedings Sixth International World Wide Web Conference*, pp. 1429–1437.

Gellersen, H.-W., and Gaedke, M., 1999, Object-oriented Web application development. *IEEE Internet Computing*, Jan.–Feb., **3**(1): 60–68.

Ginige, A., 2002, Workshop on Web Engineering: Web Engineering: Managing the complexity of Web systems development. *Proceedings 14th International Conference on Software Engineering and Knowledge Engineering*, July, pp. 72–729.

Ginige, A., and Murugesan, S., 2001, Web Engineering: An introduction. *IEEE Multimedia*, Jan.–Mar., **8**(1):. 14–18.

Goldstein, M., and Goldstein, I.F., 1978, *How We Know: An Exploration of the Scientific Process*, Plenum Press, New York.

Harper Collins Publishers, 2000, *Collins English Dictionary*.

Houghton Mifflin Company, 1994, *The American Heritage Concise Dictionary*, 3rd ed.

Kitchenham, B., Pickard, L., and Pfleeger, S.L., 1995, Case studies for method and tool evaluation. *IEEE Software*, **12**(4): 52–62.

Kitchenham, B.A., Pfleeger, S.L., Pickard, L.M., Jones, P.W., Hoaglin, D.C., El Emam, K., and Rosenberg, J., 2002, Preliminary guidelines for empirical research in software engineering. *IEEE Transactions on Software Engineering*, August, **28**(8): 721–734.

Lee, S.C., and Shirani, A.I., 2004, A component based methodology for Web application development. *Journal of Systems and Software*, **71**(1–2): 177–187.

Maxwell, K., 2005, What you need to know about statistics. In *Web Engineering*, E. Mendes and N. Mosley, eds., Springer, New York, pp. 365–407.

Murugesan, S., and Deshpande, Y., 2001, *Web Engineering, Managing Diversity and Complexity of Web Application Development*, Lecture Notes in Computer Science 2016, Springer-Verlag, Heidelberg.

Murugesan, S., and Deshpande, Y., 2002, Meeting the challenges of Web application development: The Web Engineering approach. In *Proceedings 24th International Conference on Software Engineering*, May, pp. 687–688.

Offutt, J., 2002, Quality attributes of Web software applications. *IEEE Software*, Mar.–Apr., **19**(2): 25–32.

Pfleeger, S.L., Jeffery, R., Curtis, B., and Kitchenham, B., 1997, Status report on software measurement. *IEEE Software*, Mar.–Apr., **14**(2): 33–43.

Porter, A.A., Siy, H.P., Toman, C.A., and Votta, L.G., 1997, An experiment to assess the cost-benefits of code inspections in large-scale software development, *TSE*, **23**(6): 329–346.

Pressman, R.S., 1998, Can Internet-based applications be engineered? *IEEE Software*, Sept.–Oct., **15**(5): 104–110.

Ricca, F., and Tonella, P., 2001, Analysis and testing of Web applications. In *Proceedings 23rd International Conference on Software Engineering*, pp. 25–34.

Taylor, M.J., McWilliam, J., Forsyth, H., and Wade, S., 2002, Methodologies and Website development: A survey of practice. *Information and Software Technology*, **44**(6): 381–391.

Wikipedia, http://en.wikipedia.org/wiki/Main_Page (accessed on 25 October 2004).

Wild, C., and Seber, G., 2000, *Chance Encounters: A First Course in Data Analysis and Inference*, John Wiley & Sons, New York.

Wohlin, C., Host, M., and Henningsson, K., 2005, Empirical research methods in Web and software Engineering. In *Web Engineering*, E. Mendes and N. Mosley, eds., Springer, New York, pp. 409–430.

Chapter 15

CONCLUSIONS

Oscar Pastor,[4] Gustavo Rossi,[1] Luis Olsina,[3] Daniel Schwabe[2]

[1]*LIFIA, Facultad de Informatica,Universidad Nacional de La Plata, (also at CONICET) Argentina,* gustavo@lifia.info.unlp.edu.ar

[2]*Departamento de Informática, PUC-Rio, Rio de Janeiro, Brazil,* dschwabe@inf.puc-rio.br

[3]*GIDIS_Web, Engineering School, Universidad Nacional de La Pampa, Calle 9 y 110, (6360) General Pico, LP, Argentina,* olsinal@ing.unlpam.edu.ar

[4]*DSIC, Valencia University of Technology, Valencia, Spain,* opastor@dsic.upv.es

Historically, software engineering's main challenge has been to provide those processes, methods, and supporting tools capable of producing quality software. Given that the Web is, nowadays, the major delivery platform, and an important development support platform, it is only natural that it should evolve toward Web Engineering.

In the chapter written by San Murugesan (Chapter 2), we have covered different issues of the Web Engineering discipline. In this introductory chapter, San outlines the evolution of the Web and the unique aspects of Web applications and discusses some of the key challenges present in Web application development. It also examines what differentiates development of Web applications from other types of software or computer application development. It then reviews the problems and limitations of current Web development practices and their implications. Finally, it outlines key elements of a Web Engineering process and discusses the role of Web Engineering in successful Web application development.

These ideas are properly complemented with Chapter 3 by Martin Gaedke and Johannes Meinecke, which extends the view of the Web as the right platform to build distributed applications, with all the particular problems of these environments.

In this context, our perception is that well-known problems that have been identified in the so-called conventional software engineering community are translated in a natural and evolutionary way into a Web Engineering context. In particular, model-based software production processes, methods, and tools are strongly required by the Web Engineering community, to make true the MDA basic statement of producing Web applications through an adequate model transformation process, from requirements to the final software product.

Several evolutionary technologies related with Web Engineering are being proposed based on the Model-based Web Application Development Process. Another strong need is to properly fit all the pieces of this apparent Web Engineering puzzle, where we can find issues related with Web services, the Semantic Web, ontologies definition and management, adaptivity, conceptual modeling of Web applications, electronic commerce, and so on. Too often, these technologies are developed and evolve independently, while a sound Web Engineering process will require all of them to be properly aligned and connected.

In this context, it is true that many approaches have already been proposed, providing answers to some of the problems related with all these issues. But more than ever, we think that we need a place where all those approaches that share common roots are presented, their particularities analyzed, and their solutions presented in the context of a common example. Such material will let readers understand commonalities and their differences between the approaches, allowing them to make informed choices that best fit their own needs and situation. This is the spirit of this book.

The Web Engineering discipline is in constant evolution, and over the last few years a set of methods to model, design, and implement Web applications has been proposed. The central core of this book is the presentation of some of these relevant approaches. Clearly, not all such approaches have been presented, but we attempted to present the ones that have been discussed the most in the literature. The methods selected cover a very relevant and big spectrum of what Web application development methods currently mean, addressing the main issues and concepts that are critical in producing good Web applications.

The list of selected methods includes UWE, IDM, WebML, Hera, WSDM, OOWS/OO-Method, and OOHDM. Each one is presented in its own chapter, covering the main aspects of the method, the models used, the expressiveness provided, and the relevant parts of their software process. This is done with the support of a common example: a popular online repository of information related to the movie industry. We have selected an example that is both well-known and easy to interpret in terms of how every method manages it. The proposed requirements included both information recovery (such as, for instance, queries on films or actors) and service execution (such as, for instance, buying tickets to see a particular film). We did that to cover contents and functionality, to make it possible to analyze how the methods deal with these aspects.

Furthermore, this Web application was designed to be used by different types of users (anonymous, registered, system administrators, etc.), each one with a particular behavior and accessing some information and functionality; adaptivity must therefore be properly faced. Thus, a distinguishing feature of this book is precisely the exercise of applying each method to the same problem.

It is our feeling that the ultimate evaluation of each method will be made by the readers. Nevertheless, a preliminary side-by-side comparison shows some noticeable points that raise interesting questions. First, all the methods use different notations to deal with similar concepts. Could a common, standard notation be used? This is an interesting open question for the audience. All the methods share some relevant points of view:

- A clear separation of concerns with respect to conceptual modeling for Web applications, focusing basically on contents, functionality, navigation, and presentation. An interesting task for any reader is to compare how each method represents and manages these different modeling perspectives.
- The modeling and code generation tools developed to support the methods (such as WebRatio, ONME, ArgoUWE, HyperDE) emphasize the use of Model-Driven Software Development strategies as the right approach. It seems that, logically, model transformation technologies are also strongly present in the Web Engineering community.

As stated before, working on the same example, the readers can understand the basic models and primitives provided by the methods and can even personally evaluate the different solutions provided by them. As a matter of fact, it has been interesting for us to verify that all of them share some common conceptual constructs, but at the same time each one orders them in different ways, emphasizing some particular aspects considered more or less relevant. A strong value of this book is in providing adequate material to allow you—our reader—to reach your own conclusions about each method and how they compare.

It has also been remarkable to see how all these approaches are in constant evolution. They are extending their expressive capabilities, to give support to the most advanced characteristics required by modern Web applications. New aspects encompass, among other things, supporting business process execution, the development of adaptive Web applications, the proper use of Semantic Web representations in the Web application construction process, the use of Web services, the multidevice-oriented development of Web applications, etc. This simply emphasizes the fact that research continues, constantly extending the material already presented here.

Model transformation is present throughout the book, and it is behind all the presented methods for Web application design and implementation. To delve further into this, the chapter written by Antonio Vallecillo et al. provides a precise view on the current status of what we call Model-Driven Web Engineering (MDWE) in the context of MDA. Apart from introducing the main concepts and mechanisms related to MDWE and MDA, it also discusses the strengths and benefits of the use of MDA for MDWE, as well as its major limitations and the challenges it currently faces for wider adoption. We think that this is the right point to anchor the ideas introduced in this book addressing the applicability of model transformation to obtain a sound Web development process.

We would like to close these conclusions by commenting on interesting aspects that we have learned during the editing process. First, it should be realized that the development of private notations to model Web applications can make the wider adoption and acceptance of these methods by the industry more difficult. Representing the same concepts in the same way would improve the understanding of the models used by the different approaches, which would help their use in practice

Standards are being continuously updated, especially in the context of Web services definition languages and Semantic Web-oriented languages. It could be argued that we should have sound solutions before having their associated standards. But looking at the methods presented, we can easily conclude that they already provide solutions that are sound enough to be incorporated in appropriate standards for Web application development methods. This is probably a task to be accomplished in the near future, not unlike the context where UML was initially proposed, as an attempt to unify the diverse set of notations for object-oriented analysis and design that were present in the mid-1990s. For instance, if the required conceptual primitives for specifying Web navigation were fixed, a proper notation to represent them in a clear way could be proposed with major agreement.

Furthermore, as we are talking about engineering, we cannot forget those aspects related to quality evaluation and with empirical Web Engineering. Luis Olsina et al. analyzed in their chapter the rationale to measure and evaluate software and Web applications or products, from different perspectives. To complement this view, there is a strong need to evaluate the software artifacts obtained with the methods presented in the book together with their associated tools. Empirical studies and techniques such as those presented in the chapter written by Emilia Mendes are very relevant when the objective is to demonstrate the quality and precision of the generated software product. We feel that this provides the proper tone to end our book.

Readers interested in such aspects will find basic information and pointers to further reading in these two final chapters.

So, this is all folks… . If you have reached this point, which could mean that you have read the whole book, first of all, congratulations!, and second and more important, we really hope we have been able to provide interesting and fruitful material that will help anyone better understand what modern Web Engineering means, and how all the ideas involved can be successfully put into practice, from both an academic and an industrial point of view.

INDEX

Printed in the United States
97316LV00002B/1-45/A

9 781846 289224